D0076033

$f(t)$	$\mathscr{L}\{f(t)\} = F(s)$
39. $\dfrac{e^{bt} - e^{at}}{t}$	$\ln \dfrac{s-a}{s-b}$
40. $\dfrac{2(1 - \cos kt)}{t}$	$\ln \dfrac{s^2 + k^2}{s^2}$
41. $\dfrac{2(1 - \cosh kt)}{t}$	$\ln \dfrac{s^2 - k^2}{s^2}$
42. $\dfrac{\sin at}{t}$	$\arctan\left(\dfrac{a}{s}\right)$
43. $\dfrac{\sin at \cos bt}{t}$	$\dfrac{1}{2}\arctan\dfrac{a+b}{s} + \dfrac{1}{2}\arctan\dfrac{a-b}{s}$
44. $\dfrac{1}{\sqrt{\pi t}}\, e^{-a^2/4t}$	$\dfrac{e^{-a\sqrt{s}}}{\sqrt{s}}$
45. $\dfrac{a}{2\sqrt{\pi t^3}} e^{-a^2/4t}$	$e^{-a\sqrt{s}}$
46. $\operatorname{erfc}\left(\dfrac{a}{2\sqrt{t}}\right)$	$\dfrac{e^{-a\sqrt{s}}}{s}$
47. $2\sqrt{\dfrac{t}{\pi}}\, e^{-a^2/4t} - a \operatorname{erfc}\left(\dfrac{a}{2\sqrt{t}}\right)$	$\dfrac{e^{-a\sqrt{s}}}{s\sqrt{s}}$
48. $e^{ab}e^{b^2t} \operatorname{erfc}\left(b\sqrt{t} + \dfrac{a}{2\sqrt{t}}\right)$	$\dfrac{e^{-a\sqrt{s}}}{\sqrt{s}(\sqrt{s} + b)}$
49. $-e^{ab}e^{b^2t} \operatorname{erfc}\left(b\sqrt{t} + \dfrac{a}{2\sqrt{t}}\right) + \operatorname{erfc}\left(\dfrac{a}{2\sqrt{t}}\right)$	$\dfrac{be^{-a\sqrt{s}}}{s(\sqrt{s} + b)}$
50. $\delta(t)$	1
51. $\delta(t - t_0)$	e^{-st_0}
52. $e^{at}f(t)$	$F(s - a)$
53. $f(t - a)\mathscr{U}(t - a)$	$e^{-as}F(s)$
54. $\mathscr{U}(t - a)$	$\dfrac{e^{-as}}{s}$
55. $f^{(n)}(t)$	$s^nF(s) - s^{(n-1)}f(0) - \cdots - f^{(n-1)}(0)$
56. $t^nf(t)$	$(-1)^n \dfrac{d^n}{ds^n}F(s)$
57. $\int_0^t f(\tau)g(t - \tau)\, d\tau$	$F(s)G(s)$

A First Course in DIFFERENTIAL EQUATIONS

A First Course in Differential Equations

THE CLASSIC 5TH EDITION

DENNIS G. ZILL
Loyola Marymount University

BROOKS/COLE
CENGAGE Learning™

Australia • Brazil • Japan • Korea • Mexico • Singapore • Spain • United Kingdom • United States

BROOKS/COLE
CENGAGE Learning™

A First Course in Differential Equations
The Classic 5th Edition
Dennis G. Zill

Publisher: **Gary Ostedt**

Marketing Team: **Karin Sandberg, Samantha Cabaluna**

Marketing Associate: **Beth Kroenke**

Assistant Editor: **Carol Benedict**

Editorial Assistant: **Daniel Thiem**

Production Editor: **Keith Faivre**

Production Service: **Lifland et al., Bookmakers**

Manucript Editor: **Gail Magin/Lifland**

Cover Design: **Roy R. Neuhaus**

Cover Photo: **The Stock Market/ M. Mastrorillo**

Interior Illustration: **Network Graphics**

Photo Researcher: **Gail Magin/Lifland**

Print Buyer: **Vena Dyer**

Typesetting: **The PRD Group**

ISBN-13: 978-0-534-37388-7

ISBN-10: 0-534-37388-7

Brooks/Cole
10 Davis Drive
Belmont Drive, CA 94002-3098
USA

Cengage Learning is a leading provider of customized learning solutions with office locations around the globe, including Singapore, the United Kingdom, Australia, Mexico, Brazil, and Japan. Locate your local office at:
international.cengage.com/region

Cengage Learning products are represented in Canada by Nelson Education, Ltd.

For your course and learning solutions, visit **academic.cengage.com**

Purchase any of our products at your local college store or at our preferred online store **www.ichapters.com**

Printed in the United States of America
9 10 11 12 11 10

CONTENTS

---

* Optional

CHAPTER 7 LAPLACE TRANSFORM 280

CHAPTER 8 SYSTEMS OF LINEAR DIFFERENTIAL EQUATIONS 335

CHAPTER 9 NUMERICAL METHODS FOR ORDINARY DIFFERENTIAL EQUATIONS 421

APPENDIXES

A WORD FROM THE PUBLISHER

Because of its continued popularity and the requests that we have received from our many friends at the community colleges, we are pleased to provide a reprint of the 5th edition of *A First Course in Differential Equations* by Dennis G. Zill. To distinguish this reprint from the 7th edition of the same text and from the other two texts in differential equations by the same author, we call this "The Classic Fifth Edition."

Gary W. Ostedt
Publisher

PREFACE TO THE FIFTH EDITION

My goal for the fifth edition was to achieve a balance between the concepts and presentation of materials that appealed to users of previous editions and the substantive changes made to strengthen and modernize the text. I feel I have achieved this balance, thus enabling the text to appeal to an even wider audience. Many of the additions and changes are the result of user and reviewer comments and suggestions. Moreover, these changes were made with the ultimate audience in mind—the students who will be using it. For this reason, solutions of every example have been read with an eye to improving their clarity. In various places I have added either further explanations where I thought they might be helpful or "guidance boxes" at crucial points in the flow of the solution.

NEW FEATURES

Some new features, that I hope students will find both interesting and motivational, have been added to the text. Essays written by mathematicians prominent in their specialty are included after Chapters 3, 4, 5, and 9. Each essay reflects the thoughts, creativity, and opinions of the individual author and is intended to enhance the material found in the preceding chapter. It is my hope that the addition of these essays will spark the interest of the students, encourage them to read mathematics, and help them to gain a realization that differential equations is not simply a dry collection of methods, facts, and formulas, but a vibrant field in which people can, and do, work.

Color inserts also have been added at intervals in the text. These pages consist of illustrations matched with photographs relating to some of the applications found in the text. I feel that these contribute to the visualization of the applications and thereby provide an added insight to students.

CHANGES IN THIS EDITION

- Section 1.2 is now devoted solely to the concept of a differential equation as a mathematical model.
- The material on the differential equation of a family of curves has been deleted. A brief discussion of this concept is now given in Section 3.1 (Orthogonal Trajectories).

- The method of undetermined coefficients is one of the more controversial topics in a course in differential equations. In the last three editions, this topic was developed from the viewpoint of using a differential annihilator as an aid in determining the correct form of a particular solution. While preparing this revision, a substantial number of reviewers indicated that the annihilator approach was too sophisticated for their students and requested a simpler rule-based approach. Other reviewers, however, desired no change. In order to satisfy each of these preferences, both approaches are presented in this edition. The instructor can now choose between undetermined coefficients based on the superposition principle for nonhomogeneous linear differential equations (Section 4.4) or those based on the concept of differential annihilators (Section 4.6). Moreover, the notion of a differential operator is now introduced in a separate section (Section 4.5). Thus, covering Section 4.4 does not preclude coverage of the otherwise useful concept of a differential operator.
- The review of power series in Section 6.2 has been greatly expanded. A discussion of the arithmetic of powers series (addition, multiplication, and division of series) has been added.
- A brief discussion of the "cover-up method" for determining coefficients in a partial fraction decomposition and a historical note on Oliver Heaviside have been added to Section 7.2.
- The discussion on the operational properties of the Laplace transform has now been divided into two sections: Section 7.3, Translation Theorems and Derivatives of Transforms, and Section 7.4, Transforms of Derivatives, Integrals, and Periodic Functions. This separation allows for a clearer, more comprehensive treatment of these topics.
- Gaussian elimination, in addition to Gauss-Jordan elimination, is now discussed in Section 8.4. The notation for indicating row operations on an augmented matrix has been improved.
- Chapter 9, "Numerical Methods for Ordinary Differential Equations," has been significantly expanded and partially rewritten. The Adams-Bashforth/Adams-Moulton multistep method has been added to Section 9.5. Section 9.6, Errors and Stability, and Section 9.8, Second-Order Boundary-Value Problems, are new to this edition.
- Chapter 10, on partial differential equations and Fourier series, has been eliminated from this edition. It was the consensus of users that this material was unnecessary in a beginning course. The topics: Fourier series, partial differential equations, and solutions of boundary-value problems by separation of variables, integral transforms, and numerical methods, are covered in detail in the expanded version of this text, *Differential Equations with Boundary-Value Problems,* Third Edition.
- New problems, applications, illustrations, remarks, and historical footnotes, have been added throughout the text.

Supplements Available

For Instructors

Complete Solutions Manual, Warren S. Wright and Carol D. Wright, Loyola Marymount University. This manual contains complete, worked-out solutions to every problem in the text.

For Students

Student Solutions Manual, Warren S. Wright and Carol D. Wright. This manual provides solutions to every third problem in each exercise set.

Dennis G. Zill
Los Angeles

A First Course in
DIFFERENTIAL EQUATIONS

1 INTRODUCTION TO DIFFERENTIAL EQUATIONS

INTRODUCTION The words *differential* and *equations* certainly suggest solving some kind of equation that contains derivatives. So it is; in fact, the preceding sentence tells the complete story about the course that you are about to begin. But before you start solving anything, you must learn some of the basic definitions and terminology of the subject. This is what Section 1.1 is all about. Section 1.2 is intended to be motivational. Why should you, an erstwhile scientist or engineer, study this subject? The answer is simple: Differential equations are the mathematical backbone of many areas of science and engineering. Hence, in Section 1.2 we examine, albeit briefly, how differential equations arise from attempts to formulate, or describe, certain physical systems in terms of mathematics.

1.1 BASIC DEFINITIONS AND TERMINOLOGY

- *Ordinary differential equation* • *Partial differential equation*
- *Order of an equation* • *Linear equation* • *Nonlinear equation* • *Solution*
- *Trivial solution* • *Explicit and implicit solutions* • *n-parameter family of solutions*
- *Particular solution* • *Singular solution* • *General solution*

In calculus you learned that given a function $y = f(x)$, the derivative

$$\frac{dy}{dx} = f'(x)$$

is itself a function of x and is found by some appropriate rule. For example, if $y = e^{x^2}$, then

$$\frac{dy}{dx} = 2xe^{x^2} \quad \text{or} \quad \frac{dy}{dx} = 2xy. \tag{1}$$

The problem that we face in this course is not this: Given a function $y = f(x)$, find its derivative. Rather, our problem is this: If we are given an equation such as $dy/dx = 2xy$, we must somehow find a function $y = f(x)$ that satisfies the equation. In a word, we wish to *solve* differential equations.

DEFINITION 1.1 Differential Equation

An equation containing the derivatives or differentials of one or more dependent variables, with respect to one or more independent variables, is said to be a **differential equation** (DE).

Differential equations are classified according to **type, order,** and **linearity.**

Classification by Type If an equation contains only ordinary derivatives of one or more dependent variables, with respect to a single independent variable, it is then said to be an **ordinary differential equation** (ODE). For example,

$$(y - x)\,dx + 4x\,dy = 0, \quad \frac{du}{dx} - \frac{dv}{dx} = x, \quad \frac{d^2y}{dx^2} - 2\frac{dy}{dx} + 6y = 0$$

are ordinary differential equations. An equation involving the partial derivatives of one or more dependent variables of two or more independent variables is called a **partial differential equation** (PDE). For example,

$$\frac{\partial u}{\partial y} = -\frac{\partial v}{\partial x}, \quad x\frac{\partial u}{\partial x} + y\frac{\partial u}{\partial y} = u, \quad \frac{\partial^2 u}{\partial x^2} = \frac{\partial^2 u}{\partial t^2} - 2\frac{\partial u}{\partial t}$$

are partial differential equations.

Classification by Order The order of the highest-order derivative in a differential equation is called the **order of the equation.** For example,

$$\underset{\text{second-order}}{\overset{\downarrow}{\frac{d^2y}{dx^2}}} + 5\left(\underset{\text{first-order}}{\overset{\downarrow}{\frac{dy}{dx}}}\right)^3 - 4y = e^x$$

is a second-order ordinary differential equation. Since the differential equation $(y - x)\,dx + 4x\,dy = 0$ can be put into the form

$$4x\frac{dy}{dx} + y = x$$

by dividing by the differential dx, it is a first-order ordinary differential equation. The equation

$$a^2\frac{\partial^4 u}{\partial x^4} + \frac{\partial^2 u}{\partial t^2} = 0$$

is a fourth-order partial differential equation.

Although partial differential equations are very important, their study demands a good foundation in the theory of ordinary differential equations. Consequently, in the discussion that follows we shall confine our attention to ordinary differential equations.

A general nth-order, ordinary differential equation is often represented by the symbolism

$$F\left(x, y, \frac{dy}{dx}, \ldots, \frac{d^ny}{dx^n}\right) = 0. \tag{2}$$

The following is a special case of (2).

Classification as Linear or Nonlinear A differential equation is said to be **linear** if it can be written in the form

$$a_n(x)\frac{d^ny}{dx^n} + a_{n-1}(x)\frac{d^{n-1}y}{dx^{n-1}} + \cdots + a_1(x)\frac{dy}{dx} + a_0(x)y = g(x).$$

It should be observed that linear differential equations are characterized by two properties:

(*i*) The dependent variable y and all its derivatives are of the first degree; that is, the power of each term involving y is 1.
(*ii*) Each coefficient depends on only the independent variable x.

An equation that is not linear is said to be **nonlinear.** The equations

$$x\,dy + y\,dx = 0, \quad y'' - 2y' + y = 0, \quad \text{and} \quad x^3\frac{d^3y}{dx^3} + 3x\frac{dy}{dx} + 5y = e^x$$

are linear first-, second-, and third-order ordinary differential equations, respectively. On the other hand,

$$\underset{\substack{\text{coefficient depends}\\ \text{on } y\\ \downarrow}}{y}y'' - 2y' = x \quad \text{and} \quad \frac{d^3y}{dx^3} + \underset{\substack{\text{power not 1}\\ \downarrow}}{y^2} = 0$$

are nonlinear second- and third-order ordinary differential equations, respectively.

Solutions As mentioned before, our goal in this course is to solve, or find solutions of, differential equations.

DEFINITION 1.2 Solution of a Differential Equation

Any function f defined on some interval I, which when substituted into a differential equation reduces the equation to an identity, is said to be a **solution** of the equation on the interval.

In other words, a solution of an ordinary differential equation

$$F(x, y, y', \ldots, y^{(n)}) = 0$$

is a function f that possesses at least n derivatives and *satisfies* the equation; that is,

$$F(x, f(x), f'(x), \ldots, f^{(n)}(x)) = 0$$

for every x in the interval I. The precise form of the interval I is purposely left vague in Definition 1.2. Depending on the context of the discussion, I could represent an open interval (a, b), a closed interval $[a, b]$, an infinite interval $(0, \infty)$, and so on.

EXAMPLE 1 Verification of a Solution

Verify that $y = x^4/16$ is a solution of the nonlinear equation

$$\frac{dy}{dx} = xy^{1/2}$$

on the interval $(-\infty, \infty)$.

Solution One way of verifying that the given function is a solution is to write the differential equation as $dy/dx - xy^{1/2} = 0$ and then see, after substituting, whether the sum $dy/dx - xy^{1/2}$ is zero for every x in the interval. Using

$$\frac{dy}{dx} = 4\frac{x^3}{16} = \frac{x^3}{4} \quad \text{and} \quad y^{1/2} = \left(\frac{x^4}{16}\right)^{1/2} = \frac{x^2}{4},$$

we see that

$$\frac{dy}{dx} - xy^{1/2} = \frac{x^3}{4} - x\left(\frac{x^4}{16}\right)^{1/2} = \frac{x^3}{4} - \frac{x^3}{4} = 0$$

for every real number. ■

EXAMPLE 2 Verification of a Solution

The function $y = xe^x$ is a solution of the linear equation

$$y'' - 2y' + y = 0$$

on $(-\infty, \infty)$. To see this, we compute

$$y' = xe^x + e^x \quad \text{and} \quad y'' = xe^x + 2e^x.$$

Observe

$$y'' - 2y' + y = (xe^x + 2e^x) - 2(xe^x + e^x) + xe^x = 0$$

for every real number. ■

Notice that in Examples 1 and 2 the constant function $y = 0$, for $-\infty < x < \infty$, also satisfies the given differential equation. A solution of a differential equation that is identically zero on an interval I is often referred to as a **trivial solution.**

Not every differential equation that we write necessarily has a solution.

EXAMPLE 3 Some Special DEs

(a) The first-order differential equations

$$\left(\frac{dy}{dx}\right)^2 + 1 = 0 \quad \text{and} \quad (y')^2 + y^2 + 4 = 0$$

possess no real solutions. Why?

(b) The second-order equation $(y'')^2 + 10y^4 = 0$ possesses only one real solution. What is it? ■

Explicit and Implicit Solutions You should be familiar with the notions of explicit and implicit functions from your study of calculus. Similarly, solutions of differential equations are distinguished as either explicit solutions or implicit solutions. A solution of an ordinary differential equation (2) that can be written in the form $y = f(x)$ is said to be an **explicit solution.** We saw in our initial discussion that $y = e^{x^2}$ is an explicit solution of $dy/dx = 2xy$. In Examples 1 and 2, $y = x^4/16$ and $y = xe^x$ are explicit solutions of $dy/dx = xy^{1/2}$ and $y'' - 2y' + y = 0$, respectively. A relation $G(x, y) = 0$ is said to be an **implicit solution** of an ordinary differential equation (2) on an interval I provided it defines one or more explicit solutions on I.

EXAMPLE 4 **An Implicit Solution**

For $-2 < x < 2$ the relation $x^2 + y^2 - 4 = 0$ is an implicit solution of the differential equation

$$\frac{dy}{dx} = -\frac{x}{y}.$$

By implicit differentiation it follows that

$$\frac{d}{dx}(x^2) + \frac{d}{dx}(y^2) - \frac{d}{dx}(4) = 0$$

$$2x + 2y\frac{dy}{dx} = 0 \quad \text{or} \quad \frac{dy}{dx} = -\frac{x}{y}.$$

The relation $x^2 + y^2 - 4 = 0$ in Example 4 defines two explicit differentiable functions: $y = \sqrt{4 - x^2}$ and $y = -\sqrt{4 - x^2}$ on the interval $(-2, 2)$. Also note that any relation of the form $x^2 + y^2 - c = 0$ *formally* satisfies $dy/dx = -x/y$ for any constant c. However, it is naturally understood that the relation should always make sense in the real number system; thus we cannot say that $x^2 + y^2 + 1 = 0$ determines a solution of the differential equation.

Since the distinction between an explicit and an implicit solution should be intuitively clear, we shall not belabor the issue by always saying "here is an explicit (implicit) solution."

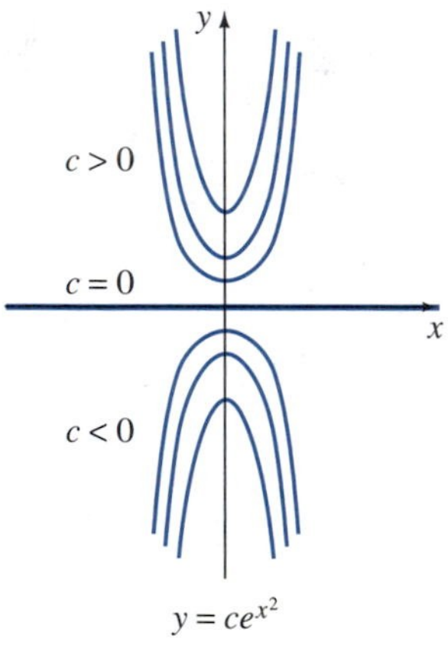

Figure 1.1

Number of Solutions You should become accustomed to the fact that a given differential equation usually possesses an *infinite* number of solutions. By direct substitution, we can prove that any curve—that is, function—in the one-parameter family $y = ce^{x^2}$, where c is any arbitrary constant, also satisfies (1). As indicated in Figure 1.1, the trivial solution is a member of this family of solutions corresponding to $c = 0$. In Example 2, tracing back through the work reveals that $y = cxe^x$ is a family of solutions of the given differential equation.

EXAMPLE 5 **An Infinite Number of Solutions**

For any value of c the function $y = c/x + 1$ is a solution of the first-order differential equation

$$x\frac{dy}{dx} + y = 1$$

on the interval $(0, \infty)$. We have

$$\frac{dy}{dx} = c\frac{d}{dx}(x^{-1}) + \frac{d}{dx}(1) = -cx^{-2} = -\frac{c}{x^2}$$

so

$$x\frac{dy}{dx} + y = x\left(-\frac{c}{x^2}\right) + \left(\frac{c}{x} + 1\right) = 1.$$

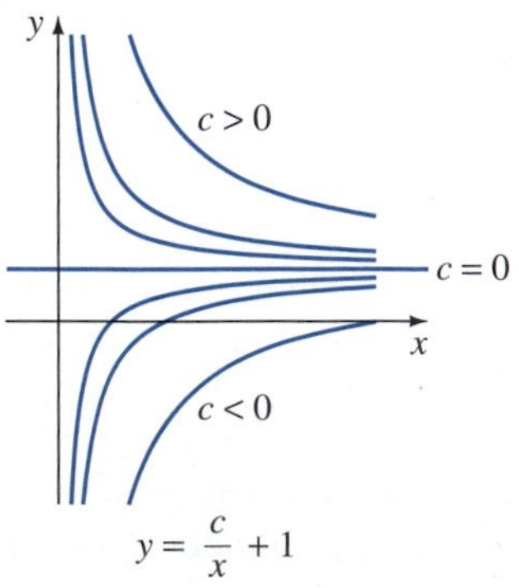

Figure 1.2

By choosing c to be any real number, we can generate an infinite number of solutions. In particular, for $c = 0$ we obtain a constant solution $y = 1$. See Figure 1.2.

In Example 5, $y = c/x + 1$ is a solution of the differential equation on any interval not containing the origin. The function is not differentiable at $x = 0$.

Under some circumstances when we add two solutions of a differential equation, we get another solution.

EXAMPLE 6 An Infinite Number of Solutions

(a) The functions $y = c_1 \cos 4x$ and $y = c_2 \sin 4x$, where c_1 and c_2 are arbitrary constants, are solutions of the differential equation

$$y'' + 16y = 0.$$

For $y = c_1 \cos 4x$, the first and second derivatives are

$$y' = -4c_1 \sin 4x \quad \text{and} \quad y'' = -16c_1 \cos 4x,$$

and so

$$y'' + 16y = -16c_1 \cos 4x + 16(c_1 \cos 4x) = 0.$$

Similarly, for $y = c_2 \sin 4x$,

$$y'' + 16y = -16c_2 \sin 4x + 16(c_2 \sin 4x) = 0.$$

(b) The sum of the solutions in part (a), $y = c_1 \cos 4x + c_2 \sin 4x$, can also be shown to be a solution of $y'' + 16y = 0$. ■

EXAMPLE 7 An Infinite Number of Solutions

You should be able to show that

$$y = e^x, \quad y = e^{-x}, \quad y = c_1e^x, \quad y = c_2e^{-x}, \quad \text{and} \quad y = c_1e^x + c_2e^{-x}$$

are all solutions of the linear second-order differential equation

$$y'' - y = 0.$$

Note that $y = c_1e^x$ is a solution for any choice of c_1, but $y = e^x + c_1$, $c_1 \neq 0$ does *not* satisfy the equation since, for this latter family of functions, we would get $y'' - y = -c_1$. ■

The next example shows that a solution of a differential equation can be a piecewise-defined function.

EXAMPLE 8 A Piecewise-Defined Solution

Any function in the one-parameter family $y = cx^4$ is a solution of the differential equation

$$xy' - 4y = 0.$$

We have $xy' - 4y = x(4cx^3) - 4cx^4 = 0$. The piecewise-defined function

$$y = \begin{cases} -x^4, & x < 0 \\ x^4, & x \geq 0 \end{cases}$$

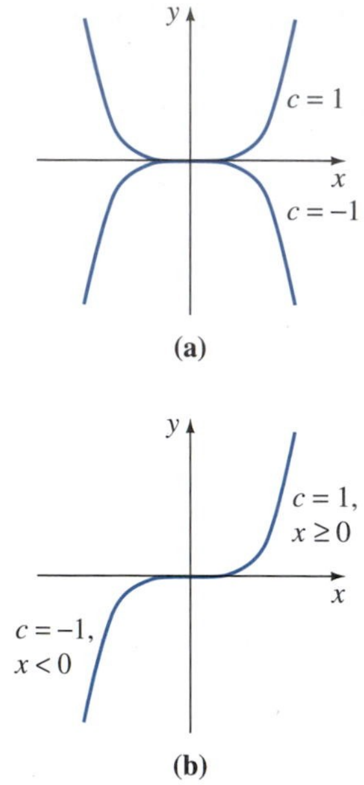

Figure 1.3

is also a solution. Observe that this function cannot be obtained from $y = cx^4$ by a single selection of the parameter c. See Figure 1.3(b).

Further Terminology The study of differential equations is similar to integral calculus. When evaluating an antiderivative or indefinite integral, we utilize a single constant of integration. In like manner, when solving a first-order differential equation $F(x, y, y') = 0$, we usually obtain a family of curves or functions $G(x, y, c) = 0$ containing one arbitrary parameter such that each member of the family is a solution of the differential equation. In fact, when solving an nth-order equation $F(x, y, y', \ldots, y^{(n)}) = 0$, where $y^{(n)}$ means $d^n y/dx^n$, we expect an ***n*-parameter family of solutions** $G(x, y, c_1, \ldots, c_n) = 0$.

A solution of a differential equation that is free of arbitrary parameters is called a **particular solution.** One way of obtaining a particular solution is to choose specific values of the parameter(s) in a family of solutions. For example, it is readily seen that $y = ce^x$ is a one-parameter family of solutions of the simple first-order equation $y' = y$. For $c = 0, -2$, and 5, we get the particular solutions $y = 0$, $y = -2e^x$, and $y = 5e^x$, respectively.

Sometimes a differential equation possesses a solution that cannot be obtained by specializing the parameters in a family of solutions. Such a solution is called a **singular solution.**

EXAMPLE 9 A One-Parameter Family of Solutions

In Section 2.2 we shall prove that a one-parameter family of solutions of $y' = xy^{1/2}$ is given by $y = (\frac{1}{4}x^2 + c)^2$. When $c = 0$, the resulting particular solution is $y = \frac{1}{16}x^4$. In this case the trivial solution $y = 0$ is a singular solution of the equation, since it cannot be obtained from the family for any choice of the parameter c.

If *every* solution of $F(x, y, y', \ldots, y^{(n)}) = 0$ on an interval I can be obtained from $G(x, y, c_1, \ldots, c_n) = 0$ by appropriate choices of the c_i, $i = 1, 2, \ldots, n$, we then say that the n-parameter family is the **general,** or **complete,** solution of the differential equation.

Remarks There are two schools of thought concerning the concept of a "general solution." An alternative viewpoint holds that a general solution of an nth-order differential equation is a family of solutions containing n essential* parameters. Period! In other words, the family is not required to contain all solutions of the differential equation on some interval. The difference in these opinions is really a distinction between the solutions to linear and nonlinear equations. In solving linear differential equations we

* We won't try to define this concept. But roughly it means: Don't play games with the constants. Certainly $y = x + c_1 + c_2$ represents a family of solutions of $y' = 1$. If we rename $c_1 + c_2$ as c, the family has *essentially* one constant: $y = x + c$. You should verify that $y = c_1 + \ln c_2 x$ is a solution of $x^2y'' + xy' = 0$ on the interval $(0, \infty)$ for any choice of c_1 and $c_2 > 0$. Are c_1 and c_2 essential parameters?

shall impose relatively simple restrictions on the coefficients; with these restrictions one can always be assured not only that a solution does exist on an interval but also that a family of solutions indeed yields all possible solutions.

Another fact deserves mention at this time. Nonlinear equations, with the exception of some first-order equations, are usually difficult or *impossible* to solve in terms of familiar elementary functions such as algebraic functions, exponential and logarithmic functions, and trigonometric and inverse trigonometric functions. Furthermore, if we happen to have a family of solutions for a nonlinear equation, it is not obvious when this family constitutes a general solution. On a practical level, then, the designation "general solution" is applied only to linear differential equations.

EXERCISES 1.1

Answers to odd-numbered problems begin on page A-1.

In Problems 1–10 state whether the given differential equations are linear or nonlinear. Give the order of each equation.

1. $(1 - x)y'' - 4xy' + 5y = \cos x$

2. $x\dfrac{d^3y}{dx^3} - 2\left(\dfrac{dy}{dx}\right)^4 + y = 0$

3. $yy' + 2y = 1 + x^2$

4. $x^2\,dy + (y - xy - xe^x)\,dx = 0$

5. $x^3y^{(4)} - x^2y'' + 4xy' - 3y = 0$

6. $\dfrac{d^2y}{dx^2} + 9y = \sin y$

7. $\dfrac{dy}{dx} = \sqrt{1 + \left(\dfrac{d^2y}{dx^2}\right)^2}$

8. $\dfrac{d^2r}{dt^2} = -\dfrac{k}{r^2}$

9. $(\sin x)y''' - (\cos x)y' = 2$

10. $(1 - y^2)\,dx + x\,dy = 0$

In Problems 11–40 verify that the indicated function is a solution of the given differential equation. Where appropriate, c_1 and c_2 denote constants.

11. $2y' + y = 0; \quad y = e^{-x/2}$

12. $y' + 4y = 32; \quad y = 8$

13. $\dfrac{dy}{dx} - 2y = e^{3x}; \quad y = e^{3x} + 10e^{2x}$

14. $\dfrac{dy}{dt} + 20y = 24; \quad y = \frac{6}{5} - \frac{6}{5}e^{-20t}$

15. $y' = 25 + y^2; \quad y = 5\tan 5x$

16. $\dfrac{dy}{dx} = \sqrt{\dfrac{y}{x}}; \quad y = (\sqrt{x} + c_1)^2, x > 0, c_1 > 0$

17. $y' + y = \sin x; \quad y = \frac{1}{2}\sin x - \frac{1}{2}\cos x + 10e^{-x}$

18. $2xy\,dx + (x^2 + 2y)\,dy = 0; \quad x^2y + y^2 = c_1$

19. $x^2\,dy + 2xy\,dx = 0; \quad y = -\dfrac{1}{x^2}$

20. $(y')^3 + xy' = y; \quad y = x + 1$

21. $y = 2xy' + y(y')^2; \quad y^2 = c_1(x + \frac{1}{4}c_1)$

22. $y' = 2\sqrt{|y|}; \quad y = x|x|$

23. $y' - \dfrac{1}{x}y = 1; \quad y = x \ln x, x > 0$

24. $\dfrac{dP}{dt} = P(a - bP); \quad P = \dfrac{ac_1 e^{at}}{1 + bc_1 e^{at}}$

25. $\dfrac{dX}{dt} = (2 - X)(1 - X); \quad \ln \dfrac{2 - X}{1 - X} = t$

26. $y' + 2xy = 1; \quad y = e^{-x^2}\int_0^x e^{t^2}\,dt + c_1 e^{-x^2}$

27. $(x^2 + y^2)\,dx + (x^2 - xy)\,dy = 0; \quad c_1(x + y)^2 = xe^{y/x}$

28. $y'' + y' - 12y = 0; \quad y = c_1 e^{3x} + c_2 e^{-4x}$

29. $y'' - 6y' + 13y = 0; \quad y = e^{3x} \cos 2x$

30. $\dfrac{d^2y}{dx^2} - 4\dfrac{dy}{dx} + 4y = 0; \quad y = e^{2x} + xe^{2x}$

31. $y'' = y; \quad y = \cosh x + \sinh x$

32. $y'' + 25y = 0; \quad y = c_1 \cos 5x$

33. $y'' + (y')^2 = 0; \quad y = \ln|x + c_1| + c_2$

34. $y'' + y = \tan x; \quad y = -\cos x \ln(\sec x + \tan x)$

35. $x\dfrac{d^2y}{dx^2} + 2\dfrac{dy}{dx} = 0; \quad y = c_1 + c_2 x^{-1}$

36. $x^2y'' - xy' + 2y = 0; \quad y = x\cos(\ln x), x > 0$

37. $x^2y'' - 3xy' + 4y = 0; \quad y = x^2 + x^2 \ln x, x > 0$

38. $y''' - y'' + 9y' - 9y = 0; \quad y = c_1 \sin 3x + c_2 \cos 3x + 4e^x$

39. $y''' - 3y'' + 3y' - y = 0; \quad y = x^2 e^x$

40. $x^3\dfrac{d^3y}{dx^3} + 2x^2\dfrac{d^2y}{dx^2} - x\dfrac{dy}{dx} + y = 12x^2; \quad y = c_1x + c_2 x \ln x + 4x^2, x > 0$

In Problems 41 and 42 verify that the indicated piecewise-defined function is a solution of the given differential equation.

41. $xy' - 2y = 0; \quad y = \begin{cases} -x^2, & x < 0 \\ x^2, & x \geq 0 \end{cases}$

42. $(y')^2 = 9xy; \quad y = \begin{cases} 0, & x < 0 \\ x^3, & x \geq 0 \end{cases}$

43. Verify that a one-parameter family of solutions for

$$y = xy' + (y')^2 \quad \text{is} \quad y = cx + c^2.$$

Determine a value of k such that $y = kx^2$ is a singular solution of the differential equation.

44. Verify that a one-parameter family of solutions for

$$y = xy' + \sqrt{1 + (y')^2} \quad \text{is} \quad y = cx + \sqrt{1 + c^2}.$$

Show that the relation $x^2 + y^2 = 1$ defines a singular solution of the equation on the interval $(-1, 1)$.

45. A one-parameter family of solutions for

$$y' = y^2 - 1 \quad \text{is} \quad y = \frac{1 + ce^{2x}}{1 - ce^{2x}}.$$

By inspection,* determine a singular solution of the differential equation.

46. On page 6 we saw that $y = \sqrt{4 - x^2}$ and $y = -\sqrt{4 - x^2}$ are solutions of $dy/dx = -x/y$ on the interval $(-2, 2)$. Explain why

$$y = \begin{cases} \sqrt{4 - x^2}, & -2 < x < 0 \\ -\sqrt{4 - x^2}, & 0 \le x < 2 \end{cases}$$

is not a solution of the differential equation on the interval.

In Problems 47 and 48 find values of m so that $y = e^{mx}$ is a solution of each differential equation.

47. $y'' - 5y' + 6y = 0$

48. $y'' + 10y' + 25y = 0$

In Problems 49 and 50 find values of m so that $y = x^m$ is a solution of each differential equation.

49. $x^2y'' - y = 0$

50. $x^2y'' + 6xy' + 4y = 0$

51. Show that $y_1 = x^2$ and $y_2 = x^3$ are both solutions of

$$x^2y'' - 4xy' + 6y = 0.$$

Are the constant multiples c_1y_1 and c_2y_2, with c_1 and c_2 arbitrary, also solutions? Is the sum $y_1 + y_2$ a solution?

52. Show that $y_1 = 2x + 2$ and $y_2 = -x^2/2$ are both solutions of

$$y = xy' + \tfrac{1}{2}(y')^2.$$

Are the constant multiples c_1y_1 and c_2y_2, with c_1 and c_2 arbitrary, also solutions? Is the sum $y_1 + y_2$ a solution?

53. By inspection determine, if possible, a real solution of the given differential equation.

(a) $\left|\dfrac{dy}{dx}\right| + |y| = 0$ **(b)** $\left|\dfrac{dy}{dx}\right| + |y| + 1 = 0$ **(c)** $\left|\dfrac{dy}{dx}\right| + |y| = 1$

1.2 SOME MATHEMATICAL MODELS

• *Mathematical model* • *State of the system*

In science, engineering, economics, and even psychology, we often wish to describe or **model** the behavior of some system or phenomenon in mathematical terms. This descriptions starts with

(*i*) identifying the variables that are responsible for changing the system and
(*ii*) a set of reasonable assumptions about the system.

---

* Translated, this means take a good guess and see if it works.

These assumptions also include any empirical laws that are applicable to the system. The mathematical construct of all these assumptions, or the **mathematical model** of the system, is in many instances a differential equation or a system of differential equations. We expect a reasonable mathematical model of a system to have a solution that is consistent with the known behavior of the system.

A mathematical model of a *physical* system often involves the variable time. The solution of the model then gives the **state of the system;** in other words, for appropriate values of time t, the values of the dependent variable (or variables) describe the system in the past, present, and future.

Falling Body The mathematical description of a body falling vertically under the influence of gravity leads to a simple second-order differential equation. The solution of this equation gives the position of the body relative to the ground.

EXAMPLE 1 Freely Falling Body

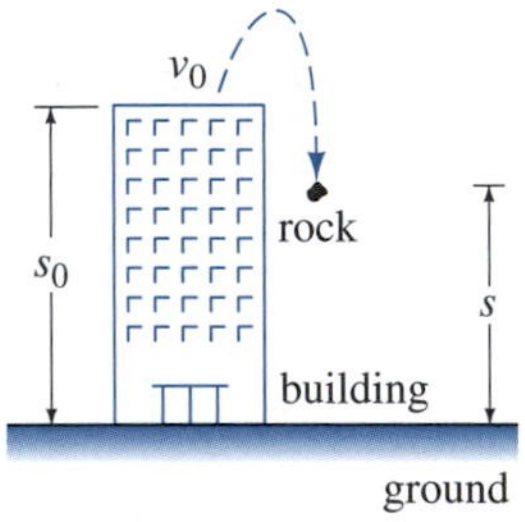

Figure 1.4

It is well known that free-falling objects close to the surface of the earth accelerate at a constant rate g. Acceleration is the derivative of the velocity, and this, in turn, is the derivative of the distance s. Suppose a rock is tossed upward from the roof of a building as illustrated in Figure 1.4. If we assume that the upward direction is positive, then the mathematical statement

$$\frac{d^2s}{dt^2} = -g$$

is the differential equation that governs the vertical distance that the body travels. The minus sign is used because the weight of the body is a force directed opposite to the positive direction.

If we suppose further that the height of the building is s_0 and the initial velocity of the rock is v_0, then we must find a solution of the differential equation

$$\frac{d^2s}{dt^2} = -g, \quad 0 < t < t_1$$

that also satisfies the side conditions $s(0) = s_0$ and $s'(0) = v_0$. Here $t = 0$ is the initial time the rock leaves the roof of the building, and t_1 is the elapsed time when the rock hits the ground. Since the rock is thrown upward in the positive direction, it is naturally assumed that $v_0 > 0$.

Note that this formulation of the problem ignores other forces such as air resistance acting on the body. ■

Spring-Mass System When Newton's second law of motion is combined with Hooke's law, we can derive a differential equation governing the motion of a mass attached to a spring.

EXAMPLE 2 Vibrations of a Mass on a Spring

To find the vertical displacement $x(t)$ of a mass attached to a spring, we use two different empirical laws: Newton's second law of motion and Hooke's law. The

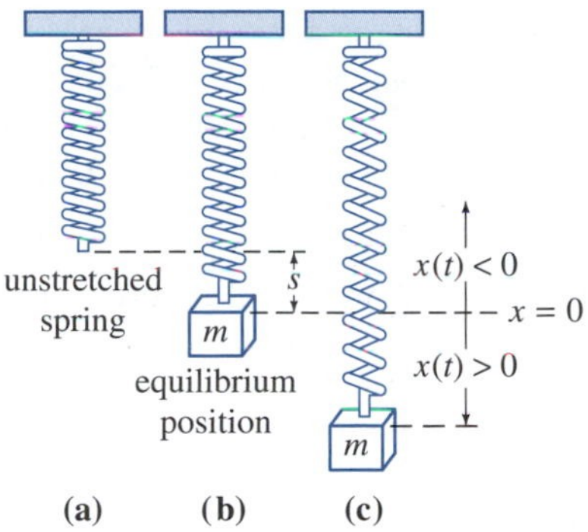

Figure 1.5

former law states that the net force acting on the system in motion is $F = ma$, where m is the mass and a is acceleration. Hooke's law states that the restoring force of a stretched spring is proportional to the elongation $s + x$; that is, the restoring force is $k(s + x)$, where $k > 0$ is a constant. As shown in Figure 1.5(b), s is the elongation of the spring after the mass has been attached and the system hangs at rest in the *equilibrium position.* When the system is in motion, the variable x represents a directed distance of the mass beyond the equilibrium position. In Chapter 5 we shall prove that when the system is in motion, the *net force* acting on the mass is simply $F = -kx$. Thus, in the absence of damping and other external forces that might be impressed on the system, the differential equation of the vertical motion through the center of gravity of the mass can be obtained by equating

$$m\frac{d^2x}{dt^2} = -kx.$$

Here the minus sign means that the restoring force of the spring acts opposite to the direction of motion—that is, toward the equilibrium position. In practice this second-order differential equation is often written as

$$\frac{d^2x}{dt^2} + \omega^2 x = 0, \tag{1}$$

where $\omega^2 = k/m$.

Units A word is in order regarding the system of units that is used in describing dynamic problems such as those illustrated in the last two examples. Three commonly used systems of units are summarized in Table 1.1. In each system the basic unit of time is the second.

Table 1.1 Units of Measurement

Quantity	Engineering system*	SI system†	cgs
Force	pound (lb)	newton (N)	dyne
Mass	slug	kilogram (kg)	gram (g)
Distance	foot (ft)	meter (m)	centimeter (cm)
Acceleration of gravity g (approximate)	32 ft/s^2	9.8 m/s^2	980 cm/s^2

* Also known as the English gravitational system or British engineering system.
† International system of units. SI is the abbreviation for *Systeme International.*

The gravitational *force* exerted by the earth on a body of mass m is called its *weight* W. In the absence of air resistance, the only force acting on a freely falling body is its weight. Hence, from Newton's second law of motion, it follows that mass m and weight W are related by

$$W = mg.$$

For example, in the engineering system a mass of $\frac{1}{4}$ slug corresponds to an 8-lb weight. Since $m = W/g$, a 64-lb weight corresponds to a mass of $\frac{64}{32} = 2$ slugs. In the cgs system a weight of 2450 dynes has a mass of $\frac{2450}{980} = 2.5$ grams. In the SI system a weight of 50 newtons has a mass of $50/9.8 = 5.1$ kilograms. We note that

$$1 \text{ newton} = 10^5 \text{ dynes} = 0.2247 \text{ pound}.$$

In the next example we derive the differential equation that describes the motion of a *simple pendulum.*

Simple Pendulum Any object that swings back and forth is called a **physical pendulum.** The simple pendulum is a special case of the physical pendulum and consists of a rod to which a mass is attached at one end. In describing the motion of a **simple pendulum,** we make the simplifying assumptions that the mass of the rod is negligible and that no external damping forces act on the system (such as air resistance).

EXAMPLE 3 Simple Pendulum

A mass m having weight W is suspended from the end of a rod of constant length l. For motion in a vertical plane, we would like to determine the displacement angle θ, measured from the vertical, as a function of time t (we consider $\theta > 0$ to the right of OP and $\theta < 0$ to left of OP). Recall that an arc s of a circle of radius l is related to the central angle θ through the formula $s = l\theta$. Hence the angular acceleration is

$$a = \frac{d^2s}{dt^2} = l\frac{d^2\theta}{dt^2}.$$

From Newton's second law we then have

$$F = ma = ml\frac{d^2\theta}{dt^2}.$$

Figure 1.6

From Figure 1.6 we see that the tangential component of the force due to the weight W is $mg \sin \theta$. We equate the two different formulations of the tangential force to obtain

$$ml\frac{d^2\theta}{dt^2} = -mg \sin \theta \quad \text{or} \quad \frac{d^2\theta}{dt^2} + \frac{g}{l}\sin \theta = 0. \tag{2}$$

Because of the presence of $\sin \theta$, the differential equation (2) is nonlinear. It is also known that this equation cannot be solved in terms of elementary functions. Hence a further simplifying assumption is made. If the angular displacements θ are not too large, we can use the approximation $\sin \theta \approx \theta$,* so (2) can be replaced with the linear second-order differential equation

$$\frac{d^2\theta}{dt^2} + \frac{g}{l}\theta = 0. \tag{3}$$

* For small values of θ (in radians), powers θ^3 and higher can be ignored in the Maclaurin series $\sin \theta = \theta - \theta^3/3! + \cdots$, and so we get $\sin \theta \approx \theta$. Use a calculator and compare the values of sin(0.05) and sin(0.005) with 0.05 and 0.005.

If we set $\omega^2 = g/l$, (3) has the exact same structure as the differential equation (1) governing the free vibrations of a weight on a spring. The fact that one basic differential equation can describe many diverse physical or even social/economic phenomena is a common occurrence in the study of applicable mathematics. ■

Rotating String We encounter equation (1) again in the analysis of a rotating string.

EXAMPLE 4 Shape of Rotating String

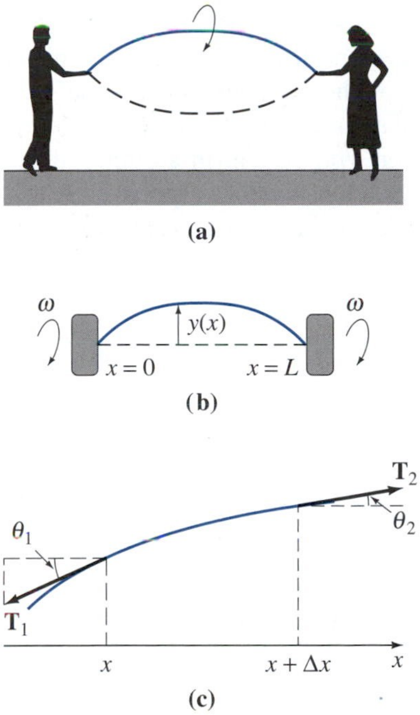

Figure 1.7

Suppose a string of length L with constant linear density ρ (mass per unit length) is stretched along the x-axis and fixed at $x = 0$ and $x = L$. Suppose the string is then rotated about that axis at a constant angular speed ω. This is analogous to two persons holding a jump rope and then twirling it in a synchronous manner. See Figure 1.7(a). We want to find the differential equation that defines the shape $y(x)$ of the string, or the **deflection curve** away from its initial position. See Figure 1.7(b). To do this, consider a portion of the string on the interval $[x, x + \Delta x]$, where Δx is small. If the magnitude T of the tension $\mathbf{T}$ acting tangential to the string is constant along the string, then the desired differential equation can be obtained by equating two different formulations of the net force acting on the string on the interval $[x, x + \Delta x]$. First, we see from Figure 1.7(c) that the net vertical force is

$$F = T\sin\theta_2 - T\sin\theta_1. \tag{4}$$

When the angles θ_1 and θ_2 (measured in radians) are small, we have

$$\sin\theta_2 \approx \tan\theta_2 \approx y'(x + \Delta x) \quad \text{and} \quad \sin\theta_1 \approx \tan\theta_1 \approx y'(x),$$

and so (4) becomes

$$F = T[y'(x + \Delta x) - y'(x)]. \tag{5}$$

Second, the net force is also given by Newton's second law $F = ma$. Here the mass of string on the interval is $m = \rho\,\Delta x$; the centripetal acceleration of a point rotating with angular speed ω in a circle of radius r is $a = r\omega^2$. With Δx small we take $r = y$. Thus another formulation of the net force is

$$F \approx -(\rho\,\Delta x)y\omega^2, \tag{6}$$

where the minus sign comes from the fact that the acceleration points in the direction opposite to the positive y direction. Now by equating (5) and (6), we have

$$T[y'(x + \Delta x) - y'(x)] \approx -(\rho\,\Delta x)y\omega^2 \quad \text{or} \quad T\frac{y'(x + \Delta x) - y'(x)}{\Delta x} \approx -\rho\omega^2 y. \tag{7}$$

For Δx close to zero, $[y'(x + \Delta x) - y'(x)]/\Delta x \approx d^2y/dx^2$, so the last expression in (7) gives

$$T\frac{d^2y}{dx^2} = -\rho\omega^2 y \quad \text{or} \quad T\frac{d^2y}{dx^2} + \rho\omega^2 y = 0. \tag{8}$$

Since the string is anchored at $x = 0$ and $x = L$, we expect that the solution $y(x)$

of the last equation would also satisfy the side conditions $y(0) = 0$ and $y(L) = 0$. ■

Dividing the last equation in (8) by T gives

$$\frac{d^2y}{dx^2} + \frac{\rho\omega^2}{T}y = 0,$$

which is analogous to (1) and (3). If the magnitude T of the tension is not constant throughout the interval $[0, L]$, then it can be shown that the differential equation for the deflection curve of the string is

$$\frac{d}{dx}\left[T(x)\frac{dy}{dx}\right] + \rho\omega^2 y = 0. \qquad \textbf{(9)}$$

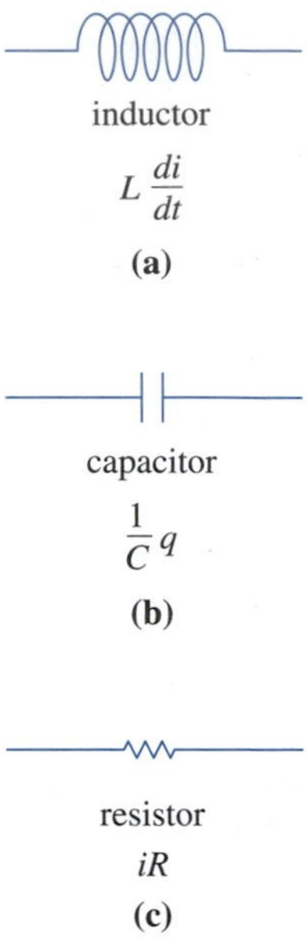

Figure 1.8

Series Circuits According to Kirchhoff's second law, the impressed voltage $E(t)$ on a closed loop must equal the sum of the voltage drops in the loop. Figure 1.8 shows the symbols and the formulas for the respective voltage drops across an inductor, a capacitor, and a resistor. The current in a circuit after a switch is closed is denoted by $i(t)$; the charge on a capacitor at time t is denoted by $q(t)$. The letters L, C, and R are constants known as inductance, capacitance, and resistance, respectively.

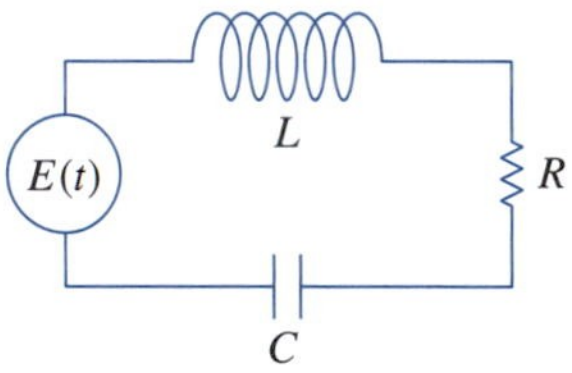

Figure 1.9

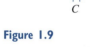

EXAMPLE 5 Charge on a Capacitor

Consider the single-loop series circuit containing an inductor, resistor, and capacitor shown in Figure 1.9. A second-order differential equation for the charge $q(t)$ on the capacitor can be obtained by adding the voltage drops

$$\text{inductor} = L\frac{di}{dt} = L\frac{d^2q}{dt^2} \quad \leftarrow$$

$$\text{resistor} = iR = R\frac{dq}{dt} \quad \leftarrow$$

current i related to charge q by $i = dq/dt$

$$\text{capacitor} = \frac{1}{C}q$$

and equating the sum to the impressed voltage $E(t)$:

$$L\frac{d^2q}{dt^2} + R\frac{dq}{dt} + \frac{1}{C}q = E(t). \qquad \textbf{(10)}$$ ■

Table 1.2 Units of Measurement

Quantity	Unit
Impressed voltage, or emf	volt (V)
Inductance L	henry (H)
Capacitance C	farad (F)
Resistance R	ohm (Ω)
Charge q	coulomb (C)
Current i	ampere (A)

In Example 5 the side conditions $q(0)$ and $q'(0)$ represent the charge on the capacitor and the current in the circuit, respectively, at time $t = 0$. Also, the impressed voltage $E(t)$ is said to be an **electromotive force, or emf.** An emf, as well as a charge on a capacitor, causes the current in a circuit to flow. Table 1.2 shows the basic units of measurement used in circuit analysis.

Newton's Law of Cooling According to Newton's empirical law of cooling, the rate at which a body cools is proportional to the difference between the tem-

ILLUSTRATED APPLICATIONS

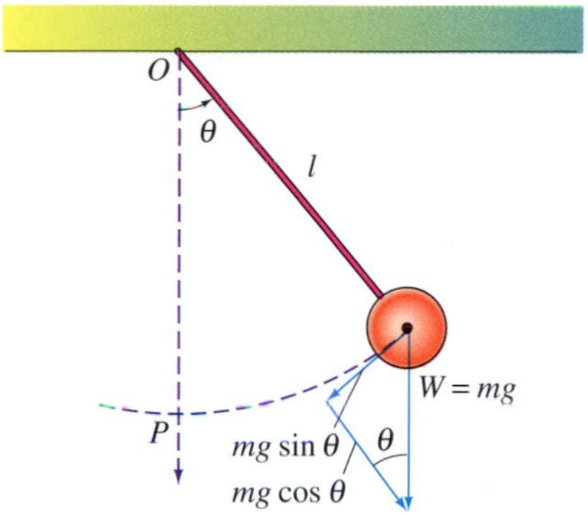

The swinging bob in a grandfather's clock and a child's swing are examples of pendulums. The displacement angle θ of a simple plane pendulum of length l is determined from the nonlinear second-order differential equation $d^2\theta/dt^2 + (g/l)\sin\theta = 0$. When the displacements of the pendulum are not too large, we can use the replacement $\sin\theta \approx \theta$ and thus approximate θ by solving the linear equation $d^2\theta/dt^2 + (g/l)\theta = 0$. See pages 14 and 212.

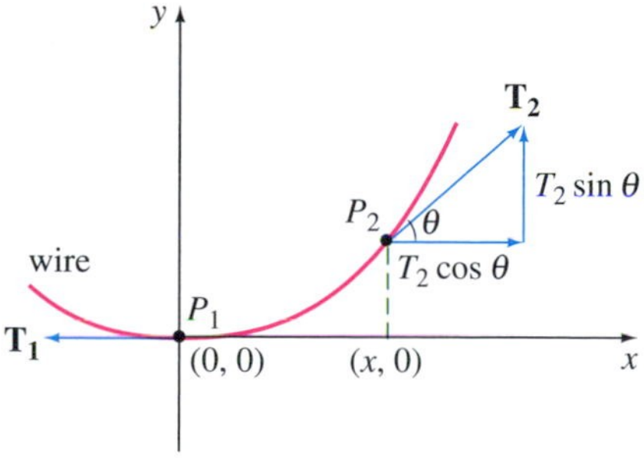

To determine the shape of a wire hanging under its own weight, such as a wire strung between two telephone poles, we must solve the nonlinear differential equation

$$d^2y/dx^2 + (w/T)\sqrt{1 + (dy/dx)^2} = 0.$$

It turns out that the wire has essentially the shape of the graph of a hyperbolic cosine. This graph of a hyperbolic cosine is called a *catenary*. The famous Gateway Arch in St. Louis has the shape of an inverted catenary. See pages 17–18.

Under some circumstances a body moving through the air encounters a resistance that is proportional to its velocity v. In general, air resistance is directly proportional to a positive power of the velocity of the body—the faster the body travels, the greater the resistance. For bodies moving at high speeds, such as projectiles or freely falling skydivers, air resistance is often taken to be proportional to v^2. See Problem 27, page 91 and Problem 8, pages 104–105.

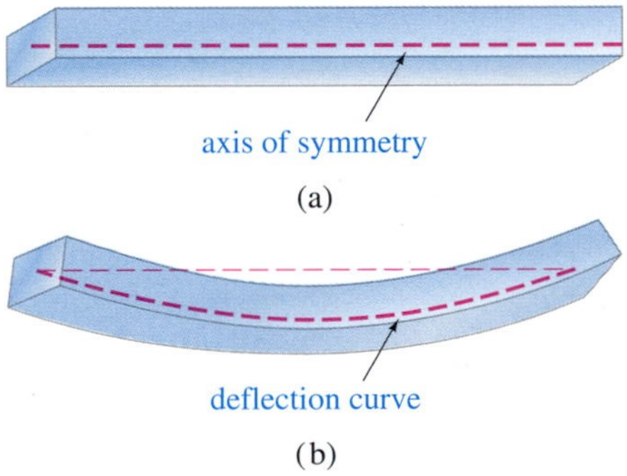

Forces acting on beams and girders cause them to distort or deform. This deformation, or deflection, $y(x)$ is described by the fourth-order differential equation $EIy^{(4)} = w(x)$. A beam that is clamped or embedded at one end and free at the other is called a *cantilever beam*. A diving board, an outstretched arm, and an airplane wing are common examples of such beams; but even flagpoles, skyscrapers, and the Washington Monument act as cantilever beams. See pages 19–20, 323, 327, and 331.

Any phenomenon modeled by the simple differential equation $dx/dt = kx$ grows ($k > 0$) or decays ($k < 0$) exponentially. The growth in the population P of bacteria, insects, or even humans can often be predicted over short time periods by the exponential solution $P(t) = ce^{kt}$. The study of substances decaying through radioactivity led to the discovery of carbon dating, which is a means of dating fossils or even a mummy. See Section 3.2 and the essay at the end of Chapter 3.

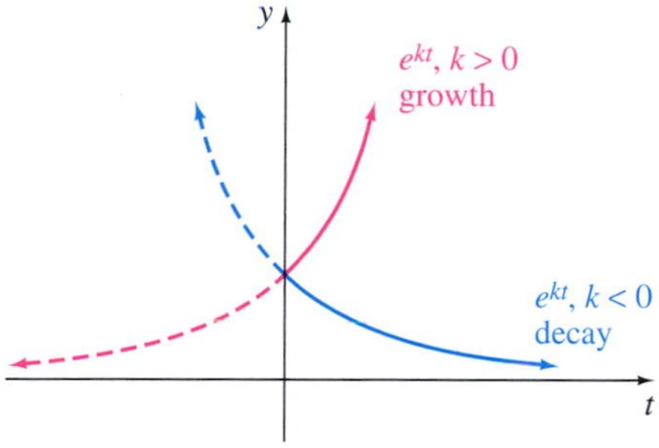

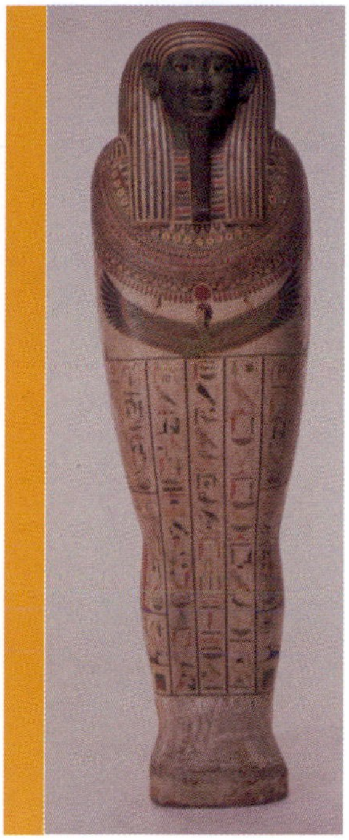

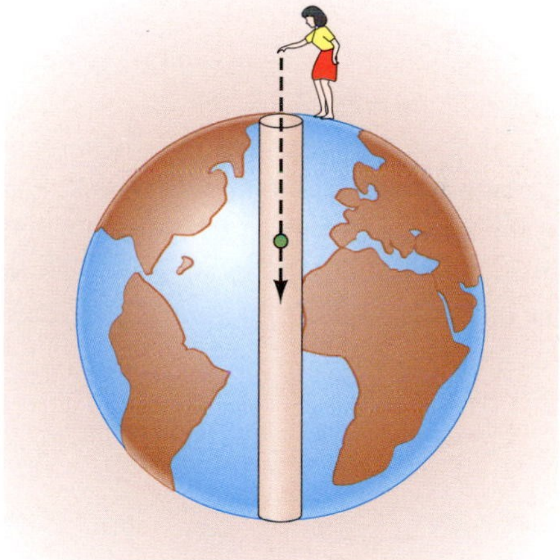

Imagine that a shaft is drilled from one side of the earth to the opposite side through its center, and a heavy object, such as a bowling ball, is dropped into the shaft. What would be the motion of the object? Would it drop straight through? Would it stop at the center? In Problem 20 of Exercise 1.2 you are asked to construct a mathematical model for the motion of the object. You can then solve this differential equation with the techniques discussed in Section 4.2.

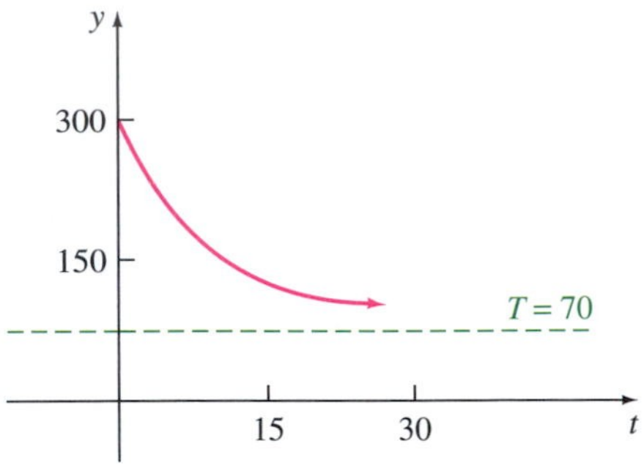

Before taking the first sip of coffee, one usually waits a short time for the liquid to cool. A forgotten cup of coffee is almost undrinkable when it has cooled to room temperature. An empirical law of cooling, attributed to Isaac Newton, asserts that the rate at which the temperature of a body cools is proportional to the difference between the temperature of the body and the temperature of the surrounding medium. The preceding sentence is a verbal description of a differential equation. See pages 16–17 and 85.

Suppose that a cell is suspended in a solution containing a solute, such as potassium, with a constant concentration and that the solute is absorbed into the cell through its permeable membrane. A mathematical model for the concentration $C(t)$ of the solute inside the cell at time t can be found using Fick's principle of passive diffusion. See Problem 30, Exercises 3.2.

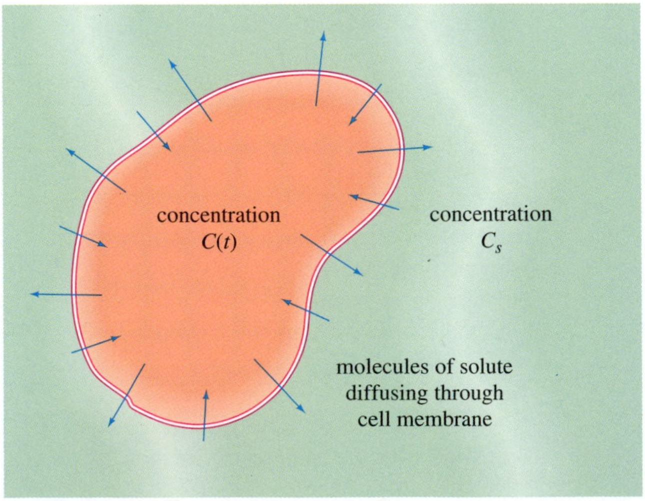

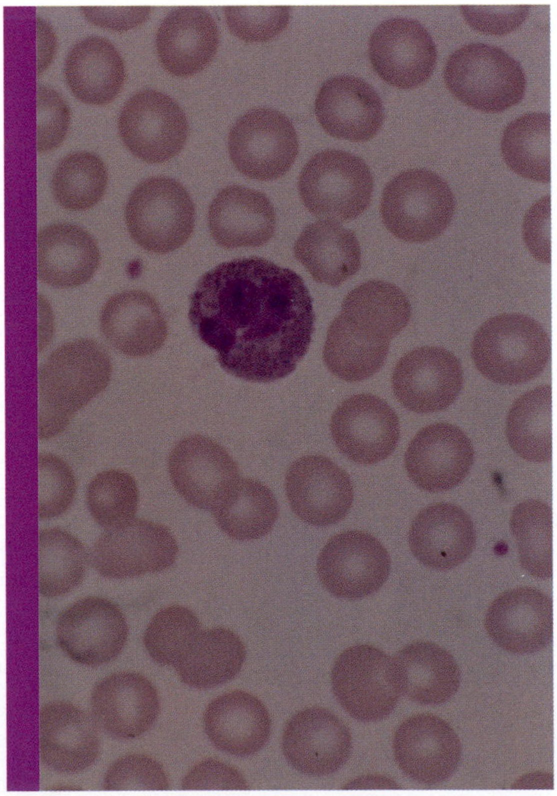

perature of the body and the temperature of the surrounding medium, the so-called ambient temperature.

EXAMPLE 6 Cooling of a Body

Suppose $T(t)$ denotes the temperature of a body at time t and T_m is the constant temperature of the surrounding medium. If dT/dt represents the rate at which a body cools, then Newton's law of cooling translates into the mathematical statement

$$\frac{dT}{dt} \propto T - T_m \quad \text{or} \quad \frac{dT}{dt} = k(T - T_m), \tag{11}$$

where k is a constant of proportionality. Since the body is assumed to be cooling, we must have $T > T_m$ and so it stands to reason that $k < 0$. ■

Suspended Wire Suppose a suspended wire hangs under its own weight. As Figure 1.10(a) shows, a physical model for this could be a long telephone wire strung between two posts. As in Example 4, our goal in the example that follows is to determine the differential equation that describes the shape that the hanging wire assumes.

EXAMPLE 7 Shape of a Hanging Wire

Let us examine only a portion of the wire between the lowest point P_1 and any arbitrary point P_2. See Figure 1.10(b). Three forces are acting on the wire: the weight of the segment P_1P_2 and the tensions $\mathbf{T}_1$ and $\mathbf{T}_2$ in the wire at P_1 and P_2, respectively. If w is the linear density (measured, say, in lb/ft) and s is the length of the segment P_1P_2, its weight is necessarily ws.

Now the tension $\mathbf{T}_2$ resolves into horizontal and vertical components (scalar quantities) $T_2 \cos\theta$ and $T_2 \sin\theta$. Because of equilibrium we can write

$$|\mathbf{T}_1| = T_1 = T_2 \cos\theta \quad \text{and} \quad ws = T_2 \sin\theta.$$

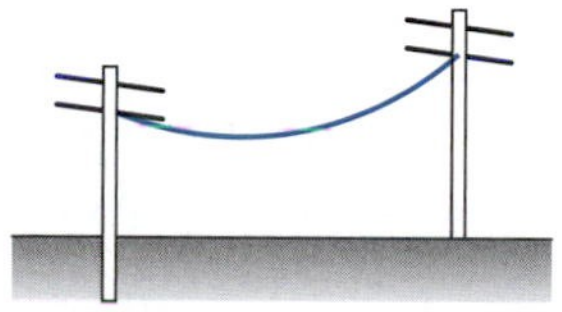

(a)

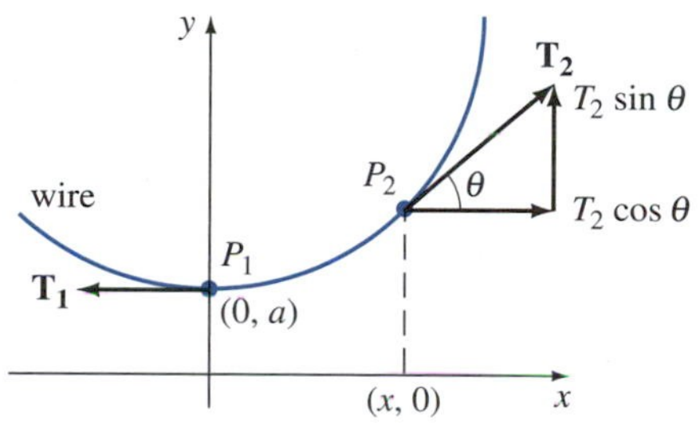

(b)

Figure 1.10

Dividing the last two equations, we find

$$\tan\theta = \frac{ws}{T_1} \quad \text{or} \quad \frac{dy}{dx} = \frac{ws}{T_1}. \tag{12}$$

Now since the length of the arc between points P_1 and P_2 is

$$s = \int_0^x \sqrt{1 + \left(\frac{dy}{dx}\right)^2}\, dx,$$

it follows from one form of the fundamental theorem of calculus that

$$\frac{ds}{dx} = \sqrt{1 + \left(\frac{dy}{dx}\right)^2}. \tag{13}$$

Differentiating (12) with respect to x and using (13) lead to

$$\frac{d^2y}{dx^2} = \frac{w}{T_1}\frac{ds}{dx} \quad \text{or} \quad \frac{d^2y}{dx^2} = \frac{w}{T_1}\sqrt{1 + \left(\frac{dy}{dx}\right)^2}. \tag{14}$$

One might conclude from Figure 1.10 that the shape the hanging wire assumes is parabolic. However, this is not the case; a wire or heavy rope hanging under only its own weight takes the shape of a hyperbolic cosine. See Problem 12, Exercises 3.3. Recall that the graph of the hyperbolic cosine is called a **catenary,** which stems from the Latin word *catena* meaning "chain." The Romans used the catena as a dog leash. Probably the most graphic example of the shape of a catenary is the 630-ft-high Gateway Arch in St. Louis, Missouri.

Discharge Through an Orifice In hydrodynamics, Torricelli's theorem states that the speed v of efflux of water through a sharp-edged orifice at the bottom of a tank filled to a depth h is the same as the speed that a body (in this case a drop of water) would acquire in falling freely from a height h:

$$v = \sqrt{2gh},$$

where g is the acceleration due to gravity. This last expression comes from equating the kinetic energy $\frac{1}{2}mv^2$ with the potential energy mgh and solving for v.

EXAMPLE 8 Depth of Water in a Draining Tank

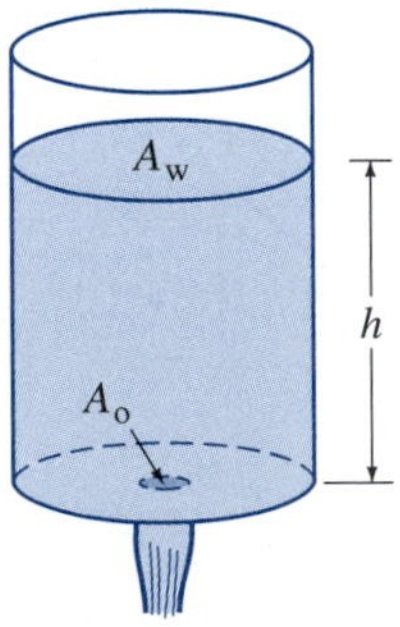

Figure 1.11

Suppose a tank filled with water is allowed to discharge through an orifice under the influence of gravity. We would like to find the depth h of water remaining in the tank at time t.

Consider the tank shown in Figure 1.11. If the area of the orifice is A_o (in ft^2) and the speed of the water leaving the tank is $v = \sqrt{2gh}$ (in ft/s), then the volume of water leaving the tank per second is $A_o\sqrt{2gh}$ (in ft^3/s). Thus, if $V(t)$ denotes the volume of water in the tank at time t, then

$$\frac{dV}{dt} = -A_o\sqrt{2gh}, \tag{15}$$

where the minus sign indicates that V is decreasing. Note here that we are ignoring the possibility of friction at the orifice, which might reduce the rate of flow there.

Now if the tank is such that the volume of water in it at time t can be written as $V(t) = A_w h$, where A_w (in ft^2) is the *constant* area of the upper surface of the water (see Figure 1.11), then $dV/dt = A_w(dh/dt)$. Substituting this last expression into (15) gives us the desired differential equation for the height h of the water:

$$\frac{dh}{dt} = -\frac{A_0}{A_w}\sqrt{2gh}. \tag{16}$$

It is interesting to observe that (16) remains valid even when A_w is not constant. In this case we must express the upper surface area of the water as a function of h: $A_w = A(h)$. See Problem 9 in Exercises 1.2 and Problem 19 in the Chapter 1 Review Exercises.

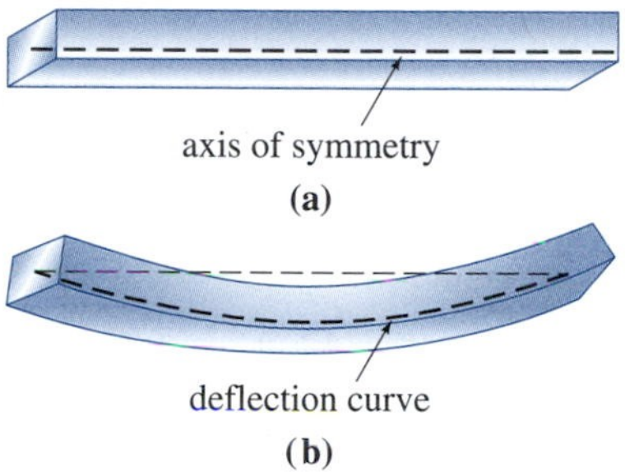

Figure 1.12

Deflection of Beams In engineering an important problem is to determine the static deflection of an elastic beam caused by its weight or by an external load. We assume that the beam is homogeneous and has uniform cross sections along its length. Let L denote the length of the beam. In the absence of any load on the beam (including its weight), a curve joining the centroids of all its cross sections is a straight line called the **axis of symmetry.** See Figure 1.12(a). If a load is applied to the beam in a vertical plane containing the axis of symmetry, then, as shown in Figure 1.12(b), the beam undergoes a distortion and the curve connecting the centroids of all cross sections is called the **deflection curve** or **elastic curve.** In the next example we derive the differential equation of the deflection curve. This derivation uses principles from elasticity and a concept from calculus called curvature.

EXAMPLE 9 Deflection of a Cantilever Beam

For the sake of illustration let us consider a cantilever beam embedded at its left end and free at its right end. As shown in Figure 1.13, we let the embedded end of the beam coincide with $x = 0$ and its free end with $x = L$. The x-axis coincides with the axis of symmetry, and the deflection $y(x)$ is measured from this axis and is considered positive if downward. Now in the theory of elasticity it is shown that the bending moment $M(x)$ at a point x along the beam is related to the load per unit length $w(x)$ by the equation

$$\frac{d^2M}{dx^2} = w(x). \tag{17}$$

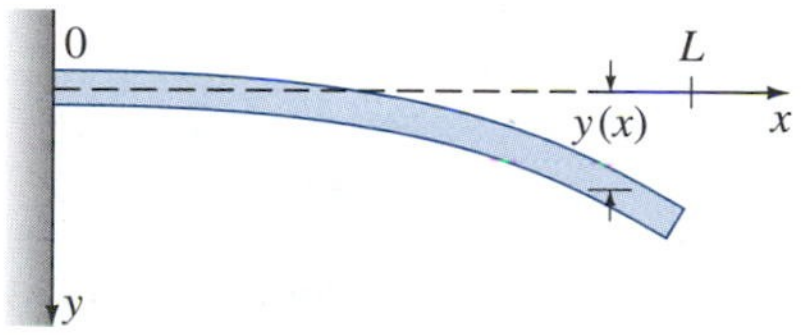

Figure 1.13

In addition, the bending moment $M(x)$ is proportional to the curvature κ of the elastic curve:

$$M(x) = EI\kappa, \tag{18}$$

where E and I are constants; E is Young's modulus of elasticity of the material of the beam and I is the moment of inertia of a cross-section of the beam (about an axis known as the neutral axis). The product EI is called the flexural rigidity of the beam.

Now, from calculus, curvature is given by

$$\kappa = \frac{y''}{[1 + (y')^2]^{3/2}}.$$

When the deflection $y(x)$ is small, the slope $y' \approx 0$ and so $[1 + (y')^2]^{3/2} \approx 1$. If we let $\kappa = y''$, equation (18) becomes $M = EIy''$. The second derivative of this last expression is

$$\frac{d^2M}{dx^2} = EI\frac{d^2}{dx^2}y'' = EI\frac{d^4y}{dx^4}. \tag{19}$$

Using the given result in (17) to replace d^2M/dx^2 in (19), we see that the deflection $y(x)$ satisfies the fourth-order differential equation

$$EI\frac{d^4y}{dx^4} = w(x). \tag{20}$$

As we shall see later on, it is extremely important to note any side conditions that accompany a differential equation in the mathematical description of a physical phenomenon. For the cantilever beam in Example 9, in addition to satisfying (20) we would expect the deflection $y(x)$ to satisfy the following conditions at the ends of the beam:

- $y(0) = 0$ since there is no deflection at the embedded end.
- $y'(0) = 0$ since the deflection curve is tangent to the x-axis at the embedded end.
- $y''(L) = 0$ since the bending moment is zero at a free end.
- $y'''(L) = 0$ since the shear force is zero at a free end.

The function $F(x) = dM/dx = EI\,(d^3y/dx^3)$ is called the shear force.

Population Growth In the next examples we examine some mathematical models of biological growth.

EXAMPLE 10 Population Growth

It seems plausible to expect that the rate at which a population P expands is proportional to the population that is present at time t. Roughly put, the more people there are, the more there are going to be. Thus one model for population growth is given by the differential equation

$$\frac{dP}{dt} = kP, \tag{21}$$

where k is a constant of proportionality. Since we also expect the population to expand, we must have $dP/dt > 0$ and thus $k > 0$. ■

EXAMPLE 11 Spread of a Disease

In the spread of a contagious disease, for example a flu virus, it is reasonable to assume that the rate, dx/dt, at which the disease spreads is proportional not only to the number of people, $x(t)$, who have contracted the disease but also to the number of people, $y(t)$, who have not yet been exposed; that is,

$$\frac{dx}{dt} = kxy, \tag{22}$$

where k is the usual constant of proportionality. If one infected person is introduced into a fixed population of n people, then x and y are related by

$$x + y = n + 1. \tag{23}$$

Using (23) to eliminate y in (22), we get

$$\frac{dx}{dt} = kx(n + 1 - x). \tag{24}$$

The obvious side condition accompanying equation (24) is $x(0) = 1$. ■

Logistic Equation The nonlinear first-order equation (24) is a special case of a more general equation

$$\frac{dP}{dt} = P(a - bP), \quad a \text{ and } b \text{ constants}, \tag{25}$$

known as the **logistic equation.** See Section 3.3. The solution of this equation is very important in ecological, sociological, and even managerial sciences.

Continuous Compounding of Interest It is very common for savings institutions to advertise that interest is being compounded daily, by the hour, or even by the minute. Of course, there is no reason to stop there; interest could as well be compounded every second, every half-second, every microsecond, and so on. That is to say, interest could be compounded continuously.

EXAMPLE 12 Continuous Compound Interest

When interest is compounded continuously, the rate at which an amount of money S grows is proportional to the amount of money present at time t; that is,

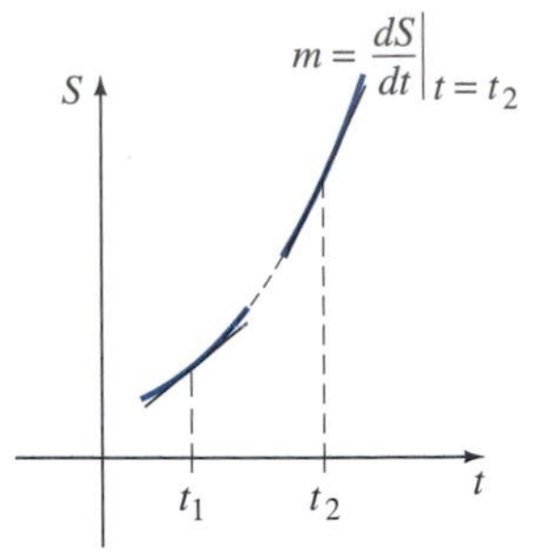

Figure 1.14

$$\frac{dS}{dt} = rS, \tag{26}$$

where r is the annual rate of interest. This mathematical description is analogous to the population growth of Example 10. The rate of growth is large when the amount of money present in the account is also large. Translated geometrically, this means the tangent line is steep when S is large. See Figure 1.14.

The definition of a derivative provides an interesting derivation of (26). Suppose $S(t)$ is the amount of money accrued in a savings account after t years when the annual rate of interest r is compounded continuously. If h denotes an increment in time, then the interest obtained in the time span $(t + h) - t$ is the difference in the amounts accrued:

$$S(t + h) - S(t). \tag{27}$$

Since interest is given by (rate) $\times$ (time) $\times$ (principal), we can approximate the interest earned in this same time period by either

$$rhS(t) \quad \text{or} \quad rhS(t + h).$$

Intuitively we see that $rhS(t)$ and $rhS(t + h)$ are lower and upper bounds, respectively, for the actual interest (27); that is,

$$rhS(t) \le S(t + h) - S(t) \le rhS(t + h)$$

or

$$rS(t) \le \frac{S(t + h) - S(t)}{h} \le rS(t + h). \tag{28}$$

Taking the limit of (28) as $h \to 0$, we get

$$rS(t) \le \lim_{h\to 0} \frac{S(t + h) - S(t)}{h} \le rS(t),$$

and so it must follow that

$$\lim_{h\to 0} \frac{S(t + h) - S(t)}{h} = rS(t) \quad \text{or} \quad \frac{dS}{dt} = rS.$$

EXERCISES 1.2

Answers to odd-numbered problems begin on page A-1.

In Problems 1–22 derive the differential equation(s) describing the given physical situation.

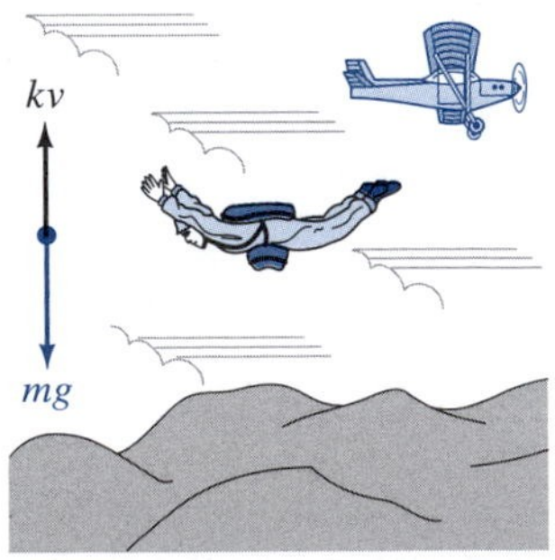

Figure 1.15

1. Under some circumstances a falling body B of mass m, such as the skydiver shown in Figure 1.15, encounters air resistance proportional to its instantaneous velocity v. Use Newton's second law to find the differential equation for the velocity v of the body at time t. Recall that acceleration is $a = dv/dt$. Assume in this case that the positive direction is downward.

2. What is the differential equation for the velocity v of a body of mass m falling vertically downward through a medium (such as water) that offers a resistance proportional to the square of its instantaneous velocity? Assume the positive direction is downward.

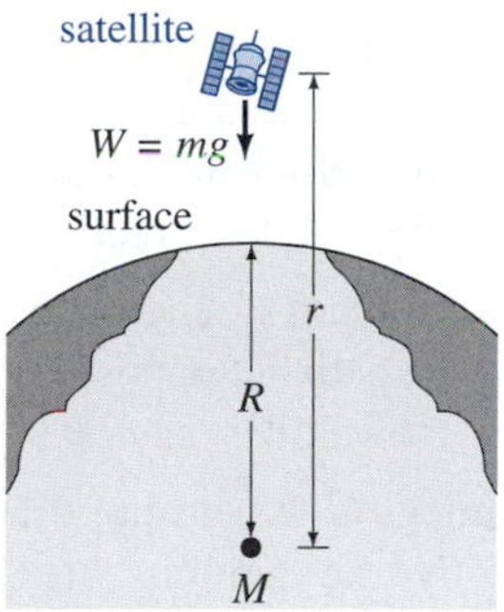

Figure 1.16

3. By Newton's universal law of gravitation the free-fall acceleration a of a body, such as the satellite shown in Figure 1.16, falling a great distance to the surface is *not* the constant g. Rather, the acceleration a is inversely proportional to the square of the distance r from the center of the earth, $a = k/r^2$, where k is the constant of proportionality.

(a) Use the fact that at the surface of the earth $r = R$ and $a = g$ to determine the constant of proportionality k.

(b) Use Newton's second law and part (a) to find a differential equation for the distance r.

(c) Use the chain rule in the form

$$\frac{d^2r}{dt^2} = \frac{dv}{dt} = \frac{dv}{dr}\frac{dr}{dt}$$

to express the differential equation in part (b) as a differential equation involving v and dv/dr.

4. (a) Use part (b) of Problem 3 to find the differential equation for r if the resistance to the falling satellite is proportional to its instantaneous velocity.

(b) Near the surface of the earth, use the approximation $R \approx r$ to show that the differential equation in part (a) reduces to the equation derived in Problem 1.

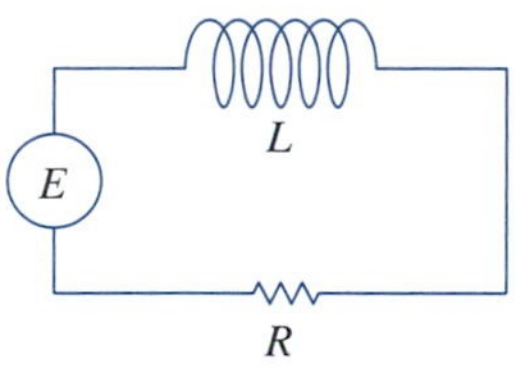

Figure 1.17

5. A series circuit contains a resistor and an inductor as shown in Figure 1.17. Determine the differential equation for the current $i(t)$ if the resistance is R, the inductance is L, and the impressed voltage is $E(t)$.

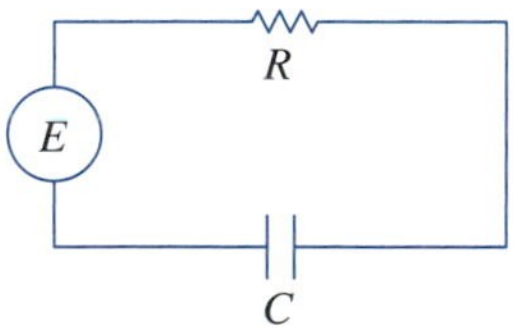

Figure 1.18

6. A series circuit contains a resistor and a capacitor as shown in Figure 1.18. Determine the differential equation for the charge $q(t)$ on the capacitor if the resistance is R, the capacitance is C, and the impressed voltage is $E(t)$.

7. Suppose a tank is discharging water through a circular orifice of cross-sectional area A_o at its bottom. It has been shown experimentally that when friction at the orifice is taken into consideration, the volume of water leaving the tank per second is approximately $0.6A_o\sqrt{2gh}$. Find the differential equation for the height h of water at time t for the cubical tank in Figure 1.19. The radius of the orifice is 2 in. and $g = 32$ ft/s^2.

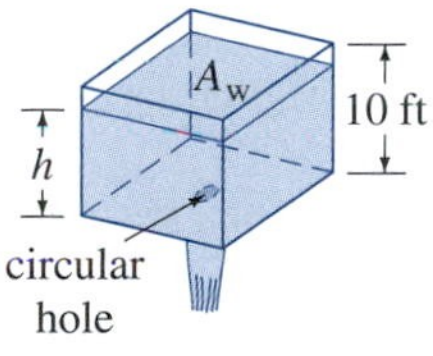

Figure 1.19

8. A tank in the form of a right circular cylinder of radius 2 ft and height 10 ft is standing on end. The tank is initially full of water, and water leaks from a circular hole of radius $\frac{1}{2}$ in. at its bottom. Use the information in Problem 7 to obtain the differential equation for the height h of the water at time t.

9. A water tank has the shape of a hemisphere with radius 5 ft. Water leaks out of a circular hole of radius 1 in. at its flat bottom. Use the information in Problem 7 to obtain the differential equation for the height h of the water at time t.

10. The rate at which a radioactive substance decays is proportional to the amount $A(t)$ of the substance remaining at time t. Determine the differential equation for the amount $A(t)$.

11. A drug is infused into a patient's bloodstream at a constant rate of r grams per second. Simultaneously, the drug is removed at a rate proportional to the amount $x(t)$ of the drug present at time t. Determine the differential equation governing the amount $x(t)$.

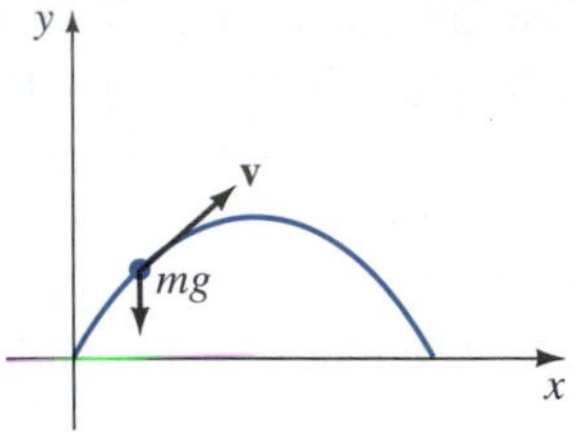

Figure 1.20

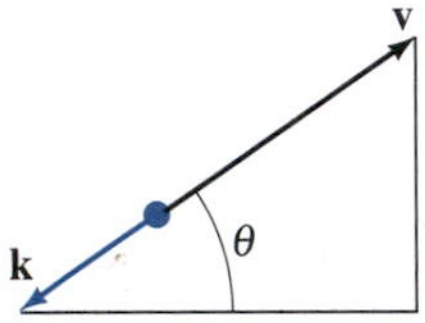

Figure 1.21

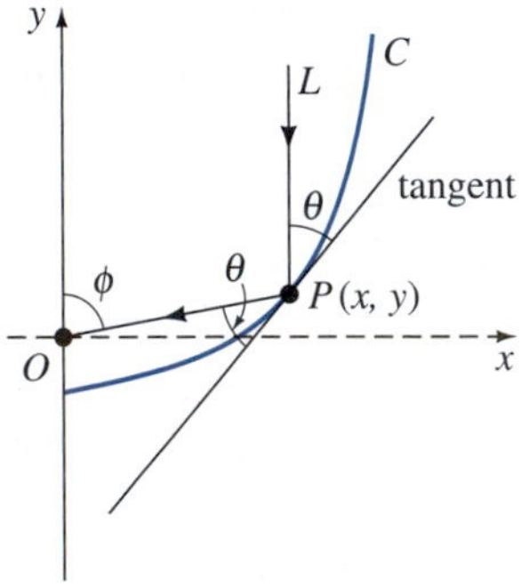

Figure 1.22

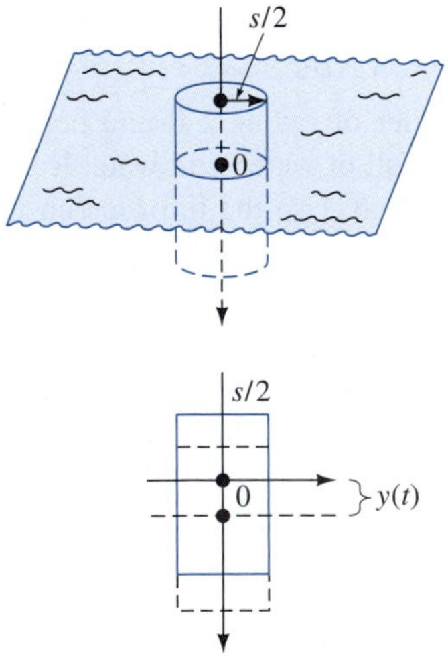

Figure 1.23

12. A projectile shot from a gun has weight $w = mg$ and velocity $\mathbf{v}$ tangent to its path of motion. Ignoring air resistance and all other forces except its weight, find the system of differential equations that describes the motion. See Figure 1.20. [*Hint*: Use Newton's second law in the x and y direction.]

13. Determine the equations of motion if the projectile in Problem 12 encounters a retarding force $\mathbf{k}$ (of magnitude k) acting tangent to the path but opposite to the motion. See Figure 1.21. [*Hint*: $\mathbf{k}$ is a multiple of the velocity, say $c\mathbf{v}$.]

14. Two chemicals A and B react to form a new chemical C. Assuming that the concentrations of both A and B decrease by the amount of C formed, find the differential equation governing the concentration $x(t)$ of the chemical C if the rate at which the chemical reaction takes place is proportional to the product of the remaining concentrations of A and B.

15. Light strikes a plane curve C in such a manner that all beams L parallel to the y-axis are reflected to a single point O. Determine the differential equation for the function $y = f(x)$ describing the shape of the curve. (The fact that the angle of incidence is equal to the angle of reflection is a principle of optics.) [*Hint*: Inspection of Figure 1.22 shows that the inclination of the tangent line from the horizontal at $P(x, y)$ is $\pi/2 - \theta$ and that we can write $\phi = 2\theta$. (Why?) Also, don't be afraid to use a trigonometric identity.]

16. A cylindrical barrel s feet in diameter of weight w lb is floating in water. After an initial depression, the barrel exhibits an up-and-down bobbing motion along a vertical line. Using Figure 1.23(b), determine the differential equation for the vertical displacements $y(t)$ if the origin is taken to be on the vertical axis at the surface of the water when the barrel is at rest. Use Archimedes' principle that the buoyancy, or upward force of the water on the barrel, is equal to the weight of the water displaced and the fact that the density of water is 62.4 lb/ft^3. Assume that the downward direction is positive. Ignore the resistance of the water.

17. A rocket is shot vertically upward from the surface of the earth. After all its fuel has been expended, the mass of the rocket is a constant m. Use Newton's second law of motion and the fact that the force of gravity varies inversely as the square of the distance to find the differential equation for distance y from the earth's center to the rocket at time t after burnout. State appropriate conditions at $t = 0$ associated with this differential equation.

18. Newton's second law $F = ma$ can be written $F = \dfrac{d}{dt}(mv)$. When the mass of an object is variable, this latter formulation is used. The mass $m(t)$ of a rocket launched upward changes as its fuel is consumed.* If $v(t)$ denotes its velocity at any time, it can be shown that

$$-mg = m\frac{dv}{dt} - V\frac{dm}{dt}, \tag{29}$$

* It is assumed that the total mass,

mass of vehicle + *mass of fuel* + *mass of exhaust gases,*

is constant. In this case $m(t)$ = *mass of vehicle* + *mass of fuel.*

where V is the constant velocity of the exhaust gases relative to the rocket. Use (29) to find the differential equation for v if it is known that $m(t) = m_0 - at$ and $V = -b$, where m_0, a, and b are constants.

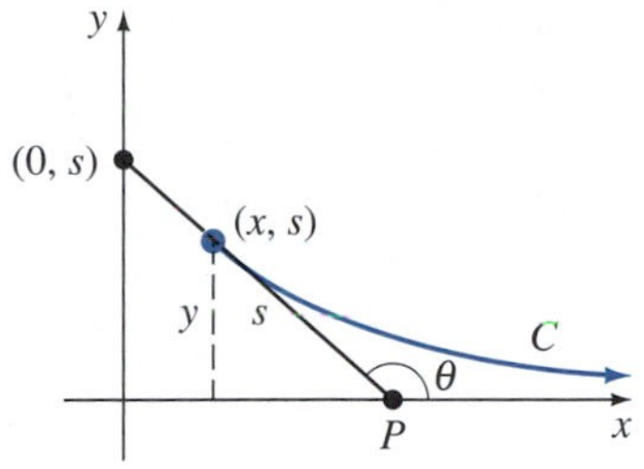

Figure 1.24

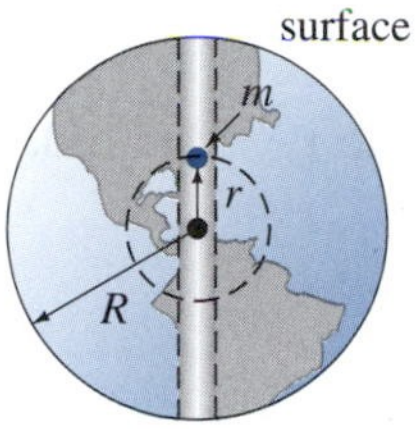

Figure 1.25

19. A person P, starting at the origin, moves in the direction of the positive x-axis, pulling a weight along the curve C (called a **tractrix**) as shown in Figure 1.24. The weight, initially located on the y-axis at $(0, s)$, is pulled by a rope of constant length s, which is kept taut throughout the motion. Find the differential equation of the path of motion. [*Hint*: The rope is always tangent to C; consider the angle of inclination θ as shown in the figure.]

20. Suppose a hole is drilled through the center of the earth. A body with mass m is dropped into the hole. Let the distance from the center of the earth to the mass at time t be denoted by r. See Figure 1.25.

(a) Let M denote the mass of the earth and M_r denote the mass of that portion of the earth within a sphere of radius r. The gravitational force on m is $F = -kM_r m/r^2$, where the minus sign indicates that the force is one of attraction. Use this fact to show that

$$F = -k\frac{mM}{R^3}r.$$

[*Hint:* Assume that the earth is homogeneous—that is, has a constant density δ. Use mass = density × volume.]

(b) Use Newton's second law and the result in part (a) to derive the differential equation

$$\frac{d^2r}{dt^2} + \omega^2 r = 0,$$

where $\omega^2 = kM/R^3 = g/R$.

21. In the theory of learning, the rate at which a subject is memorized is assumed to be proportional to the amount that is left to be memorized. If M denotes the total amount that is to be memorized and $A(t)$ the amount memorized in time t, find the differential equation for A.

22. In Problem 21 assume that the amount of material forgotten is proportional to the amount memorized in time t. What is the differential equation for A when forgetfulness is taken into account?

CHAPTER 1 REVIEW

We classify a differential equation by its type, **ordinary** or **partial;** by its **order;** and by whether it is **linear** or **nonlinear.**

A **solution** of a differential equation is any function having a sufficient number of derivatives that satisfies the equation identically on some interval.

When solving an nth-order ordinary differential equation, we expect to find an n-parameter family of solutions. A **particular solution** is any solution free of arbitrary parameters that satisfies the differential equation. A **singular solution** is any solution that cannot be obtained from an n-parameter family of solutions by assigning values to the parameters. When an n-parameter family of solutions gives every solution of a differential equation on some interval, it is called a **general,** or **complete,** solution.

In the analysis of physical problems, many differential equations can be obtained by equating two different empirical formulations of the same situation. For example, a differential equation of motion can sometimes be obtained by simply equating Newton's second law of motion with the net forces acting on a body.

CHAPTER 1 REVIEW EXERCISES

Answers to odd-numbered problems begin on page A-2.

In Problems 1–4 classify the given differential equation as to type and order. Classify the ordinary differential equations as to linearity.

1. $(2xy - y^2)\,dx + e^x\,dy = 0$

2. $(\sin xy)y''' + 4xy' = 0$

3. $\dfrac{\partial^2 u}{\partial x^2} + \dfrac{\partial^2 u}{\partial y^2} = u$

4. $x^2\dfrac{d^2y}{dx^2} - 3x\dfrac{dy}{dx} + y = x^2$

In Problems 5–8 verify that the indicated function is a solution of the given differential equation.

5. $y' + 2xy = 2 + x^2 + y^2;\quad y = x + \tan x$

6. $x^2y'' + xy' + y = 0;\quad y = c_1\cos(\ln x) + c_2\sin(\ln x),\ x > 0$

7. $y''' - 2y'' - y' + 2y = 6;\quad y = c_1e^x + c_2e^{-x} + c_3e^{2x} + 3$

8. $y^{(4)} - 16y = 0;\quad y = \sin 2x + \cosh 2x$

In Problems 9–16 determine by inspection at least one solution for the given differential equation.

9. $y' = 2x$

10. $\dfrac{dy}{dx} = 5y$

11. $y'' = 1$

12. $y' = y^3 - 8$

13. $y'' = y'$

14. $2y\dfrac{dy}{dx} = 1$

15. $y'' = -y$

16. $y'' = y$

17. Determine an interval on which $y^2 - 2y = x^2 - x - 1$ defines a solution of $2(y - 1)\,dy + (1 - 2x)\,dx = 0$.

18. Explain why the differential equation

$$\left(\frac{dy}{dx}\right)^2 = \frac{4 - y^2}{4 - x^2}$$

possesses no real solutions for $|x| < 2$, $|y| > 2$. Are there other regions in the xy-plane for which the equation has no solutions?

19. The conical tank shown in Figure 1.26 loses water out of an orifice at its bottom. If the cross-sectional area of the orifice is $\frac{1}{4}$ ft^2, find the differential

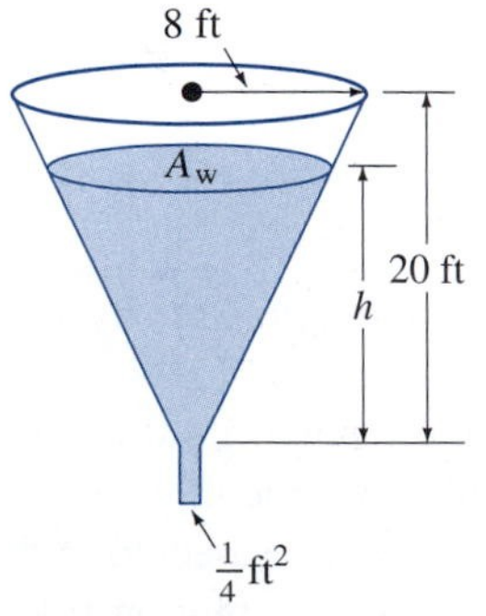

Figure 1.26

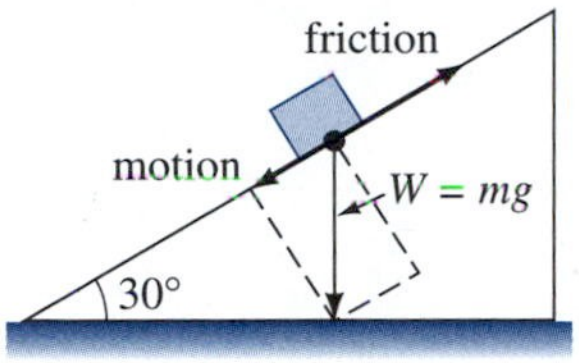

Figure 1.27

equation representing the height of the water h at time t. Ignore friction at the orifice.

20. A weight of 96 lb slides down an incline making a 30° angle with the horizontal. If the coefficient of sliding friction is μ, determine the differential equation for the velocity $v(t)$ of the weight at time t. Use the fact that the force of friction opposing the motion is μN, where N is the normal component of the weight. See Figure 1.27.

2 FIRST-ORDER DIFFERENTIAL EQUATIONS

Introduction

We are now in a position to solve some differential equations. We begin with first-order differential equations.

If a first-order differential equation can be solved, we shall see that the technique or method for solving it depends on what kind of first-order equation it is. Over the years mathematicians struggled to solve many specialized kinds of equations. Thus there are many methods of solution; what works for one kind of first-order equation does not necessarily apply to another kind of equation. Although we consider solution methods for seven classical types of equations in this chapter, our focus is on four types of equations. Some of these four types are important in applications.

2.1 PRELIMINARY THEORY

• *Initial-value problem* • *Initial condition* • *Existence of a solution* • *Uniqueness of a solution*

Initial-Value Problem We are often interested in solving a first-order differential equation*

$$\frac{dy}{dx} = f(x, y) \tag{1}$$

subject to a side condition $y(x_0) = y_0$, where x_0 is a number in an interval I and y_0 is an arbitrary real number. The problem

$$\begin{aligned} &\textit{Solve:} && \frac{dy}{dx} = f(x, y) \\ &\textit{Subject to:} && y(x_0) = y_0 \end{aligned} \tag{2}$$

is called an **initial-value problem** (IVP). The side condition is known as an **initial condition.** In geometric terms we are seeking at least one solution of the differential equation defined on some interval I such that the graph of the solution passes through a prescribed point (x_0, y_0). See Figure 2.1.

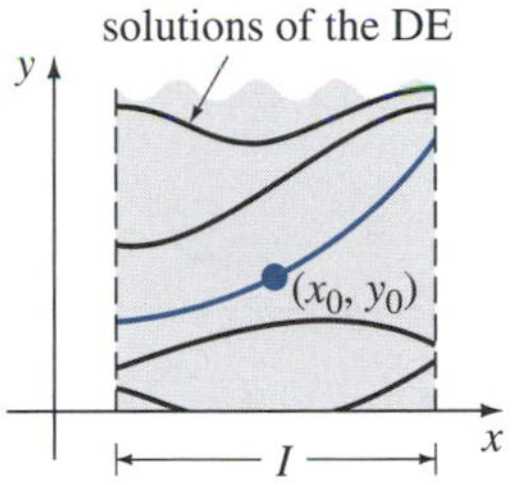

Figure 2.1

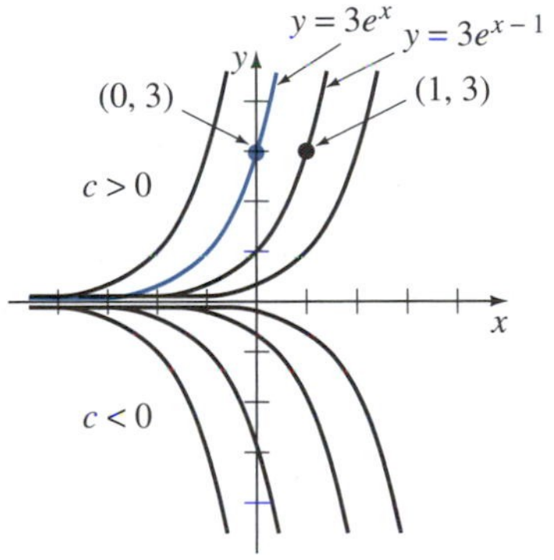

Figure 2.2

EXAMPLE 1 First-Order IVPs

We have seen (page 8) that $y = ce^x$ is a one-parameter family of solutions for $y' = y$ on the interval $(-\infty, \infty)$. If we specify, say, $y(0) = 3$, then substituting $x = 0$, $y = 3$ in the family yields $3 = ce^0 = c$. Thus, as shown in Figure 2.2, the function $y = 3e^x$ is a solution of the initial-value problem

$$y' = y, \quad y(0) = 3.$$

Had we demanded that a solution of $y' = y$ pass through the point (1, 3) rather than (0, 3), then $y(1) = 3$ would yield $c = 3e^{-1}$ and so $y = 3e^{x-1}$. The graph of this function is also indicated in Figure 2.2. ■

Two fundamental questions arise in considering an initial-value problem such as (2):

Does a solution of the problem *exist*?
If a solution exists, is it *unique*?

In other words, does the differential equation $dy/dx = f(x, y)$ possess a solution whose graph passes through (x_0, y_0), and if it does, is there precisely *one* such solution?

As the next example shows, the answer to the second question is sometimes no.

* In this text we assume that a differential equation $F(x, y, y', \ldots, y^{(n)}) = 0$ can be solved for the highest order derivative: $y^{(n)} = f(x, y, y', \ldots, y^{(n-1)})$. There are exceptions.

EXAMPLE 2 An IVP Can Have Several Solutions

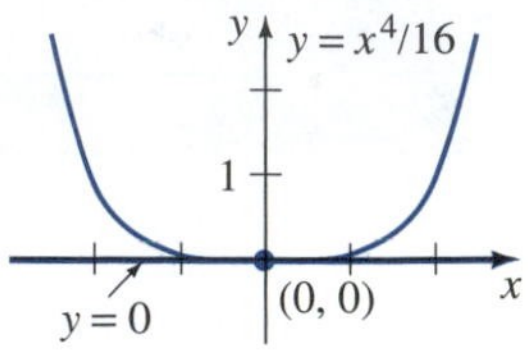

Figure 2.3

You should verify that each of the functions $y = 0$ and $y = x^4/16$ satisfies the differential equation and initial condition in the problem

$$\frac{dy}{dx} = xy^{1/2}, \quad y(0) = 0.$$

As illustrated in Figure 2.3, the graphs of both functions pass through the same point (0, 0).

It is often desirable to know before tackling an initial-value problem whether a solution exists and, when it does, whether it is the only solution of the problem. The following theorem due to Picard* gives conditions that are sufficient to guarantee that a solution exists and that it is also a unique solution.

THEOREM 2.1 Existence of a Unique Solution

Let R be a rectangular region in the xy-plane defined by $a \le x \le b$, $c \le y \le d$ that contains the point (x_0, y_0) in its interior. If $f(x, y)$ and $\partial f/\partial y$ are continuous on R, then there exist an interval I centered at x_0 and a unique function $y(x)$ defined on I satisfying the initial-value problem (2).

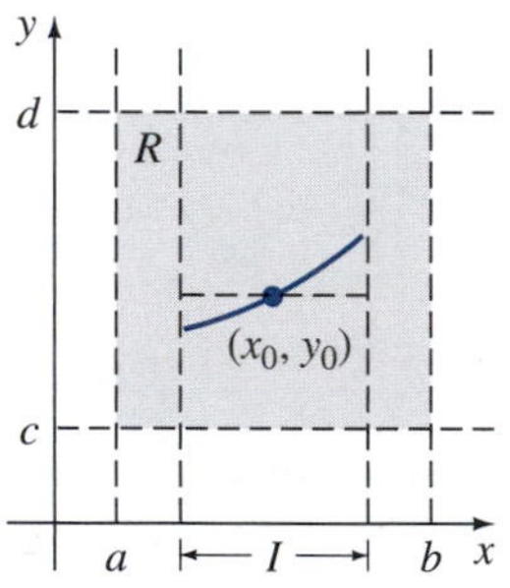

Figure 2.4

The foregoing is one of the most popular existence and uniqueness theorems for first-order differential equations because the criteria of continuity of $f(x, y)$ and $\partial f/\partial y$ are relatively easy to check. In general, it is not always possible to find a specific interval I on which a solution is defined without actually solving the differential equation (see Problem 16). The geometry of Theorem 2.1 is illustrated in Figure 2.4.

EXAMPLE 3 Example 2 Revisited

We saw in Example 2 that the differential equation

$$\frac{dy}{dx} = xy^{1/2}$$

possesses at least two solutions whose graphs pass through (0, 0). The functions

$$f(x, y) = xy^{1/2} \quad \text{and} \quad \frac{\partial f}{\partial y} = \frac{x}{2y^{1/2}}$$

* CHARLES EMILE PICARD (1856–1941) Picard was one of the prominent French mathematicians of the latter nineteenth and early twentieth centuries. He made significant contributions to the fields of differential equations and complex variables. In 1899 Picard lectured at Clark University in Worcester, Massachusetts.

are continuous in the upper half-plane defined by $y > 0$. We conclude from Theorem 2.1 that through any point (x_0, y_0), $y_0 > 0$ (say, for example, (0, 1)), there is some interval around x_0 on which the given differential equation has a unique solution. ■

EXAMPLE 4 Using Theorem 2.1

Theorem 2.1 guarantees that there exists an interval about $x = 0$ on which $y = 3e^x$ is the only solution of the initial-value problem of Example 1:

$$y' = y, \quad y(0) = 3.$$

This follows from the fact that $f(x, y) = y$ and $\partial f/\partial y = 1$ are continuous throughout the entire xy-plane. It can be further shown that this interval is $(-\infty, \infty)$. ■

EXAMPLE 5 Using Theorem 2.1

For

$$\frac{dy}{dx} = x^2 + y^2$$

we observe that $f(x, y) = x^2 + y^2$ and $\partial f/\partial y = 2y$ are continuous throughout the entire xy-plane. Therefore, through any given point (x_0, y_0) there passes one and only one solution of the differential equation. ■

Remarks (*i*) You should be aware of the distinction between a solution *existing* and *exhibiting* a solution. Clearly if we find a solution by exhibiting it, we can say that it exists, but on the other hand a solution can exist and we may not be able to display it. From Example 5 we know that a solution of the problem $dy/dx = x^2 + y^2$, $y(0) = 1$ exists on some interval around $x = 0$ and is unique. However, the equation cannot be solved in terms of elementary functions; we can approximate the solution using the methods in Chapter 9.
(*ii*) The conditions stated in Theorem 2.1 are *sufficient* but not *necessary.* When $f(x, y)$ and $\partial f/\partial y$ are continuous on a rectangular region R, it must always follow that there exists a unique solution of (2) when (x_0, y_0) is a point interior to R. However, if the conditions stated in the hypothesis of the theorem do not hold, then the initial-value problem (2) may still have (a) no solution, (b) more than one solution, or (c) a unique solution. Furthermore, the continuity condition on $\partial f/\partial y$ can be relaxed somewhat without changing the conclusion of the theorem. This results in a stronger theorem but is, unfortunately, not as easy to apply as Theorem 2.1. Indeed, if we are not interested in uniqueness, then a famous theorem due to the Italian mathematician Guiseppe Peano states that the continuity of $f(x, y)$ on R is sufficient to guarantee the existence of at least one solution of $dy/dx = f(x, y)$ through a point (x_0, y_0) interior to R.

EXERCISES 2.1

Answers to odd-numbered problems begin on page A-2.

In Problems 1–10 determine a region of the xy-plane for which the given differential equation would have a unique solution through a point (x_0, y_0) in the region.

1. $\dfrac{dy}{dx} = y^{2/3}$

2. $\dfrac{dy}{dx} = \sqrt{xy}$

3. $x\dfrac{dy}{dx} = y$

4. $\dfrac{dy}{dx} - y = x$

5. $(4 - y^2)y' = x^2$

6. $(1 + y^3)y' = x^2$

7. $(x^2 + y^2)y' = y^2$

8. $(y - x)y' = y + x$

9. $\dfrac{dy}{dx} = x^3 \cos y$

10. $\dfrac{dy}{dx} = (x - 1)e^{y/(x-1)}$

In Problems 11 and 12 determine by inspection at least two solutions of the given initial-value problem.

11. $y' = 3y^{2/3}, \quad y(0) = 0$

12. $x\dfrac{dy}{dx} = 2y, \quad y(0) = 0$

13. By inspection determine a solution of the nonlinear differential equation $y' = y^3$ satisfying $y(0) = 0$. Is the solution unique?

14. By inspection find a solution of the initial-value problem

$$y' = |y - 1|, \quad y(0) = 1.$$

State why the conditions of Theorem 2.1 do not hold for this differential equation. Although we shall not prove it, the solution to this initial-value problem is unique.

15. Verify that $y = cx$ is a solution of the differential equation $xy' = y$ for every value of the parameter c. Find at least two solutions of the initial-value problem

$$xy' = y, \quad y(0) = 0.$$

Observe that the piecewise-defined function

$$y = \begin{cases} 0, & x < 0 \\ x, & x \geq 0 \end{cases}$$

satisfies the condition $y(0) = 0$. Is it a solution of the initial-value problem?

16. **(a)** Consider the differential equation

$$\frac{dy}{dx} = 1 + y^2.$$

Determine a region of the xy-plane for which the equation has a unique solution through a point (x_0, y_0) in the region.

(b) Formally show that $y = \tan x$ satisfies the differential equation and the condition $y(0) = 0$.

(c) Explain why $y = \tan x$ is not a solution of the initial-value problem

$$\frac{dy}{dx} = 1 + y^2, \quad y(0) = 0$$

on the interval $(-2, 2)$.

(d) Explain why $y = \tan x$ is a solution of the initial-value problem in part (c) on the interval $(-1, 1)$.

In Problems 17–20 determine whether Theorem 2.1 guarantees that the differential equation $y' = \sqrt{y^2 - 9}$ possesses a unique solution through the given point.

17. $(1, 4)$

18. $(5, 3)$

19. $(2, -3)$

20. $(-1, 1)$

2.2 SEPARABLE VARIABLES

• *Separation of variables* • *Losing a solution*

Note to the Student In solving a differential equation you will often have to utilize, say, integration by parts, partial fractions, or possibly a substitution. It will be worth a few minutes of your time to review some techniques of integration.

We begin our study of the methodology of solving first-order equations with the simplest of all differential equations.

If $g(x)$ is a continuous function, then the first-order equation

$$\frac{dy}{dx} = g(x) \tag{1}$$

can be solved by integration. The solution of (1) is

$$y = \int g(x)\,dx = G(x) + c,$$

where $G(x)$ is an antiderivative (indefinite integral) of $g(x)$.

EXAMPLE 1 Solution by Integration

Solve **(a)** $\dfrac{dy}{dx} = 1 + e^{2x}$ and **(b)** $\dfrac{dy}{dx} = \sin x$.

Solution As illustrated above, both equations can be solved by integration.

(a) $y = \int (1 + e^{2x})\,dx = x + \frac{1}{2}e^{2x} + c$

(b) $y = \int \sin x\,dx = -\cos x + c$

Equation (1), as well as its method of solution, is just a special case of the following:

DEFINITION 2.1 Separable Equation

A differential equation of the form

$$\frac{dy}{dx} = \frac{g(x)}{h(y)}$$

is said to be **separable** or to have **separable variables.**

Observe that a separable equation can be written as

$$h(y)\frac{dy}{dx} = g(x). \tag{2}$$

It is seen immediately that (2) reduces to (1) when $h(y) = 1$.

Now if $y = f(x)$ denotes a solution of (2), we must have

$$h(f(x))\, f'(x) = g(x),$$

and therefore

$$\int h(f(x))\, f'(x)\, dx = \int g(x)\, dx + c. \tag{3}$$

But $dy = f'(x)\,dx$, so (3) is the same as

$$\int h(y)\, dy = \int g(x)\, dx + c. \tag{4}$$

Method of Solution Equation (4) indicates the procedure for solving separable differential equations. A one-parameter family of solutions, usually given implicitly, is obtained by integrating both sides of $h(y)\,dy = g(x)\,dx$.

Note There is no need to use two constants in the integration of a separable equation, since $\int h(y)\,dy + c_1 = \int g(x)\,dx + c_2$ can be written $\int h(y)\,dy = \int g(x)\,dx + c_2 - c_1 = \int g(x)\,dx + c$, where c is completely arbitrary. In many instances throughout the following chapters, we shall not hesitate to relabel constants in a manner that may prove convenient for a given equation. For example, multiples of constants or combinations of constants can sometimes be replaced by one constant.

EXAMPLE 2 Solving a Separable DE

Solve $(1 + x)\,dy - y\,dx = 0$.

Solution Dividing by $(1 + x)y$, we can write $dy/y = dx/(1 + x)$, from which it follows that

$$\int \frac{dy}{y} = \int \frac{dx}{1+x}.$$

$$\begin{aligned} \ln|y| &= \ln|1+x| + c_1 \\ y &= e^{\ln|1+x|+c_1} \\ &= e^{\ln|1+x|} \cdot e^{c_1} \\ &= |1+x|e^{c_1} \\ &= \pm e^{c_1}(1+x) \quad \begin{cases} |1+x| = 1+x, \ x \geq -1 \\ |1+x| = -(1+x), \ x < -1 \end{cases} \end{aligned}$$

Relabeling $\pm e^{c_1}$ as c then gives $y = c(1 + x)$.

Alternative Solution Since each integral results in a logarithm, a judicious choice for the constant of integration is $\ln|c|$ rather than c:

$$\ln|y| = \ln|1+x| + \ln|c| \quad \text{or} \quad \ln|y| = \ln|c(1+x)|$$

so

$$y = c(1 + x).$$

Even if not *all* the indefinite integrals are logarithms, it may still be advantageous to use $\ln|c|$. However, no firm rule can be given.

EXAMPLE 3 An Initial-Value Problem

Solve the initial-value problem $\dfrac{dy}{dx} = -\dfrac{x}{y}$, $y(4) = 3$.

Solution From $y\,dy = -x\,dx$ we get

$$\int y\,dy = -\int x\,dx \quad \text{and} \quad \frac{y^2}{2} = -\frac{x^2}{2} + c_1.$$

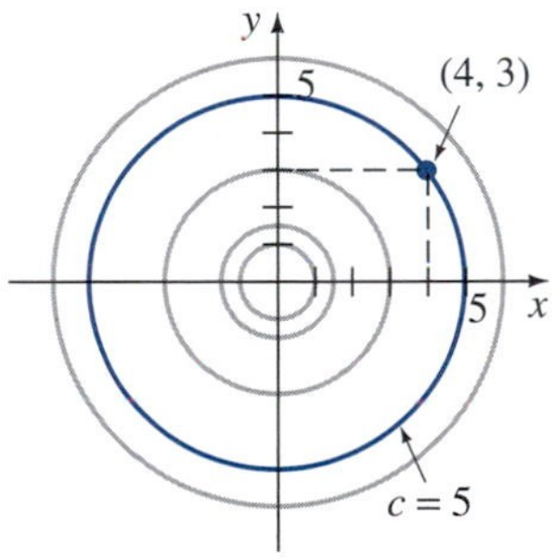

Figure 2.5

This solution can be written as $x^2 + y^2 = c^2$ by replacing the constant $2c_1$ by c^2. The solution represents a family of concentric circles.

Now when $x = 4$, $y = 3$, so $16 + 9 = 25 = c^2$. Thus the initial-value problem determines $x^2 + y^2 = 25$. In view of Theorem 2.1, we can conclude that it is the only circle of the family passing through the point (4, 3). See Figure 2.5.

EXAMPLE 4 Solving a Separable DE

Solve $xe^{-y}\sin x\,dx - y\,dy = 0$.

Solution After multiplying by e^y, we get

$$x \sin x\,dx = ye^y\,dy.$$

Integration by parts on both sides of the equality gives

$$-x\cos x + \sin x = ye^y - e^y + c.$$

EXAMPLE 5 Solving a Separable DE

Solve $xy^4\,dx + (y^2 + 2)e^{-3x}\,dy = 0$. **(5)**

Solution By multiplying the given equation by e^{3x} and dividing by y^4, we obtain

$$xe^{3x}\,dx + \frac{y^2+2}{y^4}\,dy = 0 \quad \text{or} \quad xe^{3x}\,dx + (y^{-2} + 2y^{-4})\,dy = 0. \tag{6}$$

Using integration by parts on the first term yields

$$\frac{1}{3}xe^{3x} - \frac{1}{9}e^{3x} - y^{-1} - \frac{2}{3}y^{-3} = c_1.$$

The one-parameter family of solutions can also be written as

$$e^{3x}(3x-1) = \frac{9}{y} + \frac{6}{y^3} + c, \tag{7}$$

where the constant $9c_1$ is rewritten as c.

Two points are worth mentioning at this time. First, unless it is important or convenient there is no need to try to solve an expression representing a family of solutions for y explicitly in terms of x. Equation (7) shows that this task may present more problems than just the drudgery of symbol pushing. As a consequence it is often the case that the interval over which a solution is valid is not apparent. Second, some care should be exercised when separating variables to make certain that divisors are not zero. A constant solution may sometimes get lost in the shuffle of solving the problem. In Example 5 observe that $y = 0$ is a perfectly good solution of (5) but is not a member of the set of solutions defined by (7).

EXAMPLE 6 Losing a Solution

Solve the initial-value problem $\dfrac{dy}{dx} = y^2 - 4$, $y(0) = -2$.

Solution We put the equation into the form

$$\frac{dy}{y^2-4} = dx \tag{8}$$

and use partial fractions on the left side. We have

$$\left[\frac{-\frac{1}{4}}{y+2} + \frac{\frac{1}{4}}{y-2}\right]dy = dx \tag{9}$$

so

$$-\frac{1}{4}\ln|y+2| + \frac{1}{4}\ln|y-2| = x + c_1. \tag{10}$$

Thus

$$\ln\left|\frac{y-2}{y+2}\right| = 4x + c_2 \quad \text{and} \quad \frac{y-2}{y+2} = ce^{4x},$$

where we have replaced $4c_1$ by c_2 and e^{c_2} by c. Finally, solving the last equation for y, we get

$$y = 2\frac{1+ce^{4x}}{1-ce^{4x}}. \tag{11}$$

Substituting $x = 0$, $y = -2$ leads to the dilemma

$$-2 = 2\frac{1+c}{1-c}$$
$$-1 + c = 1 + c \quad \text{or} \quad -1 = 1.$$

Let us consider the differential equation a little more carefully. The fact is, the equation

$$\frac{dy}{dx} = (y+2)(y-2)$$

is satisfied by two constant functions—namely, $y = -2$ and $y = 2$. Inspection of equations (8), (9), and (10) clearly indicates we must preclude $y = -2$ and $y = 2$ at those steps in our solution. But it is interesting to observe that we can subsequently recover the solution $y = 2$ by setting $c = 0$ in equation (11). However, there is no finite value of c that will ever yield the solution $y = -2$. This latter constant function is the only solution to the original initial-value problem. See Figure 2.6. ■

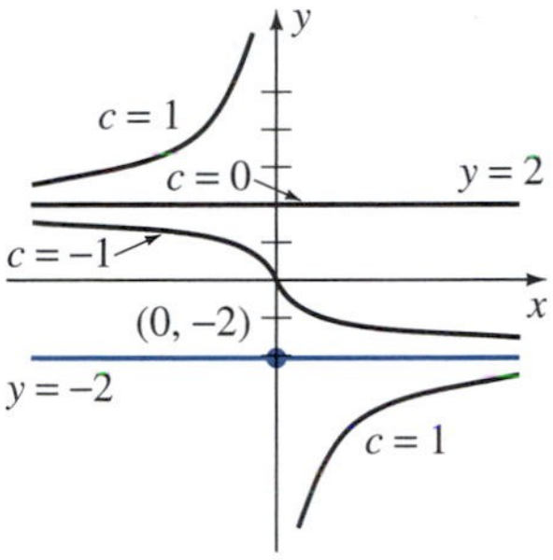

Figure 2.6

If, in Example 6, we had used $\ln|c|$ for the constant of integration, then the form of the one-parameter family of solutions would be

$$y = 2\frac{c + e^{4x}}{c - e^{4x}}. \tag{12}$$

Note that (12) reduces to $y = -2$ when $c = 0$, but now there is no finite value of c that will give the constant solution $y = 2$.

If an initial condition leads to a particular solution by finding a specific value of the parameter c in a family of solutions for a first-order differential equation, it is a natural inclination of most students (and instructors) to relax and be content. In Section 2.1 we saw, however, that a solution of an initial-value problem may not be unique. For example, the problem

$$\frac{dy}{dx} = xy^{1/2}, \quad y(0) = 0 \tag{13}$$

has at least two solutions—namely, $y = 0$ and $y = x^4/16$. We are now in a position to solve the equation. Separating variables

$$y^{-1/2}dy = x\,dx$$

and integrating gives

$$2y^{1/2} = \frac{x^2}{2} + c_1 \quad \text{or} \quad y = \left(\frac{x^2}{4} + c\right)^2.$$

When $x = 0$, $y = 0$, so necessarily $c = 0$. Therefore $y = x^4/16$. The solution $y = 0$ was lost by dividing by $y^{1/2}$. In addition, the initial-value problem (13) possesses infinitely more solutions, since for any choice of the parameter $a \geq 0$ the piecewise-defined function

$$y = \begin{cases} 0, & x < a \\ \dfrac{(x^2 - a^2)^2}{16}, & x \geq a \end{cases}$$

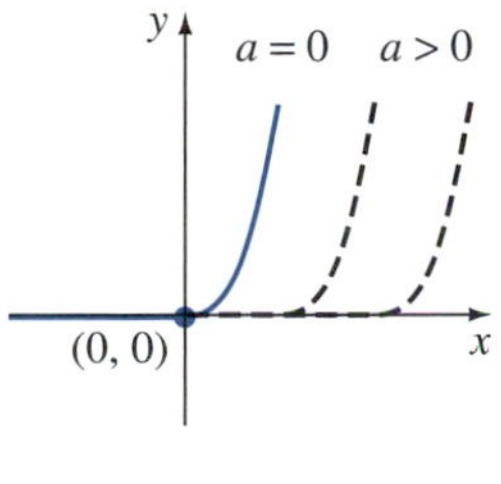

Figure 2.7

satisfies both the differential equation and the initial condition. See Figure 2.7.

Remarks We saw in some of the preceding examples that the constant in the one-parameter family of solutions for a first-order differential equation can be relabeled when convenient. Also, it can easily happen that two individuals solving the same equation arrive at dissimilar answers. For example, by separation of variables we can show that one-parameter families of solutions for $(1 + y^2)\,dx + (1 + x^2)\,dy = 0$ are

$$\arctan x + \arctan y = c \quad\text{or}\quad \arctan x + \arctan y = \arctan c \quad\text{or}\quad \frac{x + y}{1 - xy} = c.$$

As you work your way through the next several sections, bear in mind that families of solutions may be *equivalent* in the sense that one family may be obtained from another by either relabeling the constant or applying algebra and trigonometry.

EXERCISES 2.2

Answers to odd-numbered problems begin on page A-2.

In Problems 1–40 solve the given differential equation by separation of variables.

1. $\dfrac{dy}{dx} = \sin 5x$

2. $\dfrac{dy}{dx} = (x + 1)^2$

3. $dx + e^{3x}\,dy = 0$

4. $dx - x^2 dy = 0$

5. $(x + 1)\dfrac{dy}{dx} = x + 6$

6. $e^x \dfrac{dy}{dx} = 2x$

7. $xy' = 4y$

8. $\dfrac{dy}{dx} + 2xy = 0$

9. $\dfrac{dy}{dx} = \dfrac{y^3}{x^2}$

10. $\dfrac{dy}{dx} = \dfrac{y + 1}{x}$

11. $\dfrac{dx}{dy} = \dfrac{x^2y^2}{1 + x}$

12. $\dfrac{dx}{dy} = \dfrac{1 + 2y^2}{y \sin x}$

13. $\dfrac{dy}{dx} = e^{3x+2y}$

14. $e^x y \dfrac{dy}{dx} = e^{-y} + e^{-2x-y}$

15. $(4y + yx^2)\,dy - (2x + xy^2)\,dx = 0$

16. $(1 + x^2 + y^2 + x^2y^2)\,dy = y^2\,dx$

17. $2y(x + 1)\,dy = x\,dx$

18. $x^2y^2\,dy = (y + 1)\,dx$

19. $y \ln x \dfrac{dx}{dy} = \left(\dfrac{y + 1}{x}\right)^2$

20. $\dfrac{dy}{dx} = \left(\dfrac{2y + 3}{4x + 5}\right)^2$

21. $\dfrac{dS}{dr} = kS$

22. $\dfrac{dQ}{dt} = k(Q - 70)$

23. $\dfrac{dP}{dt} = P - P^2$

24. $\dfrac{dN}{dt} + N = Nte^{t+2}$

25. $\sec^2 x\,dy + \csc y\,dx = 0$

26. $\sin 3x\,dx + 2y\cos^3 3x\,dy = 0$

27. $e^y \sin 2x\, dx + \cos x(e^{2y} - y)dy = 0$

28. $\sec x\, dy = x \cot y\, dx$

29. $(e^y + 1)^2 e^{-y}\, dx + (e^x + 1)^3 e^{-x}\, dy = 0$

30. $\dfrac{y}{x}\dfrac{dy}{dx} = (1 + x^2)^{-1/2}(1 + y^2)^{1/2}$

31. $(y - yx^2)\dfrac{dy}{dx} = (y + 1)^2$

32. $2\dfrac{dy}{dx} - \dfrac{1}{y} = \dfrac{2x}{y}$

33. $\dfrac{dy}{dx} = \dfrac{xy + 3x - y - 3}{xy - 2x + 4y - 8}$

34. $\dfrac{dy}{dx} = \dfrac{xy + 2y - x - 2}{xy - 3y + x - 3}$

35. $\dfrac{dy}{dx} = \sin x\,(\cos 2y - \cos^2 y)$

36. $\sec y \dfrac{dy}{dx} + \sin(x - y) = \sin(x + y)$

37. $x\sqrt{1 - y^2}\, dx = dy$

38. $y(4 - x^2)^{1/2}\, dy = (4 + y^2)^{1/2}\, dx$

39. $(e^x + e^{-x})\dfrac{dy}{dx} = y^2$

40. $(x + \sqrt{x})\dfrac{dy}{dx} = y + \sqrt{y}$

In Problems 41–48 solve the given differential equation subject to the indicated initial condition.

41. $(e^{-y} + 1)\sin x\, dx = (1 + \cos x)dy, \quad y(0) = 0$

42. $(1 + x^4)dy + x(1 + 4y^2)dx = 0, \quad y(1) = 0$

43. $y\, dy = 4x(y^2 + 1)^{1/2}\, dx, \quad y(0) = 1$

44. $\dfrac{dy}{dt} + ty = y, \quad y(1) = 3$

45. $\dfrac{dx}{dy} = 4(x^2 + 1), \quad x\left(\dfrac{\pi}{4}\right) = 1$

46. $\dfrac{dy}{dx} = \dfrac{y^2 - 1}{x^2 - 1}, \quad y(2) = 2$

47. $x^2 y' = y - xy, \quad y(-1) = -1$

48. $y' + 2y = 1, \quad y(0) = \dfrac{5}{2}$

In Problems 49 and 50 find a solution of the given differential equation that passes through the indicated points.

49. $\dfrac{dy}{dx} - y^2 = -9$ **(a)** $(0, 0)$ **(b)** $(0, 3)$ **(c)** $(\frac{1}{3}, 1)$

50. $x\dfrac{dy}{dx} = y^2 - y$ **(a)** $(0, 1)$ **(b)** $(0, 0)$ **(c)** $(\frac{1}{2}, \frac{1}{2})$

51. Find a singular solution for the equation in Problem 37.

52. Find a singular solution for the equation in Problem 39.

Often a radical change in the solution of a differential equation corresponds to a very small change in either the initial condition or the equation itself. In Problems 53–56 compare the solutions of the given initial-value problems.

53. $\dfrac{dy}{dx} = (y - 1)^2, \quad y(0) = 1$

54. $\dfrac{dy}{dx} = (y - 1)^2, \quad y(0) = 1.01$

55. $\dfrac{dy}{dx} = (y-1)^2 + 0.01, \quad y(0) = 1$ **56.** $\dfrac{dy}{dx} = (y-1)^2 - 0.01, \quad y(0) = 1$

A differential equation of the form $dy/dx = f(ax + by + c), b \neq 0$ can always be reduced to an equation with separable variables by means of the substitution $u = ax + by + c$. Use this procedure to solve Problems 57–62.

57. $\dfrac{dy}{dx} = (x + y + 1)^2$

58. $\dfrac{dy}{dx} = \dfrac{1 - x - y}{x + y}$

59. $\dfrac{dy}{dx} = \tan^2(x + y)$

60. $\dfrac{dy}{dx} = \sin(x + y)$

61. $\dfrac{dy}{dx} = 2 + \sqrt{y - 2x + 3}$

62. $\dfrac{dy}{dx} = 1 + e^{y-x+5}$

2.3 HOMOGENEOUS EQUATIONS

• *Homogeneous function* • *Homogeneous equation*

Before considering the concept of a **homogeneous first-order differential equation** and its method of solution, we need to examine closely the nature of a **homogeneous function.** We begin with the definition of this concept.

DEFINITION 2.2 Homogeneous Function

If a function f has the property that

$$f(tx, ty) = t^n f(x, y) \tag{1}$$

for some real number n, then f is said to be a **homogeneous function of degree n.**

EXAMPLE 1 Some Homogeneous Functions

(a) $f(x, y) = x^2 - 3xy + 5y^2$

$$\begin{aligned} f(tx, ty) &= (tx)^2 - 3(tx)(ty) + 5(ty)^2 \\ &= t^2x^2 - 3t^2xy + 5t^2y^2 \\ &= t^2[x^2 - 3xy + 5y^2] = t^2 f(x, y) \end{aligned}$$

The function is homogeneous of degree two.

(b) $f(x, y) = \sqrt[3]{x^2 + y^2}$

$$f(tx, ty) = \sqrt[3]{t^2x^2 + t^2y^2} = t^{2/3}\sqrt[3]{x^2 + y^2} = t^{2/3} f(x, y)$$

The function is homogeneous of degree $\frac{2}{3}$.

(c) $f(x, y) = x^3 + y^3 + 1$

$f(tx, ty) = t^3x^3 + t^3y^3 + 1 \neq t^3f(x, y)$

since $t^3f(x, y) = t^3x^3 + t^3y^3 + t^3$. The function is not homogeneous.

(d) $f(x, y) = \dfrac{x}{2y} + 4$

$$f(tx, ty) = \frac{tx}{2ty} + 4 = \frac{x}{2y} + 4 = t^0f(x, y)$$

The function is homogeneous of degree zero. ■

As parts (c) and (d) of Example 1 show, a constant added to a function destroys homogeneity, unless the function is homogeneous of degree zero. Also, in many instances a homogeneous function can be recognized by examining the *total degree* of each term.

EXAMPLE 2 Examining the Degree of Each Term

(a) $$f(x, y) = \overbrace{6\overset{\text{degree 1}}{x}\overset{\text{degree 3}}{y^3}}^{\text{degree 4}} - \underbrace{\underset{\text{degree 2}}{x^2}\overset{\text{degree 2}}{y^2}}_{\text{degree 4}}$$

The function is homogeneous of degree four.

(b) $$f(x, y) = \left.\overset{\text{degree 2}}{x^2} - \underset{\text{degree 1}}{y}\quad\right\}\ \text{degrees not the same}$$

The function is not homogeneous since the degrees of the two terms are different. ■

If $f(x, y)$ is a homogeneous function of degree n, notice that we can write

$$f(x, y) = x^nf\left(1, \frac{y}{x}\right) \quad \text{and} \quad f(x, y) = y^nf\left(\frac{x}{y}, 1\right), \tag{2}$$

where $f(1, y/x)$ and $f(x/y, 1)$ are both homogeneous of degree zero.

EXAMPLE 3 Special Forms of Homogeneous Functions

We see that $f(x, y) = x^2 + 3xy + y^2$ is homogeneous of degree two. Thus

$$f(x, y) = x^2\left[1 + 3\left(\frac{y}{x}\right) + \left(\frac{y}{x}\right)^2\right] = x^2f\left(1, \frac{y}{x}\right)$$

$$f(x, y) = y^2\left[\left(\frac{x}{y}\right)^2 + 3\left(\frac{x}{y}\right) + 1\right] = y^2f\left(\frac{x}{y}, 1\right).$$ ■

A homogeneous first-order differential equation is defined in terms of homogeneous functions.

DEFINITION 2.3 Homogeneous Equation

A differential equation of the form

$$M(x, y)\,dx + N(x, y)\,dy = 0 \tag{3}$$

is said to be **homogeneous** if both coefficients M and N are homogeneous functions of the same degree.

In other words, $M(x, y)\,dx + N(x, y)\,dy = 0$ is homogeneous if

$$M(tx, ty) = t^n M(x, y) \quad \text{and} \quad N(tx, ty) = t^n N(x, y).$$

Method of Solution A homogeneous differential equation $M(x, y)\,dx + N(x, y)\,dy = 0$ can be solved by means of an algebraic substitution. Specifically, *either* substitution $y = ux$ or $x = vy$, where u and v are new dependent variables, *will reduce the equation to a separable first-order differential equation.* To see this, let us substitute $y = ux$ and its differential $dy = u\,dx + x\,du$ into (3):

$$M(x, ux)\,dx + N(x, ux)[u\,dx + x\,du] = 0.$$

Now, by the homogeneity property given in (2), we can write

$$x^n M(1, u)\,dx + x^n N(1, u)[u\,dx + x\,du] = 0$$

or

$$[M(1, u) + uN(1, u)]\,dx + xN(1, u)\,du = 0,$$

which gives

$$\frac{dx}{x} + \frac{N(1, u)\,du}{M(1, u) + uN(1, u)} = 0.$$

We hasten to point out that the preceding formula should not be memorized; rather, the *procedure should be worked through each time.* The proof that the substitution $x = vy$ in (3) also leads to a separable equation is left as an exercise. See Problem 45.

EXAMPLE 4 Solving a Homogeneous DE

Solve $(x^2 + y^2)\,dx + (x^2 - xy)\,dy = 0.$

Solution Both $M(x, y)$ and $N(x, y)$ are homogeneous of degree two. If we let $y = ux$, it follows that

$$(x^2 + u^2x^2)\,dx + (x^2 - ux^2)[u\,dx + x\,du] = 0$$

$$x^2(1 + u)\,dx + x^3(1 - u)\,du = 0$$

$$\frac{1 - u}{1 + u}\,du + \frac{dx}{x} = 0$$

$$\left[-1 + \frac{2}{1 + u}\right]du + \frac{dx}{x} = 0 \quad \leftarrow \text{long division}$$

After integration the last line gives

$$-u + 2\ln|1 + u| + \ln|x| = \ln|c| \quad \text{or} \quad -\frac{y}{x} + 2\ln\left|1 + \frac{y}{x}\right| + \ln|x| = \ln|c|.$$

Using the properties of logarithms, we can write the preceding solution as

$$\ln\left|\frac{(x+y)^2}{cx}\right| = \frac{y}{x}.$$

The definition of a logarithm then yields

$$(x + y)^2 = cxe^{y/x}.$$

EXAMPLE 5 Solving a Homogeneous DE

Solve $(2\sqrt{xy} - y)dx - x\,dy = 0$.

Solution The coefficients $M(x, y)$ and $N(x, y)$ are homogeneous of degree one. If $y = ux$, the differential equation becomes, after simplifying,

$$\frac{du}{2u - 2u^{1/2}} + \frac{dx}{x} = 0.$$

The integral of the first term can be evaluated by the further substitution $t = u^{1/2}$. The result is

$$\frac{dt}{t-1} + \frac{dx}{x} = 0.$$

Integrating yields $\qquad \ln|t - 1| + \ln|x| = \ln|c|$

$$\ln\left|\sqrt{\frac{y}{x}} - 1\right| + \ln|x| = \ln|c| \leftarrow t = u^{1/2} \text{ and } u = y/x$$

$$x\left(\sqrt{\frac{y}{x}} - 1\right) = c \quad \text{or} \quad \sqrt{xy} - x = c.$$

By now you may be asking, When should the substitution $x = vy$ be used? Although it can be used for every homogeneous differential equation, in practice we try $x = vy$ whenever the function $M(x, y)$ is simpler than $N(x, y)$. In solving $(x^2 + y^2)dx + (x^2 - xy)dy = 0$, for example, we know there is no appreciable difference between M and N, so either $y = ux$ or $x = vy$ could be used. Also, it could happen that after using one substitution, we encounter integrals that are difficult or impossible to evaluate in closed form; switching substitutions may result in an easier problem.

EXAMPLE 6 Solving a Homogeneous DE

Solve $2x^3y\,dx + (x^4 + y^4)\,dy = 0$.

Solution Each coefficient is a homogeneous function of degree four. Since the coefficient of dx is slightly simpler than the coefficient of dy, we try $x = vy$. After substituting, we simplify the equation

$$2v^3y^4[v\,dy + y\,dv] + (v^4y^4 + y^4)dy = 0 \quad \text{to} \quad \frac{2v^3\,dv}{3v^4 + 1} + \frac{dy}{y} = 0.$$

Integration gives

$$\frac{1}{6}\ln(3v^4 + 1) + \ln|y| = \ln|c_1| \quad \text{or} \quad 3x^4y^2 + y^6 = c,$$

where $c = c_1^6$. Now had the substitution $y = ux$ been used, then

$$\frac{dx}{x} + \frac{u^4 + 1}{u^5 + 3u}\,du = 0.$$

You are urged to reflect on how to evaluate the integral of the second term in the last equation.

A homogeneous differential equation can always be expressed in the alternative form

$$\frac{dy}{dx} = F\left(\frac{y}{x}\right).$$

To see this, suppose we write the equation $M(x, y)dx + N(x, y)dy = 0$ as $dy/dx = f(x, y)$, where

$$f(x,y) = -\frac{M(x, y)}{N(x, y)}.$$

The function $f(x, y)$ must necessarily be homogeneous of degree zero when M and N are homogeneous of degree n. From (2), it follows that

$$f(x, y) = -\frac{x^nM\left(1, \frac{y}{x}\right)}{x^nN\left(1, \frac{y}{x}\right)} = -\frac{M\left(1, \frac{y}{x}\right)}{N\left(1, \frac{y}{x}\right)}.$$

The last ratio is recognized as a function of the form $F(y/x)$. We leave it as an exercise to demonstrate that a homogeneous differential equation can also be written as $dy/dx = G(x/y)$. See Problem 47.

EXAMPLE 7 An Initial-Value Problem

Solve the initial-value problem $x\dfrac{dy}{dx} = y + xe^{y/x}$, $y(1) = 1$.

Solution By writing the equation as

$$\frac{dy}{dx} = \frac{y}{x} + e^{y/x},$$

we see that the function to the right of the equality is homogeneous of degree zero. From the form of this function we are prompted to use $u = y/x$. After differentiating $y = ux$ by the product rule and substituting, we find

$$u + x\frac{du}{dx} = u + e^u \quad \text{or} \quad e^{-u}\,du = \frac{dx}{x}.$$

Integrating and substituting $u = y/x$ gives

$$-e^{-u} + c = \ln|x| \quad \text{or} \quad -e^{-y/x} + c = \ln|x|.$$

Since $y = 1$ when $x = 1$, we get $-e^{-1} + c = 0$ or $c = e^{-1}$. Therefore the solution to the initial-value problem is

$$e^{-1} - e^{-y/x} = \ln|x|.$$

EXERCISES 2.3

Answers to odd-numbered problems begin on page A-2.

In Problems 1–10 determine whether the given function is homogeneous. If so, state the degree of homogeneity.

1. $x^3 + 2xy^2 - y^4/x$
2. $\sqrt{x + y}\,(4x + 3y)$
3. $\dfrac{x^3y - x^2y^2}{(x + 8y)^2}$
4. $\dfrac{x}{y^2 + \sqrt{x^4 + y^4}}$
5. $\cos\dfrac{x^2}{x + y}$
6. $\sin\dfrac{x}{x + y}$
7. $\ln x^2 - 2\ln y$
8. $\dfrac{\ln x^3}{\ln y^3}$
9. $(x^{-1} + y^{-1})^2$
10. $(x + y + 1)^2$

In Problems 11–30 solve the given differential equation by using an appropriate substitution.

11. $(x - y)\,dx + x\,dy = 0$
12. $(x + y)\,dx + x\,dy = 0$
13. $x\,dx + (y - 2x)\,dy = 0$
14. $y\,dx = 2(x + y)\,dy$
15. $(y^2 + yx)\,dx - x^2\,dy = 0$
16. $(y^2 + yx)\,dx + x^2\,dy = 0$
17. $\dfrac{dy}{dx} = \dfrac{y - x}{y + x}$
18. $\dfrac{dy}{dx} = \dfrac{x + 3y}{3x + y}$
19. $-y\,dx + (x + \sqrt{xy})\,dy = 0$
20. $x\dfrac{dy}{dx} - y = \sqrt{x^2 + y^2}$
21. $2x^2y\,dx = (3x^3 + y^3)\,dy$
22. $(x^4 + y^4)\,dx - 2x^3y\,dy = 0$
23. $\dfrac{dy}{dx} = \dfrac{y}{x} + \dfrac{x}{y}$
24. $\dfrac{dy}{dx} = \dfrac{y}{x} + \dfrac{x^2}{y^2} + 1$
25. $y\dfrac{dx}{dy} = x + 4ye^{-2x/y}$
26. $(x^2e^{-y/x} + y^2)\,dx = xy\,dy$
27. $\left(y + x\cot\dfrac{y}{x}\right)dx - x\,dy = 0$
28. $\dfrac{dy}{dx} = \dfrac{y}{x}\ln\dfrac{y}{x}$
29. $(x^2 + xy - y^2)\,dx + xy\,dy = 0$
30. $(x^2 + xy + 3y^2)\,dx - (x^2 + 2xy)\,dy = 0$

In Problems 31–44 solve the given differential equation subject to the indicated initial condition.

31. $xy^2\dfrac{dy}{dx} = y^3 - x^3, \quad y(1) = 2$

32. $(x^2 + 2y^2)\,dx = xy\,dy, \quad y(-1) = 1$

33. $2x^2 \dfrac{dy}{dx} = 3xy + y^2, \quad y(1) = -2$

34. $xy\,dx - x^2\,dy = y\sqrt{x^2 + y^2}\,dy, \quad y(0) = 1$

35. $(x + ye^{y/x})\,dx - xe^{y/x}\,dy = 0, \quad y(1) = 0$

36. $y\,dx + \left(y \cos\dfrac{x}{y} - x\right)dy = 0, \quad y(0) = 2$

37. $(y^2 + 3xy)\,dx = (4x^2 + xy)\,dy, \quad y(1) = 1$

38. $y^3\,dx = 2x^3\,dy - 2x^2y\,dx, \quad y(1) = \sqrt{2}$

39. $(x + \sqrt{xy})\dfrac{dy}{dx} + x - y = x^{-1/2}y^{3/2}, \quad y(1) = 1$

40. $y\,dx + x\,(\ln x - \ln y - 1)\,dy = 0, \quad y(1) = e$

41. $y^2\,dx + (x^2 + xy + y^2)\,dy = 0, \quad y(0) = 1$

42. $(\sqrt{x} + \sqrt{y})^2\,dx = x\,dy, \quad y(1) = 0$

43. $(x + \sqrt{y^2 - xy})\dfrac{dy}{dx} = y, \quad y(\tfrac{1}{2}) = 1$

44. $\dfrac{dy}{dx} - \dfrac{y}{x} = \cosh\dfrac{y}{x}, \quad y(1) = 0$

45. Suppose $M(x, y)\,dx + N(x, y)\,dy = 0$ is a homogeneous equation. Show that the substitution $x = vy$ reduces the equation to one with separable variables.

46. Suppose $M(x, y)\,dx + N(x, y)\,dy = 0$ is a homogeneous equation. Show that the substitutions $x = r\cos\theta$, $y = r\sin\theta$ reduce the equation to one with separable variables.

47. Suppose $M(x, y)\,dx + N(x, y)\,dy = 0$ is a homogeneous equation. Show that the equation has the alternative form $dy/dx = G(x/y)$.

48. If $f(x,y)$ is a homogeneous function of degree n, show that

$$x\frac{\partial f}{\partial x} + y\frac{\partial f}{\partial y} = nf.$$

2.4 EXACT EQUATIONS

• *Exact differential* • *Exact equation* • *Integrating factor*

Although the simple equation

$$y\,dx + x\,dy = 0$$

is both separable and homogeneous, we should recognize that it is also equivalent to the differential of the product of x and y; that is,

$$y\,dx + x\,dy = d(xy) = 0.$$

By integrating we immediately obtain the implicit solution $xy = c$.

From calculus you might remember that if $z = f(x, y)$ is a function having continuous first partial derivatives in a region R of the xy-plane, then its *total differential* is

$$dz = \frac{\partial f}{\partial x}\,dx + \frac{\partial f}{\partial y}\,dy. \tag{1}$$

Now if $f(x, y) = c$, it follows from (1) that

$$\frac{\partial f}{\partial x}\,dx + \frac{\partial f}{\partial y}\,dy = 0. \tag{2}$$

In other words, given a family of curves $f(x, y) = c$, we can generate a first-order differential equation by computing the total differential.

EXAMPLE 1 Differential Yields a DE

If $x^2 - 5xy + y^3 = c$, then (2) gives

$$(2x - 5y)\,dx + (-5x + 3y^2)\,dy = 0 \quad \text{or} \quad \frac{dy}{dx} = \frac{5y - 2x}{-5x + 3y^2}.$$

For our purposes, it is more important to turn the problem around; namely, given an equation such as

$$\frac{dy}{dx} = \frac{5y - 2x}{-5x + 3y^2}, \tag{3}$$

can we identify the equation as being equivalent to the statement

$$d(x^2 - 5xy + y^3) = 0?$$

Notice that equation (3) is neither separable nor homogeneous.

DEFINITION 2.4 Exact Equation

A differential expression

$$M(x, y)\,dx + N(x, y)\,dy$$

is an **exact differential** in a region R of the xy-plane if it corresponds to the total differential of some function $f(x, y)$. A differential equation of the form

$$M(x, y)\,dx + N(x, y)\,dy = 0$$

is said to be an **exact equation** if the expression on the left side is an exact differential.

EXAMPLE 2 An Exact DE

The equation $x^2y^3\,dx + x^3y^2\,dy = 0$ is exact since it is recognized that

$$d\left(\tfrac{1}{3}x^3y^3\right) = x^2y^3\,dx + x^3y^2\,dy.$$

The following theorem is a test for an exact differential.

THEOREM 2.2 **Criterion for an Exact Differential**

Let $M(x, y)$ and $N(x, y)$ be continuous and have continuous first partial derivatives in a rectangular region R defined by $a < x < b, c < y < d$. Then a necessary and sufficient condition that

$$M(x, y)\, dx + N(x, y)\, dy$$

be an exact differential is

$$\frac{\partial M}{\partial y} = \frac{\partial N}{\partial x}. \tag{4}$$

Proof of the Necessity For simplicity let us assume that $M(x, y)$ and $N(x, y)$ have continuous first partial derivatives for all (x, y). Now if the expression $M(x, y)\, dx + N(x, y)\, dy$ is exact, there exists some function f for which

$$M(x, y)\, dx + N(x, y)\, dy = \frac{\partial f}{\partial x}\, dx + \frac{\partial f}{\partial y}\, dy$$

for all (x, y) in R. Therefore

$$M(x, y) = \frac{\partial f}{\partial x}, \quad N(x, y) = \frac{\partial f}{\partial y},$$

and $$\frac{\partial M}{\partial y} = \frac{\partial}{\partial y}\left(\frac{\partial f}{\partial x}\right) = \frac{\partial^2 f}{\partial y\, \partial x} = \frac{\partial}{\partial x}\left(\frac{\partial f}{\partial y}\right) = \frac{\partial N}{\partial x}.$$

The equality of the mixed partials is a consequence of the continuity of the first partial derivatives of $M(x, y)$ and $N(x, y)$. ■

The sufficiency part of Theorem 2.2 consists of showing that there exists a function f for which $\partial f/\partial x = M(x, y)$ and $\partial f/\partial y = N(x, y)$ whenever (4) holds. The construction of the function f actually reflects a basic procedure for solving exact equations.

Method of Solution Given the equation

$$M(x, y)\, dx + N(x, y)\, dy = 0, \tag{5}$$

first show that $$\frac{\partial M}{\partial y} = \frac{\partial N}{\partial x}.$$

Then assume that $$\frac{\partial f}{\partial x} = M(x, y)$$

so we can find f by integrating $M(x, y)$ with respect to x, while holding y constant. We write

$$f(x, y) = \int M(x, y)\, dx + g(y), \tag{6}$$

where the arbitrary function $g(y)$ is the "constant" of integration. Now differentiate (6) with respect to y and assume $\partial f/\partial y = N(x, y)$:

$$\frac{\partial f}{\partial y} = \frac{\partial}{\partial y}\int M(x, y)\,dx + g'(y) = N(x, y).$$

This gives
$$g'(y) = N(x, y) - \frac{\partial}{\partial y}\int M(x, y)\,dx. \tag{7}$$

Finally, integrate (7) with respect to y and substitute the result in (6). The solution of the equation is $f(x, y) = c$.

Note Some observations are in order. First, it is important to realize that the expression $N(x, y) - (\partial/\partial y)\int M(x, y)\,dx$ in (7) is independent of x, since

$$\frac{\partial}{\partial x}\left[N(x, y) - \frac{\partial}{\partial y}\int M(x, y)\,dx\right] = \frac{\partial N}{\partial x} - \frac{\partial}{\partial y}\left(\frac{\partial}{\partial x}\int M(x, y)\,dx\right)$$

$$= \frac{\partial N}{\partial x} - \frac{\partial M}{\partial y} = 0.$$

Second, we could just as well start the foregoing procedure with the assumption that $\partial f/\partial y = N(x, y)$. After integrating N with respect to y and then differentiating that result, we find the analogues of (6) and (7) would be, respectively,

$$f(x, y) = \int N(x, y)\,dy + h(x) \quad \text{and} \quad h'(x) = M(x, y) - \frac{\partial}{\partial x}\int N(x, y)\,dy.$$

In either case *none of these formulas should be memorized.* Also, when testing an equation for exactness, make sure it is of form (5). Often a differential equation is written $G(x, y)\,dx = H(x, y)\,dy$. In this case write the equation as $G(x, y)\,dx - H(x, y)\,dy = 0$ and then identify $M(x, y) = G(x, y)$ and $N(x, y) = -H(x, y)$.

EXAMPLE 3 Solving an Exact DE

Solve $2xy\,dx + (x^2 - 1)\,dy = 0$.

Solution With $M(x, y) = 2xy$ and $N(x, y) = x^2 - 1$, we have

$$\frac{\partial M}{\partial y} = 2x = \frac{\partial N}{\partial x}.$$

Thus the equation is exact, and so, by Theorem 2.2, there exists a function $f(x, y)$ such that

$$\frac{\partial f}{\partial x} = 2xy \quad \text{and} \quad \frac{\partial f}{\partial y} = x^2 - 1.$$

From the first of these equations we obtain, after integrating,

$$f(x, y) = x^2 y + g(y).$$

Taking the partial derivative of the last expression with respect to y and setting the result equal to $N(x, y)$ gives

$$\frac{\partial f}{\partial y} = x^2 + g'(y) = x^2 - 1. \quad \leftarrow N(x, y)$$

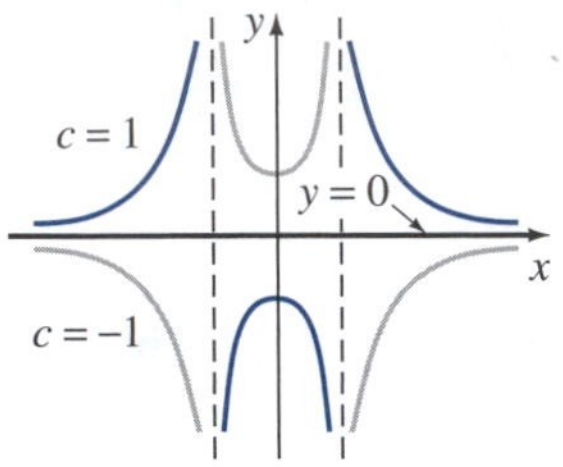

Figure 2.8

It follows that $g'(y) = -1$ and $g(y) = -y$.

The constant of integration need not be included in the preceding line since the solution is $f(x, y) = c$. Some of the family of curves $x^2y - y = c$ are given in Figure 2.8.

Note The solution of the equation is *not* $f(x, y) = x^2y - y$. Rather it is $f(x, y) = c$ or $f(x, y) = 0$ if a constant is used in the integration of $g'(y)$. Observe that the equation could also be solved by separation of variables.

EXAMPLE 4 Solving an Exact DE

Solve $(e^{2y} - y\cos xy)\,dx + (2xe^{2y} - x\cos xy + 2y)\,dy = 0$.

Solution The equation is neither separable nor homogeneous but is exact since

$$\frac{\partial M}{\partial y} = 2e^{2y} + xy\sin xy - \cos xy = \frac{\partial N}{\partial x}.$$

Hence a function $f(x, y)$ exists for which

$$M(x, y) = \frac{\partial f}{\partial x} \quad \text{and} \quad N(x, y) = \frac{\partial f}{\partial y}.$$

Now for variety we shall start with the assumption that $\partial f/\partial y = N(x, y)$;

that is,
$$\frac{\partial f}{\partial y} = 2xe^{2y} - x\cos xy + 2y$$

$$f(x, y) = 2x\int e^{2y}dy - x\int \cos xy\,dy + 2\int y\,dy.$$

Remember, the reason x can come out in front of the symbol $\int$ is that in the integration with respect to y, x is treated as an ordinary constant. It follows that

$$f(x, y) = xe^{2y} - \sin xy + y^2 + h(x)$$

$$\frac{\partial f}{\partial x} = e^{2y} - y\cos xy + h'(x) = e^{2y} - y\cos xy, \quad \leftarrow M(x, y)$$

so
$$h'(x) = 0 \quad \text{and} \quad h(x) = c.$$

Hence a one-parameter family of solutions is given by

$$xe^{2y} - \sin xy + y^2 + c = 0.$$

EXAMPLE 5 An Initial-Value Problem

Solve the initial-value problem

$$(\cos x\sin x - xy^2)\,dx + y(1 - x^2)\,dy = 0, \quad y(0) = 2.$$

Solution The equation is exact since

$$\frac{\partial M}{\partial y} = -2xy = \frac{\partial N}{\partial x}.$$

Now
$$\frac{\partial f}{\partial y} = y(1 - x^2)$$
$$f(x, y) = \frac{y^2}{2}(1 - x^2) + h(x)$$
$$\frac{\partial f}{\partial x} = -xy^2 + h'(x) = \cos x \sin x - xy^2.$$

The last equation implies

$$h'(x) = \cos x \sin x \quad \text{or} \quad h(x) = -\int(\cos x)(-\sin x\, dx) = -\frac{1}{2}\cos^2 x.$$

Thus
$$\frac{y^2}{2}(1 - x^2) - \frac{1}{2}\cos^2 x = c_1 \quad \text{or} \quad y^2(1 - x^2) - \cos^2 x = c,$$

where $2c_1$ has been replaced by c. The initial condition $y = 2$ when $x = 0$ demands that $4(1) - \cos^2(0) = c$ or that $c = 3$. Thus a solution of the problem is

$$y^2(1 - x^2) - \cos^2 x = 3.$$

Integrating Factor It is sometimes possible to convert a nonexact differential equation into an exact equation by multiplying it by a function $\mu(x, y)$ called an **integrating factor.** However, the resulting exact equation

$$\mu M(x, y)\, dx + \mu N(x, y)\, dy = 0$$

may not be equivalent to the original equation in the sense that a solution of one is also a solution of the other. It is possible for a solution to be lost or gained as a result of the multiplication.

EXAMPLE 6 Using an Integrating Factor

Solve $(x + y)\, dx + x \ln x\, dy = 0$, using $\mu(x, y) = \dfrac{1}{x}$, on $(0, \infty)$.

Solution Let $M(x, y) = x + y$ and $N(x, y) = x \ln x$ so $\partial M/\partial y = 1$ and $\partial N/\partial x = 1 + \ln x$. The given equation is not exact. However, if we multiply the equation by $\mu(x, y) = 1/x$, we obtain

$$\left(1 + \frac{y}{x}\right) dx + \ln x\, dy = 0.$$

From this latter form we make the identifications:

$$M(x, y) = 1 + \frac{y}{x}, \quad N(x, y) = \ln x, \quad \frac{\partial M}{\partial y} = \frac{1}{x} = \frac{\partial N}{\partial x}.$$

Therefore the second differential equation is exact. It follows that

$$\frac{\partial f}{\partial x} = 1 + \frac{y}{x} = M(x, y)$$
$$f(x, y) = x + y \ln x + g(y)$$
$$\frac{\partial f}{\partial y} = 0 + \ln x + g'(y) = \ln x$$

and so $$g'(y) = 0 \quad \text{and} \quad g(y) = c.$$

Hence $f(x, y) = x + y \ln x + c$. It is readily verified that

$$x + y \ln x + c = 0$$

is a solution of both equations on $(0, \infty)$.

EXERCISES 2.4

Answers to odd-numbered problems begin on page A-3.

In Problems 1–24 determine whether the given equation is exact. If it is exact, solve.

1. $(2x - 1)dx + (3y + 7)dy = 0$
2. $(2x + y)dx - (x + 6y)dy = 0$
3. $(5x + 4y)dx + (4x - 8y^3)dy = 0$
4. $(\sin y - y \sin x)dx + (\cos x + x \cos y - y)dy = 0$
5. $(2y^2x - 3)dx + (2yx^2 + 4)dy = 0$
6. $\left(2y - \dfrac{1}{x} + \cos 3x\right)\dfrac{dy}{dx} + \dfrac{y}{x^2} - 4x^3 + 3y \sin 3x = 0$
7. $(x + y)(x - y)dx + x(x - 2y)dy = 0$
8. $\left(1 + \ln x + \dfrac{y}{x}\right)dx = (1 - \ln x)\, dy$
9. $(y^3 - y^2 \sin x - x)dx + (3xy^2 + 2y \cos x)dy = 0$
10. $(x^3 + y^3)dx + 3xy^2 dy = 0$
11. $(y \ln y - e^{-xy})\, dx + \left(\dfrac{1}{y} + x \ln y\right)dy = 0$
12. $\dfrac{2x}{y}\, dx - \dfrac{x^2}{y^2}\, dy = 0$
13. $x \dfrac{dy}{dx} = 2xe^x - y + 6x^2$
14. $(3x^2y + e^y)dx + (x^3 + xe^y - 2y)dy = 0$
15. $\left(1 - \dfrac{3}{x} + y\right)dx + \left(1 - \dfrac{3}{y} + x\right)dy = 0$
16. $(e^y + 2xy \cosh x)y' + xy^2 \sinh x + y^2 \cosh x = 0$
17. $\left(x^2y^3 - \dfrac{1}{1 + 9x^2}\right)\dfrac{dx}{dy} + x^3y^2 = 0$
18. $(5y - 2x)y' - 2y = 0$

19. $(\tan x - \sin x \sin y)\,dx + \cos x \cos y\,dy = 0$

20. $(3x \cos 3x + \sin 3x - 3)\,dx + (2y + 5)\,dy = 0$

21. $(1 - 2x^2 - 2y)\dfrac{dy}{dx} = 4x^3 + 4xy$

22. $(2y \sin x \cos x - y + 2y^2 e^{xy^2})\,dx = (x - \sin^2 x - 4xye^{xy^2})\,dy$

23. $(4x^3y - 15x^2 - y)\,dx + (x^4 + 3y^2 - x)\,dy = 0$

24. $\left(\dfrac{1}{x} + \dfrac{1}{x^2} - \dfrac{y}{x^2 + y^2}\right)dx + \left(ye^y + \dfrac{x}{x^2 + y^2}\right)dy = 0$

In Problems 25–30 solve the given differential equation subject to the indicated initial condition.

25. $(x + y)^2\,dx + (2xy + x^2 - 1)\,dy = 0,\quad y(1) = 1$

26. $(e^x + y)\,dx + (2 + x + ye^y)\,dy = 0,\quad y(0) = 1$

27. $(4y + 2x - 5)\,dx + (6y + 4x - 1)\,dy = 0,\quad y(-1) = 2$

28. $\left(\dfrac{3y^2 - x^2}{y^5}\right)\dfrac{dy}{dx} + \dfrac{x}{2y^4} = 0,\quad y(1) = 1$

29. $(y^2 \cos x - 3x^2y - 2x)\,dx + (2y \sin x - x^3 + \ln y)\,dy = 0,\quad y(0) = e$

30. $\left(\dfrac{1}{1 + y^2} + \cos x - 2xy\right)\dfrac{dy}{dx} = y(y + \sin x),\quad y(0) = 1$

In Problems 31–34 find the value of k so that the given differential equation is exact.

31. $(y^3 + kxy^4 - 2x)\,dx + (3xy^2 + 20x^2y^3)\,dy = 0$

32. $(2x - y \sin xy + ky^4)\,dx - (20xy^3 + x \sin xy)\,dy = 0$

33. $(2xy^2 + ye^x)\,dx + (2x^2y + ke^x - 1)\,dy = 0$

34. $(6xy^3 + \cos y)\,dx + (kx^2y^2 - x \sin y)\,dy = 0$

35. Determine a function $M(x, y)$ so that the following differential equation is exact:

$$M(x,y)\,dx + \left(xe^{xy} + 2xy + \frac{1}{x}\right)dy = 0.$$

36. Determine a function $N(x, y)$ so that the following differential equation is exact:

$$\left(y^{1/2}x^{-1/2} + \frac{x}{x^2 + y}\right)dx + N(x, y)\,dy = 0.$$

In Problems 37–42 solve the given differential equation by verifying that the indicated function $\mu(x,y)$ is an integrating factor.

37. $6xy\,dx + (4y + 9x^2)\,dy = 0,\quad \mu(x, y) = y^2$

38. $-y^2\,dx + (x^2 + xy)\,dy = 0,\quad \mu(x, y) = 1/x^2y$

39. $(-xy \sin x + 2y \cos x)\,dx + 2x \cos x\,dy = 0,\quad \mu(x, y) = xy$

40. $y(x + y + 1)\,dx + (x + 2y)\,dy = 0,\quad \mu(x, y) = e^x$

41. $(2y^2 + 3x)\,dx + 2xy\,dy = 0,\quad \mu(x, y) = x$

42. $(x^2 + 2xy - y^2)\,dx + (y^2 + 2xy - x^2)\,dy = 0,\quad \mu(x, y) = (x + y)^{-2}$

43. Show that any separable first-order differential equation in the form $h(y)\,dy - g(x)\,dx = 0$ is also exact.

2.5 LINEAR EQUATIONS

• Linear equation • Integrating factor • General solution

In Chapter 1 we defined the general form of a linear differential equation of order n to be

$$a_n(x)\frac{d^ny}{dx^n} + a_{n-1}(x)\frac{d^{n-1}y}{dx^{n-1}} + \cdots + a_1(x)\frac{dy}{dx} + a_0(x)y = g(x).$$

We remind you that linearity means that all coefficients are functions of x only and that y and all its derivatives are raised to the first power. Now when $n = 1$, we obtain a linear first-order equation.

DEFINITION 2.5 Linear Equation

A differential equation of the form

$$a_1(x)\frac{dy}{dx} + a_0(x)y = g(x)$$

is said to be a **linear equation.**

Dividing by the lead coefficient $a_1(x)$ gives a more useful form, the **standard form,** of a linear equation:

$$\frac{dy}{dx} + P(x)y = f(x). \qquad \textbf{(1)}$$

We seek a solution of (1) on an interval I for which the functions $P(x)$ and $f(x)$ are continuous. In the discussion that follows, we tacitly assume that (1) has a solution.

Integrating Factor Using differentials, we can rewrite equation (1) as

$$dy + [P(x)y - f(x)]\,dx = 0. \qquad \textbf{(2)}$$

Linear equations possess the pleasant property that a function $\mu(x)$ can always be found such that the multiple of (2),

$$\mu(x)dy + \mu(x)[P(x)y - f(x)]dx = 0, \qquad \textbf{(3)}$$

is an exact differential equation. By Theorem 2.2 we know that the left side of equation (3) is an exact differential if

$$\frac{\partial}{\partial x}\mu(x) = \frac{\partial}{\partial y}\mu(x)[P(x)y - f(x)] \qquad \textbf{(4)}$$

or

$$\frac{d\mu}{dx} = \mu P(x).$$

This is a separable equation from which we can determine $\mu(x)$. We have

$$\frac{d\mu}{\mu} = P(x)dx$$

$$\ln|\mu| = \int P(x)\,dx \tag{5}$$

so

$$\mu(x) = e^{\int P(x)dx}. \tag{6}$$

The function $\mu(x)$ defined in (6) is an **integrating factor** for the linear equation. Note that we need not use a constant of integration in (5) since (3) is unaffected by a constant multiple. Also, $\mu(x) \neq 0$ for every x in I and is continuous and differentiable.

It is interesting to observe that equation (3) is still an exact differential equation even when $f(x) = 0$. In fact, $f(x)$ plays no part in determining $\mu(x)$ since we see from (4) that $(\partial/\partial y)\mu(x)f(x) = 0$. Thus both

$$e^{\int P(x)dx}\,dy + e^{\int P(x)dx}[P(x)y - f(x)]\,dx$$

and

$$e^{\int P(x)dx}\,dy + e^{\int P(x)dx}\,P(x)y\,dx$$

are exact differentials. We now write (3) in the form

$$e^{\int P(x)dx}\,dy + e^{\int P(x)dx}\,P(x)y\,dx = e^{\int P(x)dx}\,f(x)\,dx$$

and recognize that we can write this equation as

$$d[e^{\int P(x)dx}y] = e^{\int P(x)dx}f(x)\,dx.$$

Integrating the last equation gives

$$e^{\int P(x)dx}y = \int e^{\int P(x)dx}f(x)\,dx + c$$

or

$$y = e^{-\int P(x)dx}\int e^{\int P(x)dx}f(x)\,dx + ce^{-\int P(x)dx}. \tag{7}$$

In other words, if (1) has a solution, it must be of form (7). Conversely, it is a straightforward matter to verify that (7) constitutes a one-parameter family of solutions of equation (1).

Summary of the Method No attempt should be made to memorize the formula given in (7). The procedure should be followed each time, so for convenience we summarize the results.

SOLVING A LINEAR FIRST-ORDER EQUATION

(*i*) To solve a linear first-order equation first put it into the form (1); that is, make the coefficient of dy/dx unity.

(*ii*) Identify $P(x)$ and find the integrating factor

$$e^{\int P(x)dx}.$$

(*iii*) Multiply the equation obtained in step (*i*) by the integrating factor:

$$e^{\int P(x)dx}\frac{dy}{dx} + P(x)e^{\int P(x)dx}y = e^{\int P(x)dx}f(x).$$

(*iv*) The left side of the equation in step (*iii*) is the derivative of the product of the integrating factor and the dependent variable y; that is,

$$\frac{d}{dx}\left[e^{\int P(x)\,dx}y\right] = e^{\int P(x)\,dx}f(x).$$

(*v*) Integrate both sides of the equation found in step (*iv*).

EXAMPLE 1 Solving a Linear DE

Solve $x\dfrac{dy}{dx} - 4y = x^6e^x$.

Solution Write the equation as

$$\frac{dy}{dx} - \frac{4}{x}y = x^5e^x \tag{8}$$

by dividing by x. Since $P(x) = -4/x$, the integrating factor is

$$e^{-4\int dx/x} = e^{-4\ln|x|} = e^{\ln x^{-4}} = x^{-4}.$$

Here we have used the basic identity $b^{\log_b N} = N$, $N > 0$. Now if we multiply (8) by this term,

$$x^{-4}\frac{dy}{dx} - 4x^{-5}y = xe^x, \tag{9}$$

we obtain

$$\frac{d}{dx}\left[x^{-4}y\right] = xe^x.* \tag{10}$$

It follows from integration by parts that

$$x^{-4}y = xe^x - e^x + c \quad \text{or} \quad y = x^5e^x - x^4e^x + cx^4.$$

EXAMPLE 2 Solving a Linear DE

Solve $\dfrac{dy}{dx} - 3y = 0$.

Solution The equation is already in form (1). Hence the integrating factor is

$$e^{\int(-3)\,dx} = e^{-3x}.$$

Therefore

$$e^{-3x}\frac{dy}{dx} - 3e^{-3x}y = 0$$

$$\frac{d}{dx}\left[e^{-3x}y\right] = 0.$$

---

* You should perform the indicated differentiations a few times in order to be convinced that all equations such as (8), (9), and (10) are formally equivalent.

Integrating, we have $$e^{-3x}y = c$$

and so $$y = ce^{3x}.$$

General Solution If it is assumed that $P(x)$ and $f(x)$ are continuous on an interval I and x_0 is any point in the interval, then it follows from Theorem 2.1 that there exists only one solution of the initial-value problem

$$\frac{dy}{dx} + P(x)y = f(x), \quad y(x_0) = y_0. \tag{11}$$

But we saw earlier that (1) possesses a family of solutions and that every solution of the equation on the interval I is of form (7). Thus, obtaining the solution of (11) is a simple matter of finding an appropriate value of c in (7). Consequently, we are justified in calling (7) the **general solution** of the differential equation. In retrospect, you should recall that in several instances we found singular solutions of nonlinear equations. This cannot happen in the case of a linear equation when proper attention is paid to solving the equation over a common interval on which $P(x)$ and $f(x)$ are continuous.

EXAMPLE 3 General Solution

Find the general solution of $(x^2 + 9)\dfrac{dy}{dx} + xy = 0.$

Solution We write $$\frac{dy}{dx} + \frac{x}{x^2 + 9}y = 0.$$

The function $P(x) = x/(x^2 + 9)$ is continuous on $(-\infty, \infty)$. Now the integrating factor for the equation is

$$e^{\int x\,dx/(x^2+9)} = e^{\frac{1}{2}\int 2x\,dx/(x^2+9)} = e^{\frac{1}{2}\ln(x^2+9)} = \sqrt{x^2 + 9}$$

so
$$\sqrt{x^2 + 9}\,\frac{dy}{dx} + \frac{x}{\sqrt{x^2 + 9}}y = 0$$

$$\frac{d}{dx}\left[\sqrt{x^2 + 9}\,y\right] = 0$$

$$\sqrt{x^2 + 9}\,y = c.$$

Hence the general solution on the interval is

$$y = \frac{c}{\sqrt{x^2 + 9}}.$$

EXAMPLE 4 An Initial-Value Problem

Solve the initial-value problem $\dfrac{dy}{dx} + 2xy = x,\ y(0) = -3.$

Solution The functions $P(x) = 2x$ and $f(x) = x$ are continuous on $(-\infty, \infty)$. The integrating factor is

$$e^{2\int x\,dx} = e^{x^2}$$

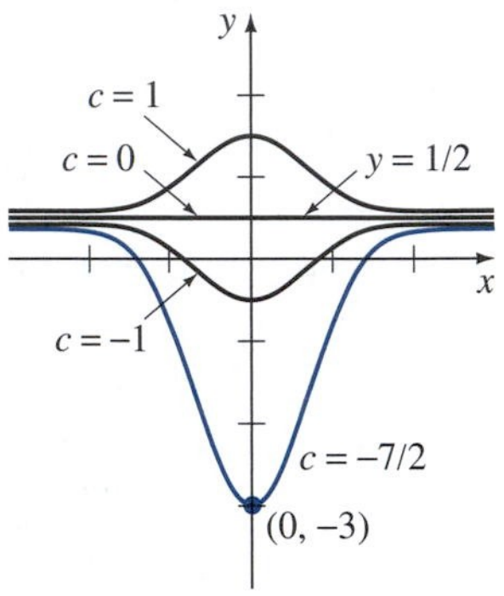

Figure 2.9

so
$$e^{x^2}\frac{dy}{dx} + 2xe^{x^2}y = xe^{x^2}$$
$$\frac{d}{dx}\left[e^{x^2}y\right] = xe^{x^2}$$
$$e^{x^2}y = \int xe^{x^2}dx = \frac{1}{2}e^{x^2} + c.$$

Thus the general solution of the differential equation is

$$y = \frac{1}{2} + ce^{-x^2}.$$

The condition $y(0) = -3$ gives $c = -\frac{7}{2}$, and hence the solution of the initial-value problem on the interval is

$$y = \frac{1}{2} - \frac{7}{2}e^{-x^2}.$$

See Figure 2.9. ■

EXAMPLE 5 An Initial-Value Problem

Solve the initial-value problem $x\dfrac{dy}{dx} + y = 2x,\ y(1) = 0.$

Solution Write the given equation as

$$\frac{dy}{dx} + \frac{1}{x}y = 2$$

and observe that $P(x) = 1/x$ is continuous on any interval not containing the origin. In view of the initial condition, we solve the problem on the interval $(0, \infty)$.

The integrating factor is

$$e^{\int dx/x} = e^{\ln|x|} = x$$

and so
$$\frac{d}{dx}[xy] = 2x$$

gives
$$xy = x^2 + c.$$

The general solution of the equation is

$$y = x + \frac{c}{x}. \tag{12}$$

But $y(1) = 0$ implies $c = -1$. Hence we obtain

$$y = x - \frac{1}{x}, \quad 0 < x < \infty. \tag{13}$$

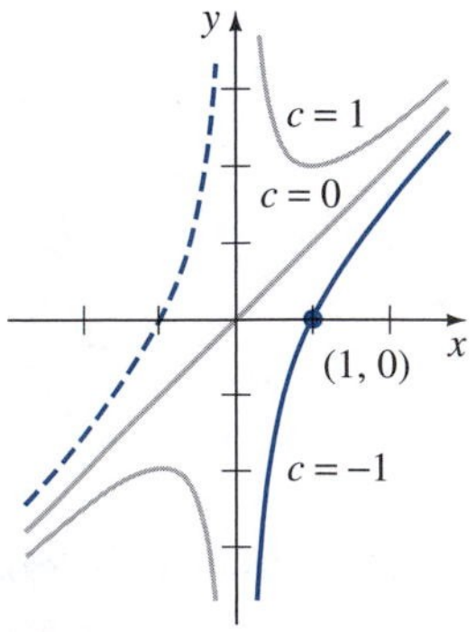

Figure 2.10

Considered as a one-parameter family of curves, the graph of (12) is given in Figure 2.10. The solution (13) of the initial-value problem is indicated by the colored portion of the graph. ■

EXAMPLE 6 An Initial-Value Problem

Solve the initial-value problem $\dfrac{dy}{dx} = \dfrac{1}{x + y^2}$, $y(-2) = 0$.

Solution The given differential equation is not separable, homogeneous, exact, or linear in the variable y. However, if we take the reciprocal, then

$$\frac{dx}{dy} = x + y^2 \quad \text{or} \quad \frac{dx}{dy} - x = y^2.$$

This latter equation is *linear in* x, so the corresponding integrating factor is $e^{-\int dy} = e^{-y}$. Hence,

$$\frac{d}{dy}[e^{-y}x] = y^2e^{-y} \quad \text{and} \quad e^{-y}x = \int y^2e^{-y}\,dy.$$

Integrating by parts twice then gives

$$\begin{aligned} e^{-y}x &= -y^2e^{-y} - 2ye^{-y} - 2e^{-y} + c \\ x &= -y^2 - 2y - 2 + ce^{y}. \end{aligned}$$

When $x = -2$, $y = 0$, we find $c = 0$ and so

$$x = -y^2 - 2y - 2.$$

The next example illustrates a way of solving (1) when the function f is discontinuous.

EXAMPLE 7 A Discontinuous $f(x)$

Find a continuous solution satisfying

$$\frac{dy}{dx} + y = f(x), \quad \text{where} \quad f(x) = \begin{cases} 1, & 0 \le x \le 1 \\ 0, & x > 1 \end{cases}$$

and the initial condition $y(0) = 0$.

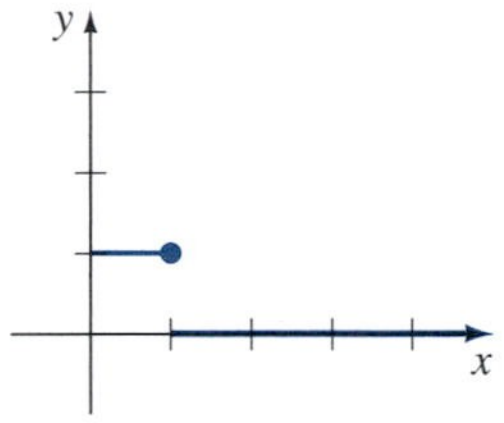

Figure 2.11

Solution From Figure 2.11 we see that f is discontinuous at $x = 1$. Consequently, we solve the problem in two parts. For $0 \le x \le 1$ we have

$$\frac{dy}{dx} + y = 1 \quad \text{or} \quad \frac{d}{dx}[e^{x}y] = e^{x}.$$

The last equation yields $y = 1 + c_1e^{-x}$. Since $y(0) = 0$, we must have $c_1 = -1$, and therefore

$$y = 1 - e^{-x}, \quad 0 \le x \le 1.$$

For $x > 1$ we then have $$\frac{dy}{dx} + y = 0,$$

which leads to $y = c_2e^{-x}$. Hence we can write

$$y = \begin{cases} 1 - e^{-x}, & 0 \le x \le 1 \\ c_2e^{-x}, & x > 1 \end{cases}$$

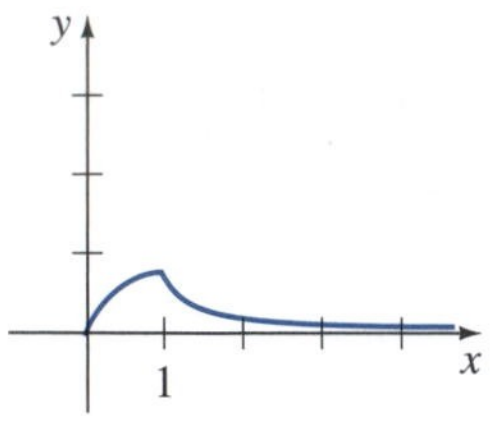

Figure 2.12

Now in order that y be a continuous function, we certainly want $\lim_{x\to 1^+} y(x) = y(1)$. This latter requirement is equivalent to $c_2e^{-1} = 1 - e^{-1}$ or $c_2 = e - 1$. As Figure 2.12 shows, the function

$$y = \begin{cases} 1 - e^{-x}, & 0 \le x \le 1 \\ (e-1)e^{-x}, & x > 1 \end{cases}$$

is continuous but not differentiable at $x = 1$.

Remarks Formula (7), representing the general solution of (1), actually consists of the sum of two solutions. We define

$$y = y_c + y_p, \tag{14}$$

where

$$y_c = ce^{-\int P(x)dx} \quad \text{and} \quad y_p = e^{-\int P(x)dx}\int e^{\int P(x)dx} f(x)\, dx.$$

The function y_c is readily shown to be the general solution of $y' + P(x)y = 0$, whereas y_p is a particular solution of $y' + P(x)y = f(x)$. As we shall see in Chapter 4, the additivity property of solutions (14) to form a general solution is an intrinsic property of linear equations of any order.

EXERCISES 2.5

Answers to odd-numbered problems begin on page A-3.

In Problems 1–40 find the general solution of the given differential equation. State an interval on which the general solution is defined.

1. $\dfrac{dy}{dx} = 5y$

2. $\dfrac{dy}{dx} + 2y = 0$

3. $3\dfrac{dy}{dx} + 12y = 4$

4. $x\dfrac{dy}{dx} + 2y = 3$

5. $\dfrac{dy}{dx} + y = e^{3x}$

6. $\dfrac{dy}{dx} = y + e^x$

7. $y' + 3x^2y = x^2$

8. $y' + 2xy = x^3$

9. $x^2y' + xy = 1$

10. $y' = 2y + x^2 + 5$

11. $(x + 4y^2)\,dy + 2y\,dx = 0$

12. $\dfrac{dx}{dy} = x + y$

13. $x\,dy = (x \sin x - y)dx$

14. $(1 + x^2)\,dy + (xy + x^3 + x)\,dx = 0$

15. $(1 + e^x)\dfrac{dy}{dx} + e^xy = 0$

16. $(1 - x^3)\dfrac{dy}{dx} = 3x^2y$

17. $\cos x \dfrac{dy}{dx} + y \sin x = 1$

18. $\dfrac{dy}{dx} + y \cot x = 2 \cos x$

19. $x \dfrac{dy}{dx} + 4y = x^3 - x$

20. $(1 + x)y' - xy = x + x^2$

21. $x^2y' + x(x + 2)y = e^x$

22. $xy' + (1 + x)y = e^{-x} \sin 2x$

23. $\cos^2 x \sin x \, dy + (y \cos^3 x - 1)\, dx = 0$

24. $(1 - \cos x)\, dy + (2y \sin x - \tan x)\, dx = 0$

25. $y\, dx + (xy + 2x - ye^y)\, dy = 0$

26. $(x^2 + x)\, dy = (x^5 + 3xy + 3y)\, dx$

27. $x \dfrac{dy}{dx} + (3x + 1)y = e^{-3x}$

28. $(x + 1) \dfrac{dy}{dx} + (x + 2)y = 2xe^{-x}$

29. $y\, dx - 4(x + y^6)\, dy = 0$

30. $xy' + 2y = e^x + \ln x$

31. $\dfrac{dy}{dx} + y = \dfrac{1 - e^{-2x}}{e^x + e^{-x}}$

32. $\dfrac{dy}{dx} - y = \sinh x$

33. $y\, dx + (x + 2xy^2 - 2y)\, dy = 0$

34. $y\, dx = (ye^y - 2x)\, dy$

35. $\dfrac{dr}{d\theta} + r \sec \theta = \cos \theta$

36. $\dfrac{dP}{dt} + 2tP = P + 4t - 2$

37. $(x + 2)^2 \dfrac{dy}{dx} = 5 - 8y - 4xy$

38. $(x^2 - 1) \dfrac{dy}{dx} + 2y = (x + 1)^2$

39. $y' = (10 - y) \cosh x$

40. $dx = (3e^y - 2x)\, dy$

In Problems 41–54 solve the given differential equation subject to the indicated initial condition.

41. $\dfrac{dy}{dx} + 5y = 20, \quad y(0) = 2$

42. $y' = 2y + x(e^{3x} - e^{2x}), \quad y(0) = 2$

43. $L \dfrac{di}{dt} + Ri = E; \quad L, R,$ and E constants, $i(0) = i_0$

44. $y \dfrac{dx}{dy} - x = 2y^2, \quad y(1) = 5$

45. $y' + (\tan x)y = \cos^2 x, \quad y(0) = -1$

46. $\dfrac{dQ}{dx} = 5x^4 Q, \quad Q(0) = -7$

47. $\dfrac{dT}{dt} = k(T - 50); \quad k$ a constant, $T(0) = 200$

48. $x\, dy + (xy + 2y - 2e^{-x})\, dx = 0, \quad y(1) = 0$

49. $(x + 1) \dfrac{dy}{dx} + y = \ln x, \quad y(1) = 10$

50. $xy' + y = e^x, \quad y(1) = 2$

51. $x(x - 2)y' + 2y = 0, \quad y(3) = 6$

52. $\sin x \dfrac{dy}{dx} + (\cos x)y = 0, \quad y\left(-\dfrac{\pi}{2}\right) = 1$

53. $\dfrac{dy}{dx} = \dfrac{y}{y - x}, \quad y(5) = 2$

54. $\cos^2 x \dfrac{dy}{dx} + y = 1, \quad y(0) = -3$

In Problems 55–58 find a continuous solution satisfying each differential equation and the given initial condition.

55. $\dfrac{dy}{dx} + 2y = f(x), \quad f(x) = \begin{cases} 1, & 0 \le x \le 3 \\ 0, & x > 3 \end{cases}, \quad y(0) = 0$

56. $\dfrac{dy}{dx} + y = f(x), \quad f(x) = \begin{cases} 1, & 0 \le x \le 1 \\ -1, & x > 1 \end{cases}, \quad y(0) = 1$

57. $\dfrac{dy}{dx} + 2xy = f(x), \quad f(x) = \begin{cases} x, & 0 \le x < 1 \\ 0, & x \ge 1 \end{cases}, \quad y(0) = 2$

58. $(1 + x^2)\dfrac{dy}{dx} + 2xy = f(x), \quad f(x) = \begin{cases} x, & 0 \le x < 1 \\ -x, & x \ge 1 \end{cases}, \quad y(0) = 0$

2.6* EQUATIONS OF BERNOULLI, RICATTI, AND CLAIRAUT†

• *Bernoulli's equation* • *Ricatti's equation* • *Clairaut's equation*

In this section we are not going to study any one particular type of differential equation. Rather, we are going to consider three classical equations that in some instances can be transformed into equations we have already studied.

Bernoulli's Equation The differential equation

$$\frac{dy}{dx} + P(x)y = f(x)y^n, \tag{1}$$

where n is any real number, is called **Bernoulli's equation.** For $n = 0$ and $n = 1$, equation (1) is linear in y. Now for $y \neq 0$, (1) can be written as

$$y^{-n}\frac{dy}{dx} + P(x)y^{1-n} = f(x). \tag{2}$$

---

* This section is an optional section.

† JAKOB BERNOULLI (1654–1705) The Bernoullis were a Swiss family of scholars whose contributions to mathematics, physics, astronomy, and history spanned the sixteenth to the twentieth centuries. Jakob, the elder of the two sons of the patriarch Jacques Bernoulli, made many contributions to the then-new fields of calculus and probability. Originally the second of the two major divisions of calculus was called *calculus summatorius*. In 1696, at Jakob Bernoulli's suggestion, its name was changed to *calculus integralis* or, as we know it today, integral calculus.

JACOBO FRANCESCO RICATTI (1676–1754) An Italian count, Ricatti was also a mathematician and philosopher.

ALEXIS CLAUDE CLAIRAUT (1713–1765) Born in Paris in 1713, Clairaut was a child prodigy who wrote his first book on mathematics at the age of eleven. He was among the first to discover singular solutions of differential equations. Like many mathematicians of his era, Clairaut was also a physicist and astronomer.

If we let $w = y^{1-n}$, $n \neq 0$, $n \neq 1$, then

$$\frac{dw}{dx} = (1 - n)y^{-n}\frac{dy}{dx}.$$

With these substitutions, (2) can be simplified to the linear equation

$$\frac{dw}{dx} + (1 - n)\,P(x)w = (1 - n)\,f(x). \tag{3}$$

Solving (3) for w and using $y^{1-n} = w$ leads to a solution of (1).

EXAMPLE 1 Solving a Bernoulli DE

Solve $\dfrac{dy}{dx} + \dfrac{1}{x}y = xy^2$.

Solution From (1) we identify $P(x) = 1/x$, $f(x) = x$, and $n = 2$. Thus the substitution $w = y^{-1}$ gives

$$\frac{dw}{dx} - \frac{1}{x}w = -x.$$

The integrating factor for this linear equation on, say, $(0, \infty)$ is

$$e^{-\int dx/x} = e^{-\ln|x|} = e^{\ln|x|^{-1}} = x^{-1}.$$

Hence

$$\frac{d}{dx}\left[x^{-1}w\right] = -1.$$

Integrating this latter form, we get

$$x^{-1}w = -x + c \quad \text{or} \quad w = -x^2 + cx.$$

Since $w = y^{-1}$, we obtain $y = 1/w$ or

$$y = \frac{1}{-x^2 + cx}.$$

For $n > 0$ note that the trivial solution $y = 0$ is a solution of (1). In Example 1, $y = 0$ is a singular solution of the given equation.

Ricatti's Equation The nonlinear differential equation

$$\frac{dy}{dx} = P(x) + Q(x)y + R(x)y^2 \tag{4}$$

is called **Ricatti's equation.** If y_1 is a *known* particular solution of (4), then the substitutions

$$y = y_1 + u \quad \text{and} \quad \frac{dy}{dx} = \frac{dy_1}{dx} + \frac{du}{dx}$$

in (4) lead to the following differential equation for u:

$$\frac{du}{dx} - (Q + 2y_1R)u = Ru^2. \tag{5}$$

Since (5) is a Bernoulli equation with $n = 2$, it can in turn be reduced to the linear equation

$$\frac{dw}{dx} + (Q + 2y_1R)w = -R \tag{6}$$

by the substitution $w = u^{-1}$. See Problems 25 and 26.

As Example 2 illustrates, in many cases a solution of a Ricatti equation cannot be expressed in terms of elementary functions.

EXAMPLE 2 Solving a Ricatti DE

Solve $\dfrac{dy}{dx} = 2 - 2xy + y^2$.

Solution It is easily verified that a particular solution of this equation is $y_1 = 2x$. From (4) we make the identifications $P(x) = 2$, $Q(x) = -2x$, and $R(x) = 1$ and then solve the linear equation (6):

$$\frac{dw}{dx} + (-2x + 4x)w = -1 \quad \text{or} \quad \frac{dw}{dx} + 2xw = -1.$$

The integrating factor for the last equation is e^{x^2}, and so

$$\frac{d}{dx}[e^{x^2}w] = -e^{x^2}.$$

Now the integral $\int_{x_0}^{x} e^{t^2}\,dt$ cannot be expressed in terms of elementary functions.* Thus we write

$$e^{x^2}w = -\int_{x_0}^{x} e^{t^2}\,dt + c \quad \text{or} \quad e^{x^2}\left(\frac{1}{u}\right) = -\int_{x_0}^{x} e^{t^2}\,dt + c,$$

so

$$u = \frac{e^{x^2}}{c - \int_{x_0}^{x} e^{t^2}\,dt}.$$

A solution of the equation is then $y = 2x + u$.

Clairaut's Equation It is left as an exercise to show that a solution of **Clairaut's equation**

$$y = xy' + f(y') \tag{7}$$

is the family of straight lines $y = cx + f(c)$, where c is an arbitrary constant. See Problem 29. Furthermore (7) may also possess a solution in parametric form:

$$x = -f'(t), \quad y = f(t) - tf'(t). \tag{8}$$

---

* When an integral $\int f(x)\,dx$ cannot be evaluated in terms of elementary functions, it is customary to write $\int_{x_0}^{x} f(t)\,dt$, where x_0 is a constant. When an initial condition is specified, it is imperative that this form be used.

This last solution is a singular solution since, if $f''(t) \neq 0$, it cannot be obtained from the family of solutions $y = cx + f(c)$.

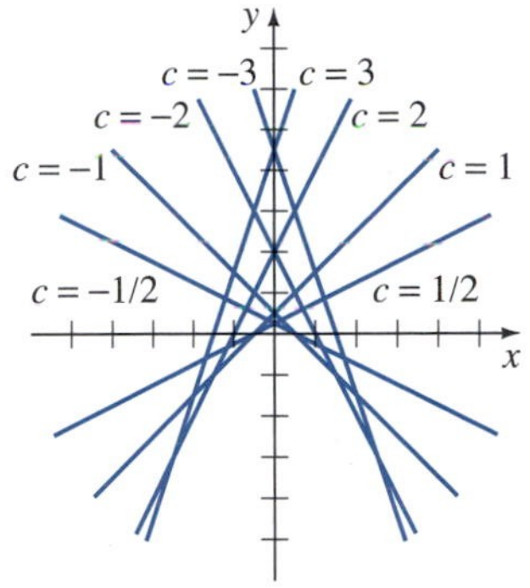

Figure 2.13

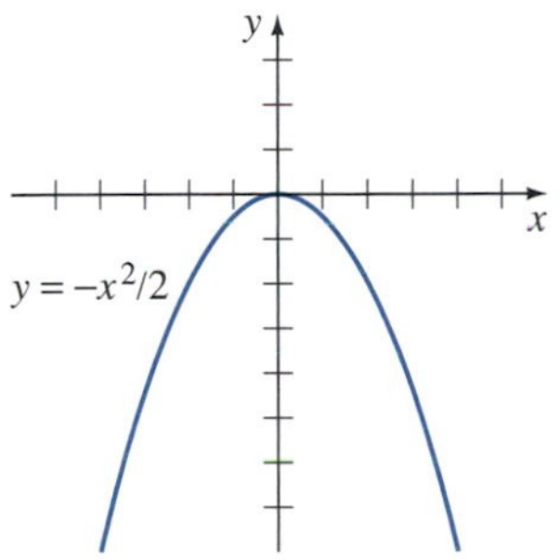

Figure 2.14

EXAMPLE 3 Solving a Clairaut DE

Solve $y = xy' + \frac{1}{2}(y')^2$.

Solution We first make the identification $f(y') = \frac{1}{2}(y')^2$ so that $f(t) = \frac{1}{2}t^2$. It follows from the preceding discussion that a family of solutions is

$$y = cx + \frac{1}{2}c^2. \tag{9}$$

The graph of this family is given in Figure 2.13. Since $f'(t) = t$, a singular solution is obtained from (8):

$$x = -t, \quad y = \frac{1}{2}t^2 - t \cdot t = -\frac{1}{2}t^2.$$

After eliminating the parameter, we find this latter solution is the same as $y = -\frac{1}{2}x^2$. One can readily see that this function is not part of the family (9). See Figure 2.14.

EXERCISES 2.6

Answers to odd-numbered problems begin on page A-3.

In Problems 1–6 solve the given Bernoulli equation.

1. $x\dfrac{dy}{dx} + y = \dfrac{1}{y^2}$

2. $\dfrac{dy}{dx} - y = e^x y^2$

3. $\dfrac{dy}{dx} = y(xy^3 - 1)$

4. $x\dfrac{dy}{dx} - (1 + x)y = xy^2$

5. $x^2\dfrac{dy}{dx} + y^2 = xy$

6. $3(1 + x^2)\dfrac{dy}{dx} = 2xy(y^3 - 1)$

In Problems 7–10 solve the given differential equation subject to the indicated initial condition.

7. $x^2\dfrac{dy}{dx} - 2xy = 3y^4, \quad y(1) = \dfrac{1}{2}$

8. $y^{1/2}\dfrac{dy}{dx} + y^{3/2} = 1, \quad y(0) = 4$

9. $xy(1 + xy^2)\dfrac{dy}{dx} = 1, \quad y(1) = 0$

10. $2\dfrac{dy}{dx} = \dfrac{y}{x} - \dfrac{x}{y^2}, \quad y(1) = 1$

In Problems 11–16 solve the given Ricatti equation; y_1 is a known solution of the equation.

11. $\dfrac{dy}{dx} = -2 - y + y^2, \quad y_1 = 2$

12. $\dfrac{dy}{dx} = 1 - x - y + xy^2, \quad y_1 = 1$

13. $\dfrac{dy}{dx} = -\dfrac{4}{x^2} - \dfrac{1}{x}y + y^2, \quad y_1 = \dfrac{2}{x}$

14. $\dfrac{dy}{dx} = 2x^2 + \dfrac{1}{x}y - 2y^2, \quad y_1 = x$

15. $\dfrac{dy}{dx} = e^{2x} + (1 + 2e^x)y + y^2, \quad y_1 = -e^x$

16. $\dfrac{dy}{dx} = \sec^2 x - (\tan x)y + y^2, \quad y_1 = \tan x$

17. Solve $\dfrac{dy}{dx} = 6 + 5y + y^2$.

18. Solve $\dfrac{dy}{dx} = 9 + 6y + y^2$.

In Problems 19–24 solve the given Clairaut equation. Obtain a singular solution.

19. $y = xy' + 1 - \ln y'$

20. $y = xy' + (y')^{-2}$

21. $y = x\dfrac{dy}{dx} - \left(\dfrac{dy}{dx}\right)^3$

22. $y = (x + 4)y' + (y')^2$

23. $xy' - y = e^{y'}$

24. $y - xy' = \ln y'$

25. Show that if y_1 is a solution of (4), then the substitution $y = y_1 + u$ in (4) yields (5).

26. Show that (5) can be reduced to (6) by means of the substitution $w = u^{-1}$.

27. When $R(x) = -1$, the Ricatti equation can be written as $y' + y^2 - Q(x)y - P(x) = 0$. Show that the substitution $y = w'/w$ leads to the linear second-order equation $w'' - Q(x)w' - P(x)w = 0$. (When Q and P are also constants, there is little difficulty in solving equations of this type.)

28. An alternative definition of Clairaut's equation is any equation of the form $F(y - xy', y') = 0$.

(a) Show that a family of solutions of the latter equation is

$$F(y - cx, c) = 0.$$

(b) Use the result of part (a) to solve

$$(xy' - y)^3 = (y')^2 + 5.$$

29. Show that $y = cx + f(c)$, where c is an arbitrary constant, is a solution of (7).

30. Show that (8) is a solution of (7). [*Hint:* Differentiate both sides of (7) with respect to x and consider two cases. Use parametric differentiation to show that

$$\frac{dy}{dx} = \frac{dy/dt}{dx/dt} = t, \quad f''(t) \neq 0.$$

Note that since the slope of $y = cx + f(c)$ is constant, the singular solution cannot be obtained from it.]

2.7* SUBSTITUTIONS

In the preceding sections, we saw that sometimes a differential equation could be transformed by means of a substitution into a form that could then be solved by one of the standard methods. An equation may look different from any of those that we have just studied, but through a judicious change of variables perhaps an apparently difficult problem may be readily solved. Although we can give no firm rules on what, *if any*, substitution to use, a working axiom might be "Try something!" It sometimes pays to be clever.

EXAMPLE 1 DE Reduced to Separable Variables

The differential equation

$$y(1 + 2xy)\,dx + x(1 - 2xy)\,dy = 0$$

is not separable, homogeneous, exact, linear, or Bernoulli. However, if we stare at the equation long enough, we might be prompted to try the substitution

$$u = 2xy \quad \text{or} \quad y = \frac{u}{2x}.$$

Since

$$dy = \frac{x\,du - u\,dx}{2x^2},$$

the equation becomes, after we simplify,

$$2u^2\,dx + (1 - u)x\,du = 0.$$

We recognize the last equation as separable, and so from

$$2\frac{dx}{x} + \frac{1 - u}{u^2}\,du = 0$$

we obtain

$$2\ln|x| - u^{-1} - \ln|u| = c \quad \text{or} \quad \ln\left|\frac{x}{2y}\right| = c + \frac{1}{2xy}$$

$$\frac{x}{2y} = c_1 e^{1/2xy} \quad \text{or} \quad x = 2yc_1 e^{1/2xy},$$

where e^c was replaced by c_1. We can also replace $2c_1$ by c_2 if desired. ■

Notice that the differential equation in Example 1 possesses the trivial solution $y = 0$ but that this function is not included in the one-parameter family of solutions.

EXAMPLE 2 DE Reduced to a Linear Equation

Solve $2xy\dfrac{dy}{dx} + 2y^2 = 3x - 6$.

---

* This section is an optional section.

Solution The presence of the term $2y\dfrac{dy}{dx}$ prompts us to try $u = y^2$ since

$$\frac{du}{dx} = 2y\frac{dy}{dx}.$$

Now

$$x\frac{du}{dx} + 2u = 3x - 6$$

has the linear form

$$\frac{du}{dx} + \frac{2}{x}u = 3 - \frac{6}{x},$$

so multiplication by the integrating factor $e^{\int(2/x)\,dx} = e^{\ln x^2} = x^2$ gives

$$\frac{d}{dx}[x^2u] = 3x^2 - 6x$$

$$x^2u = x^3 - 3x^2 + c \quad \text{or} \quad x^2y^2 = x^3 - 3x^2 + c.$$

EXAMPLE 3 DE Reduced to Separable Variables

Solve $x\dfrac{dy}{dx} - y = \dfrac{x^3}{y}e^{y/x}$.

Solution If we let $u = y/x$, the differential equation can be simplified to

$$ue^{-u}\,du = dx.$$

Integrating by parts and replacing $-c$ by c_1 then yields

$$-ue^{-u} - e^{-u} = x + c \quad \text{or} \quad u + 1 = (c_1 - x)e^{u}.$$

We then resubstitute $u = y/x$ and simplify:

$$y + x = x(c_1 - x)e^{y/x}.$$

Some higher-order differential equations can be reduced to first-order equations by a substitution.

EXAMPLE 4 Second-Order DE Reduced to a First-Order DE

Solve $y'' = 2x(y')^2$.

Solution If we let $u = y'$ so that $du/dx = y''$, the equation reduces to a separable form. We have

$$\frac{du}{dx} = 2xu^2 \quad \text{or} \quad \frac{du}{u^2} = 2x\,dx$$

$$\int u^{-2}\,du = \int 2x\,dx \quad \text{or} \quad -u^{-1} = x^2 + c_1^2.$$

The constant of integration is written as c_1^2 for convenience. The reason should be obvious in the next few steps. Since $u^{-1} = 1/y'$, it follows that

$$\frac{dy}{dx} = -\frac{1}{x^2 + c_1^2} \quad \text{or} \quad dy = -\frac{dx}{x^2 + c_1^2}$$

$$\int dy = -\int \frac{dx}{x^2 + c_1^2} \quad \text{or} \quad y + c_2 = -\frac{1}{c_1}\tan^{-1}\frac{x}{c_1}.$$

EXERCISES 2.7

Answers to odd-numbered problems begin on page A-3.

In Problems 1–26 solve the given differential equation by using an appropriate substitution.

1. $xe^{2y}\dfrac{dy}{dx} + e^{2y} = \dfrac{\ln x}{x}$

2. $y' + y\ln y = ye^x$

3. $y\,dx + (1 + ye^x)\,dy = 0$

4. $(2 + e^{-x/y})\,dx + 2\left(1 - \dfrac{x}{y}\right)dy = 0$

5. $\dfrac{dy}{dx} - \dfrac{4}{x}y = 2x^5e^{y/x^4}$

6. $\dfrac{dy}{dx} + x + y + 1 = (x + y)^2e^{3x}$

7. $2yy' + x^2 + y^2 + x = 0$

8. $y' = y + x(y + 1)^2 + 1$

9. $2x\csc 2y\,\dfrac{dy}{dx} = 2x - \ln(\tan y)$

10. $x^2\dfrac{dy}{dx} + 2xy = x^4y^2 + 1$

11. $x^4y^2y' + x^3y^3 = 2x^3 - 3$

12. $xe^yy' - 2e^y = x^2$

13. $y' + 1 = e^{-(x+y)}\sin x$

14. $\sin y\sinh x\,dx + \cos y\cosh x\,dy = 0$

15. $y\dfrac{dx}{dy} + 2x\ln x = xe^y$

16. $x\sin y\dfrac{dy}{dx} + \cos y = -x^2e^x$

17. $y'' + (y')^2 + 1 = 0$

18. $xy'' = y' + x(y')^2$

19. $xy'' = y' + (y')^3$

20. $x^2y'' + (y')^2 = 0$

21. $y' - xy'' - (y'')^3 = 1$

22. $y'' = 1 + (y')^2$

23. $xy'' - y' = 0$

24. $y'' + (\tan x)y' = 0$

25. $y'' + 2y(y')^3 = 0$

26. $y^2y'' = y'$

$$\left[\textit{Hint: } \text{Let } u = y' \text{ so that } y'' = \frac{du}{dx} = \frac{du}{dy}\frac{dy}{dx} = \frac{du}{dy}u.\right]$$

27. In calculus the curvature of a curve whose equation is $y = f(x)$ is defined to be the number

$$\kappa = \frac{y''}{[1 + (y')^2]^{3/2}}.$$

Determine a function for which $\kappa = 1$. [*Hint:* For simplicity ignore constants of integration. Also consider a trigonometric substitution.]

2.8* PICARD'S METHOD

- *Successive approximations* • *Picard's method of iteration*

The initial-value problem

$$y' = f(x, y), \quad y(x_0) = y_0, \tag{1}$$

first considered in Section 2.1, can be written in an alternative manner. Let f be continuous in a region containing the point (x_0, y_0). By integrating both sides of the differential equation with respect to x, we get

$$y(x) = c + \int_{x_0}^{x} f(t, y(t))\, dt.$$

Now $$y(x_0) = c + \int_{x_0}^{x_0} f(t, y(t))\, dt = c$$

implies $c = y_0$. Thus

$$y(x) = y_0 + \int_{x_0}^{x} f(t, y(t))\, dt. \tag{2}$$

Conversely, if we start with (2), we can obtain (1). In other words, the integral equation (2) and the initial-value problem (1) are equivalent. We now try to solve (2) by a *method of successive approximations.*

Suppose $y_0(x)$ is an arbitrary continuous function that represents a guess or approximation to the solution of (2). Since $f(x, y_0(x))$ is a known function depending solely on x, it can be integrated. With $y(t)$ replaced by $y_0(t)$, the right-hand side of (2) defines another function, which we write as

$$y_1(x) = y_0 + \int_{x_0}^{x} f(t, y_0(t))\, dt.$$

It is hoped that this new function is a better approximation to the solution. When we repeat the procedure, yet another function is given by

$$y_2(x) = y_0 + \int_{x_0}^{x} f(t, y_1(t))\, dt.$$

In this manner we obtain a sequence of functions $y_1(x), y_2(x), y_3(x), \ldots$ whose nth term is defined by the relation

$$y_n(x) = y_0 + \int_{x_0}^{x} f(t, y_{n-1}(t))\, dt, \quad n = 1, 2, 3, \ldots. \tag{3}$$

In the application of (3), it is common practice to choose the initial function as $y_0(x) = y_0$. The repetitive use of formula (3) is known as **Picard's method of iteration.**

---

* This section is an optional section.

EXAMPLE 1 Using Picard's Method

Consider the problem $y' = y - 1, y(0) = 2$. Use Picard's method to find the approximations y_1, y_2, y_3, y_4.

Solution If we identify $x_0 = 0$, $y_0(x) = 2$, and $f(t, y_{n-1}(t)) = y_{n-1}(t) - 1$, equation (3) becomes

$$y_n(x) = 2 + \int_0^x (y_{n-1}(t) - 1)\, dt, \quad n = 1, 2, 3, \ldots.$$

Iterating this last expression then gives

$$y_1(x) = 2 + \int_0^x 1 \cdot dt = 2 + x$$

$$y_2(x) = 2 + \int_0^x (1 + t)\, dt = 2 + x + \frac{x^2}{2}$$

$$y_3(x) = 2 + \int_0^x \left(1 + t + \frac{t^2}{2}\right) dt = 2 + x + \frac{x^2}{2} + \frac{x^3}{2 \cdot 3}$$

$$y_4(x) = 2 + \int_0^x \left(1 + t + \frac{t^2}{2} + \frac{t^3}{2 \cdot 3}\right) dt$$

$$= 2 + x + \frac{x^2}{2} + \frac{x^3}{2 \cdot 3} + \frac{x^4}{2 \cdot 3 \cdot 4}.$$

By induction it can be shown in Example 1 that the nth term of the sequence of approximations is

$$y_n(x) = 2 + x + \frac{x^2}{2!} + \frac{x^3}{3!} + \cdots + \frac{x^n}{n!} = 1 + \sum_{k=0}^{n} \frac{x^k}{k!}.$$

From this latter form we recognize that the *limit* of $y_n(x)$ as $n \to \infty$ is $y(x) = 1 + e^x$. It should come as no surprise to note that the function is an exact solution of the given initial-value problem.

You should not be deceived by the relative ease with which the iterates $y_n(x)$ were obtained in the last example. In general, the integration involved in generating each $y_n(x)$ can become complicated very quickly. Nor, for that matter, is it always apparent that the sequence $\{y_n(x)\}$ converges to a nice explicit function. Thus it is fair to ask at this point: Is Picard's method a practical means of solving a first-order equation $y' = f(x, y)$ subject to $y(x_0) = y_0$? In most cases the answer is no. One might ask further, in the spirit of a scientist/engineer: What *is* it good for? The answer is not bound to please: Picard's method of iteration is a theoretical tool used in the consideration of the existence and uniqueness of solutions of differential equations. Under certain conditions on $f(x, y)$ it can be shown that as $n \to \infty$, the sequence $\{y_n(x)\}$ defined by (3) converges to a function $y(x)$ that satisfies the integral equation (2) and hence the initial-value problem (1). Indeed, it is precisely Picard's method of successive approximations that is used in proving Picard's theorem of Section 2.1. However, the proof of Theorem 2.1 uses concepts from advanced calculus and is not presented here.

Our purpose in introducing this topic is twofold: for you to gain an appreciation for the potential of the procedure and obtain at least a nodding acquaintance with an iterative technique. In Chapter 9 we shall consider other methods for approximating solutions of differential equations that also utilize iteration.

EXERCISES 2.8

Answers to odd-numbered problems begin on page A-4.

In Problems 1–6 use Picard's method to find y_1, y_2, y_3, y_4. Determine the limit of the sequence $\{y_n(x)\}$ as $n \to \infty$.

1. $y' = -y, \quad y(0) = 1$

2. $y' = x + y, \quad y(0) = 1$

3. $y' = 2xy, \quad y(0) = 1$

4. $y' + 2xy = x, \quad y(0) = 0$

5. $y' + y^2 = 0, \quad y(0) = 0$

6. $y' = 2e^x - y, \quad y(0) = 1$

7. **(a)** Use Picard's method to find y_1, y_2, y_3 for the problem

$$y' = 1 + y^2, \quad y(0) = 0.$$

(b) Solve the initial-value problem in part (a) by one of the methods of this chapter.

(c) Compare the results of parts (a) and (b).

8. In Picard's method the initial choice $y_0(x) = y_0$ is not necessary. Rework Problem 3 with **(a)** $y_0(x) = k$ a constant and $k \neq 1$ and **(b)** $y_0(x) = x$.

CHAPTER 2 REVIEW

An **initial-value problem** consists of finding a solution of

$$\frac{dy}{dx} = f(x, y), \quad y(x_0) = y_0$$

on an interval I containing x_0. If $f(x,y)$ and $\partial f/\partial y$ are continuous in a rectangular region of the xy-plane with (x_0, y_0) in its interior, then we are guaranteed that there exists an interval around x_0 on which the problem has a unique solution.

The method of solution for a first-order differential equation depends on an appropriate classification of the equation. We summarize five cases.

An equation is **separable** if it can be put into the form $h(y)\,dy = g(x)\,dx$. The solution results from integrating both sides of the equation.

If $M(x, y)$ and $N(x, y)$ are **homogeneous functions** of the same degree, then $M(x, y)\,dx + N(x, y)\,dy = 0$ can be reduced to an equation with separable variables by either the substitution $y = ux$ or the substitution $x = vy$. The choice of substitution usually depends on which coefficient is simpler.

The differential equation $M(x, y)\,dx + N(x, y)\,dy = 0$ is said to be **exact** if the form $M(x, y)\,dx + N(x, y)\,dy$ is an exact differential. When $M(x, y)$ and $N(x, y)$ are continuous and have continuous first partial derivatives, then $\partial M/\partial y = \partial N/\partial x$ is a necessary and sufficient condition that $M(x, y)\,dx + N(x, y)\,dy$ be exact. This means there exists some function $f(x, y)$ for which $M(x, y) = \partial f/\partial x$ and $N(x, y) = \partial f/\partial y$. The method of solution for an exact equation starts with integrating either of these latter expressions.

If a first-order equation can be put into the form $dy/dx + P(x)y = f(x)$, it is said to be **linear** in the variable y. We solve the equation by first finding the **integrating factor** $e^{\int P(x)\,dx}$, multiplying both sides of the equation by this factor, and then integrating both sides of

$$\frac{d}{dx}\left[e^{\int P(x)\,dx}y\right] = e^{\int P(x)\,dx}f(x).$$

Bernoulli's equation is $dy/dx + P(x)y = f(x)y^n$, where n is any real number. When $n \neq 0$ and $n \neq 1$, Bernoulli's equation can be reduced to a linear equation by the substitution $w = y^{1-n}$.

In certain circumstances a differential equation can be reduced to one of the familiar forms by an appropriate **substitution,** or **change, of variables.** Of course we already know that this is the procedure when solving a homogeneous or a Bernoulli equation. In the general context, no rule on when to use a substitution can be given.

By converting an initial-value problem to an equivalent integral equation, **Picard's method of iteration** provides one way of obtaining an approximation to the solution of the problem.

CHAPTER 2 REVIEW EXERCISES

Answers to odd-numbered problems begin on page A-4.

Answer Problems 1–4 without referring back to the text. Fill in the blank or answer true or false.

1. The differential equation $y' = 1/(25 - x^2 - y^2)$ has a unique solution through any point (x_0, y_0) in the region(s) defined by ______.
2. The initial-value problem $xy' = 3y$, $y(0) = 0$ has the solutions $y = x^3$ and ______.
3. The initial-value problem $y' = y^{1/2}$, $y(0) = 0$ has no solution since $\partial f/\partial y$ is discontinuous on the line $y = 0$. ______
4. There exists an interval centered at 2 on which the unique solution of the initial-value problem $y' = (y - 1)^3$, $y(2) = 1$ is $y = 1$. ______
5. Without solving, classify each of the following equations as to whether it is separable, homogeneous, exact, linear, Bernoulli, Ricatti, or Clairaut.

(a) $\dfrac{dy}{dx} = \dfrac{1}{y - x}$ **(b)** $\dfrac{dy}{dx} = \dfrac{x - y}{x}$

(c) $\left(\dfrac{dy}{dx}\right)^2 + 2y = 2x\dfrac{dy}{dx}$ **(d)** $\dfrac{dy}{dx} = \dfrac{1}{x(x - y)}$

(e) $\dfrac{dy}{dx} = \dfrac{y^2 + y}{x^2 + x}$ **(f)** $\dfrac{dy}{dx} = 4 + 5y + y^2$

(g) $y\,dx = (y - xy^2)\,dy$ **(h)** $x\dfrac{dy}{dx} = ye^{x/y} - x$

(i) $xyy' + y^2 = 2x$ **(j)** $2xyy' + y^2 = 2x^2$

(k) $y\,dx + x\,dy = 0$

(l) $\left(x^2 + \dfrac{2y}{x}\right) dx = (3 - \ln x^2)\,dy$

(m) $\dfrac{dy}{dx} = \dfrac{x}{y} + \dfrac{y}{x} + 1$ **(n)** $\dfrac{y}{x^2}\dfrac{dy}{dx} + e^{2x^3+y^2} = 0$

(o) $y = xy' + (y' - 3)^2$ **(p)** $y' + 5y^2 = 3x^4 - 2xy$

6. Solve $(y^2 + 1)\,dx = y \sec^2 x\,dy$.

7. Solve $\dfrac{y}{x}\dfrac{dy}{dx} = \dfrac{e^x}{\ln y}$ subject to $y(1) = 1$.

8. Solve $y(\ln x - \ln y)\,dx = (x \ln x - x \ln y - y)\,dy$.

9. Solve $xyy' = 3y^2 + x^2$ subject to $y(-1) = 2$.

10. Solve $(6x + 1)y^2 \dfrac{dy}{dx} + 3x^2 + 2y^3 = 0$.

11. Solve $ye^{xy}\dfrac{dx}{dy} + xe^{xy} = 12y^2$ subject to $y(0) = -1$.

12. Solve $x\,dy + (xy + y - x^2 - 2x)\,dx = 0$.

13. Solve $(x^2 + 4)\dfrac{dy}{dx} = 2x - 8xy$ subject to $y(0) = -1$.

14. Solve $(2x + y)y' = 1$.

15. Solve $x\dfrac{dy}{dx} + 4y = x^4 y^2$ subject to $y(1) = 1$.

16. Solve $-xy' + y = (y' + 1)^2$ subject to $y(0) = 0$.

In Problems 17 and 18 solve the given differential equation by means of a substitution.

17. $\dfrac{dy}{dx} + xy^3 \sec\dfrac{1}{y^2} = 0$ **18.** $y'' = x - y'$

19. Use the Picard method to find approximations y_1 and y_2 for

$$y' = x^2 + y^2, \quad y(0) = 1.$$

20. Solve $y' + 2y = 4$, $y(0) = 3$ by one of the usual methods. Solve the same problem by Picard's method and compare the results.

3 APPLICATIONS OF FIRST-ORDER DIFFERENTIAL EQUATIONS

INTRODUCTION

In Section 1.2 we saw that a differential equation used to describe the behavior of some real-life system, whether physical, sociological, or even economic, is called a **mathematical model.** Mathematical models for phenomena, such as radioactive decay, population growth, chemical reactions, cooling of bodies, velocity of a falling body, rate of memorization, or current in a series circuit, are often first-order differential equations.

In this chapter we are concerned with solving some of the more commonly occurring linear and nonlinear first-order differential equations that arise in applications.

3.1 ORTHOGONAL TRAJECTORIES

• Differential equation of a family of curves • Orthogonal trajectories

Differential Equation of a Family of Curves At the end of Section 1.1 we expressed the expectation or, better, the hope that an nth-order ordinary differential will yield an n-parameter family of solutions. On the other hand, suppose we turn the problem around: Starting with an n-parameter family of curves, can we find an associated nth-order differential equation that is entirely free of arbitrary parameters and represents the given family? In most cases the answer is yes.

In the discussion that follows we are interested in finding the differential equation $dy/dx = f(x, y)$ of a one-parameter family of curves.

EXAMPLE 1 DE of a Family of Curves

Find the differential equation of the family $y = c_1x^3$.

Solution Differentiation gives

$$\frac{dy}{dx} = 3c_1x^2.$$

We can eliminate the parameter c_1 from the equation by using $c_1 = y/x^3$ obtained from the first equation:

$$\frac{dy}{dx} = 3\left(\frac{y}{x^3}\right)x^2 \quad \text{or} \quad \frac{dy}{dx} = 3\frac{y}{x}.$$

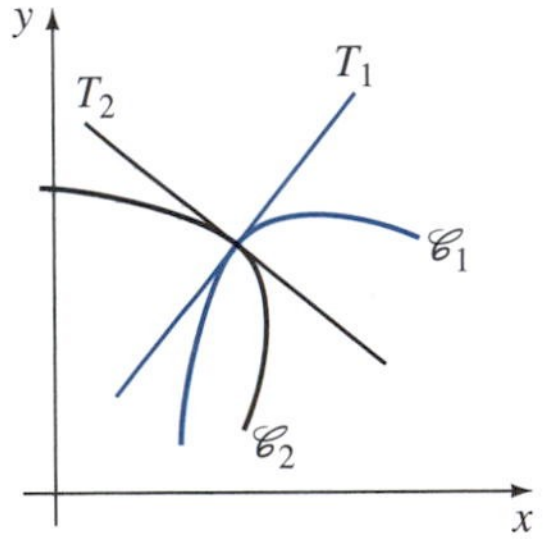

Figure 3.1

Orthogonal Curves Recall from your study of analytic geometry that two lines L_1 and L_2, which are not parallel to the coordinate axes, are perpendicular if and only if their respective slopes satisfy the relationship $m_1m_2 = -1$. For this reason, the graphs of $y = (-1/2)x + 1$ and $y = 2x + 4$ are obviously perpendicular. In general, two curves $\mathscr{C}_1$ and $\mathscr{C}_2$ are said to be **orthogonal** at a point if and only if their tangent lines T_1 and T_2 are perpendicular at the point of intersection. See Figure 3.1. Except for the case when T_1 and T_2 are parallel to the coordinate axes, this means the slopes of the tangents are negative reciprocals of each other.

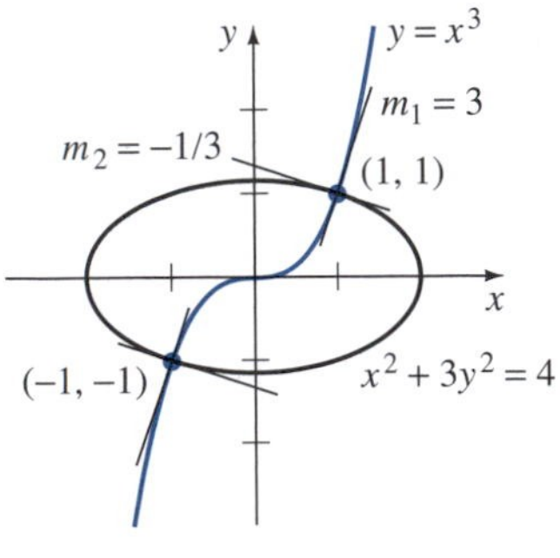

Figure 3.2

EXAMPLE 2 Orthogonal Curves

Show that the curves $\mathscr{C}_1$ and $\mathscr{C}_2$ defined by $y = x^3$ and $x^2 + 3y^2 = 4$ are orthogonal at their point(s) of intersection.

Solution In Figure 3.2 it is seen that the points of intersection of the graphs are $(1, 1)$ and $(-1, -1)$. Now the slope of the tangent line to $y = x^3$ at any point is $dy/dx = 3x^2$, so

$$\left.\frac{dy}{dx}\right|_{x=1} = \left.\frac{dy}{dx}\right|_{x=-1} = 3.$$

We use implicit differentiation to obtain dy/dx for the second curve:

$$2x + 6y\frac{dy}{dx} = 0 \quad \text{or} \quad \frac{dy}{dx} = -\frac{x}{3y}$$

and therefore
$$\left.\frac{dy}{dx}\right|_{(1,1)} = \left.\frac{dy}{dx}\right|_{(-1,-1)} = -\frac{1}{3}.$$

Thus, at either (1, 1) or $(-1, -1)$ we have

$$\left(\frac{dy}{dx}\right)_{\mathscr{C}_1} \cdot \left(\frac{dy}{dx}\right)_{\mathscr{C}_2} = -1.$$

It is easy to show that any curve $\mathscr{C}_1$ in the family $y = c_1x^3$, $c_1 \neq 0$ is orthogonal to each curve $\mathscr{C}_2$ in the family $x^2 + 3y^2 = c_2$, $c_2 > 0$. From Example 1 we know that the differential equation of the first family is

$$\frac{dy}{dx} = 3\frac{y}{x}.$$

Implicit differentiation of $x^2 + 3y^2 = c_2$ leads to exactly the same differential equation as for $x^2 + 3y^2 = 4$ in Example 2—namely,

$$\frac{dy}{dx} = -\frac{x}{3y}.$$

Hence, at the point (x, y) on each curve,

$$\left(\frac{dy}{dx}\right)_{\mathscr{C}_1} \cdot \left(\frac{dy}{dx}\right)_{\mathscr{C}_2} = \left(\frac{3y}{x}\right)\left(-\frac{x}{3y}\right) = -1.$$

Since the slopes of the tangent lines are negative reciprocals, the curves $\mathscr{C}_1$ and $\mathscr{C}_2$ intersect each other in an orthogonal manner.

This discussion leads to the following definition.

DEFINITION 3.1 Orthogonal Trajectories

When all the curves of one family of curves $G(x, y, c_1) = 0$ intersect orthogonally all the curves of another family $H(x, y, c_2) = 0$, then the families are said to be **orthogonal trajectories** of each other.

In other words, an orthogonal trajectory is any *one* curve that intersects every curve of another family at right angles.

EXAMPLE 3 Orthogonal Trajectories

(a) The graph of $y = -\frac{1}{2}x + 1$ is an orthogonal trajectory of $y = 2x + c_1$. The families $y = -\frac{1}{2}x + c_2$ and $y = 2x + c_1$ are orthogonal trajectories.

(b) The graph of $y = 4x^3$ is an orthogonal trajectory of $x^2 + 3y^2 = c_2$. The families $y = c_1x^3$ and $x^2 + 3y^2 = c_2$ are orthogonal trajectories.

(c) In Figure 3.3 it is seen that the family of straight lines $y = c_1x$ through the origin and the family $x^2 + y^2 = c_2$ of concentric circles with center at the origin are orthogonal trajectories.

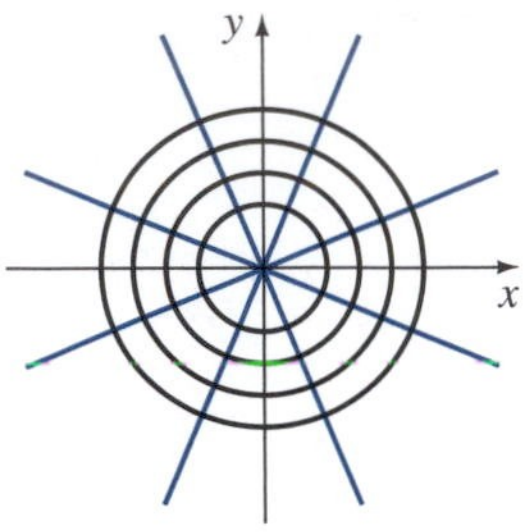

Figure 3.3

Orthogonal trajectories occur naturally in the construction of meteorological maps and in the study of electricity and magnetism. For example, in an electric field around two bodies of opposite charge, the lines of force are perpendicular to the equipotential curves (that is, curves along which the potential is constant). The lines of force are indicated in Figure 3.4 by dashed lines.

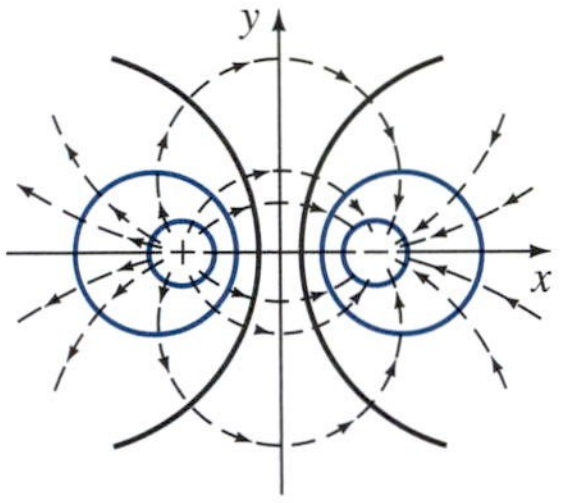

Figure 3.4

General Method To find the orthogonal trajectories of a given family of curves we first find the differential equation

$$\frac{dy}{dx} = f(x, y)$$

that describes the family. The differential equation of the second, and orthogonal, family is then

$$\frac{dy}{dx} = \frac{-1}{f(x, y)}.$$

EXAMPLE 4 Finding Orthogonal Trajectories

Find the orthogonal trajectories of the family of rectangular hyperbolas

$$y = \frac{c_1}{x}.$$

Solution The derivative of $y = c_1/x$ is

$$\frac{dy}{dx} = \frac{-c_1}{x^2}.$$

Replacing c_1 by $c_1 = xy$ yields the differential equation of the given family:

$$\frac{dy}{dx} = -\frac{y}{x}.$$

The differential equation of the orthogonal family is then

$$\frac{dy}{dx} = \frac{-1}{(-y/x)} = \frac{x}{y}.$$

We solve this last equation by separation of variables. From

$$y\,dy = x\,dx \quad \text{we get} \quad \int y\,dy = \int x\,dx$$

$$\frac{y^2}{2} = \frac{x^2}{2} + c_2' \quad \text{or} \quad y^2 - x^2 = c_2,$$

where we replaced $2c_2'$ by c_2. We recognize that this solution represents another family of hyperbolas. The graphs of both families, for various values of c_1 and c_2, are given in Figure 3.5.

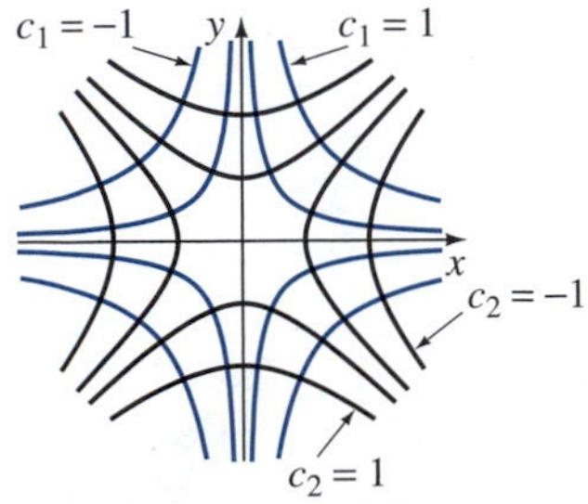

Figure 3.5

EXAMPLE 5 **Finding Orthogonal Trajectories**

Find the orthogonal trajectories of $y = \dfrac{c_1 x}{1 + x}$.

Solution From the quotient rule we find

$$\frac{dy}{dx} = \frac{c_1}{(1+x)^2} \quad \text{or} \quad \frac{dy}{dx} = \frac{y}{x(1+x)},$$

since $c_1 = y(1 + x)/x$. The differential equation of the orthogonal trajectories is then

$$\frac{dy}{dx} = -\frac{x(1+x)}{y}.$$

Again, by separating variables, we have

$$y\,dy = -x(1+x)\,dx \quad \text{and} \quad \int y\,dy = -\int (x + x^2)\,dx$$

$$\frac{y^2}{2} = -\frac{x^2}{2} - \frac{x^3}{3} + c_2' \quad \text{or} \quad 3y^2 + 3x^2 + 2x^3 = c_2.$$

Polar Curves In calculus it is shown that for a graph of a polar equation $r = f(\theta)$,

$$r\frac{d\theta}{dr} = \tan\psi,$$

where ψ is the positive counterclockwise angle between the radial line and the tangent line. See Figure 3.6. It is left as an exercise to show that two polar curves $r = f_1(\theta)$ and $r = f_2(\theta)$ are orthogonal at a point of intersection if and only if

$$(\tan\psi_1)_{\mathscr{C}_1}(\tan\psi_2)_{\mathscr{C}_2} = -1. \tag{1}$$

See Problem 42.

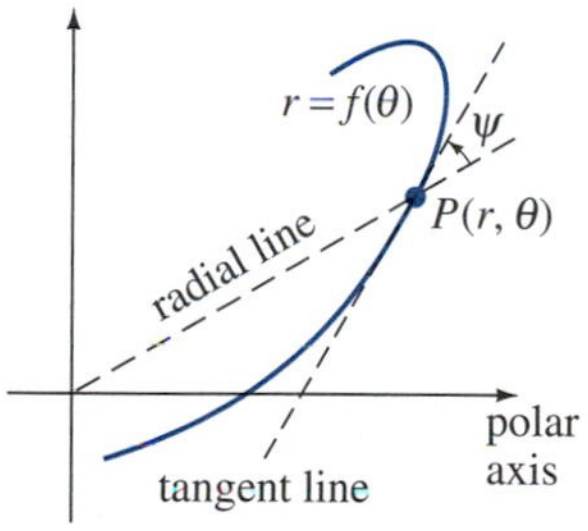

Figure 3.6

EXAMPLE 6 **Polar Orthogonal Trajectories**

Find the orthogonal trajectories of $r = c_1(1 - \sin\theta)$.

Solution For the given curve we can write

$$\frac{dr}{d\theta} = -c_1\cos\theta = \frac{-r\cos\theta}{1 - \sin\theta}$$

so

$$r\frac{d\theta}{dr} = -\frac{1 - \sin\theta}{\cos\theta} = \tan\psi_1.$$

Thus, by (1), the differential equation of the orthogonal trajectories is

$$r\frac{d\theta}{dr} = \frac{\cos\theta}{1 - \sin\theta} = \tan\psi_2.$$

Separating variables then gives

$$\frac{dr}{r} = \frac{1 - \sin\theta}{\cos\theta}\,d\theta = (\sec\theta - \tan\theta)\,d\theta$$

so

$$\ln|r| = \ln|\sec\theta + \tan\theta| + \ln|\cos\theta| + \ln c_2$$

$$= \ln|c_2(1 + \sin\theta)|.$$

Hence

$$r = c_2(1 + \sin\theta).$$

EXERCISES 3.1

Answers to odd-numbered problems begin on page A-4.

In Problems 1–26 find the orthogonal trajectories of the given family of curves.

1. $y = c_1 x$

2. $3x + 4y = c_1$

3. $y = c_1 x^2$

4. $y = (x - c_1)^2$

5. $c_1 x^2 + y^2 = 1$

6. $2x^2 + y^2 = c_1^2$

7. $y = c_1 e^{-x}$

8. $y = e^{c_1 x}$

9. $y^2 = c_1 x^3$

10. $y^a = c_1 x^b$, a and b constants

11. $y = \dfrac{x}{1 + c_1 x}$

12. $y = \dfrac{1 + c_1 x}{1 - c_1 x}$

13. $2x^2 + y^2 = 4c_1 x$

14. $x^2 + y^2 = 2c_1 x$

15. $y^3 + 3x^2 y = c_1$

16. $y^2 - x^2 = c_1 x^3$

17. $y = \dfrac{c_1}{1 + x^2}$

18. $y = \dfrac{1}{c_1 + x}$

19. $4y + x^2 + 1 + c_1 e^{2y} = 0$

20. $y = -x - 1 + c_1 e^x$

21. $y = \dfrac{1}{\ln c_1 x}$

22. $y = \ln(\tan x + c_1)$

23. $\sinh y = c_1 x$

24. $y = c_1 \sin x$

25. $x^{1/3} + y^{1/3} = c_1$

26. $x^a + y^a = c_1$, $a \neq 2$

27. Find the member of the orthogonal trajectories for $x + y = c_1 e^y$ that passes through (0, 5).

28. Find the member of the orthogonal trajectories for $3xy^2 = 2 + 3c_1 x$ that passes through (0, 10).

In Problems 29–34 find the orthogonal trajectories of the given polar curves.

29. $r = 2c_1 \cos\theta$

30. $r = c_1(1 + \cos\theta)$

31. $r^2 = c_1 \sin 2\theta$

32. $r = \dfrac{c_1}{1 + \cos\theta}$

33. $r = c_1 \sec\theta$

34. $r = c_1 e^{\theta}$

35. A family of curves that intersects a given family of curves at a specified constant angle $\alpha \neq \pi/2$ is said to be an isogonal family. The two families are said to be **isogonal trajectories** of each other. If $dy/dx = f(x, y)$ is the differential equation of the given family, show that the differential equation of the isogonal family is

$$\frac{dy}{dx} = \frac{f(x, y) \pm \tan \alpha}{1 \mp f(x, y)\tan \alpha}.$$

In Problems 36–38 use the results of Problem 35 to find the isogonal family that intersects the one-parameter family of straight lines $y = c_1x$ at the given angle.

36. $\alpha = 45°$ **37.** $\alpha = 60°$ **38.** $\alpha = 30°$

A family of curves can be **self-orthogonal** in the sense that a member of the orthogonal trajectories is also a member of the original family. In Problems 39 and 40 show that the given family of curves is self-orthogonal.

39. parabolas $y^2 = c_1 (2x + c_1)$

40. confocal conics $\dfrac{x^2}{c_1 + 1} + \dfrac{y^2}{c_1} = 1$

41. Verify that the orthogonal trajectories of the family of curves given by the parametric equations $x = c_1e^t \cos t$, $y = c_1e^t \sin t$ are

$$x = c_2e^{-t} \cos t, \quad y = c_2e^{-t} \sin t.$$

[*Hint:* $dy/dx = (dy/dt)/(dx/dt)$.]

42. Show that two polar curves $r = f_1(\theta)$ and $r = f_2(\theta)$ are orthogonal at a point of intersection if and only if

$$(\tan \psi_1)_{\mathscr{C}_1}(\tan \psi_2)_{\mathscr{C}_2} = -1.$$

3.2 APPLICATIONS OF LINEAR EQUATIONS

- *Exponential growth and decay* • *Half-life* • *Carbon dating* • *Response*
- *Transient term* • *Steady-state term*

Growth and Decay The initial-value problem

$$\frac{dx}{dt} = kx, \quad x(t_0) = x_0, \tag{1}$$

where k is a constant of proportionality, occurs in many physical theories involving either **growth** or **decay.** For example, in biology it is often observed that the rate at which certain bacteria grow is proportional to the number of bacteria present at time t. Over short intervals of time, the population of small animals, such as rodents, can be predicted fairly accurately by the solution of (1). In physics an initial-value problem such as (1) provides a model for approximating the remaining amount of a substance that is disintegrating, or decaying, through radioactivity. The differential equation in (1) could also determine the temperature of a cooling body. In chemistry the amount of a substance remaining during certain reactions is also described by (1).

The constant of proportionality k in (1) is either positive or negative and can be determined from the solution of the problem using a subsequent measurement of x at a time $t_1 > t_0$.

EXAMPLE 1 Bacterial Growth

A culture initially has N_0 number of bacteria. At $t = 1$ hour the number of bacteria is measured to be $\frac{3}{2}N_0$. If the rate of growth is proportional to the number of bacteria present, determine the time necessary for the number of bacteria to triple.

Solution We first solve the differential equation

$$\frac{dN}{dt} = kN \tag{2}$$

subject to $N(0) = N_0$. Then we use the empirical condition $N(1) = \frac{3}{2}N_0$ to determine the constant of proportionality k.

Now (2) is both separable and linear. When it is put into the form

$$\frac{dN}{dt} - kN = 0,$$

we can see by inspection that the integrating factor is e^{-kt}. Multiplying both sides of the equation by this term gives immediately

$$\frac{d}{dt}[e^{-kt}N] = 0.$$

Integrating both sides of the last equation yields

$$e^{-kt}N = c \quad \text{or} \quad N(t) = ce^{kt}.$$

At $t = 0$ it follows that $N_0 = ce^0 = c$ and so $N(t) = N_0e^{kt}$. At $t = 1$ we have

$$\frac{3}{2}N_0 = N_0e^k \quad \text{or} \quad e^k = \frac{3}{2},$$

from which we get, to four decimal places, $k = \ln\frac{3}{2} = 0.4055$. Thus

$$N(t) = N_0e^{0.4055t}.$$

To find the time at which the bacteria have tripled we solve

$$3N_0 = N_0e^{0.4055t}$$

for t. It follows from this equation that $0.4055t = \ln 3$ and so

$$t = \frac{\ln 3}{0.4055} \approx 2.71 \text{ hours}.$$

See Figure 3.7.

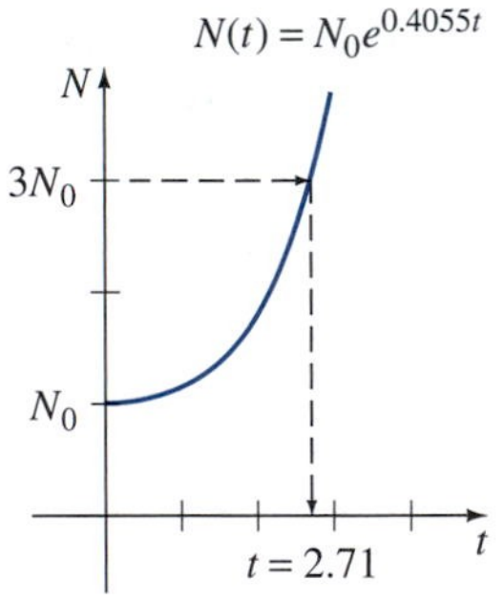

Figure 3.7

Note We can write the function $N(t)$ obtained in Example 1 in an alternative form. From the laws of exponents,

$$N(t) = N_0(e^k)^t = N_0\left(\frac{3}{2}\right)^t$$

since $e^k = \frac{3}{2}$. This latter solution provides a convenient method for computing $N(t)$ for small positive integral values of t. It also clearly shows the influence of the subsequent experimental observation at $t = 1$ on the solution for all time.

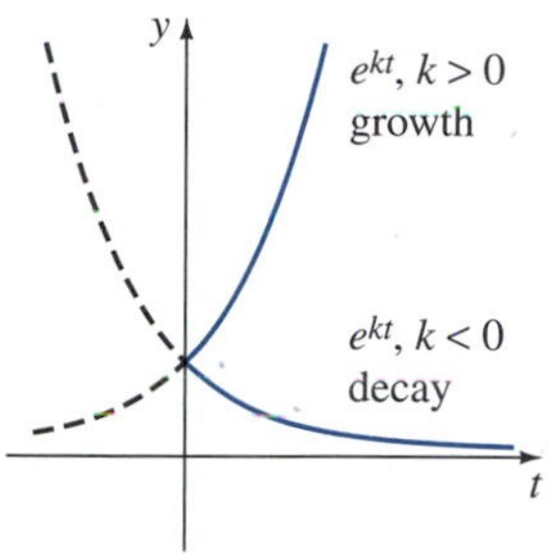

Figure 3.8

We notice, too, that the actual number of bacteria present at time $t = 0$ is quite irrelevant in finding the time required to triple the number in the culture. The necessary time to triple, say, 100 or 10,000 bacteria is still approximately 2.71 hours.

As shown in Figure 3.8, the exponential function e^{kt} increases as t increases for $k > 0$, and decreases as t increases for $k < 0$. Thus problems describing growth, such as population, bacteria, or even capital, are characterized by a positive value of k, whereas problems involving decay, as in radioactive disintegration, yield a negative k value.

Half-Life In physics the **half-life** is a measure of the stability of a radioactive substance. The half-life is simply the time it takes for one-half of the atoms in an initial amount A_0 to disintegrate, or transmute, into the atoms of another element. The longer the half-life of a substance, the more stable it is. For example, the half-life of highly radioactive radium, Ra-226, is about 1700 years. In 1700 years one-half of a given quantity of Ra-226 is transmuted into radon, Rn-222. The most commonly occurring uranium isotope, U-238, has a half-life of approximately 4,500,000,000 years. In about 4.5 billion years, one-half of a quantity of U-238 is transmuted into lead, Pb-206.

EXAMPLE 2 Half-Life of a Radioactive Substance

A breeder reactor converts the relatively stable uranium 238 into the isotope plutonium 239. After 15 years it is determined that 0.043% of the initial amount A_0 of the plutonium has disintegrated. Find the half-life of this isotope if the rate of disintegration is proportional to the amount remaining at time t.

Solution Let $A(t)$ denote the amount of plutonium remaining at time t. As in Example 1, the solution of the initial-value problem

$$\frac{dA}{dt} = kA, \quad A(0) = A_0$$

is $A(t) = A_0e^{kt}$. If 0.043% of the atoms of A_0 have disintegrated, then 99.957% of the substance remains. To find k we use $0.99957A_0 = A(15)$; that is,

$$0.99957A_0 = A_0e^{15k}.$$

Solving for k then gives $k = \frac{1}{15}\ln(0.99957) = -0.00002867$. Hence

$$A(t) = A_0e^{-0.00002867t}.$$

Now the half-life is the corresponding value of time for which $A(t) = \frac{1}{2}A_0$. Solving

$$\frac{A_0}{2} = A_0e^{-0.00002867t} \quad \text{or} \quad \frac{1}{2} = e^{-0.00002867t}$$

for t gives $-0.00002867t = \ln\frac{1}{2} = -\ln 2$ or

$$t = \frac{\ln 2}{0.00002867} \approx 24{,}180 \text{ years.}$$

Carbon Dating About 1950 the chemist Willard Libby devised a method of using radioactive carbon as a means of determining the approximate ages of fossils. The theory of **carbon dating** is based on the fact that the isotope carbon 14 is produced in the atmosphere by the action of cosmic radiation on nitrogen. The ratio of the amount of C-14 to ordinary carbon in the atmosphere appears to be a constant, and as a consequence the proportionate amount of the isotope present in all living organisms is the same as that in the atmosphere. When an organism dies, the absorption of C-14, by either breathing or eating, ceases. Thus, by comparing the proportionate amount of C-14 present, say, in a fossil with the constant ratio found in the atmosphere, it is possible to obtain a reasonable estimation of its age. The method is based on the knowledge that the half-life of the radioactive C-14 is approximately 5600 years. For his work Libby won the Nobel Prize for chemistry in 1960. Libby's method has been used to date wooden furniture in Egyptian tombs and the woven flax wrappings of the Dead Sea scrolls.

EXAMPLE 3 Age of a Fossil

A fossilized bone is found to contain $\frac{1}{1000}$ the original amount of C-14. Determine the age of the fossil.

Solution The starting point is again $A(t) = A_0e^{kt}$. To determine the value of k we use the fact that $\frac{1}{2}A_0 = A(5600)$ or $\frac{1}{2}A_0 = A_0e^{5600k}$. Hence we have $5600k = \ln\frac{1}{2}$ or $k = \frac{-1}{5600}\ln 2 = -0.00012378$. Therefore

$$A(t) = A_0e^{-0.00012378t}.$$

When $A(t) = \frac{1}{1000}A_0$, we have $\frac{1}{1000}A_0 = A_0e^{-0.00012378t}$. Solving the last equation for t then yields

$$t = \frac{\ln 1000}{0.00012378} \approx 55{,}800 \text{ years.}$$

The date found in Example 3 is really at the border of accuracy for this method. The usual carbon 14 technique is limited to about 9 half-lives of the isotope, or about 50,000 years. One reason is that the chemical analysis needed to obtain an accurate measurement of the remaining C-14 becomes somewhat formidable around the point of $A_0/1000$. Also, this analysis demands the destruction of a rather large sample of the specimen. If this measurement is accomplished indirectly, based on the actual radioactivity of the specimen, then it is very difficult to distinguish between the radiation from the fossil and the normal background radiation. But in recent developments, the use of a particle accelerator has enabled scientists to separate the C-14 from the stable C-12 directly. By computing the precise value of the ratio of C-14 to C-12, the accuracy of this method can be extended to 70,000–100,000 years. Other isotopic techniques such as using potassium 40 and argon 40 can give dates of several million years. Nonisotopic methods based on the use of amino acids are also sometimes possible.

Cooling Newton's law of cooling states that the rate at which the temperature $T(t)$ changes in a cooling body is proportional to the difference between the temperature in the body and the constant temperature T_m of the surrounding medium—that is,

$$\frac{dT}{dt} = k(T - T_m), \tag{3}$$

where k is a constant of proportionality.

EXAMPLE 4 Newton's Law of Cooling

When a cake is removed from a baking oven, its temperature is measured at 300°F. Three minutes later its temperature is 200°F. How long will it take to cool off to a room temperature of 70°F?

Solution In (3) we make the identification $T_m = 70$. We must then solve the initial-value problem

$$\frac{dT}{dt} = k(T - 70), \quad T(0) = 300 \tag{4}$$

and determine the value of k so that $T(3) = 200$.

Equation (4) is both linear and separable. Separating variables, we find that

$$\frac{dT}{T - 70} = k\,dt \quad \text{yields} \quad \ln|T - 70| = kt + c_1$$

$$T - 70 = c_2 e^{kt} \qquad \text{or} \qquad T = 70 + c_2 e^{kt}.$$

When $t = 0$, $T = 300$, so $300 = 70 + c_2$ gives $c_2 = 230$ and therefore $T = 70 + 230e^{kt}$.

From $T(3) = 200$ we find that $e^{3k} = \frac{13}{23}$ or $k = \frac{1}{3}\ln\frac{13}{23} = -0.19018$. Thus

$$T(t) = 70 + 230e^{-0.19018t}. \tag{5}$$

We note that (5) furnishes no finite solution to $T(t) = 70$ since $\lim_{t\to\infty} T(t) = 70$. Yet intuitively we expect the cake to reach the room temperature after a reasonably long period of time. How long is long? Of course, we should not be disturbed by the fact that the model (4) does not quite live up to our physical intuition. Parts (a) and (b) of Figure 3.9 clearly show that the cake will be approximately at room temperature in about one-half hour. ■

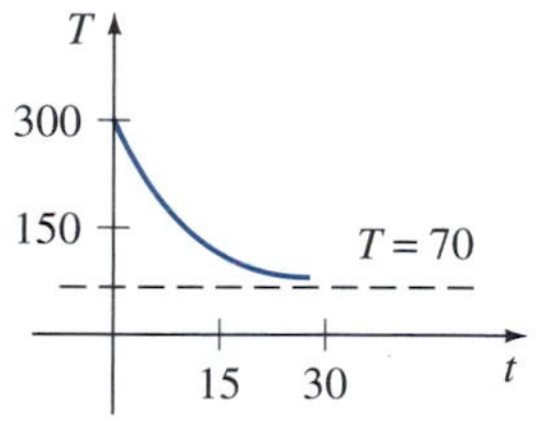

(a)

$T(t)$	t (minutes)
75°	20.1
74°	21.3
73°	22.8
72°	24.9
71°	28.6
70.5°	32.3

(b)

Figure 3.9

Series Circuits In a series circuit containing only a resistor and an inductor, Kirchhoff's second law states that the sum of the voltage drop across the inductor ($L(di/dt)$) and the voltage drop across the resistor (iR) is the same as the impressed voltage ($E(t)$) on the circuit. See Figure 3.10.

Thus we obtain the linear differential equation for the current $i(t)$,

$$L\frac{di}{dt} + Ri = E(t), \tag{6}$$

where L and R are constants known as the inductance and the resistance, respectively. The current $i(t)$ is sometimes called the **response** of the system.

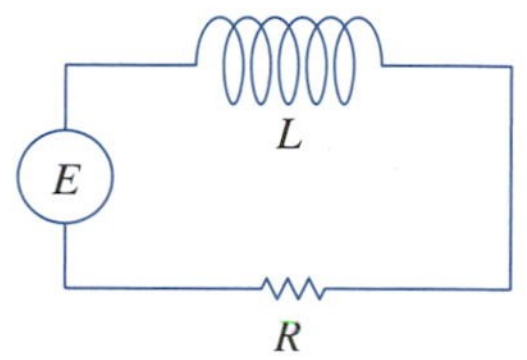

Figure 3.10 L-R series circuit

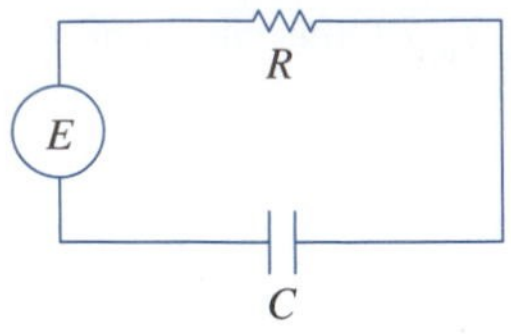

Figure 3.11 R-C series circuit

The voltage drop across a capacitor with capacitance C is given by $q(t)/C$, where q is the charge on the capacitor. Hence, for the series circuit shown in Figure 3.11, Kirchhoff's second law gives

$$Ri + \frac{1}{C}q = E(t). \tag{7}$$

But current i and charge q are related by $i = dq/dt$, so (7) becomes the linear differential equation

$$R\frac{dq}{dt} + \frac{1}{C}q = E(t). \tag{8}$$

EXAMPLE 5 Current in a Series Circuit

A 12-volt battery is connected to a series circuit in which the inductance is $\frac{1}{2}$ henry and the resistance is 10 ohms. Determine the current i if the initial current is zero.

Solution From (6) we see that we must solve

$$\frac{1}{2}\frac{di}{dt} + 10i = 12$$

subject to $i(0) = 0$. First, we multiply the differential equation by 2 and read off the integrating factor e^{20t}. We then obtain

$$\frac{d}{dt}\left[e^{20t}i\right] = 24e^{20t}$$

$$e^{20t}i = \frac{24}{20}e^{20t} + c \quad \text{or} \quad i = \frac{6}{5} + ce^{-20t}.$$

Now $i(0) = 0$ implies $0 = \frac{6}{5} + c$ or $c = -\frac{6}{5}$. Therefore the response is

$$i(t) = \frac{6}{5} - \frac{6}{5}e^{-20t}.$$

From (7) of Section 2.5 we can write a general solution of (6):

$$i(t) = \frac{e^{-(R/L)t}}{L}\int e^{(R/L)t}E(t)\,dt + ce^{-(R/L)t}. \tag{9}$$

In particular, when $E(t) = E_0$ is a constant, (9) becomes

$$i(t) = \frac{E_0}{R} + ce^{-(R/L)t}. \tag{10}$$

Note that as $t \to \infty$, the second term in equation (10) approaches zero. Such a term is usually called a **transient term;** any remaining terms are called the **steady-state** part of the solution. In this case E_0/R is also called the **steady-state current;** for large values of time it then appears that the current in the circuit is simply governed by Ohm's law ($E = iR$).

Mixture Problem The mixing of two fluids sometimes gives rise to a linear first-order differential equation. In the next example we consider the mixture of two salt solutions with different concentrations.

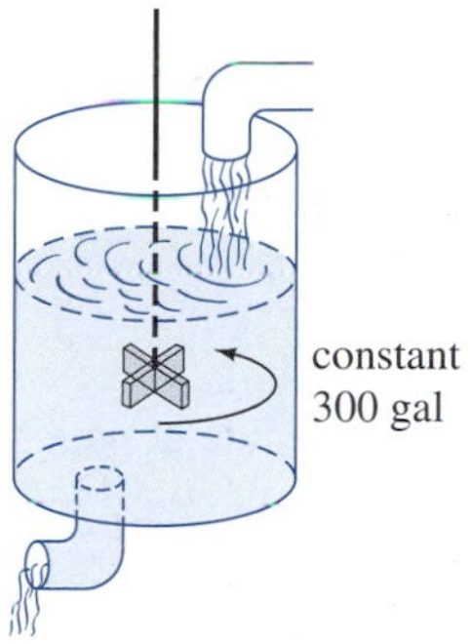

Figure 3.12

EXAMPLE 6 Mixture of Two Salt Solutions

Initially 50 pounds of salt is dissolved in a large tank holding 300 gallons of water. A brine solution is pumped into the tank at a rate of 3 gallons per minute, and the well-stirred solution is then pumped out at the same rate. See Figure 3.12. If the concentration of the solution entering is 2 pounds per gallon, determine the amount of salt in the tank at time t. How much salt is present after 50 minutes? after a long time?

Solution Let $A(t)$ be the amount of salt (in pounds) in the tank at any time. For problems of this sort, the net rate at which $A(t)$ changes is given by

$$\frac{dA}{dt} = \left(\begin{array}{c}\textit{rate of}\\ \textit{substance entering}\end{array}\right) - \left(\begin{array}{c}\textit{rate of}\\ \textit{substance leaving}\end{array}\right) = R_1 - R_2. \quad \textbf{(11)}$$

Now the rate at which the salt enters the tank is, in pounds per minute,

$$R_1 = (3 \text{ gal/min}) \cdot (2 \text{ lb/gal}) = 6 \text{ lb/min},$$

whereas the rate at which salt is leaving is

$$R_2 = (3 \text{ gal/min}) \cdot \left(\frac{A}{300} \text{ lb/gal}\right) = \frac{A}{100} \text{ lb/min}.$$

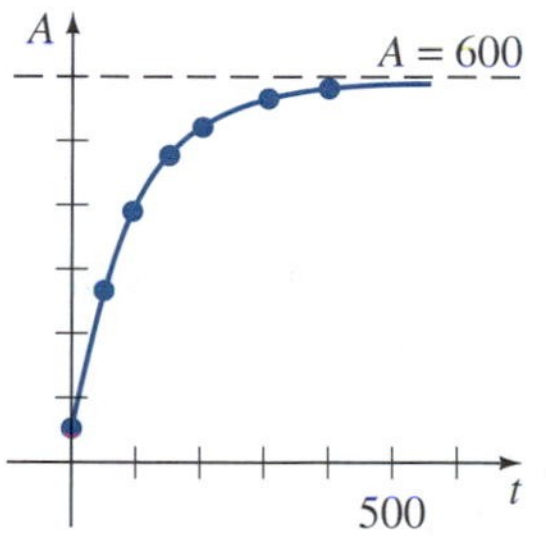

(a)

t (minutes)	A (lb)
50	266.41
100	397.67
150	477.27
200	525.57
300	572.62
400	589.93

(b)

Figure 3.13

Thus equation (11) becomes

$$\frac{dA}{dt} = 6 - \frac{A}{100}, \quad \textbf{(12)}$$

which we solve subject to the initial condition $A(0) = 50$.

Since the integrating factor is $e^{t/100}$, we can write (12) as

$$\frac{d}{dt}\left[e^{t/100}A\right] = 6e^{t/100}$$

and therefore

$$e^{t/100}A = 600e^{t/100} + c \quad \text{or} \quad A = 600 + ce^{-t/100}. \quad \textbf{(13)}$$

When $t = 0$, $A = 50$, so we find that $c = -550$. Finally, we obtain

$$A(t) = 600 - 550e^{-t/100}. \quad \textbf{(14)}$$

At $t = 50$ we find that $A(50) = 266.41$ pounds. Also, as $t \to \infty$ it is seen from (14) and Figure 3.13 that $A \to 600$. Of course this is what we would expect; over a long period of time the number of pounds of salt in the solution must be

$$(300 \text{ gal})(2 \text{ lb/gal}) = 600 \text{ lb}.$$

In Example 6 we assumed that the rate at which the solution was pumped in was the same as the rate at which the solution was pumped out. However, this need not be the case; the mixed brine solution could be pumped out at a rate faster or slower than the rate at which the other solution is pumped in. The resulting differential equation in this latter situation is linear with a variable coefficient.

EXAMPLE 7 Mixture of Two Salt Solutions

If the well-stirred solution in Example 6 is pumped out at a slower rate of 2 gallons per minute, then the solution is *accumulating* at a rate of

$$(3 - 2)\ \text{gal/min} = 1\ \text{gal/min}.$$

After t minutes there are $300 + t$ gallons of brine in the tank. The rate at which the salt is leaving is then

$$R_2 = (2\ \text{gal/min}) \cdot \left(\frac{A}{300 + t}\ \text{lb/gal}\right) = \frac{2A}{300 + t}\ \text{lb/min}.$$

Hence equation (11) becomes

$$\frac{dA}{dt} = 6 - \frac{2A}{300 + t} \quad \text{or} \quad \frac{dA}{dt} + \frac{2A}{300 + t} = 6.$$

Finding the integrating factor and solving the last equation, we get

$$A(t) = 2(300 + t) + c(300 + t)^{-2}.$$

The initial condition $A(0) = 50$ yields $c = -4.95 \times 10^7$ and so

$$A(t) = 2(300 + t) - (4.95 \times 10^7)(300 + t)^{-2}.$$

Remarks Consider the differential equation in Example 1 that describes the growth of bacteria. The solution $N(t) = N_0 e^{0.4055t}$ of the initial-value problem $dN/dt = kN$, $N(t_0) = N_0$ is of course a continuous function. But in the example we are talking about a population of bacteria and so common sense dictates that N take on only positive integer values. Moreover, the population does not necessarily grow continuously (that is, every second, every microsecond, and so on), as predicted by the function $N(t) = N_0 e^{0.4055t}$; there may be time intervals $[t_1, t_2]$ over which there is no growth at all. Perhaps, then, the graph shown in Figure 3.14(a) gives a more

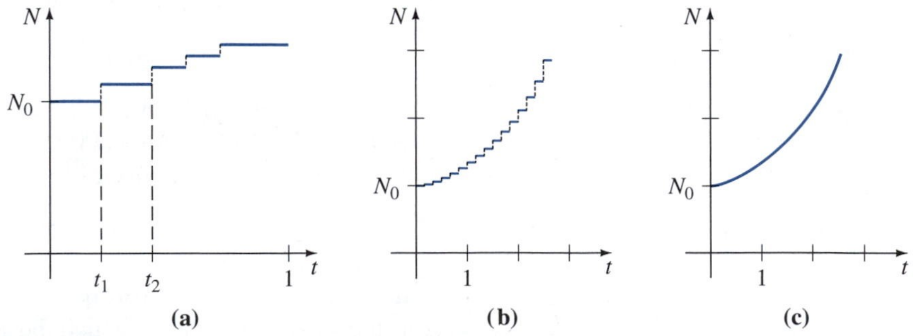

Figure 3.14

realistic description of N than that given by the graph of an exponential function. The point is that in many instances a mathematical model describes a system in only approximate terms. It is often more convenient than accurate to use a continuous function to describe a discrete phenomenon. However, for some purposes we may be satisfied if our model describes the system fairly accurately when viewed macroscopically in time as in Figures 3.14(b) and (c) rather than microscopically.

EXERCISES 3.2

Answers to odd-numbered problems begin on page A-5.

1. The population of a certain community is known to increase at a rate proportional to the number of people present at time t. If the population has doubled in 5 years, how long will it take to triple? to quadruple?

2. Suppose it is known that the population of the community in Problem 1 is 10,000 after 3 years. What was the initial population? What will be the population in 10 years?

3. The population of a town grows at a rate proportional to the population at time t. Its initial population of 500 increases by 15% in 10 years. What will be the population in 30 years?

4. The population of bacteria in a culture grows at a rate proportional to the number of bacteria present at time t. After 3 hours it is observed that there are 400 bacteria present. After 10 hours there are 2000 bacteria present. What is the initial number of bacteria?

5. The radioactive isotope of lead, Pb-209, decays at a rate proportional to the amount present at time t and has a half-life of 3.3 hours. If 1 gram of lead is present initially, how long will it take for 90% of the lead to decay?

6. Initially there were 100 milligrams of a radioactive substance present. After 6 hours the mass had decreased by 3%. If the rate of decay is proportional to the amount of the substance present at time t, find the amount remaining after 24 hours.

7. Determine the half-life of the radioactive substance described in Problem 6.

8. Show that the half-life of a radioactive substance is, in general,

$$t = \frac{(t_2 - t_1)\ln 2}{\ln(A_1/A_2)},$$

where $A_1 = A(t_1)$ and $A_2 = A(t_2)$, $t_1 < t_2$.

9. When a vertical beam of light passes through a transparent substance, the rate at which its intensity I decreases is proportional to $I(t)$, where t represents the thickness of the medium (in feet). In clear seawater the intensity 3 feet below the surface is 25% of the initial intensity I_0 of the incident beam. What is the intensity of the beam 15 feet below the surface?

10. When interest is compounded continuously, the amount of money S increases at a rate proportional to the amount present at time t: $dS/dt = rS$, where r is the annual rate of interest (see (26) of Section 1.2).

(**a**) Find the amount of money accrued at the end of 5 years when \$5000 is deposited in a savings account drawing $5\frac{3}{4}\%$ annual interest compounded continuously.

(**b**) In how many years will the initial sum deposited be doubled?

(**c**) Use a hand calculator to compare the number obtained in part (a) with the value

$$S = 5000\left(1 + \frac{0.0575}{4}\right)^{5(4)}.$$

This value represents the amount accrued when interest is compounded quarterly.

11. In a piece of burned wood, or charcoal, it was found that 85.5% of the C-14 had decayed. Use the information in Example 3 to determine the approximate age of the wood. (It is precisely these data that archaeologists used to date prehistoric paintings in a cave in Lascaux, France.)

12. A thermometer is taken from an inside room to the outside where the air temperature is 5°F. After 1 minute the thermometer reads 55°F, and after 5 minutes the reading is 30°F. What was the initial temperature of the room?

13. A thermometer is removed from a room where the air temperature is 70°F to the outside where the temperature is 10°F. After $\frac{1}{2}$ minute the thermometer reads 50°F. What is the reading at $t = 1$ minute? How long will it take for the thermometer to reach 15°F?

14. Formula (3) also holds when an object absorbs heat from the surrounding medium. If a small metal bar whose initial temperature is 20°C is dropped into a container of boiling water, how long will it take for the bar to reach 90°C if it is known that its temperature increased 2° in 1 second? How long will it take the bar to reach 98°C?

15. A 30-volt electromotive force is applied to an L-R series circuit in which the inductance is 0.1 henry and the resistance is 50 ohms. Find the current $i(t)$ if $i(0) = 0$. Determine the current as $t \to \infty$.

16. Solve equation (6) under the assumption that $E(t) = E_0 \sin \omega t$ and $i(0) = i_0$.

17. A 100-volt electromotive force is applied to an R-C series circuit in which the resistance is 200 ohms and the capacitance is 10^{-4} farad. Find the charge $q(t)$ on the capacitor if $q(0) = 0$. Find the current $i(t)$.

18. A 200-volt electromotive force is applied to an R-C series circuit in which the resistance is 1000 ohms and the capacitance is 5×10^{-6} farad. Find the charge $q(t)$ on the capacitor if $i(0) = 0.4$. Determine the charge and current at $t = 0.005$ second. Determine the charge as $t \to \infty$.

19. An electromotive force

$$E(t) = \begin{cases} 120, & 0 \le t \le 20 \\ 0, & t > 20 \end{cases}$$

is applied to an L-R series circuit in which the inductance is 20 henry and the resistance is 2 ohms. Find the current $i(t)$ if $i(0) = 0$.

20. Suppose an R-C series circuit has a variable resistor. If the resistance at time t is given by $R = k_1 + k_2 t$, where $k_1 > 0$ and $k_2 > 0$ are known constants, then (8) becomes

$$(k_1 + k_2 t)\frac{dq}{dt} + \frac{1}{C}q = E(t).$$

Show that if $E(t) = E_0$ and $q(0) = q_0$, then

$$q(t) = E_0 C + (q_0 - E_0 C)\left(\frac{k_1}{k_1 + k_2 t}\right)^{1/Ck_2}.$$

21. A tank contains 200 liters of fluid in which 30 g of salt is dissolved. Brine containing 1 g of salt per liter is then pumped into the tank at a rate of 4 liters per minute; the well-mixed solution is pumped out at the same rate. Find the number of grams of salt $A(t)$ in the tank at time t.

22. Solve Problem 21 assuming pure water is pumped into the tank.

23. A large tank is filled with 500 gallons of pure water. Brine containing 2 lb of salt per gallon is pumped into the tank at a rate of 5 gallons per minute. The well-mixed solution is pumped out at the same rate. Find the number of pounds of salt $A(t)$ in the tank at time t.

24. Solve Problem 23 under the assumption that the solution is pumped out at a faster rate of 10 gallons per minute. When is the tank empty?

25. A large tank is partially filled with 100 gallons of fluid in which 10 lb of salt is dissolved. Brine containing $\frac{1}{2}$ lb of salt per gallon is pumped into the tank at a rate of 6 gallons per minute. The well-mixed solution is then pumped out at a slower rate of 4 gallons per minute. Find the number of pounds of salt in the tank after 30 minutes.

26. Beer containing 6% alcohol per gallon is pumped into a vat that initially contains 400 gallons of beer at 3% alcohol. The rate at which the beer is pumped in is 3 gallons per minute, whereas the mixed liquid is pumped out at a rate of 4 gallons per minute. Find the number of gallons of alcohol $A(t)$ in the tank at time t. What is the percentage of alcohol in the tank after 60 minutes? When is the tank empty?

Miscellaneous Applications

27. The differential equation governing the velocity v of a falling mass m subjected to air resistance proportional to the instantaneous velocity is

$$m\frac{dv}{dt} = mg - kv,$$

where k is a positive constant of proportionality.

(a) Solve the equation subject to the initial condition $v(0) = v_0$.

(b) Determine the limiting, or terminal, velocity of the weight.

(c) If distance s is related to velocity $ds/dt = v$, find an explicit expression for s if it is further known that $s(0) = s_0$.

28. The rate at which a drug disseminates into the bloodstream is governed by the differential equation

$$\frac{dX}{dt} = A - BX,$$

where A and B are positive constants. The function $X(t)$ describes the concentration of the drug in the bloodstream at time t. Find the limiting value of X as $t \to \infty$. At what time is the concentration one-half this limiting value? Assume that $X(0) = 0$.

29. A heart pacemaker, as shown in Figure 3.15, consists of a battery, a capacitor, and the heart as a resistor. When the switch S is at P the capacitor charges; when S is at Q the capacitor discharges, sending an electrical stimulus to the heart. During this time the voltage E applied to the heart is given by

$$\frac{dE}{dt} = -\frac{1}{RC}E, \quad t_1 < t < t_2,$$

where R and C are constants. Determine $E(t)$ if $E(t_1) = E_0$. (Of course, the opening and closing of the switch are periodic in time, to simulate the natural heartbeat.)

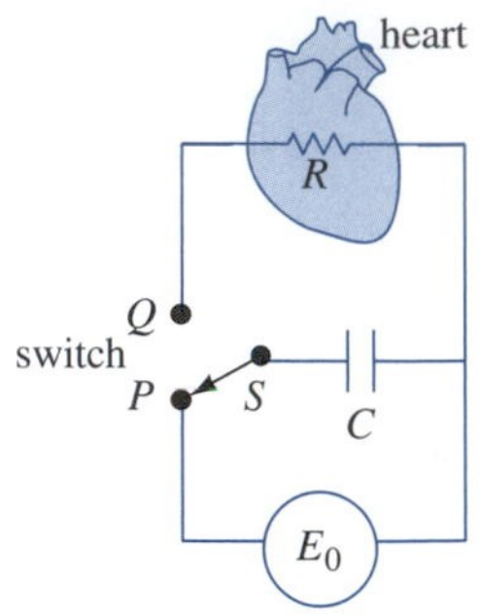

Figure 3.15

30. Suppose a cell is suspended in a solution containing a solute of constant concentration C_s. Suppose further that the cell has constant volume V and that the area of its permeable membrane is the constant A. By Fick's law the rate of change of its mass m is directly proportional to the area A and the difference $C_s - C(t)$, where $C(t)$ is the concentration of the solute inside the cell at time t. Find $C(t)$ if $m = VC(t)$ and $C(0) = C_0$. See Figure 3.16.

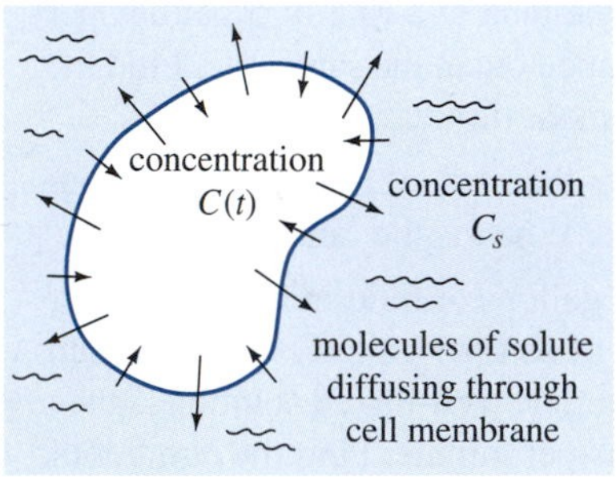

Figure 3.16

31. In one model of the changing population $P(t)$ of a community, it is assumed that

$$\frac{dP}{dt} = \frac{dB}{dt} - \frac{dD}{dt},$$

where dB/dt and dD/dt are the birth and death rates, respectively.

(**a**) Solve for $P(t)$ if

$$\frac{dB}{dt} = k_1P \quad \text{and} \quad \frac{dD}{dt} = k_2P.$$

(**b**) Analyze the cases $k_1 > k_2$, $k_1 = k_2$, and $k_1 < k_2$.

32. The differential equation

$$\frac{dP}{dt} = (k \cos t)P,$$

where k is a positive constant, is often used as a model of a population that undergoes yearly seasonal fluctuations. Solve for $P(t)$ and graph the solution. Assume $P(0) = P_0$.

33. In polar coordinates the angular momentum of a moving body of mass m is defined to be $L = mr^2(d\theta/dt)$. Assume that the coordinates of the body are (r_1, θ_1) and (r_2, θ_2) at times $t = a$ and $t = b$, $a < b$, respectively. If L is constant, show that the area A swept out by r is $A = L\,(b - a)/2m$. When the sun is taken to be at the origin, this proves **Kepler's second law of planetary motion:** The radius vector joining the sun sweeps out equal areas in equal intervals of time. See Figure 3.17.

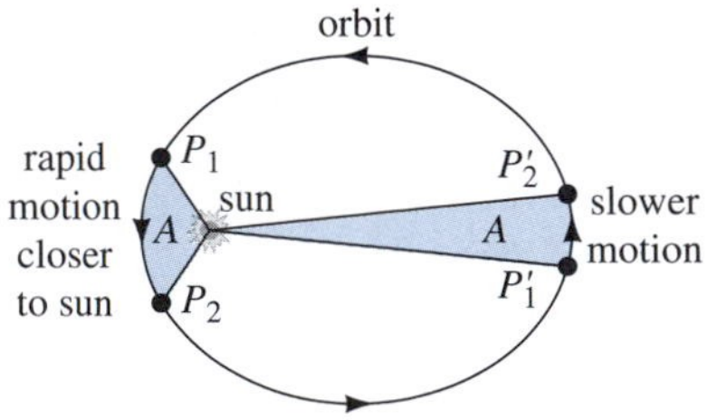

Figure 3.17

34. When forgetfulness is taken into account, the rate of memorization of a subject is given by

$$\frac{dA}{dt} = k_1(M - A) - k_2A,$$

where $k_1 > 0$, $k_2 > 0$, $A(t)$ is the amount of material memorized in time t, M is the total amount to be memorized, and $M - A$ is the amount remaining to be memorized. Solve for $A(t)$ and graph the solution. Assume $A(0) = 0$. Find the limiting value of A as $t \to \infty$ and interpret the result.

3.3 APPLICATIONS OF NONLINEAR EQUATIONS

• *Logistic equation* • *Chemical reactions* • *Escape velocity*

We have seen that if a population P is described by

$$\frac{dP}{dt} = kP, \quad k > 0, \tag{1}$$

then $P(t)$ exhibits unbounded exponential growth. In many instances this differential equation provides an unrealistic model of the growth of a population; that is, what is actually observed differs substantially from what is predicted.

Around 1840 the Belgian mathematician-biologist P. F. Verhulst was concerned with mathematical formulations for predicting the human populations of various countries. One of the equations he studied was

$$\frac{dP}{dt} = P(a - bP), \tag{2}$$

where a and b are positive constants. Equation (2) came to be known as the **logistic equation,** and its solution is called the **logistic function** (the graph of which is naturally called a logistic curve).

Equation (1) does not provide a very accurate model for population growth when the population itself is very large. Overcrowded conditions with the resulting detrimental effects on the environment, such as pollution and excessive and competitive demands for food and fuel, can have an inhibitive effect on the population growth. If a, $a > 0$ is a constant average birth rate, let us assume that the average death rate is proportional to the population $P(t)$ at time t. Thus, if $(1/P)(dP/dt)$ is the rate of growth per individual in a population, then

$$\frac{1}{P}\frac{dP}{dt} = \begin{pmatrix} \textit{average} \\ \textit{birth rate} \end{pmatrix} - \begin{pmatrix} \textit{average} \\ \textit{death rate} \end{pmatrix} = a - bP, \tag{3}$$

where b is a positive constant of proportionality. Cross multiplying (3) by P immediately gives (2).

As we shall now see, the solution of (2) is bounded as $t \to \infty$. If we rewrite (2) as $dP/dt = aP - bP^2$, the term $-bP^2$, $b > 0$ can be interpreted as an "inhibition" or "competition" term. Also, in most applications, the positive constant a is much larger than the constant b.

Logistic curves have proved to be quite accurate in predicting the growth patterns, in a limited space, of certain types of bacteria, protozoa, water fleas (*Daphnia*), and fruit flies (*Drosophila*). We have already seen equation (2) in the form $dx/dt = kx(n + 1 - x)$, $k > 0$. This differential equation provides a reasonable model for describing the spread of an epidemic brought about by initially introducing an infected individual into a static population. The solution

$x(t)$ represents the number of individuals infected with the disease at time t (see Example 11, Section 1.2). Sociologists and even business analysts have borrowed this latter model to study the spread of information and the impact of advertising in certain centers of population.

Solution One method for solving equation (2) is separation of variables.* By using partial fractions, we can write

$$\frac{dP}{P(a - bP)} = dt \quad \text{as} \quad \left(\frac{1/a}{P} + \frac{b/a}{a - bP}\right) dP = dt.$$

Integrating, we get

$$\frac{1}{a}\ln|P| - \frac{1}{a}\ln|a - bP| = t + c \qquad \text{or} \qquad \ln\left|\frac{P}{a - bP}\right| = at + ac$$

$$\frac{P}{a - bP} = c_1 e^{at}. \tag{4}$$

It follows from the last equation that

$$P(t) = \frac{ac_1e^{at}}{1 + bc_1e^{at}} = \frac{ac_1}{bc_1 + e^{-at}} \tag{5}$$

Now if we are given the initial condition $P(0) = P_0$, $P_0 \neq a/b$,† equation (4) implies $c_1 = P_0/(a - bP_0)$. Substituting this value in (5) and simplifying then gives

$$P(t) = \frac{aP_0}{bP_0 + (a - bP_0)e^{-at}}. \tag{6}$$

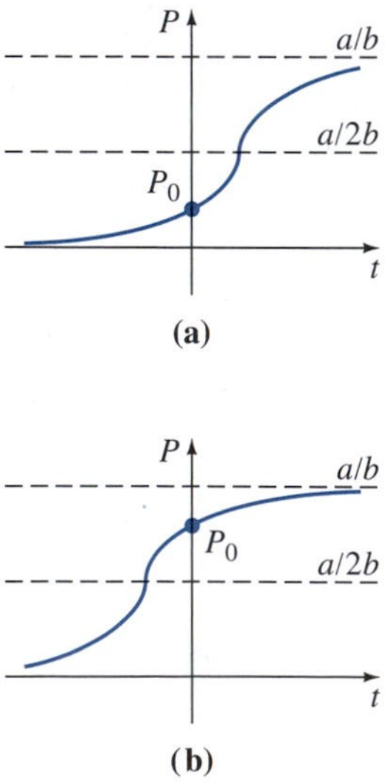

Figure 3.18

Graphs of *P*(*t*) The basic shape of the graph of the logistic function $P(t)$ can be obtained without too much effort. Although the variable t usually represents time and we are seldom concerned with applications in which $t < 0$, it is nonetheless of some interest to include this interval when displaying the various graphs of P. From (6) we see that

$$P(t) \to \frac{aP_0}{bP_0} = \frac{a}{b} \text{ as } t \to \infty \quad \text{and} \quad P(t) \to 0 \text{ as } t \to -\infty.$$

Now differentiating (2) by the product rule gives

$$\begin{aligned} \frac{d^2P}{dt^2} &= P\left(-b\frac{dP}{dt}\right) + (a - bP)\frac{dP}{dt} \\ &= \frac{dP}{dt}(a - 2bP) \\ &= P(a - bP)(a - 2bP) \\ &= 2b^2P\left(P - \frac{a}{b}\right)\left(P - \frac{a}{2b}\right). \end{aligned} \tag{7}$$

From calculus recall that the points where $d^2P/dt^2 = 0$ are possible points of inflection, but $P = 0$ and $P = a/b$ can obviously be ruled out. Hence $P = a/2b$

---

* In the form $dP/dt - aP = -bP^2$ you might recognize the logistic equation as a special case of Bernoulli's equation (see Section 2.6).

† Notice that $P = a/b$ is a singular solution of equation (2).

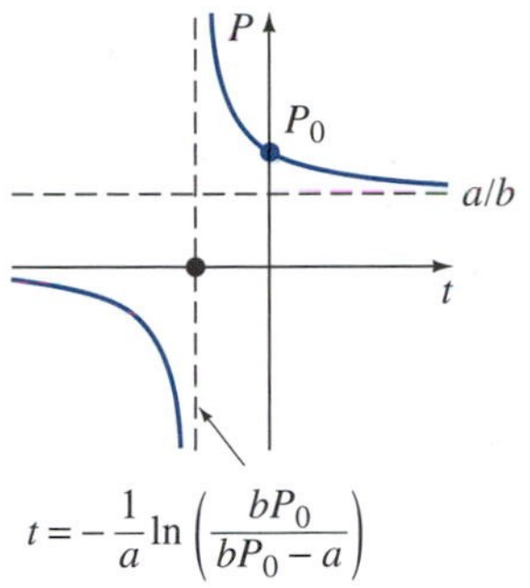

Figure 3.19

is the only possible ordinate value at which the concavity of the graph can change. For $0 < P < a/2b$ it follows from (7) that $P'' > 0$, and $a/2b < P < a/b$ implies $P'' < 0$. Thus, as we read from left to right, the graph changes from concave up to concave down at the point corresponding to $P = a/2b$. When the initial value satisfies $0 < P_0 < a/2b$, the graph of $P(t)$ assumes the shape of an S, as we see in Figure 3.18(a). For $a/2b < P_0 < a/b$ the graph is still S-shaped but the point of inflection occurs at a negative value of t, as shown in Figure 3.18(b).

If $P_0 > a/b$, equation (7) shows that $P'' > 0$ for all t in the domain of $P(t)$ for which $P > 0$. When $P < 0$, equation (7) implies $P'' < 0$. However, $P = 0$ is not a point of inflection since, whenever $a - bP_0 < 0$, an inspection of (6) reveals a vertical asymptote at

$$t = -\frac{1}{a}\ln\left(\frac{bP_0}{bP_0 - a}\right).$$

The graph of $P(t)$ in this case is given in Figure 3.19.

EXAMPLE 1 Spread of a Flu Virus

Suppose a student carrying a flu virus returns to an isolated college campus of 1000 students. If it is assumed that the rate at which the virus spreads is proportional not only to the number x of infected students but also to the number of students not infected, determine the number of infected students after 6 days if it is further observed that after 4 days $x(4) = 50$.

Solution Assuming that no one leaves the campus throughout the duration of the disease, we must solve the initial-value problem

$$\frac{dx}{dt} = kx(1000 - x), \quad x(0) = 1.$$

By making the identifications $a = 1000k$ and $b = k$, we have immediately from (6) that

$$x(t) = \frac{1000k}{k + 999ke^{-1000kt}} = \frac{1000}{1 + 999e^{-1000kt}}. \tag{8}$$

Now, using the information $x(4) = 50$, we determine k from

$$50 = \frac{1000}{1 + 999e^{-4000k}}.$$

We find that $k = \frac{-1}{4000}\ln\frac{19}{999} = 0.0009906$. Thus (8) becomes

$$x(t) = \frac{1000}{1 + 999e^{-0.9906t}}.$$

Finally,

$$x(6) = \frac{1000}{1 + 999e^{-5.9436}} = 276 \text{ students}.$$

Additional calculated values of $x(t)$ are given in the table in Figure 3.20.

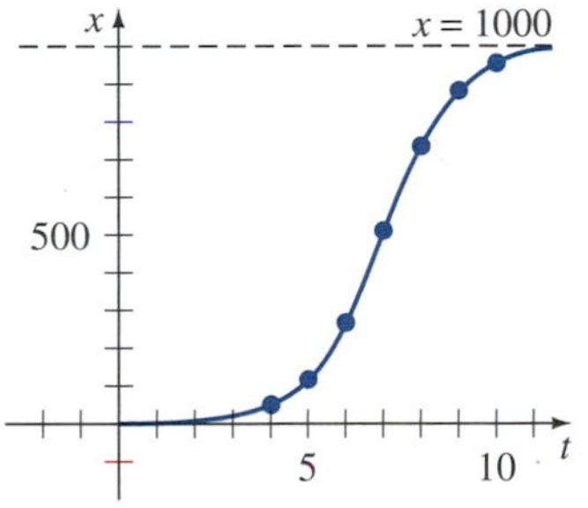

(a)

t (days)	x (number infected)
4	50 (observed)
5	124
6	276
7	507
8	735
9	882
10	953

(b)

Figure 3.20

Gompertz Curves A modification of the logistic equation is

$$\frac{dP}{dt} = P(a - b \ln P), \tag{9}$$

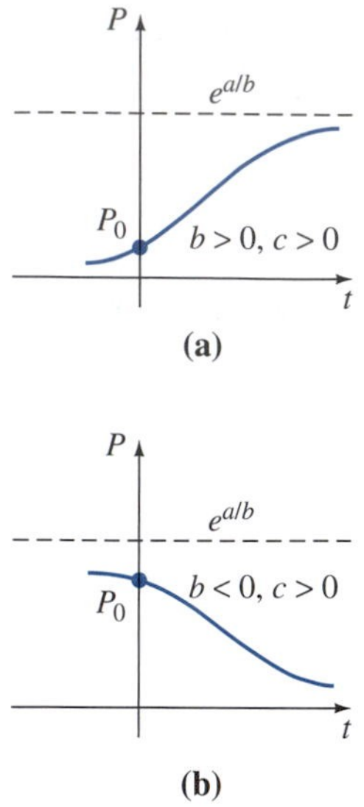

Figure 3.21

where a and b are constants. It is readily shown by separation of variables (see Problem 5) that a solution of (9) is

$$P(t) = e^{a/b}e^{-ce^{-bt}}, \tag{10}$$

where c is an arbitrary constant. We note that when $b > 0$, $P \to e^{a/b}$ as $t \to \infty$, whereas for $b < 0$, $c > 0$, $P \to 0$ as $t \to \infty$. The graph of the function (10), called a **Gompertz curve,** is quite similar to the graph of the logistic function. Figure 3.21 shows two possibilities for the graph of $P(t)$.

Functions such as (10) are encountered, for example, in studies of the growth or decline of certain populations, in the growth of solid tumors, in actuarial predictions, and in the study of growth of revenue in the sale of a commercial product.

Chemical Reactions The disintegration of a radioactive substance, governed by equation (1) of the preceding section, is said to be a **first-order reaction.** In chemistry a few reactions follow the same empirical law: If the molecules of a substance A decompose into smaller molecules, it is a natural assumption that the rate at which this decomposition takes place is proportional to the amount of the first substance that has not undergone conversion; that is, if $X(t)$ is the amount of substance A remaining at time t, then

$$\frac{dX}{dt} = kX,$$

where k is negative, since X is decreasing. An example of a first-order chemical reaction is the conversion of t-butyl chloride into t-butyl alcohol:

$$(CH_3)_3CCl + NaOH \to (CH_3)_3COH + NaCl.$$

Only the concentration of the t-butyl chloride controls the rate of reaction.

Now in the reaction

$$CH_3Cl + NaOH \to CH_3OH + NaCl,$$

for every molecule of methyl chloride one molecule of sodium hydroxide is consumed, thus forming one molecule of methyl alcohol and one molecule of sodium chloride. In this case the rate at which the reaction proceeds is proportional to the product of the remaining concentrations of CH_3Cl and of NaOH. If X denotes the amount of CH_3OH formed and α and β are the given amounts of the first two chemicals A and B, then the instantaneous amounts not converted to chemical C are $\alpha - X$ and $\beta - X$, respectively. Hence the rate of formation of C is given by

$$\frac{dX}{dt} = k(\alpha - X)(\beta - X), \tag{11}$$

where k is a constant of proportionality. A reaction described by equation (11) is said to be of **second-order.**

EXAMPLE 2 Second-Order Chemical Reaction

A compound C is formed when two chemicals A and B are combined. The resulting reaction between the two chemicals is such that for each gram of A, 4 grams of B are used. It is observed that 30 grams of the compound C are formed in 10 minutes. Determine the amount of C at any time if the rate of the reaction is proportional to the amounts of A and B remaining and if initially there are 50 grams of A and 32 grams of B. How much of the compound C is present at 15 minutes? Interpret the solution as $t \to \infty$.

Solution Let $X(t)$ denote the number of grams of the compound C present at time t. Clearly $X(0) = 0$ and $X(10) = 30$.

Now for example, if there are 2 grams of compound C, we must have used, say, a grams of A and b grams of B so

$$a + b = 2 \quad \text{and} \quad b = 4a.$$

Thus we must use $a = \frac{2}{5} = 2\frac{1}{5}$ grams of chemical A and $b = \frac{8}{5} = 2\frac{4}{5}$ grams of B. In general, for X grams of C we must use

$$\frac{X}{5} \text{ grams of } A \quad \text{and} \quad \frac{4}{5}X \text{ grams of } B.$$

The amounts of A and B remaining at time t are then

$$50 - \frac{X}{5} \quad \text{and} \quad 32 - \frac{4}{5}X,$$

respectively.

Now we know that the rate at which chemical C is formed satisfies

$$\frac{dX}{dt} \propto \left(50 - \frac{X}{5}\right)\left(32 - \frac{4}{5}X\right).$$

To simplify the subsequent algebra, we factor $\frac{1}{5}$ from the first term and $\frac{4}{5}$ from the second, and then introduce the constant of proportionality:

$$\frac{dX}{dt} = k(250 - X)(40 - X).$$

By separation of variables and partial fractions, we can write

$$\frac{dX}{(250 - X)(40 - X)} = k\,dt$$

$$-\frac{1/210}{250 - X}dX + \frac{1/210}{40 - X}dX = k\,dt$$

$$\ln\left|\frac{250 - X}{40 - X}\right| = 210kt + c_1$$

$$\frac{250 - X}{40 - X} = c_2e^{210kt}. \tag{12}$$

When $t = 0$, $X = 0$, so it follows at this point that $c_2 = \frac{25}{4}$. Using $X = 30$ at $t = 10$, we find that $210k = \frac{1}{10}\ln\frac{88}{25} = 0.1258$. With this information we solve (12) for X:

$$X(t) = 1000\frac{1 - e^{-0.1258t}}{25 - 4e^{-0.1258t}}. \tag{13}$$

The behavior of X as a function of time is displayed in Figure 3.22. It is clear from the accompanying table and equation (13) that $X \to 40$ as $t \to \infty$. This means there are 40 grams of compound C formed, leaving

$$50 - \tfrac{1}{5}(40) = 42 \text{ g of chemical } A \quad \text{and} \quad 32 - \tfrac{4}{5}(40) = 0 \text{ g of chemical } B.$$

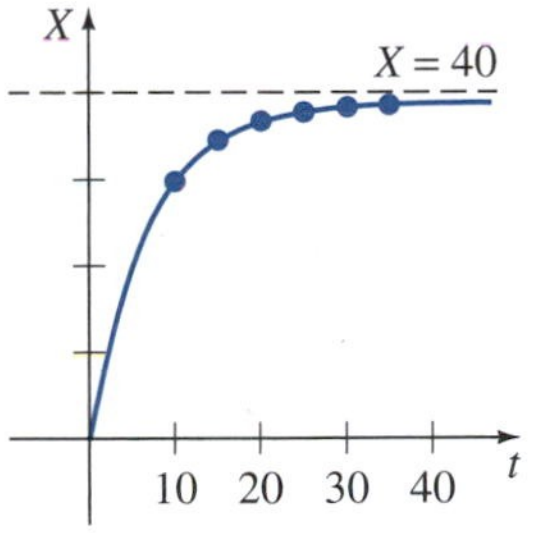

(a)

t (minutes)	X (grams)
10	30 (measured)
15	34.78
20	37.25
25	38.54
30	39.22
35	39.59

(b)

Figure 3.22

Law of Mass Action The preceding example can be generalized in the following manner. Suppose that a grams of substance A are combined with b grams of substance B. If there are M parts of A and N parts of B formed in the compound, then the amounts of substances A and B remaining at time t are, respectively,

$$a - \frac{M}{M+N}X \quad \text{and} \quad b - \frac{N}{M+N}X.$$

Thus

$$\frac{dX}{dt} \propto \left[a - \frac{M}{M+N}X\right]\left[b - \frac{N}{M+N}X\right]. \tag{14}$$

Proceeding as before, if we factor out $M/(M+N)$ from the first term and $N/(M+N)$ from the second term, the resulting differential equation is the same as (11):

$$\frac{dX}{dt} = k(\alpha - X)(\beta - X), \tag{15}$$

where

$$\alpha = \frac{a(M+N)}{M} \quad \text{and} \quad \beta = \frac{b(M+N)}{N}.$$

Chemists refer to reactions described by equation (15) as the **law of mass action.**

When $\alpha \neq \beta$, it is readily shown (see Problem 9) that a solution of (15) is

$$\frac{1}{\alpha - \beta}\ln\left|\frac{\alpha - X}{\beta - X}\right| = kt + c. \tag{16}$$

When we assume the natural initial condition $X(0) = 0$, equation (16) yields the explicit solution

$$X(t) = \frac{\alpha\beta[1 - e^{(\alpha-\beta)kt}]}{\beta - \alpha e^{(\alpha-\beta)kt}}. \tag{17}$$

Without loss of generality we assume in (17) that $\beta > \alpha$ or $\alpha - \beta < 0$. Since $X(t)$ is an increasing function, we expect $k > 0$, and so it follows immediately from (17) that $X \rightarrow \alpha$ as $t \rightarrow \infty$.

Escape Velocity In Example 1 of Section 1.2 we saw that the differential equation of a free-falling object of mass m near the surface of the earth is given by

$$m\frac{d^2s}{dt^2} = -mg \quad \text{or simply} \quad \frac{d^2s}{dt^2} = -g,$$

where s represents the distance from the surface of the earth to the object and the positive direction is considered to be upward. In other words, the underlying assumption here is that the distance s to the object is small when compared with the radius R of the earth; put yet another way, the distance y from the center of the earth to the object is approximately the same as R. If, on the other hand, the distance y to an object, such as a rocket or a space probe, is large compared to R, then we combine Newton's second law of motion and his universal law of gravitation to derive a differential equation in the variable y. The solution of this differential equation can be used to determine the minimum velocity, the so-called **escape velocity,** needed by a rocket to break free of the earth's gravitational attraction.

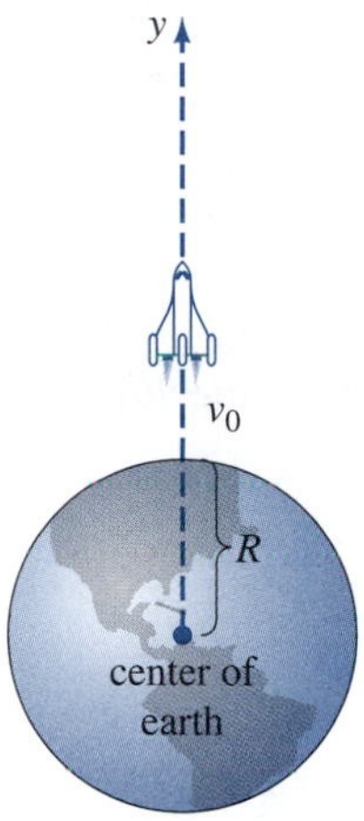

Figure 3.23

EXAMPLE 3 Velocity of a Rocket

A rocket is shot vertically upward from the ground as shown in Figure 3.23. If the positive direction is upward and air resistance is ignored, then the differential equation of motion after fuel burnout is

$$m\frac{d^2y}{dt^2} = -k\frac{mM}{y^2} \quad \text{or} \quad \frac{d^2y}{dt^2} = -k\frac{M}{y^2}, \tag{18}$$

where k is a constant of proportionality, y is the distance from the center of the earth to the rocket, M is the mass of the earth, and m is the mass of the rocket. To determine the constant k, we use the fact that, when $y = R$,

$$k\frac{mM}{R^2} = mg \quad \text{or} \quad k = \frac{gR^2}{M}.$$

Thus the last equation in (18) becomes

$$\frac{d^2y}{dt^2} = -g\frac{R^2}{y^2}. \tag{19}$$

Although this is not a first-order equation, if we write the acceleration as

$$\frac{d^2y}{dt^2} = \frac{dv}{dt} = \frac{dv}{dy}\frac{dy}{dt} = v\frac{dv}{dy},$$

then (19) becomes first-order in v; that is,

$$v\frac{dv}{dy} = -g\frac{R^2}{y^2}.$$

This last equation can be solved by separation of variables. From

$$\int v\,dv = -gR^2\int y^{-2}dy \quad \text{we get} \quad \frac{v^2}{2} = g\frac{R^2}{y} + c. \tag{20}$$

If we assume that the velocity is $v = v_0$ at burnout and that $y \approx R$ at that instant, we can obtain the (approximate) value of c. From (20) we find $c = -gR + v_0^2/2$. Substituting this value in (20) and multiplying the resulting equation by 2 yields

$$v^2 = 2g\frac{R^2}{y} - 2gR + v_0^2. \tag{21}$$

You might object, correctly, that in Example 3 we have not really solved the original equation for y. Actually the solution (21) gives quite a bit of information. Now that we have done the hard part, we leave the actual determination of the escape velocity from the earth as an exercise. See Problem 11.

EXERCISES 3.3

Answers to odd-numbered problems begin on page A-5.

1. The number of supermarkets $C(t)$ throughout the country that are using a computerized checkout system is described by the initial-value problem

$$\frac{dC}{dt} = C(1 - 0.0005C), \quad C(0) = 1,$$

where $t > 0$. How many supermarkets are using the computerized method when $t = 10$? How many companies are estimated to adopt the new procedure over a long period of time?

2. The number of people $N(t)$ in a community who are exposed to a particular advertisement is governed by the logistic equation. Initially $N(0) = 500$, and it is observed that $N(1) = 1000$. If it is predicted that the limiting number of people in the community who will see the advertisement is 50,000, determine $N(t)$ at time t.

3. The population $P(t)$ at time t in a suburb of a large city is governed by the initial-value problem

$$\frac{dP}{dt} = P(10^{-1} - 10^{-7}P), \quad P(0) = 5000,$$

where t is measured in months. What is the limiting value of the population? At what time will the population be equal to one-half of this limiting value?

4. Find a solution of the **modified logistic equation**

$$\frac{dP}{dt} = P(a - bP)(1 - cP^{-1}), \quad a, b, c > 0.$$

5. (**a**) Solve equation (9).

(**b**) Determine the value of c in equation (10) if $P(0) = P_0$.

6. Assuming $0 < P_0 < e^{a/b}$ and $a > 0$, use equation (9) to find the ordinate of the point of inflection for a Gompertz curve.

7. Two chemicals A and B are combined to form a chemical C. The rate or velocity of the reaction is proportional to the product of the instantaneous amounts of A and B not converted to chemical C. Initially there are 40 grams of A and 50 grams of B, and for each gram of B, 2 grams of A are used. It is observed that 10 grams of C are formed in 5 minutes. How much is formed in 20 minutes? What is the limiting amount of C after a long time? How much of chemicals A and B remains after a long time?

8. Solve Problem 7 if 100 grams of chemical A are present initially. At what time is chemical C half-formed?

9. Obtain a solution of the equation

$$\frac{dX}{dt} = k(\alpha - X)(\beta - X)$$

governing second-order reactions in the two cases $\alpha \neq \beta$ and $\alpha = \beta$.

10. In a third-order chemical reaction the number of grams X of a compound obtained by combining three chemicals is governed by

$$\frac{dX}{dt} = k(\alpha - X)(\beta - X)(\gamma - X).$$

Solve the equation under the assumption $\alpha \neq \beta \neq \gamma$.

11. (**a**) Use equation (21) to show that the escape velocity of the rocket is given by $v_0 = \sqrt{2gR}$. [*Hint*: Take $y \to \infty$ in (21) and assume $v > 0$ for all times t.]

(**b**) The result in part (a) holds for any body in the solar system. Use the values $g = 32\ \text{ft/s}^2$ and $R = 4000$ miles to show that the escape velocity from the earth is (approximately) $v_0 = 25{,}000$ mi/h.

(**c**) Find the escape velocity from the moon if the acceleration of gravity is $0.165g$ and $R = 1080$ miles.

Miscellaneous Applications

12. In Example 7 of Section 1.2 we saw that the differential equation describing the shape of a wire of constant linear density w hanging under its own weight is

$$\frac{d^2y}{dx^2} = \frac{w}{T_1}\sqrt{1 + \left(\frac{dy}{dx}\right)^2},$$

where T_1 is the horizontal tension in the wire at its lowest point. Using the substitution $p = dy/dx$, solve this equation subject to the initial conditions $y(0) = 1$, $y'(0) = 0$.

13. An equation similar to that given in Problem 12 is

$$x\frac{d^2y}{dx^2} = \frac{v_1}{v_2}\sqrt{1 + \left(\frac{dy}{dx}\right)^2}.$$

In this case the equation arises in the study of the shape of the path that a pursuer, traveling at a speed v_2, must take in order to intercept a prey traveling at speed v_1. Use the same substitution as in Problem 12 and the initial conditions $y(1) = 0$, $y'(1) = 0$ to solve the equation. Consider the two cases $v_1 = v_2$ and $v_1 \neq v_2$.

14. According to **Stefan's law** of radiation, the rate of change of temperature from a body at absolute temperature T is

$$\frac{dT}{dt} = k(T^4 - T_{\text{m}}^4),$$

where T_{m} is the absolute temperature of the surrounding medium. Find a solution of this differential equation. It can be shown that when $T - T_{\text{m}}$ is small compared to T_{m}, this particular equation is closely approximated by Newton's law of cooling (Equation (3), Section 3.2).

15. The height h of water that is flowing through an orifice at the bottom of a cylindrical tank is given by

$$\frac{dh}{dt} = -\frac{A_{\text{o}}}{A_{\text{w}}}\sqrt{2gh}, \quad g = 32\ \text{ft/s}^2,$$

where A_{w} and A_{o} are the cross-sectional areas of the water and orifice, respectively (see Example 8, Section 1.2). Solve the equation if the initial height of the water is 20 ft and $A_{\text{w}} = 50\ \text{ft}^2$ and $A_{\text{o}} = \frac{1}{4}\ \text{ft}^2$. At what time is the tank empty?

16. The nonlinear differential equation

$$\left(\frac{dr}{dt}\right)^2 = \frac{2\mu}{r} + 2h,$$

where μ and h are nonnegative constants, arises in the study of the two-body problem of celestial mechanics. Here the variable r represents the distance between the two masses. Solve the equation in the two cases $h = 0$ and $h > 0$.

17. Solve the differential equation of the **tractrix**

$$\frac{dy}{dx} = -\frac{y}{\sqrt{s^2 - y^2}}$$

(see Problem 19, Exercises 1.2). Assume that the initial point on the y-axis is (0, 10) and the length of rope is $s = 10$ ft.

18. A body of mass m falling through a viscous medium encounters a resisting force proportional to the square of its instantaneous velocity. In this situation the differential equation for the velocity $v(t)$ at time t is

$$m\frac{dv}{dt} = mg - kv^2,$$

where k is a positive constant of proportionality. Solve the equation subject to $v(0) = v_0$. What is the limiting velocity of the falling body?

19. The differential equation

$$x\left(\frac{dx}{dy}\right)^2 + 2y\frac{dx}{dy} = x,$$

where $x = x(y)$, occurs in the study of optics. The equation describes the type of plane curve that will reflect all incoming light rays to the same point (see Problem 15, Exercises 1.2). Show that the curve must be a parabola. [*Hint*: Use the substitution $w = x^2$ and then re-examine Section 2.6.]

20. Solve the equation in Problem 19 with the aid of the quadratic formula.

21. The equations of Lotka and Volterra*

$$\frac{dy}{dt} = y(\alpha - \beta x)$$

$$\frac{dx}{dt} = x(-\gamma + \delta y),$$

---

*A. J. LOTKA (1880–1949) Lotka, born in Austria, was an American biomathematician.

VITO VOLTERRA (1860–1940) Born in Ancona, Italy, Vito Volterra showed an early aptitude for mathematics. He studied calculus on his own initiative and investigated problems in gravitation at the age of twelve. Although his education was a constant financial struggle, Volterra quickly attained prominence as a scientist and mathematician. He was also an active politician and was appointed Senator of the Kingdom of Italy in 1905. Volterra became interested in the applications of mathematics to ecology in the mid-1920s and formulated this system of differential equations in an attempt to explain the variations in the fish population in the Mediterranean as a result of predator–prey interactions. (Lotka, working independently, arrived at the same system of equations and published the result in 1925 in his text *Elements of Physical Biology*.) Through his research into mathematical models of population, Volterra established the groundwork for a field of mathematics known as integral equations. A man of principle, Volterra refused to sign a loyalty oath to the fascist regime of Benito Mussolini and eventually resigned his chair of mathematics at the University of Rome and all his memberships in Italian scientific societies.

where α, β, γ, and δ are positive constants, occur in the analysis of the biological balance of two species of animals such as predators and prey (for example, foxes and rabbits). Here $x(t)$ and $y(t)$ denote the populations of the two species at time t. Although no explicit solutions of the system exist, solutions can be found relating the two populations at any time. Divide the first equation by the second and solve the resulting nonlinear first-order differential equation.

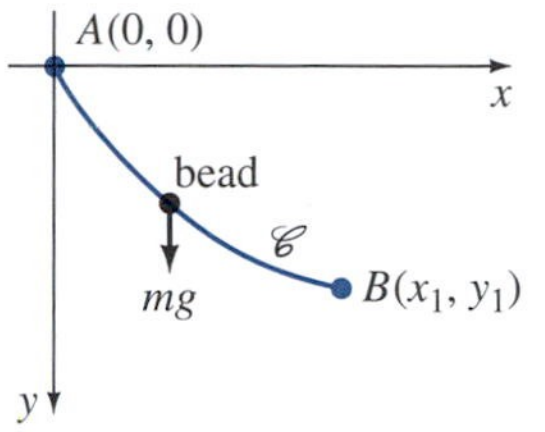

Figure 3.24

22. A classical problem in the calculus of variations is to find the shape of a curve $\mathscr{C}$ such that a bead, under the influence of gravity, will slide from $A(0, 0)$ to $B(x_1, y_1)$ in the least time. See Figure 3.24. It can be shown that the differential equation for the shape of the path is $y[1 + (y')^2] = k$, where k is a constant. First solve for dx in terms of y and dy, and then use the substitution $y = k \sin^2\theta$ to obtain the parametric form of the solution. The curve $\mathscr{C}$ turns out to be a cycloid.

23. The initial-value problem describing the motion of a simple pendulum released from rest from an angle $\theta_0 > 0$ is

$$\frac{d^2\theta}{dt^2} + \frac{g}{l}\sin\theta = 0, \quad \theta(0) = \theta_0, \quad \left.\frac{d\theta}{dt}\right|_{t=0} = 0.$$

(**a**) Obtain the first-order equation

$$\left(\frac{d\theta}{dt}\right)^2 = \frac{2g}{l}(\cos\theta - \cos\theta_0).$$

[*Hint*: Multiply the given equation by $2\,d\theta/dt$.]

(**b**) Use the equation in part (a) to show that the period of motion is

$$T = 2\sqrt{\frac{2l}{g}}\int_0^{\theta_0}\frac{d\theta}{\sqrt{\cos\theta - \cos\theta_0}}.$$

CHAPTER 3 REVIEW

If every curve in a one-parameter family of curves $G(x, y, c_1) = 0$ is orthogonal to every curve in a second one-parameter family $H(x, y, c_2) = 0$, we say that the two families are **orthogonal trajectories.** Two curves are orthogonal if their tangent lines are perpendicular at a point of intersection. When given a family, we find its differential equation $dy/dx = f(x, y)$ by differentiating the equation $G(x, y, c_1) = 0$ and eliminating the parameter c_1. The differential equation of the second and orthogonal family is then $dy/dx = -1/f(x, y)$. We solve this latter equation by the methods of Chapter 2.

In the mathematical analysis of population growth, radioactive decay, or chemical mixtures, we often encounter **linear** differential equations such as

$$\frac{dx}{dt} = kx \quad \text{and} \quad \frac{dx}{dt} = a + bx$$

or **nonlinear** differential equations such as

$$\frac{dx}{dt} = x(a - bx) \quad \text{and} \quad \frac{dx}{dt} = k(\alpha - x)(\beta - x).$$

You should be able to solve these particular equations without hesitation. It is never a good idea simply to memorize solutions of differential equations.

CHAPTER 3 REVIEW EXERCISES

Answers to odd-numbered problems begin on page A-6.

1. Find the orthogonal trajectories of the family of curves $y(x^3 + c_1) = 3$.
2. Find the orthogonal trajectory to the family $y = 4x + 1 + c_1e^{4x}$ passing through the point $(0, 0)$.
3. Find the orthogonal trajectories of the family of parabolas opening in the y direction with vertex at $(1, 2)$.
4. Show that if a population expands at a rate proportional to the number of people present at any time, then the doubling time of the population is $T = (\ln 2)/k$, where k is the positive growth rate. This is known as the **Law of Malthus.**
5. In March 1976 the world population reached 4 billion. A popular news magazine has predicted that with an average yearly growth rate of 1.8%, the world population will be 8 billion in 45 years. How does this value compare with that predicted by the model that says the rate of increase is proportional to the population at time t?
6. Air containing 0.06% carbon dioxide is pumped into a room whose volume is 8000 ft^3. The rate at which the air is pumped in is 2000 ft^3/min, and the circulated air is then pumped out at the same rate. If there is an initial concentration of 0.2% carbon dioxide, determine the subsequent amount in the room at any time. What is the concentration at 10 minutes? What is the steady-state or equilibrium concentration of carbon dioxide?
7. The populations of two species of animals are described by the nonlinear system of first-order differential equations

$$\frac{dx}{dt} = k_1x(\alpha - x), \quad \frac{dy}{dt} = k_2xy.$$

   Solve for x and y in terms of t.

8. A projectile is shot vertically into the air with an initial velocity of v_0 ft/s. Assuming that air resistance is proportional to the square of the instantaneous velocity, the motion is described by this pair of differential equations:

$$m\frac{dv}{dt} = -mg - kv^2, \quad k > 0,$$

   positive y-axis up, origin at ground level so that $v = v_0$ at $y = 0$, and

$$m\frac{dv}{dt} = mg - kv^2, \quad k > 0,$$

   positive y-axis down, origin at the maximum height so that $v = 0$ at $y = h$. The first and second equations describe the motion of the projectile when rising and falling, respectively.

   (a) Determine the limiting, or terminal, velocity of the falling projectile. Compare this terminal velocity with that obtained in Problem 27 in Exercises 3.2.

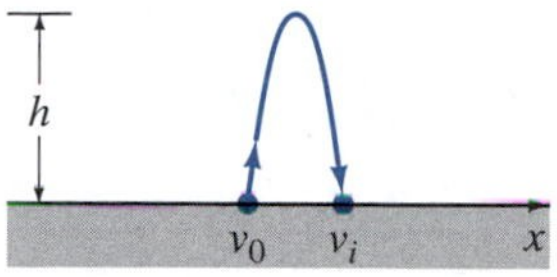

Figure 3.25

(b) Prove that the impact velocity v_i of the projectile is less than the initial velocity v_0. It can also be shown that the time t_1 needed to attain its maximum height h is less than the time t_2 that it takes to fall from this height. See Figure 3.25.

9. Consider Newton's law of cooling $dT/dt = k(T - T_m)$, $k < 0$, where the temperature of the surrounding medium T_m changes with time. Suppose the initial temperature of a body is T_1 and the initial temperature of the surrounding medium is T_2 and $T_m = T_2 + B(T_1 - T)$, where $B > 0$ is a constant.

(a) Find the temperature of the body at any time t.

(b) What is the limiting value of the temperature as $t \to \infty$?

(c) What is the limiting value of T_m as $t \to \infty$?

10. An L-R series circuit has a variable inductor with the inductance defined by

$$L = \begin{cases} 1 - \dfrac{t}{10}, & 0 \le t < 10 \\ 0, & t \ge 10 \end{cases}$$

Find the current $i(t)$ if the resistance is 0.2 ohm, the impressed voltage is $E(t) = 4$, and $i(0) = 0$. Graph $i(t)$.

Essay

by
Michael Olinick
Department of Mathematics and Computer Sciences, Middlebury College

POPULATION DYNAMICS

"The fact that ecology is essentially a mathematical subject is becoming ever more widely accepted," writes Evelyn C. Pielou [2]. "Ecologists everywhere are attempting to formulate and solve their problems by mathematical reasoning." Historically, the first and perhaps most important branch of mathematical ecology is the investigation of population dynamics: how populations grow and decline. First-order differential equations have been a critically important tool in these studies.

Many attempts to model population growth begin with the assumption that the rate of population growth is dependent on the size of the population. If P represents the population at time t, then these models all have the form

$$\frac{dP}{dt} = f(P),$$

where f is some function of the population level P. How should f be selected?

The central figure in the history of population is the Reverend Thomas Robert Malthus (1766–1834). Malthus was an honors graduate in mathematics at Cambridge University, an ordained minister in the Church of England, and a professor of history and political economy. In a seminal work, *An Essay on the Principle of Population* [3], Malthus argued that the appropriate form of $f(P)$, at least when the population is small, should be a constant multiple of P; that is,

$$\frac{dP}{dt} = rP,$$

where r is a constant.

As we have seen, this model yields **exponential growth,** since the solution of the differential equation is

$$P(t) = P_0e^{rt}.$$

One characteristic of exponential growth is **constant doubling time:** It takes exactly the same amount of time, $(\ln 2)/r$, for the population to double in size from P_0 to $2P_0$ regardless of the size of P_0. Another way to examine exponential growth is to examine the population at successive time units:

$$P(0), P(1), P(2), P(3), \ldots, P(k), P(k+1), \ldots.$$

Since $P(k) = P_0e^{rk} = P_0(e^r)^k$, these populations form a **geometric sequence**

$$a, ac, ac^2, ac^3, ac^4, ac^5, \ldots$$

with initial term $a = P_0$ and constant ratio $c = e^r$. Malthus, in fact, begins his famous essay with the observation "In taking a view of animated nature, we cannot fail to be struck with a prodigious power of increase in plants and animals ... whether they increase slowly or rapidly, their natural tendency must be to increase in a geometrical ratio, that is, by multiplication; and at whatever rate they are increasing during any one period, if no further obstacles be opposed to them, they must proceed in a geometrical progression." After carefully examining figures collected during the first censuses of the United States and looking at data from other countries, Malthus concluded that "the natural progress of population" was exponential in nature with a doubling time of about 25 years for humans.

Since this model asserts that there is no limit to the number of individuals in this population, it is clear that the exponential model is not a completely realistic picture. The exponential model *may* be a realistic one for the growth of some populations over relatively short time intervals. The population of the United States during the period from 1790 to 1860 grew at such an exponential pace, with an annual growth rate of about 3%.

It's instructive to examine the actual U.S. census figures alongside those given by an exponential growth model. Table 3.1 shows the actual and predicted populations in millions. The "error" column displays the difference between the actual and predicted numbers. The final column, "% error," is derived from the ratio of the error to the actual population. The "actual population" is that reported by the U.S. Census Bureau. The "predicted population" is generated from the equation

$$P(t) = 3.929e^{0.029655t}.$$

From the table we see that the exponential growth model fits the actual census data quite closely for the 70-year period beginning in 1790; the largest error is less than 2%. "Reality" and the model's predictions begin to diverge by 1870; the model

Table 3.1

Year	Actual population	Predicted population	Error	% Error
1790	3.929	3.929	0.000	0.00
1800	5.308	5.285	0.023	0.43
1810	7.24	7.110	0.130	1.80
1820	9.638	9.564	0.074	0.76
1830	12.866	12.866	0.000	0.00
1840	17.069	17.307	−0.238	−1.40
1850	23.192	23.282	−0.090	−0.39
1860	31.433	31.319	0.114	0.36
1870	38.558	42.131	−3.573	−9.27
1880	50.156	56.675	−6.519	−13.00
1890	62.948	76.240	−13.292	−21.12

had no way of predicting the Civil War, which raged from 1861 to 1865 and during which more than half a million American young men lost their lives. We have cut off the table in 1890. The disparity between the actual and predicted populations becomes enormously large in the twentieth century, reaching a staggering 495% error by 1990! One of the consequences we can draw from this model is that if the U.S. population had continued to grow at the same rate it had grown in the first half of the nineteenth century, we would have today a nation six times more populous than the one we do have. That's worth contemplating next time you're stuck in a traffic jam or waiting to check out at your local supermarket!

There are many generalizations of this model. In the **logistic growth model,** first developed by the Belgian mathematician Pierre-Francois Verhulst (1804–1849), we assume that r is not a constant, but a variable that *decreases* in a simple linear fashion as the population increases. Thus we can represent r as $a - bP$, where a and b are positive constants. This yields the model $dP/dt = (a - bP)P$, which has a solution of the form

$$P(t) = \frac{K}{1 + e^{d-at}}.$$

Although Verhulst attempted to test his model on actual population data, he was frustrated by the inaccurate census information available in the early 1840s when he carried out his studies. Because the existing data on population were too inadequate to form any effective test of the logistic model at the time, Verhulst's work lay forgotten for nearly 80 years. It was rediscovered independently by two American scientists working at Johns Hopkins University, Raymond Pearl and Lowell J. Reed.

In 1920 Pearl and Reed [1] examined how closely the U.S. population growth curve followed a logistic curve. Using data from the censuses of 1790, 1850, and 1910 to find values for K, d, and a, they found that the logistic equation

$$P(t) = \frac{197.274}{1 + e^{3.896 - 0.031t}}$$

matched well the actual population figures for the 120-year period beginning in 1790. In fact, the logistic model gives an excellent portrayal of the changes in the U.S. population from 1790 through 1950. Table 3.2 shows the comparison between the predictions of this logistic model and U.S. census data.

We have excellent agreement between the model's predictions and the observed population between 1790 and 1950. The largest error is about 3.5%. While the model predicts a leveling off of U.S. inhabitants that should continue after mid-century, the actual data show the "baby boom" of the 1950s, which increased the population by almost 30 million in 10 years.

It is remarkable that a relatively simple model such as the logistic one can give such accurate results for a period of 160 years. The particular logistic curve calculated by Pearl and Reed could have been derived as early as 1911, when the results of the 1910 census were published. Their equation could have been used for 40 years to give accurate population projections that would have been useful for government planning.

How has the population of the United States grown in the last half of the twentieth century? Can a simple model explain the observed changes and give us a reasonable prediction for the near future? As populations become large, we often observe a slowing down of the rate of increase. Various reasons for this have been given, the principal one being competition for limited resources.

Table 3.2

Year	Predicted population	Census population	Error	% Error
1790	3.929	3.929	0.000	0.00
1800	5.336	5.308	0.028	0.53
1810	7.228	7.240	−0.012	−0.17
1820	9.757	9.638	0.119	1.23
1830	13.109	12.866	0.243	1.89
1840	17.506	17.069	0.437	2.56
1850	23.192	23.192	0.000	0.00
1860	30.412	31.433	−1.021	−3.25
1870	39.372	38.558	0.814	2.11
1880	50.177	50.156	0.021	0.04
1890	62.769	62.948	−0.179	−0.28
1900	76.870	75.996	0.874	1.15
1910	91.972	91.972	0.000	0.00
1920	107.395	105.711	1.684	1.59
1930	122.398	122.775	−0.377	−0.31
1940	136.318	131.669	4.649	3.53
1950	148.678	150.697	−2.019	−1.34
1960	159.230	179.323	−20.093	−11.20
1970	167.944	203.185	−35.241	−17.34
1980	174.941	226.546	−51.605	−22.78
1990	180.437	248.710	−68.273	−27.45

Let's look at one of the simplest mathematical assumptions we can make: The growth rate is **inversely** proportional to the population. Our mathematical model looks like

$$\frac{dP}{dt} = \frac{b}{P} \quad \text{with} \quad P(0) = P_0,$$

where b is a positive constant.

The differential equation is easily solved by separating the variables and integrating. We obtain

$$P(t) = \sqrt{2bt + P_0^2}.$$

If b is positive, this model predicts that the population will continue to increase without bound.

If the population at a later time t_1 is observed to be P_1, then we can determine the value of the parameter b as

$$b = \frac{P_1^2 - P_0^2}{2t_1}.$$

As a test of this simple model, we will look at the U.S. census figures. Taking $t = 0$ to correspond with the 1950 figure of 150.697 million and $t = 1$ for the 1960 census of 179.323 million, we obtain a value of 4723.58 for b. Thus our fitted model has the form

$$P(t) = \sqrt{9447.16t + (150.697)^2},$$

where each unit increase in t represents 10 years.

Table 3.3

Year	Actual pop.	Pred. pop.	% Error
1970	203.302	203.970	0.33
1980	226.505	225.945	0.24
1990	248.710	245.964	1.10

With these values, in Table 3.3 we can compare the predictions of the model with the best estimates of the U.S. Census Bureau for the years 1970, 1980, and 1990. (Here both actual and predicted populations are given in millions; the percentage error is obtained by comparing the absolute value of the difference between actual and predicted values to the actual population level.)

We observe that this simple model yields surprisingly accurate results for a least one "real world" set of data. Our value of $t = 0$ corresponds to the April 1, 1950, census date. The U.S. Census Bureau makes population projections for the first day of July. A recent projection for July 1, 2000, is 268.266 million. Using the corresponding value for t of 5.025, our model predicts a population of 264.918 million. The difference is about 1.25%.

More accurate models of the growth of the U.S. population might be obtained by refining the logistic models in several different ways. The function $f(P)$—chosen to be quadratic in the logistic model—might be taken to be a polynomial of higher degree so that higher-order effects of the size of the population on the growth rate could be included. Additional factors might be attached to the differential equation to incorporate the concept that the rate of change of population is a function not only of population but of time as well; that is, you could examine models of the form

$$\frac{dP}{dt} = f(P, t).$$

We should note that the **Gompertz model** can also be viewed as a generalization of the exponential model $dP/dt = rP$, with the constant r replaced by a variable $r(t)$ which itself decreases at a constant percentage rate; that is,

$$\frac{dr}{dt} = \alpha r,$$

where α is a negative constant. This gives r as $r_0 e^{\alpha t}$, so the Gompertz model takes the form

$$\frac{dP}{dt} = r_0 e^{\alpha t} P.$$

Generalizations of the Gompertz model include approaches where $r(t)$ is taken to be a polynomial in t of a fixed degree.

Demographers are using increasingly complex and sophisticated mathematical models of both deterministic and probabilistic character to study changes in population growth in the past and to make projections about the future. Although there are many different approaches to the creation of such models, most of them are extensions or generalizations of a first-order nonlinear differential equation.

REFERENCES

1. Pearl, Raymond, and Lowell Reed. "On the Rate of Growth of the United States Population Since 1790 and Its Mathematical Representation." *Proceedings of the National Academy of Sciences* **6** (1920): 275–288.
2. Pielou, E. C. *An Introduction to Mathematical Ecology*, second edition. New York: Wiley, 1977.
3. Malthus, Thomas R. *An Essay on the Principle of Population and a Summary View of the Principle of Population*. Baltimore: Penguin, 1970.

4 LINEAR DIFFERENTIAL EQUATIONS OF HIGHER ORDER

Introduction

We turn now to the solution of differential equations of order two or higher. Although we can solve some *nonlinear* first-order equations by the techniques considered in Chapter 2, nonlinear equations of higher order generally defy solution. This does not mean a nonlinear equation has no solution, but rather there are no rules or methods whereby its solution can be exhibited in terms of elementary or other kinds of functions. As a consequence, in attempting to solve higher-order equations, we shall confine our attention to *linear* equations.

We begin this chapter by first examining the underlying theory of linear equations. As we did in Section 2.5, we set conditions on the differential equation under which we can obtain its *general solution.* Recall that a general solution contains all solutions of the equation on some interval. For the remainder of the chapter we then develop methods for obtaining a general solution for a linear equation with *constant coefficients.* We shall see that our ability to solve an nth-order linear differential equation with constant coefficients hinges on our ability to solve an nth-degree polynomial equation. The method of solution of linear equations with *nonconstant coefficients* will be taken up in Chapter 6.

4.1 PRELIMINARY THEORY

- *Initial-value problem* • *Initial conditions* • *Boundary-value problem*
- *Boundary conditions* • *Linear dependence* • *Linear independence*
- *Wronskian* • *Homogeneous equation* • *Nonhomogeneous equation*
- *Superposition principle* • *Fundamental set of solutions* • *General solution*
- *Particular solution* • *Complementary function*

We begin the discussion of higher-order differential equations, as we did that of first-order equations, with the notion of an initial-value problem. However, we confine our attention to linear differential equations.

4.1.1 INITIAL-VALUE AND BOUNDARY-VALUE PROBLEMS

Initial-Value Problem For a linear nth-order differential equation, the problem

$$\textit{Solve:} \qquad a_n(x)\frac{d^ny}{dx^n} + a_{n-1}(x)\frac{d^{n-1}y}{dx^{n-1}} + \cdots + a_1(x)\frac{dy}{dx} + a_0(x)y = g(x) \tag{1}$$

$$\textit{Subject to:} \quad y(x_0) = y_0, \quad y'(x_0) = y'_0, \quad \ldots, \quad y^{(n-1)}(x_0) = y_0^{(n-1)},$$

where $y_0, y'_0, \ldots, y_0^{(n-1)}$ are arbitrary constants, is called an **initial-value problem** (IVP). The specified values $y(x_0) = y_0, y'(x_0) = y'_0, \ldots, y^{(n-1)}(x_0) = y_0^{(n-1)}$ are called **initial conditions.** We seek a solution on some interval I containing x_0.

In the case of a linear second-order equation, a solution of the initial-value problem

$$a_2(x)\frac{d^2y}{dx^2} + a_1(x)\frac{dy}{dx} + a_0(x)y = g(x), \quad y(x_0) = y_0, \quad y'(x_0) = y'_0$$

is a function satisfying the differential equation on I whose graph passes through (x_0, y_0) such that the slope of the curve at the point is the number y'_0. See Figure 4.1.

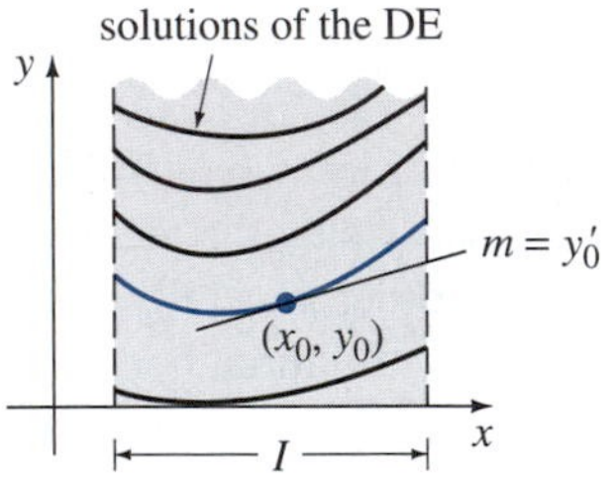

Figure 4.1

The next theorem gives sufficient conditions for the existence of a unique solution to (1).

THEOREM 4.1 Existence of a Unique Solution

> Let $a_n(x), a_{n-1}(x), \ldots, a_1(x), a_0(x)$, and $g(x)$ be continuous on an interval I and let $a_n(x) \neq 0$ for every x in this interval. If $x = x_0$ is any point in this interval, then a solution $y(x)$ of the initial-value problem (1) exists on the interval and is unique.

EXAMPLE 1 Solution of an IVP

You should verify that the function $y = 3e^{2x} + e^{-2x} - 3x$ is a solution of the initial-value problem

$$y'' - 4y = 12x, \quad y(0) = 4, \quad y'(0) = 1.$$

Now the differential equation is linear, the coefficients as well as $g(x) = 12x$ are continuous, and $a_2(x) = 1 \neq 0$ on any interval containing $x = 0$. We conclude from Theorem 4.1 that the given function is the unique solution. ■

EXAMPLE 2 Trivial Solution of an IVP

The initial-value problem

$$3y''' + 5y'' - y' + 7y = 0, \quad y(1) = 0, \quad y'(1) = 0, \quad y''(1) = 0$$

possesses the trivial solution $y = 0$. Since the third-order equation is linear with constant coefficients, it follows that all the conditions of Theorem 4.1 are fulfilled. Hence $y = 0$ is the *only* solution on any interval containing $x = 1$. ■

EXAMPLE 3 Solution of an IVP

The function $y = \frac{1}{4}\sin 4x$ is a solution of the initial-value problem

$$y'' + 16y = 0, \quad y(0) = 0, \quad y'(0) = 1.$$

It follows from Theorem 4.1 that on any interval containing $x = 0$ the solution is unique. ■

The requirements in Theorem 4.1 that $a_i(x)$, $i = 0, 1, 2, \ldots, n$ be continuous and $a_n(x) \neq 0$ for every x in I are both important. Specifically, if $a_n(x) = 0$ for some x in the interval, then the solution of a linear initial-value problem may not be unique or even exist.

EXAMPLE 4 Family of Solutions of an IVP

Verify that the function $y = cx^2 + x + 3$ is a solution of the initial-value problem

$$x^2y'' - 2xy' + 2y = 6, \quad y(0) = 3, \quad y'(0) = 1$$

on the interval $(-\infty, \infty)$ for any choice of the parameter c.

Solution Since $y' = 2cx + 1$ and $y'' = 2c$, it follows that

$$\begin{aligned} x^2y'' - 2xy' + 2y &= x^2(2c) - 2x(2cx + 1) + 2(cx^2 + x + 3) \\ &= 2cx^2 - 4cx^2 - 2x + 2cx^2 + 2x + 6 = 6. \end{aligned}$$

Also $\quad y(0) = c(0)^2 + 0 + 3 = 3 \quad$ and $\quad y'(0) = 2c(0) + 1 = 1.$ ■

Although the differential equation in Example 4 is linear and the coefficients and $g(x) = 6$ are continuous everywhere, the obvious difficulties are that $a_2(x) = x^2$ is zero at $x = 0$ and that the initial conditions are also imposed at $x = 0$.

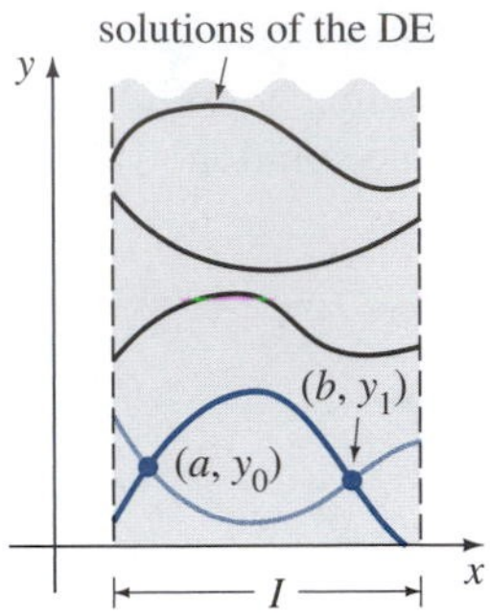

Figure 4.2

Boundary-Value Problem Another type of problem consists of solving a differential equation of order two or greater in which the dependent variable y or its derivatives are specified at *different points.* A problem such as

$$\textit{Solve:} \qquad a_2(x)\frac{d^2y}{dx^2} + a_1(x)\frac{dy}{dx} + a_0(x)y = g(x)$$

$$\textit{Subject to:} \quad y(a) = y_0, \quad y(b) = y_1$$

is called a **two-point boundary-value problem** or simply a **boundary-value problem** (BVP). The specified values $y(a) = y_0$ and $y(b) = y_1$ are called **boundary conditions.** A solution of the foregoing problem is a function satisfying the differential equation on some interval I, containing a and b, whose graph passes through the two points (a, y_0) and (b, y_1). See Figure 4.2.

EXAMPLE 5 Solution of a BVP

You should verify that on the interval $(0, \infty)$ the function $y = 3x^2 - 6x + 3$ satisfies both the differential equation and the boundary conditions of the two-point boundary-value problem

$$x^2y'' - 2xy' + 2y = 6, \quad y(1) = 0, \quad y(2) = 3.$$

For a second-order differential equation, other pairs of boundary conditions could be

$$y'(a) = y'_0, \quad y(b) = y_1,$$
$$y(a) = y_0, \quad y'(b) = y'_1,$$
or
$$y'(a) = y'_0, \quad y'(b) = y'_1,$$

where y_0, y'_0, y_1, and y'_1 denote arbitrary constants. These three pairs of conditions are just special cases of the general boundary conditions

$$\alpha_1 y(a) + \beta_1 y'(a) = \gamma_1$$
$$\alpha_2 y(b) + \beta_2 y'(b) = \gamma_2.$$

The next examples show that even when the conditions of Theorem 4.1 are fulfilled, a boundary-value problem may have

(*i*) several solutions (as suggested in Figure 4.2),
(*ii*) a unique solution, or
(*iii*) no solution at all.

EXAMPLE 6 Family of Solutions of a BVP

In Example 6 of Section 1.1, we saw that a two-parameter family of solutions for the differential equation $y'' + 16y = 0$ is

$$y = c_1 \cos 4x + c_2 \sin 4x.$$

Suppose we now wish to determine that solution of the equation that further satisfies the boundary conditions

$$y(0) = 0, \quad y\left(\frac{\pi}{2}\right) = 0.$$

Observe that the first condition

$$0 = c_1 \cos 0 + c_2 \sin 0$$

implies $c_1 = 0$, so $y = c_2 \sin 4x$. But when $x = \pi/2$, we have

$$0 = c_2 \sin 2\pi.$$

Since $\sin 2\pi = 0$, this latter condition is satisfied for any choice of c_2, so it follows that a solution of the problem

$$y'' + 16y = 0, \quad y(0) = 0, \quad y\left(\frac{\pi}{2}\right) = 0$$

is the one-parameter family

$$y = c_2 \sin 4x.$$

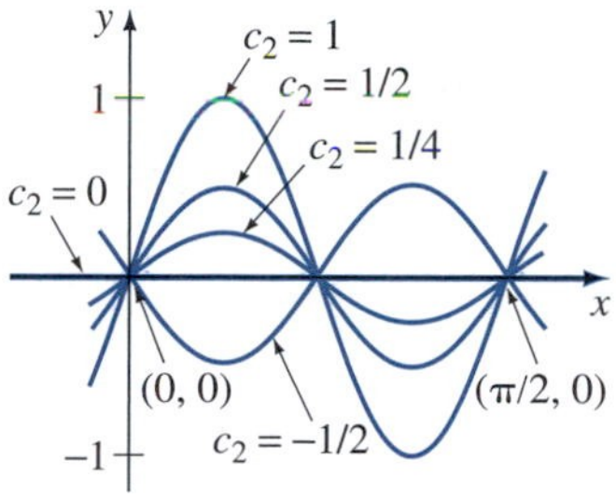

Figure 4.3

As Figure 4.3 shows, there are an infinite number of functions satisfying the differential equation whose graphs pass through the two points (0, 0) and $(\pi/2, 0)$.

If the boundary conditions were $y(0) = 0$ and $y(\pi/8) = 0$, then necessarily c_1 and c_2 would both equal zero. Thus $y = 0$ would be a solution of this new boundary-value problem. In fact, as we shall see later on in this section, it is the only solution. ■

EXAMPLE 7 BVP with No Solution

The boundary-value problem

$$y'' + 16y = 0, \quad y(0) = 0, \quad y\left(\frac{\pi}{2}\right) = 1$$

has no solution in the family $y = c_1 \cos 4x + c_2 \sin 4x$. As in Example 6, the condition $y(0) = 0$ still implies that $c_1 = 0$. Thus $y = c_2 \sin 4x$, so when $x = \pi/2$ we obtain the contradiction $1 = c_2 \cdot 0 = 0$. ■

Boundary-value problems are often encountered in the applications of partial differential equations.

4.1.2 Linear Dependence and Linear Independence

The next two concepts are basic to the study of linear differential equations.

DEFINITION 4.1 Linear Dependence

A set of functions $f_1(x), f_2(x), \ldots, f_n(x)$ is said to be **linearly dependent** on an interval I if there exist constants $c_1, c_2, \ldots, c_n$, not all zero, such that

$$c_1 f_1(x) + c_2 f_2(x) + \cdots + c_n f_n(x) = 0$$

for every x in the interval.

DEFINITION 4.2 Linear Independence

A set of functions $f_1(x), f_2(x), \ldots, f_n(x)$ is said to be **linearly independent** on an interval I if it is not linearly dependent on the interval.

In other words, a set of functions is linearly independent on an interval if the only constants for which

$$c_1 f_1(x) + c_2 f_2(x) + \cdots + c_n f_n(x) = 0$$

for every x in the interval are $c_1 = c_2 = \cdots = c_n = 0$.

It is easy to understand these definitions in the case of a set of two functions $f_1(x)$ and $f_2(x)$. If the set of functions is linearly dependent on an interval, then there exist constants c_1 and c_2 that are not both zero such that for every x in the interval,

$$c_1 f_1(x) + c_2 f_2(x) = 0.$$

Therefore, if we assume that $c_1 \neq 0$, it follows that

$$f_1(x) = -\frac{c_2}{c_1} f_2(x);$$

that is, *if a set of two functions is linearly dependent, then one is simply a constant multiple of the other.* Conversely, if $f_1(x) = c_2 f_2(x)$ for some constant c_2, then

$$(-1) \cdot f_1(x) + c_2 f_2(x) = 0$$

for every x on some interval. Hence the functions are linearly dependent since at least one of the constants (namely, $c_1 = -1$) is not zero. We conclude that *a set of two functions is linearly independent when neither is a constant multiple of the other* on an interval.

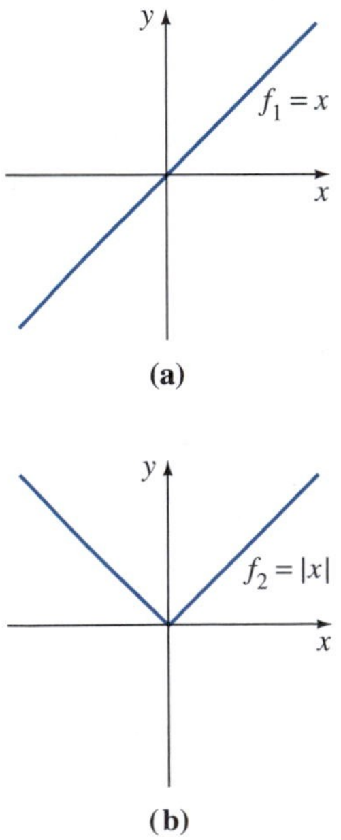

Figure 4.4

EXAMPLE 8 Linearly Dependent Functions

The set of functions $f_1(x) = \sin 2x$ and $f_2(x) = \sin x \cos x$ is linearly dependent on the interval $(-\infty, \infty)$, since

$$c_1 \sin 2x + c_2 \sin x \cos x = 0$$

is satisfied for every real x if we choose $c_1 = \frac{1}{2}$ and $c_2 = -1$. (Recall the trigonometric identity $\sin 2x = 2 \sin x \cos x$.) ■

EXAMPLE 9 Linearly Independent Functions

The set of functions $f_1(x) = x$ and $f_2(x) = |x|$ is linearly independent on the interval $(-\infty, \infty)$. Inspection of Figure 4.4 should convince you that neither function is a constant multiple of the other. Thus in order to have $c_1 f_1(x) + c_2 f_2(x) = 0$ for every real x, we must choose $c_1 = 0$ and $c_2 = 0$. ■

In the consideration of linear dependence or linear independence, the interval on which the functions are defined is important. The set of functions $f_1(x) = x$ and $f_2(x) = |x|$ in Example 9 is linearly dependent on the interval $(0, \infty)$, since

$$c_1x + c_2|x| = c_1x + c_2x = 0$$

is satisfied for any nonzero choice of c_1 and c_2 for which $c_1 = -c_2$.

EXAMPLE 10 Linearly Dependent Functions

The set of functions $f_1(x) = \cos^2 x$, $f_2(x) = \sin^2 x$, $f_3(x) = \sec^2 x$, $f_4(x) = \tan^2 x$ is linearly dependent on the interval $(-\pi/2, \pi/2)$, since

$$c_1 \cos^2 x + c_2 \sin^2 x + c_3 \sec^2 x + c_4 \tan^2 x = 0$$

when $c_1 = c_2 = 1, c_3 = -1, c_4 = 1$. We note that $\cos^2 x + \sin^2 x = 1$ and $1 + \tan^2 x = \sec^2 x$.

A set of functions $f_1(x), f_2(x), \ldots, f_n(x)$ is linearly dependent on an interval if at least one function can be expressed as a linear combination of the remaining functions.

EXAMPLE 11 Linearly Dependent Functions

The set of functions $f_1(x) = \sqrt{x} + 5$, $f_2(x) = \sqrt{x} + 5x$, $f_3(x) = x - 1$, $f_4(x) = x^2$ is linearly dependent on the interval $(0, \infty)$, since f_2 can be written as a linear combination of f_1, f_3, and f_4. Observe that

$$f_2(x) = 1 \cdot f_1(x) + 5 \cdot f_3(x) + 0 \cdot f_4(x)$$

for every x in the interval $(0, \infty)$.

Wronskian The following theorem provides a sufficient condition for the linear independence of n functions on an interval. Each function is assumed to be differentiable at least $n - 1$ times.

THEOREM 4.2 Criterion for Linearly Independent Functions

Suppose $f_1(x), f_2(x), \ldots, f_n(x)$ possess at least $n - 1$ derivatives. If the determinant

$$\begin{vmatrix} f_1 & f_2 & \cdots & f_n \\ f_1' & f_2' & \cdots & f_n' \\ \vdots & & & \vdots \\ f_1^{(n-1)} & f_2^{(n-1)} & \cdots & f_n^{(n-1)} \end{vmatrix}$$

is not zero for at least one point in the interval I, then the set of functions $f_1(x), f_2(x), \ldots, f_n(x)$ is linearly independent on the interval.

The determinant in the preceding theorem is denoted by

$$W(f_1(x), f_2(x), \ldots, f_n(x))$$

and is called the **Wronskian*** of the functions.

Proof We prove Theorem 4.2 by contradiction for the case when $n = 2$. Assume that $W(f_1(x_0), f_2(x_0)) \neq 0$ for a fixed x_0 in the interval I and that $f_1(x)$ and $f_2(x)$ are linearly dependent on the interval. The fact that the set of functions is linearly dependent means there exist constants c_1 and c_2, not both zero, for which

$$c_1 f_1(x) + c_2 f_2(x) = 0$$

for every x in I. Differentiating this combination then gives

$$c_1 f_1'(x) + c_2 f_2'(x) = 0.$$

Thus we obtain the system of linear equations

$$\begin{aligned} c_1 f_1(x) + c_2 f_2(x) &= 0 \\ c_1 f_1'(x) + c_2 f_2'(x) &= 0. \end{aligned} \tag{2}$$

But the linear dependence of f_1 and f_2 implies that (2) possesses a nontrivial solution for each x in the interval. Hence

$$W(f_1(x), f_2(x)) = \begin{vmatrix} f_1(x) & f_2(x) \\ f_1'(x) & f_2'(x) \end{vmatrix} = 0$$

for every x in I.† This contradicts the assumption that $W(f_1(x_0), f_2(x_0)) \neq 0$. We conclude that the set of functions f_1 and f_2 is linearly independent. ■

COROLLARY

If $f_1(x), f_2(x), \ldots, f_n(x)$ possess at least $n - 1$ derivatives and are linearly dependent on I, then

$$W(f_1(x), f_2(x), \ldots, f_n(x)) = 0$$

for every x in the interval.

EXAMPLE 12 Zero Wronskian

The set of functions $f_1(x) = \sin^2 x$ and $f_2(x) = 1 - \cos 2x$ is linearly dependent on $(-\infty, \infty)$. (Why?) By the preceding corollary, $W(\sin^2 x, 1 - \cos 2x) = 0$ for every real number. To see this, we observe

---

* JOSEF MARIA HOËNE WRONSKI (1778–1853) Born in Poland and educated in Germany, Wronski lived most of his life in France. More a philosopher than a mathematician, he believed that absolute truth could be attained through mathematics. Wronski's only noteworthy contribution to mathematics was the above determinant. Always an eccentric, he eventually went insane.

† See Appendix III for a review of determinants.

$$\begin{aligned} W(\sin^2 x, 1 - \cos 2x) &= \begin{vmatrix} \sin^2 x & 1 - \cos 2x \\ 2 \sin x \cos x & 2 \sin 2x \end{vmatrix} \\ &= 2 \sin^2 x \sin 2x - 2 \sin x \cos x \\ &\quad + 2 \sin x \cos x \cos 2x \\ &= \sin 2x[2 \sin^2 x - 1 + \cos 2x] \\ &= \sin 2x[2 \sin^2 x - 1 + \cos^2 x - \sin^2 x] \\ &= \sin 2x[\sin^2 x + \cos^2 x - 1] = 0. \end{aligned}$$

Here we have used the trigonometric identities $\sin 2x = 2 \sin x \cos x$, $\cos 2x = \cos^2 x - \sin^2 x$, and $\sin^2 x + \cos^2 x = 1$. ■

EXAMPLE 13 Nonzero Wronskian

For $f_1(x) = e^{m_1 x}$, $f_2(x) = e^{m_2 x}$, $m_1 \neq m_2$,

$$W(e^{m_1 x}, e^{m_2 x}) = \begin{vmatrix} e^{m_1 x} & e^{m_2 x} \\ m_1 e^{m_1 x} & m_2 e^{m_2 x} \end{vmatrix} = (m_2 - m_1) e^{(m_1 + m_2)x} \neq 0$$

for every real value of x. Thus the set of functions f_1 and f_2 is linearly independent on any interval of the x-axis. ■

EXAMPLE 14 Nonzero Wronskian

If α and β are real numbers, $\beta \neq 0$, then $y_1 = e^{\alpha x} \cos \beta x$ and $y_2 = e^{\alpha x} \sin \beta x$ are linearly independent on any interval of the x-axis, since

$$\begin{aligned} W(e^{\alpha x} \cos \beta x, e^{\alpha x} \sin \beta x) &= \begin{vmatrix} e^{\alpha x} \cos \beta x & e^{\alpha x} \sin \beta x \\ -\beta e^{\alpha x} \sin \beta x + \alpha e^{\alpha x} \cos \beta x & \beta e^{\alpha x} \cos \beta x + \alpha e^{\alpha x} \sin \beta x \end{vmatrix} \\ &= \beta e^{2\alpha x}(\cos^2 \beta x + \sin^2 \beta x) = \beta e^{2\alpha x} \neq 0. \end{aligned}$$

Notice when $\alpha = 0$ that $\cos \beta x$ and $\sin \beta x$, $\beta \neq 0$ are also linearly independent on any interval of the x-axis. ■

EXAMPLE 15 Nonzero Wronskian

The set of functions $f_1(x) = e^x$, $f_2(x) = xe^x$, and $f_3(x) = x^2 e^x$ is linearly independent on any interval of the x-axis, since

$$W(e^x, xe^x, x^2 e^x) = \begin{vmatrix} e^x & xe^x & x^2 e^x \\ e^x & xe^x + e^x & x^2 e^x + 2xe^x \\ e^x & xe^x + 2e^x & x^2 e^x + 4xe^x + 2e^x \end{vmatrix} = 2e^{3x}$$

is not zero for any real value of x. ■

EXAMPLE 16 Wronskian Cannot Be Used

In Example 9 we saw that the set of functions $f_1(x) = x$ and $f_2(x) = |x|$ is linearly independent on $(-\infty, \infty)$; however, we cannot compute the Wronskian since f_2 is not differentiable at $x = 0$. ■

We leave it as an exercise to show that a set of functions $f_1(x), f_2(x), \ldots, f_n(x)$ could be linearly independent on some interval and yet have a vanishing Wronskian. See Problem 30. In other words, if $W(f_1(x), f_2(x), \ldots, f_n(x)) = 0$ for every x in an interval, it does not necessarily mean that the functions are linearly dependent.

4.1.3 SOLUTIONS OF LINEAR EQUATIONS

Homogeneous Equations A linear nth-order differential equation of the form

$$a_n(x)\frac{d^ny}{dx^n} + a_{n-1}(x)\frac{d^{n-1}y}{dx^{n-1}} + \cdots + a_1(x)\frac{dy}{dx} + a_0(x)y = 0 \qquad \textbf{(3)}$$

is said to be **homogeneous,** whereas

$$a_n(x)\frac{d^ny}{dx^n} + a_{n-1}(x)\frac{d^{n-1}y}{dx^{n-1}} + \cdots + a_1(x)\frac{dy}{dx} + a_0(x)y = g(x), \qquad \textbf{(4)}$$

$g(x)$ not identically zero, is said to be **nonhomogeneous.**

The word *homogeneous* in this context does not refer to coefficients that are homogeneous functions. See Section 2.3.

EXAMPLE 17 Homogeneous/Nonhomogeneous

(a) The equation $2y'' + 3y' - 5y = 0$ is a homogeneous linear second-order ordinary differential equation.
(b) The equation $x^3y''' - 2xy'' + 5y' + 6y = e^x$ is a nonhomogeneous linear third-order ordinary differential equation. ■

We shall see in the latter part of this section, as well as in the subsequent sections of this chapter, that in order to solve a nonhomogeneous equation (4), we must first solve the associated homogeneous equation (3).

Note To avoid needless repetition throughout the remainder of this text, we shall, as a matter of course, make the following important assumptions when giving definitions and proving theorems about the linear equations (3) and (4). On some common interval I,

- the coefficients $a_i(x)$, $i = 0, 1, \ldots, n$ are continuous;
- the right-hand member $g(x)$ is continuous; and
- $a_n(x) \neq 0$ for every x in the interval.

Superposition Principle In the next theorem we see that the sum, or **superposition,** of two or more solutions of a homogeneous linear differential equation is also a solution.

THEOREM 4.3 Superposition Principle—Homogeneous Equations

Let $y_1, y_2, \ldots, y_k$ be solutions of the homogeneous linear nth-order differential equation (3) on an interval I. Then the linear combination

$$y = c_1y_1(x) + c_2y_2(x) + \cdots + c_ky_k(x), \qquad \textbf{(5)}$$

where the c_i, $i = 1, 2, \ldots, k$ are arbitrary constants, is also a solution on the interval.

Proof We prove the case when $n = k = 2$. Let $y_1(x)$ and $y_2(x)$ be solutions of

$$a_2(x)y'' + a_1(x)y' + a_0(x)y = 0.$$

If we define $y = c_1y_1(x) + c_2y_2(x)$, then

$$\begin{aligned} &a_2(x)[c_1y_1'' + c_2y_2''] + a_1(x)[c_1y_1' + c_2y_2'] + a_0(x)[c_1y_1 + c_2y_2] \\ &= c_1\underbrace{[a_2(x)y_1'' + a_1(x)y_1' + a_0(x)y_1]}_{\text{zero}} + c_2\underbrace{[a_2(x)y_2'' + a_1(x)y_2' + a_0(x)y_2]}_{\text{zero}} \\ &= c_1 \cdot 0 + c_2 \cdot 0 = 0. \end{aligned}$$

■

COROLLARIES TO THEOREM 4.3

(A) A constant multiple $y = c_1y_1(x)$ of a solution $y_1(x)$ of a homogeneous linear differential equation is also a solution.

(B) A homogeneous linear differential equation always possesses the trivial solution $y = 0$.

The superposition principle defined by (5) and its special case given in Corollary (A) are properties that nonlinear differential equations, in general, do not possess. See Problems 31 and 32.

EXAMPLE 18 Superposition—Homogeneous DE

The functions $y_1 = x^2$ and $y_2 = x^2 \ln x$ are both solutions of the homogeneous third-order equation

$$x^3y''' - 2xy' + 4y = 0$$

on the interval $(0, \infty)$. By the superposition principle the linear combination

$$y = c_1x^2 + c_2x^2 \ln x$$

is also a solution of the equation on the interval. ■

EXAMPLE 19 Superposition—Homogeneous DE

The functions $y_1 = e^x$, $y_2 = e^{2x}$, and $y_3 = e^{3x}$ all satisfy the homogeneous equation

$$\frac{d^3y}{dx^3} - 6\frac{d^2y}{dx^2} + 11\frac{dy}{dx} - 6y = 0$$

on $(-\infty, \infty)$. By Theorem 4.3 another solution is

$$y = c_1e^x + c_2e^{2x} + c_3e^{3x}.$$

■

EXAMPLE 20 **Constant Multiple Is a Solution**

The function $y = x^2$ is a solution of the homogeneous linear equation

$$x^2y'' - 3xy' + 4y = 0$$

on $(0, \infty)$. Hence $y = cx^2$ is also a solution. For various values of c we see that $y = 3x^2$, $y = ex^2$, $y = 0, \ldots$ are all solutions of the equation on the interval. ■

Linearly Independent Solutions We are interested in determining when a set of n solutions $y_1, y_2, \ldots, y_n$ of the homogeneous differential equation (3) is linearly independent. Surprisingly, the nonvanishing of the Wronskian of a set of n such solutions on an interval I is both necessary and sufficient for linear independence.

THEOREM 4.4 **Criterion for Linearly Independent Solutions**

Let $y_1, y_2, \ldots, y_n$ be n solutions of the homogeneous linear nth-order differential equation (3) on an interval I. Then the set of solutions is **linearly independent** on I if and only if

$$W(y_1, y_2, \ldots, y_n) \neq 0$$

for every x in the interval.

Proof We prove Theorem 4.4 for the case when $n = 2$. First, if $W(y_1, y_2) \neq 0$ for every x in I, it follows immediately from Theorem 4.2 that y_1 and y_2 are linearly independent. Next, we must show that if y_1 and y_2 are linearly independent solutions of a homogeneous linear second-order differential equation, then $W(y_1, y_2) \neq 0$ for every x in I. To show this, let us suppose y_1 and y_2 are linearly independent and there is some fixed x_0 in I for which $W(y_1(x_0), y_2(x_0)) = 0$. Hence there must exist c_1 and c_2, not both zero, such that

$$\begin{aligned} c_1y_1(x_0) + c_2y_2(x_0) &= 0 \\ c_1y_1'(x_0) + c_2y_2'(x_0) &= 0. \end{aligned} \tag{6}$$

If we define $$y(x) = c_1y_1(x) + c_2y_2(x),$$

then in view of (6), $y(x)$ must also satisfy

$$y(x_0) = 0, \quad y'(x_0) = 0. \tag{7}$$

But the identically zero function satisfies both the differential equation and the initial conditions (7), and thus by Theorem 4.1 it is the only solution. In other words, $y = 0$ or $c_1y_1(x) + c_2y_2(x) = 0$ for every x in I. This contradicts the assumption that y_1 and y_2 are linearly independent on the interval. ■

From the foregoing discussion we conclude that when $y_1, y_2, \ldots, y_n$ are n solutions of (3) on an interval I, the Wronskian either is identically zero or is never zero on the interval.

DEFINITION 4.3 **Fundamental Set of Solutions**

Any linearly independent set $y_1, y_2, \ldots, y_n$ of n solutions of the homogeneous linear nth-order differential equation (3) on an interval I is said to be a **fundamental set of solutions** on the interval.

THEOREM 4.5 **Existence of Constants**

Let $y_1, y_2, \ldots, y_n$ be a fundamental set of solutions of the homogeneous linear nth-order differential equation (3) on an interval I. Then for any solution $Y(x)$ of (3) on I, constants $C_1, C_2, \ldots, C_n$ can be found so that

$$Y = C_1 y_1(x) + C_2 y_2(x) + \cdots + C_n y_n(x).$$

Proof We prove the case when $n = 2$. Let Y be a solution, and let y_1 and y_2 be linearly independent solutions of

$$a_2(x)y'' + a_1(x)y' + a_0(x)y = 0$$

on an interval I. Suppose $x = t$ is a point in this interval for which $W(y_1(t), y_2(t)) \neq 0$. Suppose also that the values of $Y(t)$ and $Y'(t)$ are given by $Y(t) = k_1$, $Y'(t) = k_2$. If we now examine the system of equations

$$\begin{aligned} C_1 y_1(t) + C_2 y_2(t) &= k_1 \\ C_1 y_1'(t) + C_2 y_2'(t) &= k_2, \end{aligned}$$

it follows that we can determine C_1 and C_2 uniquely provided the determinant of the coefficients satisfies

$$\begin{vmatrix} y_1(t) & y_2(t) \\ y_1'(t) & y_2'(t) \end{vmatrix} \neq 0.$$

But this latter determinant is simply the Wronskian evaluated at $x = t$, and by assumption $W \neq 0$. If we now define the function

$$G(x) = C_1 y_1(x) + C_2 y_2(x),$$

we observe the following:

(*i*) $G(x)$ satisfies the differential equation since it is the superposition of two known solutions y_1 and y_2.

(*ii*) $G(x)$ satisfies the initial conditions

$$G(t) = C_1 y_1(t) + C_2 y_2(t) = k_1, \quad G'(t) = C_1 y_1'(t) + C_2 y_2'(t) = k_2.$$

(*iii*) $Y(x)$ satisfies the *same* linear equation and the *same* initial conditions.

Since the solution of this linear initial-value problem is unique (Theorem 4.1), we have $Y(x) = G(x)$ or

$$Y(x) = C_1 y_1(x) + C_2 y_2(x). \qquad \blacksquare$$

The basic question of whether a fundamental set of solutions exists for a linear equation is answered in the next theorem.

THEOREM 4.6 Existence of a Fundamental Set

There exists a fundamental set of solutions for the homogeneous linear nth-order differential equation (3) on an interval I.

The proof of this result follows from Theorem 4.1. The justification of Theorem 4.6 in the special case of second-order equations is left as an exercise.

Since we have shown that any solution of (3) is obtained from a linear combination of functions in a fundamental set of solutions, we are able to make the following definition.

DEFINITION 4.4 General Solution—Homogeneous Equations

Let $y_1, y_2, \ldots, y_n$ be a fundamental set of solutions of the homogeneous linear nth-order differential equation (3) on an interval I. The **general solution** of the equation on the interval is defined to be

$$y = c_1 y_1(x) + c_2 y_2(x) + \cdots + c_n y_n(x),$$

where the c_i, $i = 1, 2, \ldots, n$ are arbitrary constants.

Recall that the general solution as defined in Section 1.1 is also called the **complete solution** of the differential equation.

EXAMPLE 21 General Solution—Homogeneous DE

The second-order equation $y'' - 9y = 0$ possesses two solutions

$$y_1 = e^{3x} \quad \text{and} \quad y_2 = e^{-3x}.$$

Since $$W(e^{3x}, e^{-3x}) = \begin{vmatrix} e^{3x} & e^{-3x} \\ 3e^{3x} & -3e^{-3x} \end{vmatrix} = -6 \neq 0$$

for every value of x, y_1 and y_2 form a fundamental set of solutions on $(-\infty, \infty)$. The general solution of the differential equation on the interval is

$$y = c_1 e^{3x} + c_2 e^{-3x}.$$

EXAMPLE 22 Solution Obtained from General Solution

You should verify that the function $y = 4 \sinh 3x - 5e^{-3x}$ also satisfies the differential equation in Example 21. By choosing $c_1 = 2$, $c_2 = -7$ in the general solution $y = c_1 e^{3x} + c_2 e^{-3x}$, we obtain

$$\begin{aligned} y &= 2e^{3x} - 7e^{-3x} = 2e^{3x} - 2e^{-3x} - 5e^{-3x} \\ &= 4\left(\frac{e^{3x} - e^{-3x}}{2}\right) - 5e^{-3x} \\ &= 4 \sinh 3x - 5e^{-3x}. \end{aligned}$$

EXAMPLE 23 General Solution—Homogeneous DE

The functions $y_1 = e^x$, $y_2 = e^{2x}$, and $y_3 = e^{3x}$ satisfy the third-order equation

$$\frac{d^3y}{dx^3} - 6\frac{d^2y}{dx^2} + 11\frac{dy}{dx} - 6y = 0.$$

Since $$W(e^x, e^{2x}, e^{3x}) = \begin{vmatrix} e^x & e^{2x} & e^{3x} \\ e^x & 2e^{2x} & 3e^{3x} \\ e^x & 4e^{2x} & 9e^{3x} \end{vmatrix} = 2e^{6x} \neq 0$$

for every real value of x, y_1, y_2, and y_3 form a fundamental set of solutions on $(-\infty, \infty)$. We conclude that

$$y = c_1 e^x + c_2 e^{2x} + c_3 e^{3x}$$

is the general solution of the differential equation on the interval.

Nonhomogeneous Equations We now turn our attention to defining the general solution of a **nonhomogeneous** linear equation. Any function y_p, free of arbitrary parameters, that satisfies (4) is said to be a **particular solution** of the equation (sometimes also called a **particular integral**).

EXAMPLE 24 Particular Solutions of Nonhomogeneous DEs

(a) A particular solution of

$$y'' + 9y = 27$$

is $y_p = 3$, since $y_p'' = 0$ and $0 + 9y_p = 9(3) = 27$.

(b) $y_p = x^3 - x$ is a particular solution of

$$x^2y'' + 2xy' - 8y = 4x^3 + 6x,$$

since $y_p' = 3x^2 - 1$, $y_p'' = 6x$, and

$$x^2y_p'' + 2xy_p' - 8y_p = x^2(6x) + 2x(3x^2 - 1) - 8(x^3 - x) = 4x^3 + 6x.$$ ■

THEOREM 4.7 A Solution of Equation (4)

Let $y_1, y_2, \ldots, y_k$ be solutions of the homogeneous linear nth-order differential equation (3) on an interval I, and let y_p be any solution of the nonhomogeneous equation (4) on the same interval. Then

$$y = c_1y_1(x) + c_2y_2(x) + \cdots + c_ky_k(x) + y_p(x)$$

is also a solution of the nonhomogeneous equation on the interval for any constants $c_1, c_2, \ldots, c_k$.

We can now prove the following analogue of Theorem 4.5 for nonhomogeneous differential equations.

THEOREM 4.8 Existence of Constants

Let y_p be a given solution of the nonhomogeneous linear nth-order differential equation (4) on an interval I, and let $y_1, y_2, \ldots, y_n$ be a fundamental set of solutions of the associated homogeneous equation (3) on the interval. Then for any solution $Y(x)$ of (4) on I, constants $C_1, C_2, \ldots, C_n$ can be found so that

$$Y = C_1y_1(x) + C_2y_2(x) + \cdots + C_ny_n(x) + y_p(x).$$

Proof We prove the case when $n = 2$. Suppose Y and y_p are both solutions of $a_2(x)y'' + a_1(x)y' + a_0(x)y = g(x)$. If we define a function u by $u(x) = Y(x) - y_p(x)$, then

$$\begin{aligned}
&a_2(x)u'' + a_1(x)u' + a_0(x)u \\
&\quad = a_2(x)[Y'' - y_p''] + a_1(x)[Y' - y_p'] + a_0(x)[Y - y_p] \\
&\quad = a_2(x)Y'' + a_1(x)Y' + a_0(x)Y - [a_2(x)y_p'' + a_1(x)y_p' + a_0(x)y_p] \\
&\quad = g(x) - g(x) = 0.
\end{aligned}$$

Therefore, in view of Definition 4.4 and Theorem 4.5, we can write

$$\begin{aligned}
u(x) &= C_1y_1(x) + C_2y_2(x) \\
Y(x) - y_p(x) &= C_1y_1(x) + C_2y_2(x)
\end{aligned}$$

or

$$Y(x) = C_1y_1(x) + C_2y_2(x) + y_p(x).$$ ■

Thus we come to the last definition of this section.

DEFINITION 4.5 General Solution—Nonhomogeneous Equations

Let y_p be a given solution of the nonhomogeneous linear nth-order differential equation (4) on an interval I, and let

$$y_c = c_1y_1(x) + c_2y_2(x) + \cdots + c_ny_n(x)$$

denote the general solution of the associated homogeneous equation (3) on the interval. The **general solution** of the nonhomogeneous equation on the interval is defined to be

$$y = c_1y_1(x) + c_2y_2(x) + \cdots + c_ny_n(x) + y_p(x) = y_c(x) + y_p(x).$$

Complementary Function In Definition 4.5 the linear combination

$$y_c(x) = c_1y_1(x) + c_2y_2(x) + \cdots + c_ny_n(x),$$

which is the general solution of (3), is called the **complementary function** for equation (4). In other words, the general solution of a nonhomogeneous linear differential equation is

y = complementary function + any particular solution.

EXAMPLE 25 General Solution—Nonhomogeneous DE

By substitution, the function $y_p = -\frac{11}{12} - \frac{1}{2}x$ is readily shown to be a particular solution of the nonhomogeneous equation

$$\frac{d^3y}{dx^3} - 6\frac{d^2y}{dx^2} + 11\frac{dy}{dx} - 6y = 3x. \tag{8}$$

In order to write the general solution of (8), we must also be able to solve the associated homogeneous equation

$$\frac{d^3y}{dx^3} - 6\frac{d^2y}{dx^2} + 11\frac{dy}{dx} - 6y = 0.$$

But in Example 23 we saw that the general solution of this latter equation on the interval $(-\infty, \infty)$ was

$$y_c = c_1e^x + c_2e^{2x} + c_3e^{3x}.$$

Hence the general solution of (8) on the interval is

$$y = y_c + y_p = c_1e^x + c_2e^{2x} + c_3e^{3x} - \frac{11}{12} - \frac{1}{2}x.$$

Another Superposition Principle The last theorem of this discussion will be useful in Section 4.4 when we consider a method for finding particular solutions of nonhomogeneous equations.

THEOREM 4.9 Superposition Principle—Nonhomogeneous Equations

Let $y_{p_1}, y_{p_2}, \ldots, y_{p_k}$ be k particular solutions of the linear nth-order differential equation (4) on an interval I corresponding, in turn, to k distinct functions $g_1, g_2, \ldots, g_k$. That is, suppose y_{p_i} denotes a particular solution of the corresponding differential equation

$$a_n(x)y^{(n)} + a_{n-1}(x)y^{(n-1)} + \cdots + a_1(x)y' + a_0(x)y = g_i(x),$$

where $i = 1, 2, \ldots, k$. Then

$$y_p = y_{p_1}(x) + y_{p_2}(x) + \cdots + y_{p_k}(x)$$

is a particular solution of

$$a_n(x)y^{(n)} + a_{n-1}(x)y^{(n-1)} + \cdots + a_1(x)y' + a_0(x)y$$
$$= g_1(x) + g_2(x) + \cdots + g_k(x).$$

We leave the proof of this result when $k = 2$ as an exercise. See Problem 50.

EXAMPLE 26 Superposition—Nonhomogeneous DE

You should verify that

$y_{p_1} = -4x^2$ is a particular solution of $y'' - 3y' + 4y = -16x^2 + 24x - 8$,
$y_{p_2} = e^{2x}$ is a particular solution of $y'' - 3y' + 4y = 2e^{2x}$,
$y_{p_3} = xe^{x}$ is a particular solution of $y'' - 3y' + 4y = 2xe^{x} - e^{x}$.

It follows from Theorem 4.9 that the superposition of y_{p_1}, y_{p_2}, and y_{p_3},

$$y = y_{p_1} + y_{p_2} + y_{p_3} = -4x^2 + e^{2x} + xe^{x},$$

is a solution of

$$y'' - 3y' + 4y = \underbrace{-16x^2 + 24x - 8}_{g_1(x)} + \underbrace{2e^{2x}}_{g_2(x)} + \underbrace{2xe^{x} - e^{x}}_{g_3(x)}.$$

Before we actually start solving homogeneous and nonhomogeneous linear differential equations, we need the one additional bit of theory presented in the next section.

Remarks A physical system that changes with time and whose mathematical model is a linear differential equation

$$a_n(t)y^{(n)} + a_{n-1}(t)y^{(n-1)} + \cdots + a_1(t)y' + a_0(t)y = g(t)$$

is said to be a **linear system.** The values of the variables $y(t)$, $y'(t)$, . . ., $y^{(n-1)}(t)$ at a specific time t_0 describe the state of the system. The function

g is variously called the **input function, forcing function,** or **excitation function.** A solution $y(t)$ of the differential equation is said to be the **output** or **response** of the system. The dependence of the output on the input is illustrated in Figure 4.5.

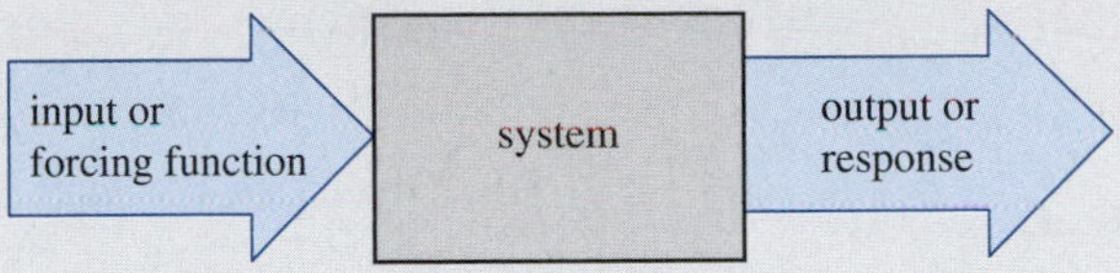

Figure 4.5

In order for a physical system to be a linear system, it is necessary that the superposition principle (Theorem 4.9) hold in the system; that is, the response of the system to a superposition of inputs is a superposition of outputs.

EXERCISES 4.1

Answers to odd-numbered problems begin on page A-6.

4.1.1 Initial-Value and Boundary-Value Problems

1. Given that $y = c_1e^x + c_2e^{-x}$ is a two-parameter family of solutions of $y'' - y = 0$ on the interval $(-\infty, \infty)$, find a member of the family satisfying the initial conditions $y(0) = 0$, $y'(0) = 1$.
2. Find a solution of the differential equation in Problem 1 satisfying the boundary conditions $y(0) = 0$, $y(1) = 1$.
3. Given that $y = c_1e^{4x} + c_2e^{-x}$ is a two-parameter family of solutions of $y'' - 3y' - 4y = 0$ on the interval $(-\infty, \infty)$, find a member of the family satisfying the initial conditions $y(0) = 1$, $y'(0) = 2$.
4. Given that $y = c_1 + c_2 \cos x + c_3 \sin x$ is a three-parameter family of solutions of $y''' + y' = 0$ on the interval $(-\infty, \infty)$, find a member of the family satisfying the initial conditions $y(\pi) = 0$, $y'(\pi) = 2$, $y''(\pi) = -1$.
5. Given that $y = c_1x + c_2x \ln x$ is a two-parameter family of solutions of $x^2y'' - xy' + y = 0$ on the interval $(-\infty, \infty)$, find a member of the family satisfying the initial conditions $y(1) = 3$, $y'(1) = -1$.
6. Given that $y = c_1 + c_2x^2$ is a two-parameter family of solutions of $xy'' - y' = 0$ on the interval $(-\infty, \infty)$, show that constants c_1 and c_2 cannot be found so that a member of the family satisfies the initial conditions $y(0) = 0$, $y'(0) = 1$. Explain why this does not violate Theorem 4.1.
7. Find two members of the family of solutions of $xy'' - y' = 0$ given in Problem 6 satisfying the initial conditions $y(0) = 0$, $y'(0) = 0$.
8. Find a member of the family of solutions of $xy'' - y' = 0$ given in Prob-

lem 6 satisfying the boundary conditions $y(0) = 1, y'(1) = 6$. Does Theorem 4.1 guarantee that this solution is unique?

9. Given that $y = c_1e^x \cos x + c_2e^x \sin x$ is a two-parameter family of solutions of $y'' - 2y' + 2y = 0$ on the interval $(-\infty, \infty)$, determine whether a member of the family can be found that satisfies the conditions

(a) $y(0) = 1, \quad y'(0) = 0$ **(b)** $y(0) = 1, \quad y(\pi) = -1$

(c) $y(0) = 1, \quad y(\pi/2) = 1$ **(d)** $y(0) = 0, \quad y(\pi) = 0$

10. Given that $y = c_1x^2 + c_2x^4 + 3$ is a two-parameter family of solutions of $x^2y'' - 5xy' + 8y = 24$ on the interval $(-\infty, \infty)$, determine whether a member of the family can be found that satisfies the conditions

(a) $y(-1) = 0, \quad y(1) = 4$ **(b)** $y(0) = 1, \quad y(1) = 2$

(c) $y(0) = 3, \quad y(1) = 0$ **(d)** $y(1) = 3, \quad y(2) = 15$

In Problems 11 and 12 find an interval around $x = 0$ for which the given initial-value problem has a unique solution.

11. $(x - 2)y'' + 3y = x; \quad y(0) = 0, y'(0) = 1$

12. $y'' + (\tan x)y = e^x; \quad y(0) = 1, y'(0) = 0$

13. Given that $y = c_1 \cos \lambda x + c_2 \sin \lambda x$ is a family of solutions of the differential equation $y'' + \lambda^2 y = 0$, determine the values of the parameter λ for which the boundary-value problem

$$y'' + \lambda^2 y = 0, \quad y(0) = 0, \quad y(\pi) = 0$$

has nontrivial solutions.

14. Determine the values of the parameter λ for which the boundary-value problem

$$y'' + \lambda^2 y = 0, \quad y(0) = 0, \quad y(5) = 0$$

has nontrivial solutions. See Problem 13.

4.1.2 Linear Dependence and Linear Independence

In Problems 15–22 determine whether the given functions are linearly independent or dependent on $(-\infty, \infty)$.

15. $f_1(x) = x, \quad f_2(x) = x^2, \quad f_3(x) = 4x - 3x^2$

16. $f_1(x) = 0, \quad f_2(x) = x, \quad f_3(x) = e^x$

17. $f_1(x) = 5, \quad f_2(x) = \cos^2 x, \quad f_3(x) = \sin^2 x$

18. $f_1(x) = \cos 2x, \quad f_2(x) = 1, \quad f_3(x) = \cos^2 x$

19. $f_1(x) = x, \quad f_2(x) = x - 1, \quad f_3(x) = x + 3$

20. $f_1(x) = 2 + x, \quad f_2(x) = 2 + |x|$

21. $f_1(x) = 1 + x, \quad f_2(x) = x, \quad f_3(x) = x^2$

22. $f_1(x) = e^x, \quad f_2(x) = e^{-x}, \quad f_3(x) = \sinh x$

In Problems 23–28 show by computing the Wronskian that the given functions are linearly independent on the indicated interval.

23. $x^{1/2}, x^2; \quad (0, \infty)$ **24.** $1 + x, x^3; \quad (-\infty, \infty)$

25. $\sin x, \csc x; \quad (0, \pi)$

26. $\tan x, \cot x; \quad (0, \pi/2)$

27. $e^x, e^{-x}, e^{4x}; \quad (-\infty, \infty)$

28. $x, x \ln x, x^2 \ln x; \quad (0, \infty)$

29. Observe that for the functions $f_1(x) = 2$ and $f_2(x) = e^x$,

$$1 \cdot f_1(0) - 2 \cdot f_2(0) = 0.$$

Does this imply that f_1 and f_2 are linearly dependent on any interval containing $x = 0$?

30. (a) Show graphically that $f_1(x) = x^2$ and $f_2(x) = x|x|$ are linearly independent on $(-\infty, \infty)$.

(b) Show that $W(f_1(x), f_2(x)) = 0$ for every real number.

4.1.3 Solutions of Linear Equations

31. (a) Verify that $y = 1/x$ is a solution of the nonlinear differential equation $y'' = 2y^3$ on the interval $(0, \infty)$.

(b) Show that the constant multiple $y = c/x$ is not a solution of the equation when $c \neq 0, \pm 1$.

32. (a) Verify that $y_1 = 1$ and $y_2 = \ln x$ are solutions of the nonlinear differential equation $y'' + (y')^2 = 0$ on the interval $(0, \infty)$.

(b) Is $y_1 + y_2$ a solution of the equation? Is $c_1 y_1 + c_2 y_2$, c_1 and c_2 arbitrary, a solution of the equation?

In Problems 33–40 verify that the given functions form a fundamental set of solutions of the differential equation on the indicated interval. Form the general solution.

33. $y'' - y' - 12y = 0; \quad e^{-3x}, e^{4x}, (-\infty, \infty)$

34. $y'' - 4y = 0; \quad \cosh 2x, \sinh 2x, (-\infty, \infty)$

35. $y'' - 2y' + 5y = 0; \quad e^x \cos 2x, e^x \sin 2x, (-\infty, \infty)$

36. $4y'' - 4y' + y = 0; \quad e^{x/2}, xe^{x/2}, (-\infty, \infty)$

37. $x^2y'' - 6xy' + 12y = 0; \quad x^3, x^4, (0, \infty)$

38. $x^2y'' + xy' + y = 0; \quad \cos(\ln x), \sin(\ln x), (0, \infty)$

39. $x^3y''' + 6x^2y'' + 4xy' - 4y = 0; \quad x, x^{-2}, x^{-2}\ln x, (0, \infty)$

40. $y^{(4)} + y'' = 0; \quad 1, x, \cos x, \sin x, (-\infty, \infty)$

In Problems 41–44 verify that the given two-parameter family of functions is the general solution of the nonhomogeneous differential equation on the indicated interval.

41. $y'' - 7y' + 10y = 24e^x$
$y = c_1e^{2x} + c_2e^{5x} + 6e^x, (-\infty, \infty)$

42. $y'' + y = \sec x$
$y = c_1 \cos x + c_2 \sin x + x \sin x + (\cos x)\ln(\cos x), (-\pi/2, \pi/2)$

43. $y'' - 4y' + 4y = 2e^{2x} + 4x - 12$
$y = c_1e^{2x} + c_2xe^{2x} + x^2e^{2x} + x - 2, (-\infty, \infty)$

44. $2x^2y'' + 5xy' + y = x^2 - x$
$y = c_1x^{-1/2} + c_2x^{-1} + \frac{1}{15}x^2 - \frac{1}{6}x, (0, \infty)$

45. **(a)** Verify that $y_1 = x^3$ and $y_2 = |x|^3$ are linearly independent solutions of the differential equation $x^2y'' - 4xy' + 6y = 0$ on $(-\infty, \infty)$.

(b) Show that $W(y_1, y_2) = 0$ for every real number.

(c) Does the result of part (b) violate Theorem 4.4?

(d) Verify that $Y_1 = x^3$ and $Y_2 = x^2$ are also linearly independent solutions of the differential equation on the interval $(-\infty, \infty)$.

(e) Find a solution of the equation satisfying $y(0) = 0$, $y'(0) = 0$.

(f) By the superposition principle both linear combinations

$$y = c_1y_1 + c_2y_2 \quad \text{and} \quad y = c_1Y_1 + c_2Y_2$$

are solutions of the differential equation. Is one, both, or neither the general solution of the differential equation on $(-\infty, \infty)$?

46. Consider the second-order differential equation

$$a_2(x)y'' + a_1(x)y' + a_0(x)y = 0, \tag{9}$$

where $a_2(x)$, $a_1(x)$, and $a_0(x)$ are continuous on an interval I and $a_2(x) \neq 0$ for every x in the interval. From Theorem 4.1 there exists only one solution y_1 of the equation satisfying $y(x_0) = 1$ and $y'(x_0) = 0$, where x_0 is a point in I. Similarly there exists a unique solution y_2 of the equation satisfying $y(x_0) = 0$ and $y'(x_0) = 1$. Show that y_1 and y_2 form a fundamental set of solutions of the differential equation on the interval I.

47. Let y_1 and y_2 be two solutions of (9).

(a) If $W(y_1, y_2)$ is the Wronskian of y_1 and y_2, show that

$$a_2(x)\frac{dW}{dx} + a_1(x)W = 0.$$

(b) Derive **Abel's formula,***

$$W = ce^{-\int[a_1(x)/a_2(x)]\,dx},$$

where c is a constant.

(c) Using an alternative form of Abel's formula,

$$W = ce^{-\int_{x_0}^{x}[a_1(t)/a_2(t)]\,dt},$$

for x_0 in I, show that

$$W(y_1, y_2) = W(x_0)e^{-\int_{x_0}^{x}[a_1(t)/a_2(t)]\,dt}.$$

---

* NIELS HENRIK ABEL (1802–1829) Abel was a brilliant Norwegian mathematician whose tragic death at age 26 due to tuberculosis was an inestimable loss for mathematics. His greatest achievement was the solution of a problem that baffled mathematicians for centuries: he showed that a general fifth-degree polynomial equation cannot be solved algebraically—that is, in terms of radicals. Abel's contemporary, the Frenchman Evariste Galois, then proved that it was impossible to solve *any* general polynomial equation of degree greater than four in an algebraic manner. Galois is another tragic figure in the history of mathematics; a political activist, he was killed in a duel at the age of 22.

(d) Show that if $W(x_0) = 0$, then $W = 0$ for every x in I, whereas if $W(x_0) \neq 0$, then $W \neq 0$ for every x in the interval.

In Problems 48 and 49 use the results of Problem 47.

48. If y_1 and y_2 are two solutions of

$$(1 - x^2)y'' - 2xy' + n(n + 1)y = 0$$

on $(-1, 1)$, show that $W(y_1, y_2) = c/(1 - x^2)$, where c is a constant.

49. In Chapter 6 we shall see that the solutions y_1 and y_2 of $xy'' + y' + xy = 0$, $0 < x < \infty$ are infinite series. Suppose we consider initial conditions

$$y_1(x_0) = k_1, \quad y_1'(x_0) = k_2$$

and
$$y_2(x_0) = k_3, \quad y_2'(x_0) = k_4$$

for $x_0 > 0$. Show that

$$W(y_1, y_2) = \frac{(k_1k_4 - k_2k_3)x_0}{x}.$$

50. Suppose the mathematical model of a linear system is given by

$$a_2(t)\frac{d^2y}{dt^2} + a_1(t)\frac{dy}{dt} + a_0(t)y = E(t).$$

If y_1 is a response of the system to an input $E_1(t)$ and y_2 is a response of the same system to an input $E_2(t)$, show that $y_1 + y_2$ is a response of the system to the input $E_1(t) + E_2(t)$.

4.2 CONSTRUCTING A SECOND SOLUTION FROM A KNOWN SOLUTION

- *Reduction of order*

Reduction of Order It is one of the more interesting as well as important facts in the study of linear *second-order* differential equations that we can construct a second solution from a *known* solution. Suppose $y_1(x)$ is a nonzero solution of the equation

$$a_2(x)y'' + a_1(x)y' + a_0(x)y = 0. \tag{1}$$

We assume, as we did in the preceding section, that the coefficients in (1) are continuous and $a_2(x) \neq 0$ for every x in some interval I. The process we use to find a second solution $y_2(x)$ consists of **reducing the order** of equation (1) to a first-order equation. For example, it is easily verified that $y_1 = e^x$ satisfies the differential equation $y'' - y = 0$. If we try to determine a solution of the form $y = u(x)e^x$, then

$$y' = ue^x + e^xu', \quad y'' = ue^x + 2e^xu' + e^xu''$$

and so
$$y'' - y = e^x(u'' + 2u') = 0.$$

Since $e^x \neq 0$, this last equation requires that $u'' + 2u' = 0$.

If we let $w = u'$, then the latter equation is recognized as a linear first-order equation $w' + 2w = 0$. Using the integrating factor e^{2x}, we can write

$$\frac{d}{dx}[e^{2x}w] = 0$$
$$w = c_1e^{-2x} \quad \text{or} \quad u' = c_1e^{-2x}.$$

Thus

$$u = -\frac{c_1}{2}e^{-2x} + c_2 \quad \text{and} \quad y = u(x)e^x = -\frac{c_1}{2}e^{-x} + c_2e^x.$$

By picking $c_2 = 0$ and $c_1 = -2$, we obtain the second solution $y_2 = e^{-x}$. Since $W(e^x, e^{-x}) \neq 0$ for every x, the solutions are linearly independent on $(-\infty, \infty)$ and thus the expression for y is actually the general solution of the given equation.

EXAMPLE 1 A Second Solution by Reduction of Order

Given that $y_1 = x^3$ is a solution of $x^2y'' - 6y = 0$, use reduction of order to find a second solution on the interval $(0, \infty)$.

Solution Define $y = u(x)x^3$ so that

$$y' = 3x^2u + x^3u', \quad y'' = x^3u'' + 6x^2u' + 6xu$$

and
$$x^2y'' - 6y = x^2(x^3u'' + 6x^2u' + 6xu) - 6ux^3$$
$$= x^5u'' + 6x^4u' = 0$$

provided $u(x)$ is a solution of

$$x^5u'' + 6x^4u' = 0 \quad \text{or} \quad u'' + \frac{6}{x}u' = 0.$$

If $w = u'$, we obtain the linear first-order equation

$$w' + \frac{6}{x}w = 0,$$

which possesses the integrating factor $e^{6\int dx/x} = e^{6\ln x} = x^6$. Now

$$\frac{d}{dx}[x^6w] = 0 \quad \text{gives} \quad x^6w = c_1.$$

Therefore $w = u' = \dfrac{c_1}{x^6}$, and thus

$$u = -\frac{c_1}{5x^5} + c_2 \quad \text{and} \quad y = u(x)x^3 = -\frac{c_1}{5x^2} + c_2x^3.$$

Choosing $c_2 = 0$ and $c_1 = -5$ yields the second solution $y_2 = 1/x^2$. ■

General Case Suppose we divide by $a_2(x)$ in order to put equation (1) in the standard form

$$y'' + P(x)y' + Q(x)y = 0, \tag{2}$$

where $P(x)$ and $Q(x)$ are continuous on some interval I. Let us suppose further that $y_1(x)$ is a known solution of (2) on I and that $y_1(x) \neq 0$ for every x in the interval. If we define $y = u(x)y_1(x)$, it follows that

$$y' = uy_1' + y_1u', \quad y'' = uy_1'' + 2y_1'u' + y_1u''$$
$$y'' + Py' + Qy = u\underbrace{[y_1'' + Py_1' + Qy_1]}_{\text{zero}} + y_1u'' + (2y_1' + Py_1)u' = 0.$$

This implies that we must have

$$y_1u'' + (2y_1' + Py_1)u' = 0 \quad \text{or} \quad y_1w' + (2y_1' + Py_1)w' = 0, \tag{3}$$

where we have let $w = u'$. Observe that equation (3) is both linear and separable. Applying the latter technique, we obtain

$$\frac{dw}{w} + 2\frac{y_1'}{y_1}dx + P\,dx = 0$$

$$\ln|w| + 2\ln|y_1| = -\int P\,dx + c \quad \text{or} \quad \ln|wy_1^2| = -\int P\,dx + c$$

$$wy_1^2 = c_1e^{-\int P\,dx} \quad \text{or} \quad w = u' = c_1\frac{e^{-\int P\,dx}}{y_1^2}.$$

Integrating again gives $u = c_1\displaystyle\int \frac{e^{-\int P\,dx}}{y_1^2}\,dx + c_2$ and therefore

$$y = u(x)y_1(x) = c_1y_1(x)\int \frac{e^{-\int P(x)\,dx}}{y_1^2(x)}\,dx + c_2y_1(x).$$

By choosing $c_2 = 0$ and $c_1 = 1$, we find that a second solution of equation (2) is

$$y_2 = y_1(x)\int \frac{e^{-\int P(x)\,dx}}{y_1^2(x)}\,dx. \tag{4}$$

It makes a good review exercise in differentiation to start with formula (4) and actually verify that equation (2) is satisfied.

Now $y_1(x)$ and $y_2(x)$ are linearly independent, since

$$W(y_1(x), y_2(x)) = \begin{vmatrix} y_1 & y_1\displaystyle\int \frac{e^{-\int P\,dx}}{y_1^2}\,dx \\ y_1' & \dfrac{e^{-\int P\,dx}}{y_1}\,dx + y_1'\displaystyle\int \frac{e^{-\int P\,dx}}{y_1^2}\,dx \end{vmatrix} = e^{-\int P\,dx}$$

is not zero on any interval on which $y_1(x)$ is not zero.*

EXAMPLE 2 A Second Solution by Formula (4)

The function $y_1 = x^2$ is a solution of $x^2y'' - 3xy' + 4y = 0$. Find the general solution on the interval $(0, \infty)$.

* Alternatively, if $y_2 = u(x)y_1$, then $W(y_1, y_2) = u'(y_1)^2 \neq 0$, since $y_1 \neq 0$ for every x on some interval. If $u' = 0$, then $u =$ constant.

Solution Since the equation has the alternative form

$$y'' - \frac{3}{x}y' + \frac{4}{x^2}y = 0,$$

we find from (4)

$$y_2 = x^2 \int \frac{e^{3\int dx/x}}{x^4}\,dx \qquad \leftarrow e^{3\int dx/x} = e^{\ln x^3} = x^3$$

$$= x^2 \int \frac{dx}{x} = x^2 \ln x.$$

The general solution on $(0, \infty)$ is given by $y = c_1 y_1 + c_2 y_2$; that is,

$$y = c_1 x^2 + c_2 x^2 \ln x.$$

EXAMPLE 3 A Second Solution by Formula (4)

It can be verified that $y_1 = \dfrac{\sin x}{\sqrt{x}}$ is a solution of $x^2y'' - xy' + (x^2 - \frac{1}{4})y = 0$ on $(0, \pi)$. Find a second solution.

Solution First put the equation into the form

$$y'' + \frac{1}{x}y' + \left(1 - \frac{1}{4x^2}\right)y = 0.$$

Then from (4) we have

$$y_2 = \frac{\sin x}{\sqrt{x}} \int \frac{e^{-\int dx/x}}{\left(\dfrac{\sin x}{\sqrt{x}}\right)^2}\,dx \qquad \leftarrow e^{-\int dx/x} = e^{\ln x^{-1}} = x^{-1}$$

$$= \frac{\sin x}{\sqrt{x}} \int \csc^2 x\,dx$$

$$= \frac{\sin x}{\sqrt{x}}(-\cot x) = -\frac{\cos x}{\sqrt{x}}.$$

Since the differential equation is homogeneous, we can disregard the negative sign and take the second solution to be $y_2 = (\cos x)/\sqrt{x}$.

Observe that $y_1(x)$ and $y_2(x)$ of Example 3 are linearly independent solutions of the given differential equation on the larger interval $(0, \infty)$.

Remarks We have derived and illustrated how to use (4) because you will see this formula again in the next section and in Section 6.1. We use (4) simply to save time in obtaining a desired result. Your instructor will tell you whether you should memorize (4) or whether you should know the first principles of reduction of order.

EXERCISES 4.2

Answers to odd-numbered problems begin on page A-7.

In Problems 1–30 find a second solution of each differential equation. Use reduction of order or formula (4) as instructed. Assume an appropriate interval of validity.

1. $y'' + 5y' = 0;\quad y_1 = 1$
2. $y'' - y' = 0;\quad y_1 = 1$
3. $y'' - 4y' + 4y = 0;\quad y_1 = e^{2x}$
4. $y'' + 2y' + y = 0;\quad y_1 = xe^{-x}$
5. $y'' + 16y = 0;\quad y_1 = \cos 4x$
6. $y'' + 9y = 0;\quad y_1 = \sin 3x$
7. $y'' - y = 0;\quad y_1 = \cosh x$
8. $y'' - 25y = 0;\quad y_1 = e^{5x}$
9. $9y'' - 12y' + 4y = 0;\quad y_1 = e^{2x/3}$
10. $6y'' + y' - y = 0;\quad y_1 = e^{x/3}$
11. $x^2y'' - 7xy' + 16y = 0;\quad y_1 = x^4$
12. $x^2y'' + 2xy' - 6y = 0;\quad y_1 = x^2$
13. $xy'' + y' = 0;\quad y_1 = \ln x$
14. $4x^2y'' + y = 0;\quad y_1 = x^{1/2}\ln x$
15. $(1 - 2x - x^2)y'' + 2(1 + x)y' - 2y = 0;\quad y_1 = x + 1$
16. $(1 - x^2)y'' - 2xy' = 0;\quad y_1 = 1$
17. $x^2y'' - xy' + 2y = 0;\quad y_1 = x\sin(\ln x)$
18. $x^2y'' - 3xy' + 5y = 0;\quad y_1 = x^2\cos(\ln x)$
19. $(1 + 2x)y'' + 4xy' - 4y = 0;\quad y_1 = e^{-2x}$
20. $(1 + x)y'' + xy' - y = 0;\quad y_1 = x$
21. $x^2y'' - xy' + y = 0;\quad y_1 = x$
22. $x^2y'' - 20y = 0;\quad y_1 = x^{-4}$
23. $x^2y'' - 5xy' + 9y = 0;\quad y_1 = x^3\ln x$
24. $x^2y'' + xy' + y = 0;\quad y_1 = \cos(\ln x)$
25. $x^2y'' - 4xy' + 6y = 0;\quad y_1 = x^2 + x^3$
26. $x^2y'' - 7xy' - 20y = 0;\quad y_1 = x^{10}$
27. $(3x + 1)y'' - (9x + 6)y' + 9y = 0;\quad y_1 = e^{3x}$
28. $xy'' - (x + 1)y' + y = 0;\quad y_1 = e^x$
29. $y'' - 3(\tan x)y' = 0;\quad y_1 = 1$
30. $xy'' - (2 + x)y' = 0;\quad y_1 = 1$

In Problems 31–34 use the method of reduction of order to find a solution of the given nonhomogeneous equation. The indicated function $y_1(x)$ is a solution of the associated homogeneous equation. Determine a second solution of the homogeneous equation and a particular solution of the nonhomogeneous equation.

31. $y'' - 4y = 2;\quad y_1 = e^{-2x}$
32. $y'' + y' = 1;\quad y_1 = 1$
33. $y'' - 3y' + 2y = 5e^{3x};\quad y_1 = e^x$
34. $y'' - 4y' + 3y = x;\quad y_1 = e^x$
35. Verify by direct substitution that formula (4) satisfies equation (2).

4.3 HOMOGENEOUS LINEAR EQUATIONS WITH CONSTANT COEFFICIENTS

• Auxiliary equation • Second-order equations • Euler's formula • Higher-order equations

We have seen that the linear first-order equation $dy/dx + ay = 0$, where a is a constant, has the exponential solution $y = c_1e^{-ax}$ on $(-\infty, \infty)$. Therefore, it is natural to seek to determine whether exponential solutions exist on $(-\infty, \infty)$ for higher-order equations such as

$$a_n y^{(n)} + a_{n-1}y^{(n-1)} + \cdots + a_2 y'' + a_1 y' + a_0 y = 0, \tag{1}$$

where the a_i, $i = 0, 1, \ldots, n$ are constants. The surprising fact is that *all* solutions of (1) are exponential functions or constructed out of exponential functions. We begin by considering the special case of the second-order equation

$$ay'' + by' + cy = 0. \tag{2}$$

Auxiliary Equation If we try a solution of the form $y = e^{mx}$, then $y' = me^{mx}$ and $y'' = m^2e^{mx}$, so equation (2) becomes

$$am^2e^{mx} + bme^{mx} + ce^{mx} = 0 \quad \text{or} \quad e^{mx}(am^2 + bm + c) = 0.$$

Because e^{mx} is never zero for real values of x, it is apparent that the only way that this exponential function can satisfy the differential equation is if we choose m so that it is a root of the quadratic equation

$$am^2 + bm + c = 0. \tag{3}$$

This latter equation is called the **auxiliary equation,** or **characteristic equation,** of the differential equation (2). We consider three cases—namely, the solutions of the auxiliary equation corresponding to distinct real roots, real but equal roots, and a conjugate pair of complex roots.

Case I: Distinct Real Roots Under the assumption that the auxiliary equation (3) has two unequal real roots m_1 and m_2, we find two solutions $y_1 = e^{m_1x}$ and $y_2 = e^{m_2x}$. We have seen that these functions are linearly independent on $(-\infty, \infty)$ (see Example 13, Section 4.1) and hence form a fundamental set. It follows that the general solution of (2) on this interval is

$$y = c_1e^{m_1x} + c_2e^{m_2x}. \tag{4}$$

Case II: Repeated Real Roots When $m_1 = m_2$, we necessarily obtain only one exponential solution $y_1 = e^{m_1x}$. However, it follows immediately from the discussion of Section 4.2 that a second solution is

$$y_2 = e^{m_1x}\int \frac{e^{-(b/a)x}}{e^{2m_1x}}\,dx. \tag{5}$$

But from the quadratic formula, we have $m_1 = -b/2a$ since the only way to have $m_1 = m_2$ is to have $b^2 - 4ac = 0$. In view of the fact that $2m_1 = -b/a$, (5) becomes

$$y_2 = e^{m_1 x} \int \frac{e^{2m_1 x}}{e^{2m_1 x}}\, dx = e^{m_1 x} \int dx = xe^{m_1 x}.$$

The general solution of (2) is then

$$y = c_1 e^{m_1 x} + c_2 x e^{m_1 x}. \tag{6}$$

Case III: Conjugate Complex Roots If m_1 and m_2 are complex, then we can write

$$m_1 = \alpha + i\beta \quad \text{and} \quad m_2 = \alpha - i\beta,$$

where α and $\beta > 0$ are real and $i^2 = -1$. Formally there is no difference between this case and Case I, and hence

$$y = C_1 e^{(\alpha + i\beta)x} + C_2 e^{(\alpha - i\beta)x}.$$

However, in practice we prefer to work with real functions instead of complex exponentials. To this end we use Euler's formula:*

$$e^{i\theta} = \cos\theta + i\sin\theta,$$

where θ is any real number. It follows from this formula that

$$e^{i\beta x} = \cos\beta x + i\sin\beta x \quad \text{and} \quad e^{-i\beta x} = \cos\beta x - i\sin\beta x, \tag{7}$$

where we have used $\cos(-\beta x) = \cos\beta x$ and $\sin(-\beta x) = -\sin\beta x$. Note that by first adding and then subtracting the two equations in (7), we obtain, respectively,

$$e^{i\beta x} + e^{-i\beta x} = 2\cos\beta x \quad \text{and} \quad e^{i\beta x} - e^{-i\beta x} = 2i\sin\beta x.$$

Since $y = C_1 e^{(\alpha + i\beta)x} + C_2 e^{(\alpha - i\beta)x}$ is a solution of (2) for any choice of the constants C_1 and C_2, the choices $C_1 = C_2 = 1$ and $C_1 = 1$, $C_2 = -1$ give, in turn, two solutions:

$$y_1 = e^{(\alpha + i\beta)x} + e^{(\alpha - i\beta)x} \quad \text{and} \quad y_2 = e^{(\alpha + i\beta)x} - e^{(\alpha - i\beta)x}.$$

---

* **LEONHARD EULER** (1707–1783) A man with a prodigious memory and phenomenal powers of concentration, Euler had almost universal interests; he was a theologian, physicist, astronomer, linguist, physiologist, classical scholar, and, primarily, mathematician. Euler is considered to be a true genius of the era. In mathematics, he made lasting contributions to algebra, trigonometry, analytic geometry, calculus, calculus of variations, differential equations, complex variables, number theory, and topology. The volume of his mathematical output did not seem to be affected by the distractions of thirteen children or the fact that he was totally blind for the last seventeen years of his life. Euler wrote over 700 papers and 32 books on mathematics and was responsible for introducing many of the symbols (such as e, π, and $i = \sqrt{-1}$) and notations (such as $f(x)$, Σ, sin *x*, and cos *x*) that are still used. Euler was born in Basel, Switzerland, on April 15, 1707, and died of a stroke in St. Petersburg on September 18, 1783, while serving in the court of the Russian empress Catherine the Great.

See Appendix IV for a review of complex numbers and a derivation of Euler's formula.

But $$y_1 = e^{\alpha x}(e^{i\beta x} + e^{-i\beta x}) = 2e^{\alpha x}\cos \beta x$$

and $$y_2 = e^{\alpha x}(e^{i\beta x} - e^{-i\beta x}) = 2ie^{\alpha x}\sin \beta x.$$

Hence, from Corollary (A) of Theorem 4.3 the last two results show that the *real* functions $e^{\alpha x}\cos \beta x$ and $e^{\alpha x}\sin \beta x$ are solutions of (2). Moreover, from Example 14 of Section 4.1 we have $W(e^{\alpha x}\cos \beta x, e^{\alpha x}\sin \beta x) = \beta e^{2\alpha x} \neq 0, \beta > 0$, and so we can conclude that $e^{\alpha x}\cos \beta x$ and $e^{\alpha x}\sin \beta x$ themselves form a fundamental set of solutions of the differential equation on $(-\infty, \infty)$. By the superposition principle, the general solution is

$$y = c_1 e^{\alpha x}\cos \beta x + c_2 e^{\alpha x}\sin \beta x = e^{\alpha x}(c_1 \cos \beta x + c_2 \sin \beta x). \tag{8}$$

EXAMPLE 1 Second-Order DEs

Solve the following differential equations:

(a) $2y'' - 5y' - 3y = 0$ **(b)** $y'' - 10y' + 25y = 0$ **(c)** $y'' + y' + y = 0$

Solution

(a) $2m^2 - 5m - 3 = (2m + 1)(m - 3) = 0, m_1 = -\frac{1}{2}, m_2 = 3,$
$$y = c_1 e^{-x/2} + c_2 e^{3x}$$

(b) $m^2 - 10m + 25 = (m - 5)^2 = 0, m_1 = m_2 = 5,$
$$y = c_1 e^{5x} + c_2 x e^{5x}$$

(c) $m^2 + m + 1 = 0, m_1 = -\dfrac{1}{2} + \dfrac{\sqrt{3}}{2}i, m_2 = -\dfrac{1}{2} - \dfrac{\sqrt{3}}{2}i,$
$$y = e^{-x/2}\left(c_1 \cos \frac{\sqrt{3}}{2}x + c_2 \sin \frac{\sqrt{3}}{2}x\right)$$

EXAMPLE 2 An Initial-Value Problem

Solve the initial-value problem $y'' - 4y' + 13y = 0, y(0) = -1, y'(0) = 2.$

Solution The roots of the auxiliary equation $m^2 - 4m + 13 = 0$ are $m_1 = 2 + 3i$ and $m_2 = 2 - 3i$, so

$$y = e^{2x}(c_1 \cos 3x + c_2 \sin 3x).$$

The condition $y(0) = -1$ implies

$$-1 = e^0(c_1 \cos 0 + c_2 \sin 0) = c_1,$$

from which we can write

$$y = e^{2x}(-\cos 3x + c_2 \sin 3x).$$

Differentiating this latter expression yields

$$y' = e^{2x}(3 \sin 3x + 3c_2 \cos 3x) + 2e^{2x}(-\cos 3x + c_2 \sin 3x)$$

and using $y'(0) = 2$ gives $3c_2 - 2 = 2$, and so $c_2 = \frac{4}{3}$. Hence

$$y = e^{2x}\left(-\cos 3x + \frac{4}{3}\sin 3x\right).$$

EXAMPLE 3 Two Second-Order DEs Worth Knowing

The two equations

$$y'' + k^2y = 0 \tag{9}$$

$$y'' - k^2y = 0 \tag{10}$$

are frequently encountered in the study of applied mathematics. For the former differential equation, the auxiliary equation $m^2 + k^2 = 0$ has the roots $m_1 = ki$ and $m_2 = -ki$. It follows from (8) that the general solution of (9) is

$$y = c_1 \cos kx + c_2 \sin kx. \tag{11}$$

The differential equation (10) has the auxiliary equation $m^2 - k^2 = 0$, with real roots $m_1 = k$ and $m_2 = -k$, so its general solution is

$$y = c_1e^{kx} + c_2e^{-kx}. \tag{12}$$

Notice that if we choose $c_1 = c_2 = \frac{1}{2}$ in (12), then

$$y = \frac{e^{kx} + e^{-kx}}{2} = \cosh kx$$

is also a solution of (10). Furthermore, if $c_1 = \frac{1}{2}$, $c_2 = -\frac{1}{2}$, then (12) becomes

$$y = \frac{e^{kx} - e^{-kx}}{2} = \sinh kx.$$

Since $\cosh kx$ and $\sinh kx$ are linearly independent on any interval of the x-axis, they form a fundamental set. Thus an alternative form for the general solution of (10) is

$$y = c_1 \cosh kx + c_2 \sinh kx. \tag{13}$$

Higher-Order Equations In general, to solve an nth-order differential equation

$$a_ny^{(n)} + a_{n-1}y^{(n-1)} + \cdots + a_2y'' + a_1y' + a_0y = 0, \tag{14}$$

where the a_i, $i = 0, 1, \ldots, n$ are real constants, we must solve an nth-degree polynominal equation

$$a_nm^n + a_{n-1}m^{n-1} + \cdots + a_2m^2 + a_1m + a_0 = 0. \tag{15}$$

If all the roots of (15) are real and distinct, then the general solution of (14) is

$$y = c_1e^{m_1x} + c_2e^{m_2x} + \cdots + c_ne^{m_nx}. \tag{16}$$

It is somewhat harder to summarize the analogues of Cases II and III because the roots of an auxiliary equation of degree greater than two can occur in many combinations. For example, a fifth-degree equation could have five distinct real roots, or three distinct real and two complex roots, or one real and four complex roots, or five real but equal roots, or five real roots but two of them equal, and so on. When m_1 is a root of multiplicity k of an nth-degree auxiliary equation (that is, k roots are equal to m_1), it can be shown that the linearly independent solutions are

$$e^{m_1x},\ xe^{m_1x},\ x^2e^{m_1x},\ \ldots,\ x^{k-1}e^{m_1x}$$

and the general solution must contain the linear combination

$$c_1e^{m_1x} + c_2xe^{m_1x} + c_3x^2e^{m_1x} + \cdots + c_kx^{k-1}e^{m_1x}.$$

Lastly, it should be remembered that when the coefficients are real, complex roots of an auxiliary equation always appear in conjugate pairs. Thus, for example, a cubic polynomial equation can have at most two complex roots.

EXAMPLE 4 Third-Order DE

Solve $y''' + 3y'' - 4y = 0$.

Solution It should be apparent from inspection of

$$m^3 + 3m^2 - 4 = 0$$

that one root is $m_1 = 1$. Now if we divide $m^3 + 3m^2 - 4$ by $m - 1$, we find that

$$m^3 + 3m^2 - 4 = (m - 1)(m^2 + 4m + 4) = (m - 1)(m + 2)^2,$$

and so the other roots are $m_2 = m_3 = -2$. Thus the general solution is

$$y = c_1e^{x} + c_2e^{-2x} + c_3xe^{-2x}.$$

Of course, the most difficult aspect of solving constant-coefficient equations is finding the roots of auxiliary equations of degree greater than two. As illustrated in Example 4, one way to solve an equation is to guess a root m_1. If we have found one root m_1, we then know from the **factor theorem** that $m - m_1$ is a factor of the polynomial. By dividing the polynomial by $m - m_1$, we obtain the factorization $(m - m_1)Q(m)$. We then try to find the roots of the quotient $Q(m)$. The algebraic technique of **synthetic division** is also very helpful in finding **rational** roots of polynomial equations. Specifically, if $m_1 = p/q$ is a **rational real root** (p and q integers, p/q in lowest terms) of an auxiliary equation

$$a_nm^n + \cdots + a_1m + a_0 = 0$$

with integer coefficients, then p is a factor of a_0 and q is a factor of a_n. Thus to determine whether a polynomial equation has rational roots we need only examine all ratios of each factor of a_0 to each factor of a_n. In this manner we construct a list of all the possible rational roots of the equation. We test each of these numbers by synthetic division. If the remainder is zero, the number m_1 being tested is a root of the equation and thus $m - m_1$ is a factor of the polynomial.

The next example illustrates this method.

EXAMPLE 5 Using Synthetic Division

Solve $3y''' + 5y'' + 10y' - 4y = 0$.

Solution The auxiliary equation is

$$3m^3 + 5m^2 + 10m - 4 = 0.$$

All the factors of $a_0 = -4$ and $a_n = 3$ are

$$p: \pm 1, \pm 2, \pm 4 \quad \text{and} \quad q: \pm 1, \pm 3,$$

respectively. Therefore the possible rational roots of the auxiliary equation are

$$\frac{p}{q}: -1, 1, -2, 2, -4, 4, -\frac{1}{3}, \frac{1}{3}, -\frac{2}{3}, \frac{2}{3}, -\frac{4}{3}, \frac{4}{3}.$$

Testing each of these numbers in turn by synthetic division, we eventually find

coefficients of auxiliary equation

$\frac{1}{3}$	3	5	10	−4	
		1	2	4	
	3	6	12	0	remainder

Consequently $m_1 = \frac{1}{3}$ is a root. Furthermore, you should verify that it is the *only* rational root. The numbers highlighted in the preceding division are the coefficients of the quotient. Thus the auxiliary equation can be written as

$$(m - \tfrac{1}{3})(3m^2 + 6m + 12) = 0 \quad \text{or} \quad (3m - 1)(m^2 + 2m + 4) = 0.$$

Solving $m^2 + 2m + 4 = 0$ by the quadratic formula leads to the complex roots $m_2 = -1 + \sqrt{3}\,i$ and $m_3 = -1 - \sqrt{3}\,i$. Thus the general solution of the differential equation is

$$y = c_1 e^{x/3} + e^{-x}(c_2 \cos \sqrt{3}x + c_3 \sin \sqrt{3}x).$$

EXAMPLE 6 Fourth-Order DE

Solve $\dfrac{d^4y}{dx^4} + 2\dfrac{d^2y}{dx^2} + y = 0$.

Solution The auxiliary equation

$$m^4 + 2m^2 + 1 = (m^2 + 1)^2 = 0$$

has roots $m_1 = m_3 = i$ and $m_2 = m_4 = -i$. Thus from Case II the solution is

$$y = C_1 e^{ix} + C_2 e^{-ix} + C_3 x e^{ix} + C_4 x e^{-ix}.$$

By Euler's formula the grouping $C_1 e^{ix} + C_2 e^{-ix}$ can be rewritten as

$$c_1 \cos x + c_2 \sin x$$

after a relabeling of constants. Similarly, $x(C_3 e^{ix} + C_4 e^{-ix})$ can be expressed as $x(c_3 \cos x + c_4 \sin x)$. Hence the general solution is

$$y = c_1 \cos x + c_2 \sin x + c_3 x \cos x + c_4 x \sin x.$$

Example 6 illustrates a special case when the auxiliary equation has repeated complex roots. In general, if $m_1 = \alpha + i\beta$, $\beta > 0$ is a complex root of multiplicity k of an auxiliary equation with real coefficients, then its conjugate $m_2 = \alpha - i\beta$ is also a root of multiplicity k. From the $2k$ complex-valued solutions

$$e^{(\alpha+i\beta)x}, \quad xe^{(\alpha+i\beta)x}, \quad x^2e^{(\alpha+i\beta)x}, \quad \ldots, \quad x^{k-1}e^{(\alpha+i\beta)x}$$
$$e^{(\alpha-i\beta)x}, \quad xe^{(\alpha-i\beta)x}, \quad x^2e^{(\alpha-i\beta)x}, \quad \ldots, \quad x^{k-1}e^{(\alpha-i\beta)x}$$

we conclude, with the aid of Euler's formula, that the general solution of the corresponding differential equation must then contain a linear combination of the $2k$ real linearly independent solutions

$$e^{\alpha x}\cos\beta x, \quad xe^{\alpha x}\cos\beta x, \quad x^2e^{\alpha x}\cos\beta x, \quad \ldots, \quad x^{k-1}e^{\alpha x}\cos\beta x$$
$$e^{\alpha x}\sin\beta x, \quad xe^{\alpha x}\sin\beta x, \quad x^2e^{\alpha x}\sin\beta x, \quad \ldots, \quad x^{k-1}e^{\alpha x}\sin\beta x.$$

In Example 6 we identify $k = 2$, $\alpha = 0$, and $\beta = 1$.

EXERCISES 4.3

Answers to odd-numbered problems begin on page A-8.

In Problems 1–36 find the general solution of the given differential equation.

1. $4y'' + y' = 0$

2. $2y'' - 5y' = 0$

3. $y'' - 36y = 0$

4. $y'' - 8y = 0$

5. $y'' + 9y = 0$

6. $3y'' + y = 0$

7. $y'' - y' - 6y = 0$

8. $y'' - 3y' + 2y = 0$

9. $\dfrac{d^2y}{dx^2} + 8\dfrac{dy}{dx} + 16y = 0$

10. $\dfrac{d^2y}{dx^2} - 10\dfrac{dy}{dx} + 25y = 0$

11. $y'' + 3y' - 5y = 0$

12. $y'' + 4y' - y = 0$

13. $12y'' - 5y' - 2y = 0$

14. $8y'' + 2y' - y = 0$

15. $y'' - 4y' + 5y = 0$

16. $2y'' - 3y' + 4y = 0$

17. $3y'' + 2y' + y = 0$

18. $2y'' + 2y' + y = 0$

19. $y''' - 4y'' - 5y' = 0$

20. $4y''' + 4y'' + y' = 0$

21. $y''' - y = 0$

22. $y''' + 5y'' = 0$

23. $y''' - 5y'' + 3y' + 9y = 0$

24. $y''' + 3y'' - 4y' - 12y = 0$

25. $y''' + y'' - 2y = 0$

26. $y''' - y'' - 4y = 0$

27. $y''' + 3y'' + 3y' + y = 0$

28. $y''' - 6y'' + 12y' - 8y = 0$

29. $\dfrac{d^4y}{dx^4} + \dfrac{d^3y}{dx^3} + \dfrac{d^2y}{dx^2} = 0$

30. $\dfrac{d^4y}{dx^4} - 2\dfrac{d^2y}{dx^2} + y = 0$

31. $16\dfrac{d^4y}{dx^4} + 24\dfrac{d^2y}{dx^2} + 9y = 0$

32. $\dfrac{d^4y}{dx^4} - 7\dfrac{d^2y}{dx^2} - 18y = 0$

33. $\dfrac{d^5y}{dx^5} - 16\dfrac{dy}{dx} = 0$

34. $\dfrac{d^5y}{dx^5} - 2\dfrac{d^4y}{dx^4} + 17\dfrac{d^3y}{dx^3} = 0$

35. $\dfrac{d^5y}{dx^5} + 5\dfrac{d^4y}{dx^4} - 2\dfrac{d^3y}{dx^3} - 10\dfrac{d^2y}{dx^2} + \dfrac{dy}{dx} + 5y = 0$

36. $2\dfrac{d^5y}{dx^5} - 7\dfrac{d^4y}{dx^4} + 12\dfrac{d^3y}{dx^3} + 8\dfrac{d^2y}{dx^2} = 0$

In Problems 37–52 solve the given differential equation subject to the indicated initial conditions.

37. $y'' + 16y = 0, \quad y(0) = 2, y'(0) = -2$

38. $y'' - y = 0, \quad y(0) = y'(0) = 1$

39. $y'' + 6y' + 5y = 0, \quad y(0) = 0, y'(0) = 3$

40. $y'' - 8y' + 17y = 0, \quad y(0) = 4, y'(0) = -1$

41. $2y'' - 2y' + y = 0, \quad y(0) = -1, y'(0) = 0$

42. $y'' - 2y' + y = 0, \quad y(0) = 5, y'(0) = 10$

43. $y'' + y' + 2y = 0, \quad y(0) = y'(0) = 0$

44. $4y'' - 4y' - 3y = 0, \quad y(0) = 1, y'(0) = 5$

45. $y'' - 3y' + 2y = 0, \quad y(1) = 0, y'(1) = 1$

46. $y'' + y = 0, \quad y\left(\dfrac{\pi}{3}\right) = 0, y'\left(\dfrac{\pi}{3}\right) = 2$

47. $y''' + 12y'' + 36y' = 0, \quad y(0) = 0, y'(0) = 1, y''(0) = -7$

48. $y''' + 2y'' - 5y' - 6y = 0, \quad y(0) = y'(0) = 0, y''(0) = 1$

49. $y''' - 8y = 0, \quad y(0) = 0, y'(0) = -1, y''(0) = 0$

50. $\dfrac{d^4y}{dx^4} = 0, \quad y(0) = 2, y'(0) = 3, y''(0) = 4, y'''(0) = 5$

51. $\dfrac{d^4y}{dx^4} - 3\dfrac{d^3y}{dx^3} + 3\dfrac{d^2y}{dx^2} - \dfrac{dy}{dx} = 0, \quad y(0) = y'(0) = 0, y''(0) = y'''(0) = 1$

52. $\dfrac{d^4y}{dx^4} - y = 0, \quad y(0) = y'(0) = y''(0) = 0, y'''(0) = 1$

In Problems 53–56 solve the given differential equation subject to the indicated boundary conditions.

53. $y'' - 10y' + 25y = 0, \quad y(0) = 1, y(1) = 0$

54. $y'' + 4y = 0, \quad y(0) = 0, y(\pi) = 0$

55. $y'' + y = 0, \quad y'(0) = 0, y'\left(\dfrac{\pi}{2}\right) = 2$

56. $y'' - y = 0, \quad y(0) = 1, y'(1) = 0$

57. The roots of an auxiliary equation are $m_1 = 4, m_2 = m_3 = -5$. What is the corresponding differential equation?

58. The roots of an auxiliary equation are $m_1 = -\frac{1}{2}$, $m_2 = 3 + i$, $m_3 = 3 - i$. What is the corresponding differential equation?

In Problems 59 and 60 find the general solution of the given equation if it is known that y_1 is a solution.

59. $y''' - 9y'' + 25y' - 17y = 0; \quad y_1 = e^x$

60. $y''' + 6y'' + y' - 34y = 0; \quad y_1 = e^{-4x}\cos x$

In Problems 61–64 determine a homogeneous linear differential equation with constant coefficients having the given solutions.

61. $4e^{6x}, 3e^{-3x}$

62. $10\cos 4x, -5\sin 4x$

63. $3, 2x, -e^{7x}$

64. $8 \sinh 3x$, $12 \cosh 3x$

65. Use the facts

$$i = \left(\frac{\sqrt{2}}{2} + \frac{\sqrt{2}}{2}i\right)^2 \quad \text{and} \quad -i = \left(\frac{\sqrt{2}}{2} - \frac{\sqrt{2}}{2}i\right)^2$$

to solve the differential equation

$$\frac{d^4y}{dx^4} + y = 0.$$

[*Hint*: Write the auxiliary equation $m^4 + 1 = 0$ as $(m^2 + 1)^2 - 2m^2 = 0$. See what happens when you factor.]

4.4 UNDETERMINED COEFFICIENTS—SUPERPOSITION APPROACH

• *Undetermined coefficients* • *Particular solution* • *Synthetic division*

Note to the Instructor In this section the method of undetermined coefficients is developed from the viewpoint of the superposition principle for nonhomogeneous differential equations (Theorem 4.9). In Section 4.6 an entirely different approach to this method will be presented, one utilizing the concept of differential annihilator operators. Take your pick.

To obtain the general solution of a nonhomogeneous linear differential equation we must do two things:

(*i*) Find the complementary function y_c.
(*ii*) Find *any* particular solution y_p of the nonhomogeneous equation.

Recall from the discussion of Section 4.1 that a particular solution is any function, free of arbitrary constants, that satisfies the differential equation identically. The general solution of a nonhomogeneous equation on an interval is then $y = y_c + y_p$.

As in the Section 4.3 we begin with second-order equations, but in this case nonhomogeneous equations of the form

$$ay'' + by' + cy = g(x), \tag{1}$$

where a, b, and c are constants. Although the **method of undetermined coefficients** presented in this section is *not* limited to second-order equations, it *is* limited to nonhomogeneous linear equations in which

- coefficients are constant and
- $g(x)$ is a constant k, a polynomial function, an exponential function $e^{\alpha x}$, $\sin \beta x$, $\cos \beta x$, or finite sums and products of these functions.

Note Strictly speaking, $g(x) = k$ (a constant) is a polynomial function. Since a constant function is probably not the first thing that comes to mind when you think of polynomial functions, for emphasis we continue to use the redundancy "constant functions, polynomials, ..."

The following are some examples of the types of input functions $g(x)$ that are appropriate for this discussion:

$$g(x) = 10, \quad g(x) = x^2 - 5x, \quad g(x) = 15x - 6 + 8e^{-4x},$$
$$g(x) = \sin 3x - 5x \cos 2x, \quad g(x) = e^x \cos x - (3x^2 - 1)e^{-x},$$

and so on. That is, $g(x)$ is a linear combination of functions of the type

$$k \text{ (constant)}, \quad x^n, \quad x^n e^{\alpha x}, \quad x^n e^{\alpha x} \cos \beta x, \quad \text{and} \quad x^n e^{\alpha x} \sin \beta x,$$

where n is a nonnegative integer and α and β are real numbers. The method of undetermined coefficients is not applicable to equations of form (1) when

$$g(x) = \ln x, \quad g(x) = \frac{1}{x}, \quad g(x) = \tan x, \quad g(x) = \sin^{-1} x,$$

and so on. Differential equations with this latter kind of input function will be considered in Section 4.7.

The set of functions that consists of constants, polynomials, exponentials $e^{\alpha x}$, sines, and cosines has the remarkable property that derivatives of their sums and products are again sums and products of constants, polynomials, exponentials $e^{\alpha x}$, sines, and cosines. Since the linear combination of derivatives $ay_p'' + by_p' + cy_p$ must be identically equal to $g(x)$, it seems reasonable to assume then that y_p *has the same form as* $g(x)$. This assumption could be better characterized as an educated conjecture or guess. The next two examples illustrate the basic method.

EXAMPLE 1 General Solution

Solve $y'' + 4y' - 2y = 2x^2 - 3x + 6$. **(2)**

Solution **Step 1.** We first solve the associated homogeneous equation $y'' + 4y' - 2y = 0$. From the quadratic formula we find that the roots of the auxiliary equation $m^2 + 4m - 2 = 0$ are $m_1 = -2 - \sqrt{6}$ and $m_2 = -2 + \sqrt{6}$. Hence the complementary function is

$$y_c = c_1 e^{-(2+\sqrt{6})x} + c_2 e^{(-2+\sqrt{6})x}.$$

Step 2. Now since the input function $g(x)$ is a quadratic polynomial, let us assume a particular solution that is also in the form of a quadratic polynomial:

$$y_p = Ax^2 + Bx + C.$$

We seek to determine *specific* coefficients A, B, and C for which y_p is a solution of (2). Substituting y_p and the derivatives

$$y_p' = 2Ax + B \quad \text{and} \quad y_p'' = 2A$$

into the given differential equation (2), we get

$$y_p'' + 4y_p' - 2y_p = 2A + 8Ax + 4B - 2Ax^2 - 2Bx - 2C = 2x^2 - 3x + 6.$$

Since the last equation is supposed to be an identity, the coefficients of like powers of x must be equal:

equal

$$\boxed{-2A}\,x^2 + \boxed{8A - 2B}\,x + \boxed{2A + 4B - 2C} = 2x^2 - 3x + 6.$$

That is,

$$-2A = 2, \quad 8A - 2B = -3, \quad 2A + 4B - 2C = 6.$$

Solving this system of equations leads to the values $A = -1$, $B = -\frac{5}{2}$, and $C = -9$. Thus a particular solution is

$$y_p = -x^2 - \frac{5}{2}x - 9.$$

Step 3. The general solution of the given equation is

$$y = y_c + y_p = c_1 e^{-(2+\sqrt{6})x} + c_2 e^{(-2+\sqrt{6})x} - x^2 - \frac{5}{2}x - 9.$$

EXAMPLE 2 Particular Solution

Find a particular solution of $y'' - y' + y = 2 \sin 3x$.

Solution A natural first guess for a particular solution would be $A \sin 3x$. But since successive differentiations of $\sin 3x$ produce $\sin 3x$ *and* $\cos 3x$, we are prompted instead to assume a particular solution that includes both of these terms:

$$y_p = A \cos 3x + B \sin 3x.$$

Differentiating y_p and substituting the results into the differential equation gives, after regrouping,

$$y_p'' - y_p' + y_p = (-8A - 3B)\cos 3x + (3A - 8B)\sin 3x = 2 \sin 3x$$

or

equal

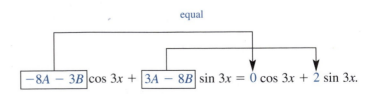

$$\boxed{-8A - 3B}\cos 3x + \boxed{3A - 8B}\sin 3x = 0 \cos 3x + 2 \sin 3x.$$

From the resulting system of equations

$$-8A - 3B = 0, \quad 3A - 8B = 2$$

we get $A = \frac{6}{73}$ and $B = -\frac{16}{73}$. A particular solution of the equation is

$$y_p = \frac{6}{73}\cos 3x - \frac{16}{73}\sin 3x.$$

As we mentioned, the form that we assume for the particular solution y_p is an educated guess; it is not a blind guess. This educated guess must take into consideration not only the types of functions that make up $g(x)$ but also, as we shall see in Example 4, the functions that make up the complementary function y_c. ■

EXAMPLE 3 Forming y_p by Superposition

Solve

$$y'' - 2y' - 3y = 4x - 5 + 6xe^{2x}. \tag{3}$$

Solution **Step 1.** First, the solution of the associated homogeneous equation $y'' - 2y' - 3y = 0$ is found to be $y_c = c_1e^{-x} + c_2e^{3x}$.

Step 2. Next, the presence of $4x - 5$ in $g(x)$ suggests that the particular solution includes a linear polynomial. Furthermore, since the derivative of the product xe^{2x} produces $2xe^{2x}$ and e^{2x}, we also assume that the particular solution includes both xe^{2x} and e^{2x}. In other words, g is the sum of two basic kinds of functions:

$$g(x) = g_1(x) + g_2(x) = \textit{polynomial} + \textit{exponentials}.$$

Correspondingly, the superposition principle for nonhomogeneous equations (Theorem 4.9) suggests that we seek a particular solution

$$y_p = y_{p_1} + y_{p_2},$$

where $y_{p_1} = Ax + B$ and $y_{p_2} = Cxe^{2x} + De^{2x}$. Substituting

$$y_p = Ax + B + Cxe^{2x} + De^{2x}$$

into the given equation (3) and grouping like terms gives

$$y_p'' - 2y_p' - 3y_p = -3Ax - 2A - 3B - 3Cxe^{2x} + (2C - 3D)e^{2x} = 4x - 5 + 6xe^{2x}. \tag{4}$$

From this identity we obtain the system of four equations in four unknowns:

$$-3A = 4, \quad -2A - 3B = -5, \quad -3C = 6, \quad 2C - 3D = 0.$$

The last equation in this system results from the interpretation that the coefficient of e^{2x} in the right member of (4) is zero. Solving, we find $A = -\frac{4}{3}$, $B = -\frac{23}{9}$, $C = -2$, and $D = -\frac{4}{3}$. Consequently,

$$y_p = -\frac{4}{3}x + \frac{23}{9} - 2xe^{2x} - \frac{4}{3}e^{2x}.$$

Step 3. The general solution of the equation is

$$y = c_1e^{-x} + c_2e^{3x} - \frac{4}{3}x + \frac{23}{9} - \left(2x + \frac{4}{3}\right)e^{2x}.$$ ■

In light of the superposition principle (Theorem 4.9) we can also approach Example 3 from the viewpoint of solving two simpler problems. You should verify that substituting

$$y_{p_1} = Ax + B \quad \text{into} \quad y'' - 2y' - 3y = 4x - 5$$

and

$$y_{p_2} = Cxe^{2x} + De^{2x} \quad \text{into} \quad y'' - 2y' - 3y = 6xe^{2x}$$

yields in turn $y_{p_1} = -\frac{4}{3}x + \frac{23}{9}$ and $y_{p_2} = -(2x + \frac{4}{3})e^{2x}$. A particular solution of (3) is then $y_p = y_{p_1} + y_{p_2}$.

The next example illustrates that sometimes the "obvious" assumption for the form of y_p is not a correct assumption.

EXAMPLE 4 A Glitch in the Method

Find a particular solution of $y'' - 5y' + 4y = 8e^x$.

Solution Differentiation of e^x produces no new functions. Thus, proceeding as we did in the earlier examples, we can reasonably assume a particular solution of the form

$$y_p = Ae^x.$$

But in this case substitution of this expression into the differential equation yields the contradictory statement

$$0 = 8e^x,$$

and so we have clearly made the wrong guess for y_p.

The difficulty here is apparent upon examining the complementary function $y_c = c_1e^x + c_2e^{4x}$. Observe that our assumption Ae^x is already present in y_c. This means that e^x is a solution of the associated homogeneous differential equation, and a constant multiple Ae^x when substituted into the differential equation necessarily produces zero.

What then should be the form of y_p? Inspired by Case II of Section 4.3, let's see whether we can find a particular solution of the form

$$y_p = Axe^x.$$

Using $y_p' = Axe^x + Ae^x$ and $y_p'' = Axe^x + 2Ae^x$, we obtain

$$y_p'' - 5y_p' + 4y_p = Axe^x + 2Ae^x - 5Axe^x - 5Ae^x + 4Axe^x = 8e^x$$

or

$$-3Ae^x = 8e^x.$$

From this last equation we see that the value of A is now determined as $A = -\frac{8}{3}$. Therefore

$$y_p = -\frac{8}{3}xe^x$$

must be a particular solution of the given equation.

The difference in the procedures used in Examples 1–3 and in Example 4 suggests that we consider two cases. The first case reflects the situation in Examples 1–3.

Case I *No function in the assumed particular solution is a solution of the associated homogeneous differential equation.*

Table 4.1 illustrates some specific examples of $g(x)$ in (1) along with the corresponding form of the particular solution. We are, of course, taking for granted that no function in the assumed particular solution y_p is duplicated by a function in the complementary function y_c.

Table 4.1 Trial Particular Solutions

$g(x)$	Form of y_p
1. 1 (any constant)	A
2. $5x + 7$	$Ax + B$
3. $3x^2 - 2$	$Ax^2 + Bx + C$
4. $x^3 - x + 1$	$Ax^3 + Bx^2 + Cx + D$
5. $\sin 4x$	$A \cos 4x + B \sin 4x$
6. $\cos 4x$	$A \cos 4x + B \sin 4x$
7. e^{5x}	Ae^{5x}
8. $(9x - 2)e^{5x}$	$(Ax + B)e^{5x}$
9. x^2e^{5x}	$(Ax^2 + Bx + C)e^{5x}$
10. $e^{3x}\sin 4x$	$Ae^{3x}\cos 4x + Be^{3x}\sin 4x$
11. $5x^2\sin 4x$	$(Ax^2 + Bx + C)\cos 4x + (Dx^2 + Ex + F)\sin 4x$
12. $xe^{3x}\cos 4x$	$(Ax + B)e^{3x}\cos 4x + (Cx + D)e^{3x}\sin 4x$

EXAMPLE 5 Forms of Particular Solutions—Case I

Determine the form of a particular solution of

(a) $y'' - 8y' + 25y = 5x^3e^{-x} - 7e^{-x}$ and **(b)** $y'' + 4y = x \cos x$.

Solution

(a) We can write

$$g(x) = (5x^3 - 7)e^{-x}.$$

Using entry 9 in the table as a model, we assume a particular solution of the form

$$y_p = (Ax^3 + Bx^2 + Cx + D)e^{-x}.$$

Note that there is no duplication between the terms in y_p and the terms in the complementary function $y_c = e^{4x}(c_1 \cos 3x + c_2 \sin 3x)$.

(b) The function $g(x) = x \cos x$ is similar to entry 11 in the table, except of course we use a linear rather than a quadratic polynomial and $\cos x$ and $\sin x$ instead of $\cos 4x$ and $\sin 4x$ in the form of y_p:

$$y_p = (Ax + B)\cos x + (Cx + D)\sin x.$$

Again observe that there is no duplication of terms between y_p and $y_c = c_1 \cos 2x + c_2 \sin 2x$. ■

If $g(x)$ consists of a sum of, say, m terms of the kind listed in the table, then (as in Example 3) the assumption for a particular solution y_p consists of the sum of the trial forms $y_{p_1}, y_{p_2}, \ldots, y_{p_m}$ corresponding to these terms:

$$y_p = y_{p_1} + y_{p_2} + \cdots + y_{p_m}.$$

Put another way:

The form of y_p is a linear combination of all linearly independent functions that are generated by repeated differentiations of $g(x)$.

EXAMPLE 6 Forming y_p by Superposition—Case I

Determine the form of a particular solution of

$$y'' - 9y' + 14y = 3x^2 - 5\sin 2x + 7xe^{6x}.$$

Solution

Corresponding to $3x^2$, we assume $\quad y_{p_1} = Ax^2 + Bx + C.$

Corresponding to $-5 \sin 2x$, we assume $\quad y_{p_2} = D\cos 2x + E \sin 2x.$

Corresponding to $7xe^{6x}$, we assume $\quad y_{p_3} = (Fx + G)e^{6x}.$

The assumption for the particular solution is then

$$y_p = y_{p_1} + y_{p_2} + y_{p_3} = Ax^2 + Bx + C + D\cos 2x + E\sin 2x + (Fx + G)e^{6x}.$$

No term in this assumption duplicates a term in $y_c = c_1e^{2x} + c_2e^{7x}$. ■

Case II *A function in the assumed particular solution is also a solution of the associated homogeneous differential equation.*

The next example is similar to Example 4.

EXAMPLE 7 Particular Solution—Case II

Find a particular solution of $y'' - 2y' + y = e^x$.

Solution The complementary function is $y_c = c_1e^x + c_2xe^x$. As in Example 4, the assumption $y_p = Ae^x$ will fail since it is apparent from y_c that e^x is a solution of the associated homogeneous equation $y'' - 2y' + y = 0$. Moreover, we will not be able to find a particular solution of the form $y_p = Axe^x$ since the term xe^x is also duplicated in y_c. We next try

$$y_p = Ax^2e^x.$$

Substituting into the given differential equation yields

$$2Ae^x = e^x \quad \text{and so} \quad A = \frac{1}{2}.$$

Thus a particular solution is $y_p = \frac{1}{2}x^2e^x$. ■

Suppose again that $g(x)$ consists of m terms of the kind given in the table and suppose further that the usual assumption for a particular solution is

$$y_p = y_{p_1} + y_{p_2} + \cdots + y_{p_m},$$

where the y_{p_i}, $i = 1, 2, \ldots, m$ are the trial particular solution forms corresponding to these terms. Under the circumstance described in Case II, we can make up the following **general rule:**

> *If any y_{p_i} contains terms that duplicate terms in y_c, then that y_{p_i} must be multiplied by x^n, where n is the smallest positive integer that eliminates that duplication.*

EXAMPLE 8 An Initial-Value Problem

Solve the initial-value problem $y'' + y = 4x + 10\sin x$, $y(\pi) = 0$, $y'(\pi) = 2$.

Solution The solution of the associated homogeneous equation $y'' + y = 0$ is

$$y_c = c_1 \cos x + c_2 \sin x.$$

Now since $g(x)$ is the sum of a linear polynomial and a sine function, our normal assumption for y_p from entries 2 and 5 of the trial solutions table would be the sum of $y_{p_1} = Ax + B$ and $y_{p_2} = C\cos x + D\sin x$:

$$y_p = Ax + B + C\cos x + D\sin x. \tag{5}$$

But there is an obvious duplication of the terms $\cos x$ and $\sin x$ in this assumed form and two terms in the complementary function. This duplication can be eliminated by simply multiplying y_{p_2} by x. Instead of (5) we now use

$$y_p = Ax + B + Cx\cos x + Dx\sin x. \tag{6}$$

Differentiating this expression and substituting the results into the differential equation gives

$$y_p'' + y_p = Ax + B - 2C\sin x + 2D\cos x = 4x + 10\sin x,$$

and so

$$A = 4, \quad B = 0, \quad -2C = 10, \quad 2D = 0.$$

The solutions of the system are immediate: $A = 4, B = 0, C = -5$, and $D = 0$. Therefore from (6) we obtain

$$y_p = 4x - 5x\cos x.$$

The general solution of the given equation is

$$y = y_c + y_p = c_1\cos x + c_2\sin x + 4x - 5x\cos x.$$

We now apply the prescribed initial conditions to the general solution of the equation. First, $y(\pi) = c_1\cos\pi + c_2\sin\pi + 4\pi - 5\pi\cos\pi = 0$ yields $c_1 = 9\pi$, since $\cos\pi = -1$ and $\sin\pi = 0$. Next, from the derivative

$$y' = -9\pi\sin x + c_2\cos x + 4 + 5x\sin x - 5\cos x$$

and

$$y'(\pi) = -9\pi\sin\pi + c_2\cos\pi + 4 + 5\pi\sin\pi - 5\cos\pi = 2$$

we find $c_2 = 7$. The solution of the initial-value problem is then

$$y = 9\pi\cos x + 7\sin x + 4x - 5x\cos x.$$

EXAMPLE 9 Particular Solution—Case II

Solve $y'' - 6y' + 9y = 6x^2 + 2 - 12e^{3x}$.

Solution The complementary function is

$$y_c = c_1e^{3x} + c_2xe^{3x},$$

and based on entries 3 and 7 of the table, the usual assumption for a particular solution would be

$$y_p = \underbrace{Ax^2 + Bx + C}_{y_{p_1}} + \underbrace{De^{3x}}_{y_{p_2}}.$$

Inspection of these functions shows that the one term in y_{p_2} is duplicated in y_c. If we multiply y_{p_2} by x, we note that the term xe^{3x} is still part of y_c. But multiplying y_{p_2} by x^2 eliminates all duplications. Thus the operative form of a particular solution is

$$y_p = Ax^2 + Bx + C + Dx^2e^{3x}.$$

Differentiating this last form, substituting into the differential equation, and collecting like terms gives

$$y_p'' - 6y_p' + 9y_p = 9Ax^2 + (-12A + 9B)x + 2A - 6B + 9C + 2De^{3x} = 6x^2 + 2 - 12e^{3x}.$$

It follows from this identity that $A = \frac{2}{3}$, $B = \frac{8}{9}$, $C = \frac{2}{3}$, and $D = -6$. Hence the general solution $y = y_c + y_p$ is

$$y = c_1e^{3x} + c_2xe^{3x} + \frac{2}{3}x^2 + \frac{8}{9}x + \frac{2}{3} - 6x^2e^{3x}.$$

Higher-Order Equations The method of undetermined coefficients given here is not restricted to second-order equations but can be used as well with higher-order equations

$$a_ny^{(n)} + a_{n-1}y^{(n-1)} + \cdots + a_1y' + a_0y = g(x)$$

with constant coefficients. It is only necessary that $g(x)$ consist of the proper kinds of functions discussed above.

EXAMPLE 10 Third-Order DE

Solve $y''' + y'' = e^x \cos x$.

Solution From the characteristic equation $m^3 + m^2 = 0$ we find $m_1 = m_2 = 0$ and $m_3 = -1$. Hence the complementary solution of the equation is $y_c = c_1 + c_2x + c_3e^{-x}$. With $g(x) = e^x \cos x$ we see from entry 10 of the table of trial particular solutions that we should assume

$$y_p = Ae^x \cos x + Be^x \sin x.$$

Since there are no functions in y_p that duplicate functions in the complementary solution, we proceed in the usual manner. From

$$y_p''' + y_p'' = (-2A + 4B)e^x \cos x + (-4A - 2B)e^x \sin x = e^x \cos x$$

we get $$-2A + 4B = 1, \quad -4A - 2B = 0.$$

This system gives $A = -\frac{1}{10}$ and $B = \frac{1}{5}$, so a particular solution is

$$y_p = -\frac{1}{10}e^x \cos x + \frac{1}{5}e^x \sin x.$$

The general solution of the equation is

$$y = y_c + y_p = c_1 + c_2x + c_3e^{-x} - \frac{1}{10}e^x \cos x + \frac{1}{5}e^x \sin x.$$

EXAMPLE 11 Fourth-Order DE

Determine the form of a particular solution of $y^{(4)} + y''' = 1 - e^{-x}$.

Solution Comparing the complementary function

$$y_c = c_1 + c_2x + c_3x^2 + c_4e^{-x}$$

with our normal assumption for a particular solution

$$y_p = \underbrace{A}_{y_{p_1}} + \underbrace{Be^{-x}}_{y_{p_2}},$$

we see that the duplications between y_c and y_p are eliminated when y_{p_1} is multiplied by x^3 and y_{p_2} is multiplied by x. Thus the correct assumption for a particular solution is

$$y_p = Ax^3 + Bxe^{-x}.$$

EXERCISES 4.4

Answers to odd-numbered problems begin on page A-8.

In Problems 1–26 solve the given differential equation by the method of undetermined coefficients.

1. $y'' + 3y' + 2y = 6$
2. $4y'' + 9y = 15$
3. $y'' - 10y' + 25y = 30x + 3$
4. $y'' + y' - 6y = 2x$
5. $\frac{1}{4}y'' + y' + y = x^2 - 2x$
6. $y'' - 8y' + 20y = 100x^2 - 26xe^x$
7. $y'' + 3y = -48x^2e^{3x}$
8. $4y'' - 4y' - 3y = \cos 2x$
9. $y'' - y' = -3$
10. $y'' + 2y' = 2x + 5 - e^{-2x}$
11. $y'' - y' + \frac{1}{4}y = 3 + e^{x/2}$
12. $y'' - 16y = 2e^{4x}$
13. $y'' + 4y = 3 \sin 2x$
14. $y'' + 4y = (x^2 - 3)\sin 2x$
15. $y'' + y = 2x \sin x$
16. $y'' - 5y' = 2x^3 - 4x^2 - x + 6$
17. $y'' - 2y' + 5y = e^x \cos 2x$
18. $y'' - 2y' + 2y = e^{2x}(\cos x - 3 \sin x)$
19. $y'' + 2y' + y = \sin x + 3 \cos 2x$

20. $y'' + 2y' - 24y = 16 - (x + 2)e^{4x}$

21. $y''' - 6y'' = 3 - \cos x$

22. $y''' - 2y'' - 4y' + 8y = 6xe^{2x}$

23. $y''' - 3y'' + 3y' - y = x - 4e^{x}$

24. $y''' - y'' - 4y' + 4y = 5 - e^{x} + e^{2x}$

25. $y^{(4)} + 2y'' + y = (x - 1)^2$

26. $y^{(4)} - y'' = 4x + 2xe^{-x}$

In Problems 27 and 28 use a trigonometric identity as an aid in finding a particular solution of the given differential equation.

27. $y'' + y = 8\sin^2 x$

28. $y'' + y = \sin x \cos 2x$

In Problems 29–40 solve the given differential equation subject to the indicated initial conditions.

29. $y'' + 4y = -2,\ y\left(\frac{\pi}{8}\right) = \frac{1}{2},\ y'\left(\frac{\pi}{8}\right) = 2$

30. $2y'' + 3y' - 2y = 14x^2 - 4x - 11,\ y(0) = 0,\ y'(0) = 0$

31. $5y'' + y' = -6x,\ y(0) = 0,\ y'(0) = -10$

32. $y'' + 4y' + 4y = (3 + x)e^{-2x},\ y(0) = 2,\ y'(0) = 5$

33. $y'' + 4y' + 5y = 35e^{-4x},\ y(0) = -3,\ y'(0) = 1$

34. $y'' - y = \cosh x,\ y(0) = 2,\ y'(0) = 12$

35. $\dfrac{d^2x}{dt^2} + \omega^2 x = F_0 \sin \omega t,\ x(0) = 0,\ x'(0) = 0$

36. $\dfrac{d^2x}{dt^2} + \omega^2 x = F_0 \cos \gamma t,\ x(0) = 0,\ x'(0) = 0$

37. $y'' + y = \cos x - \sin 2x,\ y\left(\frac{\pi}{2}\right) = 0,\ y'\left(\frac{\pi}{2}\right) = 0$

38. $y'' - 2y' - 3y = 2\cos^2 x,\ y(0) = -\frac{1}{3},\ y'(0) = 0$

39. $y''' - 2y'' + y' = 2 - 24e^{x} + 40e^{5x},\ y(0) = \frac{1}{2},\ y'(0) = \frac{5}{2},\ y''(0) = -\frac{9}{2}$

40. $y''' + 8y = 2x - 5 + 8e^{-2x},\ y(0) = -5,\ y'(0) = 3,\ y''(0) = -4$

In Problems 41 and 42 solve the given differential equation subject to the indicated boundary conditions.

41. $y'' + y = x^2 + 1,\ y(0) = 5,\ y(1) = 0$

42. $y'' - 2y' + 2y = 2x - 2,\ y(0) = 0,\ y(\pi) = \pi$

43. In applications the input function $g(x)$ is often discontinuous. Solve the initial-value problem $y'' + 4y = g(x),\ y(0) = 1,\ y'(0) = 2$, where

$$g(x) = \begin{cases} \sin x, & 0 \le x \le \pi/2 \\ 0, & x > \pi/2 \end{cases}$$

[*Hint*: Solve the problem on the two intervals and then find a solution so that y and y' are continuous at $x = \pi/2$.]

4.5 DIFFERENTIAL OPERATORS

• Differential operator • nth-order linear differential operator • Annihilator operator

In calculus, differentiation is often denoted by the capital letter D; that is,

$$\frac{dy}{dx} = Dy.$$

The symbol D is called a **differential operator,** and it transforms a differentiable function into another function; for example,

$$D(e^{4x}) = 4e^{4x}, \quad D(5x^3 - 6x^2) = 15x^2 - 12x, \quad D(\cos 2x) = -2\sin 2x.$$

The differential operator D also possesses a linearity property; D operating on a linear combination of two differentiable functions is the same as the linear combination of D operating on the individual functions. In symbols, this means

$$D\{af(x) + bg(x)\} = aDf(x) + bDg(x), \tag{1}$$

where a and b are constants. Because of (1) we say that D is a **linear differential operator.**

Higher-Order Derivatives Higher-order derivatives can be expressed in terms of D in a natural manner:

$$\frac{d}{dx}\left(\frac{dy}{dx}\right) = \frac{d^2y}{dx^2} = D(Dy) = D^2y \quad \text{and in general} \quad \frac{d^ny}{dx^n} = D^ny,$$

where y represents a sufficiently differentiable function. Polynomial expressions involving D such as

$$D + 3, \quad D^2 + 3D - 4, \quad \text{and} \quad 5D^3 - 6D^2 + 4D + 9$$

are also linear differential operators.

Differential Equations Any linear differential equation can be expressed in terms of the D notation. For example, a second-order differential equation with constant coefficients $ay'' + by' + cy = g(x)$ can be written as

$$aD^2y + bDy + cy = g(x) \quad \text{or} \quad (aD^2 + bD + c)y = g(x).$$

If we further define $L = aD^2 + bD + c$, then the last equation can be written compactly as

$$L(y) = g(x).$$

The operator $L = aD^2 + bD + c$ is said to be a **second-order linear differential operator with constant coefficients.**

EXAMPLE 1 Second-Order Differential Operator

The differential equation $y'' + y' + 2y = 5x - 3$ can be written as

$$(D^2 + D + 2)y = 5x - 3.$$

An ***n*th-order linear differential operator**

$$L = a_n D^n + a_{n-1} D^{n-1} + \cdots + a_1 D + a_0$$

with constant coefficients can be factored whenever the characteristic polynomial $a_n m^n + a_{n-1} m^{n-1} + \cdots + a_1 m + a_0$ factors.* For example, if we treat D as an algebraic quantity, then $D^2 + 5D + 6$ can be factored as $(D + 2)(D + 3)$ or as $(D + 3)(D + 2)$. In other words, for a function $y = f(x)$ possessing a second derivative

$$(D^2 + 5D + 6)y = (D + 2)(D + 3)y = (D + 3)(D + 2)y.$$

To see why this is so, let $w = (D + 3)y = y' + 3y$. Then

$$(D + 2)w = Dw + 2w = \underbrace{(y'' + 3y')}_{Dw} + \underbrace{(2y' + 6y)}_{2w} = y'' + 5y' + 6y.$$

Similarly, if we let $w = (D + 2)y = y' + 2y$, then

$$(D + 3)w = Dw + 3w = \underbrace{(y'' + 2y')}_{Dw} + \underbrace{(3y' + 6y)}_{3w} = y'' + 5y' + 6y.$$

This illustrates a general property:

Factors of a linear differential operator with constant coefficients commute.

EXAMPLE 2 Factoring a Differential Operator

(a) The operator $D^2 - 1$ can be written as

$$(D + 1)(D - 1) \quad \text{or} \quad (D - 1)(D + 1).$$

(b) The operator $D^2 + D + 2$ in Example 1 does not factor with real numbers.

EXAMPLE 3 Factoring a Differential Operator

The differential equation $y'' + 4y' + 4y = 0$ is written as

$$(D^2 + 4D + 4)y = 0 \quad \text{or} \quad (D + 2)(D + 2)y = 0 \quad \text{or} \quad (D + 2)^2 y = 0.$$

Annihilator Operator If L is a linear differential operator with constant coefficients and $y = f(x)$ is a sufficiently differentiable function such that

$$L(y) = 0,$$

then L is said to be an **annihilator** of the function. For example, if $y = k$ (a constant), then $Dk = 0$. Also, $D^2 x = 0$, $D^3 x^2 = 0$, and so on.

* If one is willing to use complex numbers, then a differential operator with constant coefficients can *always* be factored. We are, however, primarily concerned with writing differential equations in operator form with real coefficients.

The differential operator D^n annihilates each of the functions **(2)**

$$1, x, x^2, \dots, x^{n-1}.$$

As an immediate consequence of (2) and the fact that differentiation can be done term by term, a polynomial

$$c_0 + c_1x + \cdots + c_{n-1}x^{n-1}$$

can be annihilated by finding an operator that annihilates the highest power of x.

EXAMPLE 4 Annihilator Operator

Find a differential operator that annihilates $1 - 5x^2 + 8x^3$.

Solution From (2) we know that $D^4x^3 = 0$, and so it follows that

$$D^4(1 - 5x^2 + 8x^3) = 0.$$

Note The functions that are annihilated by an nth-order linear differential operator L are simply those functions that can be obtained from the general solution of the homogeneous differential equation $L(y) = 0$.

The differential operator $(D - \alpha)^n$ annihilates each of the functions **(3)**

$$e^{\alpha x}, xe^{\alpha x}, x^2e^{\alpha x}, \dots, x^{n-1}e^{\alpha x}.$$

To see this, note that the auxiliary equation of the homogeneous equation $(D - \alpha)^n y = 0$ is $(m - \alpha)^n = 0$. Since α is a root of multiplicity n, the general solution is

$$y = c_1e^{\alpha x} + c_2xe^{\alpha x} + \cdots + c_nx^{n-1}e^{\alpha x}. \quad \textbf{(4)}$$

EXAMPLE 5 Annihilator Operator

Find an annihilator operator for **(a)** e^{5x} and **(b)** $4e^{2x} - 6xe^{2x}$.

Solution

(a) From (3) with $\alpha = 5$ and $n = 1$, we see that

$$(D - 5)e^{5x} = 0.$$

(b) From (3) and (4) with $\alpha = 2$ and $n = 2$, we have

$$(D - 2)^2(4e^{2x} - 6xe^{2x}) = 0.$$

When α and β are real numbers, the quadratic formula reveals that $[m^2 - 2\alpha m + (\alpha^2 + \beta^2)]^n = 0$ has complex roots $\alpha + i\beta, \alpha - i\beta$, both of multiplicity n. From the discussion at the end of Section 4.3, we have the next result.

The differential operator $[D^2 - 2\alpha D + (\alpha^2 + \beta^2)]^n$ annihilates each of the functions

$$e^{\alpha x}\cos\beta x, xe^{\alpha x}\cos\beta x, x^2e^{\alpha x}\cos\beta x, \dots, x^{n-1}e^{\alpha x}\cos\beta x,$$
$$e^{\alpha x}\sin\beta x, xe^{\alpha x}\sin\beta x, x^2e^{\alpha x}\sin\beta x, \dots, x^{n-1}e^{\alpha x}\sin\beta x. \tag{5}$$

EXAMPLE 6 Using (5)

From (5), with $\alpha = -1, \beta = 2$, and $n = 1$, we see that

$$(D^2 + 2D + 5)e^{-x}\cos 2x = 0 \quad \text{and} \quad (D^2 + 2D + 5)e^{-x}\sin 2x = 0.$$

Since $y_1(x) = e^{-x}\cos 2x$ and $y_2(x) = e^{-x}\sin 2x$ are the two linearly independent functions in the general solution of $(D^2 + 2D + 5)y = 0$, the linear operator $D^2 + 2D + 5$ also annihilates any linear combination of these functions, such as $5e^{-x}\cos 2x - 9e^{-x}\sin 2x$.

EXAMPLE 7 Using (5)

From (5), with $\alpha = 0, \beta = 1$, and $n = 2$, it is seen that the differential operator $(D^2 + 1)^2$ or $D^4 + 2D^2 + 1$ annihilates $\cos x, x\cos x, \sin x$, and $x\sin x$. Moreover, $(D^2 + 1)^2$ annihilates any linear combination of these functions.

When $\alpha = 0$ and $n = 1$, a special case of (5) is

$$(D^2 + \beta^2)\begin{cases}\cos\beta x\\ \sin\beta x\end{cases} = 0. \tag{6}$$

We are often interested in annihilating the sum of two or more functions. As seen in Examples 4–7, if L is a linear differential operator such that $L(y_1) = 0$ and $L(y_2) = 0$, then L annihilates the linear combination $c_1y_1(x) + c_2y_2(x)$. This is a direct consequence of Theorem 4.3. Let us now suppose that L_1 and L_2 are linear differential operators with constant coefficients such that L_1 annihilates $y_1(x)$ and L_2 annihilates $y_2(x)$ but $L_1(y_2) \neq 0$ and $L_2(y_1) \neq 0$. Then the *product* of differential operators L_1L_2 annihilates the sum $c_1y_1(x) + c_2y_2(x)$. We can easily demonstrate this using linearity and the fact that $L_1L_2 = L_2L_1$:

$$\begin{aligned} L_1L_2(y_1 + y_2) &= L_1L_2(y_1) + L_1L_2(y_2) \\ &= L_2L_1(y_1) + L_1L_2(y_2) \\ &= L_2[\underbrace{L_1(y_1)}_{\text{zero}}] + L_1[\underbrace{L_2(y_2)}_{\text{zero}}] = 0. \end{aligned} \tag{7}$$

EXAMPLE 8 Using (2) and (6)

Find a differential operator that annihilates $7 - x + 6 \sin 3x$.

Solution From (2) and (6) we have, respectively,

$$D^2(7 - x) = 0 \quad \text{and} \quad (D^2 + 9)\sin 3x = 0.$$

It follows from (7) that the operator $D^2(D^2 + 9)$ annihilates the given linear combination.

EXAMPLE 9 Using (3)

Find a differential operator that annihilates $e^{-3x} + xe^x$.

Solution From (3),

$$(D + 3)e^{-3x} = 0 \quad \text{and} \quad (D - 1)^2 xe^x = 0.$$

Hence the product of the two operations $(D + 3)(D - 1)^2$ annihilates the given linear combination.

Remarks The differential operator that annihilates a function is not unique. For example, we know that $D - 5$ annihilates e^{5x}, but so do differential operators of higher order such as $(D - 5)(D + 1)$ and $(D - 5)D^2$. (Verify this.) When we seek a differential annihilator for a function $y = f(x)$, we want the operator of lowest possible order that does the job.

EXERCISES 4.5

Answers to odd-numbered problems begin on page A-8.

In Problems 1–6 write the given differential equation in the form $L(y) = g(x)$, where L is a differential operator with constant coefficients.

1. $\dfrac{dy}{dx} + 5y = 9 \sin x$

2. $4\dfrac{dy}{dx} + 8y = x + 3$

3. $3y'' - 5y' + y = e^x$

4. $y''' - 2y'' + 7y' - 6y = 1 - \sin x$

5. $y''' - 4y'' + 5y' = 4x$

6. $y^{(4)} - 2y'' + y = e^{-3x} + e^{2x}$

In Problems 7–16, if possible, factor the given differential operator.

7. $9D^2 - 4$

8. $D^2 - 5$

9. $D^2 - 4D - 12$

10. $2D^2 - 3D - 2$

11. $D^3 + 10D^2 + 25D$

12. $D^3 + 4D$

13. $D^3 + 2D^2 - 13D + 10$

14. $D^3 + 4D^2 + 3D$

15. $D^4 + 8D$

16. $D^4 - 8D^2 + 16$

In Problems 17–20 verify that the given differential operator annihilates the indicated function.

17. D^4; $y = 10x^3 - 2x$

18. $2D - 1$; $y = 4e^{x/2}$

19. $(D - 2)(D + 5)$; $y = 4e^{2x}$

20. $D^2 + 64$; $y = 2\cos 8x - 5\sin 8x$

In Problems 21–32 find a differential operator that annihilates the given function.

21. $1 + 6x - 2x^3$

22. $x^3(1 - 5x)$

23. $1 + 7e^{2x}$

24. $x + 3xe^{6x}$

25. $\cos 2x$

26. $1 + \sin x$

27. $13x + 9x^2 - \sin 4x$

28. $8x - \sin x + 10\cos 5x$

29. $e^{-x} + 2xe^x - x^2e^x$

30. $(2 - e^x)^2$

31. $3 + e^x\cos 2x$

32. $e^{-x}\sin x - e^{2x}\cos x$

In Problems 33–40 find linearly independent functions that are annihilated by the given differential operator.

33. D^5

34. $D^2 + 4D$

35. $(D - 6)(2D + 3)$

36. $D^2 - 9D - 36$

37. $D^2 + 5$

38. $D^2 - 6D + 10$

39. $D^3 - 10D^2 + 25D$

40. $D^2(D - 5)(D - 7)$

4.6 UNDETERMINED COEFFICIENTS— ANNIHILATOR APPROACH

• *Method of undetermined coefficients* • *Particular solution*

To obtain the general solution of a nonhomogeneous linear differential equation we must do two things:

(*i*) Find the complementary function y_c.
(*ii*) Find *any* particular solution y_p of the nonhomogeneous equation.

Recall from the discussion of Section 4.1 that a particular solution is any function, free of arbitrary constants, that satisfies the differential equation identically. The general solution of a nonhomogeneous equation on an interval is then $y = y_c + y_p$.

If L denotes a linear differential operator of the form $a_n D^n + a_{n-1}D^{n-1} + \cdots + a_1D + a_0$, then a nonhomogeneous linear differential equation can be written simply as

$$L(y) = g(x). \tag{1}$$

The **method of undetermined coefficients** presented in this section is limited to nonhomogeneous linear equations in which

- coefficients are constant and
- $g(x)$ is a constant k, a polynomial function, an exponential function $e^{\alpha x}$, $\sin \beta x$, $\cos \beta x$, or finite sums and products of these functions.

Note Strictly speaking, $g(x) = k$ (a constant) is a polynomial function. Since a constant function is probably not the first thing that comes to mind when you think of polynomial functions, for emphasis we continue to use the redundancy "constant functions, polynomials, ..."

The following are some examples of the types of input functions $g(x)$ that are appropriate for this discussion:

$$g(x) = 10, \quad g(x) = x^2 - 5x, \quad g(x) = 15x - 6 + 8e^{4x},$$

$$g(x) = \sin 3x - 5x \cos 2x, \quad g(x) = e^x \cos x - (3x^2 - 1)e^{-x},$$

and so on. In other words, $g(x)$ is a linear combination of functions of the form

$$k \text{ (constant)}, \quad x^m, \quad x^m e^{\alpha x}, \quad x^m e^{\alpha x} \cos \beta x, \quad \text{and} \quad x^m e^{\alpha x} \sin \beta x,$$

where m is a nonnegative integer and α and β are real numbers. The method of undetermined coefficients is not applicable to equations of the form (1) when

$$g(x) = \ln x, \quad g(x) = \frac{1}{x}, \quad g(x) = \tan x, \quad g(x) = \sin^{-1} x,$$

and so on. Differential equations with this latter kind of input function will be considered in Section 4.7.

As we saw in Section 4.5, a linear combination of functions of the type k, x^m, $x^m e^{\alpha x}$, $x^m e^{\alpha x} \cos \beta x$, and $x^m e^{\alpha x} \sin \beta x$ is precisely that kind of function that can be annihilated by an operator L_1 (of lowest order) consisting of a product of operators such as D^n, $(D - \alpha)^n$, and $(D^2 - 2\alpha D + \alpha^2 + \beta^2)^n$. Applying L_1 to both members of (1) yields

$$L_1 L(y) = L_1(g(x)) = 0. \tag{2}$$

By solving the *homogeneous higher-order* equation $L_1 L(y) = 0$, we can discover the form of a particular solution y_p for the original nonhomogeneous equation $L(y) = g(x)$.

The next several examples illustrate the method. The general solution of each equation is defined on the interval $(-\infty, \infty)$.

EXAMPLE 1 Using (2) of Section 4.5

Solve
$$\frac{d^2y}{dx^2} + 3\frac{dy}{dx} + 2y = 4x^2. \tag{3}$$

Solution **Step 1.** We first solve the homogeneous equation

$$\frac{d^2y}{dx^2} + 3\frac{dy}{dx} + 2y = 0.$$

From the auxiliary equation $m^2 + 3m + 2 = (m + 1)(m + 2) = 0$, we find $m_1 = -1, m_2 = -2$, and so the complementary function is

$$y_c = c_1e^{-x} + c_2e^{-2x}.$$

Step 2. In view of (2) of Section 4.5, (3) can be rendered homogeneous by taking three derivatives of each side of the equation. In other words,

$$D^3(D^2 + 3D + 2)y = 4D^3x^2 = 0, \tag{4}$$

since $D^3x^2 = 0$. The auxiliary equation of (4),

$$m^3(m^2 + 3m + 2) = 0 \qquad \text{or} \qquad m^3(m + 1)(m + 2) = 0,$$

has roots 0, 0, 0, -1, and -2. Thus its general solution must be

$$y = c_1 + c_2x + c_3x^2 + \boxed{c_4e^{-x} + c_5e^{-2x}}. \tag{5}$$

The terms in the box in (5) constitute the complementary function of the original equation (3). We can then argue that a particular solution y_p of (3) should also satisfy equation (4). This means that the terms remaining in (5) must be the basic structure of y_p:

$$y_p = A + Bx + Cx^2, \tag{6}$$

where, for convenience, we have replaced c_1, c_2, and c_3 by A, B, and C, respectively. For (6) to be a particular solution of (3), it is necessary to find *specific* coefficients A, B, and C. Differentiating (6), we have

$$y_p' = B + 2Cx, \quad y_p'' = 2C,$$

and substitution into (3) then gives

$$y_p'' + 3y_p' + 2y_p = 2C + 3B + 6Cx + 2A + 2Bx + 2Cx^2 = 4x^2.$$

Since the last equation is supposed to be an identity, the coefficients of like powers of x must be equal:

equal

$$\boxed{2C}\,x^2 + \boxed{2B + 6C}\,x + \boxed{2A + 3B + 2C} = 4x^2 + 0x + 0.$$

That is,

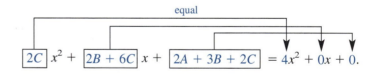

$$2C = 4, \quad 2B + 6C = 0, \quad 2A + 3B + 2C = 0. \tag{7}$$

Solving (7) gives $A = 7$, $B = -6$, and $C = 2$. Thus $y_p = 7 - 6x + 2x^2$.

Step 3. The general solution of (3) is $y = y_c + y_p$ or

$$y = c_1e^{-x} + c_2e^{-2x} + 7 - 6x + 2x^2.$$

EXAMPLE 2 Using (3) and (6) of Section 4.5

Solve

$$y'' - 3y' = 8e^{3x} + 4\sin x. \tag{8}$$

Solution **Step 1.** The auxiliary equation for the homogeneous equation $y'' - 3y' = 0$ is $m(m - 3) = 0$, and so

$$y_c = c_1 + c_2e^{3x}.$$

Step 2. Now since $(D - 3)e^{3x} = 0$ and $(D^2 + 1)\sin x = 0$, we apply the differential operator $(D - 3)(D^2 + 1)$ to both sides of (8):

$$(D - 3)(D^2 + 1)(D^2 - 3D)y = 0. \tag{9}$$

The auxiliary equation of (9) is

$$(m - 3)(m^2 + 1)(m^2 - 3m) = 0 \quad \text{or} \quad m(m - 3)^2(m^2 + 1) = 0.$$

Thus
$$y = \boxed{c_1 + c_2e^{3x}} + c_3xe^{3x} + c_4\cos x + c_5\sin x. \tag{10}$$

After excluding the linear combination of terms in the box that corresponds to y_c, we arrive at the form of y_p:

$$y_p = Axe^{3x} + B\cos x + C\sin x.$$

Substituting y_p in (8) and simplifying yields

$$\begin{aligned} y_p'' - 3y_p' &= 3Ae^{3x} + (-B - 3C)\cos x + (3B - C)\sin x \\ &= 8e^{3x} + 4\sin x. \end{aligned}$$

Equating coefficients gives

$$3A = 8, \quad -B - 3C = 0, \quad 3B - C = 4.$$

We find $A = \frac{8}{3}$, $B = \frac{6}{5}$, $C = -\frac{2}{5}$, and consequently

$$y_p = \frac{8}{3}xe^{3x} + \frac{6}{5}\cos x - \frac{2}{5}\sin x.$$

Step 3. The general solution of (8) is then

$$y = c_1 + c_2e^{3x} + \frac{8}{3}xe^{3x} + \frac{6}{5}\cos x - \frac{2}{5}\sin x.$$

EXAMPLE 3 Using (2) and (3) of Section 4.5

Solve
$$y'' + 8y = 5x + 2e^{-x}. \tag{11}$$

Solution From (2) and (3) of Section 4.5 we know that $D^2x = 0$ and $(D + 1)e^{-x} = 0$, respectively. Hence we apply $D^2(D + 1)$ to (11):

$$D^2(D + 1)(D^2 + 8)y = 0.$$

It is seen that

$$y = \boxed{c_1\cos 2\sqrt{2}x + c_2\sin 2\sqrt{2}x} + c_3 + c_4x + c_5e^{-x}$$

and so
$$y_p = A + Bx + Ce^{-x}.$$

Substituting y_p into (11) yields

$$y_p'' + 8y_p = 8A + 8Bx + 9Ce^{-x} = 5x + 2e^{-x}.$$

This implies $A = 0, B = \frac{5}{8}$, and $C = \frac{2}{9}$, so the general solution of (1) is

$$y = c_1\cos 2\sqrt{2}x + c_2\sin 2\sqrt{2}x + \frac{5}{8}x + \frac{2}{9}e^{-x}.$$

EXAMPLE 4 Using (5) of Section 4.5

Solve
$$y'' + y = x\cos x - \cos x. \tag{12}$$

Solution In Example 7 of Section 4.5 we saw that $x\cos x$ and $\cos x$ are annihilated by the operator $(D^2 + 1)^2$. Thus

$$(D^2 + 1)^2(D^2 + 1)y = 0 \quad \text{or} \quad (D^2 + 1)^3 y = 0.$$

Since i and $-i$ are both complex roots of multiplicity 3 of the auxiliary equation of the last differential equation, we conclude

$$y = c_1\cos x + c_2\sin x + c_3 x\cos x + c_4 x\sin x + c_5 x^2\cos x + c_6 x^2\sin x.$$

We substitute

$$y_p = Ax\cos x + Bx\sin x + Cx^2\cos x + Ex^2\sin x$$

into (12) and simplify:

$$\begin{aligned} y_p'' + y_p &= 4Ex\cos x - 4Cx\sin x + (2B + 2C)\cos x + (-2A + 2E)\sin x \\ &= x\cos x - \cos x. \end{aligned}$$

Equating coefficients gives the equations

$$4E = 1, \quad -4C = 0, \quad 2B + 2C = -1, \quad -2A + 2E = 0,$$

from which we find $E = \frac{1}{4}$, $C = 0$, $B = -\frac{1}{2}$, and $A = \frac{1}{4}$. Hence the general solution of (12) is

$$y = c_1\cos x + c_2\sin x + \frac{1}{4}x\cos x - \frac{1}{2}x\sin x + \frac{1}{4}x^2\sin x.$$

EXAMPLE 5 Form of a Particular Solution

Determine the form of a particular solution for

$$y'' - 2y' + y = 10e^{-2x}\cos x. \tag{13}$$

Solution The complementary function for the given equation is $y_c = c_1 e^x + c_2 x e^x$.

Now from (5) of Section 4.5, with $\alpha = -2$, $\beta = 1$, and $n = 1$, we know that

$$(D^2 + 4D + 5)e^{-2x}\cos x = 0.$$

Applying the operator $D^2 + 4D + 5$ to (13) gives

$$(D^2 + 4D + 5)(D^2 - 2D + 1)y = 0. \tag{14}$$

Since the roots of the auxiliary equation of (14) are $-2 - i$, $-2 + i$, 1, and 1,

$$y = c_1 e^x + c_2 x e^x + c_3 e^{-2x}\cos x + c_4 e^{-2x}\sin x.$$

Hence a particular solution of (13) can be found with the form

$$y_p = Ae^{-2x}\cos x + Be^{-2x}\sin x.$$

EXAMPLE 6 Form of a Particular Solution

Determine the form of a particular solution for

$$y''' - 4y'' + 4y' = 5x^2 - 6x + 4x^2e^{2x} + 3e^{5x}. \tag{15}$$

Solution Observe that

$$D^3(5x^2 - 6x) = 0, \quad (D - 2)^3x^2e^{2x} = 0, \quad \text{and} \quad (D - 5)e^{5x} = 0.$$

Therefore $D^3(D - 2)^3(D - 5)$ applied to (15) gives

$$D^3(D - 2)^3(D - 5)(D^3 - 4D^2 + 4D)y = 0$$

or

$$D^4(D - 2)^5(D - 5)y = 0.$$

The roots of the auxiliary equation for the last differential equation are easily seen to be 0, 0, 0, 0, 2, 2, 2, 2, 2, and 5. Hence

$$y = c_1 + c_2x + c_3x^2 + c_4x^3 + c_5e^{2x} + c_6xe^{2x} + c_7x^2e^{2x} + c_8x^3e^{2x} + c_9x^4e^{2x} + c_{10}e^{5x}. \quad \textbf{(16)}$$

Since the linear combination $c_1 + c_5e^{2x} + c_6xe^{2x}$ can be taken as the complementary function of (15), the remaining terms in (16) give the form of a particular solution of the differential equation:

$$y_p = Ax + Bx^2 + Cx^3 + Ex^2e^{2x} + Fx^3e^{2x} + Gx^4e^{2x} + He^{5x}.$$

Summary of the Method For your convenience the method of undetermined coefficients is summarized here.

UNDETERMINED COEFFICIENTS—ANNIHILATOR APPROACH

The differential equation $L(y) = g(x)$ has constant coefficients and the function $g(x)$ consists of finite sums and products of constants, polynomials, exponential functions $e^{\alpha x}$, sines, and cosines.

(i) Find the complementary solution y_c for the homogeneous equation $L(y) = 0$.

(ii) Operate on both sides of the nonhomogeneous equation $L(y) = g(x)$ with a differential operator L_1 that annihilates the function $g(x)$.

(iii) Find the general solution of the higher-order homogeneous differential equation $L_1L(y) = 0$.

(iv) Delete all those terms from the solution in step *(iii)* that are duplicated in the complementary solution y_c found in step *(i)*. Form a linear combination y_p of the terms that remain. This is the form of a particular solution of $L(y) = g(x)$.

(v) Substitute y_p found in step *(iv)* into $L(y) = g(x)$. Match coefficients of the various functions on each side of the equality and solve the resulting system of equations for the unknown coefficients in y_p.

(vi) With the particular solution found in step *(v)*, form the general solution $y = y_c + y_p$ of the given differential equation.

EXERCISES 4.6

Answers to odd-numbered problems begin on page A-8.

In Problems 1–32 solve the given differential equation by the method of undetermined coefficients.

1. $y'' - 9y = 54$

2. $2y'' - 7y' + 5y = -29$

3. $y'' + y' = 3$

4. $y''' + 2y'' + y' = 10$

5. $y'' + 4y' + 4y = 2x + 6$

6. $y'' + 3y' = 4x - 5$

7. $y''' + y'' = 8x^2$

8. $y'' - 2y' + y = x^3 + 4x$

9. $y'' - y' - 12y = e^{4x}$

10. $y'' + 2y' + 2y = 5e^{6x}$

11. $y'' - 2y' - 3y = 4e^x - 9$

12. $y'' + 6y' + 8y = 3e^{-2x} + 2x$

13. $y'' + 25y = 6 \sin x$

14. $y'' + 4y = 4 \cos x + 3 \sin x - 8$

15. $y'' + 6y' + 9y = -xe^{4x}$

16. $y'' + 3y' - 10y = x(e^x + 1)$

17. $y'' - y = x^2e^x + 5$

18. $y'' + 2y' + y = x^2e^{-x}$

19. $y'' - 2y' + 5y = e^x \sin x$

20. $y'' + y' + \frac{1}{4}y = e^x(\sin 3x - \cos 3x)$

21. $y'' + 25y = 20 \sin 5x$

22. $y'' + y = 4 \cos x - \sin x$

23. $y'' + y' + y = x \sin x$

24. $y'' + 4y = \cos^2 x$

25. $y''' + 8y'' = -6x^2 + 9x + 2$

26. $y''' - y'' + y' - y = xe^x - e^{-x} + 7$

27. $y''' - 3y'' + 3y' - y = e^x - x + 16$

28. $2y''' - 3y'' - 3y' + 2y = (e^x + e^{-x})^2$

29. $y^{(4)} - 2y''' + y'' = e^x + 1$

30. $y^{(4)} - 4y'' = 5x^2 - e^{2x}$

31. $16y^{(4)} - y = e^{x/2}$

32. $y^{(4)} - 5y'' + 4y = 2 \cosh x - 6$

In Problems 33–40 solve the given differential equation subject to the indicated initial conditions.

33. $y'' - 64y = 16, \quad y(0) = 1, \ y'(0) = 0$

34. $y'' + y' = x, \quad y(0) = 1, \ y'(0) = 0$

35. $y'' - 5y' = x - 2, \quad y(0) = 0, \ y'(0) = 2$

36. $y'' + 5y' - 6y = 10e^{2x}, \quad y(0) = 1, \ y'(0) = 1$

37. $y'' + y = 8 \cos 2x - 4 \sin x, \quad y(\pi/2) = -1, \ y'(\pi/2) = 0$

38. $y''' - 2y'' + y' = xe^x + 5, \quad y(0) = 2, \ y'(0) = 2, \ y''(0) = -1$

39. $y'' - 4y' + 8y = x^3, \quad y(0) = 2, \ y'(0) = 4$

40. $y^{(4)} - y''' = x + e^x, \quad y(0) = 0, \ y'(0) = 0, \ y''(0) = 0, \ y'''(0) = 0$

In Problems 41 and 42 determine the form of a particular solution for the given differential equation.

41. $y'' - y = e^x(2 + 3x \cos 2x)$

42. $y'' + y' = 9 - e^{-x} + x^2 \sin x$

43. Show that the operator $(xD - 1)(D + 4)$ is not the same as the operator $(D + 4)(xD - 1)$.

44. Prove that the differential equation

$$a_n y^{(n)} + a_{n-1} y^{(n-1)} + \cdots + a_1 y' + a_0 y = k,$$

k a constant, $a_0 \neq 0$, has the particular solution $y_p = k/a_0$.

4.7 VARIATION OF PARAMETERS

• Variation of parameters • Wronskian • Particular solution

Linear First-Order Equation Revisited In Chapter 2 we saw that the general solution of the linear first-order differential equation

$$\frac{dy}{dx} + P(x)y = f(x), \tag{1}$$

where $P(x)$ and $f(x)$ are continuous on an interval I, is

$$y = e^{-\int P(x)\,dx}\int e^{\int P(x)\,dx} f(x)\,dx + c_1 e^{-\int P(x)\,dx}. \tag{2}$$

Now (2) has the form $y = y_c + y_p$, where $y_c = c_1 e^{-\int P(x)\,dx}$ is a solution of

$$\frac{dy}{dx} + P(x)y = 0 \tag{3}$$

and

$$y_p = e^{-\int P(x)\,dx}\int e^{\int P(x)\,dx} f(x)\,dx \tag{4}$$

is a particular solution of (1). As a means of motivating an additional method for solving nonhomogeneous linear equations of higher order, we rederive (4) by a method known as **variation of parameters.** The basic procedure is essentially that used in Section 4.2.

Suppose y_1 is a known solution of (3); that is,

$$\frac{dy_1}{dx} + P(x)y_1 = 0.$$

We proved in Section 2.5 that $y_1 = e^{-\int P(x)\,dx}$ is a solution, and since the differential equation is linear, its general solution is $y = c_1 y_1(x)$. Variation of parameters consists of finding a function u_1 such that

$$y_p = u_1(x)y_1(x)$$

is a particular solution of (1). In other words, we replace the parameter c_1 by a variable u_1.

Substituting $y_p = u_1 y_1$ into (1) gives

$$\frac{d}{dx}[u_1 y_1] + P(x)u_1 y_1 = f(x)$$

$$u_1\frac{dy_1}{dx} + y_1\frac{du_1}{dx} + P(x)u_1 y_1 = f(x)$$

$$u_1\underbrace{\left[\frac{dy_1}{dx} + P(x)y_1\right]}_{\text{zero}} + y_1\frac{du_1}{dx} = f(x),$$

so

$$y_1\frac{du_1}{dx} = f(x).$$

By separating variables, we find that

$$du_1 = \frac{f(x)}{y_1(x)}\,dx \quad\text{and}\quad u_1 = \int \frac{f(x)}{y_1(x)}\,dx,$$

from which it follows that

$$y = u_1 y_1 = y_1 \int \frac{f(x)}{y_1(x)}\, dx.$$

From the definition of y_1 we see that the last result is identical to (4).

Second-Order Equations To adapt the foregoing procedure to a linear second-order differential equation

$$a_2(x)y'' + a_1(x)y' + a_0(x)y = g(x), \tag{5}$$

we put (5) in the standard form

$$y'' + P(x)y' + Q(x)y = f(x) \tag{6}$$

by dividing through by $a_2(x)$. Here we assume that $P(x)$, $Q(x)$, and $f(x)$ are continuous on some interval I. Equation (6) is the analogue of (1). As we know, when $P(x)$ and $Q(x)$ are constants, we have absolutely no difficulty in writing y_c.

Suppose y_1 and y_2 form a fundamental set of solutions on I of the associated homogeneous form of (6); that is,

$$y_1'' + P(x)y_1' + Q(x)y_1 = 0 \quad \text{and} \quad y_2'' + P(x)y_2' + Q(x)y_2 = 0.$$

Now we ask: Can two functions u_1 and u_2 be found so that

$$y_p = u_1(x)y_1(x) + u_2(x)y_2(x)$$

is a particular solution of (6)? Notice that our assumption for y_p is the same as $y_c = c_1y_1 + c_2y_2$, but we have replaced c_1 and c_2 by the "variable parameters" u_1 and u_2. Because we seek to determine two unknown functions, reason dictates that we need two equations. As in the introductory discussion leading to the discovery of (4), one of these equations results from substituting $y_p = u_1y_1 + u_2y_2$ into the given differential equation (6). The other equation that we impose is

$$y_1u_1' + y_2u_2' = 0. \tag{7}$$

This equation is an assumption that is made to simplify the first derivative and, thereby, the second derivative of y_p. Using the product rule to differentiate y_p, we get

$$y_p' = u_1y_1' + y_1u_1' + u_2y_2' + y_2u_2' = u_1y_1' + u_2y_2' + \overbrace{y_1u_1' + y_2u_2'}^{\text{zero from (7)}} \tag{8}$$

and so $\quad y_p' = u_1y_1' + u_2y_2'.$

Continuing, we find $\quad y_p'' = u_1y_1'' + y_1'u_1' + u_2y_2'' + y_2'u_2'.$

Substitution of these results in (6) yields

$$\begin{aligned} y_p'' + Py_p' + Qy_p &= u_1y_1'' + y_1'u_1' + u_2y_2'' + y_2'u_2' + Pu_1y_1' + Pu_2y_2' + Qu_1y_1 + Qu_2y_2 \\ &= u_1[\underbrace{y_1'' + Py_1' + Qy_1}_{\text{zero}}] + u_2[\underbrace{y_2'' + Py_2' + Qy_2}_{\text{zero}}] + y_1'u_1' + y_2'u_2' = f(x). \end{aligned}$$

In other words, u_1 and u_2 must be functions that also satisfy the condition

$$y_1'u_1' + y_2'u_2' = f(x). \tag{9}$$

Equations (7) and (9) constitute a linear system of equations for determining the derivatives u_1' and u_2'. By Cramer's rule,* the solution of

$$\begin{aligned} y_1u_1' + y_2u_2' &= 0 \\ y_1'u_1' + y_2'u_2' &= f(x) \end{aligned}$$

* See Appendix III for a review of Cramer's rule.

can be expressed in terms of determinants:

$$u_1' = \frac{W_1}{W} \quad \text{and} \quad u_2' = \frac{W_2}{W}, \tag{10}$$

where

$$W = \begin{vmatrix} y_1 & y_2 \\ y_1' & y_2' \end{vmatrix}, \quad W_1 = \begin{vmatrix} 0 & y_2 \\ f(x) & y_2' \end{vmatrix}, \quad \text{and} \quad W_2 = \begin{vmatrix} y_1 & 0 \\ y_1' & f(x) \end{vmatrix}. \tag{11}$$

The determinant W is recognized as the Wronskian of y_1 and y_2. By linear independence of y_1 and y_2 on I, we know that $W(y_1(x), y_2(x)) \neq 0$ for every x in the interval.

Summary of the Method Usually it is not a good idea to memorize formulas in lieu of understanding a procedure. However, the foregoing procedure is too long and complicated to use each time we wish to solve a differential equation. In this case it is more efficient to simply use the formulas in (10). Thus, to solve $a_2y'' + a_1y' + a_0y = g(x)$, first find the complementary function $y_c = c_1y_1 + c_2y_2$ and then compute the Wronskian

$$W = \begin{vmatrix} y_1 & y_2 \\ y_1' & y_2' \end{vmatrix}.$$

By dividing by a_2, we put the equation into the form $y'' + Py' + Qy = f(x)$ to determine $f(x)$. We find u_1 and u_2 by integrating $u_1' = W_1/W$ and $u_2' = W_2/W$, where W_1 and W_2 are defined in (11). A particular solution is $y_p = u_1y_1 + u_2y_2$. The general solution of the equation is then $y = y_c + y_p$.

EXAMPLE 1 General Solution

Solve $y'' - 4y' + 4y = (x + 1)e^{2x}$.

Solution Since the auxiliary equation is $m^2 - 4m + 4 = (m - 2)^2 = 0$, we have $y_c = c_1e^{2x} + c_2xe^{2x}$. Identifying $y_1 = e^{2x}$ and $y_2 = xe^{2x}$, we next compute the Wronskian

$$W(e^{2x}, xe^{2x}) = \begin{vmatrix} e^{2x} & xe^{2x} \\ 2e^{2x} & 2xe^{2x} + e^{2x} \end{vmatrix} = e^{4x}.$$

Since the given differential equation is already in form (6) (that is, the coefficient of y'' is 1), we identify $f(x) = (x + 1)e^{2x}$. From (11) we obtain

$$W_1 = \begin{vmatrix} 0 & xe^{2x} \\ (x+1)e^{2x} & 2xe^{2x} + e^{2x} \end{vmatrix} = -(x + 1)xe^{4x}$$

$$W_2 = \begin{vmatrix} e^{2x} & 0 \\ 2e^{2x} & (x+1)e^{2x} \end{vmatrix} = (x + 1)e^{4x}$$

and so, from (10),

$$u_1' = -\frac{(x + 1)xe^{4x}}{e^{4x}} = -x^2 - x, \qquad u_2' = \frac{(x + 1)e^{4x}}{e^{4x}} = x + 1.$$

It follows that

$$u_1 = -\frac{x^3}{3} - \frac{x^2}{2} \quad \text{and} \quad u_2 = \frac{x^2}{2} + x.$$

Therefore
$$y_p = \left(-\frac{x^3}{3} - \frac{x^2}{2}\right)e^{2x} + \left(\frac{x^2}{2} + x\right)xe^{2x}$$
$$= \left(\frac{x^3}{6} + \frac{x^2}{2}\right)e^{2x}.$$

Hence
$$y = y_c + y_p = c_1e^{2x} + c_2xe^{2x} + \left(\frac{x^3}{6} + \frac{x^2}{2}\right)e^{2x}.$$

EXAMPLE 2 General Solution

Solve $4y'' + 36y = \csc 3x$.

Solution We first put the equation in the standard form (6) by dividing by 4:

$$y'' + 9y = \frac{1}{4}\csc 3x.$$

Since the roots of the auxiliary equation $m^2 + 9 = 0$ are $m_1 = 3i$ and $m_2 = -3i$, the complementary function is $y_c = c_1 \cos 3x + c_2 \sin 3x$. Using $y_1 = \cos 3x, y_2 = \sin 3x$, and $f(x) = \frac{1}{4}\csc 3x$, we find

$$W(\cos 3x, \sin 3x) = \begin{vmatrix} \cos 3x & \sin 3x \\ -3\sin 3x & 3\cos 3x \end{vmatrix} = 3$$
$$W_1 = \begin{vmatrix} 0 & \sin 3x \\ \frac{1}{4}\csc 3x & 3\cos 3x \end{vmatrix} = -\frac{1}{4} \quad \leftarrow \csc 3x = \frac{1}{\sin 3x}$$
$$W_2 = \begin{vmatrix} \cos 3x & 0 \\ -3\sin 3x & \frac{1}{4}\csc 3x \end{vmatrix} = \frac{1}{4}\frac{\cos 3x}{\sin 3x}.$$

Integrating

$$u_1' = \frac{W_1}{W} = -\frac{1}{12} \quad \text{and} \quad u_2' = \frac{W_2}{W} = \frac{1}{12}\frac{\cos 3x}{\sin 3x}$$

gives
$$u_1 = -\frac{1}{12}x \quad \text{and} \quad u_2 = \frac{1}{36}\ln|\sin 3x|.$$

Thus a particular solution is

$$y_p = -\frac{1}{12}x\cos 3x + \frac{1}{36}(\sin 3x)\ln|\sin 3x|.$$

The general solution of the equation is

$$y = y_c + y_p = c_1\cos 3x + c_2\sin 3x - \frac{1}{12}x\cos 3x + \frac{1}{36}(\sin 3x)\ln|\sin 3x|. \quad \textbf{(12)}$$

Equation (12) represents the general solution of the differential equation on, say, the interval $(0, \pi/6)$.

Constants of Integration When computing the indefinite integrals of u_1' and u_2', we need not introduce any constants. This is because

$$\begin{aligned} y = y_c + y_p &= c_1y_1 + c_2y_2 + (u_1 + a_1)y_1 + (u_2 + b_1)y_2 \\ &= (c_1 + a_1)y_1 + (c_2 + b_1)y_2 + u_1y_1 + u_2y_2 \\ &= C_1y_1 + C_2y_2 + u_1y_1 + u_2y_2. \end{aligned}$$

EXAMPLE 3 Using Nonelementary Integrals

Solve $y'' - y = \dfrac{1}{x}$.

Solution The auxiliary equation $m^2 - 1 = 0$ yields $m_1 = -1$ and $m_2 = 1$. Therefore $y_c = c_1e^x + c_2e^{-x}$ and

$$W(e^x, e^{-x}) = \begin{vmatrix} e^x & e^{-x} \\ e^x & -e^{-x} \end{vmatrix} = -2$$

$$u_1' = -\frac{e^{-x}(1/x)}{-2}, \quad u_1 = \frac{1}{2}\int_{x_0}^{x} \frac{e^{-t}}{t}\,dt$$

$$u_2' = \frac{e^x(1/x)}{-2}, \quad u_2 = -\frac{1}{2}\int_{x_0}^{x} \frac{e^{t}}{t}\,dt.$$

It is well known that the integrals defining u_1 and u_2 cannot be expressed in terms of elementary functions. Hence we write

$$y_p = \frac{1}{2}e^x\int_{x_0}^{x} \frac{e^{-t}}{t}\,dt - \frac{1}{2}e^{-x}\int_{x_0}^{x} \frac{e^{t}}{t}\,dt,$$

and so

$$y = y_c + y_p = c_1e^x + c_2e^{-x} + \frac{1}{2}e^x\int_{x_0}^{x} \frac{e^{-t}}{t}\,dt - \frac{1}{2}e^{-x}\int_{x_0}^{x} \frac{e^{t}}{t}\,dt.$$

In Example 3 we can integrate on any interval $x_0 \le t \le x$ not containing the origin.

Higher-Order Equations The method we have just examined for nonhomogeneous second-order differential equations can be generalized to nth-order linear equations that have been put into the form

$$y^{(n)} + P_{n-1}(x)y^{(n-1)} + \cdots + P_1(x)y' + P_0(x)y = f(x). \qquad \textbf{(13)}$$

If $y = c_1y_1 + c_2y_2 + \cdots + c_ny_n$ is the complementary function for (13), then a particular solution is

$$y_p = u_1(x)y_1(x) + u_2(x)y_2(x) + \cdots + u_n(x)y_n(x),$$

where the u_k', $k = 1, 2, \ldots, n$ are determined by the n equations

$$\begin{aligned} y_1u_1' + y_2u_2' + \cdots + y_nu_n' &= 0 \\ y_1'u_1' + y_2'u_2' + \cdots + y_n'u_n' &= 0 \\ \vdots \qquad\qquad & \quad \vdots \\ y_1^{(n-1)}u_1' + y_2^{(n-1)}u_2' + \cdots + y_n^{(n-1)}u_n' &= f(x). \end{aligned}$$

The first $n - 1$ equations in this system, like (7), are assumptions made to simplify the first $n - 1$ derivatives of y_p. The last equation of the system results from substituting the nth derivative of y_p and the simplified lower derivatives into (13). In this case, Cramer's rule gives

$$u'_k = \frac{W_k}{W}, \quad k = 1, 2, \ldots, n,$$

where W is the Wronskian of $y_1, y_2, \ldots, y_n$, and W_k is the determinant obtained by replacing the kth column of the Wronskian by the column whose entries are $0, 0, \ldots, 0, f(x)$. When $n = 2$, we get (10) and (11).

Remarks (*i*) Variation of parameters has a distinct advantage over the method of undetermined coefficients in that it will *always* yield a particular solution y_p, provided the related homogeneous equation can be solved. The present method is not limited to a function $f(x)$, which is a combination of the four types of functions listed on page 146. Also, variation of parameters, unlike undetermined coefficients, is applicable to differential equations with variable coefficients.

In the problems that follow do not hesitate to simplify the form of y_p. Depending on how the antiderivatives of u'_1 and u'_2 are found, you may not obtain the same y_p as given in the answer section. For example, in Problem 3, both $y_p = \frac{1}{2}\sin x - \frac{1}{2}x\cos x$ and $y_p = \frac{1}{4}\sin x - \frac{1}{2}x\cos x$ are valid answers. In either case, the general solution $y = y_c + y_p$ simplifies to $y = c_1\cos x + c_2\sin x - \frac{1}{2}x\cos x$. Why?
(*ii*) In Problems 25–28 you are asked to solve initial-value problems. Be sure to apply the initial conditions to the general solution $y = y_c + y_p$. Students often make the mistake of applying the initial conditions to only the complementary function y_c since it is that part of the solution that contains the constants. Review Example 8 in Section 4.4 for the correct procedure.

EXERCISES 4.7

Answers to odd-numbered problems begin on page A-9.

In Problems 1–24 solve each differential equation by variation of parameters. State an interval on which the general solution is defined.

1. $y'' + y = \sec x$

2. $y'' + y = \tan x$

3. $y'' + y = \sin x$

4. $y'' + y = \sec x \tan x$

5. $y'' + y = \cos^2 x$

6. $y'' + y = \sec^2 x$

7. $y'' - y = \cosh x$

8. $y'' - y = \sinh 2x$

9. $y'' - 4y = e^{2x}/x$

10. $y'' - 9y = 9x/e^{3x}$

11. $y'' + 3y' + 2y = 1/(1 + e^x)$

12. $y'' - 3y' + 2y = e^{3x}/(1 + e^x)$

13. $y'' + 3y' + 2y = \sin e^x$

14. $y'' - 2y' + y = e^x \arctan x$

15. $y'' - 2y' + y = e^x/(1 + x^2)$

16. $y'' - 2y' + 2y = e^x \sec x$

17. $y'' + 2y' + y = e^{-x}\ln x$

18. $y'' + 10y' + 25y = e^{-10x}/x^2$

19. $3y'' - 6y' + 30y = e^x \tan 3x$

20. $4y'' - 4y' + y = e^{x/2}\sqrt{1 - x^2}$

21. $y''' + y' = \tan x$

22. $y''' + 4y' = \sec 2x$

23. $y''' - 2y'' - y' + 2y = e^{3x}$

24. $2y''' - 6y'' = x^2$

In Problems 25–28 solve each differential equation by variation of parameters subject to the initial conditions $y(0) = 1$, $y'(0) = 0$.

25. $4y'' - y = xe^{x/2}$

26. $2y'' + y' - y = x + 1$

27. $y'' + 2y' - 8y = 2e^{-2x} - e^{-x}$

28. $y'' - 4y' + 4y = (12x^2 - 6x)e^{2x}$

29. Given that $y_1 = x$ and $y_2 = x \ln x$, form a fundamental set of solutions of $x^2y'' - xy' + y = 0$ on $(0, \infty)$. Find the general solution of

$$x^2y'' - xy' + y = 4x \ln x.$$

30. Given that $y_1 = x^2$ and $y_2 = x^3$, form a fundamental set of solutions of $x^2y'' - 4xy' + 6y = 0$ on $(0, \infty)$. Find the general solution of

$$x^2y'' - 4xy' + 6y = \frac{1}{x}.$$

31. Given that $y_1 = x^{-1/2} \cos x$ and $y_2 = x^{-1/2} \sin x$, form a fundamental set of solutions of $x^2y'' + xy' + (x^2 - \frac{1}{4})y = 0$ on $(0, \infty)$. Find the general solution of

$$x^2y'' + xy' + (x^2 - \tfrac{1}{4})y = x^{3/2}.$$

32. Given that $y_1 = \cos(\ln x)$ and $y_2 = \sin(\ln x)$ are known linearly independent solutions of $x^2y'' + xy' + y = 0$ on $(0, \infty)$:

(a) Find a particular solution of

$$x^2y'' + xy' + y = \sec(\ln x).$$

(b) Give the general solution of the equation and state an interval of validity. [*Hint*: It is *not* $(0, \infty)$. Why?]

33. **(a)** Use undetermined coefficients to find a particular solution of

$$y'' + 2y' + y = 4x^2 - 3.$$

(b) Use variation of parameters to find a particular solution of

$$y'' + 2y' + y = \frac{e^{-x}}{x}.$$

(c) Use the superposition principle (Theorem 4.9) to find a particular solution of

$$y'' + 2y' + y = 4x^2 - 3 + \frac{e^{-x}}{x}.$$

34. Use the method outlined in Problem 33 to find a particular solution of

$$y'' + y = 2x - e^{3x} + \cot x.$$

CHAPTER 4 REVIEW

We summarize the important results of this chapter for **linear second-order** differential equations.

The equation

$$a_2(x)y'' + a_1(x)y' + a_0(x)y = 0 \tag{1}$$

is said to be **homogeneous,** whereas

$$a_2(x)y'' + a_1(x)y' + a_0(x)y = g(x), \tag{2}$$

$g(x)$ not identically zero, is **nonhomogeneous.** In the consideration of the linear equations (1) and (2), we assume that $a_2(x), a_1(x), a_0(x)$, and $g(x)$ are continuous on an interval I and that $a_2(x) \neq 0$ for every x in the interval. Under these assumptions there exists a unique solution of (2) satisfying the **initial conditions** $y(x_0) = y_0, y'(x_0) = y'_0$, where x_0 is a point in I.

The **Wronskian** of two differentiable functions $f_1(x)$ and $f_2(x)$ is the determinant

$$W(f_1(x), f_2(x)) = \begin{vmatrix} f_1(x) & f_2(x) \\ f'_1(x) & f'_2(x) \end{vmatrix}.$$

When $W \neq 0$ for at least one point in an interval, the set of functions is **linearly independent** on the interval. If the set of functions is **linearly dependent** on the interval, then $W = 0$ for every x in the interval.

In solving the homogeneous equation (1), we want linearly independent solutions. A necessary and sufficient condition that two solutions y_1 and y_2 are linearly independent on I is that $W(y_1, y_2) \neq 0$ for every x in I. We say y_1 and y_2 form a **fundamental set** on I when they are linearly independent solutions of (1) on the interval. For *any* two solutions y_1 and y_2, the **superposition principle** states that the linear combination $c_1y_1 + c_2y_2$ is also a solution of (1). When y_1 and y_2 form a fundamental set, the function $y = c_1y_1 + c_2y_2$ is called the **general solution** of (1). The **general solution** of (2) is $y = y_c + y_p$, where y_c is the **complementary function,** or general solution, of (1) and y_p is any **particular solution** of (2).

To solve (1) in the case $ay'' + by' + cy = 0, a, b$, and c constants, we first solve the **auxiliary equation** $am^2 + bm + c = 0$. There are three forms of the general solution depending on the three possible ways in which the roots of the auxiliary equation can occur (see Table 4.2).

Table 4.2

Roots	General solution
m_1 and m_2: real and distinct	$y = c_1e^{m_1x} + c_2e^{m_2x}$
m_1 and m_2: real but $m_1 = m_2$	$y = c_1e^{m_1x} + c_2xe^{m_1x}$
m_1 and m_2: complex $m_1 = \alpha + i\beta, \quad m_2 = \alpha - i\beta$	$y = e^{\alpha x}(c_1 \cos \beta x + c_2 \sin \beta x)$

To solve a nonhomogeneous differential equation we use either the method of **undetermined coefficients** or **variation of parameters** to find a particular solution y_p. The former procedure is limited to differential equations $ay'' + by' + cy = g(x)$, where a, b, and c are constants and $g(x)$ is a constant, a polynomial, $e^{\alpha x}$, $\cos \beta x$, $\sin \beta x$, or finite sums and products of these functions.

CHAPTER 4 REVIEW EXERCISES

Answers to odd-numbered problems begin on page A-9.

Answer Problems 1–10 without referring back to the text. Fill in the blank or answer true or false. In some cases there can be more than one correct answer.

1. The only solution of $y'' + x^2y = 0$, $y(0) = 0$, $y'(0) = 0$ is ______.
2. If two differentiable functions $f_1(x)$ and $f_2(x)$ are linearly independent on an interval, then $W(f_1(x), f_2(x)) \neq 0$ for at least one point in the interval. ______
3. Two functions $f_1(x)$ and $f_2(x)$ are linearly independent on an interval if one is not a constant multiple of the other. ______
4. The functions $f_1(x) = x^2$, $f_2(x) = 1 - x^2$, and $f_3(x) = 2 + x^2$ are linearly ______ on the interval $(-\infty, \infty)$.
5. The functions $f_1(x) = x^2$ and $f_2(x) = x|x|$ are linearly independent on the interval ______, whereas they are linearly dependent on the interval ______.
6. Two solutions y_1 and y_2 of $y'' + y' + y = 0$ are linearly dependent if $W(y_1, y_2) = 0$ for every real value of x. ______
7. A constant multiple of a solution of a differential equation is also a solution. ______
8. A fundamental set of two solutions of $(x - 2)y'' + y = 0$ exists on any interval not containing the point ______.
9. For the method of undetermined coefficients, the assumed form of the particular solution y_p for $y'' - y = 1 + e^x$ is ______.
10. A differential operator that annihilates $e^{2x}(x + \sin x)$ is ______.

In Problems 11 and 12 find a second solution for the differential equation given that $y_1(x)$ is a known solution.

11. $y'' + 4y = 0, \quad y_1 = \cos 2x$
12. $xy'' - 2(x + 1)y' + (x + 2)y = 0, \quad y_1 = e^x$

In Problems 13–18 find the general solution of each differential equation.

13. $y'' - 2y' - 2y = 0$
14. $2y'' + 2y' + 3y = 0$
15. $y''' + 10y'' + 25y' = 0$
16. $2y''' + 9y'' + 12y' + 5y = 0$
17. $3y''' + 10y'' + 15y' + 4y = 0$
18. $2\dfrac{d^4y}{dx^4} + 3\dfrac{d^3y}{dx^3} + 2\dfrac{d^2y}{dx^2} + 6\dfrac{dy}{dx} - 4y = 0$

In Problems 19–22 solve each differential equation by the method of undetermined coefficients.

19. $y'' - 3y' + 5y = 4x^3 - 2x$
20. $y'' - 2y' + y = x^2e^x$
21. $y''' - 5y'' + 6y' = 2\sin x + 8$
22. $y''' - y'' = 6$

In Problems 23 and 24 solve the given differential equation subject to the indicated conditions.

23. $y'' - 2y' + 2y = 0, \quad y(\pi/2) = 0, y(\pi) = -1$

24. $y'' - y = x + \sin x, \quad y(0) = 2, y'(0) = 3$

In Problems 25 and 26 solve each differential equation by the method of variation of parameters.

25. $y'' - 2y' + 2y = e^x \tan x$

26. $y'' - y = 2e^x/(e^x + e^{-x})$

In Problems 27 and 28 solve the given differential equation subject to the indicated conditions.

27. $(2D^3 - 13D^2 + 24D - 9)y = 36, \quad y(0) = -4, y'(0) = 0, y''(0) = \frac{5}{2}$

28. $y'' + y = \sec^3 x, \quad y(0) = 1, y'(0) = \frac{1}{2}$

Essay

by
John H. Hubbard
Department of Mathematics, Cornell University

CHAOS

A famous science fiction story tells of a politician who, soon after winning an election, takes a trip in a time travel machine back to the days of the dinosaurs. While there, and though much admonished not to disturb anything, he bends a blade of grass. When he returns to his real present, he finds that in this modified world he has lost the election.

This is what mathematicians mean when they say that a system exhibits chaos: Tiny changes in the initial state of a system can decisively affect the outcome. One speaks of the butterfly effect. Can the fluttering of the wings of a butterfly in Japan have a decisive effect on the weather in the United States a month later?

Most people would probably dismiss the suggestion as preposterous without a second thought. But I think it is true, at least if the time span of one month is increased to six months, and I propose here to give some quantitative reasons for my conclusions.

It is not obvious how you would go about justifying the reality of the butterfly effect. We do not have a time machine available; we cannot go back six weeks, catch a butterfly (while disturbing nothing else, whatever that means), and then come back and observe the consequences. We need to take a more devious tack.

To help you follow the argument, I will describe a "toy model" that clearly exhibits "butterfly behavior," and within which the notions that I will introduce can be understood.

Consider the (purely mathematical) system in which, at each tick of the clock, an angle is doubled. A state of the system is an angle, and it evolves by doubling again and again. In symbols, you could describe the system as a sequence of angles,

$$\theta_0, \theta_1, \dots,$$

where θ_0 is the initial state of the system and $\theta_{n+1} = 2\theta_n$.

This system exhibits butterfly behavior. If θ_0 is perturbed by one-billionth of a turn, then the state after 30 ticks of the clock is completely unknown. Indeed, the uncertainty in our knowledge of the state of the system doubles at each tick; after 30 ticks, our uncertainty has grown to

$$\frac{2^{30}}{1{,}000{,}000{,}000},$$

which is slightly greater than 1. Our uncertainty is now more than a turn; we don't know anything anymore.

The example above brings in a key notion in all descriptions of chaos entropy. This is essentially the rate at which information dissipates. There are many ways of describing this rate with precision, and they go by various names (for example, Lyapunov exponents), but for the purposes of this article, I will stick with the doubling time: **the amount of time it takes for a small uncertainty to double.**

How would we estimate this time for the system formed by the weather? Of course, we cannot "choose two initial states for the weather an epsilon apart and measure how fast they diverge," but we can do something like it. We can go back over history and find times when the weather was very similar. Then we can see how long it took for the weather pattern to diverge. This has been done and leads to a doubling time of about $2\frac{1}{2}$ days.

You can also go to the meteorology department of a big research institution (in practice, the Institut de Météorologie de l'université de Paris–VI) and ask what the doubling time is for their best computer models. You get much the same figure.

The next question that must be faced is this: What corresponds to the number one-billionth above? What proportion of the size of the system (the atmosphere) is our disturbance (one butterfly)? One (perhaps contestable) way of estimating this is simply to measure the ratio of the masses. We might guess that a butterfly weighs 1 gram (rather a heavy butterfly), and it turns out that the atmosphere has a mass of about 5×10^{21} grams.

The atmospheric pressure is very nearly 1 kg/cm^2, meaning that there is 1 kg of air above every square centimeter of the earth. The area of a sphere of radius r is $4\pi r^2$, and the radius of the earth is about 6000 km. Thus the mass of the atmosphere is about

$$1000 \times 4 \times \pi \times (6 \times 10^8)^2 \text{ grams.}$$

So a butterfly isn't one-billionth the size of the system; it is more nearly a thousand-billion-billionth.

Since 5×10^{21} is approximately 2^{72}, it should take about 72 doubling periods for the effects of an individual butterfly to induce perturbations on a global scale.

One consequence of this analysis is that long-term weather prediction should be completely impossible. It is inconceivable that anyone could ever know the state of the atmosphere to within the effect of one butterfly, or even on a scale a thousand billion times larger. Perturbations on that scale decisively affect the global weather within a month.

Physicists, chemists, astronomers, and mathematicians have now shown that a vast collection of systems appear to display "chaos," in the sense that they are expanding and have a doubling time for errors.

One example is due to the meteorologist E. Lorenz, whose discovery may be credited with starting this entire line of thought. In 1961 he was running a simulation of the weather. The computers of the time were primitive, so the data had to be drastically simplified for any computation to be possible. He observed various behaviors from his model, apparently quite satisfactory, until one day he decided to examine something he had already computed over a longer time. He typed in what he thought were the original initial conditions, went off to get a cup of coffee, and when he came back he found that his new "weather" was unrelated to the previous run of his model.

He eventually realized that the difference was due to the fact that he had entered the initial conditions with fewer decimals than in the initial run. After further simplification of his model, Lorenz found that the following set of three coupled differential equations exhibits the same sort of chaotic behavior:

$$\begin{aligned} x' &= 10(y - x) \\ y' &= 28x - y - xz \\ z' &= \frac{8}{3}z + xy. \end{aligned}$$

Much further work on these equations and others in R^3 have shown that chaotic behavior and fractal attractors are common.

The presence of chaos plays havoc with predictions, but it is sometimes useful; sometimes chaos can be translated into control.

NASA is not able to build rockets with enough fuel to give the speeds required to go huge distances. So they delicately bounced the spacecraft off Venus, stealing a bit of Venus's potential energy to give the spacecraft the fantastic speeds required. Only because the tiny variations in the trajectory due to the guidance rockets can be amplified to enormous variations in speed and trajectory is such a scheme feasible. But imagine how difficult this makes the long-term predictions about the orbits of comets.

The presence of chaos also has philosophical consequences, for instance, about the conflict of determinism and free will. How can humans have free will if the universe is completely governed by deterministic laws?

Well, if the equations exhibit chaos, then it follows that you cannot know that they are deterministic, however long you observe the system. If you were to observe a sequence of "doubled angles," each with 15 decimal places, you could never know whether you are observing a sequence of exact doubles of some angle or the same system perturbed at a scale smaller than 10^{-15} (such as round-off error).

Similarly, if the brain were not following exactly the laws of physics but were perturbed (by spiritual forces, free will, god) at a scale unmeasurable without drastically affecting the system (I doubt anyone reacts the same way with electrodes in his brain), then you could never know it, but over a time scale of perhaps 4 seconds, you could decisively alter all decisions. Making precise sense of the above would require knowing something about the doubling time of the brain. Introspection tells me this should be perhaps 0.1 second, the time of an elementary realization. Perhaps some day neurology will give a more credible estimate. If this is roughly accurate, then 4 seconds is 40 doubling times later, and $2^{40} \approx 10^{12}$ is approximately the number of neurons in the brain.

Personally, I do not think that there is a god tinkering with the laws of physics in my brain, but this is not knowable. Chaos prevents us from knowing such things.

More generally, although I am as firm a scientist as you could hope to find, I think that the world is essentially incomprehensible, with all sorts of tiny events having enormous consequences that no one can hope to foresee or understand.

If you find preposterous the idea that your smallest actions probably influence the entire world after a short time, consider the following. The time span used is much longer, about 1200 years, but I think the implications are much the same. Who were your ancestors in the year 800 A.D.? You have two parents, four grandparents, and so on. The number of ancestors should be 2^n in the nth generation. The year 800 is 50–60 generations ago, leading to $2^{50} \approx 10^{15}$ ancestors at that time. Of course this is a million times larger than the population of the world today, and a much larger multiple of the population at that time. This shows among other things that there must have been a lot of inbreeding.

Now think about it from the point of view of a woman in the year 800. In order for the population to be sustained, and even grow, she had to have at least an average of 2 children, 4 grandchildren, and so on. Of course, her line might become extinct, but if that were to happen, then it must happen soon, after fewer than three generations. The probability of having no descendants can be evaluated over the short term and might be $\frac{1}{3}$ or $\frac{1}{2}$. And once the descendance has become large, it tends to become the entire population quite rapidly (although the influence of inbreeding slows this down considerably).

If you put these two points of view together, you will see that essentially everyone of European ancestry is descended from every person alive in Europe in the year 800 who has any descendants at all. For instance, we are all descended from Charlemagne, but also from about half of the peasants alive at the time. If just one of those couples had had one child fewer, *everyone* would be genetically different.

Of course, this doesn't make sense; long ago the paths of history would have diverged so that no detailed comparisons are possible. But certainly the smallest variation in the sexual behavior of anyone at that time would certainly have affected the world in incalculable ways.

5 APPLICATIONS OF SECOND-ORDER DIFFERENTIAL EQUATIONS: VIBRATIONAL MODELS

INTRODUCTION

A single differential equation can serve as a mathematical model for many different phenomena. Forms of the linear second-order differential equation $ay'' + by' + cy = f(t)$ appear in the analysis of problems in physics, engineering, chemistry, and biology.

In this chapter our primary focus is on one application: the motion of a mass attached to a spring. We shall see what the individual terms ay'', by', cy, and $f(t)$ of the differential equation $ay'' + by' + cy = f(t)$ mean in the context of this vibrational system. We shall also see that, except for terminology and physical interpretation of the terms ay'', by', cy, and $f(t)$, the mathematics of a series circuit is identical to that of a vibrating spring-mass system. Our goal, of course, is not to study all possible applications but to acquaint you with the mathematical procedures that are common to these problems.

5.1 SIMPLE HARMONIC MOTION

• *Free motion* • *Simple harmonic motion* • *Free undamped motion* • *Period* • *Frequency* • *Equation of motion* • *Amplitude* • *Phase angle*

Hooke's Law Suppose, as in Figure 5.1(b), a mass m_1 is attached to a flexible spring suspended from a rigid support. When m_1 is replaced with a different mass m_2, the amount of stretch, or elongation, of the spring will of course be different.

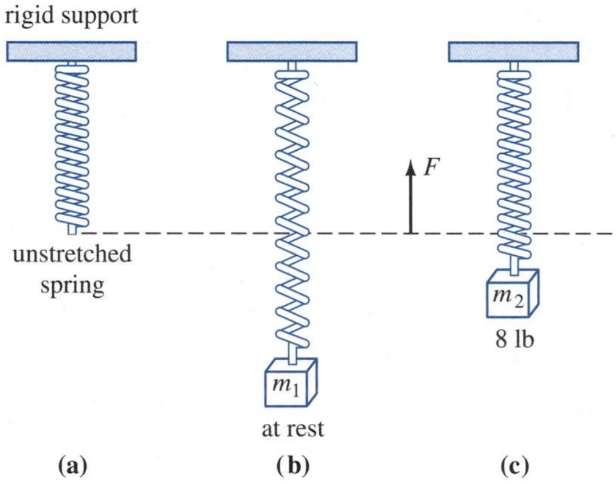

Figure 5.1

By Hooke's law,* the spring itself exerts a restoring force F opposite to the direction of elongation and proportional to the amount of elongation s. Simply stated, $F = ks$, where k is a constant of proportionality. Although masses with different weights stretch a spring by different amounts, the spring is essentially characterized by the number k. For example, if a mass weighing 10 lb stretches a spring by $\frac{1}{2}$ ft, then $10 = k(\frac{1}{2})$ implies $k = 20$ lb/ft. Necessarily then a mass weighing 8 lb stretches the same spring $\frac{2}{5}$ ft.

Newton's Second Law After a mass m is attached to a spring, it stretches the spring by an amount s and attains a position of equilibrium at which its weight

* ROBERT HOOKE (1635–1703) An English physicist and inventor, Hooke published this law in 1658. The idea of attaching a spring to a balance wheel, causing oscillatory motion that enabled a clock to mark units of time, is usually attributed to Hooke. The concept of the balance spring led to the invention of the pocket watch by Christian Huygens in 1674. Hooke accused Huygens of stealing his invention. Irascible and contentious, Hooke charged many of his colleagues, notably Isaac Newton, with plagiarism.

W is balanced by the restoring force ks. Recall from Section 1.2 that weight is defined by $W = mg$, where the mass is measured in slugs, kilograms, or grams and $g = 32$ ft/s^2, 9.8 m/s^2, or 980 cm/s^2, respectively. As indicated in Figure 5.2(b), the condition of equilibrium is $mg = ks$ or $mg - ks = 0$. If the mass is now displaced by an amount x from its equilibrium position and released, the net force F in this dynamic case is given by **Newton's second law of motion** $F = ma$, where a is the acceleration d^2x/dt^2. Assuming that there are no retarding forces acting on the system and assuming that the mass vibrates free of other external influencing forces—**free motion**—we can equate F to the resultant force of the weight and the restoring force:

$$m\frac{d^2x}{dt^2} = -k(s + x) + mg = -kx + \underbrace{mg - ks}_{\text{zero}} = -kx. \tag{1}$$

The negative sign in (1) indicates that the restoring force of the spring acts opposite to the direction of motion. Furthermore, we shall adopt the convention that displacements measured *below* the equilibrium position are positive. See Figure 5.3.

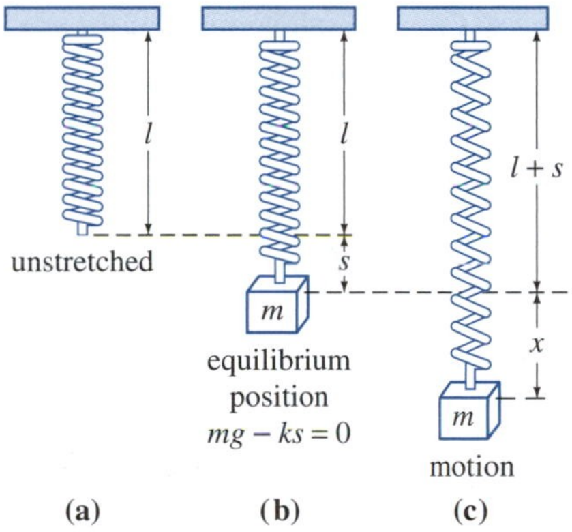

Figure 5.2

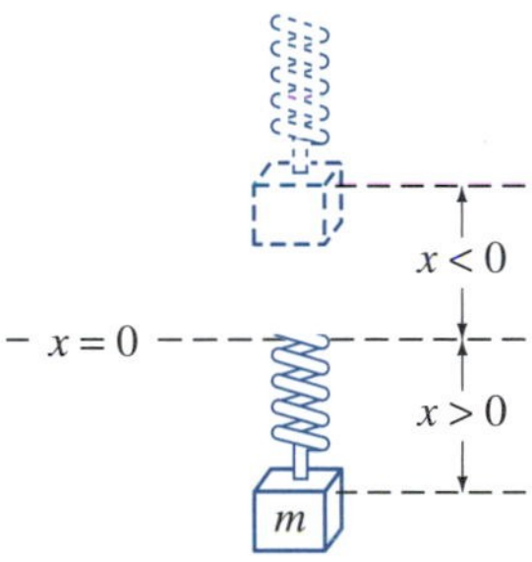

Figure 5.3

Differential Equation of Free Undamped Motion By dividing (1) by the mass m, we obtain the second-order differential equation

$$\frac{d^2x}{dt^2} + \frac{k}{m}x = 0 \tag{2}$$

or

$$\frac{d^2x}{dt^2} + \omega^2 x = 0, \tag{3}$$

where $\omega^2 = k/m$. Equation (3) is said to describe **simple harmonic motion,** or **free undamped motion.** There are two obvious initial conditions associated with (3):

$$x(0) = \alpha, \quad x'(0) = \beta, \tag{4}$$

representing the amount of initial displacement and the initial velocity, respectively. For example, if $\alpha > 0, \beta < 0$, the mass starts from a point *below* the equilibrium position with an imparted *upward* velocity. If $\alpha < 0, \beta = 0$, the mass is released from *rest* from a point $|\alpha|$ units *above* the equilibrium position, and so on.

Solution and Equation of Motion To solve equation (3) we note that the solutions of the auxiliary equation $m^2 + \omega^2 = 0$ are the complex numbers $m_1 = \omega i$, $m_2 = -\omega i$. Thus from (8) of Section 4.3, we find the general solution of (3) to be

$$x(t) = c_1 \cos \omega t + c_2 \sin \omega t. \tag{5}$$

The **period** of free vibrations described by (5) is $T = 2\pi/\omega$, and the **frequency** is $f = 1/T = \omega/2\pi$.* For example, for $x(t) = 2 \cos 3t - 4 \sin 3t$, the period is $2\pi/3$ and the frequency is $3/2\pi$. The former number means that the graph of $x(t)$ repeats every $2\pi/3$ units; the latter number means that there are 3 cycles of the graph every 2π units or, equivalently, the mass undergoes $3/2\pi$ complete vibrations per unit time. In addition, it can be shown that the period $2\pi/\omega$ is the time interval between two successive maxima of $x(t)$. Keep in mind that a maximum of $x(t)$ is a positive displacement corresponding to the mass attaining a maximum distance *below* the equilibrium position, whereas a minimum of $x(t)$ is a negative displacement corresponding to the mass attaining a maximum height *above* the equilibrium position. We shall refer to either case as an **extreme displacement** of the mass. Finally, when the initial conditions (4) are used to determine the constants c_1 and c_2 in (5), we say that the resulting particular solution is the **equation of motion.**

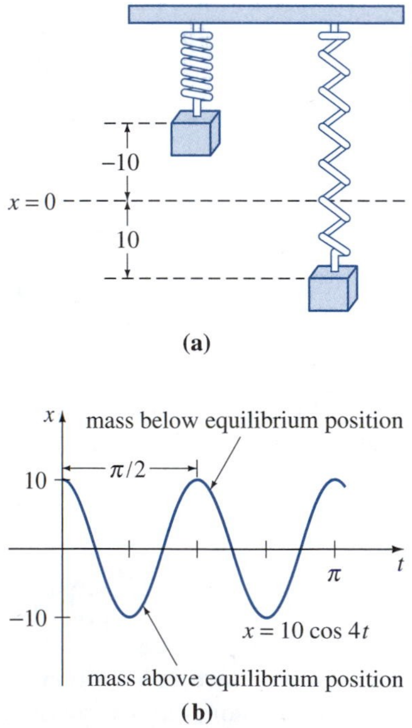

Figure 5.4

EXAMPLE 1 Interpretation of an IVP

Solve and interpret the initial-value problem

$$\frac{d^2x}{dt^2} + 16x = 0, \quad x(0) = 10, \quad x'(0) = 0.$$

Solution The problem is equivalent to pulling a mass on a spring down 10 units below the equilibrium position, holding it until $t = 0$, and then releasing it from rest. Applying the initial conditions to the solution

$$x(t) = c_1 \cos 4t + c_2 \sin 4t$$

gives $x(0) = 10 = c_1 \cdot 1 + c_2 \cdot 0$, so $c_1 = 10$ and hence

$$x(t) = 10 \cos 4t + c_2 \sin 4t$$

$$\frac{dx}{dt} = -40 \sin 4t + 4c_2 \cos 4t.$$

Now $x'(0) = 0 = 4c_2 \cdot 1$ implies that $c_2 = 0$; therefore, the equation of motion is $x(t) = 10 \cos 4t$.

The solution clearly shows that once the system is set in motion, it stays in motion with the mass bouncing back and forth 10 units on either side of the

---

* Sometimes the number ω is called the *circular frequency* of vibrations. For free undamped motion the numbers $2\pi/\omega$ and $\omega/2\pi$ are also referred to as the natural period and natural frequency, respectively.

equilibrium position $x = 0$. As shown in Figure 5.4(b), the period of oscillation is $2\pi/4 = \pi/2$ seconds.

EXAMPLE 2 Free Undamped Motion

A mass weighing 2 lb stretches a spring 6 inches. At $t = 0$ the mass is released from a point 8 inches below the equilibrium position with an upward velocity of $\frac{4}{3}$ ft/s. Determine the function $x(t)$ that describes the subsequent free motion.

Solution Since we are using the engineering system of units, the measurements given in terms of inches must be converted to feet: 6 inches $= \frac{1}{2}$ foot; 8 inches $= \frac{2}{3}$ foot. In addition we must convert the units of weight given in pounds to units of mass. From $m = W/g$ we have $m = \frac{2}{32} = \frac{1}{16}$ slug. Also, from Hooke's law, $2 = k(\frac{1}{2})$ implies $k = 4$ lb/ft. Hence the analogues of (1) and (2) are, respectively,

$$\frac{1}{16}\frac{d^2x}{dt^2} = -4x \quad \text{and} \quad \frac{d^2x}{dt^2} + 64x = 0.$$

The initial displacement and initial velocity are given by

$$x(0) = \frac{2}{3}, \quad x'(0) = -\frac{4}{3},$$

where the negative sign in the last condition is a consequence of the fact that the mass is given an initial velocity in the negative or upward direction.

Now $\omega^2 = 64$ or $\omega = 8$, so the general solution of the differential equation is

$$x(t) = c_1 \cos 8t + c_2 \sin 8t. \tag{6}$$

Applying the initial condition $x(0) = \frac{2}{3}$ to (6), we first find $\frac{2}{3} = c_1 \cdot 1 + c_2 \cdot 0$, so $c_1 = \frac{2}{3}$. Then applying $x'(0) = -\frac{4}{3}$ to

$$x'(t) = -\frac{16}{3}\sin 8t + 8c_2 \cos 8t$$

gives $-\frac{4}{3} = -\frac{16}{3} \cdot 0 + 8c_2 \cdot 1$ and $c_2 = -\frac{1}{6}$. Thus the equation of motion is

$$x(t) = \frac{2}{3}\cos 8t - \frac{1}{6}\sin 8t. \tag{7}$$

Note The distinction between weight and mass is often blurred. Thus one often speaks of both the motion of a mass on a spring and the motion of a weight on a spring.

Alternative Form of $x(t)$ When $c_1 \neq 0$ and $c_2 \neq 0$, the actual **amplitude** A of free vibrations is not obvious from inspection of equation (5). For example, although the mass in Example 2 is initially displaced $\frac{2}{3}$ ft beyond the equilibrium position, the amplitude of vibrations is a number larger than $\frac{2}{3}$. Hence it is often convenient to convert a solution of form (5) to the simpler form

$$x(t) = A\sin(\omega t + \phi), \tag{8}$$

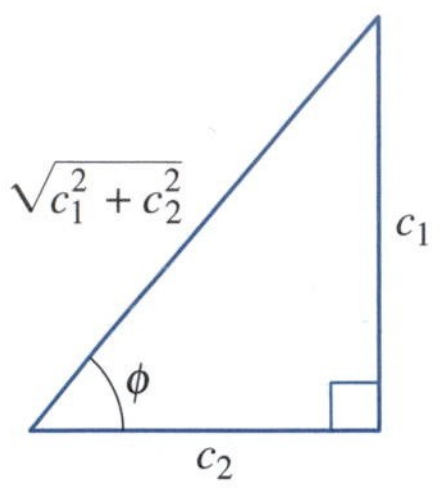

Figure 5.5

where $A = \sqrt{c_1^2 + c_2^2}$ and ϕ is a **phase angle** defined by

$$\left.\begin{aligned} \sin\phi &= \frac{c_1}{A} \\ \cos\phi &= \frac{c_2}{A} \end{aligned}\right\} \tan\phi = \frac{c_1}{c_2} \tag{9}$$

To verify this we expand (8) by the addition formula for the sine function:

$$A \sin\omega t \cos\phi + A \cos\omega t \sin\phi = (A \sin\phi)\cos\omega t + (A\cos\phi)\sin\omega t. \tag{10}$$

It follows from Figure 5.5 that if ϕ is defined by

$$\sin\phi = \frac{c_1}{\sqrt{c_1^2 + c_2^2}} = \frac{c_1}{A}, \quad \cos\phi = \frac{c_2}{\sqrt{c_1^2 + c_2^2}} = \frac{c_2}{A},$$

then (10) becomes

$$A\frac{c_1}{A}\cos\omega t + A\frac{c_2}{A}\sin\omega t = c_1\cos\omega t + c_2\sin\omega t = x(t).$$

EXAMPLE 3 Writing Solution (7) in Form (8)

In view of the foregoing discussion, we can write the solution (7) in Example 2,

$$x(t) = \frac{2}{3}\cos 8t - \frac{1}{6}\sin 8t, \quad \text{alternatively as} \quad x(t) = A\sin(8t + \phi).$$

The amplitude is given by

$$A = \sqrt{\left(\frac{2}{3}\right)^2 + \left(-\frac{1}{6}\right)^2} = \frac{\sqrt{17}}{6} \approx 0.69 \text{ ft}.$$

One should exercise some care when finding the phase angle ϕ defined by (9). In this case

$$\tan\phi = \frac{\frac{2}{3}}{-\frac{1}{6}} = -4$$

and a scientific hand calculator would give

$$\tan^{-1}(-4) = -1.326 \text{ radians.*}$$

But this angle is located in the fourth quadrant and therefore contradicts the fact that $\sin\phi > 0$ and $\cos\phi < 0$ (recall that $c_1 > 0$ and $c_2 < 0$). Hence we must take ϕ to be the second-quadrant angle

$$\phi = \pi + (-1.326) = 1.816 \text{ radians}.$$

Thus we have

$$x(t) = \frac{\sqrt{17}}{6}\sin(8t + 1.816). \tag{11}$$

* The range of the inverse tangent is $-\pi/2 < \tan^{-1}x < \pi/2$.

Form (8) is very useful since it is easy to find the values of time for which the graph of $x(t)$ crosses the positive t-axis (the line $x = 0$). We observe that $\sin(\omega t + \phi) = 0$ when $\omega t + \phi = n\pi$, where n is a nonnegative integer.

EXAMPLE 4 Times at the Equilibrium Position

For motion described by $x(t) = (\sqrt{17}/6)\sin(8t + 1.816)$, find the first value of time for which the mass passes through the equilibrium position heading downward.

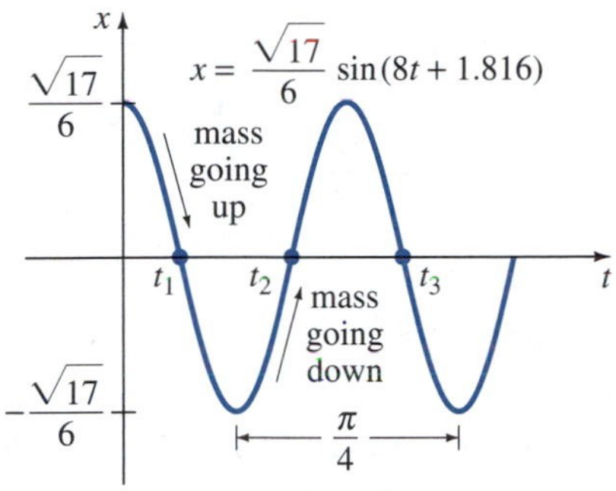

Figure 5.6

Solution The values $t_1, t_2, t_3, \ldots$ for which $\sin(8t + 1.816) = 0$ are determined from

$$8t_1 + 1.816 = \pi, \quad 8t_2 + 1.816 = 2\pi, \quad 8t_3 + 1.816 = 3\pi, \quad \ldots.$$

We find $t_1 = 0.166$, $t_2 = 0.558$, $t_3 = 0.951, \ldots$, respectively.

Figure 5.6 shows that the mass passes through $x = 0$ heading downward (namely, toward $x > 0$) the first time at $t_2 = 0.558$ second.

EXERCISES 5.1

Answers to odd-numbered problems begin on page A-9.

In Problems 1 and 2 state in words a possible physical interpretation of the given initial-value problem.

1. $\frac{4}{32}x'' + 3x = 0$
$x(0) = -3, x'(0) = -2$

2. $\frac{1}{16}x'' + 4x = 0$
$x(0) = 0.7, x'(0) = 0$

In Problems 3–8 write the solution of the given initial-value problem in form (8).

3. $x'' + 25x = 0$
$x(0) = -2, x'(0) = 10$

4. $\frac{1}{2}x'' + 8x = 0$
$x(0) = 1, x'(0) = -2$

5. $x'' + 2x = 0$
$x(0) = -1, x'(0) = -2\sqrt{2}$

6. $\frac{1}{4}x'' + 16x = 0$
$x(0) = 4, x'(0) = 16$

7. $0.1x'' + 10x = 0$
$x(0) = 1, x'(0) = 1$

8. $x'' + x = 0$
$x(0) = -4, x'(0) = 3$

9. The period of free undamped oscillations of a mass on a spring is $\pi/4$ second. If the spring constant is 16 lb/ft, what is the numerical value of the weight?

10. A spring is suspended from a ceiling. When a mass weighing 60 lb is attached, the spring is stretched $\frac{1}{2}$ ft. The mass is removed and a person grabs the end of the spring and proceeds to bounce up and down with a period of 1 second. How much does the person weigh?

11. A 4-lb weight is attached to a spring whose spring constant is 16 lb/ft. What is the period of simple harmonic motion?

12. A 20-kg mass is attached to a spring. If the frequency of simple harmonic motion is $2/\pi$ vibrations/second, what is the spring constant k? What is the frequency of simple harmonic motion if the original mass is replaced with an 80-kg mass?

13. A 24-lb weight, attached to the end of a spring, stretches it 4 in. Find the equation of motion if the weight is released from rest from a point 3 in. above the equilibrium position.

14. Determine the equation of motion if the weight in Problem 13 is released from the equilibrium position with an initial downward velocity of 2 ft/s.

15. A 20-lb weight stretches a spring 6 in. The weight is released from rest 6 in. below the equilibrium position.

(**a**) Find the position of the weight at $t = \pi/12, \pi/8, \pi/6, \pi/4, 9\pi/32$ seconds.

(**b**) What is the velocity of the weight when $t = 3\pi/16$ second? In which direction is the weight heading at this instant?

(**c**) At what times does the weight pass through the equilibrium position?

16. A force of 400 newtons stretches a spring 2 m. A mass of 50 kg is attached to the end of the spring and released from the equilibrium position with an upward velocity of 10 m/s. Find the equation of motion.

17. Another spring whose constant is 20 N/m is suspended from the same rigid support but parallel to the spring-mass system in Problem 16. A mass of 20 kg is attached to the second spring and both masses are released from the equilibrium position with an upward velocity of 10 m/s.

(**a**) Which mass exhibits the greater amplitude of motion?

(**b**) Which mass is moving faster at $t = \pi/4$ second? at $\pi/2$ seconds?

(**c**) At what times are the two masses in the same position? Where are the masses at these times? In which directions are they moving?

18. A 32-lb weight stretches a spring 2 ft. Determine the amplitude and period of motion if the weight is released 1 ft above the equilibrium position with an initial upward velocity of 2 ft/s. How many complete vibrations will the weight have completed at the end of 4π seconds?

19. An 8-lb weight attached to a spring exhibits simple harmonic motion. Determine the equation of motion if the spring constant is 1 lb/ft and if the weight is released 6 in. below the equilibrium position with a downward velocity of $\frac{3}{2}$ ft/s. Express the solution in form (8).

20. A mass weighing 10 lb stretches a spring $\frac{1}{4}$ ft. This mass is removed and replaced with a mass of 1.6 slugs, which is released $\frac{1}{3}$ ft above the equilibrium position with a downward velocity of $\frac{5}{4}$ ft/s. Express the solution in form (8). At what times does the mass attain a displacement below the equilibrium position numerically equal to one-half the amplitude?

21. A 64-lb weight attached to the end of a spring stretches it 0.32 ft. From a position 8 in. above the equilibrium position the weight is given a downward velocity of 5 ft/s.

(**a**) Find the equation of motion.

(**b**) What are the amplitude and period of motion?

(c) How many complete vibrations will the weight have completed at the end of 3π seconds?

(d) At what time does the weight pass through the equilibrium position heading downward for the second time?

(e) At what time does the weight attain its extreme displacement on either side of the equilibrium position?

(f) What is the position of the weight at $t = 3$ seconds?

(g) What is the instantaneous velocity at $t = 3$ seconds?

(h) What is the acceleration at $t = 3$ seconds?

(i) What is the instantaneous velocity at the times when the weight passes through the equilibrium position?

(j) At what times is the weight 5 in. below the equilibrium position?

(k) At what times is the weight 5 in. below the equilibrium position heading in the upward direction?

22. A mass of 1 slug is suspended from a spring whose characteristic spring constant is 9 lb/ft. Initially the mass starts from a point 1 ft above the equilibrium position with an upward velocity of $\sqrt{3}$ ft/s. Find the times for which the mass is heading downward at velocity of 3 ft/s.

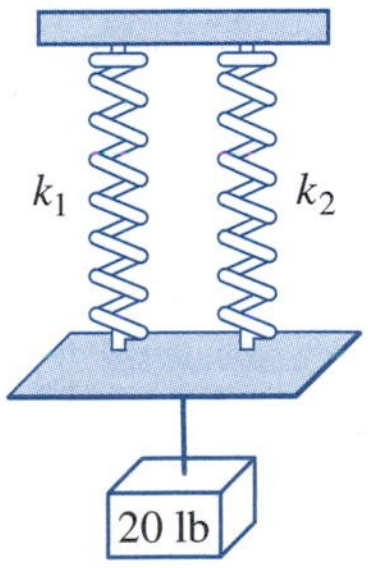

Figure 5.7

23. Under some circumstances when two parallel springs, with constants k_1 and k_2, support a single weight W, the **effective spring constant** of the system is given by $k = 4k_1k_2/(k_1 + k_2)$.* A 20-lb weight stretches one spring 6 in. and another spring 2 in. The springs are attached to a common rigid support and then to a metal plate. As shown in Figure 5.7, the 20-lb weight is attached to the center of the plate in the double spring arrangement. Determine the effective spring constant of this system. Find the equation of motion if the weight is released from the equilibrium position with a downward velocity of 2 ft/s.

24. A certain weight stretches one spring $\frac{1}{3}$ ft and another spring $\frac{1}{2}$ ft. The two springs are attached to a common rigid support in a manner indicated in Problem 23 and Figure 5.7. The first weight is set aside, and an 8-lb weight is attached to the double spring arrangement and the system is set in motion. If the period of motion is $\pi/15$ second, determine the numerical value of the first weight.

25. If x_0 and v_0 are the initial position and velocity, respectively, of a weight exhibiting simple harmonic motion, show that the amplitude of vibrations is

$$A = \sqrt{x_0^2 + \left(\frac{v_0}{\omega}\right)^2}.$$

26. Show that any linear combination $x(t) = c_1 \cos \omega t + c_2 \sin \omega t$ can also be written in the form

$$x(t) = A \cos(\omega t + \phi),$$

where

$$A = \sqrt{c_1^2 + c_2^2}, \quad \sin \phi = -\frac{c_2}{A}, \quad \cos \phi = \frac{c_1}{A}.$$

---

* If the two springs have the same natural length and if the plate is guided either by rails or by an external force so that the extensions of both springs are always the same, then the effective spring constant is simply $k = k_1 + k_2$.

27. Express the solution of Problem 3 in the form of the cosine function given in Problem 26.

28. Show that when a weight attached to a spring exhibits simple harmonic motion, the maximum value of the speed (that is, $|v(t)|$) occurs when the weight is passing through the equilibrium position.

29. A weight attached to a spring exhibits simple harmonic motion. Show that the maximum acceleration of the weight occurs at an extreme displacement and has the magnitude $4\pi^2 A/T^2$, where A is the amplitude and T is the period of free vibrations.

30. Use (8) to prove that the time interval between two successive maxima of $x(t)$ is $2\pi/\omega$.

5.2 DAMPED MOTION

• Free damped motion • Overdamping • Critical damping • Underdamping

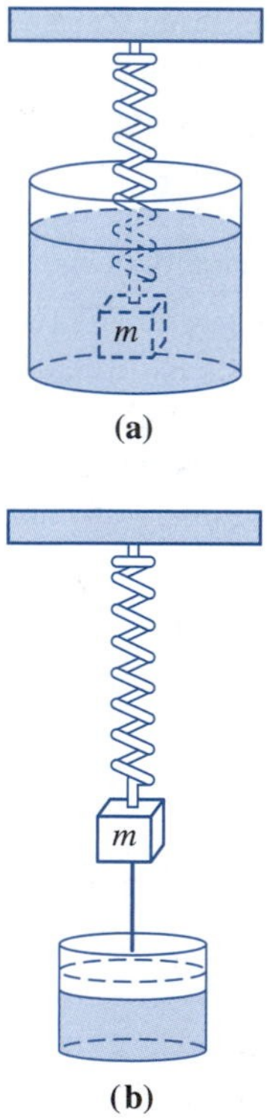

Figure 5.8

The discussion of free harmonic motion is somewhat unrealistic since the motion described by equation (2) of Section 5.1 assumes that no retarding forces are acting on the moving mass. Unless the mass is suspended in a perfect vacuum, there will be at least a resisting force due to the surrounding medium. For example, as Figure 5.8 shows, the mass m could be suspended in a viscous medium or connected to a dashpot damping device.

Differential Equation of Motion with Damping In the study of mechanics, damping forces acting on a body are considered to be proportional to a power of the instantaneous velocity. In particular, we shall assume throughout the subsequent discussion that this force is given by a constant multiple of dx/dt.* When no other external forces are impressed on the system, it follows from Newton's second law that

$$m\frac{d^2x}{dt^2} = -kx - \beta\frac{dx}{dt}, \tag{1}$$

where β is a positive *damping constant* and the negative sign is a consequence of the fact that the damping force acts in a direction opposite to the motion.

Dividing (1) by the mass m, we find the differential equation of **free damped motion** is

$$\frac{d^2x}{dt^2} + \frac{\beta}{m}\frac{dx}{dt} + \frac{k}{m}x = 0 \tag{2}$$

or

$$\frac{d^2x}{dt^2} + 2\lambda\frac{dx}{dt} + \omega^2 x = 0. \tag{3}$$

* In many instances, such as problems in hydrodynamics, the damping force is proportional to $(dx/dt)^2$.

In equation (3) we make the identifications

$$2\lambda = \frac{\beta}{m}, \quad \omega^2 = \frac{k}{m}. \tag{4}$$

The symbol 2λ is used only for algebraic convenience since the auxiliary equation is $m^2 + 2\lambda m + \omega^2 = 0$ and the corresponding roots are then

$$m_1 = -\lambda + \sqrt{\lambda^2 - \omega^2}, \quad m_2 = -\lambda - \sqrt{\lambda^2 - \omega^2}.$$

We can now distinguish three possible cases depending on the algebraic sign of $\lambda^2 - \omega^2$. Since each solution contains the *damping factor* $e^{-\lambda t}$, $\lambda > 0$, the displacements of the mass become negligible for large time.

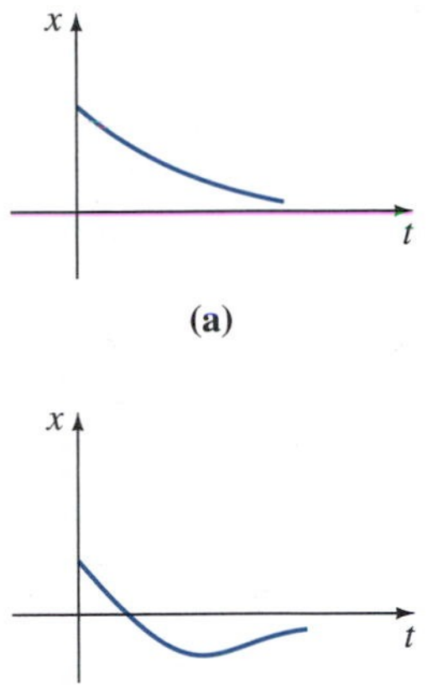

(a)

(b)

Figure 5.9

Case I $\lambda^2 - \omega^2 > 0$. In this situation the system is said to be **overdamped,** since the damping coefficient β is large when compared to the spring constant k. The corresponding solution of (3) is

$$x(t) = c_1 e^{m_1 t} + c_2 e^{m_2 t}$$

or

$$x(t) = e^{-\lambda t}\left(c_1 e^{\sqrt{\lambda^2 - \omega^2}\,t} + c_2 e^{-\sqrt{\lambda^2 - \omega^2}\,t}\right). \tag{5}$$

This equation represents a smooth and nonoscillatory motion. Figure 5.9 shows two possible graphs of $x(t)$.

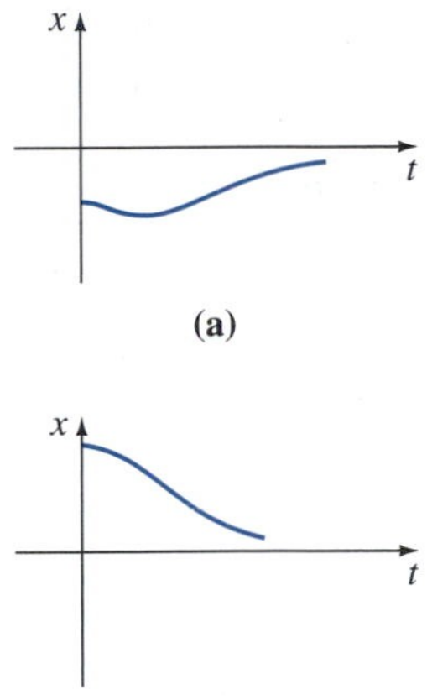

(a)

(b)

Figure 5.10

Case II $\lambda^2 - \omega^2 = 0$. The system is said to be **critically damped,** since any slight decrease in the damping force would result in oscillatory motion. The general solution of (3) is

$$x(t) = c_1 e^{m_1 t} + c_2 t e^{m_1 t}$$

or

$$x(t) = e^{-\lambda t}(c_1 + c_2 t). \tag{6}$$

Some graphs of typical motion are given in Figure 5.10. Notice that the motion is quite similar to that of an overdamped system. It is also apparent from (6) that the mass can pass through the equilibrium position at most one time.*

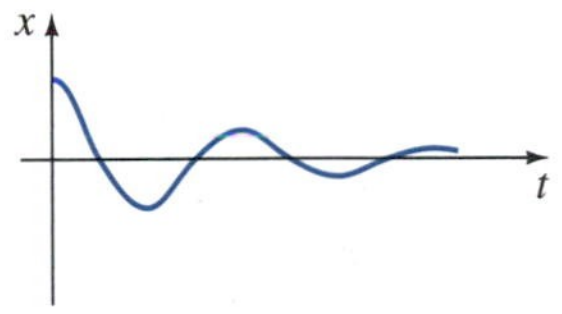

Figure 5.11

Case III $\lambda^2 - \omega^2 < 0$. In this case the system is said to be **underdamped,** since the damping coefficient is small compared to the spring constant. The roots m_1 and m_2 are now complex:

$$m_1 = -\lambda + \sqrt{\omega^2 - \lambda^2}\,i, \quad m_2 = -\lambda - \sqrt{\omega^2 - \lambda^2}\,i,$$

and so the general solution of equation (3) is

$$x(t) = e^{-\lambda t}\left(c_1 \cos\sqrt{\omega^2 - \lambda^2}\,t + c_2 \sin\sqrt{\omega^2 - \lambda^2}\,t\right). \tag{7}$$

As indicated in Figure 5.11, the motion described by (7) is oscillatory; but because of the coefficient $e^{-\lambda t}$, the amplitudes of vibration $\to 0$ as $t \to \infty$.

EXAMPLE 1 Overdamped Motion

It is readily verified that the solution of the initial-value problem

$$\frac{d^2x}{dt^2} + 5\frac{dx}{dt} + 4x = 0, \quad x(0) = 1, \quad x'(0) = 1$$

* An examination of the derivatives of (5) and (6) would show that these functions can have at most one relative maximum or one relative minimum for $t > 0$.

is
$$x(t) = \frac{5}{3}e^{-t} - \frac{2}{3}e^{-4t}. \qquad \textbf{(8)}$$

The problem can be interpreted as representing the overdamped motion of a mass on a spring. The mass starts from a position 1 unit *below* the equilibrium position with a *downward* velocity of 1 ft/s.

To graph $x(t)$ we find the value of t for which the function has an extremum—that is, the value of time for which the first derivative (velocity) is zero. Differentiating (8) gives

$$x'(t) = -\frac{5}{3}e^{-t} + \frac{8}{3}e^{-4t}$$

so $x'(t) = 0$ implies

$$e^{3t} = \frac{8}{5} \quad \text{or} \quad t = \frac{1}{3}\ln\frac{8}{5} = 0.157.$$

It follows from the first derivative test, as well as our physical intuition, that $x(0.157) = 1.069$ ft is actually a maximum. In other words, the mass attains an extreme displacement of 1.069 ft below the equilibrium position.

We should also check to see whether the graph crosses the t-axis—that is, whether the mass passes through the equilibrium position. This cannot happen in this instance since the equation $x(t) = 0$, or $e^{3t} = \frac{2}{5}$ has the physically irrelevant solution $t = \frac{1}{3}\ln\frac{2}{5} = -0.305$.

The graph of $x(t)$, along with some other pertinent data, is given in Figure 5.12.

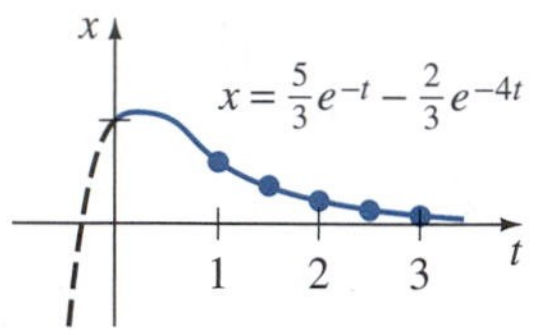

(a)

t	$x(t)$
1	0.601
1.5	0.370
2	0.225
2.5	0.137
3	0.083

(b)

Figure 5.12

EXAMPLE 2 Critically Damped Motion

An 8-lb weight stretches a spring 2 ft. Assuming a damping force numerically equal to two times the instantaneous velocity acts on the system, determine the equation of motion if the weight is released from the equilibrium position with an upward velocity of 3 ft/s.

Solution From Hooke's law we have $8 = k(2)$ or $k = 4$ lb/ft, and from $m = W/g$, $m = \frac{8}{32} = \frac{1}{4}$ slug. Thus the differential equation of motion is

$$\frac{1}{4}\frac{d^2x}{dt^2} = -4x - 2\frac{dx}{dt} \quad \text{or} \quad \frac{d^2x}{dt^2} + 8\frac{dx}{dt} + 16x = 0. \qquad \textbf{(9)}$$

The initial conditions are

$$x(0) = 0, \quad x'(0) = -3.$$

Now the auxiliary equation for (9) is

$$m^2 + 8m + 16 = (m + 4)^2 = 0$$

so $m_1 = m_2 = -4$. Hence the system is critically damped and

$$x(t) = c_1e^{-4t} + c_2te^{-4t}. \qquad \textbf{(10)}$$

The initial condition $x(0) = 0$ immediately demands that $c_1 = 0$, whereas using $x'(0) = -3$ gives $c_2 = -3$. Thus the equation of motion is

$$x(t) = -3te^{-4t}. \tag{11}$$

To graph $x(t)$ we proceed as in Example 1:

$$x'(t) = -3(-4te^{-4t} + e^{-4t}) = -3e^{-4t}(1 - 4t).$$

Clearly $x'(t) = 0$ when $t = \frac{1}{4}$. The corresponding extreme displacement is

$$x(\tfrac{1}{4}) = -3(\tfrac{1}{4})e^{-1} = -0.276 \text{ ft}.$$

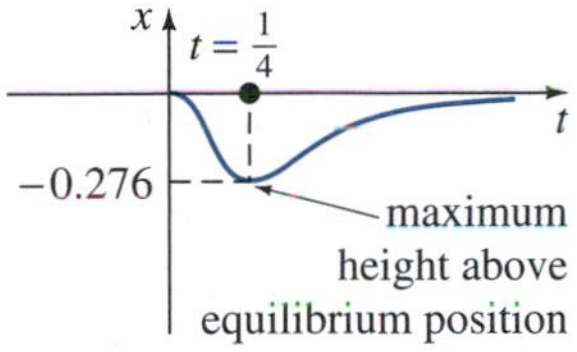

Figure 5.13

As shown in Figure 5.13, we interpret this value to mean that the weight reaches a maximum height of 0.276 ft above the equilibrium position.

EXAMPLE 3 Underdamped Motion

A 16-lb weight is attached to a 5-ft-long spring. At equilibrium the spring measures 8.2 ft. If the weight is pushed up and released from rest at a point 2 ft above the equilibrium position, find the displacements $x(t)$ if it is further known that the surrounding medium offers a resistance numerically equal to the instantaneous velocity.

Solution The elongation of the spring after the weight is attached is $8.2 - 5 = 3.2$ ft, so it follows from Hooke's law that $16 = k(3.2)$ or $k = 5$ lb/ft. In addition, $m = \frac{16}{32} = \frac{1}{2}$ slug, so the differential equation is given by

$$\frac{1}{2}\frac{d^2x}{dt^2} = -5x - \frac{dx}{dt} \quad \text{or} \quad \frac{d^2x}{dt^2} + 2\frac{dx}{dt} + 10x = 0. \tag{12}$$

This latter equation is solved subject to the conditions

$$x(0) = -2, \quad x'(0) = 0.$$

Proceeding, we find that the roots of $m^2 + 2m + 10 = 0$ are $m_1 = -1 + 3i$ and $m_2 = -1 - 3i$, which then implies the system is underdamped and

$$x(t) = e^{-t}(c_1 \cos 3t + c_2 \sin 3t). \tag{13}$$

Now $x(0) = -2 = c_1$, so

$$x(t) = e^{-t}(-2 \cos 3t + c_2 \sin 3t)$$
$$x'(t) = e^{-t}(6 \sin 3t + 3c_2 \cos 3t) - e^{-t}(-2 \cos 3t + c_2 \sin 3t).$$

Then $x'(0) = 0 = 3c_2 + 2$ gives $c_2 = -\frac{2}{3}$. Thus we finally obtain

$$x(t) = e^{-t}\left(-2 \cos 3t - \frac{2}{3} \sin 3t\right). \tag{14}$$

Alternative Form of the Solution In a manner identical to the procedure used in Section 5.1, we can write any solution

$$x(t) = e^{-\lambda t}(c_1 \cos \sqrt{\omega^2 - \lambda^2}\, t + c_2 \sin \sqrt{\omega^2 - \lambda^2}\, t)$$

in the alternative form

$$x(t) = Ae^{-\lambda t} \sin(\sqrt{\omega^2 - \lambda^2}t + \phi), \quad (15)$$

where $A = \sqrt{c_1^2 + c_2^2}$ and the phase angle ϕ is determined from the equations

$$\sin \phi = \frac{c_1}{A}, \quad \cos \phi = \frac{c_2}{A}, \quad \tan \phi = \frac{c_1}{c_2}.$$

The coefficient $Ae^{-\lambda t}$ is sometimes called the **damped amplitude** of vibrations. Because (15) is not a periodic function, the number $2\pi/\sqrt{\omega^2 - \lambda^2}$ is called the **quasi period** and $\sqrt{\omega^2 - \lambda^2}/2\pi$ is the **quasi frequency.** The quasi period is the time interval between two successive maxima of $x(t)$.

To graph an equation such as (15), we first find the intercepts $t_1, t_2, \ldots, t_k, \ldots$; that is, for some integer n

$$\sqrt{\omega^2 - \lambda^2}t + \phi = n\pi \quad \text{or} \quad t = \frac{n\pi - \phi}{\sqrt{\omega^2 - \lambda^2}}. \quad (16)$$

In addition we note that $|x(t)| \leq Ae^{-\lambda t}$ since

$$|\sin(\sqrt{\omega^2 - \lambda^2}t + \phi)| \leq 1.$$

Indeed, the graph of (15) touches the graphs of $\pm Ae^{-\lambda t}$ at the values $t_1^*, t_2^*, \ldots, t_k^*, \ldots$ for which

$$\sin(\sqrt{\omega^2 - \lambda^2}t + \phi) = \pm 1.$$

This means $\sqrt{\omega^2 - \lambda^2}t + \phi$ must be an odd multiple of $\pi/2$; that is,

$$\sqrt{\omega^2 - \lambda^2}t + \phi = (2n + 1)\frac{\pi}{2} \quad \text{or} \quad t = \frac{(2n + 1)\pi/2 - \phi}{\sqrt{\omega^2 - \lambda^2}}. \quad (17)$$

For example, if we are asked to graph $x(t) = e^{-0.5t} \sin(2t - \pi/3)$, we find the intercepts on the *positive* t-axis by solving

$$2t_1 - \frac{\pi}{3} = 0, \quad 2t_2 - \frac{\pi}{3} = \pi, \quad 2t_3 - \frac{\pi}{3} = 2\pi, \quad \ldots,$$

which gives, respectively,

$$t_1 = \frac{\pi}{6}, \quad t_2 = \frac{4\pi}{6}, \quad t_3 = \frac{7\pi}{6}, \quad \ldots.$$

Notice that even though $x(t)$ is not periodic, the difference between the successive roots is $t_k - t_{k-1} = \pi/2$ units or one-half the quasi period of $2\pi/2 = \pi$ seconds. Also $\sin(2t - \pi/3) = \pm 1$ at the solutions of

$$2t_1^* - \frac{\pi}{3} = \frac{\pi}{2}, \quad 2t_2^* - \frac{\pi}{3} = \frac{3\pi}{2}, \quad 2t_3^* - \frac{\pi}{3} = \frac{5\pi}{2}, \quad \ldots$$

or

$$t_1^* = \frac{5\pi}{12}, \quad t_2^* = \frac{11\pi}{12}, \quad t_3^* = \frac{17\pi}{12}, \quad \ldots.$$

It is readily shown that the difference between the successive t_k^* values is also $\pi/2$.* The graph of $x(t)$ is given in Figure 5.14.

---

* We note that the values of t for which the graph of $x(t)$ touches the exponential graphs are *not* the values for which the function attains its relative extrema.

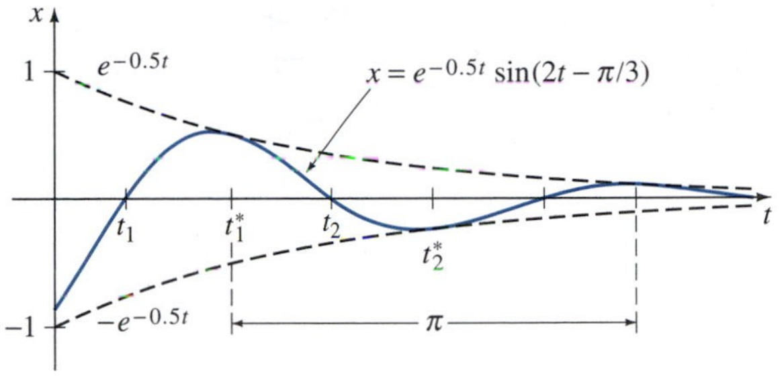

Figure 5.14

EXAMPLE 4 Writing a Solution (14) in Form (15)

Using (15), we can write the solution of the initial-value problem

$$\frac{d^2x}{dt^2} + 2\frac{dx}{dt} + 10x = 0, \quad x(0) = -2, \quad x'(0) = 0$$

in Example 3 in the form $x(t) = Ae^{-t}\sin(3t + \phi)$.

From (14) we have $c_1 = -2$, $c_2 = -\frac{2}{3}$, so

$$A = \sqrt{4 + \frac{4}{9}} = \frac{2}{3}\sqrt{10}$$

$$\tan\phi = \frac{-2}{-\frac{2}{3}} = 3 \quad \text{and} \quad \tan^{-1}(3) = 1.249 \text{ radians.}$$

But since $\sin\phi < 0$ and $\cos\phi < 0$, we take ϕ to be the third-quadrant angle $\phi = \pi + 1.249 = 4.391$ radians. Hence

$$x(t) = \frac{2}{3}\sqrt{10}e^{-t}\sin(3t + 4.391).$$

The graph of this function is given in Figure 5.15. The values of t_k and t_k^* given in the accompanying table are the intercepts and the points at which the graph of $x(t)$ touches the graphs of $\pm\frac{2}{3}\sqrt{10}e^{-t}$, respectively. In this example the quasi period is $2\pi/3$ seconds, and so the difference between the successive t_k (and the successive t_k^*) is $\pi/3$ units. ■

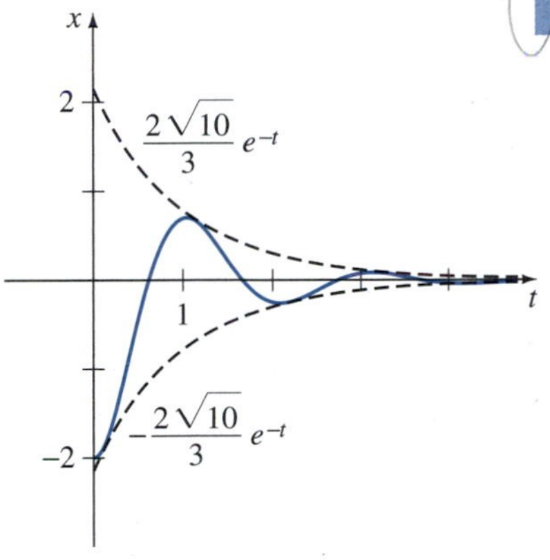

(a)

k	t_k	t_k^*	$x(t_k^*)$
1	0.631	1.154	0.665
2	1.678	2.202	−0.233
3	2.725	3.249	0.082
4	3.772	4.296	−0.029

(b)

Figure 5.15

EXERCISES 5.2

Answers to odd-numbered problems begin on page A-10.

In Problems 1 and 2 give a possible physical interpretation of the given initial-value problem.

1. $\frac{1}{16}x'' + 2x' + x = 0$

$x(0) = 0, x'(0) = -1.5$

2. $\frac{16}{32}x'' + x' + 2x = 0$

$x(0) = -2, x'(0) = 1$

In Problems 3–6 the given figure represents the graph of an equation of motion for a mass on a spring. The spring-mass system is damped. Use the graph to determine

(**a**) whether the initial displacement of the mass is above or below the equilibrium position and

(**b**) whether the mass is initially released from rest, heading downward, or heading upward.

3.

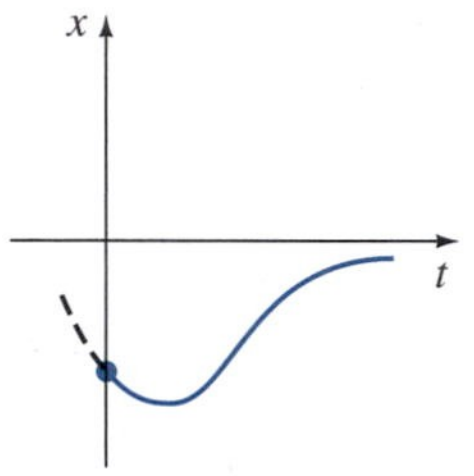

Figure 5.16

4.

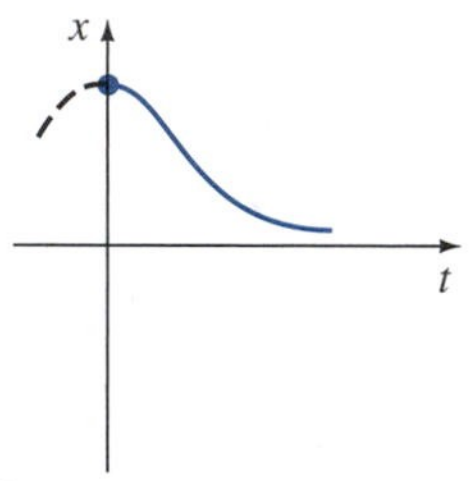

Figure 5.17

5.

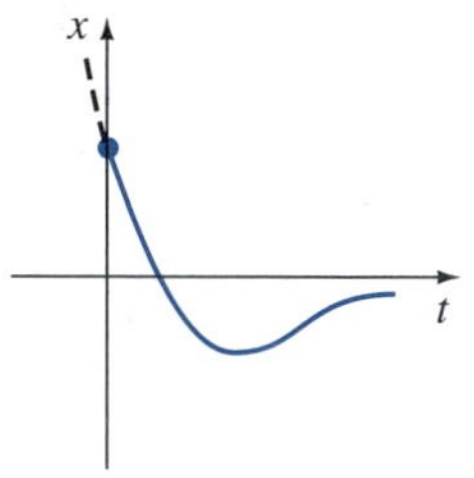

Figure 5.18

6.

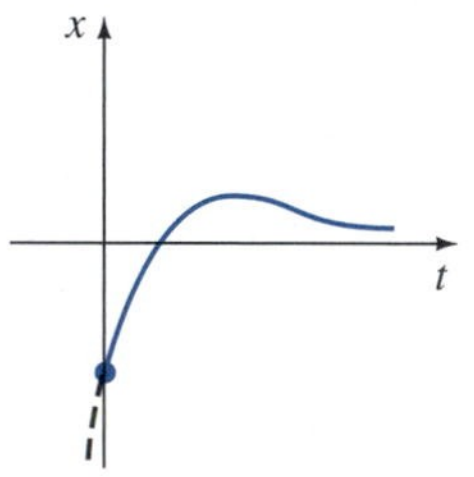

Figure 5.19

7. A 4-lb weight is attached to a spring whose constant is 2 lb/ft. The medium offers a resistance to the motion of the weight numerically equal to the instantaneous velocity. If the weight is released from a point 1 ft above the equilibrium position with a downward velocity of 8 ft/s, determine the time at which the weight passes through the equilibrium position. Find the time at which the weight attains its extreme displacement from the equilibrium position. What is the position of the weight at this instant?

8. A 4-ft spring measures 8 ft long after an 8-lb weight is attached to it. The medium through which the weight moves offers a resistance numerically equal to $\sqrt{2}$ times the instantaneous velocity. Find the equation of motion if the weight is released from the equilibrium position with a downward velocity of 5 ft/s. Find the time at which the weight attains its extreme displacement from the equilibrium position. What is the position of the weight at this instant?

9. A 1-kg mass is attached to a spring whose constant is 16 N/m and the entire system is then submerged in a liquid that imparts a damping force numerically equal to 10 times the instantaneous velocity. Determine the equations of motion if

(**a**) the weight is released from rest 1 m below the equilibrium position.

(**b**) the weight is released 1 m below the equilibrium position with an upward velocity of 12 m/s.

10. In parts (a) and (b) of Problem 9 determine whether the weight passes through the equilibrium position. In each case find the time at which the weight attains its extreme displacement from the equilibrium position. What is the position of the weight at this instant?

11. A force of 2 lb stretches a spring 1 ft. A 3.2-lb weight is attached to the spring and the system is then immersed in a medium that imparts a damping force numerically equal to 0.4 times the instantaneous velocity.

(**a**) Find the equation of motion if the weight is released from rest 1 ft above the equilibrium position.

(**b**) Express the equation of motion in the form given in (15).

(**c**) Find the first time at which the weight passes through the equilibrium position heading upward.

12. After a 10-lb weight is attached to a 5-ft spring, the spring measures 7 ft long. The 10-lb weight is removed and replaced with an 8-lb weight and the entire system is placed in a medium offering a resistance numerically equal to the instantaneous velocity.

(**a**) Find the equation of motion if the weight is released $\frac{1}{2}$ ft below the equilibrium position with a downward velocity of 1 ft/s.

(**b**) Express the equation of motion in the form given in (15).

(**c**) Find the times at which the weight passes through the equilibrium position heading downward.

(**d**) Graph the equation of motion.

13. A 10-lb weight attached to a spring stretches it 2 ft. The weight is attached to a dashpot damping device that offers a resistance numerically equal to β $(\beta > 0)$ times the instantaneous velocity. Determine the values of the damping constant β so that the subsequent motion is (**a**) overdamped, (**b**) critically damped, and (**c**) underdamped.

14. A 24-lb weight stretches a spring 4 ft. The subsequent motion takes place in a medium offering a resistance numerically equal to β $(\beta > 0)$ times the instantaneous velocity. If the weight starts from the equilibrium position with an upward velocity of 2 ft/s, show that if $\beta > 3\sqrt{2}$, the equation of motion is

$$x(t) = \frac{-3}{\sqrt{\beta^2 - 18}} e^{-2\beta t/3} \sinh \frac{2}{3} \sqrt{\beta^2 - 18}\, t.$$

15. A mass of 40 g stretches a spring 10 cm. A damping device imparts a resistance to motion numerically equal to 560 (measured in dynes/(cm/s)) times the instantaneous velocity. Find the equation of motion if the mass is released from the equilibrium position with a downward velocity of 2 cm/s.

16. Find the equation of motion for the mass in Problem 15 if the damping constant is doubled.

17. A mass of 1 slug is attached to a spring whose constant is 9 lb/ft. The medium offers a resistance to the motion numerically equal to 6 times the instantaneous velocity. The mass is released from a point 8 in. above the equilibrium position with a downward velocity of v_0 ft/s. Determine the values of v_0 such that the mass will subsequently pass through the equilibrium position.

18. The quasi period of an underdamped, vibrating 1-slug mass on a spring is $\pi/2$ seconds. If the spring constant is 25 lb/ft, find the damping constant β.

19. In the case of underdamped motion show that the difference in times between two successive positive maxima of the equation of motion is $2\pi/\sqrt{\omega^2 - \lambda^2}$.

20. Use (16) to show that the time interval between successive intercepts of (15) is one-half the quasi period.

21. Use (17) to show that the time interval between successive values of t for which the graph of (15) touches the graphs of $\pm Ae^{-\lambda t}$ is one-half the quasi period.

22. Use equation (17) to show that the intercepts of the graph of $x(t) = Ae^{-\lambda t}\sin(\sqrt{\omega^2-\lambda^2}\, t + \phi)$ are halfway between the values of t for which the graph of $x(t)$ touches the graphs of $\pm Ae^{-\lambda t}$. The values of t for which $x(t)$ is a maximum or minimum are not located halfway between the intercepts of the graph of $x(t)$. Verify this last statement by considering the function $x(t) = e^{-t}\sin(t + \pi/4)$.

23. In the case of underdamped motion show that the ratio between two consecutive maximum (or minimum) displacements x_n and x_{n+2} is the constant

$$\frac{x_n}{x_{n+2}} = e^{2\pi\lambda/\sqrt{\omega^2 - \lambda^2}}.$$

The number $\delta = \ln(x_n/x_{n+2}) = 2\pi\lambda/\sqrt{\omega^2 - \lambda^2}$ is called the **logarithmic decrement.**

24. The logarithmic decrement defined in Problem 23 is an indicator of the rate at which the motion is damped out.

(a) Describe the motion of an underdamped system if δ is a very small positive number.

(b) Compute the logarithmic decrement for the motion described in Problem 12.

5.3 FORCED MOTION

• *Forced motion* • *Transient term* • *Transient solution* • *Steady-state solution* • *Pure resonance* • *Resonance* • *Resonance curve*

With Damping Suppose we now take into consideration an external force $f(t)$ acting on a vibrating mass on a spring. For example, $f(t)$ could represent a driving force causing an oscillatory vertical motion of the support of the spring. See Figure 5.20. The inclusion of $f(t)$ in the formulation of Newton's second law gives the differential equation of **forced motion**

$$m\frac{d^2x}{dt^2} = -kx - \beta\frac{dx}{dt} + f(t), \qquad (1)$$

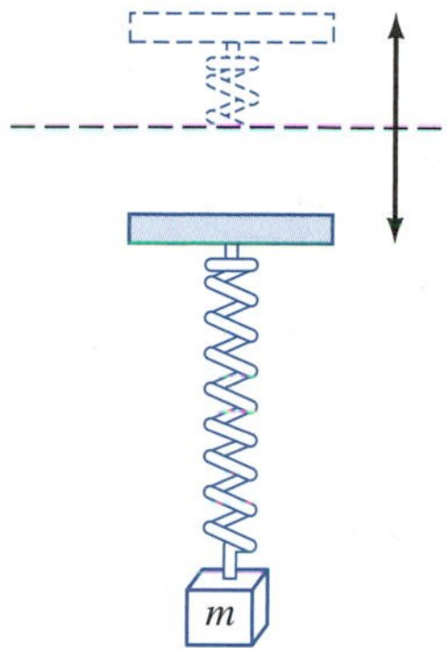

Figure 5.20

$$\frac{d^2x}{dt^2} + \frac{\beta}{m}\frac{dx}{dt} + \frac{k}{m}x = \frac{f(t)}{m}, \tag{2}$$

or

$$\frac{d^2x}{dt^2} + 2\lambda\frac{dx}{dt} + \omega^2 x = F(t), \tag{3}$$

where $F(t) = f(t)/m$ and, as in the preceding section, $2\lambda = \beta/m$, $\omega^2 = k/m$. To solve the latter nonhomogeneous equation we can use either the method of undetermined coefficients or variation of parameters.

EXAMPLE 1 Interpretation of an IVP

Interpret and solve the initial-value problem

$$\frac{1}{5}\frac{d^2x}{dt^2} + 1.2\frac{dx}{dt} + 2x = 5\cos 4t, \quad x(0) = \frac{1}{2}, \quad x'(0) = 0. \tag{4}$$

Solution We can interpret the problem to represent a vibrational system consisting of a mass ($m = \frac{1}{5}$ slug or kilogram) attached to a spring ($k = 2$ lb/ft or N/m). The mass is released from rest $\frac{1}{2}$ unit (foot or meter) below the equilibrium position. The motion is damped ($\beta = 1.2$) and is being driven by an external periodic ($T = \pi/2$ seconds) force beginning at $t = 0$. Intuitively we would expect that even with damping the system will remain in motion until such time as the forcing function is "turned off," in which case the amplitudes diminish. However, as the problem is given, $f(t) = 5\cos 4t$ will remain "on" forever.

We first multiply (4) by 5 and solve the associated homogeneous equation

$$\frac{dx^2}{dt^2} + 6\frac{dx}{dt} + 10x = 0$$

by the usual methods. Since $m_1 = -3 + i$, $m_2 = -3 - i$, it follows that

$$x_c(t) = e^{-3t}(c_1 \cos t + c_2 \sin t).$$

Using the method of undetermined coefficients, we assume a particular solution of the form $x_p(t) = A\cos 4t + B\sin 4t$. Now

$$x_p' = -4A\sin 4t + 4B\cos 4t \quad \text{and} \quad x_p'' = -16A\cos 4t - 16B\sin 4t,$$

so

$$\begin{aligned} x_p'' + 6x_p' + 10x_p &= (-6A + 24B)\cos 4t + (-24A - 6B)\sin 4t \\ &= 25\cos 4t. \end{aligned}$$

The resulting system of equations

$$-6A + 24B = 25, \quad -24A - 6B = 0$$

yields $A = -\frac{25}{102}$ and $B = \frac{50}{51}$. It follows that

$$x(t) = e^{-3t}(c_1\cos t + c_2\sin t) - \frac{25}{102}\cos 4t + \frac{50}{51}\sin 4t. \tag{5}$$

When we set $t = 0$ in the above equation, we obtain $c_1 = \frac{38}{51}$. By differentiating the expression and then setting $t = 0$, we also find that $c_2 = -\frac{86}{51}$. Therefore the equation of motion is

$$x(t) = e^{-3t}\left(\frac{38}{51}\cos t - \frac{86}{51}\sin t\right) - \frac{25}{102}\cos 4t + \frac{50}{51}\sin 4t. \qquad (6)$$

Transient and Steady-State Terms Notice that the complementary function

$$x_c(t) = e^{-3t}\left(\frac{38}{51}\cos t - \frac{86}{51}\sin t\right)$$

in Example 1 possesses the property that

$$\lim_{t\to\infty} x_c(t) = 0.$$

Since $x_c(t)$ becomes negligible (namely, $\to 0$) as $t \to \infty$, it is said to be a **transient term,** or **transient solution.** Thus for large time the displacements of the weight in the preceding problem are closely approximated by the particular solution $x_p(t)$. This latter function is also called the **steady-state solution.** When F is a periodic function, such as $F(t) = F_0 \sin \gamma t$ or $F(t) = F_0 \cos \gamma t$, the general solution of (3) consists of

$$x(t) = \textit{transient} + \textit{steady-state}.$$

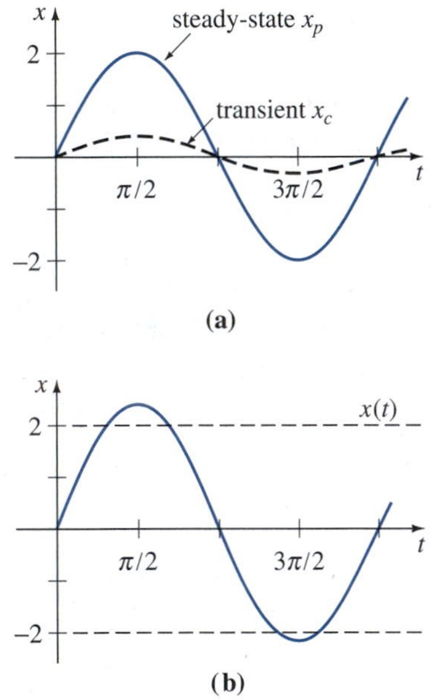

Figure 5.21

EXAMPLE 2 Transient and Steady-State Solutions

The solution to the initial-value problem

$$\frac{d^2x}{dt^2} + 2\frac{dx}{dt} + 2x = 4\cos t + 2\sin t, \quad x(0) = 0, \quad x'(0) = 3$$

is readily shown to be

$$x = x_c + x_p = \underbrace{e^{-t}\sin t}_{\text{transient}} + \underbrace{2\sin t}_{\text{steady-state}}.$$

Inspection of Figure 5.21 shows that the effect of the transient term on the solution is, in this case, negligible for about $t > 2\pi$.

Without Damping With a periodic impressed force and no damping force, there is no transient term in the solution of a problem. Also, we shall see that a periodic impressed force with a frequency near or the same as the frequency of free undamped vibrations can cause a severe problem in any oscillatory mechanical system.

EXAMPLE 3 Forced Undamped Motion

Solve the initial-value problem

$$\frac{d^2x}{dt^2} + \omega^2 x = F_0 \sin \gamma t, \quad x(0) = 0, \quad x'(0) = 0, \qquad (7)$$

where F_0 is a constant.

Solution The complementary function is $x_c(t) = c_1 \cos \omega t + c_2 \sin \omega t$. To obtain a particular solution we assume $x_p(t) = A \cos \gamma t + B \sin \gamma t$, so

$$x_p' = -A\gamma \sin \gamma t + B\gamma \cos \gamma t \quad \text{and} \quad x_p'' = -A\gamma^2 \cos \gamma t - B\gamma^2 \sin \gamma t$$

$$\begin{aligned} x_p'' + \omega^2 x_p &= A(\omega^2 - \gamma^2)\cos \gamma t + B(\omega^2 - \gamma^2)\sin \gamma t \\ &= F_0 \sin \gamma t. \end{aligned}$$

It follows that $A(\omega^2 - \gamma^2) = 0$, $B(\omega^2 - \gamma^2) = F_0$, and so

$$A = 0, \quad B = \frac{F_0}{\omega^2 - \gamma^2} \quad (\gamma \neq \omega).$$

Therefore

$$x_p(t) = \frac{F_0}{\omega^2 - \gamma^2} \sin \gamma t.$$

Applying the given initial conditions to the general solution

$$x(t) = c_1 \cos \omega t + c_2 \sin \omega t + \frac{F_0}{\omega^2 - \gamma^2} \sin \gamma t$$

yields $c_1 = 0$ and $c_2 = -\gamma F_0/\omega(\omega^2 - \gamma^2)$. Thus the solution is

$$x(t) = \frac{F_0}{\omega(\omega^2 - \gamma^2)}(-\gamma \sin \omega t + \omega \sin \gamma t), \quad \gamma \neq \omega. \tag{8}$$

Pure Resonance Although equation (8) is not defined for $\gamma = \omega$, it is interesting to observe that its limiting value as $\gamma \to \omega$ can be obtained by applying L'Hôpital's rule. This limiting process is analogous to "tuning in" the frequency of the driving force ($\gamma/2\pi$) to the frequency of free vibrations ($\omega/2\pi$). Intuitively we expect that over a length of time we should be able to substantially increase the amplitudes of vibration.* For $\gamma = \omega$ we define the solution to be

$$\begin{aligned} x(t) = \lim_{\gamma \to \omega} F_0 \frac{-\gamma \sin \omega t + \omega \sin \gamma t}{\omega(\omega^2 - \gamma^2)} &= F_0 \lim_{\gamma \to \omega} \frac{\dfrac{d}{d\gamma}(-\gamma \sin \omega t + \omega \sin \gamma t)}{\dfrac{d}{d\gamma}(\omega^3 - \omega\gamma^2)} \\ &= F_0 \lim_{\gamma \to \omega} \frac{-\sin \omega t + \omega t \cos \gamma t}{-2\omega\gamma} \\ &= F_0 \frac{-\sin \omega t + \omega t \cos \omega t}{-2\omega^2} \\ &= \frac{F_0}{2\omega^2} \sin \omega t - \frac{F_0}{2\omega} t \cos \omega t. \end{aligned} \tag{9}$$

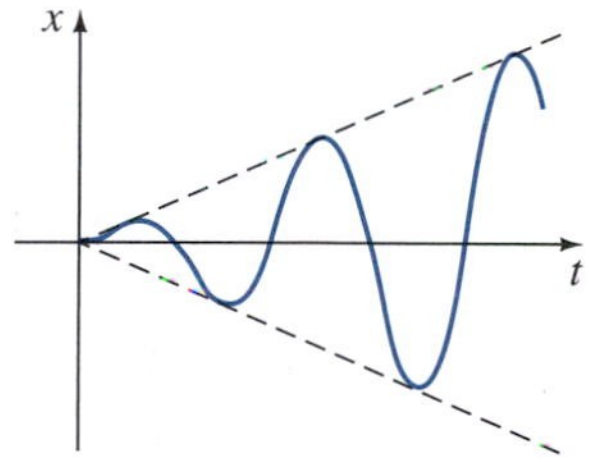

Figure 5.22

As suspected, when $t \to \infty$, the displacements become large; in fact, $|x(t_n)| \to \infty$ when $t_n = n\pi/\omega$, $n = 1, 2, \ldots$. The phenomenon we have just described is known as **pure resonance.** The graph given in Figure 5.22 shows typical motion in this case.

* If we forget about the damping effects of shock absorbers, the situation is roughly equivalent to a number of passengers jumping up and down in the back of a bus in time with the natural vertical motion caused by equally spaced faults (such as cracks) in the road. Theoretically these passengers could upset the bus—assuming they are not kicked off first!

In conclusion it should be noted that there is no actual need to use a limiting process on (8) to obtain the solution for $\gamma = \omega$. Alternatively, equation (9) follows by solving the initial-value problem

$$\frac{d^2x}{dt^2} + \omega^2 x = F_0 \sin \omega t, \quad x(0) = 0, \quad x'(0) = 0$$

directly by conventional methods.

Resonance Curve In the case of underdamped vibrations, the general solution of the differential equation

$$\frac{d^2x}{dt^2} + 2\lambda \frac{dx}{dt} + \omega^2 x = F_0 \sin \gamma t \tag{10}$$

can be shown to be

$$x(t) = Ae^{-\lambda t} \sin(\sqrt{\omega^2 - \lambda^2}\, t + \phi) + \frac{F_0}{\sqrt{(\omega^2 - \gamma^2)^2 + 4\lambda^2\gamma^2}} \sin(\gamma t + \theta), \tag{11}$$

where $A = \sqrt{c_1^2 + c_2^2}$ and the phase angles ϕ and θ are, respectively, defined by

$$\sin \phi = \frac{c_1}{A}, \quad \cos \phi = \frac{c_2}{A}$$

$$\sin \theta = \frac{-2\lambda\gamma}{\sqrt{(\omega^2 - \gamma^2)^2 + 4\lambda^2\gamma^2}}, \quad \cos \theta = \frac{\omega^2 - \gamma^2}{\sqrt{(\omega^2 - \gamma^2)^2 + 4\lambda^2\gamma^2}}.$$

EXAMPLE 4 Resonance Curve

Inspection of (11) shows that $x_c(t)$ is transient when damping is present, and hence for large values of time the solution is closely approximated by the steady-state solution

$$x_p(t) = g(\gamma) \sin(\gamma t + \theta),$$

where we define

$$g(\gamma) = \frac{F_0}{\sqrt{(\omega^2 - \gamma^2)^2 + 4\lambda^2\gamma^2}}. \tag{12}$$

Although the amplitude of x_p is bounded as $t \to \infty$, it is easily shown that the maximum oscillations will occur at the value $\gamma_1 = \sqrt{\omega^2 - 2\lambda^2}$ (see Problem 11). Thus when the frequency of the external force is $\sqrt{\omega^2 - 2\lambda^2}/2\pi$, the system is said to be in **resonance.**

In the specific case $k = 4$, $m = 1$, $F_0 = 2$, $g(\gamma)$ becomes

$$g(\gamma) = \frac{2}{\sqrt{(4 - \gamma^2)^2 + \beta^2\gamma^2}}. \tag{13}$$

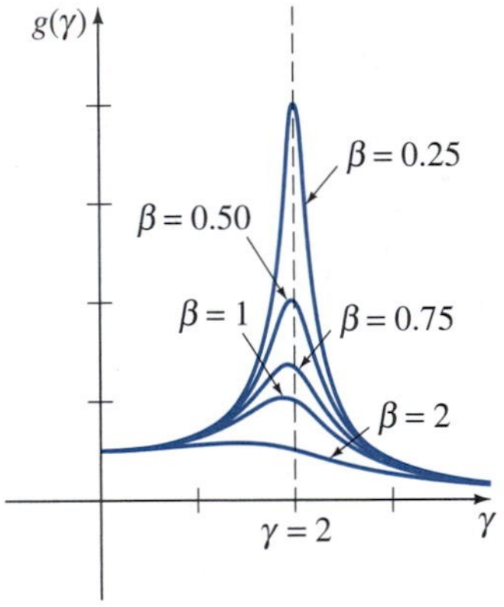

(a)

β	γ_1	$g(\gamma_1)$
2	1.41	0.58
1	1.87	1.03
0.75	1.93	1.36
0.50	1.97	2.02
0.25	1.99	4.01

(b)

Figure 5.23

Figure 5.23(a) shows the graph of (13) for various values of the damping coefficients β. This family of graphs is called the **resonance curve** of the system. Observe the behavior of the amplitudes $g(\gamma)$ as $\beta \to 0$—that is, as the system approaches pure resonance.

Remarks If a mechanical system were actually described by a function such as (9) of this section, it would necessarily fail. Large oscillations of a weight on a spring would eventually force the spring beyond its elastic limit. One might argue too that the resonating model presented in Figure 5.22 is completely unrealistic since it ignores the retarding effects of ever-present damping forces. Although it is true that pure resonance cannot occur when the smallest amount of damping is taken into consideration, large and equally destructive amplitudes of vibration (although bounded as $t \to \infty$) can occur.

If you have ever looked out an airplane window while in flight, you have probably observed that the wings on an airplane are not perfectly rigid. A reasonable amount of flex or flutter is not only tolerated but necessary to prevent the wing from snapping like a piece of peppermint stick candy. In late 1959 and early 1960 two commercial plane crashes involving a relatively new model of propjet occurred, illustrating the destructive effects of large mechanical oscillations.

The unusual aspect of these crashes was that they both happened while the planes were in mid-flight. Barring midair collisions, the safest period during any flight is when the plane has attained its cruising altitude. It is well known that a plane is most vulnerable to an accident when it is least maneuverable—namely, during either take-off or landing. So, having two planes simply fall out of the sky was not only a tragedy but an embarrassment to the aircraft industry and a thoroughly puzzling problem to aerodynamic engineers. In crashes of this sort, a structural failure of some kind is immediately suspected. After a massive technical investigation, the problem was eventually traced in each case to an outboard engine and engine housing. Roughly, it was determined that when each plane surpassed a critical speed of approximately 400 mph, a propeller and engine began to wobble, causing a gyroscopic force that could not be quelled or damped by the engine housing. This external vibrational force was then transferred to the already oscillating wing. This, in itself, need not have been destructively dangerous since aircraft wings are designed to withstand the stress of unusual and excessive forces. (In fact the particular wing in question was so incredibly strong that test engineers and pilots who were deliberately trying to snap a wing under every conceivable flight condition failed to do so.) But, unfortunately, after a short period of time during which the engine wobbled rapidly, the frequency of the impressed force actually slowed to a point at which it approached and finally coincided with the maximum frequency of wing flutter (around 3 cycles per second). The resulting resonance finally accomplished what the test engineers could not do; namely, the amplitudes of wing flutter became large enough to snap the wing. See Figure 5.24.

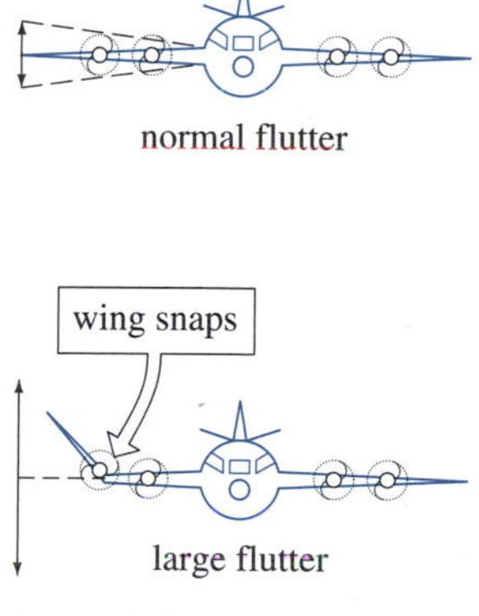

Figure 5.24

The problem was solved in two steps. All models of this particular plane were required to fly at speeds substantially below 400 mph until

Figure 5.25

each plane could be modified by considerably strengthening (or stiffening) the engine housings. A strengthened engine housing was shown to be able to impart a damping effect capable of preventing the critical resonance phenomenon even in the unlikely event of a subsequent engine wobble.*

You may be aware that soldiers usually do not march in step across bridges. The reason for breaking stride is simply to avoid any possibility of resonance occurring between the natural vibrations inherent in the bridge's structure and the frequency of the external force of a multitude of feet stomping in unison on the bridge.

Acoustic vibrations can be equally as destructive as large mechanical vibrations. In television commercials, jazz singers have inflicted destruction on the lowly wine glass. See Figure 5.25. Sounds from organs and piccolos have been known to crack windows.

> As the horns blew, the people began to shout. When they heard the signal horn, they raised a tremendous shout. The wall collapsed. ... [*Joshua 6:20*]

Did the power of acoustic resonance cause the walls of Jericho to tumble down? This is the conjecture of some contemporary scholars.

The phenomenon of resonance is not always destructive, however. For example, it is resonance of an electrical circuit that enables a radio to be tuned to a specific station.

EXERCISES 5.3

Answers to odd-numbered problems begin on page A-10.

1. A 16-lb weight stretches a spring $\frac{8}{3}$ ft. Initially the weight starts from rest 2 ft below the equilibrium position, and the subsequent motion takes place in a medium that offers a damping force numerically equal to $\frac{1}{2}$ the instantaneous velocity. Find the equation of motion if the weight is driven by an external force equal to $f(t) = 10 \cos 3t$.

2. A mass of 1 slug is attached to a spring whose constant is 5 lb/ft. Initially the mass is released 1 ft below the equilibrium position with a downward velocity of 5 ft/s, and the subsequent motion takes place in a medium that offers a damping force numerically equal to 2 times the instantaneous velocity.

(a) Find the equation of motion if the mass is driven by an external force equal to $f(t) = 12 \cos 2t + 3 \sin 2t$.

(b) Graph the transient and steady-state solutions on the same coordinate axes.

(c) Graph the equation of motion.

---

* For a fascinating nontechnical account of the investigation, see Robert J. Serling, *Loud and Clear* (New York: Dell, 1970), Chapter 5.

3. A mass of 1 slug, when attached to a spring, stretches it 2 ft and then comes to rest in the equilibrium position. Starting at $t = 0$, an external force equal to $f(t) = 8 \sin 4t$ is applied to the system. Find the equation of motion if the surrounding medium offers a damping force numerically equal to 8 times the instantaneous velocity.

4. In Problem 3 determine the equation of motion if the external force is $f(t) = e^{-t} \sin 4t$. Analyze the displacements for $t \to \infty$.

5. When a mass of 2 kilograms is attached to a spring whose constant is 32 N/m, it comes to rest in the equilibrium position. Starting at $t = 0$, a force equal to $f(t) = 68e^{-2t} \cos 4t$ is applied to the system. Find the equation of motion in the absence of damping.

6. In Problem 5 write the equation of motion in the form $x(t) = A \sin(\omega t + \phi) + Be^{-2t} \sin(4t + \theta)$. What is the amplitude of vibrations after a very long time?

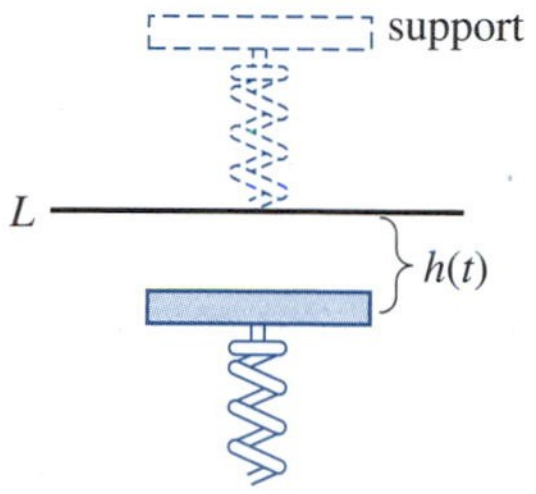

Figure 5.26

7. A mass m is attached to the end of a spring whose constant is k. After the mass reaches equilibrium, its support begins to oscillate vertically about a horizontal line L according to a formula $h(t)$. The value of h represents the distance in feet measured from L. See Figure 5.26. Determine the differential equation of motion if the entire system moves through a medium offering a damping force numerically equal to $\beta(dx/dt)$.

8. Solve the differential equation of the preceding problem if the spring is stretched 4 ft by a weight of 16 lb, and $\beta = 2$, $h(t) = 5 \cos t$, $x(0) = 0$, $x'(0) = 0$.

9. A mass of 100 g is attached to a spring whose constant is 1600 dynes/cm. After the mass reaches equilibrium, its support oscillates according to the formula $h(t) = \sin 8t$, where h represents displacement from its original position. See Problem 7 and Figure 5.26.
   - **(a)** In the absence of damping, determine the equation of motion if the mass starts from rest from the equilibrium position.
   - **(b)** At what times does the mass pass through the equilibrium position?
   - **(c)** At what times does the mass attain its extreme displacements?
   - **(d)** What are the maximum and minimum displacements?
   - **(e)** Graph the equation of motion.

10. Show that the general solution of equation (10) is given by (11).

11. **(a)** Prove that $g(\gamma)$ given in (13) of Example 4 has a maximum value at $\gamma_1 = \sqrt{\omega^2 - 2\lambda^2}$. [*Hint*: Differentiate with respect to γ.]
    (b) What is the maximum value of $g(\gamma)$ at resonance?

12. **(a)** If $k = 3$ lb/ft and $m = 1$ slug, use the information in Example 4 to show that the system is underdamped when the damping coefficient β satisfies $0 < \beta < 2\sqrt{3}$ but that resonance can occur only if $0 < \beta < \sqrt{6}$.
    (b) Construct the resonance curve of the system when $F_0 = 3$.

13. A mass of $\frac{1}{2}$ slug is suspended on a spring whose constant is 6 lb/ft. The system is set in motion in a medium offering a damping force numerically equal to twice the instantaneous velocity. Find the steady-state solution if an external force $f(t) = 40 \sin 2t$ is applied to the system starting at $t = 0$. Write this solution in the form of a constant multiple of $\sin(2t + \theta)$.

14. Verify that the mechanical system described in Problem 13 is in resonance. Show that the amplitude of the steady-state solution is the maximum value of $g(\gamma)$ described in Problem 11.

15. **(a)** Show that the solution of the initial-value problem

$$\frac{d^2x}{dt^2} + \omega^2 x = F_0 \cos \gamma t, \quad x(0) = 0, \quad x'(0) = 0$$

is
$$x(t) = \frac{F_0}{\omega^2 - \gamma^2}(\cos \gamma t - \cos \omega t).$$

(b) Evaluate $\lim_{\gamma \to \omega} \frac{F_0}{\omega^2 - \gamma^2}(\cos \gamma t - \cos \omega t)$.

16. Compare the result obtained in part (b) of Problem 15 with the solution obtained using variation of parameters when the external force is $F_0 \cos \omega t$.

In Problems 17 and 18 solve the given initial-value problem.

17. $\frac{d^2x}{dt^2} + 4x = -5 \sin 2t + 3 \cos 2t, \quad x(0) = -1, \quad x'(0) = 1$

18. $\frac{d^2x}{dt^2} + 9x = 5 \sin 3t, \quad x(0) = 2, \quad x'(0) = 0$

19. **(a)** Show that $x(t)$ given in part (a) of Problem 15 can be written in the form

$$x(t) = \frac{-2F_0}{\omega^2 - \gamma^2} \sin \frac{1}{2}(\gamma - \omega)t \sin \frac{1}{2}(\gamma + \omega)t.$$

(b) If we define $\varepsilon = \frac{1}{2}(\gamma - \omega)$, show that when ε is small an *approximate* solution is

$$x(t) = \frac{F_0}{2\varepsilon\gamma} \sin \varepsilon t \sin \gamma t.$$

When ε is small, the frequency $\gamma/2\pi$ of the impressed force is close to the frequency $\omega/2\pi$ of free vibrations. When this occurs, the motion is as indicated in Figure 5.27. Oscillations of this kind are called *beats* and are due to the fact that the frequency of $\sin \varepsilon t$ is quite small in comparison to the frequency of $\sin \gamma t$. The dashed curves, or *envelope* of the graph of $x(t)$, are obtained from the graphs of $\pm(F_0/2\varepsilon\gamma) \sin \varepsilon t$. Use a computer with various values of F_0, ε, and γ to verify the graph in Figure 5.27.

(c) Evaluate $\lim_{\varepsilon \to 0} \frac{F_0}{2\varepsilon\gamma} \sin \varepsilon t \sin \gamma t$.

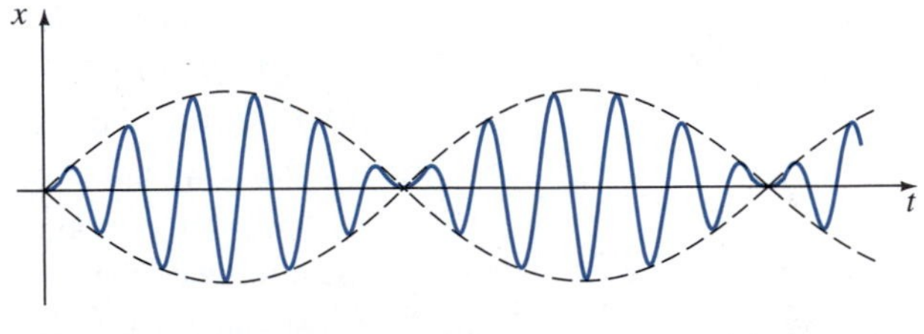

Figure 5.27

ILLUSTRATED APPLICATIONS

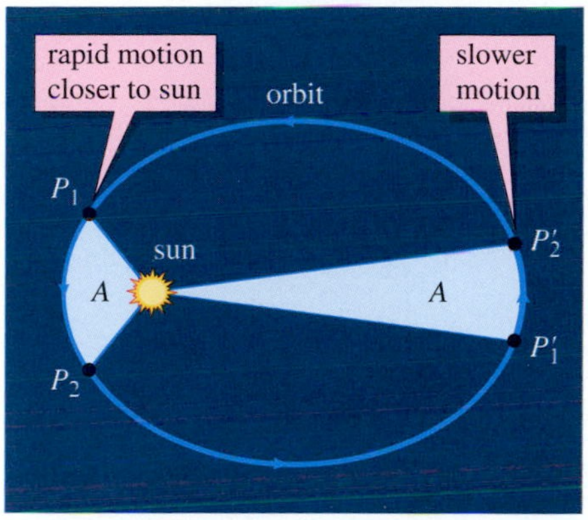

After painstaking observations, Johannes Kepler formulated and then published in 1609 three laws of planetary motion. The first law, probably his most famous, states that a planet revolves around the sun in an elliptical orbit with the sun at one focus. His second law of planetary motion states that the radius vector joining a planet with the sun sweeps out equal areas in equal intervals of time. See Problem 33, Exercises 3.2.

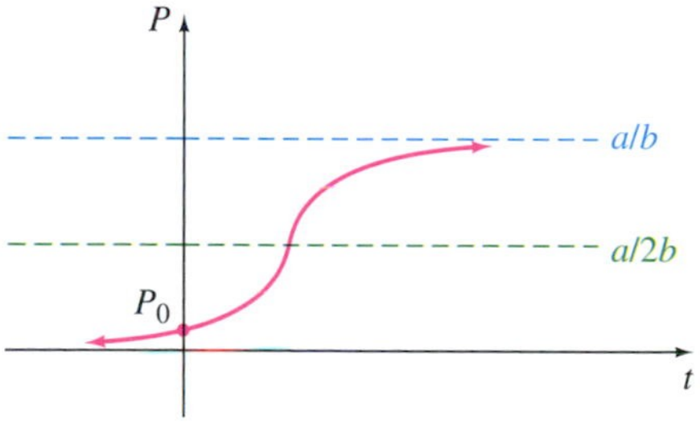

Around 1840, the Belgian mathematician P. F. Verhulst studied the nonlinear first-order differential equation $dP/dt = P(a - bP)$, called the *logistic equation,* in an attempt to model the population growth of various countries. This equation also provides a reasonable model for other growth phenomena such as the spread of flu throughout a static population, as found on the college campus illustrated here. See page 21, Section 3.3, and the essay at the end of Chapter 3.

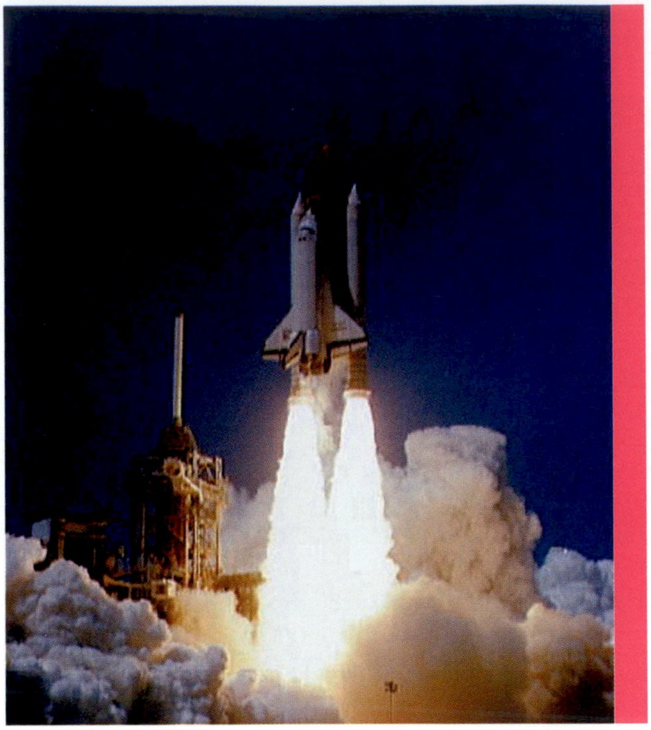

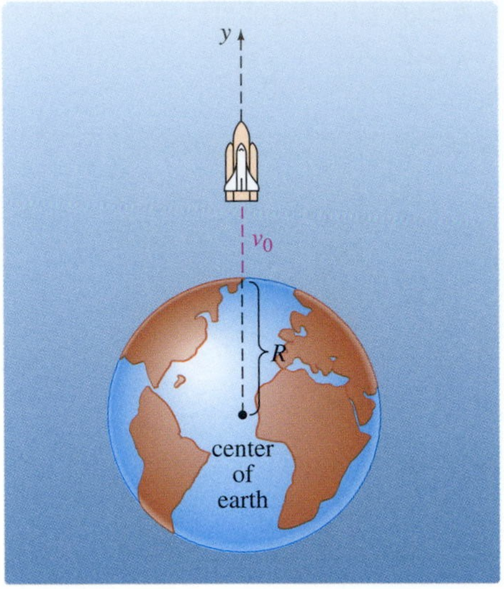

A ball or rocket projected straight up into the sky will fall back to earth under the influence of gravity unless it is given a sufficiently great velocity. The second-order differential equation $md^2y/dt^2 = -kMm/y^2$ can be used to determine the so-called "escape velocity" an object needs to break free from the gravitational attraction of any celestial body such as the earth. The escape velocity is independent of the mass of the object, so on earth the escape velocity for a ball is the same as that for a rocket. See Section 3.3.

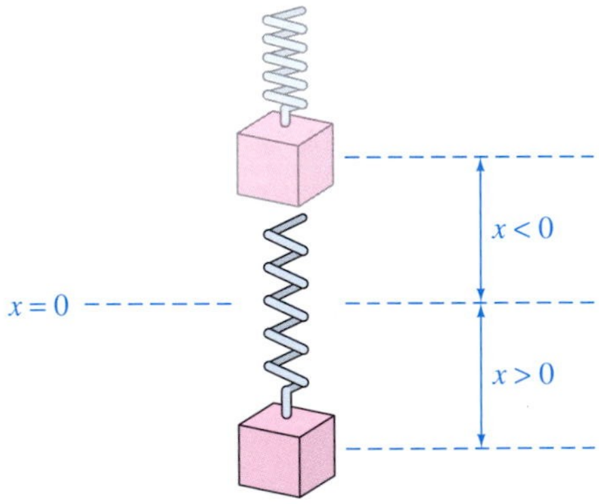

The motion of a mass attached to a spring is described by a linear second-order differential equation of the form $x'' + 2\lambda x' + \omega^2 x = f(t)$. An example of such a spring/mass system is the spring suspension system of some models of cars and trucks. See Chapter 5.

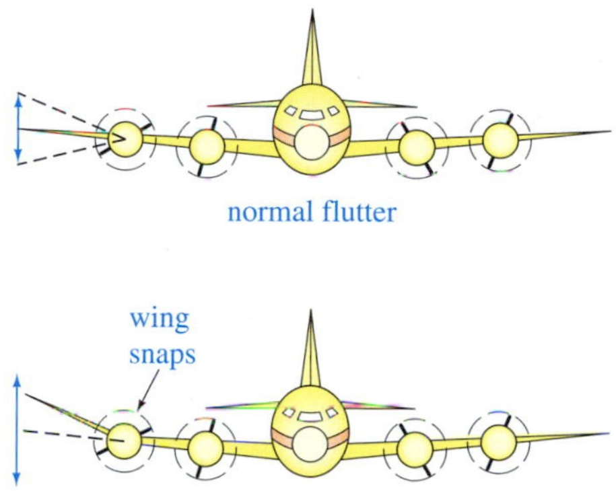

In late 1959 and early 1960, two commercial airplane crashes occurred involving the popular Lockheed Electra, a new model four-engine propjet. Braniff Flight 542 crashed near Buffalo, Texas, in September, 1959, and Northwest Flight 710 went down near Tell City, Indiana, in March, 1960. In both cases a wing had separated from the plane while at cruising altitude. These crashes illustrate the destructive effects of large mechanical oscillations or resonance. See Section 5.3.

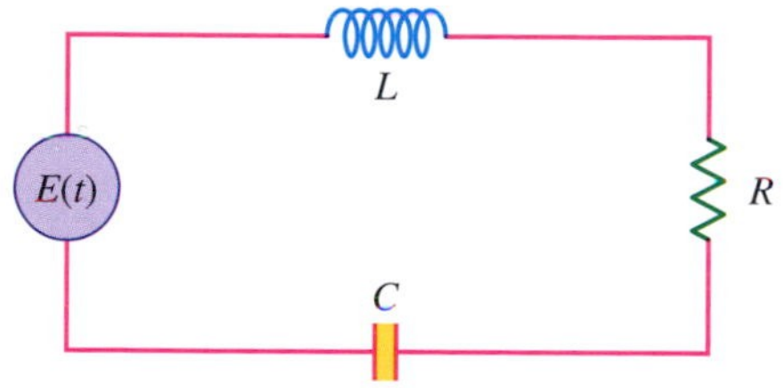

A single differential equation can serve as the mathematical model for different phenomena. The differential equation governing the charge $q(t)$ on a capacitor in an L-R-C series circuit has the same form as the differential equation governing the displacement $x(t)$ of a vibrating mass on a spring. See Sections 5.3 and 5.4.

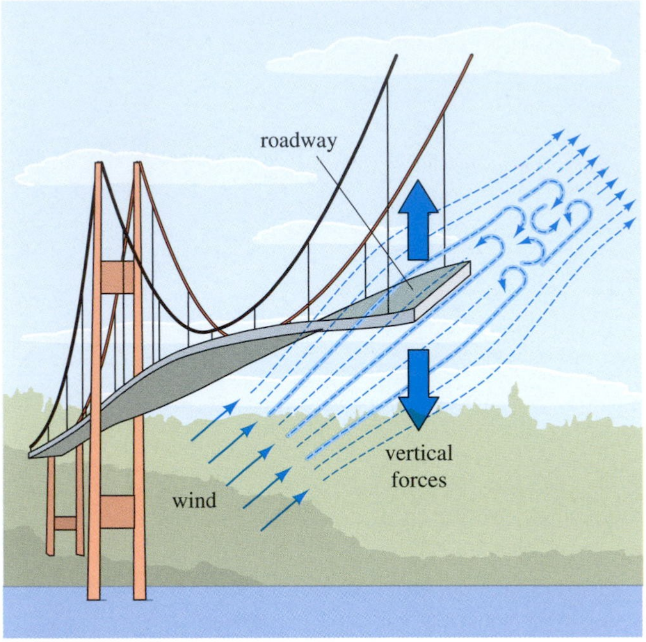

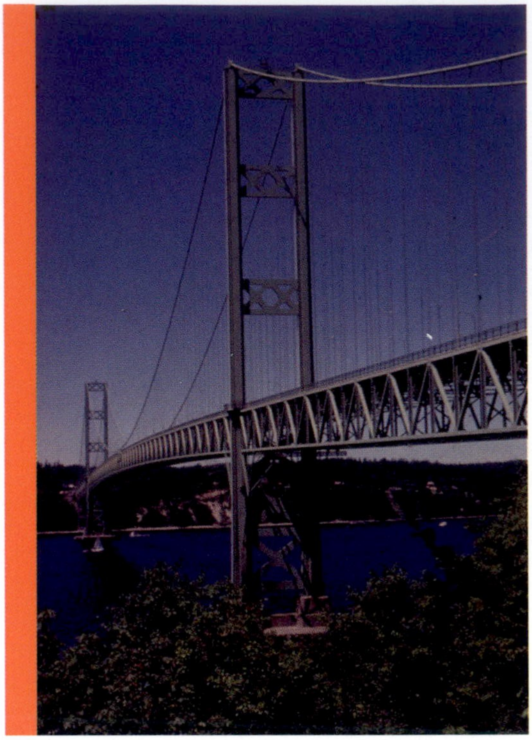

For years the 1940 collapse of the Tacoma Narrows Bridge was attributed to a resonance condition induced by wind blowing across its roadway and causing periodic vertical forces to act in the same direction as the vibration of the bridge. Recent studies suggest a different cause for the collapse. See the essay following Chapter 5. Compare the girder-stiffened roadway in this photograph, showing the reconstructed bridge, with the delicate 1940 design evident in the photographs on page 218.

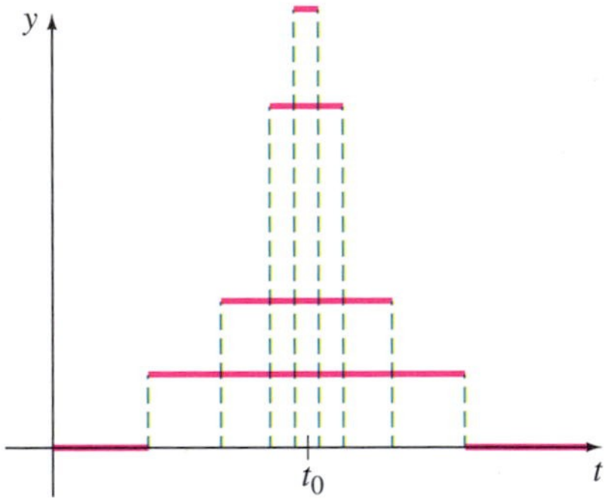

behavior of δ_a as $a \to 0$

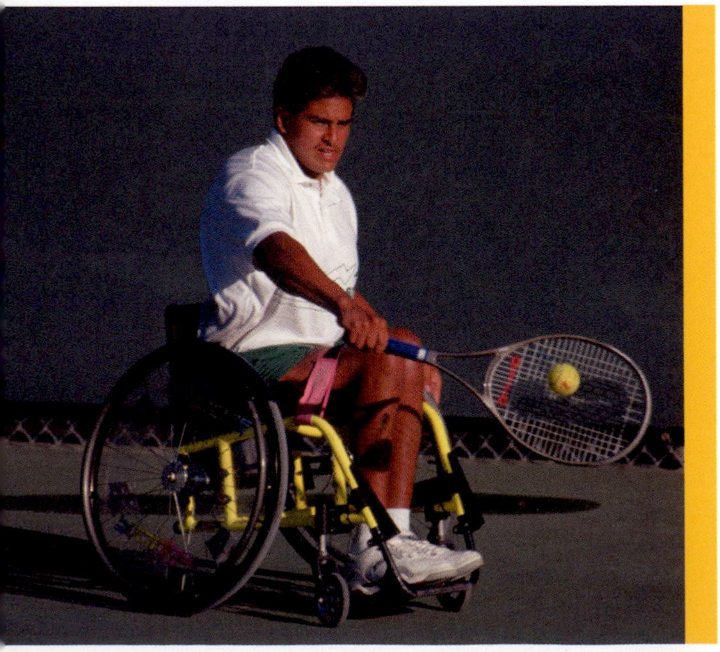

Mechanical systems are often acted upon by a sharp external force known as an impulsive force. This kind of force, which occurs in the collision of two objects, is a force of very large and approximately constant magnitude that acts on the system for a short time. A racket smashing into a tennis ball imparts a tremendous force to the ball but is in physical contact with it for only a fraction of a second. Analogous situations include a club striking a golf ball, lightning striking an airplane wing, and a proton striking the nucleus of an atom. See Section 7.6.

20. Show that the solution of the initial-value problem

$$\frac{d^2x}{dt^2} + 25x = 10\cos 7t, \quad x(0) = 0, \quad x'(0) = 0$$

is $x(t) = \frac{5}{6}\sin t \sin 6t$. Use a graphics calculator or computer to obtain the graph of $x(t)$.

5.4 ELECTRIC CIRCUITS AND OTHER ANALOGOUS SYSTEMS

- *Electrical vibrations* • *Overdamped circuit* • *Critically damped circuit*
- *Underdamped circuit* • *Steady-state current* • *Reactance* • *Impedance*
- *Extreme displacement*

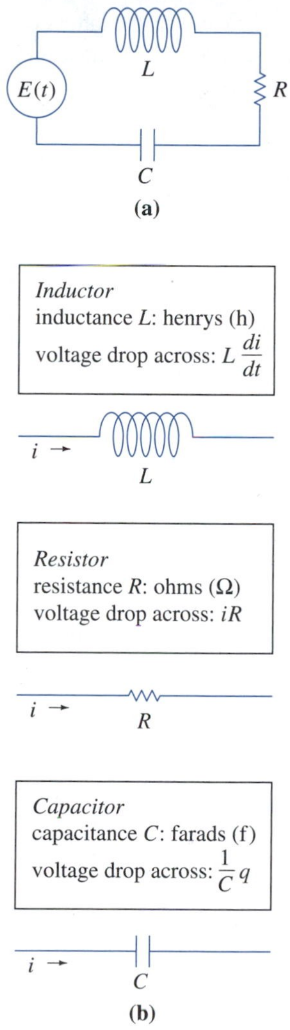

Figure 5.28

***L-R-C* Series Circuits** As mentioned in the introduction to this chapter, many different physical systems can be described by a linear second-order differential equation similar to the differential equation of forced motion with damping:

$$m\frac{d^2x}{dt^2} + \beta\frac{dx}{dt} + kx = f(t). \tag{1}$$

If $i(t)$ denotes current in an ***L-R-C* series electrical circuit** shown in Figure 5.28(a), then the voltage drops across the inductor, resistor, and capacitor are as shown in Figure 5.28(b). By Kirchhoff's second law, the sum of these voltages equals the voltage $E(t)$ impressed on the circuit; that is,

$$L\frac{di}{dt} + Ri + \frac{1}{C}q = E(t). \tag{2}$$

But the charge $q(t)$ on the capacitor is related to the current $i(t)$ by $i = dq/dt$ and so (2) becomes the linear second-order differential equation

$$L\frac{d^2q}{dt^2} + R\frac{dq}{dt} + \frac{1}{C}q = E(t). \tag{3}$$

The nomenclature used in the analysis of circuits is similar to that used to describe spring-mass systems.

If $E(t) = 0$, the **electrical vibrations** of the circuit are said to be *free*. Since the auxiliary equation for (3) is $Lm^2 + Rm + 1/C = 0$, there will be three forms of the solution when $R \neq 0$, depending on the value of the discriminant $R^2 - 4L/C$. We say that the circuit is

overdamped if $\quad R^2 - 4L/C > 0$,

critically damped if $\quad R^2 - 4L/C = 0$,

and **underdamped** if $\quad R^2 - 4L/C < 0$.

In each of these three cases, the general solution of (3) contains the factor $e^{-Rt/2L}$ and so $q(t) \to 0$ as $t \to \infty$. In the underdamped case when $q(0) = q_0$, the charge on the capacitor oscillates as it decays; in other words, the capacitor is charging and discharging as $t \to \infty$. When $E(t) = 0$ and $R = 0$, the circuit is said to be undamped and the electrical vibrations do not approach zero as t increases without bound; the response of the circuit is **simple harmonic.**

EXAMPLE 1 Simple Harmonic Response

Consider an L-C series circuit in which $E(t) = 0$. Determine the charge $q(t)$ on the capacitor for $t > 0$ if its initial charge is q_0 and if initially there is no current flowing in the circuit.

Solution In an L-C circuit there is no resistor, so from (3) we obtain

$$L\frac{d^2q}{dt^2} + \frac{1}{C}q = 0.$$

The initial conditions are $q(0) = q_0$ and $i(0) = 0$. Since $q'(t) = i(t)$, the latter condition is the same as $q'(0) = 0$. The general solution of the differential equation is

$$q(t) = c_1 \cos\frac{1}{\sqrt{LC}}t + c_2 \sin\frac{1}{\sqrt{LC}}t.$$

Now the initial conditions imply $c_1 = q_0$ and $c_2 = 0$, so

$$q(t) = q_0 \cos\frac{1}{\sqrt{LC}}t.$$

In Example 1 if we want to find the current in the circuit, we use $i(t) = q'(t)$:

$$i(t) = -\frac{q_0}{\sqrt{LC}} \sin\frac{1}{\sqrt{LC}}t.$$

EXAMPLE 2 Underdamped Series Circuit

Find the charge $q(t)$ on the capacitor in an L-R-C series circuit when $L = 0.25$ henry, $R = 10$ ohms, $C = 0.001$ farad, $E(t) = 0$, $q(0) = q_0$ coulombs, and $i(0) = 0$.

Solution Since $1/C = 1000$, equation (3) becomes

$$\frac{1}{4}q'' + 10q' + 1000q = 0 \quad \text{or} \quad q'' + 40q' + 4000q = 0.$$

Solving this homogeneous equation in the usual manner, we find that the circuit is underdamped and

$$q(t) = e^{-20t}(c_1 \cos 60t + c_2 \sin 60t).$$

Applying the initial conditions, we find $c_1 = q_0$ and $c_2 = q_0/3$. Thus the solution is given by

$$q(t) = q_0 e^{-20t}\left(\cos 60t + \frac{1}{3}\sin 60t\right).$$

The solution in Example 2 can be written as a single sine function using the method discussed in Section 5.2. From (15) of that section, we find that

$$q(t) = \frac{q_0\sqrt{10}}{3} e^{-20t} \sin(60t + 1.249).$$

When there is an impressed voltage $E(t)$ on the circuit, the electrical vibrations are said to be *forced.* Note in Example 1 that the free electrical vibrations are simple harmonic with period $2\pi/(1/\sqrt{LC}) = 2\pi\sqrt{LC}$ and frequency $1/(2\pi\sqrt{LC})$. If a periodic voltage $E(t)$ with the same frequency were impressed on the circuit, the system would be in **resonance.** In the case when $R \neq 0$, the complementary function $q_c(t)$ of (3) is called a **transient solution.** If $E(t)$ is periodic or a constant, then the particular solution $q_p(t)$ of (3) is a **steady-state solution.**

EXAMPLE 3 Steady-State Current

Find the steady-state solution $q_p(t)$ and the **steady-state current** in an L-R-C series circuit when the impressed voltage is $E(t) = E_0 \sin \gamma t$.

Solution The steady-state solution $q_p(t)$ is a particular solution of the differential equation

$$L\frac{d^2q}{dt^2} + R\frac{dq}{dt} + \frac{1}{C}q = E_0 \sin \gamma t.$$

Using the method of undetermined coefficients, we assume a particular solution of the form

$$q_p(t) = A \sin \gamma t + B \cos \gamma t. \tag{4}$$

Substituting (4) into the differential equation, simplifying, and equating coefficients gives

$$A = \frac{E_0(L\gamma - 1/C\gamma)}{-\gamma\left[L^2\gamma^2 - \dfrac{2L}{C} + \dfrac{1}{C^2\gamma^2} + R^2\right]} \quad \text{and} \quad B = \frac{E_0 R}{-\gamma\left[L^2\gamma^2 - \dfrac{2L}{C} + \dfrac{1}{C^2\gamma^2} + R^2\right]}.$$

It is convenient to express A and B in terms of some new symbols.

If $\quad X = L\gamma - \dfrac{1}{C\gamma},\quad$ then $\quad X^2 = L^2\gamma^2 - \dfrac{2L}{C} + \dfrac{1}{C^2\gamma^2}.$

If $\quad Z = \sqrt{X^2 + R^2},\quad$ then $\quad Z^2 = L^2\gamma^2 - \dfrac{2L}{C} + \dfrac{1}{C^2\gamma^2} + R^2.$

Therefore $A = E_0X/(-\gamma Z^2)$ and $B = E_0R/(-\gamma Z^2)$ and so the steady-state charge is

$$q_p(t) = -\frac{E_0X}{\gamma Z^2} \sin \gamma t - \frac{E_0R}{\gamma Z^2} \cos \gamma t.$$

Now the steady-state current is given by $i_p(t) = q_p'(t)$:

$$i_p(t) = \frac{E_0}{Z}\left(\frac{R}{Z} \sin \gamma t - \frac{X}{Z} \cos \gamma t\right). \tag{5}$$

The quantities $X = L\gamma - 1/C\gamma$ and $Z = \sqrt{X^2 + R^2}$ defined in Example 3 are called, respectively, the **reactance** and **impedance** of the circuit. Both the reactance and the impedance are measured in ohms.

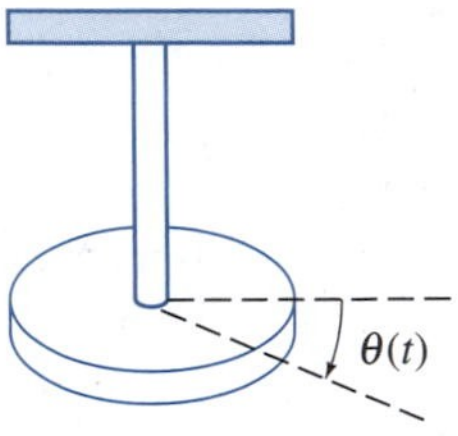

Figure 5.29

Twisted Shaft The differential equation governing the torsional motion of a weight suspended from the end of an elastic shaft is

$$I\frac{d^2\theta}{dt^2} + c\frac{d\theta}{dt} + k\theta = T(t). \tag{6}$$

As shown in Figure 5.29, the function $\theta(t)$ represents the amount of twist of the weight at any time.

By comparing equations (3) and (6) with (1), we see that, with the exception of terminology, there is absolutely no difference between the mathematics of vibrating springs, simple series circuits, and torsional vibrations. Table 5.1 gives a comparison of the analogous parts of these three kinds of systems.

Table 5.1

Mechanical	Series electrical	Torsional
m (mass)	L (inductance)	I (moment of inertia)
β (damping)	R (resistance)	c (damping)
k (spring constant)	$\frac{1}{C}$ (reciprocal of capacitance—called elastance)	k (elastic shaft constant)
$f(t)$ (applied force)	$E(t)$ (impressed voltage)	$T(t)$ (applied torque)

Simple Pendulum In Example 3 of Section 1.2 it was seen that the angular displacements θ of a simple pendulum are described by the nonlinear second-order equation

$$\frac{d^2\theta}{dt^2} + \frac{g}{l}\sin\theta = 0,$$

where l is the length of the pendulum rod. For small displacements $\sin\theta$ is replaced by θ, and the resulting differential equation

$$\frac{d^2\theta}{dt^2} + \frac{g}{l}\theta = 0 \tag{7}$$

indicates that the pendulum exhibits simple harmonic motion. Inspection of the solution (7) reveals that the period of small oscillations is given by the familiar formula from physics $T = 2\pi\sqrt{l/g}$.

EXERCISES 5.4

Answers to odd-numbered problems begin on page A-11.

In Problems 1 and 2 find the charge on the capacitor and the current in the given L-C series circuit. Assume $q(0) = 0$ and $i(0) = 0$.

1. $L = 1$ henry, $C = \frac{1}{16}$ farad, $E(t) = 60$ volts

2. $L = 5$ henrys, $C = 0.01$ farad, $E(t) = 20t$ volts

In Problems 3 and 4 without solving (3) determine whether the given L-R-C series circuit is overdamped, critically damped, or underdamped.

3. $L = 3$ henrys, $R = 10$ ohms, $C = 0.1$ farad

4. $L = 1$ henry, $R = 20$ ohms, $C = 0.01$ farad

5. Find the charge on the capacitor in an L-R-C series circuit at $t = 0.01$ second when $L = 0.05$ henry, $R = 2$ ohms, $C = 0.01$ farad, $E(t) = 0$ volts, $q(0) = 5$ coulombs, and $i(0) = 0$ amperes. Determine the first time at which the charge on the capacitor is equal to zero.

6. Find the charge on the capacitor in an L-R-C series circuit when $L = \frac{1}{4}$ henry, $R = 20$ ohms, $C = \frac{1}{300}$ farad, $E(t) = 0$ volts, $q(0) = 4$ coulombs, and $i(0) = 0$ amperes. Is the charge on the capacitor ever equal to zero?

In Problems 7 and 8 find the charge on the capacitor and the current in the given L-R-C series circuit. Find the maximum charge on the capacitor.

7. $L = \frac{5}{3}$ henrys, $R = 10$ ohms, $C = \frac{1}{30}$ farad, $E(t) = 300$ volts, $q(0) = 0$ coulombs, $i(0) = 0$ amperes

8. $L = 1$ henry, $R = 100$ ohms, $C = 0.0004$ farad, $E(t) = 30$ volts, $q(0) = 0$ coulombs, $i(0) = 2$ amperes

9. Find the steady-state charge and the steady-state current in an L-R-C series circuit when $L = 1$ henry, $R = 2$ ohms, $C = 0.25$ farad, and $E(t) = 50 \cos t$ volts.

10. Show that the amplitude of the steady-state current in the L-R-C series circuit in Example 3 is given by E_0/Z, where Z is the impedance of the circuit.

11. Show that the steady-state current in an L-R-C series circuit when $L = \frac{1}{2}$ henry, $R = 20$ ohms, $C = 0.001$ farad, and $E(t) = 100 \sin 60t$ volts is given by $i_p(t) = (4.160)\sin(60t - 0.588)$. [*Hint*: Use Problem 10.]

12. Find the steady-state current in an L-R-C series circuit when $L = \frac{1}{2}$ henry, $R = 20$ ohms, $C = 0.001$ farad, and $E(t) = 100 \sin 60t + 200 \cos 40t$ volts.

13. Find the charge on the capacitor in an L-R-C series circuit when $L = \frac{1}{2}$ henry, $R = 10$ ohms, $C = 0.01$ farad, $E(t) = 150$ volts, $q(0) = 1$ coulomb, and $i(0) = 0$ amperes. What is the charge on the capacitor after a long time?

14. Show that if L, R, C, and E_0 are constant, then the amplitude of the steady-state current in Example 3 is a maximum when $\gamma = 1/\sqrt{LC}$. What is the maximum amplitude?

15. Show that if L, R, E_0, and γ are constant, then the amplitude of the steady-state current in Example 3 is a maximum when the capacitance is $C = 1/L\gamma^2$.

16. Find the charge on the capacitor and the current in an L-C circuit when $L = 0.1$ henry, $C = 0.1$ farad, $E(t) = 100 \sin \gamma t$ volts, $q(0) = 0$ coulombs, and $i(0) = 0$ amperes.

17. Find the charge on the capacitor and the current in an L-C circuit when $E(t) = E_0 \cos \gamma t$ volts, $q(0) = q_0$ coulombs, and $i(0) = i_0$ amperes.

18. In Problem 17 find the current when the circuit is in resonance.

19. Find the equation of motion describing the small displacements $\theta(t)$ of a simple pendulum of length 2 ft released at $t = 0$ with a displacement of $\frac{1}{2}$ radian to the right of the vertical and angular velocity of $2\sqrt{3}$ ft/s to the right. What are the amplitude, period, and frequency of motion?

20. In Problem 19 at what times does the pendulum pass through its equilibrium position? At what times does the pendulum attain its extreme angular displacements on either side of its equilibrium position?

CHAPTER 5 REVIEW

When a mass is attached to a spring, it stretches to a position where the restoring force ks of the spring is balanced by the weight mg. Any subsequent motion is then measured x units (feet in the engineering system) above or below this **equilibrium position.** When the mass is above the equilibrium position, we adopt the convention that $x < 0$, whereas when the mass is below the equilibrium position, we take $x > 0$.

The differential equation of motion is obtained by equating Newton's second law $F = ma = m(d^2x/dt^2)$ with the net force acting on the mass at any time. We distinguish three cases.

Case I The equation

$$m\frac{d^2x}{dt^2} = -kx \quad \text{or} \quad \frac{d^2x}{dt^2} + \omega^2 x = 0, \tag{1}$$

where $\omega^2 = k/m$, describes the motion under the assumptions that no damping force and no external impressed forces are acting on the system. The solution of (1) is $x(t) = c_1 \cos \omega t + c_2 \sin \omega t$ and the mass is said to exhibit **simple harmonic motion.** The constants c_1 and c_2 are determined by the initial position $x(0)$ and the initial velocity $x'(0)$ of the mass.

Case II When a damping force is present, the differential equation becomes

$$m\frac{d^2x}{dt^2} = -kx - \beta\frac{dx}{dt} \quad \text{or} \quad \frac{d^2x}{dt^2} + 2\lambda\frac{dx}{dt} + \omega^2 x = 0, \tag{2}$$

where $\beta > 0, 2\lambda = \beta/m$, and $\omega^2 = k/m$. The resulting motion is said to be **overdamped, critically damped,** or **underdamped** accordingly as $\lambda^2 - \omega^2 > 0$, $\lambda^2 - \omega^2 = 0$, or $\lambda^2 - \omega^2 < 0$. The respective solutions of (2) are then

$$x(t) = c_1 e^{m_1 t} + c_2 e^{m_2 t},$$

where $m_1 = -\lambda + \sqrt{\lambda^2 - \omega^2}, m_2 = -\lambda - \sqrt{\lambda^2 - \omega^2}$;

$$x(t) = c_1 e^{m_1 t} + c_2 t e^{m_1 t},$$

where $m_1 = -\lambda$; and

$$x(t) = e^{-\lambda t}(c_1 \cos \sqrt{\omega^2 - \lambda^2}t + c_2 \sin \sqrt{\omega^2 - \lambda^2}t).$$

In each case the damping force is responsible for the displacements becoming negligible for large time; that is, $x \to 0$ as $t \to \infty$.

The motion described in Cases I and II is said to be **free motion.**

Case III When an external force is impressed on the system for $t > 0$, the differential equation becomes

$$m\frac{d^2x}{dt^2} = -kx - \beta\frac{dx}{dt} + f(t) \quad \text{or} \quad \frac{d^2x}{dt^2} + 2\lambda\frac{dx}{dt} + \omega^2 x = F(t), \qquad \textbf{(3)}$$

where λ and ω^2 are defined in Case II. The solution of the nonhomogeneous equation (3) is $x(t) = x_c + x_p$.

Since the complementary function x_c always contains the factor $e^{-\lambda t}$, it will be **transient;** that is, $x_c \to 0$ as $t \to \infty$. If $F(t)$ is periodic, then x_p will be a **steady-state solution.**

In the absence of a damping force, an impressed periodic force can cause the amplitudes of vibration to become very large. If the frequency of the external force is the same as the frequency $\omega/2\pi$ of free vibrations, we say that the system is in a state of **pure resonance.** In this case the amplitudes of vibrations become unbounded as $t \to \infty$. In the presence of a damping force, amplitudes of oscillatory motion are always bounded. However, large and potentially destructive amplitudes can occur.

When a series circuit containing an inductor, resistor, and capacitor is driven by an electromotive force $E(t)$, the resulting differential equations for the charge $q(t)$ or the current $i(t)$ are quite similar to equation (3). Hence the analysis of such circuits is the same as outlined above.

CHAPTER 5 REVIEW EXERCISES

Answers to odd-numbered problems begin on page A-11.

Answer Problems 1–9 without referring back to the text. Fill in the blank or answer true or false.

1. If a 10-lb weight stretches a spring 2.5 ft, a 32-lb weight will stretch it ________ ft.
2. The period of simple harmonic motion of an 8-lb weight attached to a spring whose constant is 6.25 lb/ft is ________ seconds.
3. The differential equation of a weight on a spring is $x'' + 16x = 0$. If the weight is released at $t = 0$ from 1 m above the equilibrium position with a downward velocity of 3 m/s, the amplitude of vibrations is ________ m.
4. Pure resonance cannot take place in the presence of a damping force. ________
5. In the presence of damping, the displacements of a weight on a spring will always approach zero as $t \to \infty$. ________
6. A weight on a spring whose motion is critically damped can possibly pass through the equilibrium position twice. ________
7. At critical damping any increase in damping will result in an ________ system.

8. When simple harmonic motion is described by $x = (\sqrt{2}/2)\sin(2t + \phi)$, the phase angle ϕ is ______ when $x(0) = -\frac{1}{2}$ and $x'(0) = 1$.

9. A 16-lb weight attached to a spring exhibits simple harmonic motion. If the frequency of oscillations is $3/2\pi$ vibrations/second, the spring constant is ______.

10. A 12-lb weight stretches a spring 2 ft. The weight is released from a point 1 ft below the equilibrium position with an upward velocity of 4 ft/s.

(a) Find the equation describing the resulting simple harmonic motion.

(b) What are the amplitude, period, and frequency of motion?

(c) At what times does the weight return to the point 1 ft below the equilibrium position?

(d) At what times does the weight pass through the equilibrium position moving upward? moving downward?

(e) What is the velocity of the weight at $t = 3\pi/16$ second?

(f) At what times is the velocity zero?

11. A force of 2 lb stretches a spring 1 ft. With one end held fixed, an 8-lb weight is attached to the other end and the system lies on a table that imparts a frictional force numerically equal to $\frac{3}{2}$ times the instantaneous velocity. Initially the weight is displaced 4 in. above the equilibrium position and released from rest. Find the equation of motion if the motion takes place along a horizontal straight line that is taken as the x-axis.

12. A 32-lb weight stretches a spring 6 in. The weight moves through a medium offering a damping force numerically equal to β times the instantaneous velocity. Determine the values of β for which the system will exhibit oscillatory motion.

13. A spring with constant $k = 2$ is suspended in a liquid that offers a damping force numerically equal to 4 times the instantaneous velocity. If a mass m is suspended from the spring, determine the values of m for which the subsequent free motion is nonoscillatory.

14. The vertical motion of a weight attached to a spring is described by the initial-value problem

$$\frac{1}{4}\frac{d^2x}{dt^2} + \frac{dx}{dt} + x = 0, \quad x(0) = 4, \quad x'(0) = 2.$$

Determine the maximum vertical displacement.

15. A 4-lb weight stretches a spring 18 in. A periodic force equal to $f(t) = \cos\gamma t + \sin\gamma t$ is impressed on the system starting at $t = 0$. In the absence of a damping force, for what value of γ will the system be in a state of pure resonance?

16. Find a particular solution for $\dfrac{d^2x}{dt^2} + 2\lambda\dfrac{dx}{dt} + \omega^2 x = A$, where A is a constant force.

17. A 4-lb weight is suspended from a spring whose constant is 3 lb/ft. The entire system is immersed in a fluid offering a damping force numerically equal to the instantaneous velocity. Beginning at $t = 0$, an external force equal to $f(t) = e^{-t}$ is impressed on the system. Determine the equation of motion if the weight is released from rest at a point 2 ft below the equilibrium position.

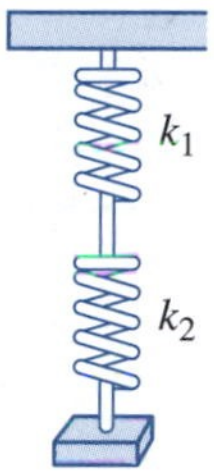

Figure 5.30

18. **(a)** Two springs are attached in series as shown in Figure 5.30. If the mass of each spring is ignored, show that the effective spring constant k is given by $1/k = 1/k_1 + 1/k_2$.

(b) A weight of W lb stretches one spring $\frac{1}{2}$ ft and stretches a different spring $\frac{1}{4}$ ft. The two springs are attached as in the figure and the weight W is then attached to the double spring. Assume that the motion is free and that there is no damping force present. Determine the equation of motion if the weight is released at a point 1 ft below the equilibrium position with a downward velocity of $\frac{2}{3}$ ft/s.

(c) Show that the maximum speed of the weight is $\frac{2}{3}\sqrt{3g + 1}$.

19. A series circuit contains an inductance of $L = 1$ henry, a capacitance of $C = 10^{-4}$ farad, and an electromotive force of $E(t) = 100 \sin 50t$ volts. Initially the charge q and current i are zero.

(a) Find the equation for the charge at time t.

(b) Find the equation for the current at time t.

(c) Find the times for which the charge on the capacitor is zero.

20. Show that the current $i(t)$ in an L-R-C series circuit satisfies the differential equation

$$L\frac{d^2i}{dt^2} + R\frac{di}{dt} + \frac{1}{C}i = E'(t),$$

where $E'(t)$ denotes the derivative of $E(t)$.

Essay

by
Gilbert N. Lewis
Department of Mathematical and Computer Sciences, Michigan Technological University

TACOMA NARROWS SUSPENSION BRIDGE COLLAPSE

As can be seen from equation (9) and Figure 5.22 of Section 5.3, the oscillations in the case of resonance become very large as time increases. In a physical system this would be catastrophic, since the ever-increasing amplitudes of oscillation would tear the system apart. Two historical examples of this have already been given (airplane wings and soldiers marching across bridges). In those examples, periodic forces of the same frequency as the natural frequency of the structures were set up in the direction of the vibrations, leading to the resonant destruction of the structures.

Another example that has, until very recently, been used to illustrate the phenomenon of resonance is the collapse of the Tacoma Narrows Bridge in the state of Washington. The bridge was completed and open to traffic in the summer of 1940. It was soon noticed that large vibrations of the roadbed were induced when wind blew across it. "Galloping Gertie," as the bridge was called, became a tourist attraction; people liked to watch it vibrate and even take an exciting roller coaster ride across it. Finally, on November 7, 1940, the entire span was shaken apart by the large vibrations and collapsed. See Figure 5.31.

Figure 5.31

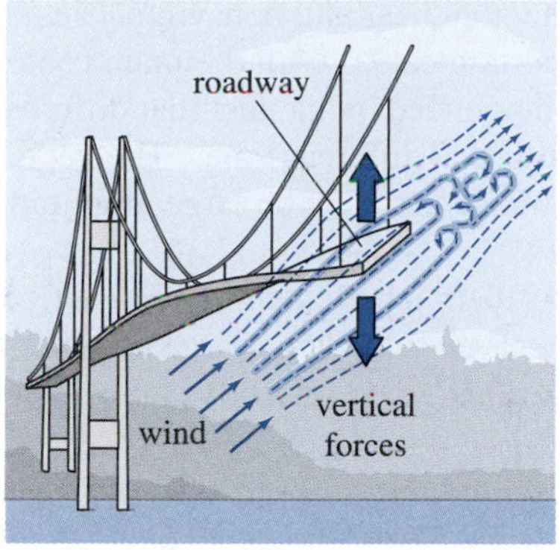

Figure 5.32

For fifty years, the popular and easily explained reason for the collapse was resonance. It was thought that, as the wind blew horizontally across the roadway, vortices (wind swirls) were shed from the downward side of the bridge alternately from above and below, thus setting up a periodic vertical force acting in the same direction as the vibration of the bridge. See Figure 5.32. It was further hypothesized that the frequency of this periodic force exactly matched the natural frequency of the bridge, thus setting up large amplitude vibrations (see equation (9)) and causing the bridge to fail. This explanation was (perhaps erroneously) attributed to the noted engineer von Karman. In his autobiography, he explained that the bridge collapse was indeed due to the von Karman vortices [4]. However, in a technical report to the Federal Works Agency, he and his coauthors concluded "it is very improbable that resonance with alternating vortices plays an important role in the oscillations of suspension bridges" [1]. Unfortunately, the resonance explanation has remained firmly entrenched in the popular and mathematical literature.

Resonance is a linear phenomenon (notice the linear differential equation (7) of Section 5.3). In addition, it is entirely dependent on an exact match between the natural frequency of the bridge (or portions thereof) and the frequency of any externally applied periodic forces. Furthermore, it requires that absolutely no damping be present in the system ($\lambda = 0$ in equation (3) of Section 5.3). It should not be surprising, therefore, that resonance was *not* the dominant factor in the collapse of the Tacoma Narrows Bridge.

If not resonance, then what? Recent research has provided an alternative explanation for the collapse of the bridge. Lazer and McKenna (see [3] for a good review article) contend that nonlinear effects, and not linear resonance, were the main factors leading to the large oscillations of the bridge [2]. There is no doubt that the wind blowing across the roadbed provided the external driving force that caused the motion. This force could even have been partly due to the shedding of vortices, as von Karman suggested. However, nonlinear interactions within the bridge and between the bridge and the external forces are a more probable culprit for the cause of the collapse.

In the linear theory, the supporting cables act as rigid elastic rods (springs). As such, the mathematical model gives rise to the linear differential equation (3), or equation (7) of Section 5.3, if no damping is present. This latter case leading to resonance is the only possible situation in the linear theory in which small external forces could lead to large amplitude vibrations. As has been mentioned, this scenario is highly unlikely.

On the other hand, nonlinear effects could yield large amplitude vibrations from small amplitude forces. They could also explain the transition from one-dimensional axial (lengthwise) oscillations to transverse (across the roadway) oscillations, which were ultimately responsible for the collapse of the bridge.

The basic idea in the Lazer–McKenna model is the following. When the vertical supporting cables are under tension (the weight of the roadbed is pulling down on them), the cables act as rigid elastic linear rods, in which case the differential equation is linear. However, once oscillations are set up in the bridge system by external forces (wind and possibly earthquakes), the cables will not always be under tension, and there will be only gravity acting downward on the roadbed. In other words, the Hooke's law term, $(k/m)x$ in equation (2) of Section 5.3, will be missing. This transition from one type of linear differential equation to another is one source of the nonlinearity. Other sources of nonlinearity in suspension bridges might include the nonlinear and nonsymmetric design of the bridge or interactions of the supporting cables with the main cables or supporting towers. This nonlinearity is compounded by the fact that different cables in the bridge system may be under tension at different times. The net result of this nonlinearity, contend Lazer and McKenna, may be large amplitude oscillations under moderate external forces.

One other aspect of nonlinear equations is their unpredictability. This might explain, for example, why the bridge was observed to undergo large amplitude oscillations under light wind conditions yet seemed to be perfectly stable under higher wind conditions.

As Lazer and McKenna point out, the nonlinear theory of suspension bridges has not been thoroughly developed yet. However, numerical simulations of their suspension bridge model agree with actual observations. It seems likely that this approach will yield a more accurate explanation of the failure of the Tacoma Narrows Suspension Bridge.

REFERENCES

1. Amann, O. H., T. von Karman, and G. B. Woodruff. *The Failure of the Tacoma Narrows Bridge.* Federal Works Agency, 1941.
2. Lazer, A. C., and P. J. McKenna. "Large-Amplitude Periodic Oscillations in Suspension Bridges: Some New Connections with Nonlinear Analysis." *SIAM Review* 32 (Dec. 1990): 537–578.
3. Peterson, I. "Rock and Roll Bridge." *Science News* 137 (1991): 344–346.
4. von Karman, T. *The Wind and Beyond, Theodore von Karman, Pioneer in Aviation and Pathfinder in Space.* Boston: Little, Brown, 1967.

6 DIFFERENTIAL EQUATIONS WITH VARIABLE COEFFICIENTS

INTRODUCTION

Up to now we have solved linear differential equations of order two or higher only in the case when the equation has constant coefficients. In applications, higher-order linear equations with nonconstant coefficients are just as important as, if not more so than, equations with constant coefficients. For example, if we wish to find the temperature or potential u in the region bounded between two concentric spheres as shown in the accompanying figure, then under some circumstances we have to solve the differential equation

$$r\frac{d^2u}{dr^2} + 2\frac{du}{dr} = 0,$$

where the variable $r > 0$ represents the radial distance measured outward from the center of the spheres. Differential equations with variable coefficients such as

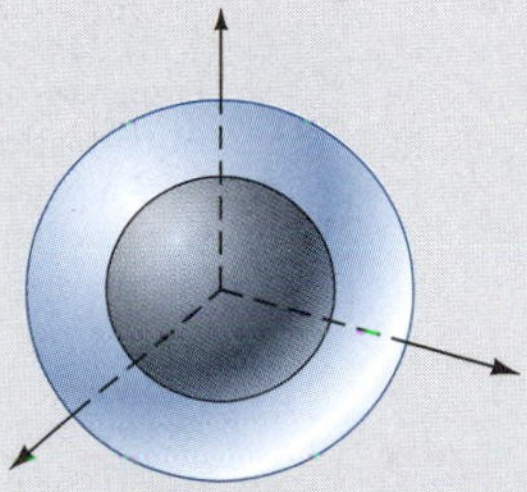

(*continued*)

$$x^2y'' + xy' + (x^2 - \nu^2)y = 0,$$
$$(1 - x^2)y'' - 2xy' + n(n + 1)y = 0,$$
and
$$y'' - 2xy' + 2ny = 0$$

occur in applications ranging from potential problems, temperature distributions, and vibrational phenomena to quantum mechanics.

The same ease with which we solved constant-coefficient differential equations does not carry over to equations with variable coefficients. In fact we cannot expect to be able to express the solution of even a simple linear equation such as $y'' - 2xy = 0$ in terms of the familiar elementary functions constructed from powers of x, sines, cosines, logarithms, and exponentials. The best we can do for the equation $y'' - 2xy = 0$ is to find a solution in the form of an infinite series. However, there is one type of variable-coefficient differential equation, called the Cauchy-Euler equation, whose general solution can always be written in terms of elementary functions. We begin the chapter with this equation.

6.1 CAUCHY-EULER EQUATION

• Cauchy-Euler equation • Auxiliary equation • Reduction to constant coefficients

Any linear differential equation of the form

$$a_nx^n \frac{d^ny}{dx^n} + a_{n-1}x^{n-1} \frac{d^{n-1}y}{dx^{n-1}} + \cdots + a_1x \frac{dy}{dx} + a_0y = g(x),$$

where $a_n, a_{n-1}, \ldots, a_0$ are constants, is said to be a **Cauchy-Euler equation,*** or **equidimensional equation.** The distinguishing characteristic of this type of differential equation is that the *degree* of each monomial coefficient matches the *order* of differentiation:

$$a_n\underset{\text{same}}{x^n} \frac{d^{\overset{\downarrow}{n}}y}{dx^n} + a_{n-1}x^{n-1} \frac{d^{n-1}y}{dx^{n-1}} + \cdots.$$

(same ↓ ↓ above x^n and d^n; same ↓ ↓ above x^{n-1} and d^{n-1})

* AUGUSTIN-LOUIS CAUCHY (1789–1857) Born during a period of upheaval in French history, Augustin-Louis Cauchy was destined to initiate a revolution of his own—in mathematics. For many original contributions, but especially for his efforts in clarifying mathematical obscurities and his incessant demand for satisfactory definitions and rigorous proofs of theorems, Cauchy is often called "the father of modern analysis." A prolific writer whose output was surpassed by only a few, Cauchy produced nearly 800 papers in astronomy, physics, and mathematics. It was he who developed the concept of convergence of an infinite series and the theory of functions of a complex variable. The same mind that was always open and inquiring in science and mathematics was narrow and unquestioning in many other areas. Outspoken and arrogant, Cauchy was passionate on political and religious issues. His stands on these issues often alienated him from his colleagues.

For the sake of discussion, we confine our attention to solving the homogeneous second-order equation

$$ax^2 \frac{d^2y}{dx^2} + bx \frac{dy}{dx} + cy = 0.$$

The solution of higher-order equations follows analogously. Also, we can solve the nonhomogeneous equation

$$ax^2 \frac{d^2y}{dx^2} + bx \frac{dy}{dx} + cy = g(x)$$

by variation of parameters once we have determined the complementary function $y_c(x)$.

Note The coefficient of d^2y/dx^2 is zero at $x = 0$. Hence in order to guarantee that the fundamental results of Theorem 4.1 are applicable to the Cauchy-Euler equation, we shall confine our attention to finding the general solution on the interval $(0, \infty)$. Solutions on the interval $(-\infty, 0)$ can be obtained by substituting $t = -x$ in the differential equation.

Method of Solution We try a solution of the form $y = x^m$, where m is to be determined. The first and second derivatives are, respectively,

$$\frac{dy}{dx} = mx^{m-1} \quad \text{and} \quad \frac{d^2y}{dx^2} = m(m-1)x^{m-2}.$$

Consequently the differential equation becomes

$$\begin{aligned} ax^2 \frac{d^2y}{dx^2} + bx \frac{dy}{dx} + cy &= ax^2 \cdot m(m-1)x^{m-2} + bx \cdot mx^{m-1} + cx^m \\ &= am(m-1)x^m + bmx^m + cx^m \\ &= x^m(am(m-1) + bm + c). \end{aligned}$$

Thus $y = x^m$ is a solution of the differential equation whenever m is a solution of the **auxiliary equation**

$$am(m-1) + bm + c = 0 \quad \text{or} \quad am^2 + (b-a)m + c = 0. \tag{1}$$

There are three different cases to be considered, depending on whether the roots of this quadratic equation are real and distinct, real and equal, or complex. In the last case, the roots appear as a conjugate pair.

Case I: Distinct Real Roots Let m_1 and m_2 denote the real roots of (1) such that $m_1 \neq m_2$. Then $y_1 = x^{m_1}$ and $y_2 = x^{m_2}$ form a fundamental set of solutions. Hence the general solution is

$$y = c_1x^{m_1} + c_2x^{m_2}. \tag{2}$$

EXAMPLE 1 Distinct Roots

Solve $x^2 \dfrac{d^2y}{dx^2} - 2x \dfrac{dy}{dx} - 4y = 0.$

Solution Rather than just memorizing equation (1), it is preferable to assume $y = x^m$ as the solution a few times in order to understand the origin and the difference between this new form of the auxiliary equation and that obtained in Chapter 4. Differentiate twice,

$$\frac{dy}{dx} = mx^{m-1}, \quad \frac{d^2y}{dx^2} = m(m-1)x^{m-2},$$

and substitute back into the differential equation,

$$\begin{aligned} x^2 \frac{d^2y}{dx^2} - 2x \frac{dy}{dx} - 4y &= x^2 \cdot m(m-1)x^{m-2} - 2x \cdot mx^{m-1} - 4x^m \\ &= x^m(m(m-1) - 2m - 4) \\ &= x^m(m^2 - 3m - 4) = 0, \end{aligned}$$

if $m^2 - 3m - 4 = 0$. Now $(m+1)(m-4) = 0$ implies $m_1 = -1$, $m_2 = 4$,

so $$y = c_1x^{-1} + c_2x^4.$$

Case II: Repeated Real Roots If the roots of (1) are repeated (that is, $m_1 = m_2$), then we obtain only one solution—namely, $y = x^{m_1}$. When the roots of the quadratic equation $am^2 + (b-a)m + c = 0$ are equal, the discriminant of the coefficients is necessarily zero. It follows from the quadratic formula that the root must be $m_1 = -(b-a)/2a$.

Now we can construct a second solution y_2, using (4) of Section 4.2. We first write the Cauchy-Euler equation in the form

$$\frac{d^2y}{dx^2} + \frac{b}{ax}\frac{dy}{dx} + \frac{c}{ax^2}y = 0$$

and make the identification $P(x) = b/ax$. Thus

$$\begin{aligned} y_2 &= x^{m_1} \int \frac{e^{-\int (b/ax)\,dx}}{(x^{m_1})^2}\,dx \\ &= x^{m_1} \int \frac{e^{-(b/a)\ln x}}{x^{2m_1}}\,dx \\ &= x^{m_1} \int x^{-b/a} \cdot x^{-2m_1}\,dx && \leftarrow e^{-(b/a)\ln x} = e^{\ln x^{-b/a}} = x^{-b/a} \\ &= x^{m_1} \int x^{-b/a} \cdot x^{(b-a)/a}\,dx && \leftarrow -2m_1 = (b-a)/a \\ &= x^{m_1} \int \frac{dx}{x} = x^{m_1} \ln x. \end{aligned}$$

The general solution is then

$$y = c_1 x^{m_1} + c_2 x^{m_1} \ln x. \tag{3}$$

EXAMPLE 2 **Repeated Roots**

Solve $4x^2 \dfrac{d^2y}{dx^2} + 8x \dfrac{dy}{dx} + y = 0.$

Solution The substitution $y = x^m$ yields

$$4x^2 \frac{d^2y}{dx^2} + 8x \frac{dy}{dx} + y = x^m(4m(m-1)8m+1) = x^m(4m^2+4m+1) = 0$$

when $4m^2 + 4m + 1 = 0$ or $(2m+1)^2 = 0$. Since $m_1 = -\frac{1}{2}$, the general solution is

$$y = c_1 x^{-1/2} + c_2 x^{-1/2} \ln x.$$

For higher-order equations, if m_1 is a root of multiplicity k, then it can be shown that

$$x^{m_1}, \quad x^{m_1} \ln x, \quad x^{m_1}(\ln x)^2, \quad \ldots, \quad x^{m_1}(\ln x)^{k-1}$$

are k linearly independent solutions. Correspondingly, the general solution of the differential equation must then contain a linear combination of these k solutions.

Case III: Conjugate Complex Roots If the roots of (1) are the conjugate pair $m_1 = \alpha + i\beta$, $m_2 = \alpha - i\beta$, where α and $\beta > 0$ are real, then a solution is

$$y = C_1 x^{\alpha + i\beta} + C_2 x^{\alpha - i\beta}.$$

But, as in the case of equations with constant coefficients, when the roots of the auxiliary equation are complex, we wish to write the solution in terms of real functions only. We note the identity

$$x^{i\beta} = (e^{\ln x})^{i\beta} = e^{i\beta \ln x},$$

which, by Euler's formula, is the same as

$$x^{i\beta} = \cos(\beta \ln x) + i \sin(\beta \ln x).$$

Similarly we have

$$x^{-i\beta} = \cos(\beta \ln x) - i \sin(\beta \ln x).$$

Adding and subtracting the last two results yields, respectively,

$$x^{i\beta} + x^{-i\beta} = 2\cos(\beta \ln x) \quad \text{and} \quad x^{i\beta} - x^{-i\beta} = 2i \sin(\beta \ln x).$$

From the fact that $y = C_1x^{\alpha+i\beta} + C_2x^{\alpha-i\beta}$ is a solution of $ax^2y'' + bxy' + cy = 0$ for any values of the constants C_1 and C_2 we see that

$$y_1 = x^\alpha(x^{i\beta} + x^{-i\beta}) \quad (C_1 = C_2 = 1)$$

$$y_2 = x^\alpha(x^{i\beta} - x^{-i\beta}) \quad (C_1 = 1, C_2 = -1)$$

and

$$y_1 = 2x^\alpha(\cos(\beta \ln x)), \qquad y_2 = 2ix^\alpha(\sin(\beta \ln x))$$

are also solutions. Since $W(x^\alpha\cos(\beta \ln x), x^\alpha\sin(\beta \ln x)) = \beta x^{2\alpha-1} \neq 0, \beta > 0$ on the interval $(0, \infty)$, we conclude that

$$y_1 = x^\alpha \cos(\beta \ln x) \quad \text{and} \quad y_2 = x^\alpha \sin(\beta \ln x)$$

constitute a fundamental set of real solutions of the differential equation. Hence, the general solution is

$$y = x^\alpha[c_1 \cos(\beta \ln x) + c_2 \sin(\beta \ln x)]. \tag{4}$$

EXAMPLE 3 An Initial-Value Problem

Solve the initial-value problem

$$x^2\frac{d^2y}{dx^2} + 3x\frac{dy}{dx} + 3y = 0, \quad y(1) = 1, \quad y'(1) = -5.$$

Solution We have

$$x^2\frac{d^2y}{dx^2} + 3x\frac{dy}{dx} + 3y = x^m(m(m-1) + 3m + 3) = x^m(m^2 + 2m + 3) = 0$$

when $m^2 + 2m + 3 = 0$. From the quadratic formula we find $m_1 = -1 + \sqrt{2}i$ and $m_2 = -1 - \sqrt{2}i$. If we make the identifications $\alpha = -1$ and $\beta = \sqrt{2}$, we see from (4) that the general solution of the differential equation is

$$y = x^{-1}[c_1 \cos(\sqrt{2} \ln x) + c_2 \sin(\sqrt{2} \ln x)].$$

By applying the conditions $y(1) = 1$, $y'(1) = -5$ to the foregoing solution, we find, in turn, $c_1 = 1$ and $c_2 = -2\sqrt{2}$. Thus the solution to the initial-value problem is

$$y = x^{-1}[\cos(\sqrt{2} \ln x) - 2\sqrt{2} \sin(\sqrt{2} \ln x)].$$

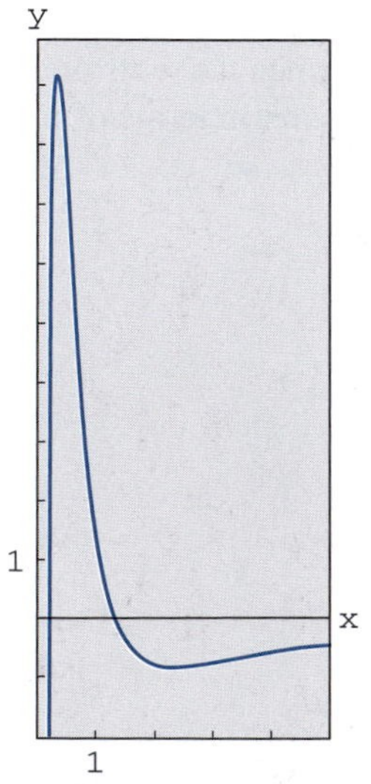

Figure 6.1

The graph of this solution, obtained with the aid of computer software, is given in Figure 6.1.

EXAMPLE 4 Third-Order Equation

Solve the third-order Cauchy-Euler equation

$$x^3\frac{d^3y}{dx^3} + 5x^2\frac{d^2y}{dx^2} + 7x\frac{dy}{dx} + 8y = 0.$$

Solution The first three derivatives of $y = x^m$ are

$$\frac{dy}{dx} = mx^{m-1}, \quad \frac{d^2y}{dx^2} = m(m-1)x^{m-2}, \quad \frac{d^3y}{dx^3} = m(m-1)(m-2)x^{m-3},$$

so the given differential equation becomes

$$\begin{aligned} x^3\frac{d^3y}{dx^3} + 5x^2\frac{d^2y}{dx^2} + 7x\frac{dy}{dx} + 8y &= x^3m(m-1)(m-2)x^{m-3} + 5x^2m(m-1)x^{m-2} + 7xmx^{m-1} + 8x^m \\ &= x^m(m(m-1)(m-2) + 5m(m-1) + 7m + 8) \\ &= x^m(m^3 + 2m^2 + 4m + 8). \end{aligned}$$

In this case we see that $y = x^m$ will be a solution of the differential equation, provided m is a root of the cubic equation

$$m^3 + 2m^2 + 4m + 8 = 0 \quad \text{or} \quad (m+2)(m^2+4) = 0.$$

The roots are $m_1 = -2$, $m_2 = 2i$, $m_3 = -2i$. Hence the general solution is

$$y = c_1x^{-2} + c_2\cos(2\ln x) + c_3\sin(2\ln x).$$

EXAMPLE 5 Using Variation of Parameters

Solve the nonhomogeneous equation

$$x^2y'' - 3xy' + 3y = 2x^4e^x.$$

Solution The substitution $y = x^m$ leads to the auxiliary equation

$$m(m-1) - 3m + 3 = 0 \quad \text{or} \quad (m-1)(m-3) = 0.$$

Thus $$y_c = c_1x + c_2x^3.$$

Before using variation of parameters to find a particular solution $y_p = u_1y_1 + u_2y_2$, recall that the formulas $u_1' = W_1/W$ and $u_2' = W_2/W$, where $W_1 = \begin{vmatrix} 0 & y_2 \\ f(x) & y_2' \end{vmatrix}$, $W_2 = \begin{vmatrix} y_1 & 0 \\ y_1' & f(x) \end{vmatrix}$, and W is the Wronskian of y_1 and y_2, were derived under the assumption that the differential equation has been put into the special form $y'' + P(x)y' + Q(x)y = f(x)$. Therefore we divide the given equation by x^2, and from

$$y'' - \frac{3}{x}y' + \frac{3}{x^2}y = 2x^2e^x$$

we make the identification $f(x) = 2x^2e^x$. Now with $y_1 = x$, $y_2 = x^3$, and

$$W = \begin{vmatrix} x & x^3 \\ 1 & 3x^2 \end{vmatrix} = 2x^3, \quad W = \begin{vmatrix} 0 & x^3 \\ 2x^2e^x & 3x^2 \end{vmatrix} = -2x^5e^x, \quad W_2 = \begin{vmatrix} x & 0 \\ 1 & 2x^2e^x \end{vmatrix} = 2x^3e^x$$

we find

$$u_1' = -\frac{2x^5e^x}{2x^3} = -x^2e^x \quad \text{and} \quad u_2' = \frac{2x^3e^x}{2x^3} = e^x.$$

The integral of the latter function is immediate, but in the case of u_1' we integrate by parts twice. The results are

$$u_1 = -x^2e^x + 2xe^x - 2e^x \quad \text{and} \quad u_2 = e^x.$$

Hence
$$\begin{aligned} y_p &= u_1y_1 + u_2y_2 \\ &= (-x^2e^x + 2xe^x - 2e^x)x + e^xx^3 = 2x^2e^x - 2xe^x. \end{aligned}$$

Finally we have $y = y_c + y_p = c_1x + c_2x^3 + 2x^2e^x - 2xe^x$.

Alternative Method of Solution Any Cauchy-Euler differential equation can be reduced to an equation with constant coefficients by means of the substitution $x = e^t$. The next example illustrates this method.

EXAMPLE 6 Reduction to Constant Coefficients

Solve $x^2\dfrac{d^2y}{dx^2} - x\dfrac{dy}{dx} + y = \ln x$.

Solution With the substitution $x = e^t$ or $t = \ln x$, it follows that

$$\frac{dy}{dx} = \frac{dy}{dt}\frac{dt}{dx} = \frac{1}{x}\frac{dy}{dt} \qquad \leftarrow \text{chain rule}$$

$$\begin{aligned} \frac{d^2y}{dx^2} &= \frac{1}{x}\frac{d}{dx}\left(\frac{dy}{dt}\right) + \frac{dy}{dt}\left(-\frac{1}{x^2}\right) \qquad \leftarrow \text{product and chain rule} \\ &= \frac{1}{x}\left(\frac{d^2y}{dt^2}\frac{1}{x}\right) + \frac{dy}{dt}\left(-\frac{1}{x^2}\right) = \frac{1}{x^2}\left(\frac{d^2y}{dt^2} - \frac{dy}{dt}\right). \end{aligned}$$

Substituting in the given differential equation and simplifying yields

$$\frac{d^2y}{dt^2} - 2\frac{dy}{dt} + y = t.$$

Since this last equation has constant coefficients, its auxiliary equation is $m^2 - 2m + 1 = 0$, or $(m - 1)^2 = 0$. Thus we obtain $y_c = c_1e^t + c_2te^t$.

By undetermined coefficients we try a particular solution of the form $y_p = A + Bt$. This assumption leads to $-2B + A + Bt = t$ so $A = 2$ and $B = 1$. Using $y = y_c + y_p$, we get

$$y = c_1e^t + c_2te^t + 2 + t,$$

and so the general solution of the original differential equation on the interval $(0, \infty)$ is

$$y = c_1x + c_2x \ln x + 2 + \ln x.$$

EXERCISES 6.1

Answers to odd-numbered problems begin on page A-11.

In Problems 1–22 solve the given differential equation.

1. $x^2y'' - 2y = 0$

2. $4x^2y'' + y = 0$

3. $xy'' + y' = 0$

4. $xy'' - y' = 0$

5. $x^2y'' + xy' + 4y = 0$

6. $x^2y'' + 5xy' + 3y = 0$

7. $x^2y'' - 3xy' - 2y = 0$

8. $x^2y'' + 3xy' - 4y = 0$

9. $25x^2y'' + 25xy' + y = 0$

10. $4x^2y'' + 4xy' - y = 0$

11. $x^2y'' + 5xy' + 4y = 0$

12. $x^2y'' + 8xy' + 6y = 0$

13. $x^2y'' - xy' + 2y = 0$

14. $x^2y'' - 7xy' + 41y = 0$

15. $3x^2y'' + 6xy' + y = 0$

16. $2x^2y'' + xy' + y = 0$

17. $x^3y''' - 6y = 0$

18. $x^3y''' + xy' - y = 0$

19. $x^3\dfrac{d^3y}{dx^3} - 2x^2\dfrac{d^2y}{dx^2} - 2x\dfrac{dy}{dx} + 8y = 0$

20. $x^3\dfrac{d^3y}{dx^3} - 2x^2\dfrac{d^2y}{dx^2} + 4x\dfrac{dy}{dx} - 4y = 0$

21. $x\dfrac{d^4y}{dx^4} + 6\dfrac{d^3y}{dx^3} = 0$

22. $x^4\dfrac{d^4y}{dx^4} + 6x^3\dfrac{d^3y}{dx^3} + 9x^2\dfrac{d^2y}{dx^2} + 3x\dfrac{dy}{dx} + y = 0$

In Problems 23–26 solve the given differential equation subject to the indicated initial conditions.

23. $x^2y'' + 3xy' = 0,\ y(1) = 0, y'(1) = 4$

24. $x^2y'' - 5xy' + 8y = 0,\ y(2) = 32, y'(2) = 0$

25. $x^2y'' + xy' + y = 0,\ y(1) = 1, y'(1) = 2$

26. $x^2y'' - 3xy' + 4y = 0,\ y(1) = 5, y'(1) = 3$

In Problems 27 and 28 solve the given differential equation subject to the indicated initial conditions. [*Hint*: Let $t = -x$.]

27. $4x^2y'' + y = 0,\ y(-1) = 2, y'(-1) = 4$

28. $x^2y'' - 4xy' + 6y = 0,\ y(-2) = 8, y'(-2) = 0$

Solve Problems 29–34 by variation of parameters.

29. $xy'' + y' = x$

30. $xy'' - 4y' = x^4$

31. $2x^2y'' + 5xy' + y = x^2 - x$

32. $x^2y'' - 2xy' + 2y = x^4e^x$

33. $x^2y'' - xy' + y = 2x$

34. $x^2y'' - 2xy' + 2y = x^3 \ln x$

In Problems 35–40 solve the given differential equation by means of the substitution $x = e^t$.

35. $x^2 \dfrac{d^2y}{dx^2} + 10x \dfrac{dy}{dx} + 8y = x^2$

36. $x^2y'' - 4xy' + 6y = \ln x^2$

37. $x^2y'' - 3xy' + 13y = 4 + 3x$

38. $2x^2y'' - 3xy' - 3y = 1 + 2x + x^2$

39. $x^2y'' + 9xy' - 20y = 5/x^3$

40. $x^3 \dfrac{d^3y}{dx^3} - 3x^2 \dfrac{d^2y}{dx^2} + 6x \dfrac{dy}{dx} - 6y = 3 + \ln x^3$

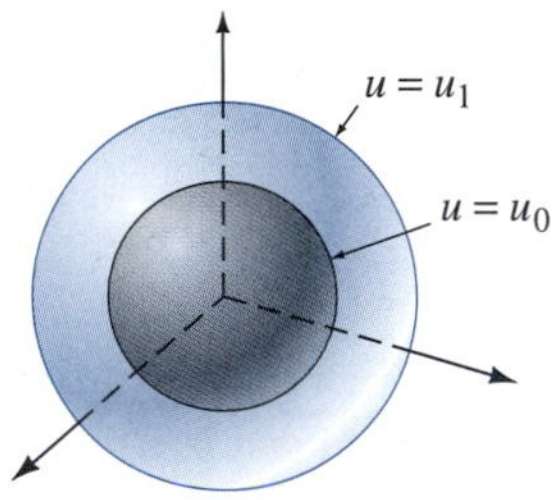

Figure 6.2

41. Consider two concentric spheres of radius $r = a$ and $r = b$, $a < b$, as shown in Figure 6.2. The temperature $u(r)$ in the region between the spheres is determined from the boundary-value problem

$$r \frac{d^2u}{dr^2} + 2 \frac{du}{dr} = 0, \quad u(a) = u_0, \quad u(b) = u_1,$$

where u_0 and u_1 are constant. Solve for $u(r)$.

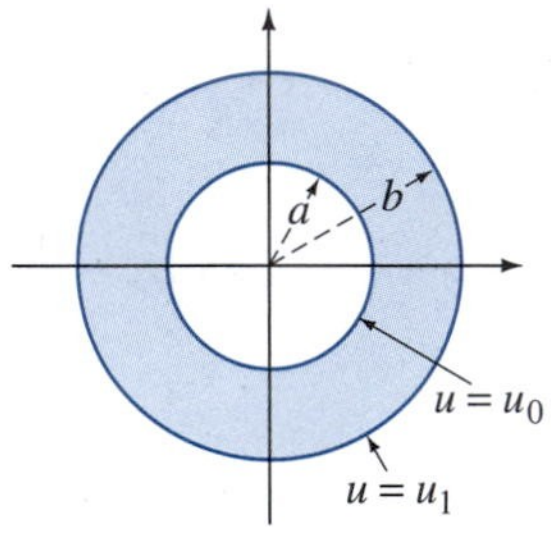

Figure 6.3

42. The temperature $u(r)$ in the circular ring shown in Figure 6.3 is determined from the boundary-value problem

$$r \frac{d^2u}{dr^2} + \frac{du}{dr} = 0, \quad u(a) = u_0, \quad u(b) = u_1,$$

where u_0 and u_1 are constant. Show that

$$u(r) = \frac{u_0 \ln(r/b) - u_1 \ln(r/a)}{\ln(a/b)}.$$

In Problems 43–45 solve the given differential equation.

43. $(x - 1)^2 \dfrac{d^2y}{dx^2} - 2(x - 1) \dfrac{dy}{dx} - 4y = 0$ [*Hint*: Let $t = x - 1$.]

44. $(3x + 4)^2y'' + 10(3x + 4)y' + 9y = 0$

45. $(x + 2)^2y'' + (x + 2)y' + y = 0$

6.2 REVIEW OF POWER SERIES; POWER SERIES SOLUTIONS

• *Power series* • *Power series solutions* • *Recurrence relation*

Review of Power Series The preceding section notwithstanding, most linear differential equations with variable coefficients cannot be solved in terms of elementary functions. A standard technique for solving higher-order linear differential equations with variable coefficients is to try to find a solution in the form of an infinite series. Often the solution can be found in the form of a power series. Because of this, it is appropriate to list some of the more important facts about power series. However, for an in-depth review of the infinite series concept you should consult a calculus text.

- **Definition of a Power Series.** A **power series** in $x - a$ is an infinite series of the form

$$\sum_{n=0}^{\infty} c_n(x-a)^n.$$

A series such as this is also said to be a **power series centered at *a*.** For example, $\sum_{n=1}^{\infty} \frac{(-1)^{n+1}}{n^2} x^n$ is a power series in x; the series is centered at zero.

- **Convergence.** For a specified value of x a power series is a series of constants. If the series equals a finite real constant for the given x, then the series is said to **converge** at x. If the series does not converge at x, it is said to **diverge** at x.

- **Interval of Convergence.** Every power series has an **interval of convergence.** The interval of convergence is the set of all numbers for which the series converges.

- **Radius of Convergence.** Every interval of convergence has a **radius of convergence** R. For a power series $\sum_{n=0}^{\infty} c_n(x-a)^n$ we have just three possibilities:

  (*i*) The series converges only at its center a. In this case $R = 0$.
  (*ii*) The series converges for all x satisfying $|x - a| < R$, where $R > 0$. The series diverges for $|x - a| > R$.
  (*iii*) The series converges for all x. In this case we write $R = \infty$.

- **Convergence at an Endpoint.** Recall that the absolute-value inequality $|x - a| < R$ is equivalent to $-R < x - a < R$, or $a - R < x < a + R$. If a power series converges for $|x - a| < R$, where $R > 0$, it may or may not converge at the endpoints of the interval $a - R < x < a + R$. Figure 6.4 shows four possible intervals of convergence.

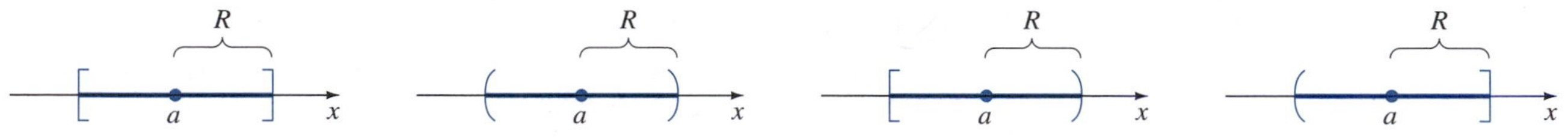

(a) $[a - R, a + R]$ Series converges at both endpoints.

(b) $(a - R, a + R)$ Series diverges at both endpoints.

(c) $[a - R, a + R)$ Series converges at $a - R$, diverges at $a + R$.

(d) $(a - R, a + R]$ Series diverges at $a - R$, converges at $a + R$.

Figure 6.4

- **Absolute Convergence.** Within its interval of convergence a power series converges absolutely. In other words, for $a - R < x < a + R$ the series of absolute values $\sum_{n=0}^{\infty} |c_n| |(x - a)^n|$ converges.

- **Finding the Interval of Convergence.** Convergence of a power series can often be determined by the **ratio test:**

$$\lim_{n\to\infty} \left| \frac{c_{n+1}}{c_n} \right| |x - a| = L.$$

The series will converge absolutely for those values of x for which $L < 1$. From this test we see that the radius of convergence is given by

$$R = \lim_{n\to\infty} \left| \frac{c_n}{c_{n+1}} \right| \tag{1}$$

provided the limit exists.

- **A Power Series Represents a Function.** A power series represents a function

$$f(x) = \sum_{n=0}^{\infty} c_n(x - a)^n = c_0 + c_1(x - a) + c_2(x - a)^2 + c_3(x - a)^3 + \cdots$$

whose domain is the interval of convergence of the series. If the series has a radius of convergence $R > 0$, then f is continuous, differentiable, and integrable on the interval $(a - R, a + R)$. Moreover, $f'(x)$ and $\int f(x)\,dx$ can be found from term-by-term differentiation and integration:

$$f'(x) = c_1 + 2c_2(x - a) + 3c_3(x - a)^2 + \cdots = \sum_{n=1}^{\infty} nc_n(x - a)^{n-1}$$

$$\int f(x)\,dx = C + c_0(x - a) + c_1\frac{(x - a)^2}{2} + c_2\frac{(x - a)^3}{3} + \cdots = C + \sum_{n=0}^{\infty} c_n \frac{(x - a)^{n+1}}{n + 1}.$$

Although the radius of convergence for both these series is R, the interval of convergence may differ from the original series in that convergence at an endpoint may be either lost by differentiation or gained through integration.

- **Series That Are Identically Zero.** If $\sum_{n=0}^{\infty} c_n(x - a)^n = 0$, $R > 0$ for all real numbers x in the interval of convergence, then $c_n = 0$ for all n.

- **Analytic at a Point.** In calculus it is seen that functions such as e^x, $\cos x$, and $\ln(x - 1)$ can be represented by power series by expansions in either Maclaurin or Taylor series. We say that a function f is **analytic at point *a*** if it can be represented by a power series in $x - a$ with a positive radius of convergence. The notion of analyticity at a point will be important in Sections 6.3 and 6.4.

- **Arithmetic of Power Series.** Power series can be combined through the operations of addition, multiplication, and division. The procedures for power series are similar to the way in which two polynomials are added, multiplied, and divided; that is, we add coefficients of like powers of x, use the distributive law and collect like terms, and perform long division. For example, if the power series $f(x) = \sum_{n=0}^{\infty} c_n x^n$ and $g(x) = \sum_{n=0}^{\infty} b_n x^n$ both converge for $|x| < R$, then

$$f(x) + g(x) = \sum_{n=0}^{\infty} (c_n + b_n)x^n$$

and

$$f(x)g(x) = c_0b_0 + (c_0b_1 + c_1b_0)x + (c_0b_2 + c_1b_1 + c_2b_0)x^2 + \cdots.$$

EXAMPLE 1 Interval of Convergence

Find the interval of convergence of the power series $\sum_{n=1}^{\infty} \frac{(x-3)^n}{2^n n}$.

Solution The power series is centered at 3. From (1), the radius of convergence is

$$R = \lim_{n\to\infty} \frac{2^{n+1}(n+1)}{2^n n} = 2.$$

The series converges absolutely for $|x - 3| < 2$, or $1 < x < 5$. At the left endpoint $x = 1$ we find that the series of constants $\sum_{n=1}^{\infty}((-1)^n/n)$ is convergent by the alternating series test. At the right endpoint $x = 5$ we find that the series is the divergent harmonic series $\sum_{n=1}^{\infty}(1/n)$. Thus the interval of convergence is $[1, 5)$.

EXAMPLE 2 Multiplying Series

Find the first four terms of a power series in x for $e^x \cos x$.

Solution From calculus the Maclaurin series for e^x and $\cos x$ are, respectively,

$$e^x = 1 + x + \frac{x^2}{2} + \frac{x^3}{6} + \frac{x^4}{24} + \cdots \quad \text{and} \quad \cos x = 1 - \frac{x^2}{2} + \frac{x^4}{24} + \cdots.$$

Multiplying out and collecting like terms yields

$$\begin{aligned} e^x \cos x &= \left(1 + x + \frac{x^2}{2} + \frac{x^3}{6} + \frac{x^4}{24} + \cdots\right)\left(1 - \frac{x^2}{2} + \frac{x^4}{24} - \cdots\right) \\ &= 1 + (1)x + \left(-\frac{1}{2} + \frac{1}{2}\right)x^2 + \left(-\frac{1}{2} + \frac{1}{6}\right)x^3 + \left(\frac{1}{24} - \frac{1}{4} + \frac{1}{24}\right)x^4 + \cdots \\ &= 1 + x - \frac{x^3}{3} - \frac{x^4}{6} + \cdots. \end{aligned}$$

In Example 2 the interval of convergence for the Maclaurin series for both e^x and $\cos x$ is $(-\infty, \infty)$. Consequently the interval of convergence for the power series for $e^x \cos x$ is also $(-\infty, \infty)$.

EXAMPLE 3 Dividing Series

Find the first four terms of a power series in x for $\sec x$.

Solution We will use the Maclaurin series for $\cos x$ given in Example 2 and then use long division. Since $\sec x = 1/\cos x$, we have

$$
\begin{array}{r}
1 + \dfrac{x^2}{2} + \dfrac{5x^4}{24} + \dfrac{61x^6}{720} + \cdots \\
\cos x = 1 - \dfrac{x^2}{2} + \dfrac{x^4}{24} - \dfrac{x^6}{720} + \cdots \overline{\big)\, 1 \qquad\qquad\qquad\qquad\qquad} \\
1 - \dfrac{x^2}{2} + \dfrac{x^4}{24} - \dfrac{x^6}{720} + \cdots \\
\hline
\dfrac{x^2}{2} - \dfrac{x^4}{24} + \dfrac{x^6}{720} + \cdots \\
\dfrac{x^2}{2} - \dfrac{x^4}{24} + \dfrac{x^6}{48} - \cdots \\
\hline
\dfrac{5x^4}{24} - \dfrac{7x^6}{360} + \cdots \\
\dfrac{5x^4}{24} - \dfrac{5x^6}{48} + \cdots \\
\hline
\dfrac{61x^6}{720} - \cdots
\end{array}
$$

Thus
$$\sec x = 1 + \frac{x^2}{2} + \frac{5x^4}{24} + \frac{61x^6}{720} + \cdots. \tag{2}$$

The interval of convergence of this series is $(-\pi/2, \pi/2)$. (Why?)

The procedures shown in Examples 2 and 3 are obviously tedious to do by hand. Problems of this sort can be done using a computer algebra system (CAS) such as *Mathematica.* When you type the command **Series[Sec[*x*], {*x*, 0, 8}]** and enter, *Mathematica* immediately gives the result obtained in (2).

For the remainder of this section, as well as this chapter, it is important that you become adept at simplifying the sum of two or more power series, each series expressed in summation (sigma) notation, to an expression with a single $\sum$. This often requires a shift of the summation indices.

EXAMPLE 4 Adding Two Series

Write $\sum_{n=1}^{\infty} 2nc_n x^{n-1} + \sum_{n=0}^{\infty} 6c_n x^{n+1}$ as one series.

Solution In order to add the series, we require that both summation indices start with the same number and that the powers of x in each series be "in phase"; that is, if one series starts with a multiple of, say, x to the first power, then we want the other series to start with the same power. By writing

series starts with x for $n = 2$ ↓ — series starts with x for $n = 0$ ↓

$$\sum_{n=1}^{\infty} 2nc_n x^{n-1} + \sum_{n=0}^{\infty} 6c_n x^{n+1} = 2 \cdot 1 \cdot c_1 x^0 + \underbrace{\sum_{n=2}^{\infty} 2nc_n x^{n-1}}_{\text{series starts with } x \text{ for } n=2} + \underbrace{\sum_{n=0}^{\infty} 6c_n x^{n+1}}_{\text{series starts with } x \text{ for } n=0}, \tag{3}$$

we have both series on the right side start with x^1. To get the same summation index we are inspired by the exponents of x; we let $k = n - 1$ in the first series and at the same time let $k = n + 1$ in the second series. Thus the right side of (3) becomes

$$2c_1 + \sum_{k=1}^{\infty} 2(k+1)c_{k+1}x^k + \sum_{k=1}^{\infty} 6c_{k-1}x^k. \tag{4}$$

Recall that the summation index is a "dummy" variable. The fact that $k = n - 1$ in one case and $k = n + 1$ in the other should cause no confusion if you keep in mind that it is the *value* of the summation index that is important. In both cases k takes on the same successive values $1, 2, 3, \dots$ for $n = 2, 3, 4, \dots$ (for $k = n - 1$) and $n = 0, 1, 2, \dots$ (for $k = n + 1$).

We are now in a position to add the series in (4) term by term:

$$\sum_{n=1}^{\infty} 2nc_n x^{n-1} + \sum_{n=0}^{\infty} 6c_n x^{n+1} = 2c_1 + \sum_{k=1}^{\infty} [2(k+1)c_{k+1} + 6c_{k-1}]x^k. \tag{5}$$

If you are not convinced, then write out a few terms on both sides of (5). ■

Power Series Solution of a Differential Equation We saw in Section 1.1 that the function $y = e^{x^2}$ is an explicit solution of the linear first-order differential equation

$$\frac{dy}{dx} - 2xy = 0. \tag{6}$$

By replacing x by x^2 in the Maclaurin series for e^x, we can write the solution of (6) as

$$y = \sum_{n=0}^{\infty} \frac{x^{2n}}{n!}.$$

This last series converges for all real values of x. In other words, knowing the solution in advance, we were able to find an infinite series solution of the differential equation.

We now propose to obtain a **power series solution** of equation (6) directly; the method of attack is similar to the technique of undetermined coefficients.

EXAMPLE 5 A Power Series Solution

Find a solution of $dy/dx - 2xy = 0$ in the form of a power series in x.

Solution If we assume that a solution of the given equation exists in the form

$$y = \sum_{n=0}^{\infty} c_n x^n, \tag{7}$$

we pose the question: Can we determine coefficients c_n for which the power series converges to a function satisfying (6)? Formal* term-by-term differentiation of (7) gives

$$\frac{dy}{dx} = \sum_{n=0}^{\infty} nc_n x^{n-1} = \sum_{n=1}^{\infty} nc_n x^{n-1}.$$

Note that since the first term in the first series (corresponding to $n = 0$) is zero, we begin the summation with $n = 1$. Using the last result and assumption (7), we find

$$\frac{dy}{dx} - 2xy = \sum_{n=1}^{\infty} nc_n x^{n-1} - \sum_{n=0}^{\infty} 2c_n x^{n+1}. \tag{8}$$

We would like to add the two series in (8). To this end we write

$$\frac{dy}{dx} - 2xy = 1 \cdot c_1 x^0 + \sum_{n=2}^{\infty} nc_n x^{n-1} - \sum_{n=0}^{\infty} 2c_n x^{n+1} \tag{9}$$

and then proceed as in Example 4 by letting $k = n - 1$ in the first series and $k = n + 1$ in the second. The right side of (9) becomes

$$c_1 + \sum_{k=1}^{\infty} (k+1)c_{k+1} x^k - \sum_{k=1}^{\infty} 2c_{k-1} x^k.$$

After we add the series termwise, it follows that

$$\frac{dy}{dx} - 2xy = c_1 + \sum_{k=1}^{\infty} [(k+1)c_{k+1} - 2c_{k-1}]x^k = 0. \tag{10}$$

Hence in order to have (10) identically zero it is necessary that the coefficients satisfy

$$c_1 = 0 \quad \text{and} \quad (k+1)c_{k+1} - 2c_{k-1} = 0, \quad k = 1, 2, 3, \ldots. \tag{11}$$

Equation (11) provides a **recurrence relation** that determines the c_k. Since $k + 1 \neq 0$ for all the indicated values of k, we can write (11) as

$$c_{k+1} = \frac{2c_{k-1}}{k+1}. \tag{12}$$

Iteration of this last formula then gives

$$k = 1, \quad c_2 = \frac{2}{2} c_0 = c_0$$

* At this point we do not know the interval of convergence.

$$k = 2, \quad c_3 = \frac{2}{3}c_1 = 0$$

$$k = 3, \quad c_4 = \frac{2}{4}c_2 = \frac{1}{2}c_0 = \frac{1}{2!}c_0$$

$$k = 4, \quad c_5 = \frac{2}{5}c_3 = 0$$

$$k = 5, \quad c_6 = \frac{2}{6}c_4 = \frac{1}{3 \cdot 2!}c_0 = \frac{1}{3!}c_0$$

$$k = 6, \quad c_7 = \frac{2}{7}c_5 = 0$$

$$k = 7, \quad c_8 = \frac{2}{8}c_6 = \frac{1}{4 \cdot 3!}c_0 = \frac{1}{4!}c_0$$

and so on. Thus from the original assumption (7), we find

$$\begin{aligned} y = \sum_{n=0}^{\infty} c_n x^n &= c_0 + c_1 x + c_2 x^2 + c_3 x^3 + c_4 x^4 + c_5 x^5 + \cdots \\ &= c_0 + 0 + c_0 x^2 + 0 + \frac{1}{2!}c_0 x^4 + 0 + \frac{1}{3!}c_0 x^6 + 0 + \cdots \\ &= c_0\left[1 + x^2 + \frac{1}{2!}x^4 + \frac{1}{3!}x^6 + \cdots\right] = c_0 \sum_{n=0}^{\infty} \frac{x^{2n}}{n!}. \end{aligned} \tag{13}$$

Since the iteration of (12) leaves c_0 completely undetermined, we have in fact found the general solution of (6). ■

The differential equation in Example 5, like the differential equation in the following example, can be easily solved by prior methods. The point of these two examples is to prepare you for the techniques considered in Sections 6.3 and 6.4.

EXAMPLE 6 A Power Series Solution

Find solutions of $4y'' + y = 0$ in the form of power series in x.

Solution If $y = \sum_{n=0}^{\infty} c_n x^n$, then

$$y' = \sum_{n=0}^{\infty} n c_n x^{n-1} = \sum_{n=1}^{\infty} n c_n x^{n-1} \quad \text{and} \quad y'' = \sum_{n=1}^{\infty} n(n-1)c_n x^{n-2} = \sum_{n=2}^{\infty} n(n-1)c_n x^{n-2}.$$

Substituting the expressions for y'' and y back into the differential equation gives

$$4y'' + y = \underbrace{\sum_{n=2}^{\infty} 4n(n-1)c_n x^{n-2} + \sum_{n=0}^{\infty} c_n x^n}_{\text{both series start with } x^0}.$$

If we substitute $k = n - 2$ in the first series and $k = n$ in the second, we get (after using, in turn, $n = k + 2$ and $n = k$)

$$\begin{aligned} 4y'' + y &= \sum_{k=0}^{\infty} 4(k+2)(k+1)c_{k+2}x^k + \sum_{k=0}^{\infty} c_k x^k \\ &= \sum_{k=0}^{\infty} [4(k+2)(k+1)c_{k+2} + c_k]x^k = 0. \end{aligned}$$

From this last identity we conclude that

$$4(k+2)(k+1)c_{k+2} + c_k = 0$$

or

$$c_{k+2} = \frac{-c_k}{4(k+2)(k+1)}, \quad k = 0, 1, 2, \ldots.$$

From iteration of this recurrence relation it follows that

$$\begin{aligned} c_2 &= \frac{-c_0}{4 \cdot 2 \cdot 1} = -\frac{c_0}{2^2 \cdot 2!} \\ c_3 &= \frac{-c_1}{4 \cdot 3 \cdot 2} = -\frac{c_1}{2^2 \cdot 3!} \\ c_4 &= \frac{-c_2}{4 \cdot 4 \cdot 3} = \frac{c_0}{2^4 \cdot 4!} \\ c_5 &= \frac{-c_3}{4 \cdot 5 \cdot 4} = \frac{c_1}{2^4 \cdot 5!} \\ c_6 &= \frac{-c_4}{4 \cdot 6 \cdot 5} = -\frac{c_0}{2^6 \cdot 6!} \\ c_7 &= \frac{-c_5}{4 \cdot 7 \cdot 6} = -\frac{c_1}{2^6 \cdot 7!} \end{aligned}$$

and so forth. This iteration leaves both c_0 and c_1 arbitrary. From the original assumption we have

$$\begin{aligned} y &= c_0 + c_1 x + c_2 x^2 + c_3 x^3 + c_4 x^4 + c_5 x^5 + c_6 x^6 + c_7 x^7 + \cdots \\ &= c_0 + c_1 x - \frac{c_0}{2^2 \cdot 2!}x^2 - \frac{c_1}{2^2 \cdot 3!}x^3 + \frac{c_0}{2^4 \cdot 4!}x^4 + \frac{c_1}{2^4 \cdot 5!}x^5 - \frac{c_0}{2^6 \cdot 6!}x^6 - \frac{c_1}{2^6 \cdot 7!}x^7 + \cdots \end{aligned}$$

or

$$\begin{aligned} y = c_0&\left[1 - \frac{1}{2^2 \cdot 2!}x^2 + \frac{1}{2^4 \cdot 4!}x^4 - \frac{1}{2^6 \cdot 6!}x^6 + \cdots\right] \\ &+ c_1\left[x - \frac{1}{2^2 \cdot 3!}x^3 + \frac{1}{2^4 \cdot 5!}x^5 - \frac{1}{2^6 \cdot 7!}x^7 + \cdots\right] \end{aligned}$$

is a general solution. When the series are written in summation notation,

$$y_1(x) = c_0 \sum_{k=0}^{\infty} \frac{(-1)^k}{(2k)!}\left(\frac{x}{2}\right)^{2k} \quad \text{and} \quad y_2(x) = 2c_1 \sum_{k=0}^{\infty} \frac{(-1)^k}{(2k+1)!}\left(\frac{x}{2}\right)^{2k+1},$$

the ratio test can be applied to show that both series converge for all x. You might also recognize the Maclaurin series as $y_1(x) = c_0 \cos(x/2)$ and $y_2(x) = 2c_1 \sin(x/2)$.

EXERCISES 6.2

Answers to odd-numbered problems begin on page A-12.

In Problems 1–10 find the interval of convergence of the given power series.

1. $\sum_{n=1}^{\infty} \frac{(-1)^n}{n} x^n$

2. $\sum_{n=1}^{\infty} \frac{x^n}{n^2}$

3. $\sum_{k=1}^{\infty} \frac{2^k}{k} x^k$

4. $\sum_{k=0}^{\infty} \frac{5^k}{k!} x^k$

5. $\sum_{n=1}^{\infty} \frac{(x-3)^n}{n^3}$

6. $\sum_{n=1}^{\infty} \frac{(x+7)^n}{\sqrt{n}}$

7. $\sum_{k=1}^{\infty} \frac{(-1)^k}{10^k}(x-5)^k$

8. $\sum_{k=1}^{\infty} \frac{k}{(k+2)^2}(x-4)^k$

9. $\sum_{k=0}^{\infty} k!2^k x^k$

10. $\sum_{k=1}^{\infty} \frac{k-1}{k^{2k}} x^k$

In Problems 11–20 find the first four terms of a power series in x for the given function. Calculate the series by hand or use a CAS as instructed.

11. $e^x \sin x$

12. $e^{-x} \cos x$

13. $\sin x \cos x$

14. $e^x \ln(1-x)$

15. $\left(x - \frac{x^3}{3} + \frac{x^5}{5} - \frac{x^7}{7} + \cdots\right)^2$

16. $\left(1 - \frac{x^2}{2} + \frac{x^4}{3} - \frac{x^6}{4} + \cdots\right)^2$

17. $\tan x$

18. $\frac{1}{e^x + e^{-x}}$

19. $\frac{1}{1 - \frac{x^2}{2} + \frac{x^4}{3} - \frac{x^6}{4} + \cdots}$

20. $\frac{1}{\left(1 - \frac{x^2}{2} + \frac{x^4}{3} - \frac{x^6}{4} + \cdots\right)^2}$

In Problems 21–30 solve each differential equation in the manner of the previous chapters and then compare the results with the solutions obtained by assuming a power series solution $y = \sum_{n=0}^{\infty} c_n x^n$.

21. $y' + y = 0$

22. $y' = 2y$

23. $y' - x^2 y = 0$

24. $y' + x^3 y = 0$

25. $(1 - x)y' - y = 0$

26. $(1 + x)y' - 2y = 0$

27. $y'' + y = 0$

28. $y'' - y = 0$

29. $y'' = y'$

30. $2y'' + y' = 0$

6.3 SOLUTIONS ABOUT ORDINARY POINTS

• *Ordinary point* • *Singular point* • *Power series solutions*

Suppose the linear second-order differential equation

$$a_2(x)y'' + a_1(x)y' + a_0(x)y = 0 \tag{1}$$

is put into the standard form

$$y'' + P(x)y' + Q(x)y = 0 \tag{2}$$

by dividing by the leading coefficient $a_2(x)$. We make the following definition.

DEFINITION 6.1 Ordinary and Singular Points

A point x_0 is said to be an **ordinary point** of the differential equation (1) if both $P(x)$ and $Q(x)$ in (2) are analytic* at x_0. A point that is not an ordinary point is said to be a **singular point** of the equation.

EXAMPLE 1 Ordinary Points of a DE

Every finite value of x is an ordinary point of

$$y'' + (e^x)y' + (\sin x)y = 0.$$

In particular we see that $x = 0$ is an ordinary point since

$$e^x = 1 + \frac{x}{1!} + \frac{x^2}{2!} + \cdots \quad \text{and} \quad \sin x = x - \frac{x^3}{3!} + \frac{x^5}{5!} - \cdots$$

converge for all finite values of x.

EXAMPLE 2 $x = 0$ Is an Ordinary Point

The differential equation $xy'' + (\sin x)y = 0$ has an ordinary point at $x = 0$ since it can be shown that $Q(x) = (\sin x)/x$ possesses the power series expansion

$$Q(x) = 1 - \frac{x^2}{3!} + \frac{x^4}{5!} - \frac{x^6}{7!} + \cdots$$

that converges for all finite values of x.

EXAMPLE 3 $x = 0$ Is a Singular Point

The differential equation $y'' + (\ln x)y = 0$ has a singular point at $x = 0$ because $Q(x) = \ln x$ possesses no power series in x.

Polynomial Coefficients Primarily we shall be concerned with the case when (1) has *polynomial* coefficients. As a consequence of Definition 6.1, we note

---

* See page 232.

that when $a_2(x)$, $a_1(x)$, and $a_0(x)$ are polynomials with *no common factors,* a point $x = x_0$ is

(*i*) an ordinary point if $a_2(x_0) \neq 0$ or (*ii*) a singular point if $a_2(x_0) = 0$.

EXAMPLE 4 Singular Points

(a) The singular points of the equation $(x^2 - 1)y'' + 2xy' + 6y = 0$ are the solutions of $x^2 - 1 = 0$ or $x = \pm 1$. All other finite values of x are ordinary points.

(b) Singular points need not be real numbers. The equation $(x^2 + 1)y'' + xy' - y = 0$ has singular points at the solutions of $x^2 + 1 = 0$—namely, $x = \pm i$. All other finite values of x, real or complex, are ordinary points.

EXAMPLE 5 $x = 0$ Is a Singular Point of a Cauchy-Euler DE

The Cauchy-Euler equation $ax^2y'' + bxy' + cy = 0$, where a, b, and c are constants, has a singular point at $x = 0$. All other finite values of x, real or complex, are ordinary points.

For our purposes ordinary points and singular points will always be finite points. It is possible for a differential equation to have, say, a singular point at infinity. (See Problem 40, Exercises 6.4.)

We state the following theorem about the existence of power series solutions without proof.

THEOREM 6.1 Existence of Power Series Solutions

If $x = x_0$ is an ordinary point of the differential equation (1), we can always find two linearly independent solutions in the form of power series centered at x_0:

$$y = \sum_{n=0}^{\infty} c_n(x - x_0)^n. \tag{3}$$

A series solution converges at least for $|x - x_0| < R$, where R is the distance from x_0 to the closest singular point (real or complex).

A solution of a differential equation of the form given in (3) is said to be a solution *about* the ordinary point x_0. The distance R given in Theorem 6.1 is the minimum value for the radius of convergence. A differential equation could have a finite singular point and yet a solution could be valid for all x; for example, the differential equation may possess a polynomial solution.

To solve a linear second-order equation such as (1) we find two sets of coefficients c_n so that we have two distinct power series $y_1(x)$ and $y_2(x)$, both expanded about the same ordinary point x_0. The procedure used to solve a second-order equation is the same as that used in Example 6 of Section 6.2; that is, we assume a solution $y = \sum_{n=0}^{\infty} c_n (x - x_0)^n$ and then determine the c_n. The general solution of the differential equation is $y = C_1 y_1(x) + C_2 y_2(x)$; in fact, it can be shown that $C_1 = c_0$ and $C_2 = c_1$, where c_0 and c_1 are arbitrary.

Note For the sake of simplicity, we assume an ordinary point is always located at $x = 0$, since, if not, the substitution $t = x - x_0$ translates the value $x = x_0$ to $t = 0$.

EXAMPLE 6 Power Series Solutions

Solve $y'' - 2xy = 0$.

Solution We see that $x = 0$ is an ordinary point of the equation. Since there are no finite singular points, Theorem 6.1 guarantees two solutions of the form $y = \sum_{n=0}^{\infty} c_n x^n$ convergent for $|x| < \infty$. Proceeding, we write

$$y' = \sum_{n=0}^{\infty} nc_n x^{n-1} = \sum_{n=1}^{\infty} nc_n x^{n-1}, \quad y'' = \sum_{n=1}^{\infty} n(n-1)c_n x^{n-2} = \sum_{n=2}^{\infty} n(n-1)c_n x^{n-2},$$

where we have used the fact that the first term in each series, corresponding to $n = 0$ and $n = 1$, respectively, is zero. Therefore

$$\begin{aligned} y'' - 2xy &= \sum_{n=2}^{\infty} n(n-1)c_n x^{n-2} - \sum_{n=0}^{\infty} 2c_n x^{n+1} \\ &= 2 \cdot 1 c_2 x^0 + \underbrace{\sum_{n=3}^{\infty} n(n-1)c_n x^{n-2} - \sum_{n=0}^{\infty} 2c_n x^{n+1}}_{\text{both series start with } x}. \end{aligned}$$

Letting $k = n - 2$ in the first series and $k = n + 1$ in the second, we have

$$\begin{aligned} y'' - 2xy &= 2c_2 + \sum_{k=1}^{\infty} (k+2)(k+1)c_{k+2} x^k - \sum_{k=1}^{\infty} 2c_{k-1} x^k \\ &= 2c_2 + \sum_{k=1}^{\infty} [(k+2)(k+1)c_{k+2} - 2c_{k-1}]x^k = 0. \end{aligned}$$

We must then have

$$2c_2 = 0 \quad \text{and} \quad (k+2)(k+1)c_{k+2} - 2c_{k-1} = 0.$$

The last expression is the same as

$$c_{k+2} = \frac{2c_{k-1}}{(k+2)(k+1)}, \quad k = 1, 2, 3, \ldots.$$

Iteration gives

$$\begin{aligned} c_3 &= \frac{2c_0}{3 \cdot 2} \\ c_4 &= \frac{2c_1}{4 \cdot 3} \\ c_5 &= \frac{2c_2}{5 \cdot 4} = 0 \end{aligned}$$

$$c_6 = \frac{2c_3}{6 \cdot 5} = \frac{2^2}{6 \cdot 5 \cdot 3 \cdot 2} c_0$$

$$c_7 = \frac{2c_4}{7 \cdot 6} = \frac{2^2}{7 \cdot 6 \cdot 4 \cdot 3} c_1$$

$$c_8 = \frac{2c_5}{8 \cdot 7} = 0$$

$$c_9 = \frac{2c_6}{9 \cdot 8} = \frac{2^3}{9 \cdot 8 \cdot 6 \cdot 5 \cdot 3 \cdot 2} c_0$$

$$c_{10} = \frac{2c_7}{10 \cdot 9} = \frac{2^3}{10 \cdot 9 \cdot 7 \cdot 6 \cdot 4 \cdot 3} c_1$$

$$c_{11} = \frac{2c_8}{11 \cdot 10} = 0$$

and so on. It should be apparent that both c_0 and c_1 are arbitrary. Now

$$\begin{aligned} y &= c_0 + c_1x + c_2x^2 + c_3x^3 + c_4x^4 + c_5x^5 + c_6x^6 + c_7x^7 + c_8x^8 \\ &\quad + c_9x^9 + c_{10}x^{10} + c_{11}x^{11} + \cdots \\ &= c_0 + c_1x + 0 + \frac{2}{3 \cdot 2} c_0x^3 + \frac{2}{4 \cdot 3} c_1x^4 + 0 + \frac{2^2}{6 \cdot 5 \cdot 3 \cdot 2} c_0x^6 \\ &\quad + \frac{2^2}{7 \cdot 6 \cdot 4 \cdot 3} c_1x^7 + 0 + \frac{2^3}{9 \cdot 8 \cdot 6 \cdot 5 \cdot 3 \cdot 2} c_0x^9 \\ &\quad + \frac{2^3}{10 \cdot 9 \cdot 7 \cdot 6 \cdot 4 \cdot 3} c_1x^{10} + 0 + \cdots \\ &= c_0\left[1 + \frac{2}{3 \cdot 2}x^3 + \frac{2^2}{6 \cdot 5 \cdot 3 \cdot 2}x^6 + \frac{2^3}{9 \cdot 8 \cdot 6 \cdot 5 \cdot 3 \cdot 2}x^9 + \cdots\right] \\ &\quad + c_1\left[x + \frac{2}{4 \cdot 3}x^4 + \frac{2^2}{7 \cdot 6 \cdot 4 \cdot 3}x^7 + \frac{2^3}{10 \cdot 9 \cdot 7 \cdot 6 \cdot 4 \cdot 3}x^{10} + \cdots\right]. \end{aligned}$$

Although the pattern of the coefficients in Example 6 should be clear, it is sometimes useful to write the solutions in terms of summation notation. By using the properties of the factorial, we can write

$$y_1(x) = c_0\left[1 + \sum_{k=1}^{\infty} \frac{2^k[1 \cdot 4 \cdot 7 \cdots (3k-2)]}{(3k)!} x^{3k}\right]$$

and
$$y_2(x) = c_1\left[x + \sum_{k=1}^{\infty} \frac{2^k[2 \cdot 5 \cdot 8 \cdots (3k-1)]}{(3k+1)!} x^{3k+1}\right].$$

In this form the ratio test can be used to show that each series converges for $|x| < \infty$.

EXAMPLE 7 Discovering a Polynomial Solution

Solve $(x^2 + 1)y'' + xy' - y = 0$.

Solution Since the singular points are $x = \pm i$, a power series solution will converge at least for $|x| < 1$.* The assumption $y = \sum_{n=0}^{\infty} c_n x^n$ leads to

$$(x^2+1)\sum_{n=2}^{\infty} n(n-1)c_n x^{n-2} + x\sum_{n=1}^{\infty} nc_n x^{n-1} - \sum_{n=0}^{\infty} c_n x^n$$

$$= \sum_{n=2}^{\infty} n(n-1)c_n x^n + \sum_{n=2}^{\infty} n(n-1)c_n x^{n-2} + \sum_{n=1}^{\infty} nc_n x^n - \sum_{n=0}^{\infty} c_n x^n$$

$$= 2c_2x^0 - c_0x^0 + 6c_3x + c_1x - c_1x + \underbrace{\sum_{n=2}^{\infty} n(n-1)c_n x^n}_{k=n}$$

$$+ \underbrace{\sum_{n=4}^{\infty} n(n-1)c_n x^{n-2}}_{k=n-2} + \underbrace{\sum_{n=2}^{\infty} nc_n x^n}_{k=n} - \underbrace{\sum_{n=2}^{\infty} c_n x^n}_{k=n}$$

$$= 2c_2 - c_0 + 6c_3x + \sum_{k=2}^{\infty}[k(k-1)c_k + (k+2)(k+1)c_{k+2} + kc_k - c_k]x^k$$

$$= 2c_2 - c_0 + 6c_3x + \sum_{k=2}^{\infty}[k(k+1)(k-1)c_k + (k+2)(k+1)c_{k+2}]x^k = 0.$$

Thus

$$2c_2 - c_0 = 0, \quad c_3 = 0$$

$$(k+1)(k-1)c_k + (k+2)(k+1)c_{k+2} = 0.$$

After dividing by $(k+2)(k+1)$, we get

$$c_2 = \frac{1}{2}c_0 \quad \text{and} \quad c_3 = 0$$

and

$$c_{k+2} = \frac{1-k}{k+2}c_k, \quad k = 2, 3, 4, \ldots.$$

Iteration of the last formula gives

$$c_4 = -\frac{1}{4}c_2 = -\frac{1}{2\cdot 4}c_0 = -\frac{1}{2^2 2!}c_0$$

$$c_5 = -\frac{2}{5}c_3 = 0$$

$$c_6 = -\frac{3}{6}c_4 = \frac{3}{2\cdot 4\cdot 6}c_0 = \frac{1\cdot 3}{2^3 3!}c_0$$

$$c_7 = -\frac{4}{7}c_5 = 0$$

$$c_8 = -\frac{5}{8}c_6 = -\frac{3\cdot 5}{2\cdot 4\cdot 6\cdot 8}c_0 = -\frac{1\cdot 3\cdot 5}{2^4 4!}c_0$$

* The **modulus** or magnitude of the complex number $x = i$ is $|x| = 1$. If $x = a + bi$ is a singular point, then $|x| = \sqrt{a^2 + b^2}$. See Appendix IV.

$$c_9 = -\frac{6}{9}c_7 = 0$$

$$c_{10} = -\frac{7}{10}c_8 = \frac{3 \cdot 5 \cdot 7}{2 \cdot 4 \cdot 6 \cdot 8 \cdot 10}c_0 = \frac{1 \cdot 3 \cdot 5 \cdot 7}{2^5 5!}c_0$$

and so on. Therefore

$$y = c_0 + c_1x + c_2x^2 + c_3x^3 + c_4x^4 + c_5x^5 + c_6x^6 + c_7x^7 + c_8x^8 + \cdots$$

$$= c_1x + c_0\left[1 + \frac{1}{2}x^2 - \frac{1}{2^2 2!}x^4 + \frac{1 \cdot 3}{2^3 3!}x^6 - \frac{1 \cdot 3 \cdot 5}{2^4 4!}x^8 + \frac{1 \cdot 3 \cdot 5 \cdot 7}{2^5 5!}x^{10} - \cdots\right].$$

The solutions are

$$y_1(x) = c_0\left[1 + \frac{1}{2}x^2 + \sum_{n=2}^{\infty}(-1)^{n-1}\frac{1 \cdot 3 \cdot 5 \cdots (2n-3)}{2^n n!}x^{2n}\right], \quad |x| < 1$$

$$y_2(x) = c_1x.$$

EXAMPLE 8 A Three-Term Recurrence Relation

If we seek a solution $y = \sum_{n=0}^{\infty} c_n x^n$ for the equation

$$y'' - (1 + x)y = 0,$$

we obtain $c_2 = c_0/2$ and the three-term recurrence relation

$$c_{k+2} = \frac{c_k + c_{k-1}}{(k+1)(k+2)}, \quad k = 1, 2, 3, \ldots.$$

To simplify the iteration we can first choose $c_0 \neq 0$, $c_1 = 0$; this yields one solution. The other solution follows from next choosing $c_0 = 0$, $c_1 \neq 0$. With the first assumption we find

$$c_2 = \frac{1}{2}c_0$$

$$c_3 = \frac{c_1 + c_0}{2 \cdot 3} = \frac{c_0}{2 \cdot 3} = \frac{1}{6}c_0$$

$$c_4 = \frac{c_2 + c_1}{3 \cdot 4} = \frac{c_0}{2 \cdot 3 \cdot 4} = \frac{1}{24}c_0$$

$$c_5 = \frac{c_3 + c_2}{4 \cdot 5} = \frac{c_0}{4 \cdot 5}\left[\frac{1}{2 \cdot 3} + \frac{1}{2}\right] = \frac{1}{30}c_0$$

and so on. Thus one solution is

$$y_1(x) = c_0\left[1 + \frac{1}{2}x^2 + \frac{1}{6}x^3 + \frac{1}{24}x^4 + \frac{1}{30}x^5 + \cdots\right].$$

Similarly if we choose $c_0 = 0$, then

$$c_2 = 0$$

$$c_3 = \frac{c_1 + c_0}{2 \cdot 3} = \frac{c_1}{2 \cdot 3} = \frac{1}{6}c_1$$

$$c_4 = \frac{c_2 + c_1}{3 \cdot 4} = \frac{c_1}{3 \cdot 4} = \frac{1}{12} c_1$$
$$c_5 = \frac{c_3 + c_2}{4 \cdot 5} = \frac{c_1}{2 \cdot 3 \cdot 4 \cdot 5} = \frac{1}{120} c_1$$

and so on. Hence another solution is

$$y_2(x) = c_1\left[x + \frac{1}{6}x^3 + \frac{1}{12}x^4 + \frac{1}{120}x^5 + \cdots\right].$$

Each series converges for all finite values of x.

Nonpolynomial Coefficients The next example illustrates how to find a power series solution about an ordinary point of a differential equation when its coefficients are not polynomials. In this example we see an application of multiplication of two power series that was discussed in Section 6.2.

EXAMPLE 9 DE with a Nonpolynomial Coefficient

Solve $y'' + (\cos x)y = 0$.

Solution Since $\cos x = 1 - \frac{x^2}{2!} + \frac{x^4}{4!} - \frac{x^6}{6!} + \cdots$, it is seen that $x = 0$ is an ordinary point. Thus the assumption $y = \sum_{n=0}^{\infty} c_n x^n$ leads to

$$\begin{aligned} y'' + (\cos x)y &= \sum_{n=2}^{\infty} n(n-1)c_n x^{n-2} + \left(1 - \frac{x^2}{2!} + \frac{x^4}{4!} - \cdots\right)\sum_{n=0}^{\infty} c_n x^n \\ &= (2c_2 + 6c_3 x + 12c_4 x^2 + 20c_5 x^3 + \cdots) \\ &\quad + \left(1 - \frac{x^2}{2} + \frac{x^4}{24} - \cdots\right)(c_0 + c_1 x + c_2 x^2 + c_3 x^3 + \cdots) \\ &= 2c_2 + c_0 + (6c_3 + c_1)x + \left(12c_4 + c_2 - \frac{1}{2}c_0\right)x^2 + \left(20c_5 + c_3 - \frac{1}{2}c_1\right)x^3 + \cdots. \end{aligned}$$

Since the last line is to be identically zero, we must have

$$\begin{aligned} 2c_2 + c_0 &= 0 \\ 6c_3 + c_1 &= 0 \\ 12c_4 + c_2 - \frac{1}{2}c_0 &= 0 \\ 20c_5 + c_3 - \frac{1}{2}c_1 &= 0 \end{aligned}$$

and so on. Since c_0 and c_1 are arbitrary, we find

$$y_1(x) = c_0\left[1 - \frac{1}{2}x^2 + \frac{1}{12}x^4 - \cdots\right] \quad \text{and} \quad y_2(x) = c_1\left[x - \frac{1}{6}x^3 + \frac{1}{30}x^5 - \cdots\right].$$

Since the differential equation has no singular points, both series converge for all finite values of x.

EXERCISES 6.3

Answers to odd-numbered problems begin on page A-12.

In Problems 1–14 for each differential equation find two linearly independent power series solutions about the ordinary point $x = 0$.

1. $y'' = xy$

2. $y'' + x^2y = 0$

3. $y'' - 2xy' + y = 0$

4. $y'' - xy' + 2y = 0$

5. $y'' + x^2y' + xy = 0$

6. $y'' + 2xy' + 2y = 0$

7. $(x - 1)y'' + y' = 0$

8. $(x + 2)y'' + xy' - y = 0$

9. $(x^2 - 1)y'' + 4xy' + 2y = 0$

10. $(x^2 + 1)y'' - 6y = 0$

11. $(x^2 + 2)y'' + 3xy' - y = 0$

12. $(x^2 - 1)y'' + xy' - y = 0$

13. $y'' - (x + 1)y' - y = 0$

14. $y'' - xy' - (x + 2)y = 0$

In Problems 15–18 use the power series method to solve the given differential equation subject to the indicated initial conditions.

15. $(x - 1)y'' - xy' + y = 0, \quad y(0) = -2, y'(0) = 6$

16. $(x + 1)y'' - (2 - x)y' + y = 0, \quad y(0) = 2, y'(0) = -1$

17. $y'' - 2xy' + 8y = 0, \quad y(0) = 3, y'(0) = 0$

18. $(x^2 + 1)y'' + 2xy' = 0, \quad y(0) = 0, y'(0) = 1$

In Problems 19–22 use the procedure illustrated in Example 9 to find two power series solutions of the given differential equation about the ordinary point $x = 0$.

19. $y'' + (\sin x)y = 0$

20. $xy'' + (\sin x)y = 0$ [*Hint*: See Example 2.]

21. $y'' + e^{-x}y = 0$

22. $y'' + e^xy' - y = 0$

In Problems 23 and 24 use the power series method to solve the nonhomogeneous equation.

23. $y'' - xy = 1$

24. $y'' - 4xy' - 4y = e^x$

25. The differential equation $y'' - 2xy' + 2ny = 0$ is known as **Hermite's equation.*** When $n \geq 0$ is an integer, Hermite's equation has a polynomial solution. Hermite polynomials have some importance in the study of quantum mechanics. Obtain the polynomial solutions corresponding to $n = 1$ and $n = 2$.

26. In the analysis of a uniform thin column of length L that is buckling under its own weight, the following boundary-value problem is encountered:

$$\theta'' + \frac{\delta g}{EI}(L - x)\theta = 0, \quad \theta(0) = 0, \quad \theta'(L) = 0.$$

Here E is Young's modulus, I is the cross-sectional moment of inertia, δ is the constant linear density, x is distance measured along the column,

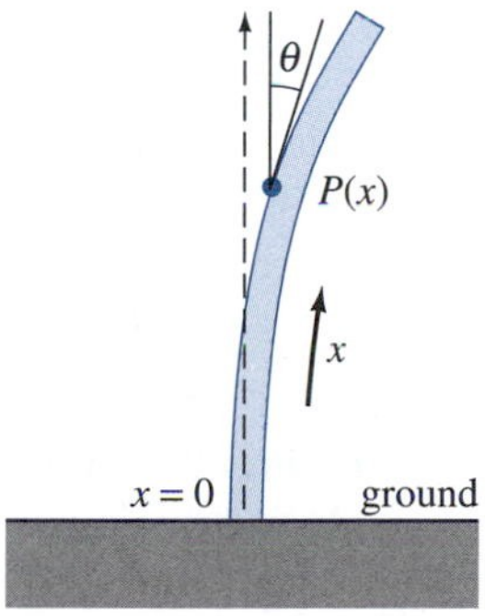

Figure 6.5

* Named after the French mathematician **Charles Hermite** (1822–1901).

and $\theta(x)$ is the angular deflection of the column from the vertical at a point $P(x)$. See Figure 6.5. Obtain a power series solution of the differential equation that satisfies the condition $\theta'(L) = 0$. For convenience define $\lambda^2 = \delta g L/EI$ and change the variable by letting $t = L - x$.

6.4 SOLUTIONS ABOUT SINGULAR POINTS

• *Regular singular point* • *Irregular singular point* • *Method of Frobenius* • *Indicial equation* • *Indicial roots*

6.4.1 REGULAR SINGULAR POINTS; METHOD OF FROBENIUS—CASE I

We saw in the preceding section that there is no basic problem in finding a power series solution of

$$a_2(x)y'' + a_1(x)y' + a_0(x)y = 0 \tag{1}$$

about an ordinary point $x = x_0$. However, when $x = x_0$ is a singular point, it is not always possible to find a solution of the form

$$y = \sum_{n=0}^{\infty} c_n(x - x_0)^n;$$

it turns out that we *may* be able to find a solution of the form

$$y = \sum_{n=0}^{\infty} c_n(x - x_0)^{n+r},$$

where r is a constant that must be determined.

Regular and Irregular Singular Points Singular points are further classified as either regular or irregular. To define these concepts we again put (1) into the standard form

$$y'' + P(x)y' + Q(x)y = 0. \tag{2}$$

DEFINITION 6.2 Regular and Irregular Singular Points

A singular point $x = x_0$ of equation (1) is said to be a **regular singular point** if both $(x - x_0)P(x)$ and $(x - x_0)^2Q(x)$ are analytic at x_0. A singular point that is not regular is said to be an **irregular singular point** of the equation.

Polynomial Coefficients In the case in which the coefficients in (1) are polynomials with no common factors, Definition 6.2 is equivalent to the following.

Let $a_2(x_0) = 0$. Form $P(x)$ and $Q(x)$ by reducing $a_1(x)/a_2(x)$ and $a_0(x)/a_2(x)$ to lowest terms, respectively. If the factor $(x - x_0)$ appears at most to the first power in the denominator of $P(x)$ and at most to the second power in the denominator of $Q(x)$, then $x = x_0$ is a regular singular point.

EXAMPLE 1 Classification of Singular Points

It should be clear that $x = -2$ and $x = 2$ are singular points of the equation

$$(x^2 - 4)^2y'' + (x - 2)y' + y = 0.$$

Dividing the equation by $(x^2 - 4)^2 = (x - 2)^2(x + 2)^2$, we find that

$$P(x) = \frac{1}{(x - 2)(x + 2)^2} \quad \text{and} \quad Q(x) = \frac{1}{(x - 2)^2(x + 2)^2}.$$

We now test $P(x)$ and $Q(x)$ at each singular point.

In order that $x = -2$ be a regular singular point, the factor $x + 2$ can appear at most to the first power in the denominator of $P(x)$ and can appear at most to the second power in the denominator of $Q(x)$. Inspection of $P(x)$ and $Q(x)$ shows that the first condition is not satisfied, and so we conclude that $x = -2$ is an irregular singular point.

In order that $x = 2$ be a regular singular point, the factor $x - 2$ can appear at most to the first power in the denominator of $P(x)$ and can appear at most to the second power in the denominator of $Q(x)$. Further inspection of $P(x)$ and $Q(x)$ shows that both these conditions are satisfied, so $x = 2$ is a regular singular point. ■

EXAMPLE 2 Classification of Singular Points

Both $x = 0$ and $x = -1$ are singular points of the differential equation

$$x^2(x + 1)^2y'' + (x^2 - 1)y' + 2y = 0.$$

Inspection of

$$P(x) = \frac{x - 1}{x^2(x + 1)} \quad \text{and} \quad Q(x) = \frac{2}{x^2(x + 1)^2}$$

shows that $x = 0$ is an irregular singular point since $(x - 0)$ appears to the second power in the denominator of $P(x)$. Note, however, that $x = -1$ is a regular singular point. ■

EXAMPLE 3 Classification of Singular Points

(a) $x = 1$ and $x = -1$ are regular singular points of

$$(1 - x^2)y'' - 2xy' + 30y = 0.$$

(b) $x = 0$ is an irregular singular point of

$$x^3y'' - 2xy' + 5y = 0$$

since
$$Q(x) = \frac{5}{x^3}.$$

(c) $x = 0$ is a regular singular point of

$$xy'' - 2xy' + 5y = 0$$

since
$$P(x) = -2 \quad \text{and} \quad Q(x) = \frac{5}{x}.$$

In part (c) of Example 3 notice that $(x - 0)$ and $(x - 0)^2$ do not even appear in the denominators of $P(x)$ and $Q(x)$, respectively. Remember, these factors can appear at most in this fashion. For a singular point $x = x_0$, any nonnegative power of $(x - x_0)$ less than one (namely, zero) and nonnegative power less than two (namely, zero and one) in the denominators of $P(x)$ and $Q(x)$, respectively, imply x_0 is a regular singular point.

Also, recall that singular points can be complex numbers. It should be apparent that both $x = 3i$ and $x = -3i$ are regular singular points of the equation $(x^2 + 9)y'' - 3xy' + (1 - x)y = 0$ since

$$P(x) = \frac{-3x}{(x - 3i)(x + 3i)} \quad \text{and} \quad Q(x) = \frac{1 - x}{(x - 3i)(x + 3i)}.$$

EXAMPLE 4 A Cauchy-Euler DE

From our discussion of the Cauchy-Euler equation in Section 6.1, we can show that $y_1 = x^2$ and $y_2 = x^2 \ln x$ are solutions of the equation $x^2y'' - 3xy' + 4y = 0$ on the interval $(0, \infty)$. If the procedure of Theorem 6.1 were attempted at the regular singular point $x = 0$ (that is, an assumed solution of the form $y = \sum_{n=0}^{\infty} c_n x^n$), we would succeed in obtaining only the solution $y_1 = x^2$. The fact that we would not obtain the second solution is not really surprising since $\ln x$ does not possess a Taylor series expansion about $x = 0$. It follows that $y_2 = x^2 \ln x$ does not have a power series in x.

EXAMPLE 5 Existence of Series Solutions

The differential equation $6x^2y'' + 5xy' + (x^2 - 1)y = 0$ has a regular singular point at $x = 0$ but does not possess *any* solution of the form $y = \sum_{n=0}^{\infty} c_n x^n$. By the procedure that we shall now consider it can be shown, however, that there exist two series solutions of the form

$$y = \sum_{n=0}^{\infty} c_n x^{n+1/2} \quad \text{and} \quad y = \sum_{n=0}^{\infty} c_n x^{n-1/3}.$$

Method of Frobenius To solve a differential equation such as (1) about a regular singular point we employ the following theorem due to Frobenius.*

THEOREM 6.2 Frobenius' Theorem

If $x = x_0$ is a regular singular point of the differential equation (1), then there exists at least one series solution of the form

$$y = (x - x_0)^r \sum_{n=0}^{\infty} c_n(x - x_0)^n = \sum_{n=0}^{\infty} c_n(x - x_0)^{n+r}, \tag{3}$$

where the number r is a constant that must be determined. The series will converge at least on some interval $0 < x - x_0 < R$.

Note the words *at least* in the second line of Theorem 6.2. This means that, in contrast to Theorem 6.1, we are *not* guaranteed two solutions of the indicated form. The **method of Frobenius** consists of identifying a regular singular point x_0, substituting $y = \sum_{n=0}^{\infty} c_n(x - x_0)^{n+r}$ into the differential equation, and determining the unknown exponent r and the coefficients c_n.

As in the preceding section, for the sake of simplicity we shall always assume $x_0 = 0$.

EXAMPLE 6 Two Series Solutions

Since $x = 0$ is a regular singular point of the differential equation

$$3xy'' + y' - y = 0, \tag{4}$$

we try a solution of the form $y = \sum_{n=0}^{\infty} c_n x^{n+r}$. Now

$$y' = \sum_{n=0}^{\infty} (n + r)c_n x^{n+r-1}, \quad y'' = \sum_{n=0}^{\infty} (n + r)(n + r - 1)c_n x^{n+r-2}$$

so

$$\begin{aligned} 3xy'' + y' - y &= 3\sum_{n=0}^{\infty} (n + r)(n + r - 1)c_n x^{n+r-1} \\ &\quad + \sum_{n=0}^{\infty} (n + r)c_n x^{n+r-1} - \sum_{n=0}^{\infty} c_n x^{n+r} \end{aligned}$$

* FERDINAND GEORG FROBENIUS (1848–1917) Although the basic idea of this series method can be traced back to Euler, the German mathematician Ferdinand Frobenius was the first to prove the result, which he published in 1878. Frobenius made many contributions to the field of analysis, but his name appears more in texts on abstract algebra than in texts on differential equations. His most significant contributions to mathematics were in the field of group theory.

$$= \sum_{n=0}^{\infty} (n + r)(3n + 3r - 2)c_n x^{n+r-1} - \sum_{n=0}^{\infty} c_n x^{n+r}$$

$$= x^r\left[r(3r - 2)c_0 x^{-1} + \underbrace{\sum_{n=1}^{\infty} (n + r)(3n + 3r - 2)c_n x^{n-1}}_{k = n - 1} - \underbrace{\sum_{n=0}^{\infty} c_n x^n}_{k = n}\right]$$

$$= x^r\left[r(3r - 2)c_0 x^{-1} + \sum_{k=0}^{\infty} [(k + r + 1)(3k + 3r + 1)c_{k+1} - c_k]x^k\right] = 0,$$

which implies $\qquad r(3r - 2)c_0 = 0$

$$(k + r + 1)(3k + 3r + 1)c_{k+1} - c_k = 0, \quad k = 0, 1, 2, \ldots. \tag{5}$$

Since nothing is gained by taking $c_0 = 0$, we must then have

$$r(3r - 2) = 0 \tag{6}$$

and
$$c_{k+1} = \frac{c_k}{(k + r + 1)(3k + 3r + 1)}, \quad k = 0, 1, 2, \ldots. \tag{7}$$

The two values of r that satisfy (6), $r_1 = \frac{2}{3}$ and $r_2 = 0$, when substituted in (7), give two different recurrence relations:

$$r_1 = \tfrac{2}{3}: \quad c_{k+1} = \frac{c_k}{(3k + 5)(k + 1)}, \quad k = 0, 1, 2, \ldots, \tag{8}$$

$$r_2 = 0: \quad c_{k+1} = \frac{c_k}{(k + 1)(3k + 1)}, \quad k = 0, 1, 2, \ldots. \tag{9}$$

Iteration of (8) gives
$$c_1 = \frac{c_0}{5 \cdot 1}$$
$$c_2 = \frac{c_1}{8 \cdot 2} = \frac{c_0}{2!5 \cdot 8}$$
$$c_3 = \frac{c_2}{11 \cdot 3} = \frac{c_0}{3!5 \cdot 8 \cdot 11}$$
$$c_4 = \frac{c_3}{14 \cdot 4} = \frac{c_0}{4!5 \cdot 8 \cdot 11 \cdot 14}$$
$$\vdots$$
$$c_n = \frac{c_0}{n!5 \cdot 8 \cdot 11 \cdots (3n + 2)}, \quad n = 1, 2, 3, \ldots,$$

whereas iteration of (9) yields

$$c_1 = \frac{c_0}{1 \cdot 1}$$
$$c_2 = \frac{c_1}{2 \cdot 4} = \frac{c_0}{2!1 \cdot 4}$$
$$c_3 = \frac{c_2}{3 \cdot 7} = \frac{c_0}{3!1 \cdot 4 \cdot 7}$$

$$c_4 = \frac{c_3}{4 \cdot 10} = \frac{c_0}{4!1 \cdot 4 \cdot 7 \cdot 10}$$

$$\vdots$$

$$c_n = \frac{c_0}{n!1 \cdot 4 \cdot 7 \cdots (3n-2)}, \quad n = 1, 2, 3, \ldots.$$

Thus we obtain two series solutions

$$y_1 = c_0 x^{2/3}\left[1 + \sum_{n=1}^{\infty} \frac{1}{n!5 \cdot 8 \cdot 11 \cdots (3n+2)} x^n\right] \tag{10}$$

and
$$y_2 = c_0 x^0\left[1 + \sum_{n=1}^{\infty} \frac{1}{n!1 \cdot 4 \cdot 7 \cdots (3n-2)} x^n\right]. \tag{11}$$

By the ratio test it can be demonstrated that both (10) and (11) converge for all finite values of x. Also it should be clear from the form of (10) and (11) that neither series is a constant multiple of the other and, therefore, $y_1(x)$ and $y_2(x)$ are linearly independent solutions on the x-axis. Hence by the superposition principle

$$y = C_1 y_1(x) + C_2 y_2(x) = C_1\left[x^{2/3} + \sum_{n=1}^{\infty} \frac{1}{n!5 \cdot 8 \cdot 11 \cdots (3n+2)} x^{n+2/3}\right] + C_2\left[1 + \sum_{n=1}^{\infty} \frac{1}{n!1 \cdot 4 \cdot 7 \cdots (3n-2)} x^n\right], \quad |x| < \infty$$

is another solution of (4). On any interval not containing the origin, this combination represents the general solution of the differential equation. ■

Although Example 6 illustrates the general procedure for using the method of Frobenius, we hasten to point out that we may not always be able to find two solutions so readily or for that matter find two solutions that are infinite series consisting entirely of powers of x.

Indicial Equation Equation (6) is called the **indicial equation** of the problem, and the values $r_1 = \frac{2}{3}$ and $r_2 = 0$ are called the **indicial roots,** or **exponents,** of the singularity. In general, if $x = 0$ is a regular singular point of (1), then the functions $xP(x)$ and $x^2Q(x)$ obtained from (2) are analytic at zero; that is, the expansions

$$\begin{aligned} xP(x) &= p_0 + p_1 x + p_2 x^2 + \cdots \\ x^2Q(x) &= q_0 + q_1 x + q_2 x^2 + \cdots \end{aligned} \tag{12}$$

are valid on intervals that have a positive radius of convergence. After we substitute $y = \sum_{n=0}^{\infty} c_n x^{n+r}$ in (1) or (2) and simplify, the indicial equation is a quadratic equation in r that results from equating the *total coefficient of the lowest power of x to zero.* It is left as an exercise to show that the general indicial equation is

$$r(r-1) + p_0 r + q_0 = 0. \tag{13}$$

See Problem 38. We then solve the latter equation for the two values of the exponents and substitute these values into a recurrence relation such as (7). Theorem 6.2 guarantees that at least one solution of the assumed series form can be found.

EXAMPLE 7 Only One Series Solution

The differential equation

$$xy'' + 3y' - y = 0 \tag{14}$$

has a regular singular point at $x = 0$. The method of Frobenius yields

$$xy'' + 3y' - y = x^r\left[r(r+2)c_0x^{-1} + \sum_{k=0}^{\infty}[(k+r+1)(k+r+3)c_{k+1} - c_k]x^k\right] = 0$$

so the indicial equation and exponents are $r(r + 2) = 0$ and $r_1 = 0$, $r_2 = -2$, respectively.

Since

$$(k + r + 1)(k + r + 3)c_{k+1} - c_k = 0, \quad k = 0, 1, 2, \ldots, \tag{15}$$

it follows that when $r_1 = 0$,

$$\begin{aligned}
c_{k+1} &= \frac{c_k}{(k+1)(k+3)} \\
c_1 &= \frac{c_0}{1 \cdot 3} \\
c_2 &= \frac{c_1}{2 \cdot 4} = \frac{2c_0}{2!4!} \\
c_3 &= \frac{c_2}{3 \cdot 5} = \frac{2c_0}{3!5!} \\
c_4 &= \frac{c_3}{4 \cdot 6} = \frac{2c_0}{4!6!} \\
&\vdots \\
c_n &= \frac{2c_0}{n!(n+2)!}, \quad n = 1, 2, 3, \ldots.
\end{aligned}$$

Thus one series solution is

$$\begin{aligned}
y_1 &= c_0x^0\left[1 + \sum_{n=1}^{\infty}\frac{2}{n!(n+2)!}x^n\right] \\
&= c_0\sum_{n=0}^{\infty}\frac{2}{n!(n+2)!}x^n, \quad |x| < \infty.
\end{aligned} \tag{16}$$

Now when $r_2 = -2$, (15) becomes

$$(k - 1)(k + 1)c_{k+1} - c_k = 0, \tag{17}$$

but note here that we *do not divide* by $(k - 1)(k + 1)$ immediately since this term is zero for $k = 1$. However, we use the recurrence relation (17) for the cases $k = 0$ and $k = 1$:

$$-1 \cdot 1c_1 - c_0 = 0 \quad \text{and} \quad 0 \cdot 2c_2 - c_1 = 0.$$

The latter equation implies that $c_1 = 0$ and so the former equation implies that $c_0 = 0$. Continuing, we find

$$c_{k+1} = \frac{c_k}{(k-1)(k+1)}, \quad k = 2, 3, 4, \ldots,$$

and so
$$c_3 = \frac{c_2}{1 \cdot 3}$$
$$c_4 = \frac{c_3}{2 \cdot 4} = \frac{2c_2}{2!4!}$$
$$c_5 = \frac{c_4}{3 \cdot 5} = \frac{2c_2}{3!5!}$$
$$\vdots$$
$$c_n = \frac{2c_2}{(n-2)!n!}, \quad n = 2, 3, 4, \ldots .$$

Thus
$$y_2 = c_2 x^{-2} \sum_{n=2}^{\infty} \frac{2}{(n-2)!n!} x^n. \tag{18}$$

However, close inspection of (18) reveals that y_2 is simply a constant multiple of (16). To see this, let $k = n - 2$ in (18). We conclude that the method of Frobenius gives only one series solution of (14).

Cases of Indicial Roots When using the method of Frobenius, we usually distinguish three cases corresponding to the nature of the indicial roots. For the sake of discussion let us suppose that r_1 and r_2 are the *real* solutions of the indicial equation and that, when appropriate, r_1 *denotes the largest root.*

Case I: Roots Not Differing by an Integer If r_1 and r_2 are distinct and do not differ by an integer, then there exist two linearly independent solutions of equation (1) of the form

$$y_1 = \sum_{n=0}^{\infty} c_n x^{n+r_1}, \quad c_0 \neq 0 \tag{19a}$$

$$y_2 = \sum_{n=0}^{\infty} b_n x^{n+r_2}, \quad b_0 \neq 0. \tag{19b}$$

EXAMPLE 8 Two Series Solutions

Solve
$$2xy'' + (1 + x)y' + y = 0. \tag{20}$$

Solution If $y = \sum_{n=0}^{\infty} c_n x^{n+r}$, then

$$\begin{aligned}
2xy'' + (1 + x)y' + y &= 2\sum_{n=0}^{\infty}(n + r)(n + r - 1)c_n x^{n+r-1} + \sum_{n=0}^{\infty}(n + r)c_n x^{n+r-1} \\
&\quad + \sum_{n=0}^{\infty}(n + r)c_n x^{n+r} + \sum_{n=0}^{\infty} c_n x^{n+r} \\
&= \sum_{n=0}^{\infty}(n + r)(2n + 2r - 1)c_n x^{n+r-1} + \sum_{n=0}^{\infty}(n + r + 1)c_n x^{n+r} \\
&= x^r\Bigg[r(2r - 1)c_0 x^{-1} + \underbrace{\sum_{n=1}^{\infty}(n + r)(2n + 2r - 1)c_n x^{n-1}}_{k = n-1} + \underbrace{\sum_{n=0}^{\infty}(n + r + 1)c_n x^n}_{k = n}\Bigg] \\
&= x^r\Bigg[r(2r - 1)c_0 x^{-1} + \sum_{k=0}^{\infty}[(k + r + 1)(2k + 2r + 1)c_{k+1} + (k + r + 1)c_k]x^k\Bigg] = 0,
\end{aligned}$$

which implies
$$r(2r - 1) = 0 \tag{21}$$

$$(k + r + 1)(2k + 2r + 1)c_{k+1} + (k + r + 1)c_k = 0, \quad k = 0, 1, 2, \ldots. \tag{22}$$

For $r_1 = \frac{1}{2}$, we can divide by $k + \frac{3}{2}$ in (22) to obtain

$$\begin{aligned}
c_{k+1} &= \frac{-c_k}{2(k+1)} \\
c_1 &= \frac{-c_0}{2 \cdot 1} \\
c_2 &= \frac{-c_1}{2 \cdot 2} = \frac{c_0}{2^2 \cdot 2!} \\
c_3 &= \frac{-c_2}{2 \cdot 3} = \frac{-c_0}{2^3 \cdot 3!} \\
&\vdots \\
c_n &= \frac{(-1)^n c_0}{2^n n!}, \quad n = 1, 2, 3, \ldots.
\end{aligned}$$

Thus we have
$$\begin{aligned}
y_1 &= c_0 x^{1/2}\left[1 + \sum_{n=1}^{\infty} \frac{(-1)^n}{2^n n!} x^n\right] \\
&= c_0 \sum_{n=0}^{\infty} \frac{(-1)^n}{2^n n!} x^{n+1/2},
\end{aligned} \tag{23}$$

which converges for $x \geq 0$. As given, the series is not meaningful for $x < 0$ because of the presence of $x^{1/2}$.

Now for $r_2 = 0$, (22) becomes

$$\begin{aligned}
c_{k+1} &= \frac{-c_k}{2k+1} \\
c_1 &= \frac{-c_0}{1} \\
c_2 &= \frac{-c_1}{3} = \frac{c_0}{1 \cdot 3} \\
c_3 &= \frac{-c_2}{5} = \frac{-c_0}{1 \cdot 3 \cdot 5} \\
c_4 &= \frac{-c_3}{7} = \frac{c_0}{1 \cdot 3 \cdot 5 \cdot 7} \\
&\vdots \\
c_n &= \frac{(-1)^n c_0}{1 \cdot 3 \cdot 5 \cdot 7 \cdots (2n-1)}, \quad n = 1, 2, 3, \ldots.
\end{aligned}$$

We conclude that a second solution to (20) is

$$y_2 = c_0\left[1 + \sum_{n=1}^{\infty} \frac{(-1)^n}{1 \cdot 3 \cdot 5 \cdot 7 \cdots (2n-1)} x^n\right], \quad |x| < \infty. \tag{24}$$

On the interval $(0, \infty)$, the general solution is $y = C_1 y_1(x) + C_2 y_2(x)$. ■

6.4.2 METHOD OF FROBENIUS—CASES II AND III

When the roots of the indicial equation differ by a positive integer, we may or may not be able to find two solutions of (1) having form (3). If not, then one solution corresponding to the smaller root contains a logarithmic term. When the exponents are equal, a second solution *always* contains a logarithm. This latter situation is analogous to the solutions of the Cauchy-Euler differential equation when the roots of the auxiliary equation are equal. We have the next two cases.

Case II: Roots Differing by a Positive Integer If $r_1 - r_2 = N$, where N is a positive integer, then there exist two linearly independent solutions of equation (1) of the form

$$y_1 = \sum_{n=0}^{\infty} c_n x^{n+r_1}, \qquad c_0 \neq 0 \tag{25a}$$

$$y_2 = Cy_1(x)\ln x + \sum_{n=0}^{\infty} b_n x^{n+r_2}, \quad b_0 \neq 0, \tag{25b}$$

where C is a constant that could be zero.

Case III: Equal Indicial Roots If $r_1 = r_2$, there always exist two linearly independent solutions of equation (1) of the form

$$y_1 = \sum_{n=0}^{\infty} c_n x^{n+r_1}, \quad c_0 \neq 0 \tag{26a}$$

$$y_2 = y_1(x)\ln x + \sum_{n=1}^{\infty} b_n x^{n+r_1}. \tag{26b}$$

EXAMPLE 9 **Two Solutions: A Polynomial and an Infinite Series**

Solve
$$xy'' + (x - 6)y' - 3y = 0. \tag{27}$$

Solution The assumption $y = \sum_{n=0}^{\infty} c_n x^{n+r}$ leads to

$$\begin{aligned} xy'' &+ (x-6)y' - 3y \\ &= \sum_{n=0}^{\infty}(n+r)(n+r-1)c_n x^{n+r-1} - 6\sum_{n=0}^{\infty}(n+r)c_n x^{n+r-1} + \sum_{n=0}^{\infty}(n+r)c_n x^{n+r} - 3\sum_{n=0}^{\infty} c_n x^{n+r} \\ &= x^r\Bigg[r(r-7)c_0x^{-1} + \underbrace{\sum_{n=1}^{\infty}(n+r)(n+r-7)c_n x^{n-1}}_{k=n-1} + \underbrace{\sum_{n=0}^{\infty}(n+r-3)c_n x^n}_{k=n}\Bigg] \\ &= x^r\Bigg[r(r-7)c_0x^{-1} + \sum_{k=0}^{\infty}[(k+r+1)(k+r-6)c_{k+1} + (k+r-3)c_k]x^k\Bigg] = 0. \end{aligned}$$

Thus $r(r-7) = 0$ so $r_1 = 7$, $r_2 = 0$, $r_1 - r_2 = 7$, and

$$(k+r+1)(k+r-6)c_{k+1} + (k+r-3)c_k = 0, \quad k = 0, 1, 2, \ldots. \tag{28}$$

For the smaller root $r_2 = 0$, (28) becomes

$$(k + 1)(k - 6)c_{k+1} + (k - 3)c_k = 0. \qquad \textbf{(29)}$$

Since $k - 6 = 0$ when $k = 6$, we do not divide by this term until $k > 6$. We find

$$\begin{aligned}
1 \cdot (-6)c_1 + (-3)c_0 &= 0 \\
2 \cdot (-5)c_2 + (-2)c_1 &= 0 \\
3 \cdot (-4)c_3 + (-1)c_2 &= 0 \\
\left.\begin{aligned}
4 \cdot (-3)c_4 + 0 \cdot c_3 &= 0 \\
5 \cdot (-2)c_5 + 1 \cdot c_4 &= 0 \\
6 \cdot (-1)c_6 + 2 \cdot c_5 &= 0 \\
7 \cdot 0c_7 + 3 \cdot c_6 &= 0
\end{aligned}\right\} &\leftarrow \text{implies } c_4 = c_5 = c_6 = 0 \text{ but } c_0 \text{ and } c_7 \text{ can be chosen arbitrarily}
\end{aligned}$$

Hence

$$\begin{aligned}
c_1 &= -\frac{1}{2}c_0 \\
c_2 &= -\frac{1}{5}c_1 = \frac{1}{10}c_0 \\
c_3 &= -\frac{1}{12}c_2 = -\frac{1}{120}c_0
\end{aligned} \qquad \textbf{(30)}$$

and, for $k \geq 7$,

$$\begin{aligned}
c_{k+1} &= \frac{-(k - 3)c_k}{(k + 1)(k - 6)} \\
c_8 &= \frac{-4}{8 \cdot 1}c_7 \\
c_9 &= \frac{-5}{9 \cdot 2}c_8 = \frac{4 \cdot 5}{2!8 \cdot 9}c_7 \\
c_{10} &= \frac{-6}{10 \cdot 3}c_9 = \frac{-4 \cdot 5 \cdot 6}{3!8 \cdot 9 \cdot 10}c_7 \\
&\vdots \\
c_n &= \frac{(-1)^{n+1}4 \cdot 5 \cdot 6 \cdots (n - 4)}{(n - 7)!8 \cdot 9 \cdot 10 \cdots n}c_7, \quad n = 8, 9, 10, \ldots.
\end{aligned} \qquad \textbf{(31)}$$

If we choose $c_7 = 0$ and $c_0 \neq 0$, we obtain the polynomial solution

$$y_1 = c_0\left[1 - \frac{1}{2}x + \frac{1}{10}x^2 - \frac{1}{120}x^3\right], \qquad \textbf{(32)}$$

but when $c_7 \neq 0$ and $c_0 = 0$, it follows that a second, though infinite series, solution is

$$\begin{aligned}
y_2 &= c_7\left[x^7 + \sum_{n=8}^{\infty} \frac{(-1)^{n+1}4 \cdot 5 \cdot 6 \cdots (n - 4)}{(n - 7)!\, 8 \cdot 9 \cdot 10 \cdots n}x^n\right] \\
&= c_7\left[x^7 + \sum_{k=1}^{\infty} \frac{(-1)^k 4 \cdot 5 \cdot 6 \cdots (k + 3)}{k!\, 8 \cdot 9 \cdot 10 \cdots (k + 7)}x^{k+7}\right], \quad |x| < \infty.
\end{aligned} \qquad \textbf{(33)}$$

Finally, the general solution of (27) on the interval $(0, \infty)$ is

$$y = C_1y_1(x) + C_2y_2(x)$$
$$= C_1\left[1 - \frac{1}{2}x + \frac{1}{10}x^2 - \frac{1}{120}x^3\right] + C_2\left[x^7 + \sum_{k=1}^{\infty}\frac{(-1)^k 4 \cdot 5 \cdot 6 \cdots (k+3)}{k!\, 8 \cdot 9 \cdot 10 \cdots (k+7)}x^{k+7}\right]. \quad ■$$

It is interesting to observe that in Example 9 the larger root $r_1 = 7$ was not used. Had we used it, we would have obtained a series solution of the form*

$$y = \sum_{n=0}^{\infty} c_n x^{n+7}, \tag{34}$$

where the c_n are defined by (28) with $r_1 = 7$:

$$c_{k+1} = \frac{-(k+4)}{(k+8)(k+1)}c_k, \quad k = 0, 1, 2, \ldots.$$

Iteration of this latter recurrence relation then would yield only *one* solution—namely, the solution given by (33) (with c_0 playing the part of c_7).

When the roots of the indicial equation differ by a positive integer, the second solution *may* contain a logarithm. In practice this is something we do not know in advance but that is determined after we have found the indicial roots and have carefully examined the recurrence relation that defines the coefficients c_n. As the foregoing example shows, we just may be lucky enough to find two solutions that involve only powers of x. On the other hand, if we fail to find a second series-type solution, we can always use the fact that

$$y_2 = y_1(x)\int \frac{e^{-\int P(x)\,dx}}{y_1^2(x)}\,dx \tag{35}$$

is also a solution of the equation $y'' + P(x)y' + Q(x)y = 0$, whenever y_1 is a known solution (see Section 4.2).

EXAMPLE 10 Finding a Second Solution of (14)

Find the general solution of $xy'' + 3y' - y = 0$.

Solution Recall from Example 7 that the method of Frobenius provides only one solution to this equation—namely,

$$y_1 = \sum_{n=0}^{\infty}\frac{2}{n!(n+2)!}x^n = 1 + \frac{1}{3}x + \frac{1}{24}x^2 + \frac{1}{360}x^3 + \cdots. \tag{36}$$

From (35) we obtain a second solution:

$$y_2 = y_1(x)\int\frac{e^{-\int(3/x)\,dx}}{y_1^2(x)}\,dx = y_1(x)\int\frac{dx}{x^3\left[1 + \frac{1}{3}x + \frac{1}{24}x^2 + \frac{1}{360}x^3 + \cdots\right]^2}$$

* Observe that both (33) and (34) start with the power x^7. In Case II it is always a good idea to work with the smaller root first.

$$= y_1(x)\int \frac{dx}{x^3\left[1 + \frac{2}{3}x + \frac{7}{36}x^2 + \frac{1}{30}x^3 + \cdots\right]} \qquad \leftarrow \text{squaring}$$

$$= y_1(x)\int \frac{1}{x^3}\left[1 - \frac{2}{3}x + \frac{1}{4}x^2 - \frac{19}{270}x^3 + \cdots\right]dx \leftarrow \text{long division}$$

$$= y_1(x)\int \left[\frac{1}{x^3} - \frac{2}{3x^2} + \frac{1}{4x} - \frac{19}{270} + \cdots\right]dx$$

$$= y_1(x)\left[-\frac{1}{2x^2} + \frac{2}{3x} + \frac{1}{4}\ln x - \frac{19}{270}x + \cdots\right]$$

or

$$y_2 = \frac{1}{4}y_1(x)\ln x + y_1(x)\left[-\frac{1}{2x^2} + \frac{2}{3x} - \frac{19}{270}x + \cdots\right]. \tag{37}$$

Hence on the interval $(0, \infty)$ the general solution is

$$y = C_1y_1(x) + C_2\left[\frac{1}{4}y_1(x)\ln x + y_1(x)\left(-\frac{1}{2x^2} + \frac{2}{3x} - \frac{19}{270}x + \cdots\right)\right], \tag{38}$$

where $y_1(x)$ is defined by (36).

Alternative Procedure There are several alternative procedures to formula (35) when the method of Frobenius fails to provide a second series solution. Although the next method is somewhat tedious, it is nonetheless straightforward. The basic idea is to assume a solution of either the form (25b) or the form (26b) and determine coefficients b_n in terms of the coefficients c_n that define the known solution $y_1(x)$.

EXAMPLE 11 **Example 10 Revisited**

The smaller of the two indicial roots for the equation $xy'' + 3y' - y = 0$ is $r_2 = -2$. From (25b) we now assume a second solution

$$y_2 = y_1 \ln x + \sum_{n=0}^{\infty} b_n x^{n-2}, \tag{39}$$

where

$$y_1 = \sum_{n=0}^{\infty} \frac{2}{n!(n+2)!}x^n. \tag{40}$$

Differentiation of (39) gives

$$y_2' = \frac{y_1}{x} + y_1'\ln x + \sum_{n=0}^{\infty}(n-2)b_nx^{n-3}$$

$$y_2'' = -\frac{y_1}{x^2} + \frac{2y_1'}{x} + y_1''\ln x + \sum_{n=0}^{\infty}(n-2)(n-3)b_nx^{n-4}$$

so

$$xy_2'' + 3y_2' - y_2 = \ln x[\underbrace{xy_1'' + 3y_1' - y_1}_{\text{zero}}] + 2y_1' + \frac{2y_1}{x} + \sum_{n=0}^{\infty}(n-2)(n-3)b_n x^{n-3}$$

$$+ 3\sum_{n=0}^{\infty}(n-2)b_n x^{n-3} - \sum_{n=0}^{\infty} b_n x^{n-2}$$

$$= 2y_1' + \frac{2y_1}{x} + \sum_{n=0}^{\infty}(n-2)nb_n x^{n-3} - \sum_{n=0}^{\infty} b_n x^{n-2}, \qquad \textbf{(41)}$$

where we have combined the first two summations and used the fact that $xy_1'' + 3y_1' - y_1 = 0$.

By differentiating (40), we can write (41) as

$$\sum_{n=0}^{\infty}\frac{4n}{n!(n+2)!}x^{n-1} + \sum_{n=0}^{\infty}\frac{4}{n!(n+2)!}x^{n-1} + \sum_{n=0}^{\infty}(n-2)nb_n x^{n-3} - \sum_{n=0}^{\infty} b_n x^{n-2}$$

$$= 0(-2)b_0x^{-3} + (-b_0 - b_1)x^{-2} + \underbrace{\sum_{n=0}^{\infty}\frac{4(n+1)}{n!(n+2)!}x^{n-1}}_{k=n} + \underbrace{\sum_{n=2}^{\infty}(n-2)nb_n x^{n-3}}_{k=n-2} - \underbrace{\sum_{n=1}^{\infty} b_n x^{n-2}}_{k=n-1}$$

$$= -(b_0 + b_1)x^{-2} + \sum_{k=0}^{\infty}\left[\frac{4(k+1)}{k!(k+2)!} + k(k+2)b_{k+2} - b_{k+1}\right]x^{k-1}. \qquad \textbf{(42)}$$

Setting (42) equal to zero then gives $b_1 = -b_0$ and

$$\frac{4(k+1)}{k!(k+2)!} + k(k+2)b_{k+2} - b_{k+1} = 0, \quad \text{for } k = 0, 1, 2, \ldots. \qquad \textbf{(43)}$$

When $k = 0$ in (43), we have $2 + 0 \cdot 2b_2 - b_1 = 0$ so $b_1 = 2$, $b_0 = -2$, but b_2 is arbitrary.

Rewriting (43) as

$$b_{k+2} = \frac{b_{k+1}}{k(k+2)} - \frac{4(k+1)}{k!(k+2)!k(k+2)} \qquad \textbf{(44)}$$

and evaluating for $k = 1, 2, \ldots$ gives

$$b_3 = \frac{b_2}{3} - \frac{4}{9}$$

$$b_4 = \frac{1}{8}b_3 - \frac{1}{32} = \frac{1}{24}b_2 - \frac{25}{288}$$

and so on. Thus we can finally write

$$y_2 = y_1 \ln x + b_0x^{-2} + b_1x^{-1} + b_2 + b_3x + \cdots$$

$$= y_1 \ln x - 2x^{-2} + 2x^{-1} + b_2 + \left(\frac{b_2}{3} - \frac{4}{9}\right)x + \cdots, \qquad \textbf{(45)}$$

where b_2 is arbitrary. ■

Equivalent Solutions At this point you may be wondering whether (37) and (45) are really equivalent. If we choose $C_2 = 4$ in (38), then

$$
\begin{aligned}
y_2 &= y_1 \ln x + y_1\left(-\frac{2}{x^2} + \frac{8}{3x} - \frac{38}{135}x + \cdots\right) \\
&= y_1 \ln x + \left(1 + \frac{1}{3}x + \frac{1}{24}x^2 + \frac{1}{360}x^3 + \cdots\right)\left(-\frac{2}{x^2} + \frac{8}{3x} - \frac{38}{135}x + \cdots\right) \\
&= y_1 \ln x - 2x^{-2} + 2x^{-1} + \frac{29}{36} - \frac{19}{108}x + \cdots, \qquad \textbf{(46)}
\end{aligned}
$$

which is precisely what we obtain from (45) if b_2 is chosen as 29/36.

The next example illustrates the case when the indicial roots are equal.

EXAMPLE 12 A General Solution

Find the general solution of

$$xy'' + y' - 4y = 0. \qquad \textbf{(47)}$$

Solution The assumption $y = \sum_{n=0}^{\infty} c_n x^{n+r}$ leads to

$$
\begin{aligned}
xy'' + y' - 4y &= \sum_{n=0}^{\infty}(n + r)(n + r - 1)c_n x^{n+r-1} + \sum_{n=0}^{\infty}(n + r)c_n x^{n+r-1} - 4\sum_{n=0}^{\infty} c_n x^{n+r} \\
&= \sum_{n=0}^{\infty}(n + r)^2 c_n x^{n+r-1} - 4\sum_{n=0}^{\infty} c_n x^{n+r} \\
&= x^r\Bigg[r^2 c_0 x^{-1} + \underbrace{\sum_{n=1}^{\infty}(n + r)^2 c_n x^{n-1}}_{k = n - 1} - 4\underbrace{\sum_{n=0}^{\infty} c_n x^n}_{k = n}\Bigg] \\
&= x^r\left[r^2 c_0 x^{-1} + \sum_{k=0}^{\infty}[(k + r + 1)^2 c_{k+1} - 4c_k]x^k\right] = 0.
\end{aligned}
$$

Therefore $r^2 = 0$, and so the indicial roots are equal: $r_1 = r_2 = 0$. Moreover, we have

$$(k + r + 1)^2 c_{k+1} - 4c_k = 0, \quad k = 0, 1, 2 \ldots. \qquad \textbf{(48)}$$

Clearly the root $r_1 = 0$ will only yield one solution corresponding to the coefficients defined by the iteration of

$$c_{k+1} = \frac{4c_k}{(k + 1)^2}, \quad k = 0, 1, 2, \ldots.$$

The result is

$$y_1 = c_0 \sum_{n=0}^{\infty} \frac{4^n}{(n!)^2} x^n, \quad |x| < \infty. \qquad \textbf{(49)}$$

To obtain the second linearly independent solution we set $c_0 = 1$ in (49) and then use (35):

$$
\begin{aligned}
y_2 = y_1(x)\int \frac{e^{-\int(1/x)\,dx}}{y_1^2(x)}\,dx &= y_1(x)\int \frac{dx}{x\left[1 + 4x + 4x^2 + \frac{16}{9}x^3 + \cdots\right]^2} \\
&= y_1(x)\int \frac{dx}{x\left[1 + 8x + 24x^2 + \frac{16}{9}x^3 + \cdots\right]}
\end{aligned}
$$

$$= y_1(x) \int \frac{1}{x}\left[1 - 8x + 40x^2 - \frac{1472}{9}x^3 + \cdots\right] dx$$

$$= y_1(x) \int \left[\frac{1}{x} - 8 + 40x - \frac{1472}{9}x^2 + \cdots\right] dx$$

$$= y_1(x)\left[\ln x - 8x + 20x^2 - \frac{1472}{27}x^3 + \cdots\right]. \tag{50}$$

Thus on the interval $(0, \infty)$ the general solution of (47) is

$$y = C_1 y_1(x) + C_2\left[y_1(x) \ln x + y_1(x)\left(-8x + 20x^2 - \frac{1472}{27}x^3 + \cdots\right)\right], \tag{51}$$

where $y_1(x)$ is defined by (49). ■

As in Case II we can also determine $y_2(x)$ of Example 12 directly from assumption (26b).

Remarks (*i*) We purposely have not considered two further complications when solving a differential equation such as (1) about a point x_0 for which $a_2(x_0) = 0$. When using (3), it is quite possible that the roots of the indicial equation could turn out to be complex numbers. When the exponents r_1 and r_2 are complex, the statement $r_1 > r_2$ is meaningless and must be replaced with $\text{Re}(r_1) > \text{Re}(r_2)$ (for example, if $r = \alpha + i\beta$, then $\text{Re}(r) = \alpha$). In particular, when the indicial equation has real coefficients, the complex roots will be a conjugate pair $r_1 = \alpha + i\beta$, $r_2 = \alpha - i\beta$, and $r_1 - r_2 = 2i\beta \neq$ integer. Thus for $x_0 = 0$ there will always exist two solutions

$$y_1 = \sum_{n=0}^{\infty} c_n x^{n+r_1} \quad \text{and} \quad y_2 = \sum_{n=0}^{\infty} b_n x^{n+r_2}.$$

Unfortunately both solutions give complex values of y for each real choice of x. This latter difficulty can be surmounted by the superposition principle. Since a combination of solutions is also a solution to the differential equation, we could form appropriate combinations of $y_1(x)$ and $y_2(x)$ to yield real solutions (see Case III of the solution of the Cauchy-Euler equation).
(*ii*) If $x = 0$ is an irregular singular point, it should be noted that we may not be able to find *any* solution of the form $y = \sum_{n=0}^{\infty} c_n x^{n+r}$.
(*iii*) In the advanced study of differential equations it is sometimes important to examine the nature of a singular point at ∞. A differential equation is said to have a singular point at ∞ if, after the substitution $z = 1/x$, the resulting equation has a singular point at $z = 0$. For example, the differential equation $y'' + xy = 0$ has no finite singular points. However, by the chain rule the substitution $z = 1/x$ transforms the equation into

$$z^5 \frac{d^2y}{dz^2} + 2z^4 \frac{dy}{dz} + y = 0.$$

(Verify this.) Inspection of $P(z) = 2/z$ and $Q(z) = 1/z^5$ shows that $z = 0$ is an irregular singular point of the equation. Hence ∞ is an irregular singular point. See Problem 40.

EXERCISES 6.4

Answers to odd-numbered problems begin on page A-13.

6.4.1 Regular Singular Points; Method of Frobenius—Case I

In Problems 1–10 determine the singular points of each differential equation. Classify each singular point as regular or irregular.

1. $x^3y'' + 4x^2y' + 3y = 0$

2. $xy'' - (x + 3)^{-2}y = 0$

3. $(x^2 - 9)^2y'' + (x + 3)y' + 2y = 0$

4. $y'' - \dfrac{1}{x}y' + \dfrac{1}{(x - 1)^3}y = 0$

5. $(x^3 + 4x)y'' - 2xy' + 6y = 0$

6. $x^2(x - 5)^2y'' + 4xy' + (x^2 - 25)y = 0$

7. $(x^2 + x - 6)y'' + (x + 3)y' + (x - 2)y = 0$

8. $x(x^2 + 1)^2y'' + y = 0$

9. $x^3(x^2 - 25)(x - 2)^2y'' + 3x(x - 2)y' + 7(x + 5)y = 0$

10. $(x^3 - 2x^2 - 3x)^2y'' + x(x - 3)^2y' - (x + 1)y = 0$

In Problems 11–22 show that the indicial roots do not differ by an integer. Use the method of Frobenius to obtain two linearly independent series solutions about the regular singular point $x_0 = 0$. Form the general solution on $(0, \infty)$.

11. $2xy'' - y' + 2y = 0$

12. $2xy'' + 5y' + xy = 0$

13. $4xy'' + \dfrac{1}{2}y' + y = 0$

14. $2x^2y'' - xy' + (x^2 + 1)y = 0$

15. $3xy'' + (2 - x)y' - y = 0$

16. $x^2y'' - \left(x - \dfrac{2}{9}\right)y = 0$

17. $2xy'' - (3 + 2x)y' + y = 0$

18. $x^2y'' + xy' + \left(x^2 - \dfrac{4}{9}\right)y = 0$

19. $9x^2y'' + 9x^2y' + 2y = 0$

20. $2x^2y'' + 3xy' + (2x - 1)y = 0$

21. $2x^2y'' - x(x - 1)y' - y = 0$

22. $x(x - 2)y'' + y' - 2y = 0$

6.4.2 Method of Frobenius—Cases II and III

In Problems 23–34 show that the indicial roots differ by an integer. Use the method of Frobenius to obtain two linearly independent series solutions about the regular singular point $x_0 = 0$. Form the general solution on $(0, \infty)$.

23. $xy'' + 2y' - xy = 0$

24. $x^2y'' + xy' + \left(x^2 - \dfrac{1}{4}\right)y = 0$

25. $x(x - 1)y'' + 3y' - 2y = 0$

26. $y'' + \dfrac{3}{x}y' - 2y = 0$

27. $xy'' + (1 - x)y' - y = 0$

28. $xy'' + y = 0$

29. $xy'' + y' + y = 0$

30. $xy'' - xy' + y = 0$

31. $x^2y'' + x(x - 1)y' + y = 0$

32. $xy'' + y' - 4xy = 0$

33. $xy'' + (x - 1)y' - 2y = 0$

34. $xy'' - y' + x^3y = 0$

In Problems 35 and 36 note that $x_0 = 0$ is an irregular singular point of each equation. In each case determine whether the method of Frobenius yields a solution.

35. $x^3y'' + y = 0$

36. $x^2y'' - y' + y = 0$

37. Solve the Cauchy-Euler equation $x^2y'' + 3xy' - 8y = 0$ by the method of Frobenius.

38. If $x = 0$ is a regular singular point, use (12) in (2) to show that (13) is the indicial equation obtained from the method of Frobenius.

39. Use (13) to find the indicial equation and exponents of

$$x^2y'' + \left(\frac{5}{3}x + x^2\right)y' - \frac{1}{3}y = 0.$$

40. **(a)** Show that the differential equation $x^2y'' - 4y = 0$ has a singular point at ∞. [*Hint*: See page 263.]

(b) Classify the singular point at ∞ as either regular or irregular.

6.5 TWO SPECIAL EQUATIONS

- *Bessel's equation* • *Parametric Bessel equation* • *Bessel functions*
- *Legendre's equation* • *Legendre polynomials*

The two equations

$$x^2y'' + xy' + (x^2 - \nu^2)y = 0 \tag{1}$$

$$(1 - x^2)y'' - 2xy' + n(n + 1)y = 0 \tag{2}$$

occur frequently in advanced studies in applied mathematics, physics, and engineering. They are called **Bessel's equation** and **Legendre's equation,** respectively.* In solving (1) we shall assume $\nu \geq 0$, whereas in (2) we shall consider only the case when n is a nonnegative integer. Since we seek series solutions of each equation about $x = 0$, we observe that the origin is a regular singular point of Bessel's equation, but it is an ordinary point of Legendre's equation.

---

* FRIEDRICH WILHELM BESSEL (1784–1846) Bessel was a German astronomer who in 1838 was the first to measure the distance to a star (61 Cygni). In 1840 he predicted the existence of a planetary mass beyond the orbit of Uranus. The planet Neptune was discovered six years later. Bessel was also the first person to calculate the orbit of Halley's comet. Although Bessel certainly studied equation (1) in his work on planetary motion, the differential equation and its solution were probably discovered by Daniel Bernoulli in his research on determining the displacements of an oscillating chain.

ADRIEN MARIE LEGENDRE (1752–1833) A French mathematician, Legendre is best remembered for spending almost forty years of his life studying and calculating elliptic integrals. However, the particular polynomial solutions of the equation that bears his name were encountered in his studies of gravitation.

6.5.1 SOLUTION OF BESSEL'S EQUATION

If we assume $y = \sum_{n=0}^{\infty} c_n x^{n+r}$, then

$$x^2y'' + xy' + (x^2 - \nu^2)y = \sum_{n=0}^{\infty} c_n(n+r)(n+r-1)x^{n+r} + \sum_{n=0}^{\infty} c_n(n+r)x^{n+r} + \sum_{n=0}^{\infty} c_n x^{n+r+2} - \nu^2 \sum_{n=0}^{\infty} c_n x^{n+r}$$

$$= c_0(r^2 - r + r - \nu^2)x^r + x^r \sum_{n=1}^{\infty} c_n[(n+r)(n+r-1) + (n+r) - \nu^2]x^n$$

$$+ x^r \sum_{n=0}^{\infty} c_n x^{n+2}$$

$$= c_0(r^2 - \nu^2)x^r + x^r \sum_{n=1}^{\infty} c_n[(n+r)^2 - \nu^2]x^n + x^r \sum_{n=0}^{\infty} c_n x^{n+2}. \quad \textbf{(3)}$$

From (3) we see that the indicial equation is $r^2 - \nu^2 = 0$, so the indicial roots are $r_1 = \nu$ and $r_2 = -\nu$. When $r_1 = \nu$, (3) becomes

$$x^\nu \sum_{n=1}^{\infty} c_n n(n+2\nu)x^n + x^\nu \sum_{n=0}^{\infty} c_n x^{n+2}$$

$$= x^\nu \Bigg[(1+2\nu)c_1 x + \underbrace{\sum_{n=2}^{\infty} c_n n(n+2\nu)x^n}_{k=n-2} + \underbrace{\sum_{n=0}^{\infty} c_n x^{n+2}}_{k=n}\Bigg]$$

$$= x^\nu \Bigg[(1+2\nu)c_1 x + \sum_{k=0}^{\infty} [(k+2)(k+2+2\nu)c_{k+2} + c_k]x^{k+2}\Bigg] = 0.$$

Therefore by the usual argument we can write

$$(1+2\nu)c_1 = 0$$

$$(k+2)(k+2+2\nu)c_{k+2} + c_k = 0$$

or

$$c_{k+2} = \frac{-c_k}{(k+2)(k+2+2\nu)}, \quad k = 0, 1, 2, \ldots. \quad \textbf{(4)}$$

The choice $c_1 = 0$ in (4) implies $c_3 = c_5 = c_7 = \cdots = 0$, so for $k = 0, 2, 4, \ldots$ we find, after letting $k + 2 = 2n$, $n = 1, 2, 3, \ldots$, that

$$c_{2n} = -\frac{c_{2n-2}}{2^2 n(n+\nu)}. \quad \textbf{(5)}$$

Thus
$$c_2 = -\frac{c_0}{2^2 \cdot 1 \cdot (1+\nu)}$$

$$c_4 = -\frac{c_2}{2^2 \cdot 2(2+\nu)} = \frac{c_0}{2^4 \cdot 1 \cdot 2(1+\nu)(2+\nu)}$$

$$c_6 = -\frac{c_4}{2^2 \cdot 3(3+\nu)} = -\frac{c_0}{2^6 \cdot 1 \cdot 2 \cdot 3(1+\nu)(2+\nu)(3+\nu)}$$

$$\vdots$$

$$c_{2n} = \frac{(-1)^n c_0}{2^{2n} n!(1+\nu)(2+\nu)\cdots(n+\nu)}, \quad n = 1, 2, 3, \ldots. \quad \textbf{(6)}$$

It is standard practice to choose c_0 to be a specific value—namely,

$$c_0 = \frac{1}{2^\nu \Gamma(1 + \nu)},$$

where $\Gamma(1 + \nu)$ is the Gamma function. See Appendix I. Since this latter function possesses the convenient property $\Gamma(1 + \alpha) = \alpha\Gamma(\alpha)$, we can reduce the indicated product in the denominator of (6) to one term. For example,

$$\Gamma(1 + \nu + 1) = (1 + \nu)\Gamma(1 + \nu)$$

$$\Gamma(1 + \nu + 2) = (2 + \nu)\Gamma(2 + \nu) = (2 + \nu)(1 + \nu)\Gamma(1 + \nu).$$

Hence we can write (6) as

$$\begin{aligned} c_{2n} &= \frac{(-1)^n}{2^{2n+\nu} n!(1 + \nu)(2 + \nu)\cdots(n + \nu)\Gamma(1 + \nu)} \\ &= \frac{(-1)^n}{2^{2n+\nu} n!\Gamma(1 + \nu + n)}, \qquad n = 0, 1, 2, \ldots. \end{aligned}$$

It follows that one solution is

$$y = \sum_{n=0}^{\infty} c_{2n} x^{2n+\nu} = \sum_{n=0}^{\infty} \frac{(-1)^n}{n!\Gamma(1 + \nu + n)} \left(\frac{x}{2}\right)^{2n+\nu}.$$

If $\nu \geq 0$, the series converges at least on the interval $[0, \infty)$.

Bessel Functions of the First Kind The foregoing series solution is usually denoted by $J_\nu(x)$:

$$J_\nu(x) = \sum_{n=0}^{\infty} \frac{(-1)^n}{n!\Gamma(1 + \nu + n)} \left(\frac{x}{2}\right)^{2n+\nu}. \tag{7}$$

Also, for the second exponent $r_2 = -\nu$, we obtain, in exactly the same manner,

$$J_{-\nu}(x) = \sum_{n=0}^{\infty} \frac{(-1)^n}{n!\Gamma(1 - \nu + n)} \left(\frac{x}{2}\right)^{2n-\nu}. \tag{8}$$

The functions $J_\nu(x)$ and $J_{-\nu}(x)$ are called **Bessel functions of the first kind** of order ν and $-\nu$, respectively. Depending on the value of ν, (8) may contain negative powers of x and hence converge on $(0, \infty)$.*

Now some care must be taken in writing the general solution of (1). When $\nu = 0$, it is apparent that (7) and (8) are the same. If $\nu > 0$ and $r_1 - r_2 = \nu - (-\nu) = 2\nu$ is not a positive integer, it follows from Case I of Section 6.4 that $J_\nu(x)$ and $J_{-\nu}(x)$ are linearly independent solutions of (1) on $(0, \infty)$, and so the general solution on the interval would be $y = c_1 J_\nu(x) + c_2 J_{-\nu}(x)$. But we also know from Case II of Section 6.4 that when $r_1 - r_2 = 2\nu$ is a positive integer, a second series solution of (1) *may* exist. In this second case we distinguish two possibilities. When $\nu = m =$ positive integer, $J_{-m}(x)$ defined by (8) and $J_m(x)$ are not linearly independent solutions. It can be shown that J_{-m} is a constant multiple of J_m (see Property (i) on page 269). In addition, $r_1 - r_2 = 2\nu$ can be a positive integer when ν is

* If we replace x by $|x|$, the series given in (7) and (8) converge for $0 < |x| < \infty$.

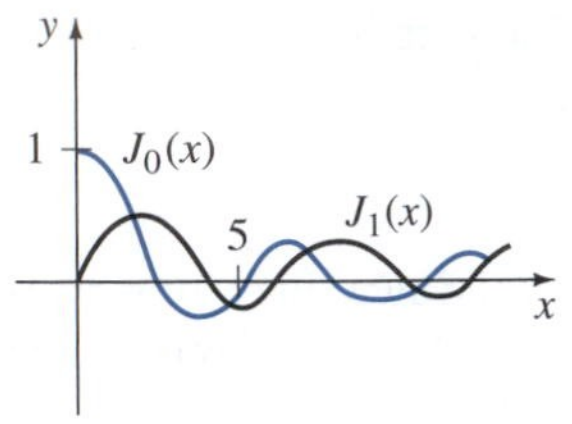

Figure 6.6

half an odd positive integer. It can be shown in this latter event that $J_\nu(x)$ and $J_{-\nu}(x)$ are linearly independent. In other words, the general solution of (1) on $(0, \infty)$ is

$$y = c_1 J_\nu(x) + c_2 J_{-\nu}(x), \quad \nu \neq \text{integer}. \tag{9}$$

The graphs of $y = J_0(x)$ and $y = J_1(x)$ are given in Figure 6.6. Observe that the graphs of J_0 and J_1 resemble damped cosine and sine graphs, respectively.*

EXAMPLE 1 General Solution: ν Not an Integer

Find the general solution of the equation $x^2y'' + xy' + (x^2 - \frac{1}{4})y = 0$ on $(0, \infty)$.

Solution We identify $\nu^2 = \frac{1}{4}$ and so $\nu = \frac{1}{2}$. From (9) we see that the general solution of the differential equation is $y = c_1 J_{1/2}(x) + c_2 J_{-1/2}(x)$. ■

Bessel Functions of the Second Kind If $\nu \neq$ integer, the function defined by the linear combination

$$Y_\nu(x) = \frac{\cos \nu\pi J_\nu(x) - J_{-\nu}(x)}{\sin \nu\pi} \tag{10}$$

and the function $J_\nu(x)$ are linearly independent solutions of (1). Thus another form of the general solution of (1) is $y = c_1 J_\nu(x) + c_2 Y_\nu(x)$, provided $\nu \neq$ integer. As $\nu \to m$, m an integer, (10) has the indeterminate form 0/0. However, it can be shown by L'Hôpital's rule that $\lim_{\nu \to m} Y_\nu(x)$ exists. Moreover, the function

$$Y_m(x) = \lim_{\nu \to m} Y_\nu(x)$$

and $J_m(x)$ are linearly independent solutions of $x^2y'' + xy' + (x^2 - m^2)y = 0$. Hence for *any* value of ν the general solution of (1) on $(0, \infty)$ can be written as

$$y = c_1 J_\nu(x) + c_2 Y_\nu(x). \tag{11}$$

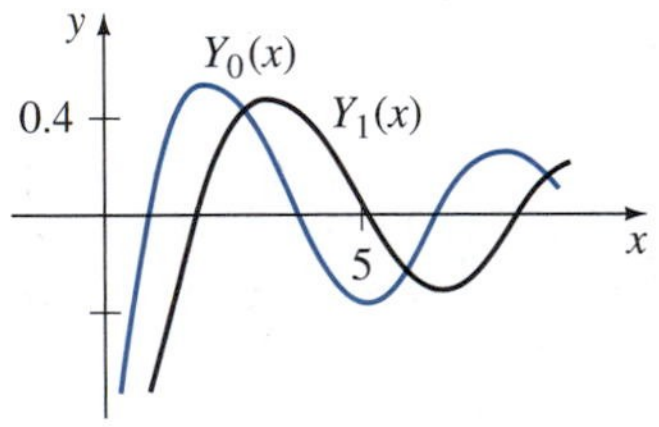

Figure 6.7

$Y_\nu(x)$ is sometimes called **Neumann's function;**† more commonly, $Y_\nu(x)$ is called the **Bessel function of the second kind** of order ν. Figure 6.7 shows the graphs of $Y_0(x)$ and $Y_1(x)$.

EXAMPLE 2 General Solution: ν an Integer

Find the general solution of the equation $x^2y'' + xy' + (x^2 - 9)y = 0$ on $(0, \infty)$.

* Bessel functions belong to a class of functions that are called "almost periodic."

† The function in (10) is also denoted by $N_\nu(x)$ in honor of the German mathematician C. Neumann (1832–1925), who investigated its properties.

Solution Since $\nu^2 = 9$, we identify $\nu = 3$. It follows from (11) that the general solution of the differential equation is $y = c_1 J_3(x) + c_2 Y_3(x)$. ■

Parametric Bessel Equation By replacing x by λx in (1) and using the chain rule, we obtain an alternative form of Bessel's equation known as the **parametric Bessel equation:**

$$x^2 y'' + xy' + (\lambda^2 x^2 - \nu^2)y = 0. \tag{12}$$

The general solution of (12) is

$$y = c_1 J_\nu(\lambda x) + c_2 Y_\nu(\lambda x). \tag{13}$$

Properties We list below a few of the more useful properties of Bessel functions of order m, $m = 0, 1, 2, \ldots$:

$$(i)\ J_{-m}(x) = (-1)^m J_m(x) \qquad (ii)\ J_m(-x) = (-1)^m J_m(x)$$

$$(iii)\ J_m(0) = \begin{cases} 0, & m > 0 \\ 1, & m = 0 \end{cases} \qquad (iv)\ \lim_{x \to 0^+} Y_m(x) = -\infty$$

Note that Property (ii) indicates $J_m(x)$ is an even function if m is an even integer and an odd function if m is an odd integer. The graphs of $Y_0(x)$ and $Y_1(x)$ in Figure 6.7 illustrate Property (iv): $Y_m(x)$ is unbounded at the origin. This last fact is not obvious from (10). It can be shown either from (10) or by the methods of Section 6.4 that for $x > 0$,

$$Y_0(x) = \frac{2}{\pi} J_0(x)\left[\gamma + \ln\frac{x}{2}\right] - \frac{2}{\pi}\sum_{k=1}^{\infty}\frac{(-1)^k}{(k!)^2}\left(1 + \frac{1}{2} + \cdots + \frac{1}{k}\right)\left(\frac{x}{2}\right)^{2k},$$

where $\gamma = 0.57721566\ldots$ is **Euler's constant.** Observe that because of the presence of the logarithmic term, $Y_0(x)$ is discontinuous at $x = 0$.

Numerical Values Some functional values of $J_0(x)$, $J_1(x)$, $Y_0(x)$, and $Y_1(x)$ for selected values of x are given in Table 6.1. The first five nonnegative zeros of $J_0(x)$, $J_1(x)$, $Y_0(x)$, and $Y_1(x)$ are given in Table 6.2.

Table 6.1 Numerical Values of J_0, J_1, Y_0, and Y_1

x	$J_0(x)$	$J_1(x)$	$Y_0(x)$	$Y_1(x)$
0	1.0000	0.0000	–	–
1	0.7652	0.4401	0.0883	−0.7812
2	0.2239	0.5767	0.5104	−0.1070
3	−0.2601	0.3391	0.3769	0.3247
4	−0.3971	−0.0660	−0.0169	0.3979
5	−0.1776	−0.3276	−0.3085	0.1479
6	0.1506	−0.2767	−0.2882	−0.1750
7	0.3001	−0.0047	−0.0259	−0.3027
8	0.1717	0.2346	0.2235	−0.1581
9	−0.0903	0.2453	0.2499	0.1043
10	−0.2459	0.0435	0.0557	0.2490
11	−0.1712	−0.1768	−0.1688	0.1637
12	0.0477	−0.2234	−0.2252	−0.0571
13	0.2069	−0.0703	−0.0782	−0.2101
14	0.1711	0.1334	0.1272	−0.1666
15	−0.0142	0.2051	0.2055	0.0211

Table 6.2 Zeros of J_0, J_1, Y_0, and Y_1

$J_0(x)$	$J_1(x)$	$Y_0(x)$	$Y_1(x)$
2.4048	0.0000	0.8936	2.1971
5.5201	3.8317	3.9577	5.4297
8.6537	7.0156	7.0861	8.5960
11.7915	10.1735	10.2223	11.7492
14.9309	13.3237	13.3611	14.8974

Recurrence Relation Recurrence formulas that relate Bessel functions of different orders are important in theory and in applications. In the next example we derive a **differential recurrence relation.**

EXAMPLE 3 Derivation Using the Series Definition

Derive the formula $xJ'_\nu(x) = \nu J_\nu(x) - xJ_{\nu+1}(x)$.

Solution It follows from (7) that

$$\begin{aligned} xJ'_\nu(x) &= \sum_{n=0}^{\infty} \frac{(-1)^n\,(2n+\nu)}{n!\Gamma(1+\nu+n)}\left(\frac{x}{2}\right)^{2n+\nu} \\ &= \nu \sum_{n=0}^{\infty} \frac{(-1)^n}{n!\Gamma(1+\nu+n)}\left(\frac{x}{2}\right)^{2n+\nu} + 2\sum_{n=0}^{\infty} \frac{(-1)^n n}{n!\Gamma(1+\nu+n)}\left(\frac{x}{2}\right)^{2n+\nu} \\ &= \nu J_\nu(x) + x \underbrace{\sum_{n=1}^{\infty} \frac{(-1)^n}{(n-1)!\Gamma(1+\nu+n)}\left(\frac{x}{2}\right)^{2n+\nu-1}}_{k=n-1} \\ &= \nu J_\nu(x) - x \sum_{k=0}^{\infty} \frac{(-1)^k}{k!\Gamma(2+\nu+k)}\left(\frac{x}{2}\right)^{2k+\nu+1} \\ &= \nu J_\nu(x) - xJ_{\nu+1}(x). \end{aligned}$$

The result in Example 3 can be written in an alternative form. Dividing $xJ'_\nu(x) - \nu J_\nu(x) = -xJ_{\nu+1}(x)$ by x gives

$$J'_\nu(x) - \frac{\nu}{x}J_\nu(x) = -J_{\nu+1}(x).$$

This last expression is recognized as a linear first-order differential equation in $J_\nu(x)$. Multiplying both sides of the equality by the integrating factor $x^{-\nu}$ then yields

$$\frac{d}{dx}[x^{-\nu}J_\nu(x)] = -x^{-\nu}J_{\nu+1}(x). \tag{14}$$

We leave it as an exercise to derive a similar formula:

$$\frac{d}{dx}[x^{\nu}J_\nu(x)] = x^{\nu}J_{\nu-1}(x). \tag{15}$$

(See Problem 20.)

When ν = half an odd integer, $J_\nu(x)$ can be expressed in terms of $\sin x$, $\cos x$, and powers of x. Such Bessel functions are called **spherical Bessel functions.**

EXAMPLE 4 Spherical Bessel Function: $\nu = \frac{1}{2}$

Find an alternative expression for $J_{1/2}(x)$. Use the fact that $\Gamma(\frac{1}{2}) = \sqrt{\pi}$.

Solution With $\nu = \frac{1}{2}$, we have from (7)

$$J_{1/2}(x) = \sum_{n=0}^{\infty} \frac{(-1)^n}{n!\Gamma(1 + \frac{1}{2} + n)} \left(\frac{x}{2}\right)^{2n+1/2}.$$

Now in view of the property $\Gamma(1 + \alpha) = \alpha\Gamma(\alpha)$ we obtain

$$n = 0, \quad \Gamma(1 + \tfrac{1}{2}) = \tfrac{1}{2}\Gamma(\tfrac{1}{2}) = \frac{1}{2}\sqrt{\pi}$$

$$n = 1, \quad \Gamma(1 + \tfrac{3}{2}) = \tfrac{3}{2}\Gamma(\tfrac{3}{2}) = \frac{3}{2^2}\sqrt{\pi}$$

$$n = 2, \quad \Gamma(1 + \tfrac{5}{2}) = \tfrac{5}{2}\Gamma(\tfrac{5}{2}) = \frac{5 \cdot 3}{2^3}\sqrt{\pi} = \frac{5 \cdot 4 \cdot 3 \cdot 2 \cdot 1}{2^3 4 \cdot 2}\sqrt{\pi} = \frac{5!}{2^5 2!}\sqrt{\pi}$$

$$n = 3, \quad \Gamma(1 + \tfrac{7}{2}) = \tfrac{7}{2}\Gamma(\tfrac{7}{2}) = \frac{7 \cdot 5!}{2^6 2!}\sqrt{\pi} = \frac{7 \cdot 6 \cdot 5!}{2^6 \cdot 6 \cdot 2!}\sqrt{\pi} = \frac{7!}{2^7 3!}\sqrt{\pi}.$$

In general,
$$\Gamma(1 + \tfrac{1}{2} + n) = \frac{(2n + 1)!}{2^{2n+1}n!}\sqrt{\pi}.$$

Hence

$$J_{1/2}(x) = \sum_{n=0}^{\infty} \frac{(-1)^n}{n!\dfrac{(2n+1)!\sqrt{\pi}}{2^{2n+1}n!}} \left(\frac{x}{2}\right)^{2n+1/2} = \sqrt{\frac{2}{\pi x}} \sum_{n=0}^{\infty} \frac{(-1)^n}{(2n + 1)!} x^{2n+1}.$$

Since the series in the last line is the Maclaurin series for $\sin x$, we have shown that

$$J_{1/2}(x) = \sqrt{\frac{2}{\pi x}} \sin x.$$

6.5.2 SOLUTION OF LEGENDRE'S EQUATION

Since $x = 0$ is an ordinary point of equation (2), we assume a solution of the form $y = \sum_{k=0}^{\infty} c_k x^k$. Therefore

$$\begin{aligned}
(1 - x^2)y'' - 2xy' + n(n + 1)y &= (1 - x^2)\sum_{k=0}^{\infty} c_k k(k - 1)x^{k-2} - 2\sum_{k=0}^{\infty} c_k k x^k + n(n + 1)\sum_{k=0}^{\infty} c_k x^k \\
&= \sum_{k=2}^{\infty} c_k k(k - 1)x^{k-2} - \sum_{k=2}^{\infty} c_k k(k - 1)x^k - 2\sum_{k=1}^{\infty} c_k k x^k + n(n + 1)\sum_{k=0}^{\infty} c_k x^k \\
&= [n(n + 1)c_0 + 2c_2]x^0 + [n(n + 1)c_1 - 2c_1 + 6c_3]x \\
&\quad + \underbrace{\sum_{k=4}^{\infty} c_k k(k - 1)x^{k-2}}_{j = k - 2} - \underbrace{\sum_{k=2}^{\infty} c_k k(k - 1)x^k}_{j = k} - 2\underbrace{\sum_{k=2}^{\infty} c_k k x^k}_{j = k} + n(n + 1)\underbrace{\sum_{k=2}^{\infty} c_k x^k}_{j = k} \\
&= [n(n + 1)c_0 + 2c_2] + [(n - 1)(n + 2)c_1 + 6c_3]x \\
&\quad + \sum_{j=2}^{\infty} [(j + 2)(j + 1)c_{j+2} + (n - j)(n + j + 1)c_j]x^j = 0
\end{aligned}$$

implies that

$$n(n+1)c_0 + 2c_2 = 0$$
$$(n-1)(n+2)c_1 + 6c_3 = 0$$
$$(j+2)(j+1)c_{j+2} + (n-j)(n+j+1)c_j = 0$$

or

$$c_2 = -\frac{n(n+1)}{2!}c_0$$
$$c_3 = -\frac{(n-1)(n+2)}{3!}c_1$$
$$c_{j+2} = -\frac{(n-j)(n+j+1)}{(j+2)(j+1)}c_j, \quad j = 2, 3, 4, \ldots. \tag{16}$$

Iterating (16) gives

$$c_4 = -\frac{(n-2)(n+3)}{4 \cdot 3}c_2 = \frac{(n-2)n(n+1)(n+3)}{4!}c_0$$
$$c_5 = -\frac{(n-3)(n+4)}{5 \cdot 4}c_3 = \frac{(n-3)(n-1)(n+2)(n+4)}{5!}c_1$$
$$c_6 = -\frac{(n-4)(n+5)}{6 \cdot 5}c_4 = -\frac{(n-4)(n-2)n(n+1)(n+3)(n+5)}{6!}c_0$$
$$c_7 = -\frac{(n-5)(n+6)}{7 \cdot 6}c_5$$
$$= -\frac{(n-5)(n-3)(n-1)(n+2)(n+4)(n+6)}{7!}c_1$$

and so on. Thus for at least $|x| < 1$, we obtain two linearly independent power series solutions.

$$y_1(x) = c_0\left[1 - \frac{n(n+1)}{2!}x^2 + \frac{(n-2)n(n+1)(n+3)}{4!}x^4 - \frac{(n-4)(n-2)n(n+1)(n+3)(n+5)}{6!}x^6 + \cdots\right]$$
$$y_2(x) = c_1\left[x - \frac{(n-1)(n+2)}{3!}x^3 + \frac{(n-3)(n-1)(n+2)(n+4)}{5!}x^5 - \frac{(n-5)(n-3)(n-1)(n+2)(n+4)(n+6)}{7!}x^7 + \cdots\right]. \tag{17}$$

Notice that if n is an even integer, the first series terminates, whereas $y_2(x)$ is an infinite series. For example, if $n = 4$, then

$$y_1(x) = c_0\left[1 - \frac{4 \cdot 5}{2!}x^2 + \frac{2 \cdot 4 \cdot 5 \cdot 7}{4!}x^4\right] = c_0\left[1 - 10x^2 + \frac{35}{3}x^4\right].$$

Similarly, when n is an odd integer, the series for $y_2(x)$ terminates with x^n; that is, *when n is a nonnegative integer we obtain an nth-degree polynomial solution* of Legendre's equation.

Since we know that a constant multiple of a solution of Legendre's equation is also a solution, it is traditional to choose specific values for c_0 and c_1, depending on whether n is an even or odd positive integer, respectively. For $n = 0$ we choose $c_0 = 1$ and for $n = 2, 4, 6, \ldots$

$$c_0 = (-1)^{n/2}\frac{1 \cdot 3 \cdots (n-1)}{2 \cdot 4 \cdots n},$$

whereas for $n = 1$ we choose $c_1 = 1$ and for $n = 3, 5, 7, \ldots$

$$c_1 = (-1)^{(n-1)/2}\frac{1 \cdot 3 \cdots n}{2 \cdot 4 \cdots (n-1)}.$$

For example, when $n = 4$, we have

$$\begin{aligned} y_1(x) &= (-1)^{4/2}\frac{1 \cdot 3}{2 \cdot 4}\left[1 - 10x^2 + \frac{35}{3}x^4\right] \\ &= \frac{3}{8} - \frac{30}{8}x^2 + \frac{35}{8}x^4 \\ &= \frac{1}{8}(35x^4 - 30x^2 + 3). \end{aligned}$$

Legendre Polynomials These specific nth-degree polynomial solutions are called **Legendre polynomials** and are denoted by $P_n(x)$. From the series for $y_1(x)$ and $y_2(x)$ and from the above choices of c_0 and c_1, we find that the first several Legendre polynomials are

$$\begin{aligned} &P_0(x) = 1, && P_1(x) = x, \\ &P_2(x) = \frac{1}{2}(3x^2 - 1), && P_3(x) = \frac{1}{2}(5x^3 - 3x), \\ &P_4(x) = \frac{1}{8}(35x^4 - 30x^2 + 3), && P_5(x) = \frac{1}{8}(63x^5 - 70x^3 + 15x). \end{aligned} \tag{18}$$

Remember, $P_0(x), P_1(x), P_2(x), P_3(x), \ldots$ are, in turn, particular solutions of the differential equations

$$\begin{aligned} n = 0, &\quad (1 - x^2)y'' - 2xy' = 0 \\ n = 1, &\quad (1 - x^2)y'' - 2xy' + 2y = 0 \\ n = 2, &\quad (1 - x^2)y'' - 2xy' + 6y = 0 \\ n = 3, &\quad (1 - x^2)y'' - 2xy' + 12y = 0 \\ &\quad \vdots \end{aligned} \tag{19}$$

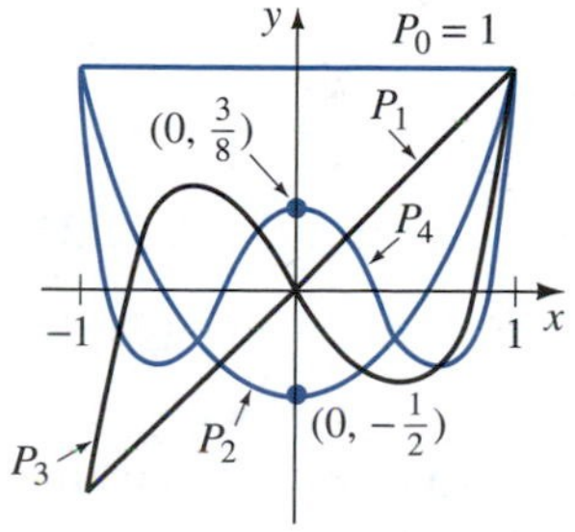

Figure 6.8

The graphs of the first four Legendre polynomials on the interval $-1 \le x \le 1$ are given in Figure 6.8.

Properties The following properties of the Legendre polynomials are apparent in (18) and Figure 6.8:

$$\begin{aligned} &(i)\ P_n(-x) = (-1)^n P_n(x) \\ &(ii)\ P_n(1) = 1 \qquad (iii)\ P_n(-1) = (-1)^n \\ &(iv)\ P_n(0) = 0,\ n \text{ odd} \qquad (v)\ P_n'(0) = 0,\ n \text{ even} \end{aligned}$$

Property (*i*) indicates that $P_n(x)$ is an even or odd function according to whether n is even or odd.

Recurrence Relation Recurrence relations that relate Legendre polynomials of different degrees are also very important in some aspects of their application. We shall derive one such relation using the formula

$$(1 - 2xt + t^2)^{-1/2} = \sum_{n=0}^{\infty} P_n(x)t^n. \tag{20}$$

The function on the left is called a **generating function** for the Legendre polynomials. Its derivation follows from the binomial theorem and is left as an exercise. (See Problem 43.)

Differentiating both sides of (20) with respect to t gives

$$(1 - 2xt + t^2)^{-3/2}(x - t) = \sum_{n=0}^{\infty} nP_n(x)t^{n-1} = \sum_{n=1}^{\infty} nP_n(x)t^{n-1},$$

so after multiplying by $1 - 2xt + t^2$, we have

$$(x - t)(1 - 2xt + t^2)^{-1/2} = (1 - 2xt + t^2)\sum_{n=1}^{\infty} nP_n(x)t^{n-1}$$

or

$$(x - t)\sum_{n=0}^{\infty} P_n(x)t^n = (1 - 2xt + t^2)\sum_{n=1}^{\infty} nP_n(x)t^{n-1}. \tag{21}$$

We multiply out and rewrite (21) as

$$\sum_{n=0}^{\infty} xP_n(x)t^n - \sum_{n=0}^{\infty} P_n(x)t^{n+1} - \sum_{n=1}^{\infty} nP_n(x)t^{n-1} + 2x\sum_{n=1}^{\infty} nP_n(x)t^n - \sum_{n=1}^{\infty} nP_n(x)t^{n+1} = 0$$

or

$$x + x^2t + \sum_{n=2}^{\infty} xP_n(x)t^n - t - \sum_{n=1}^{\infty} P_n(x)t^{n+1} - x - 2\left(\frac{3x^2 - 1}{2}\right)t$$
$$- \sum_{n=3}^{\infty} nP_n(x)t^{n-1} + 2x^2t + 2x\sum_{n=2}^{\infty} nP_n(x)t^n - \sum_{n=1}^{\infty} nP_n(x)t^{n+1} = 0.$$

Observing the appropriate cancellations, simplifying, and changing the summation indices gives

$$\sum_{k=2}^{\infty} [-(k + 1)P_{k+1}(x) + (2k + 1)xP_k(x) - kP_{k-1}(x)]t^k = 0.$$

Equating the total coefficient of t^k to zero gives the three-term recurrence relation

$$(k + 1)P_{k+1}(x) - (2k + 1)xP_k(x) + kP_{k-1}(x) = 0, \quad k = 2, 3, 4, \ldots. \tag{22}$$

This formula is also valid when $k = 1$.

In (18) we listed the first six Legendre polynomials. If, say, we wish to find $P_6(x)$, we could use (22) with $k = 5$. This relation then expresses $P_6(x)$ in terms of the known quantities $P_4(x)$ and $P_5(x)$. (See Problem 45.)

EXERCISES 6.5

Answers to odd-numbered problems begin on page A-13.

6.5.1 Solution of Bessel's Equation

In Problems 1–8 find the general solution of the given differential equation on $(0, \infty)$.

1. $x^2y'' + xy' + \left(x^2 - \frac{1}{9}\right)y = 0$
2. $x^2y'' + xy' + (x^2 - 1)y = 0$
3. $4x^2y'' + 4xy' + (4x^2 - 25)y = 0$
4. $16x^2y'' + 16xy' + (16x^2 - 1)y = 0$
5. $xy'' + y' + xy = 0$
6. $\frac{d}{dx}[xy'] + \left(x - \frac{4}{x}\right)y = 0$
7. $x^2y'' + xy' + (9x^2 - 4)y = 0$
8. $x^2y'' + xy' + \left(36x^2 - \frac{1}{4}\right)y = 0$
9. Use the change of variables $y = x^{-1/2}v(x)$ to find the general solution of the equation
$$x^2y'' + 2xy' + \lambda^2x^2y = 0, \quad x > 0.$$
10. Verify that the differential equation
$$xy'' + (1 - 2n)y' + xy = 0, \quad x > 0$$
possesses the particular solution $y = x^nJ_n(x)$.
11. Verify that the differential equation
$$xy'' + (1 + 2n)y' + xy = 0, \quad x > 0$$
possesses the particular solution $y = x^{-n}J_n(x)$.
12. Verify that the differential equation
$$x^2y'' + \left(\lambda^2x^2 - \nu^2 + \frac{1}{4}\right)y = 0, \quad x > 0$$
possesses the particular solution $y = \sqrt{x}J_\nu(\lambda x)$, where $\lambda > 0$.

In Problems 13–18 use the results of Problems 10, 11, and 12 to find a particular solution of the given differential equation on $(0, \infty)$.

13. $y'' + y = 0$
14. $xy'' - y' + xy = 0$
15. $xy'' + 3y' + xy = 0$
16. $4x^2y'' + (16x^2 + 1)y = 0$
17. $x^2y'' + (x^2 - 2)y = 0$
18. $xy'' - 5y' + xy = 0$

In Problems 19–22 derive the given recurrence relation.

19. $xJ'_\nu(x) = -\nu J_\nu(x) + xJ_{\nu-1}(x)$ [*Hint:* $2n + \nu = 2(n + \nu) - \nu$.]
20. $\frac{d}{dx}[x^\nu J_\nu(x)] = x^\nu J_{\nu-1}(x)$
21. $2\nu J_\nu(x) = xJ_{\nu+1}(x) + xJ_{\nu-1}(x)$
22. $2J'_\nu(x) = J_{\nu-1}(x) - J_{\nu+1}(x)$

In Problems 23–26 use (14) or (15) to obtain the given result.

23. $\int_0^x rJ_0(r)\,dr = xJ_1(x)$

24. $J_0'(x) = J_{-1}(x) = -J_1(x)$

25. $\int x^n J_0(x)\,dx = x^n J_1(x) + (n-1)x^{n-1}J_0(x) - (n-1)^2\int x^{n-2}J_0(x)\,dx$

26. $\int x^3 J_0(x)\,dx = x^3 J_1(x) + 2x^2 J_0(x) - 4xJ_1(x) + c$

27. Proceed as in Example 4 and express $J_{-1/2}(x)$ in terms of $\cos x$ and a power of x.

In Problems 28–33 use the recurrence relation given in Problem 21 and the results obtained in Problem 27 and Example 4 to express the given Bessel function in terms of $\sin x$, $\cos x$, and powers of x.

28. $J_{3/2}(x)$ **29.** $J_{-3/2}(x)$ **30.** $J_{5/2}(x)$

31. $J_{-5/2}(x)$ **32.** $J_{7/2}(x)$ **33.** $J_{-7/2}(x)$

34. Show that $i^{-\nu}J_\nu(ix)$, $i^2 = -1$ is a real function. The function defined by $I_\nu(x) = i^{-\nu}J_\nu(ix)$ is called a **modified Bessel function of the first kind** of order ν.

35. Find the general solution of the differential equation

$$x^2y'' + xy' - (x^2 + \nu^2)y = 0, \quad x > 0, \quad \nu \neq \text{integer}.$$

[*Hint*: $i^2x^2 = -x^2$.]

36. If $y_1 = J_0(x)$ is one solution of the zero-order Bessel equation, verify that another solution is

$$y_2 = J_0(x)\ln x + \frac{x^2}{4} - \frac{3x^4}{128} + \frac{11x^6}{13{,}824} - \cdots.$$

37. Use (8) with $\nu = m$, where m is a positive integer, and the fact that $1/\Gamma(N) = 0$, where N is a negative integer, to show that

$$J_{-m}(x) = (-1)^m J_m(x).$$

38. Use (7) with $\nu = m$, where m is a nonnegative integer, to show that

$$J_m(-x) = (-1)^m J_m(x).$$

6.5.2 Solution of Legendre's Equation

39. (**a**) Use the explicit solutions $y_1(x)$ and $y_2(x)$ of Legendre's equation and the appropriate choices of c_0 and c_1 to find the Legendre polynomials $P_6(x)$ and $P_7(x)$.

(**b**) Write the differential equations for which $P_6(x)$ and $P_7(x)$ are particular solutions.

40. Show that Legendre's equation has an alternative form

$$\frac{d}{dx}\left[(1 - x^2)\frac{dy}{dx}\right] + n(n+1)y = 0.$$

41. Show that the equation

$$\sin\theta\,\frac{d^2y}{d\theta^2} + \cos\theta\,\frac{dy}{d\theta} + n(n+1)(\sin\theta)y = 0$$

can be transformed into Legendre's equation by means of the substitution $x = \cos\theta$.

42. The general Legendre polynomial can be written as

$$P_n(x) = \sum_{k=0}^{[n/2]} \frac{(-1)^k(2n-2k)!}{2^n k!(n-k)!(n-2k)!}x^{n-2k},$$

where $[n/2]$ is the greatest integer not greater than $n/2$. Verify the results for $n = 0, 1, 2, 3, 4, 5$.

43. Use binomial series to formally show that

$$(1 - 2xt + t^2)^{-1/2} = \sum_{n=0}^{\infty} P_n(x)t^n.$$

44. Use Problem 43 to show that $P_n(1) = 1$ and $P_n(-1) = (-1)^n$.

45. Use the recurrence relation (22) and $P_0(x) = 1$, $P_1(x) = x$ to generate the next five Legendre polynomials.

46. The Legendre polynomials are also generated by **Rodrigues' formula***

$$P_n(x) = \frac{1}{2^n n!}\frac{d^n}{dx^n}(x^2 - 1)^n.$$

Verify the results for $n = 0, 1, 2, 3$.

47. Use the explicit Legendre polynomials $P_0(x)$, $P_1(x)$, $P_2(x)$, and $P_3(x)$ to evaluate $\int_{-1}^{1} P_n^2(x)\,dx$ for $n = 0, 1, 2, 3$. Generalize the results.

48. Use the explicit Legendre polynomials $P_0(x)$, $P_1(x)$, $P_2(x)$, and $P_3(x)$ to evaluate $\int_{-1}^{1} P_n(x)P_m(x)\,dx$ for $n \neq m$. Generalize the results.

49. We know that $y_1 = x$ is a solution of Legendre's equation when $n = 1$, $(1 - x^2)y'' - 2xy' + 2y = 0$. Show that a second linearly independent solution on the interval $-1 < x < 1$ is

$$y_2 = \frac{x}{2}\ln\left(\frac{1+x}{1-x}\right) - 1.$$

CHAPTER 6 REVIEW

The remarkable characteristic of a Cauchy-Euler equation is that even though it is a differential equation with variable coefficients, it can be solved in terms of elementary functions. A **second-order Cauchy-Euler equation** is any differential

* ONLINDE RODRIGUES (1794–1851) Rodrigues was a French banker and amateur mathematician. In mathematics he is remembered solely for the discovery of this one formula in 1816. In politics he is remembered as the financial backer and disciple of Count de Saint-Simon, the founder of French socialism.

equation of the form $ax^2y'' + bxy' + cy = g(x)$, where a, b, and c are constants. To solve the homogeneous equation we try a solution of the form $y = x^m$ and this in turn leads to an algebraic **auxiliary equation** $am(m - 1) + bm + c = 0$. Accordingly, when the roots are real and distinct, real and equal, or complex conjugates, the general solutions on the interval $(0, \infty)$ are, respectively,

$$y = c_1x^{m_1} + c_2x^{m_2}$$

$$y = c_1x^{m_1} + c_2x^{m_1}\ln x$$

and

$$y = x^{\alpha}[c_1\cos(\beta\ln x) + c_2\sin(\beta\ln x)].$$

We say that $x = 0$ is an **ordinary point** of the linear second-order differential equation $a_2(x)y'' + a_1(x)y' + a_0(x)y = 0$ when $a_2(0) \neq 0$ and $a_2(x)$, $a_1(x)$, $a_0(x)$ are polynomials having no common factors. Every solution has the form of a power series in x, $y = \sum_{n=0}^{\infty} c_nx^n$. To find the coefficients c_n we substitute the basic assumption into the differential equation, and after appropriate algebraic manipulations we determine a **recurrence relation** by equating to zero the combined total coefficient of x^k. Iteration of the recurrence relation yields two distinct sets of coefficients, one set containing the arbitrary coefficient c_0 and the other containing c_1. Using each set of coefficients, we form two linearly independent solutions $y_1(x)$ and $y_2(x)$. A solution is valid at least on an interval defined by $|x| < R$, where R is the distance from $x = 0$ to the closest singular point of the equation. If $a_2(0) = 0$, then $x = 0$ is a **singular point.** Singular points are classified as either **regular** or **irregular.** To determine whether $x = 0$ is a regular singular point, we examine the denominators of the rational functions P and Q that result when the equation is put into the form $y'' + P(x)y' + Q(x)y = 0$. It is understood that $a_1(x)/a_2(x)$ and $a_0(x)/a_2(x)$ are reduced to lowest terms. If x appears *at most* to the first power in the denominator of $P(x)$ and *at most* to the second power in the denominator of $Q(x)$ then $x = 0$ is a regular singular point. Around the regular singular point $x = 0$, the **method of Frobenius** guarantees that there exists *at least one* solution of the form $y = \sum_{n=0}^{\infty} c_nx^{n+r}$. The exponent r is a root of a quadratic **indicial equation.** When the indicial roots r_1 and r_2 $(r_1 > r_2)$ satisfy $r_1 - r_2 \neq$ an integer, then we can always find *two* linearly independent solutions of the assumed form. When $r_1 - r_2 =$ a positive integer, then we could *possibly* find two solutions, but when $r_1 - r_2 = 0$, or $r_1 = r_2$, we can find only one solution of the form $y = \sum_{n=0}^{\infty} c_nx^{n+r}$.

Bessel's equation $x^2y'' + xy' + (x^2 - \nu^2)y = 0$ has a regular singular point at $x = 0$, whereas $x = 0$ is an ordinary point of **Legendre's equation** $(1 - x^2)y'' - 2xy' + n(n + 1)y = 0$. The latter equation possesses a polynomial solution when n is a nonnegative integer.

CHAPTER 6 REVIEW EXERCISES

Answers to odd-numbered problems begin on page A-15.

In Problems 1–4 solve the given Cauchy-Euler equation.

1. $6x^2y'' + 5xy' - y = 0$

2. $2x^3y''' + 19x^2y'' + 39xy' + 9y = 0$

3. $x^2y'' - 4xy' + 6y = 2x^4 + x^2$

4. $x^2y'' - xy' + y = x^3$

5. Specify the ordinary points of $(x^3 - 8)y'' - 2xy' + y = 0$.

6. Specify the singular points of $(x^4 - 16)y'' + 2y = 0$.

In Problems 7–10 specify the regular and irregular singular points of the given differential equation.

7. $(x^3 - 10x^2 + 25x)y'' + y' = 0$

8. $(x^3 - 10x^2 + 25x)y'' + y = 0$

9. $x^2(x^2 - 9)^2y'' - (x^2 - 9)y' + xy = 0$

10. $x(x^2 + 1)^3y'' + y' - 8xy = 0$

In Problems 11 and 12 specify an interval around $x = 0$ for which a power series solution of the given differential equation will converge.

11. $y'' - xy' + 6y = 0$

12. $(x^2 - 4)y'' - 2xy' + 9y = 0$

In Problems 13–16 for each differential equation find two power series solutions about the ordinary point $x = 0$.

13. $y'' + xy = 0$

14. $y'' - 4y = 0$

15. $(x - 1)y'' + 3y = 0$

16. $y'' - x^2y' + xy = 0$

In Problems 17–22 find two linearly independent solutions of each equation.

17. $2x^2y'' + xy' - (x + 1)y = 0$

18. $2xy'' + y' + y = 0$

19. $x(1 - x)y'' - 2y' + y = 0$

20. $x^2y'' - xy' + (x^2 + 1)y = 0$

21. $xy'' - (2x - 1)y' + (x - 1)y = 0$

22. $x^2y'' - x^2y' + (x^2 - 2)y = 0$

23. Without referring to Section 6.5 use the method of Frobenius to obtain a solution of the Bessel equation for $\nu = 0$: $xy'' + y' + xy = 0$.

7

LAPLACE TRANSFORM

Introduction

In this chapter we examine the definition and properties of an integral known as the **Laplace transform.**

We shall see in Section 7.5 that when the Laplace transform is applied to a linear *n*th-order differential equation with constant coefficients

$$a_n \frac{d^n y}{dt^n} + a_{n-1} \frac{d^{n-1} y}{dt^{n-1}} + \cdots + a_1 \frac{dy}{dt} + a_0 y = g(t),$$

the differential equation is transformed into an *algebraic equation* that involves the conditions $y(0), y'(0), y''(0), \ldots, y^{(n-1)}(0)$. As a consequence of this property, the Laplace transform is well suited to the solution of certain kinds of initial-value problems.

Recall that in physical systems such as as a spring-mass system or a series electrical circuit, the right-hand member in the differential equations

$$m \frac{d^2 x}{dt^2} + \beta \frac{dx}{dt} + kx = f(t)$$

$$L \frac{d^2 q}{dt^2} + R \frac{dq}{dt} + \frac{1}{C} q = E(t)$$

is a driving function and represents either an external force $f(t)$ or an impressed voltage $E(t)$. In Chapter 5 we solved problems in which the functions f and E were continuous. However, piecewise continuous driving

(continued)

functions are not uncommon. For example, the impressed voltage on a circuit could be

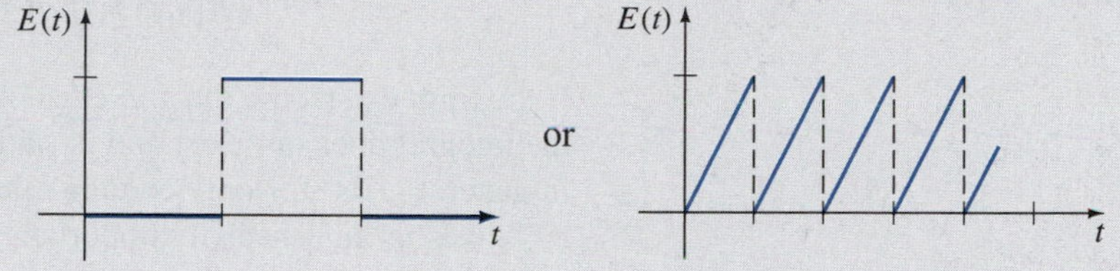

In this case, solving the differential equation of the circuit is difficult but not impossible. The Laplace transform is an invaluable aid in solving problems such as these.

7.1 LAPLACE TRANSFORM

• *Linear operation* • *Laplace transform* • *Linear transform*
• *Piecewise continuous* • *Exponential order*

In elementary calculus you learned that differentiation and integration transform a function into another function. For example, the function $f(x) = x^2$ is transformed, in turn, into a linear function, a family of cubic polynomial functions, and a constant by the operations of differentiation, indefinite integration, and definite integration:

$$\frac{d}{dx}x^2 = 2x, \quad \int x^2\,dx = \frac{x^3}{3} + c, \quad \int_0^3 x^2\,dx = 9.$$

Moreover, these three operations possess the **linearity property.** This means that for any constants α and β,

$$\frac{d}{dx}[\alpha f(x) + \beta g(x)] = \alpha\frac{d}{dx}f(x) + \beta\frac{d}{dx}g(x)$$

$$\int[\alpha f(x) + \beta g(x)]\,dx = \alpha\int f(x)\,dx + \beta\int g(x)\,dx \tag{1}$$

$$\int_a^b[\alpha f(x) + \beta g(x)]\,dx = \alpha\int_a^b f(x)\,dx + \beta\int_a^b g(x)\,dx$$

provided each derivative and integral exists.

If $f(x, y)$ is a function of two variables, then a definite integral of f with respect to one of the variables leads to a function of the other variable. For example, by holding y constant we see that $\int_1^2 2xy^2\,dx = 3y^2$. Similarly, a definite integral such as $\int_a^b K(s, t)f(t)\,dt$ transforms a function $f(t)$ into a function of the variable s. We are particularly interested in **integral transforms** of this last kind, where the interval of integration is the unbounded interval $[0, \infty)$.

Basic Definition If $f(t)$ is defined for $t \geq 0$, then the improper integral $\int_0^\infty K(s, t) f(t)\, dt$ is defined by means of a limit:

$$\int_0^\infty K(s, t) f(t)\, dt = \lim_{b\to\infty} \int_0^b K(s, t) f(t)\, dt.$$

If the limit exists, the integral exists or is convergent; if the limit does not exist, the integral does not exist and is said to be divergent. The foregoing limit will, in general, exist for only certain values of the variable s. The choice $K(s, t) = e^{-st}$ gives us an especially important integral transform.

DEFINITION 7.1 Laplace Transform

Let f be a function defined for $t \geq 0$. Then the integral

$$\mathscr{L}\{f(t)\} = \int_0^\infty e^{-st} f(t)\, dt \tag{2}$$

is said to be the **Laplace transform*** of f provided the integral converges.

When the defining integral (2) converges, the result is a function of s. In general discussion, we shall use a lowercase letter to denote the function being transformed and the corresponding capital letter to denote its Laplace transform; for example,

$$\mathscr{L}\{f(t)\} = F(s), \quad \mathscr{L}\{g(t)\} = G(s), \quad \mathscr{L}\{y(t)\} = Y(s).$$

EXAMPLE 1 Applying Definition 7.1

Evaluate $\mathscr{L}\{1\}$.

Solution

$$\mathscr{L}\{1\} = \int_0^\infty e^{-st}(1)\, dt = \lim_{b\to\infty} \int_0^b e^{-st}\, dt$$

$$= \lim_{b\to\infty} \frac{-e^{-st}}{s}\bigg|_0^b = \lim_{b\to\infty} \frac{-e^{-sb} + 1}{s}$$

$$= \frac{1}{s}$$

*PIERRE SIMON MARQUIS DE LAPLACE (1749–1827) A noted mathematician, physicist, and astronomer, Laplace was called by some of his enthusiastic contemporaries the "Newton of France." Although Laplace made use of the integral transform (2) in his work in probability theory, it is likely that the integral was first discovered by Euler. Laplace's noted treatises were *Mécanique Céleste* and *Théorie Analytique des Probabilités*. Born into a poor farming family, Laplace became a friend of Napoleon but was elevated to the nobility by Louis XVIII after the Restoration.

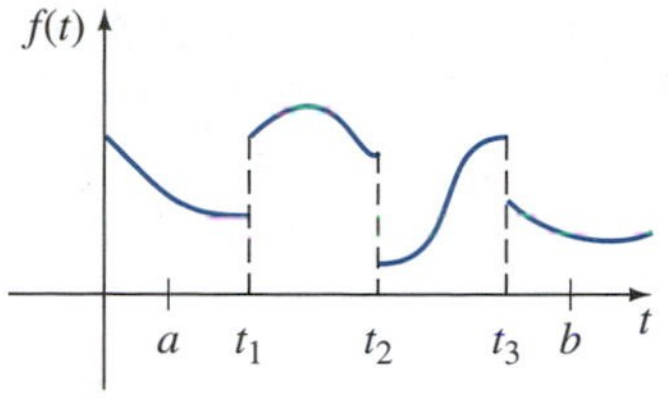

Figure 7.1

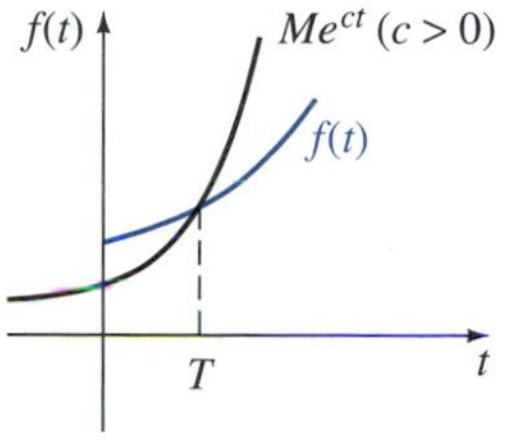

Figure 7.2

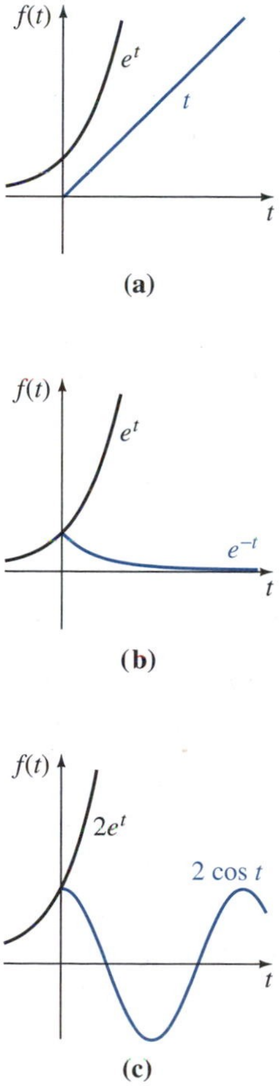

Figure 7.3

provided $s > 0$. In other words, when $s > 0$, the exponent $-sb$ is negative and $e^{-sb} \to 0$ as $b \to \infty$. When $s < 0$, the integral is divergent. ■

The use of the limit sign becomes somewhat tedious, so we shall adopt the notation $|_0^\infty$ as a shorthand to writing $\lim_{b\to\infty}(\;)|_0^b$. For example,

$$\mathscr{L}\{1\} = \int_0^\infty e^{-st}\,dt = \frac{-e^{-st}}{s}\bigg|_0^\infty = \frac{1}{s}, \quad s > 0,$$

where it is understood that at the upper limit we mean $e^{-st} \to 0$ as $t \to \infty$ for $s > 0$.

$\mathscr{L}$ a Linear Transform For a sum of functions we can write

$$\int_0^\infty e^{-st}[\alpha f(t) + \beta g(t)]\,dt = \alpha\int_0^\infty e^{-st}f(t)\,dt + \beta\int_0^\infty e^{-st}g(t)\,dt,$$

whenever both integrals converge. Hence it follows that

$$\mathscr{L}\{\alpha f(t) + \beta g(t)\} = \alpha\mathscr{L}\{f(t)\} + \beta\mathscr{L}\{g(t)\} = \alpha F(s) + \beta G(s). \tag{3}$$

Because of the property given in (3), $\mathscr{L}$ is said to be a **linear transform,** or **linear operator.**

Sufficient Conditions for Existence of $\mathscr{L}\{f(t)\}$ The integral that defines the Laplace transform does not have to converge. For example, neither $\mathscr{L}\{1/t\}$ nor $\mathscr{L}\{e^{t^2}\}$ exists. Sufficient conditions that guarantee the existence of $\mathscr{L}\{f(t)\}$ are that f be piecewise continuous on $[0, \infty)$ and that f be of exponential order for $t > T$. Recall that a function f is **piecewise continuous** on $[0, \infty)$ if, in any interval $0 \le a \le t \le b$, there are at most a finite number of points t_k, $k = 1, 2, \ldots, n$ $(t_{k-1} < t_k)$ at which f has finite discontinuities and is continuous on each open interval $t_{k-1} < t < t_k$. See Figure 7.1. The concept of **exponential order** is defined in the following manner.

DEFINITION 7.2 **Exponential Order**

> A function f is said to be of **exponential order** if there exist numbers c, $M > 0$, and $T > 0$ such that $|f(t)| \le Me^{ct}$ for $t > T$.

If, for example, f is an *increasing* function, then the condition $|f(t)| \le Me^{ct}$, $t > T$ simply states that the graph of f on the interval (T, ∞) does not grow faster than the graph of the exponential function Me^{ct}, where c is a positive constant. See Figure 7.2. The functions $f(t) = t$, $f(t) = e^{-t}$, and $f(t) = 2\cos t$ are all of exponential order for $t > 0$ since we have, respectively,

$$|t| \le e^t, \quad |e^{-t}| \le e^t, \quad |2\cos t| \le 2e^t.$$

A comparison of the graphs on the interval $(0, \infty)$ is given in Figure 7.3.

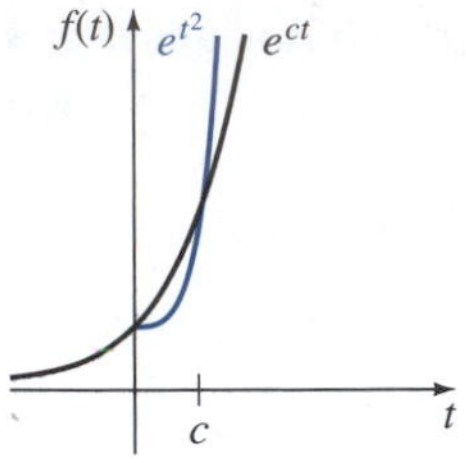

Figure 7.4

A function such as $f(t) = e^{t^2}$ is not of exponential order since, as shown in Figure 7.4, its graph grows faster than any positive linear power of e for $t > c > 0$.

A positive integral power of t is always of exponential order since, for $c > 0$,

$$|t^n| \leq Me^{ct} \quad \text{or} \quad \left|\frac{t^n}{e^{ct}}\right| \leq M \quad \text{for } t > T$$

is equivalent to showing that $\lim_{t\to\infty} t^n/e^{ct}$ is finite for $n = 1, 2, 3, \ldots$. The result follows by n applications of L'Hôpital's rule.

THEOREM 7.1 Sufficient Conditions for Existence

Let $f(t)$ be piecewise continuous on the interval $[0, \infty)$ and of exponential order for $t > T$; then $\mathscr{L}\{f(t)\}$ exists for $s > c$.

Proof $\quad \mathscr{L}\{f(t)\} = \int_0^T e^{-st}f(t)\,dt + \int_T^\infty e^{-st}f(t)\,dt = I_1 + I_2.$

The integral I_1 exists because it can be written as a sum of integrals over intervals for which $e^{-st}f(t)$ is continuous. Now

$$|I_2| \leq \int_T^\infty |e^{-st}f(t)|\,dt \leq M\int_T^\infty e^{-st}e^{ct}\,dt$$

$$= M\int_T^\infty e^{-(s-c)t}\,dt = -M\,\frac{e^{-(s-c)t}}{s-c}\bigg|_T^\infty = M\,\frac{e^{-(s-c)T}}{s-c}$$

for $s > c$. Since $\int_T^\infty Me^{-(s-c)t}\,dt$ converges, the integral $\int_T^\infty |e^{-st}f(t)|\,dt$ converges by the comparison test for improper integrals. This, in turn, implies that I_2 exists for $s > c$. The existence of I_1 and I_2 implies that $\mathscr{L}\{f(t)\} = \int_0^\infty e^{-st}f(t)\,dt$ exists for $s > c$. ■

Throughout this entire chapter we shall be concerned only with functions that are both piecewise continuous and of exponential order. We note, however, that these conditions are sufficient but not necessary for the existence of a Laplace transform. The function $f(t) = t^{-1/2}$ is not piecewise continuous on the interval $[0, \infty)$, but its Laplace transform exists. See Problem 44.

EXAMPLE 2 Applying Definition 7.1

Evaluate $\mathscr{L}\{t\}$.

Solution From Definition 7.1 we have

$$\mathscr{L}\{t\} = \int_0^\infty e^{-st}t\,dt.$$

Integrating by parts and using $\lim_{t\to\infty} te^{-st} = 0$, $s > 0$, along with the result of Example 1, we obtain

$$\mathscr{L}\{t\} = \frac{-te^{-st}}{s}\Bigg|_0^\infty + \frac{1}{s}\int_0^\infty e^{-st}\,dt = \frac{1}{s}\mathscr{L}\{1\} = \frac{1}{s}\left(\frac{1}{s}\right) = \frac{1}{s^2}.$$

EXAMPLE 3 Applying Definition 7.1

Evaluate $\mathscr{L}\{e^{-3t}\}$.

Solution From Definition 7.1 we have

$$\begin{aligned}\mathscr{L}\{e^{-3t}\} &= \int_0^\infty e^{-st}e^{-3t}\,dt \\ &= \int_0^\infty e^{-(s+3)t}\,dt \\ &= \frac{-e^{-(s+3)t}}{s+3}\Bigg|_0^\infty \\ &= \frac{1}{s+3}, \quad s > -3.\end{aligned}$$

The result follows from the fact that $\lim_{t\to\infty} e^{-(s+3)t} = 0$ for $s + 3 > 0$ or $s > -3$.

EXAMPLE 4 Applying Definition 7.1

Evaluate $\mathscr{L}\{\sin 2t\}$.

Solution From Definition 7.1 and integration by parts we have

$$\begin{aligned}\mathscr{L}\{\sin 2t\} &= \int_0^\infty e^{-st}\sin 2t\,dt \\ &= \frac{-e^{-st}\sin 2t}{s}\Bigg|_0^\infty + \frac{2}{s}\int_0^\infty e^{-st}\cos 2t\,dt \\ &= \frac{2}{s}\int_0^\infty e^{-st}\cos 2t\,dt, \quad s > 0 \\ &= \frac{2}{s}\left[\frac{-e^{-st}\cos 2t}{s}\Bigg|_0^\infty - \frac{2}{s}\int_0^\infty e^{-st}\sin 2t\,dt\right] \\ &= \frac{2}{s^2} - \frac{4}{s^2}\mathscr{L}\{\sin 2t\}.\end{aligned}$$

Laplace transform of $\sin 2t$ ↓

$\lim_{t\to\infty} e^{-st}\cos 2t = 0,\ s > 0$

Now we solve for $\mathscr{L}\{\sin 2t\}$:

$$\begin{aligned}\left[1 + \frac{4}{s^2}\right]\mathscr{L}\{\sin 2t\} &= \frac{2}{s^2} \\ \mathscr{L}\{\sin 2t\} &= \frac{2}{s^2+4}, \quad s > 0.\end{aligned}$$

EXAMPLE 5 **Using Linearity**

Evaluate $\mathscr{L}\{3t - 5 \sin 2t\}$.

Solution From Examples 2 and 4 and the linearity property of the Laplace transform we can write

$$\begin{aligned}\mathscr{L}\{3t - 5\sin 2t\} &= 3\mathscr{L}\{t\} - 5\mathscr{L}\{\sin 2t\}\\ &= 3 \cdot \frac{1}{s^2} - 5 \cdot \frac{2}{s^2+4}\\ &= \frac{-7s^2+12}{s^2(s^2+4)}, \quad s > 0.\end{aligned}$$

EXAMPLE 6 **Applying Definition 7.1**

Evaluate **(a)** $\mathscr{L}\{te^{-2t}\}$ and **(b)** $\mathscr{L}\{t^2e^{-2t}\}$.

Solution

(a) From Definition 7.1 and integration by parts we have

$$\begin{aligned}\mathscr{L}\{te^{-2t}\} &= \int_0^\infty e^{-st}(te^{-2t})\,dt\\ &= \int_0^\infty te^{-(s+2)t}\,dt\\ &= \frac{-te^{-(s+2)t}}{s+2}\bigg|_0^\infty + \frac{1}{s+2}\int_0^\infty e^{-(s+2)t}\,dt\\ &= \frac{-e^{-(s+2)t}}{(s+2)^2}\bigg|_0^\infty, \quad s > -2\\ &= \frac{1}{(s+2)^2}, \quad s > -2.\end{aligned}$$

(b) Again, integration by parts gives

$$\begin{aligned}\mathscr{L}\{t^2e^{-2t}\} &= \frac{-t^2e^{-(s+2)t}}{s+2}\bigg|_0^\infty + \frac{2}{s+2}\int_0^\infty te^{-(s+2)t}\,dt\\ &= \frac{2}{s+2}\int_0^\infty e^{-st}(te^{-2t})\,dt, \quad s > -2\\ &= \frac{2}{s+2}\mathscr{L}\{te^{-2t}\} = \frac{2}{s+2}\left[\frac{1}{(s+2)^2}\right] \quad \leftarrow \text{from part (a)}\\ &= \frac{2}{(s+2)^3}, \quad s > -2.\end{aligned}$$

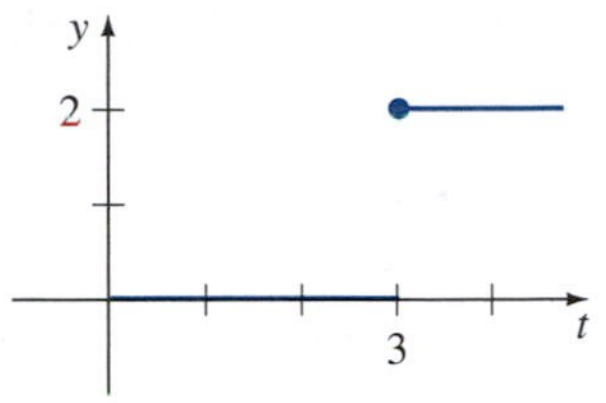

Figure 7.5

EXAMPLE 7 **Transform of a Piecewise-Defined Function**

Evaluate $\mathscr{L}\{f(t)\}$ for $f(t) = \begin{cases} 0, & 0 \le t < 3 \\ 2, & t \ge 3. \end{cases}$

Solution This piecewise continuous function is shown in Figure 7.5. Since f is defined in two pieces, $\mathscr{L}\{f(t)\}$ is expressed as the sum of two integrals:

$$\mathscr{L}\{f(t)\} = \int_0^\infty e^{-st} f(t)\, dt = \int_0^3 e^{-st}(0)\, dt + \int_3^\infty e^{-st}(2)\, dt$$

$$= -\frac{2e^{-st}}{s}\bigg|_3^\infty$$

$$= \frac{2e^{-3s}}{s}, \quad s > 0.$$

We state the generalization of some of the preceding examples by means of the next theorem. From this point on we shall also refrain from stating any restrictions on s; it is understood that s is sufficiently restricted to guarantee the convergence of the appropriate Laplace transform.

THEOREM 7.2 **Transforms of Some Basic Functions**

(a) $\mathscr{L}\{1\} = \dfrac{1}{s}$

(b) $\mathscr{L}\{t^n\} = \dfrac{n!}{s^{n+1}}, \quad n = 1, 2, 3, \ldots$

(c) $\mathscr{L}\{e^{at}\} = \dfrac{1}{s-a}$

(d) $\mathscr{L}\{\sin kt\} = \dfrac{k}{s^2 + k^2}$

(e) $\mathscr{L}\{\cos kt\} = \dfrac{s}{s^2 + k^2}$

(f) $\mathscr{L}\{\sinh kt\} = \dfrac{k}{s^2 - k^2}$

(g) $\mathscr{L}\{\cosh kt\} = \dfrac{s}{s^2 - k^2}$

Part (b) of Theorem 7.2 can be justified in the following manner: Integration by parts yields

$$\mathscr{L}\{t^n\} = \int_0^\infty e^{-st} t^n\, dt = -\frac{1}{s} e^{-st} t^n \bigg|_0^\infty + \frac{n}{s}\int_0^\infty e^{-st} t^{n-1}\, dt = \frac{n}{s}\int_0^\infty e^{-st} t^{n-1}\, dt$$

or

$$\mathscr{L}\{t^n\} = \frac{n}{s}\mathscr{L}\{t^{n-1}\}, \quad n = 1, 2, 3, \ldots.$$

Now $\mathscr{L}\{1\} = 1/s$, so it follows by recursion that

$$\mathscr{L}\{t\} = \frac{1}{s}\mathscr{L}\{1\} = \frac{2}{s^2}$$

$$\mathscr{L}\{t^2\} = \frac{2}{s}\mathscr{L}\{t\} = \frac{2}{s}\left(\frac{1}{s^2}\right) = \frac{2!}{s^3}$$

$$\mathscr{L}\{t^3\} = \frac{3}{s}\mathscr{L}\{t^2\} = \frac{3}{s}\left(\frac{2}{s^3}\right) = \frac{3!}{s^4}.$$

Although a rigorous proof requires mathematical induction, it seems reasonable to conclude from the foregoing results that in general

$$\mathscr{L}\{t^n\} = \frac{n}{s}\mathscr{L}\{t^{n-1}\} = \frac{n}{s}\left[\frac{(n-1)!}{s^n}\right] = \frac{n!}{s^{n+1}}.$$

The justifications of parts (f) and (g) of Theorem 7.2 are left to you. See Problems 33 and 34.

EXAMPLE 8 Trigonometric Identity and Linearity

Evaluate $\mathscr{L}\{\sin^2 t\}$.

Solution With the aid of a trigonometric identity, linearity, and parts (a) and (e) of Theorem 7.2, we obtain

$$\mathscr{L}\{\sin^2 t\} = \mathscr{L}\left\{\frac{1-\cos 2t}{2}\right\} = \frac{1}{2}\mathscr{L}\{1\} - \frac{1}{2}\mathscr{L}\{\cos 2t\}$$

$$= \frac{1}{2}\cdot\frac{1}{s} - \frac{1}{2}\cdot\frac{s}{s^2+4}$$

$$= \frac{2}{s(s^2+4)}.$$

EXERCISES 7.1

Answers to odd-numbered problems begin on page A-15.

In Problems 1–18 use Definition 7.1 to find $\mathscr{L}\{f(t)\}$.

1. $f(t) = \begin{cases} -1, & 0 \le t < 1 \\ 1, & t \ge 1 \end{cases}$

2. $f(t) = \begin{cases} 4, & 0 \le t < 2 \\ 0, & t \ge 2 \end{cases}$

3. $f(t) = \begin{cases} t, & 0 \le t < 1 \\ 1, & t \ge 1 \end{cases}$

4. $f(t) = \begin{cases} 2t+1, & 0 \le t < 1 \\ 0, & t \ge 1 \end{cases}$

5. $f(t) = \begin{cases} \sin t, & 0 \le t < \pi \\ 0, & t \ge \pi \end{cases}$

6. $f(t) = \begin{cases} 0, & 0 \le t < \pi/2 \\ \cos t, & t \ge \pi/2 \end{cases}$

7.

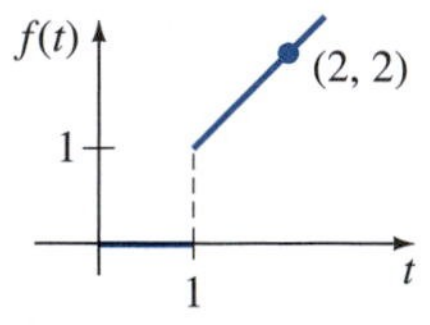

Figure 7.6

8.

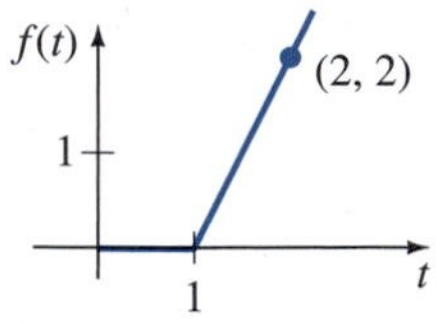

Figure 7.7

9.

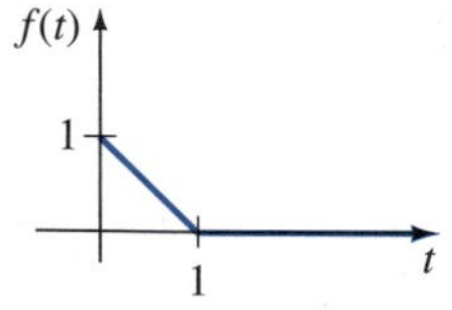

Figure 7.8

10.

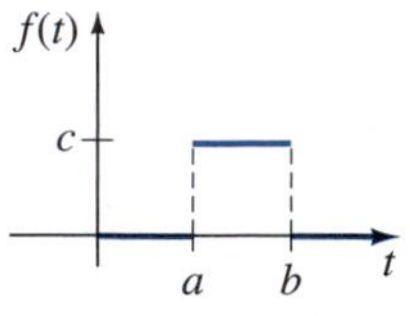

Figure 7.9

11. $f(t) = e^{t+7}$

12. $f(t) = e^{-2t-5}$

13. $f(t) = te^{4t}$

14. $f(t) = t^2e^{3t}$

15. $f(t) = e^{-t}\sin t$

16. $f(t) = e^t \cos t$

17. $f(t) = t\cos t$

18. $f(t) = t\sin t$

In Problems 19–42 use Theorem 7.2 to find $\mathscr{L}\{f(t)\}$.

19. $f(t) = 2t^4$

20. $f(t) = t^5$

21. $f(t) = 4t - 10$

22. $f(t) = 7t + 3$

23. $f(t) = t^2 + 6t - 3$

24. $f(t) = -4t^2 + 16t + 9$

25. $f(t) = (t + 1)^3$

26. $f(t) = (2t - 1)^3$

27. $f(t) = 1 + e^{4t}$

28. $f(t) = t^2 - e^{-9t} + 5$

29. $f(t) = (1 + e^{2t})^2$

30. $f(t) = (e^t - e^{-t})^2$

31. $f(t) = 4t^2 - 5\sin 3t$

32. $f(t) = \cos 5t + \sin 2t$

33. $f(t) = \sinh\ kt$

34. $f(t) = \cosh\ kt$

35. $f(t) = e^t \sinh t$

36. $f(t) = e^{-t}\cosh t$

37. $f(t) = \sin 2t \cos 2t$

38. $f(t) = \cos^2 t$

39. $f(t) = \cos t \cos 2t$ [*Hint*: Examine $\cos(t_1 \pm t_2)$.]

40. $f(t) = \sin t \sin 2t$

41. $f(t) = \sin t \cos 2t$ [*Hint*: Examine $\sin(t_1 \pm t_2)$.]

42. $f(t) = \sin^3 t$ [*Hint*: $\sin^3 t = \sin t \sin^2 t$.]

43. The **gamma function** is defined by the integral

$$\Gamma(\alpha) = \int_0^{\infty} t^{\alpha-1}e^{-t}dt, \quad \alpha > 0.$$

See Appendix I. Show that $\mathscr{L}\{t^\alpha\} = \dfrac{\Gamma(\alpha + 1)}{s^{\alpha+1}}, \alpha > -1.$

In Problems 44–46 use the result of Problem 43 to find $\mathscr{L}\{f(t)\}$.

44. $f(t) = t^{-1/2}$

45. $f(t) = t^{1/2}$

46. $f(t) = t^{3/2}$

47. Show that the function $f(t) = 1/t^2$ does not possess a Laplace transform.

[*Hint*: $\mathscr{L}\{f(t)\} = \int_0^1 e^{-st}f(t)\,dt + \int_1^{\infty} e^{-st}f(t)\,dt$. Use the definition of an improper integral to show that $\int_0^1 e^{-st}f(t)\,dt$ does not exist.]

48. Show that if the functions f and g are of exponential order for $t > T$, then the product fg is of exponential order for $t > T$.

7.2 INVERSE TRANSFORM

• *Inverse Laplace transform* • *Use of partial fractions*

In the preceding section we were concerned with the problem of transforming a function $f(t)$ into another function $F(s)$ by means of the integral $\int_0^\infty e^{-st} f(t)\,dt$. We denoted this symbolically by $\mathscr{L}\{f(t)\} = F(s)$. We now turn the problem around; namely, given $F(s)$, find the function $f(t)$ corresponding to this transform. We say $f(t)$ is the **inverse Laplace transform** of $F(s)$ and write

$$f(t) = \mathscr{L}^{-1}\{F(s)\}.$$

The analogue of Theorem 7.2 for the inverse transform is the following:

THEOREM 7.3 Some Inverse Transforms

$$\textbf{(a)}\quad 1 = \mathscr{L}^{-1}\left\{\frac{1}{s}\right\}$$

$$\textbf{(b)}\quad t^n = \mathscr{L}^{-1}\left\{\frac{n!}{s^{n+1}}\right\},\ n = 1, 2, 3, \ldots \qquad \textbf{(c)}\quad e^{at} = \mathscr{L}^{-1}\left\{\frac{1}{s-a}\right\}$$

$$\textbf{(d)}\quad \sin kt = \mathscr{L}^{-1}\left\{\frac{k}{s^2+k^2}\right\} \qquad \textbf{(e)}\quad \cos kt = \mathscr{L}^{-1}\left\{\frac{s}{s^2+k^2}\right\}$$

$$\textbf{(f)}\quad \sinh kt = \mathscr{L}^{-1}\left\{\frac{k}{s^2-k^2}\right\} \qquad \textbf{(g)}\quad \cosh kt = \mathscr{L}^{-1}\left\{\frac{s}{s^2-k^2}\right\}$$

$\mathscr{L}^{-1}$ a Linear Transform We shall assume that the inverse Laplace transform is itself a linear transform;* that is, for constants α and β,

$$\mathscr{L}^{-1}\{\alpha F(s) + \beta G(s)\} = \alpha\mathscr{L}^{-1}\{F(s)\} + \beta\mathscr{L}^{-1}\{G(s)\},$$

where F and G are the transforms of some functions f and g.

The inverse Laplace transform of a function $F(s)$ *may* not be unique. It is possible that $\mathscr{L}\{f_1(t)\} = \mathscr{L}\{f_2(t)\}$ and yet $f_1 \neq f_2$. See Problems 35 and 36. For our purposes this is not as bad as it appears. If f_1 and f_2 are piecewise continuous on $[0, \infty)$ and of exponential order for $t > 0$ and if $\mathscr{L}\{f_1(t)\} = \mathscr{L}\{f_2(t)\}$, then it can be proved that the functions f_1 and f_2 are *essentially* the same; that is, they can differ only at points of discontinuity. However, if f_1 and f_2 are continuous on $[0, \infty)$ and $\mathscr{L}\{f_1(t)\} = \mathscr{L}\{f_2(t)\}$, then $f_1 = f_2$ on the interval.

EXAMPLE 1 Applying Theorem 7.3

Evaluate $\mathscr{L}^{-1}\left\{\dfrac{1}{s^5}\right\}$.

---

* The inverse Laplace transform is actually another integral. However, evaluation of this integral demands the use of complex variables, which is beyond the scope of this text.

Solution To match the form given in part (b) of Theorem 7.3, we identify $n = 4$ and then multiply and divide by 4!. It follows that

$$\mathscr{L}^{-1}\left\{\frac{1}{s^5}\right\} = \frac{1}{4!}\mathscr{L}^{-1}\left\{\frac{4!}{s^5}\right\} = \frac{1}{24}t^4.$$

EXAMPLE 2 Applying Theorem 7.3

Evaluate $\mathscr{L}^{-1}\left\{\dfrac{1}{s^2 + 64}\right\}$.

Solution Identifying $k^2 = 64$, we multiply and divide by 8 and use part (d) of Theorem 7.3:

$$\mathscr{L}^{-1}\left\{\frac{1}{s^2 + 64}\right\} = \frac{1}{8}\mathscr{L}^{-1}\left\{\frac{8}{s^2 + 64}\right\} = \frac{1}{8}\sin 8t.$$

EXAMPLE 3 Termwise Division and Linearity

Evaluate $\mathscr{L}^{-1}\left\{\dfrac{3s + 5}{s^2 + 7}\right\}$.

Solution The given function of s can be written as two expressions by means of termwise division:

$$\frac{3s + 5}{s^2 + 7} = \frac{3s}{s^2 + 7} + \frac{5}{s^2 + 7}.$$

From the linearity property of the inverse transform and parts (e) and (d) of Theorem 7.3, we then have

$$\begin{aligned}\mathscr{L}^{-1}\left\{\frac{3s + 5}{s^2 + 7}\right\} &= 3\mathscr{L}^{-1}\left\{\frac{s}{s^2 + 7}\right\} + \frac{5}{\sqrt{7}}\mathscr{L}^{-1}\left\{\frac{\sqrt{7}}{s^2 + 7}\right\}\\ &= 3\cos\sqrt{7}t + \frac{5}{\sqrt{7}}\sin\sqrt{7}t.\end{aligned}$$

Partial Fractions The use of **partial fractions** is very important in finding inverse Laplace transforms. Here we review three basic cases of that theory. For example, the denominators of

$$(i)\ F(s) = \frac{1}{(s - 1)(s + 2)(s + 4)} \quad (ii)\ F(s) = \frac{s + 1}{s^2(s + 2)^3} \quad (iii)\ F(s) = \frac{3s - 2}{s^3(s^2 + 4)}$$

contain, respectively, only distinct linear factors, repeated linear factors, and a quadratic factor that is irreducible.*

* *Irreducible* means that the quadratic factor has no real zeros.

EXAMPLE 4 Partial Fractions: Distinct Linear Factors

Evaluate $\mathscr{L}^{-1}\left\{\dfrac{1}{(s-1)(s+2)(s+4)}\right\}$.

Solution There exist unique constants A, B, and C so that

$$\frac{1}{(s-1)(s+2)(s+4)} = \frac{A}{s-1} + \frac{B}{s+2} + \frac{C}{s+4}$$
$$= \frac{A(s+2)(s+4) + B(s-1)(s+4) + C(s-1)(s+2)}{(s-1)(s+2)(s+4)}.$$

Since the denominators are identical, the numerators are identical:

$$1 = A(s+2)(s+4) + B(s-1)(s+4) + C(s-1)(s+2).$$

By comparing coefficients of powers of s on both sides of the equality, we know that the last equation is equivalent to a system of three equations in the three unknowns A, B, and C. However, you might recall the following shortcut for determining these unknowns. If we set $s = 1$, $s = -2$, and $s = -4$, the zeros of the common denominator $(s-1)(s+2)(s+4)$, we obtain, in turn,

$$\begin{aligned} 1 &= A(3)(5), & A &= \tfrac{1}{15}, \\ 1 &= B(-3)(2), & B &= -\tfrac{1}{6}, \\ 1 &= C(-5)(-2), & C &= \tfrac{1}{10}. \end{aligned}$$

Hence we can write

$$\frac{1}{(s-1)(s+2)(s+4)} = \frac{1/15}{s-1} - \frac{1/6}{s+2} + \frac{1/10}{s+4},$$

and thus, from part (c) of Theorem 7.3,

$$\mathscr{L}^{-1}\left\{\frac{1}{(s-1)(s+2)(s+4)}\right\} = \frac{1}{15}\mathscr{L}^{-1}\left\{\frac{1}{s-1}\right\} - \frac{1}{6}\mathscr{L}^{-1}\left\{\frac{1}{s+2}\right\} + \frac{1}{10}\mathscr{L}^{-1}\left\{\frac{1}{s+4}\right\}$$
$$= \frac{1}{15}e^{t} - \frac{1}{6}e^{-2t} + \frac{1}{10}e^{-4t}.$$

EXAMPLE 5 Partial Fractions: Repeated Linear Factors

Evaluate $\mathscr{L}^{-1}\left\{\dfrac{s+1}{s^2(s+2)^3}\right\}$.

Solution Assume

$$\frac{s+1}{s^2(s+2)^3} = \frac{A}{s} + \frac{B}{s^2} + \frac{C}{s+2} + \frac{D}{(s+2)^2} + \frac{E}{(s+2)^3},$$

so that

$$s + 1 = As(s+2)^3 + B(s+2)^3 + Cs^2(s+2)^2 + Ds^2(s+2) + Es^2.$$

Setting $s = 0$ and $s = -2$ gives $B = \frac{1}{8}$ and $E = -\frac{1}{4}$, respectively. By equating the coefficients of s^4, s^3, and s, we obtain

$$0 = A + C$$
$$0 = 6A + B + 4C + D$$
$$1 = 8A + 12B,$$

from which it follows that $A = -\frac{1}{16}$, $C = \frac{1}{16}$, and $D = 0$. Hence from parts (a), (b), and (c) of Theorem 7.3,

$$\begin{aligned}\mathscr{L}^{-1}\left\{\frac{s+1}{s^2(s+2)^3}\right\} &= \mathscr{L}^{-1}\left\{-\frac{1/16}{s} + \frac{1/8}{s^2} + \frac{1/16}{s+2} - \frac{1/4}{(s+2)^3}\right\} \\ &= -\frac{1}{16}\mathscr{L}^{-1}\left\{\frac{1}{s}\right\} + \frac{1}{8}\mathscr{L}^{-1}\left\{\frac{1}{s^2}\right\} + \frac{1}{16}\mathscr{L}^{-1}\left\{\frac{1}{s+2}\right\} - \frac{1}{8}\mathscr{L}^{-1}\left\{\frac{2}{(s+2)^3}\right\} \\ &= -\frac{1}{16} + \frac{1}{8}t + \frac{1}{16}e^{-2t} - \frac{1}{8}t^2e^{-2t}.\end{aligned}$$

Here we have also used $\mathscr{L}^{-1}\{2/(s+2)^3\} = t^2e^{-2t}$ from Example 6 of Section 7.1. ■

EXAMPLE 6 Partial Fractions: Irreducible Quadratic Factor

Evaluate $\mathscr{L}^{-1}\left\{\dfrac{3s-2}{s^3(s^2+4)}\right\}$.

Solution Assume

$$\frac{3s-2}{s^3(s^2+4)} = \frac{A}{s} + \frac{B}{s^2} + \frac{C}{s^3} + \frac{Ds+E}{s^2+4}$$

so that

$$3s - 2 = As^2(s^2+4) + Bs(s^2+4) + C(s^2+4) + (Ds+E)s^3.$$

Setting $s = 0$ gives immediately $C = -\frac{1}{2}$. Now the coefficients of s^4, s^3, s^2, and s are, respectively,

$$0 = A + D, \quad 0 = B + E, \quad 0 = 4A + C, \quad 3 = 4B,$$

from which we obtain $B = \frac{3}{4}$, $E = -\frac{3}{4}$, $A = \frac{1}{8}$, and $D = -\frac{1}{8}$. Therefore from parts (a), (b), (e), and (d) of Theorem 7.3 we have

$$\begin{aligned}\mathscr{L}^{-1}\left\{\frac{3s-2}{s^3(s^2+4)}\right\} &= \mathscr{L}^{-1}\left\{\frac{1/8}{s} + \frac{3/4}{s^2} - \frac{1/2}{s^3} + \frac{-s/8 - 3/4}{s^2+4}\right\} \\ &= \frac{1}{8}\mathscr{L}^{-1}\left\{\frac{1}{s}\right\} + \frac{3}{4}\mathscr{L}^{-1}\left\{\frac{1}{s^2}\right\} - \frac{1}{4}\mathscr{L}^{-1}\left\{\frac{2}{s^3}\right\} \\ &\quad - \frac{1}{8}\mathscr{L}^{-1}\left\{\frac{s}{s^2+4}\right\} - \frac{3}{8}\mathscr{L}^{-1}\left\{\frac{2}{s^2+4}\right\} \\ &= \frac{1}{8} + \frac{3}{4}t - \frac{1}{4}t^2 - \frac{1}{8}\cos 2t - \frac{3}{8}\sin 2t.\end{aligned}$$

■

Not every arbitrary function of s is a Laplace transform of a piecewise function of exponential order.

THEOREM 7.4 **Behavior of $F(s)$ as $s \to \infty$**

Let $f(t)$ be piecewise continuous on $[0, \infty)$ and of exponential order for $t > T$; then $\lim_{s\to\infty} \mathscr{L}\{f(t)\} = 0$.

Proof Since $f(t)$ is piecewise continuous $0 \le t \le T$, it is necessarily bounded on the interval. That is, $|f(t)| \le M_1 = M_1 e^{0t}$. Also, $|f(t)| \le M_2 e^{\gamma t}$ for $t > T$. If M denotes the maximum of $\{M_1, M_2\}$ and c denotes the maximum of $\{0, \gamma\}$, then

$$|\mathscr{L}\{f(t)\}| \le \int_0^\infty e^{-st}|f(t)|\,dt \le M\int_0^\infty e^{-st} \cdot e^{ct}\,dt = -M\left.\frac{e^{-(s-c)t}}{s-c}\right|_0^\infty = \frac{M}{s-c}$$

for $s > c$. As $s \to \infty$, we have $|\mathscr{L}\{f(t)\}| \to 0$ and so $\mathscr{L}\{f(t)\} \to 0$. ■

EXAMPLE 7 **Applying Theorem 7.4**

The functions $F_1(s) = s^2$ and $F_2(s) = s/(s+1)$ are not the Laplace transforms of piecewise continuous functions of exponential order since $F_1(s) \not\to 0$ and $F_2(s) \not\to 0$ as $s \to \infty$. We say that $\mathscr{L}^{-1}\{F_1(s)\}$ and $\mathscr{L}^{-1}\{F_2(s)\}$ do not exist. ■

Remarks There is another way of determining the coefficients in a partial fraction decomposition in the special case when $\mathscr{L}\{f(t)\} = F(s)$ is a quotient of polynomials $P(s)/Q(s)$ and $Q(s)$ is a product of distinct linear factors:

$$F(s) = \frac{P(s)}{(s-r_1)(s-r_2)\cdots(s-r_n)}.$$

Let's illustrate by means of a specific example. From the theory of partial fractions we know there exist unique constants A, B, and C such that

$$\frac{s^2+4s-1}{(s-1)(s-2)(s+3)} = \frac{A}{s-1} + \frac{B}{s-2} + \frac{C}{s+3}. \tag{1}$$

Suppose we multiply both sides of this last expression by, say, $s-1$, simplify, and then set $s = 1$. Since the coefficients of B and C are zero we get

$$\left.\frac{s^2+4s-1}{(s-2)(s+3)}\right|_{s=1} = A \quad \text{or} \quad A = -1.$$

Written another way,

$$\left.\frac{s^2+4s-1}{\boxed{(s-1)}\,(s-2)(s+3)}\right|_{s=1} = A$$

where we have shaded or *covered up* the factor that canceled when the left side of (1) was multiplied by $s-1$. We *do not evaluate this covered-up factor* at $s = 1$. Now to obtain B and C we simply evaluate the left member of (1) while covering, in turn, $s - 2$ and $s + 3$:

$$\left.\frac{s^2 + 4s - 1}{(s - 1)\boxed{(s - 2)}(s + 3)}\right|_{s=2} = B \quad \text{or} \quad B = \frac{11}{5}.$$

$$\left.\frac{s^2 + 4s - 1}{(s - 1)(s - 2)\boxed{(s + 3)}}\right|_{s=-3} = C \quad \text{or} \quad C = -\frac{1}{5}.$$

Note carefully that in the calculation of C we evaluated at $s = -3$. By filling in the details in arriving at this last expression you will see why this is so. You should also verify by other means that

$$\frac{s^2 + 4s - 1}{(s - 1)(s - 2)(s + 3)} = \frac{-1}{s - 1} + \frac{11/5}{s - 2} + \frac{-1/5}{s + 3}.$$

This **cover-up method** is a simplified version of a result known as **Heaviside's expansion theorem.***

EXERCISES 7.2

Answers to odd-numbered problems begin on page A-15.

In Problems 1–34 use Theorem 7.3 to find the given inverse transform.

1. $\mathscr{L}^{-1}\left\{\dfrac{1}{s^3}\right\}$

2. $\mathscr{L}^{-1}\left\{\dfrac{1}{s^4}\right\}$

3. $\mathscr{L}^{-1}\left\{\dfrac{1}{s^2} - \dfrac{48}{s^5}\right\}$

4. $\mathscr{L}^{-1}\left\{\left(\dfrac{2}{s} - \dfrac{1}{s^3}\right)^2\right\}$

5. $\mathscr{L}^{-1}\left\{\dfrac{(s + 1)^3}{s^4}\right\}$

6. $\mathscr{L}^{-1}\left\{\dfrac{(s + 2)^2}{s^3}\right\}$

7. $\mathscr{L}^{-1}\left\{\dfrac{1}{s^2} - \dfrac{1}{s} + \dfrac{1}{s - 2}\right\}$

8. $\mathscr{L}^{-1}\left\{\dfrac{4}{s} + \dfrac{6}{s^5} - \dfrac{1}{s + 8}\right\}$

* OLIVER HEAVISIDE (1850–1925) Many of the results that we present in this chapter were devised by the English electrical engineer Oliver Heaviside and are set forth in his 1899 treatise *Electromagnetic Theory.* Heaviside originally utilized the Laplace transform as a means of solving linear differential equations with constant coefficients that arose in his investigation of transmission lines. Since many of his results lacked formal proof, the Heaviside operational calculus, as his procedures came to be called, initially met with scorn from mathematicians. A curmudgeon, Heaviside, in turn, called these "establishment" mathematicians "woodenheaded." When Heaviside, using his symbolic methods, was able to obtain answers to problems that mathematicians could not solve, their scorn turned to censure and his papers were no longer published in mathematics journals. Heaviside was also the discoverer of a layer of maximum electron density, called the Heaviside layer, in the upper atmosphere that reflects radio waves back to earth. He lived the last years of his life as a recluse in poverty, forgotten by the scientific community. He died in an unheated house in 1925.

True to their nature, mathematicians eventually seized his ideas, put them on a sound mathematical foundation, and then proceeded to generalize them into an abstract theory.

9. $\mathscr{L}^{-1}\left\{\frac{1}{4s+1}\right\}$

10. $\mathscr{L}^{-1}\left\{\frac{1}{5s-2}\right\}$

11. $\mathscr{L}^{-1}\left\{\frac{5}{s^2+49}\right\}$

12. $\mathscr{L}^{-1}\left\{\frac{10s}{s^2+16}\right\}$

13. $\mathscr{L}^{-1}\left\{\frac{4s}{4s^2+1}\right\}$

14. $\mathscr{L}^{-1}\left\{\frac{1}{4s^2+1}\right\}$

15. $\mathscr{L}^{-1}\left\{\frac{1}{s^2-16}\right\}$

16. $\mathscr{L}^{-1}\left\{\frac{10s}{s^2-25}\right\}$

17. $\mathscr{L}^{-1}\left\{\frac{2s-6}{s^2+9}\right\}$

18. $\mathscr{L}^{-1}\left\{\frac{s-1}{s^2+2}\right\}$

19. $\mathscr{L}^{-1}\left\{\frac{1}{s^2+3s}\right\}$

20. $\mathscr{L}^{-1}\left\{\frac{s+1}{s^2-4s}\right\}$

21. $\mathscr{L}^{-1}\left\{\frac{s}{s^2+2s-3}\right\}$

22. $\mathscr{L}^{-1}\left\{\frac{1}{s^2+s-20}\right\}$

23. $\mathscr{L}^{-1}\left\{\frac{0.9s}{(s-0.1)(s+0.2)}\right\}$

24. $\mathscr{L}^{-1}\left\{\frac{s-3}{(s-\sqrt{3})(s+\sqrt{3})}\right\}$

25. $\mathscr{L}^{-1}\left\{\frac{s}{(s-2)(s-3)(s-6)}\right\}$

26. $\mathscr{L}^{-1}\left\{\frac{s^2+1}{s(s-1)(s+1)(s-2)}\right\}$

27. $\mathscr{L}^{-1}\left\{\frac{2s+4}{(s-2)(s^2+4s+3)}\right\}$

28. $\mathscr{L}^{-1}\left\{\frac{s+1}{(s^2-4s)(s+5)}\right\}$

29. $\mathscr{L}^{-1}\left\{\frac{1}{s^2(s^2+4)}\right\}$

30. $\mathscr{L}^{-1}\left\{\frac{s-1}{s^2(s^2+1)}\right\}$

31. $\mathscr{L}^{-1}\left\{\frac{s}{(s^2+4)(s+2)}\right\}$

32. $\mathscr{L}^{-1}\left\{\frac{1}{s^4-9}\right\}$

33. $\mathscr{L}^{-1}\left\{\frac{1}{(s^2+1)(s^2+4)}\right\}$

34. $\mathscr{L}^{-1}\left\{\frac{6s+3}{(s^2+1)(s^2+4)}\right\}$

The inverse Laplace transform may not be unique. In Problems 35 and 36 evaluate $\mathscr{L}\{f(t)\}$.

35. $f(t) = \begin{cases} 1, & t \geq 0, t \neq 1, t \neq 2 \\ 3, & t = 1 \\ 4, & t = 2 \end{cases}$

36. $f(t) = \begin{cases} e^{3t}, & t \geq 0, t \neq 5 \\ 1, & t = 5 \end{cases}$

7.3 TRANSLATION THEOREMS AND DERIVATIVES OF A TRANSFORM

• *First translation theorem* • *Unit step function* • *Second translation theorem*

It is not convenient to use Definition 7.1 each time we wish to find the Laplace transform of a function $f(t)$. For example, the integration by parts involved in evaluating, say, $\mathscr{L}\{e^t t^2 \sin 3t\}$ is formidable to say the least. In the discussion that follows, we present several labor-saving theorems; these, in turn, enable us to build up a more extensive list of transforms without the necessity of using the definition of the Laplace transform. Indeed, we shall see that evaluating transforms such as $\mathscr{L}\{e^{4t}\cos 6t\}$, $\mathscr{L}\{t^3\sin 2t\}$, and $\mathscr{L}\{t^{10}e^{-t}\}$ is fairly straightforward, provided we know $\mathscr{L}\{\cos 6t\}$, $\mathscr{L}\{\sin 2t\}$, and $\mathscr{L}\{t^{10}\}$, respectively. Though extensive tables can be constructed and we have included a table in Appendix II, it is nonetheless a good idea to know the Laplace transforms of basic functions such as t^n, e^{at}, $\sin kt$, $\cos kt$, $\sinh kt$, and $\cosh kt$.

If we know $\mathscr{L}\{f(t)\} = F(s)$, we can compute the Laplace transform $\mathscr{L}\{e^{at}f(t)\}$ with no additional effort other than *translating,* or *shifting,* $F(s)$ to $F(s - a)$. This result is known as the **first translation theorem,** or **first shifting theorem**.

THEOREM 7.5 **First Translation Theorem**

If a is any real number, then

$$\mathscr{L}\{e^{at}f(t)\} = F(s - a),$$

where $F(s) = \mathscr{L}\{f(t)\}$.

Proof The proof is immediate, since by Definition 7.1

$$\mathscr{L}\{e^{at}f(t)\} = \int_0^\infty e^{-st}e^{at}f(t)\,dt = \int_0^\infty e^{-(s-a)t}f(t)\,dt = F(s - a). \quad \blacksquare$$

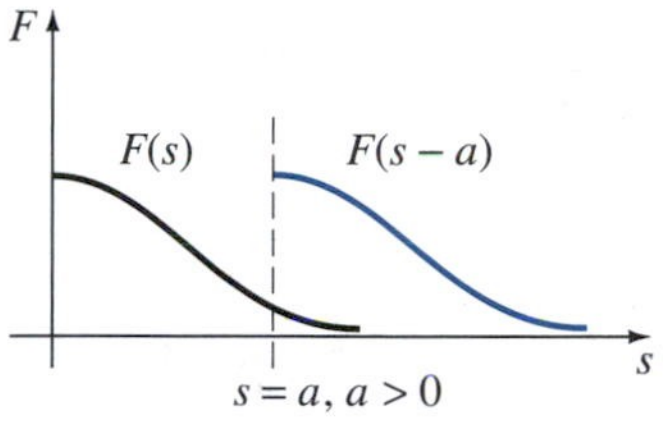

Figure 7.10

If we consider s a real variable, then the graph of $F(s - a)$ is the graph of $F(s)$ shifted on the s-axis by the amount $|a|$ units. If $a > 0$, the graph of $F(s)$ is shifted a units to the right, whereas if $a < 0$, the graph is shifted $|a|$ units to the left. See Figure 7.10.

For emphasis it is sometimes useful to use the symbolism

$$\mathscr{L}\{e^{at}f(t)\} = \mathscr{L}\{f(t)\}_{s \to s-a},$$

where $s \to s - a$ means that we replace s in $F(s)$ by $s - a$.

EXAMPLE 1 **First Translation Theorem**

Evaluate **(a)** $\mathscr{L}\{e^{5t}t^3\}$ and **(b)** $\mathscr{L}\{e^{-2t}\cos 4t\}$.

Solution The results follow from Theorem 7.5.

(a) $\mathscr{L}\{e^{5t}t^3\} = \mathscr{L}\{t^3\}_{s \to s-5}$

$$= \frac{3!}{s^4}\bigg|_{s \to s-5} = \frac{6}{(s-5)^4}.$$

(b) $\mathscr{L}\{e^{-2t}\cos 4t\} = \mathscr{L}\{\cos 4t\}_{s \to s+2}$ $\leftarrow a = -2$ so $s - a = s - (-2) = s + 2$

$$= \frac{s}{s^2 + 16}\bigg|_{s \to s+2} = \frac{s + 2}{(s + 2)^2 + 16}.$$

Inverse Form of the First Translation Theorem The inverse form of Theorem 7.5 can be written

$$\mathscr{L}^{-1}\{F(s-a)\} = \mathscr{L}^{-1}\{F(s)|_{s \to s-a}\} = e^{at}f(t), \tag{1}$$

where $f(t) = \mathscr{L}^{-1}\{F(s)\}$.

EXAMPLE 2 Completing the Square

Evaluate $\mathscr{L}^{-1}\left\{\dfrac{s}{s^2+6s+11}\right\}$.

Solution

$$\begin{aligned}
\mathscr{L}^{-1}\left\{\frac{s}{s^2+6s+11}\right\} &= \mathscr{L}^{-1}\left\{\frac{s}{(s+3)^2+2}\right\} \leftarrow \text{completion of square}\\
&= \mathscr{L}^{-1}\left\{\frac{s+3-3}{(s+3)^2+2}\right\} \leftarrow \text{adding zero in the numerator}\\
&= \mathscr{L}^{-1}\left\{\frac{s+3}{(s+3)^2+2} - \frac{3}{(s+3)^2+2}\right\} \leftarrow \text{termwise division}\\
&= \mathscr{L}^{-1}\left\{\frac{s+3}{(s+3)^2+2}\right\} - 3\mathscr{L}^{-1}\left\{\frac{1}{(s+3)^2+2}\right\}\\
&= \mathscr{L}^{-1}\left\{\frac{s}{s^2+2}\bigg|_{s\to s+3}\right\} - \frac{3}{\sqrt{2}}\mathscr{L}^{-1}\left\{\frac{\sqrt{2}}{s^2+2}\bigg|_{s\to s+3}\right\}\\
&= e^{-3t}\cos\sqrt{2}t - \frac{3}{\sqrt{2}}e^{-3t}\sin\sqrt{2}t. \leftarrow \text{from (1) and Theorem 7.3}
\end{aligned}$$

EXAMPLE 3 Completing the Square and Linearity

Evaluate $\mathscr{L}^{-1}\left\{\dfrac{1}{(s-1)^3} + \dfrac{1}{s^2+2s-8}\right\}$.

Solution Completing the square in the second denominator and using linearity yields

$$\begin{aligned}
\mathscr{L}^{-1}\left\{\frac{1}{(s-1)^3} + \frac{1}{s^2+2s-8}\right\} &= \mathscr{L}^{-1}\left\{\frac{1}{(s-1)^3} + \frac{1}{(s+1)^2-9}\right\}\\
&= \frac{1}{2!}\mathscr{L}^{-1}\left\{\frac{2!}{(s-1)^3}\right\} + \frac{1}{3}\mathscr{L}^{-1}\left\{\frac{3}{(s+1)^2-9}\right\}\\
&= \frac{1}{2!}\mathscr{L}^{-1}\left\{\frac{2!}{s^3}\bigg|_{s\to s-1}\right\} + \frac{1}{3}\mathscr{L}^{-1}\left\{\frac{3}{s^2-9}\bigg|_{s\to s+1}\right\}\\
&= \frac{1}{2}e^t t^2 + \frac{1}{3}e^{-t}\sinh 3t.
\end{aligned}$$

Unit Step Function In engineering one frequently encounters functions that can be either "on" or "off." For example, an external force acting on a mechanical system or a voltage impressed on a circuit can be turned off after a period of time. It is thus convenient to define a special function called the **unit step function.**

DEFINITION 7.3 Unit Step Function

> The function $\mathscr{U}(t-a)$ is defined to be
> $$\mathscr{U}(t-a)=\begin{cases}0, & 0\le t<a\\ 1, & t\ge a\end{cases}$$

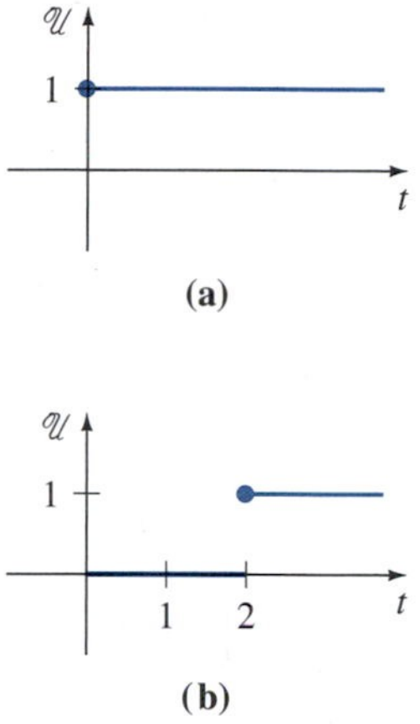

Figure 7.11

Notice that we define $\mathscr{U}(t-a)$ only on the nonnegative t-axis since this is all that we are concerned with in the study of the Laplace transform. In a broader sense, $\mathscr{U}(t-a)=0$ for $t<a$.

EXAMPLE 4 Graphs of Unit Step Functions

Graph (**a**) $\mathscr{U}(t)$ and (**b**) $\mathscr{U}(t-2)$.

Solution (**a**) $\mathscr{U}(t)=1,\quad t\ge 0$ (**b**) $\mathscr{U}(t-2)=\begin{cases}0, & 0\le t<2\\ 1, & t\ge 2\end{cases}$

The respective graphs are given in Figure 7.11. ∎

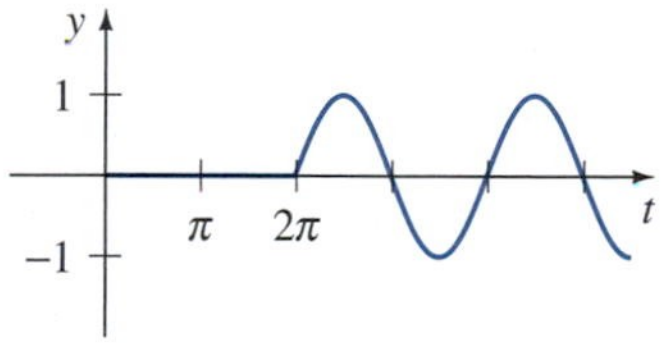

Figure 7.12

When multiplied by another function defined for $t\ge 0$, the unit step function "turns off" a portion of the graph of the function. For example, Figure 7.12 illustrates the graph of $\sin t,\ t\ge 0$ when multiplied by $\mathscr{U}(t-2\pi)$:

$$f(t)=\sin t\ \mathscr{U}(t-2\pi)=\begin{cases}0, & 0\le t<2\pi\\ \sin t, & t\ge 2\pi.\end{cases}$$

The unit step function can also be used to write piecewise-defined functions in a compact form. For instance, the piecewise-defined function

$$f(t)=\begin{cases}g(t), & 0\le t<a\\ h(t), & t\ge a\end{cases}\tag{2}$$

can be written as

$$f(t)=g(t)-g(t)\mathscr{U}(t-a)+h(t)\mathscr{U}(t-a).\tag{3}$$

To verify this, we use the definition of $\mathscr{U}(t-a)$:

$$f(t)=\begin{cases}g(t)-g(t)\cdot 0+h(t)\cdot 0, & 0\le t<a\\ g(t)-g(t)\cdot 1+h(t)\cdot 1, & t\ge a.\end{cases}$$

Similarly, a function of the type

$$f(t)=\begin{cases}0, & 0\le t<a\\ g(t), & a\le t<b\\ 0, & t\ge b\end{cases}\tag{4}$$

can be written

$$f(t) = g(t)[\mathcal{U}(t-a) - \mathcal{U}(t-b)]. \tag{5}$$

EXAMPLE 5 Voltage Expressed in Terms of a Unit Step Function

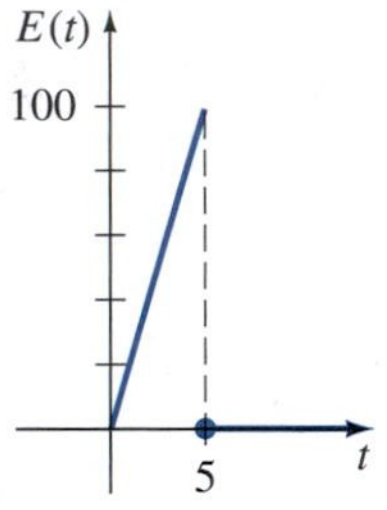

Figure 7.13

The voltage in a circuit is given by

$$E(t) = \begin{cases} 20t, & 0 \le t < 5 \\ 0, & t \ge 5 \end{cases}$$

Graph $E(t)$. Express $E(t)$ in terms of unit step functions.

Solution The graph of this piecewise-defined function is given in Figure 7.13. Now from (2) and (3) with $g(t) = 20t$ and $h(t) = 0$, we get

$$E(t) = 20t - 20t\,\mathcal{U}(t-5).$$

EXAMPLE 6 Comparison of Functions

Consider the function $y = f(t)$ defined by $f(t) = t^3$. Compare the graphs of

(a) $f(t) = t^3$ **(b)** $f(t) = t^3, \quad t \ge 0$

(c) $f(t-2), \quad t \ge 0$ **(d)** $f(t-2)\mathcal{U}(t-2), \quad t \ge 0$

Solution The respective graphs are given in Figure 7.14.

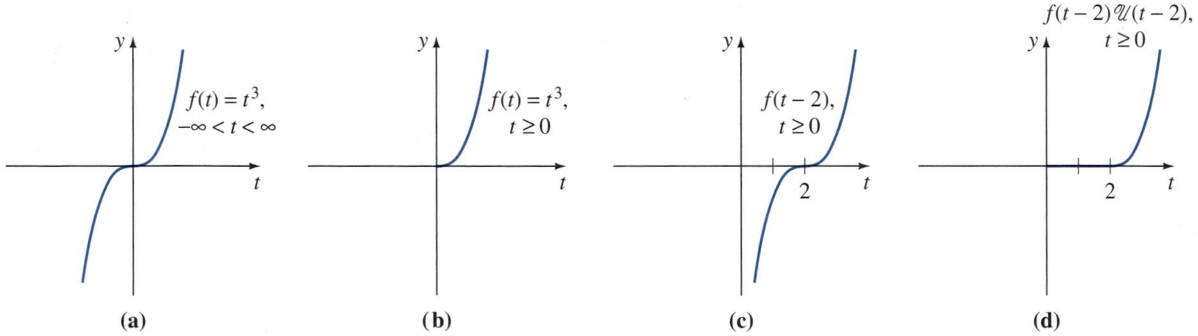

Figure 7.14

In general, if $a > 0$, then the graph of $y = f(t-a)$ is the graph of $y = f(t), t \ge 0$ shifted a units to the right on the t-axis. However, when $y = f(t-a)$ is multiplied by the unit step function $\mathcal{U}(t-a)$ in the manner illustrated in part (d) of Example 6, then the graph of the function

$$y = f(t-a)\,\mathcal{U}(t-a) \tag{6}$$

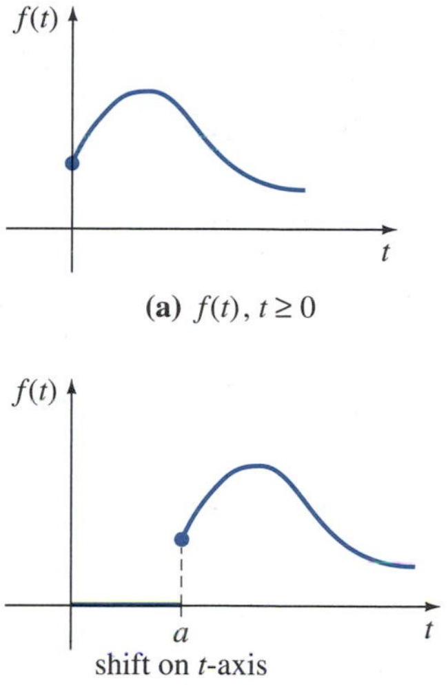

(a) $f(t),\ t \geq 0$

(b) $f(t-a)\,\mathcal{U}(t-a)$

Figure 7.15

coincides with the graph of $y = f(t - a)$ for $t \geq a$ but is identically zero for $0 \leq t < a$. See Figure 7.15.

We saw in Theorem 7.5 that an exponential multiple of $f(t)$ results in a translation or shift of the transform $F(s)$ on the s-axis. In the next theorem we see that whenever $F(s)$ is multiplied by an appropriate exponential function, the inverse transform of this product is the shifted function given in (6). This result is called the **second translation theorem**, or **second shifting theorem.**

THEOREM 7.6 Second Translation Theorem

If a is a positive constant, then

$$\mathcal{L}\{f(t-a)\mathcal{U}(t-a)\} = e^{-as}F(s),$$

where $F(s) = \mathcal{L}\{f(t)\}$.

Proof From Definition 7.1 we have

$$\mathcal{L}\{f(t-a)\mathcal{U}(t-a)\} = \int_0^\infty e^{-st} f(t-a)\mathcal{U}(t-a)\,dt$$

$$= \int_0^a e^{-st} f(t-a)\underbrace{\mathcal{U}(t-a)}_{\substack{\text{zero for} \\ 0 \leq t < a}}\,dt + \int_a^\infty e^{-st} f(t-a)\underbrace{\mathcal{U}(t-a)}_{\substack{\text{one for} \\ t \geq a}}\,dt$$

$$= \int_a^\infty e^{-st} f(t-a)\,dt.$$

Now let $v = t - a$, $dv = dt$; then

$$\mathcal{L}\{f(t-a)\mathcal{U}(t-a)\} = \int_0^\infty e^{-s(v+a)} f(v)\,dv$$

$$= e^{-as}\int_0^\infty e^{-sv} f(v)\,dv = e^{-as}\mathcal{L}\{f(t)\}. \quad \blacksquare$$

EXAMPLE 7 Second Translation Theorem

Evaluate $\mathcal{L}\{(t-2)^3\mathcal{U}(t-2)\}$.

Solution With the identification $a = 2$, it follows from Theorem 7.6 that

$$\mathcal{L}\{(t-2)^3\mathcal{U}(t-2)\} = e^{-2s}\mathcal{L}\{t^3\} = e^{-2s}\frac{3!}{s^4} = \frac{6}{s^4}e^{-2s}. \quad \blacksquare$$

We often wish to find the Laplace transform of just the unit step function. This can be found from either Definition 7.1 or Theorem 7.6. If we identify $f(t) = 1$ in Theorem 7.6, then $f(t - a) = 1$, $F(s) = \mathcal{L}\{1\} = 1/s$, and so

$$\mathcal{L}\{\mathcal{U}(t-a)\} = \frac{e^{-as}}{s}. \tag{7}$$

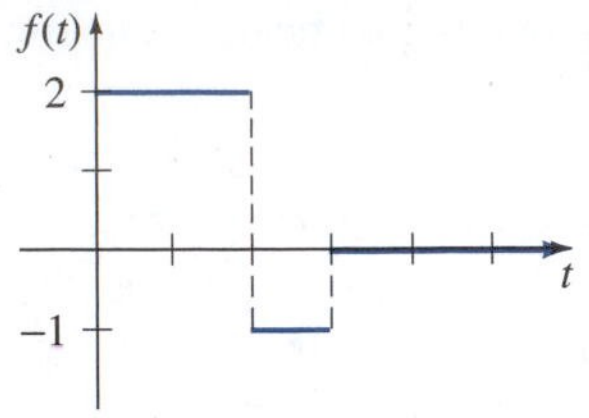

Figure 7.16

EXAMPLE 8 **Second Translation Theorem and Linearity**

Find the Laplace transform of the function shown in Figure 7.16.

Solution With the aid of the unit step function, we can write

$$f(t) = 2 - 3\mathcal{U}(t-2) + \mathcal{U}(t-3).$$

Using linearity and the result in (7) it follows that

$$\begin{aligned}\mathcal{L}\{f(t)\} &= \mathcal{L}\{2\} - 3\mathcal{L}\{\mathcal{U}(t-2)\} + \mathcal{L}\{\mathcal{U}(t-3)\}\\ &= \frac{2}{s} - 3\frac{e^{-2s}}{s} + \frac{e^{-3s}}{s}.\end{aligned}$$

EXAMPLE 9 **Second Translation Theorem**

Evaluate $\mathcal{L}\{\sin t\,\mathcal{U}(t-2\pi)\}$.

Solution With $a = 2\pi$ we have from Theorem 7.6

$$\begin{aligned}\mathcal{L}\{\sin t\,\mathcal{U}(t-2\pi)\} &= \mathcal{L}\{\sin(t-2\pi)\mathcal{U}(t-2\pi)\} \quad \leftarrow \text{sin } t \text{ has period } 2\pi\\ &= e^{-2\pi s}\mathcal{L}\{\sin t\}\\ &= \frac{e^{-2\pi s}}{s^2+1}.\end{aligned}$$

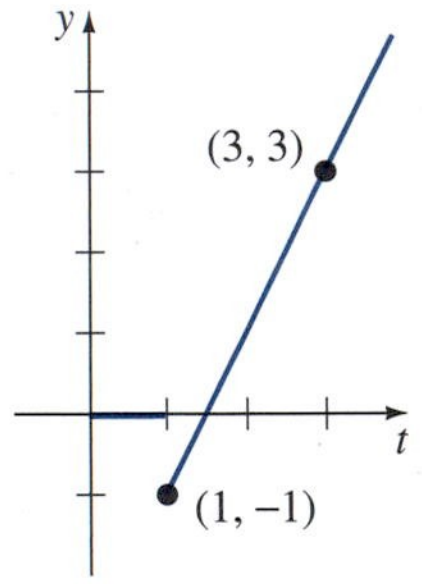

Figure 7.17

EXAMPLE 10 **Fixing Up a Function to Use Theorem 7.6**

Find the Laplace transform of the function shown in Figure 7.17.

Solution An equation of the straight line is found to be $y = 2t - 3$. To "turn off" this graph on the interval $0 \le t < 1$ we form $(2t-3)\mathcal{U}(t-1)$. Now Theorem 7.6 is not immediately applicable in the evaluation of $\mathcal{L}\{(2t-3)\mathcal{U}(t-1)\}$, since the function being transformed lacks the precise form $f(t-a)\mathcal{U}(t-a)$. However, by rewriting $2t-3$ in terms of $t-1$,

$$2t - 3 = 2(t-1) - 1,$$

we can identify $f(t-1) = 2(t-1) - 1$ and consequently $f(t) = 2t - 1$. We are now in a position to apply the second translation theorem:

$$\begin{aligned}\mathcal{L}\{(2t-3)\mathcal{U}(t-1)\} &= \mathcal{L}\{(2(t-1)-1)\mathcal{U}(t-1)\}\\ &= e^{-s}\mathcal{L}\{2t-1\}\\ &= e^{-s}\left(\frac{2}{s^2} - \frac{1}{s}\right).\end{aligned}$$

Inverse Form of the Second Translation Theorem The inverse form of Theorem 7.6 is

$$\mathcal{L}^{-1}\{e^{-as}F(s)\} = f(t-a)\mathcal{U}(t-a), \tag{8}$$

where $a > 0$ and $f(t) = \mathcal{L}^{-1}\{F(s)\}$.

EXAMPLE 11 **Inverse by Formula (8)**

Evaluate $\mathscr{L}^{-1}\left\{\dfrac{e^{-\pi s/2}}{s^2+9}\right\}$.

Solution We identify $a = \dfrac{\pi}{2}$ and $f(t) = \mathscr{L}^{-1}\left\{\dfrac{1}{s^2+9}\right\} = \dfrac{1}{3}\sin 3t$. Thus from (8),

$$\begin{aligned}\mathscr{L}^{-1}\left\{\frac{e^{-\pi s/2}}{s^2+9}\right\} &= \frac{1}{3}\mathscr{L}^{-1}\left\{\frac{3}{s^2+9}\right\}_{t\to t-\pi/2}\mathscr{U}\left(t-\frac{\pi}{2}\right)\\ &= \frac{1}{3}\sin 3\left(t-\frac{\pi}{2}\right)\mathscr{U}\left(t-\frac{\pi}{2}\right)\\ &= \frac{1}{3}\cos 3tb\,\mathscr{U}\left(t-\frac{\pi}{2}\right).\end{aligned}$$

If $F(s) = \mathscr{L}\{f(t)\}$ and if we assume that interchanging of differentiation and integration is possible, then

$$\frac{d}{ds}F(s) = \frac{d}{ds}\int_0^\infty e^{-st}f(t)\,dt = \int_0^\infty \frac{\partial}{\partial s}[e^{-st}f(t)]\,dt = -\int_0^\infty e^{-st}tf(t)\,dt = -\mathscr{L}\{tf(t)\};$$

that is,

$$\mathscr{L}\{tf(t)\} = -\frac{d}{ds}\mathscr{L}\{f(t)\}.$$

Similarly,

$$\begin{aligned}\mathscr{L}\{t^2f(t)\} &= \mathscr{L}\{t\cdot tf(t)\} = -\frac{d}{ds}\mathscr{L}\{tf(t)\}\\ &= -\frac{d}{ds}\left(-\frac{d}{ds}\mathscr{L}\{f(t)\}\right) = \frac{d^2}{ds^2}\mathscr{L}\{f(t)\}.\end{aligned}$$

The preceding two cases suggest the general result for $\mathscr{L}\{t^nf(t)\}$.

THEOREM 7.7 **Derivatives of Transforms**

For $n = 1, 2, 3, \dots,$

$$\mathscr{L}\{t^nf(t)\} = (-1)^n\frac{d^n}{ds^n}F(s),$$

where $F(s) = \mathscr{L}\{f(t)\}$.

EXAMPLE 12 **Applying Theorem 7.7**

Evaluate **(a)** $\mathscr{L}\{te^{3t}\}$, **(b)** $\mathscr{L}\{t\sin kt\}$, **(c)** $\mathscr{L}\{t^2\sin kt\}$, and **(d)** $\mathscr{L}\{te^{-t}\cos t\}$.

Solution We make use of results (c), (d), and (e) of Theorem 7.2.

(a) Note in this first example we could also use the first translation theorem. To apply Theorem 7.7 we identify $n = 1$ and $f(t) = e^{3t}$:

$$\mathscr{L}\{te^{3t}\} = -\frac{d}{ds}\mathscr{L}\{e^{3t}\} = -\frac{d}{ds}\left(\frac{1}{s-3}\right) = \frac{1}{(s-3)^2}.$$

(b)

$$\mathscr{L}\{t\sin kt\} = -\frac{d}{ds}\mathscr{L}\{\sin kt\} = -\frac{d}{ds}\left(\frac{k}{s^2+k^2}\right) = \frac{2ks}{(s^2+k^2)^2}$$

(c) With $n = 2$ in Theorem 7.7 this transform can be written

$$\mathscr{L}\{t^2\sin kt\} = \frac{d^2}{ds^2}\mathscr{L}\{\sin kt\},$$

and so by carrying out the two derivatives we obtain the result. Alternatively, we can make use of the result already obtained in part (b). Since $t^2 \sin kt = t(t\sin kt)$, we have

$$\mathscr{L}\{t^2\sin kt\} = -\frac{d}{ds}\mathscr{L}\{t\sin kt\} = -\frac{d}{ds}\left(\frac{2ks}{(s^2+k^2)^2}\right). \quad \leftarrow \text{from part (b)}$$

Differentiating and simplifying then gives

$$\mathscr{L}\{t^2\sin kt\} = \frac{6ks^2-2k^3}{(s^2+k^2)^3}.$$

(d)

$$\mathscr{L}\{te^{-t}\cos t\} = -\frac{d}{ds}\mathscr{L}\{e^{-t}\cos t\} = -\frac{d}{ds}\mathscr{L}\{\cos t\}_{s\to s+1} \quad \leftarrow \text{first translation theorem}$$

$$= -\frac{d}{ds}\left(\frac{s+1}{(s+1)^2+1}\right) = \frac{(s+1)^2-1}{[(s+1)^2+1]^2}$$

EXERCISES 7.3

Answers to odd-numbered problems begin on page A-16.

In Problems 1–44 find either $F(s)$ or $f(t)$ as indicated.

1. $\mathscr{L}\{te^{10t}\}$
2. $\mathscr{L}\{te^{-6t}\}$
3. $\mathscr{L}\{t^3e^{-2t}\}$
4. $\mathscr{L}\{t^{10}e^{-7t}\}$
5. $\mathscr{L}\{e^t\sin 3t\}$
6. $\mathscr{L}\{e^{-2t}\cos 4t\}$
7. $\mathscr{L}\{e^{5t}\sinh 3t\}$
8. $\mathscr{L}\left\{\frac{\cosh t}{e^t}\right\}$
9. $\mathscr{L}\{t(e^t+e^{2t})^2\}$
10. $\mathscr{L}\{e^{2t}(t-1)^2\}$
11. $\mathscr{L}\{e^{-t}\sin^2 t\}$
12. $\mathscr{L}\{e^t\cos^2 3t\}$

13. $\mathscr{L}^{-1}\left\{\dfrac{1}{(s+2)^3}\right\}$

14. $\mathscr{L}^{-1}\left\{\dfrac{1}{(s-1)^4}\right\}$

15. $\mathscr{L}^{-1}\left\{\dfrac{1}{s^2-6s+10}\right\}$

16. $\mathscr{L}^{-1}\left\{\dfrac{1}{s^2+2s+5}\right\}$

17. $\mathscr{L}^{-1}\left\{\dfrac{s}{s^2+4s+5}\right\}$

18. $\mathscr{L}^{-1}\left\{\dfrac{2s+5}{s^2+6s+34}\right\}$

19. $\mathscr{L}^{-1}\left\{\dfrac{s}{(s+1)^2}\right\}$

20. $\mathscr{L}^{-1}\left\{\dfrac{5s}{(s-2)^2}\right\}$

21. $\mathscr{L}^{-1}\left\{\dfrac{2s-1}{s^2(s+1)^3}\right\}$

22. $\mathscr{L}^{-1}\left\{\dfrac{(s+1)^2}{(s+2)^4}\right\}$

23. $\mathscr{L}\{(t-1)\mathscr{U}(t-1)\}$

24. $\mathscr{L}\{e^{2-t}\mathscr{U}(t-2)\}$

25. $\mathscr{L}\{t\,\mathscr{U}(t-2)\}$

26. $\mathscr{L}\{(3t+1)\mathscr{U}(t-3)\}$

27. $\mathscr{L}\{\cos 2t\,\mathscr{U}(t-\pi)\}$

28. $\mathscr{L}\left\{\sin t\,\mathscr{U}\left(t-\dfrac{\pi}{2}\right)\right\}$

29. $\mathscr{L}\{(t-1)^3e^{t-1}\mathscr{U}(t-1)\}$

30. $\mathscr{L}\{te^{t-5}\mathscr{U}(t-5)\}$

31. $\mathscr{L}^{-1}\left\{\dfrac{e^{-2s}}{s^3}\right\}$

32. $\mathscr{L}^{-1}\left\{\dfrac{(1+e^{-2s})^2}{s+2}\right\}$

33. $\mathscr{L}^{-1}\left\{\dfrac{e^{-\pi s}}{s^2+1}\right\}$

34. $\mathscr{L}^{-1}\left\{\dfrac{se^{-\pi s/2}}{s^2+4}\right\}$

35. $\mathscr{L}^{-1}\left\{\dfrac{e^{-s}}{s(s+1)}\right\}$

36. $\mathscr{L}^{-1}\left\{\dfrac{e^{-2s}}{s^2(s-1)}\right\}$

37. $\mathscr{L}\{t\cos 2t\}$

38. $\mathscr{L}\{t\sinh 3t\}$

39. $\mathscr{L}\{t^2\sinh t\}$

40. $\mathscr{L}\{t^2\cos t\}$

41. $\mathscr{L}\{te^{2t}\sin 6t\}$

42. $\mathscr{L}\{te^{-3t}\cos 3t\}$

43. $\mathscr{L}^{-1}\left\{\dfrac{s}{(s^2+1)^2}\right\}$

44. $\mathscr{L}^{-1}\left\{\dfrac{s+1}{(s^2+2s+2)^2}\right\}$

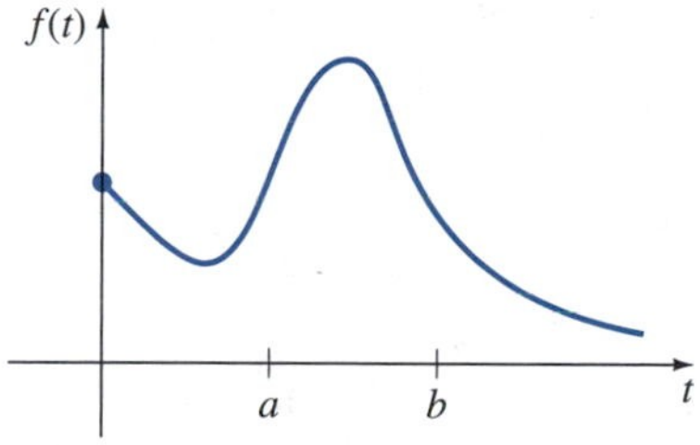

Figure 7.18

In Problems 45–50 match the given graph with one of the functions in **(a)–(f)**. The graph of $f(t)$ is given in Figure 7.18.

(a) $f(t)-f(t)\mathscr{U}(t-a)$

(b) $f(t-b)\mathscr{U}(t-b)$

(c) $f(t)\mathscr{U}(t-a)$

(d) $f(t)-f(t)\mathscr{U}(t-b)$

(e) $f(t)\mathscr{U}(t-a)-f(t)\mathscr{U}(t-b)$

(f) $f(t-a)\mathscr{U}(t-a)-f(t-a)\mathscr{U}(t-b)$

45.

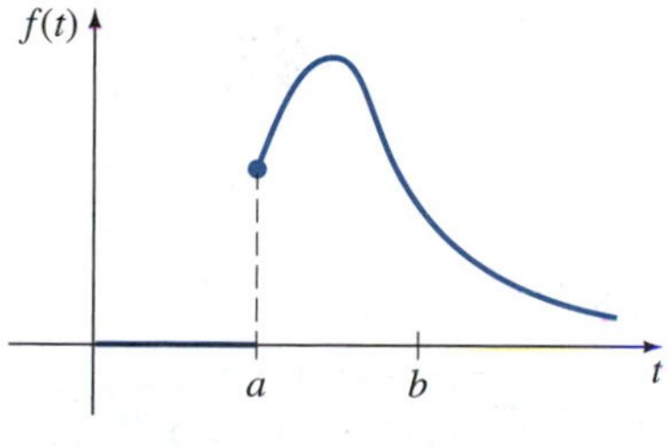

Figure 7.19

46.

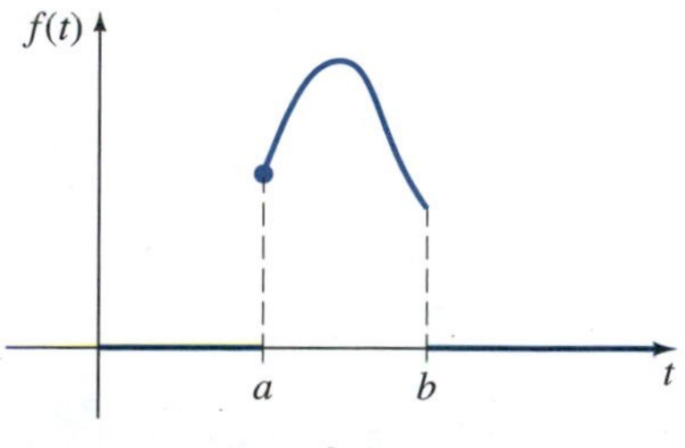

Figure 7.20

47.

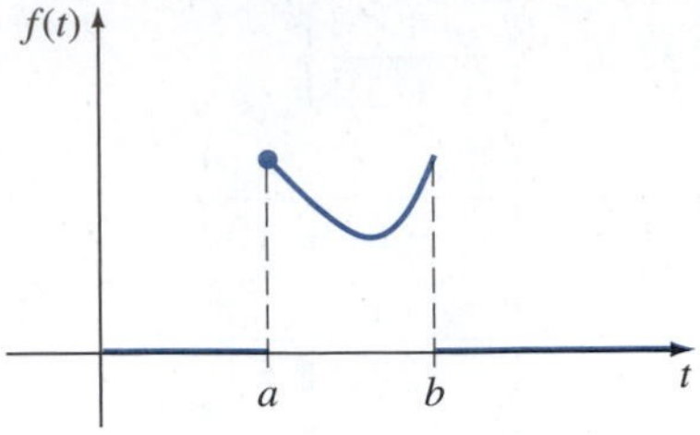

Figure 7.21

48.

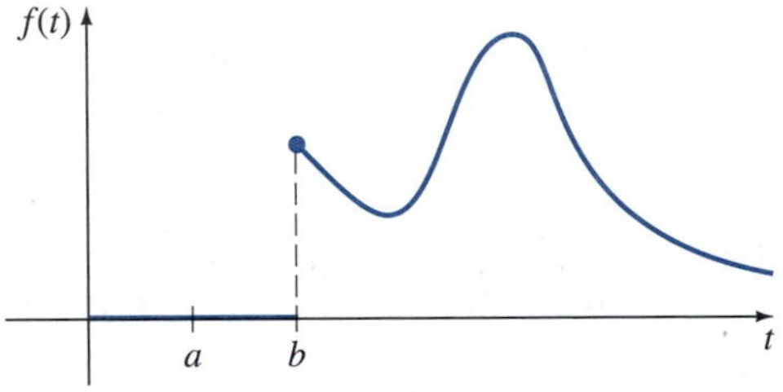

Figure 7.22

49.

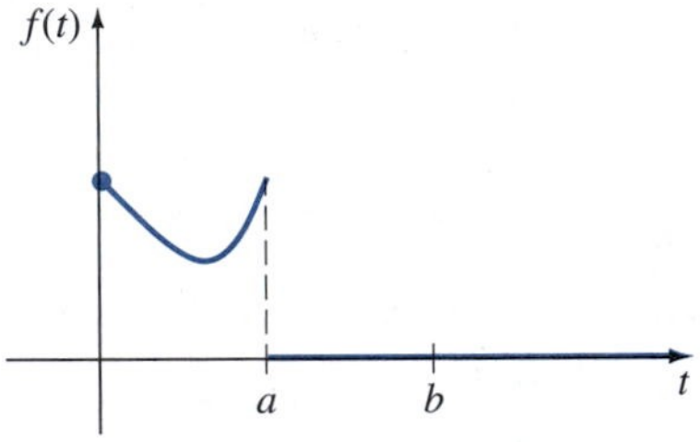

Figure 7.23

50.

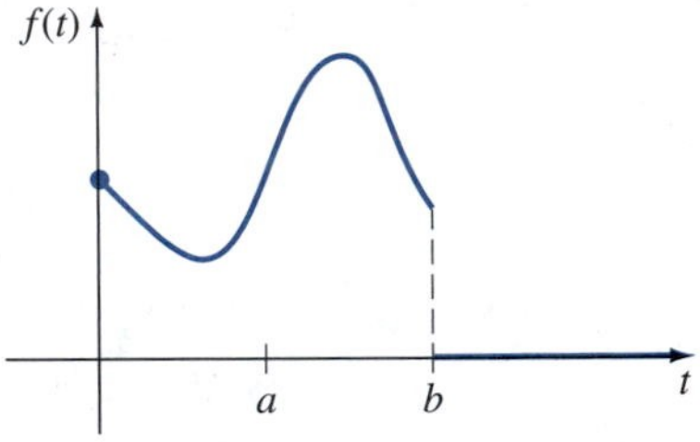

Figure 7.24

In Problems 51–58 write each function in terms of unit step functions. Find the Laplace transform of the given function.

51. $f(t) = \begin{cases} 2, & 0 \le t < 3 \\ -2, & t \ge 3 \end{cases}$

52. $f(t) = \begin{cases} 1, & 0 \le t < 4 \\ 0, & 4 \le t < 5 \\ 1, & t \ge 5 \end{cases}$

53. $f(t) = \begin{cases} 0, & 0 \le t < 1 \\ t^2, & t \ge 1 \end{cases}$

54. $f(t) = \begin{cases} 0, & 0 \le t < \dfrac{3\pi}{2} \\ \sin t, & t \ge \dfrac{3\pi}{2} \end{cases}$

55. $f(t) = \begin{cases} t, & 0 \le t < 2 \\ 0, & t \ge 2 \end{cases}$

56. $f(t) = \begin{cases} \sin t, & 0 \le t < 2\pi \\ 0, & t \ge 2\pi \end{cases}$

57.

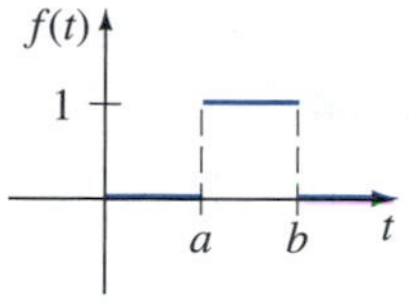

rectangular pulse

Figure 7.25

58.

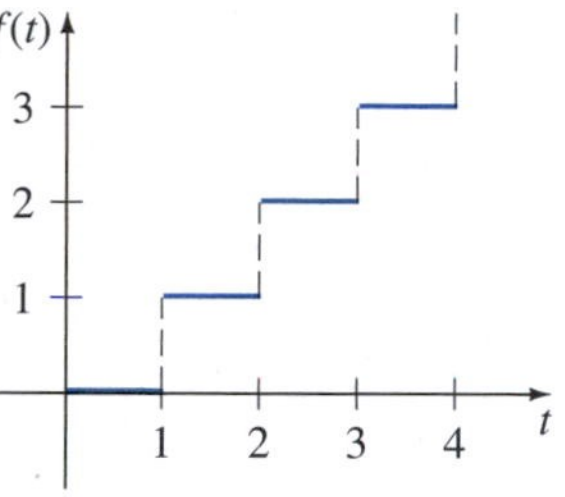

staircase function

Figure 7.26

In Problems 59 and 60 sketch the graph of the given function.

59. $f(t) = \mathscr{L}^{-1}\left\{\dfrac{1}{s^2} - \dfrac{e^{-s}}{s^2}\right\}$

60. $f(t) = \mathscr{L}^{-1}\left\{\dfrac{2}{s} - \dfrac{3e^{-s}}{s^2} + \dfrac{5e^{-2s}}{s^2}\right\}$

In Problems 61–64 use Theorem 7.7 in the form ($n = 1$)

$$f(t) = -\frac{1}{t}\mathscr{L}^{-1}\left\{\frac{d}{ds}F(s)\right\}$$

to evaluate the given inverse Laplace transform.

61. $\mathscr{L}^{-1}\left\{\ln\dfrac{s-3}{s+1}\right\}$

62. $\mathscr{L}^{-1}\left\{\ln\dfrac{s^2+1}{s^2+4}\right\}$

63. $\mathscr{L}^{-1}\left\{\dfrac{\pi}{2} - \tan^{-1}\dfrac{s}{2}\right\}$

64. $\mathscr{L}^{-1}\left\{\dfrac{1}{s} - \cot^{-1}\dfrac{4}{s}\right\}$

7.4 TRANSFORMS OF DERIVATIVES, INTEGRALS, AND PERIODIC FUNCTIONS

• *Convolution* • *Convolution theorem* • *Transform of an integral*

Our goal is to use the Laplace transform to solve certain kinds of differential equations. To that end we need to evaluate quantities such as $\mathscr{L}\{dy/dt\}$ and $\mathscr{L}\{d^2y/dt^2\}$. For example, if f' is continuous for $t \geq 0$, then integration by parts gives

$$\begin{aligned}\mathscr{L}\{f'(t)\} = \int_0^\infty e^{-st}f'(t)\,dt &= e^{-st}f(t)\Big|_0^\infty + s\int_0^\infty e^{-st}f(t)\,dt\\ &= -f(0) + s\mathscr{L}\{f(t)\}\end{aligned}$$

or

$$\mathscr{L}\{f'(t)\} = sF(s) - f(0). \tag{1}$$

Here we have assumed that $e^{-st}f(t) \to 0$ as $t \to \infty$. Similarly, the transform of the second derivative is

$$\begin{aligned}\mathscr{L}\{f''(t)\} = \int_0^\infty e^{-st}f''(t)\,dt &= e^{-st}f'(t)\Big|_0^\infty + s\int_0^\infty e^{-st}f'(t)\,dt\\ &= -f'(0) + s\mathscr{L}\{f'(t)\}\\ &= s[sF(s) - f(0)] - f'(0)\end{aligned}$$

or $$\mathscr{L}\{f''(t)\} = s^2F(s) - sf(0) - f'(0). \tag{2}$$

The results in (1) and (2) are special cases of the next theorem, which gives the Laplace transform of the nth derivative of f. The proof is omitted.

THEOREM 7.8 Transform of a Derivative

If $f(t), f'(t), \ldots, f^{(n-1)}(t)$ are continuous on $[0, \infty)$ and are of exponential order and if $f^n(t)$ is piecewise continuous on $[0, \infty)$, then

$$\mathscr{L}\{f^n(t)\} = s^nF(s) - s^{n-1}f(0) - s^{n-2}f'(0) - \cdots - f^{(n-1)}(0),$$

where $F(s) = \mathscr{L}\{f(t)\}$.

EXAMPLE 1 Applying Theorem 7.8

Observe that the sum $kt \cos kt + \sin kt$ is the derivative of $t \sin kt$. Hence

$$\begin{aligned}
\mathscr{L}\{kt \cos kt + \sin kt\} &= \mathscr{L}\left\{\frac{d}{dt}(t \sin kt)\right\} \\
&= s\mathscr{L}\{t \sin kt\} && \leftarrow \text{by (1)} \\
&= s\left(-\frac{d}{ds}\mathscr{L}\{\sin kt\}\right) && \leftarrow \text{from Theorem 7.7} \\
&= s\left(\frac{2ks}{(s^2 + k^2)^2}\right) = \frac{2ks^2}{(s^2 + k^2)^2}.
\end{aligned}$$

Convolution If functions f and g are piecewise continuous on $[0, \infty)$, then the **convolution** of f and g, denoted by $f * g$, is given by the integral

$$f * g = \int_0^t f(\tau)g(t - \tau)\, d\tau.$$

It is left as an exercise to show that $\int_0^t f(\tau)g(t - \tau)\, d\tau = \int_0^t f(t - \tau)g(\tau)\, d\tau$; that is, $f * g = g * f$. See Problem 29. This means that the convolution of two functions is commutative.

EXAMPLE 2 Convolution of Two Functions

The convolution of $f(t) = e^t$ and $g(t) = \sin t$ is

$$e^t * \sin t = \int_0^t e^\tau \sin(t - \tau)\, d\tau \tag{3}$$

$$= \frac{1}{2}(-\sin t - \cos t + e^t). \tag{4}$$

It is possible to find the Laplace transform of the convolution of two functions, such as (3), without actually evaluating the integral as we did in (4). The result that follows is known as the **convolution theorem.**

THEOREM 7.9 Convolution Theorem

Let $f(t)$ and $g(t)$ be piecewise continuous on $[0, \infty)$ and of exponential order; then

$$\mathscr{L}\{f * g\} = \mathscr{L}\{f(t)\}\mathscr{L}\{g(t)\} = F(s)G(s).$$

Proof Let $$F(s) = \mathscr{L}\{f(t)\} = \int_0^\infty e^{-s\tau} f(\tau)\, d\tau$$

and $$G(s) = \mathscr{L}\{g(t)\} = \int_0^\infty e^{-s\beta} g(\beta)\, d\beta.$$

Proceeding formally, we have

$$\begin{aligned} F(s)G(s) &= \left(\int_0^\infty e^{-s\tau} f(\tau)\, d\tau\right)\left(\int_0^\infty e^{-s\beta} g(\beta)\, d\beta\right) \\ &= \int_0^\infty \int_0^\infty e^{-s(\tau+\beta)} f(\tau) g(\beta)\, d\tau\, d\beta \\ &= \int_0^\infty f(\tau)\, d\tau \int_0^\infty e^{-s(\tau+\beta)} g(\beta)\, d\beta. \end{aligned}$$

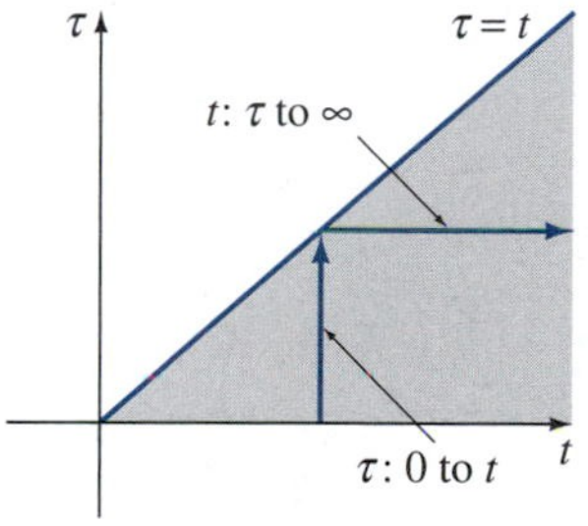

Figure 7.27

Holding τ fixed, we let $t = \tau + \beta$, $dt = d\beta$ so that

$$F(s)G(s) = \int_0^\infty f(\tau)\, d\tau \int_\tau^\infty e^{-st} g(t - \tau)\, dt.$$

In the $t\tau$-plane we are integrating over the shaded region in Figure 7.27. Since f and g are piecewise continuous on $[0, \infty)$ and of exponential order, it is possible to interchange the order of integration:

$$F(s)G(s) = \int_0^\infty e^{-st}\, dt \int_0^t f(\tau) g(t - \tau)\, d\tau = \int_0^\infty e^{-st} \left\{\int_0^t f(\tau) g(t - \tau)\, d\tau\right\} dt = \mathscr{L}\{f * g\}. \quad \blacksquare$$

When $g(t) = 1$ and $G(s) = 1/s$, the convolution theorem implies that the Laplace transform of the integral of a function f is

$$\mathscr{L}\left\{\int_0^t f(\tau)\, d\tau\right\} = \frac{F(s)}{s}. \qquad \textbf{(5)}$$

EXAMPLE 3 Transform of a Convolution

Evaluate $\mathscr{L}\left\{\int_0^t e^\tau \sin(t - \tau)\, d\tau\right\}$.

Solution With $f(t) = e^t$ and $g(t) = \sin t$, the convolution theorem states that the Laplace transform of the convolution of f and g is the product of their Laplace transforms:

$$\mathscr{L}\left\{\int_0^t e^\tau \sin(t-\tau)\,d\tau\right\} = \mathscr{L}\{e^t\}\cdot\mathscr{L}\{\sin t\}$$
$$= \frac{1}{s-1}\cdot\frac{1}{s^2+1} = \frac{1}{(s-1)(s^2+1)}.$$

Inverse Form of the Convolution Theorem The convolution theorem is sometimes useful in finding the inverse Laplace transform of a product of two Laplace transforms. From Theorem 7.9 we have

$$\mathscr{L}^{-1}\{F(s)G(s)\} = f * g. \tag{6}$$

EXAMPLE 4 Inverse Transform as a Convolution

Evaluate $\mathscr{L}^{-1}\left\{\dfrac{1}{(s-1)(s+4)}\right\}$.

Solution Partial fractions could be used, but if we identify

$$F(s) = \frac{1}{s-1} \quad\text{and}\quad G(s) = \frac{1}{s+4},$$

then $\quad \mathscr{L}^{-1}\{F(s)\} = f(t) = e^t \quad\text{and}\quad \mathscr{L}^{-1}\{G(s)\} = g(t) = e^{-4t}.$

Hence from (6) we obtain

$$\mathscr{L}^{-1}\left\{\frac{1}{(s-1)(s+4)}\right\} = \int_0^t f(\tau)g(t-\tau)\,dt = \int_0^t e^\tau e^{-4(t-\tau)}\,d\tau$$
$$= e^{-4t}\int_0^t e^{5\tau}\,d\tau$$
$$= e^{-4t}\frac{1}{5}e^{5\tau}\Big|_0^t$$
$$= \frac{1}{5}e^t - \frac{1}{5}e^{-4t}.$$

EXAMPLE 5 Inverse Transform as a Convolution

Evaluate $\mathscr{L}^{-1}\left\{\dfrac{1}{(s^2+k^2)^2}\right\}$.

Solution Let
$$F(s) = G(s) = \frac{1}{s^2 + k^2}$$
so that
$$f(t) = g(t) = \frac{1}{k}\mathscr{L}^{-1}\left\{\frac{k}{s^2 + k^2}\right\} = \frac{1}{k}\sin kt.$$
In this case (6) gives
$$\mathscr{L}^{-1}\left\{\frac{1}{(s^2 + k^2)^2}\right\} = \frac{1}{k^2}\int_0^t \sin k\tau \sin k(t - \tau)\,d\tau. \tag{7}$$
Now recall from trigonometry that
$$\cos(A + B) = \cos A \cos B - \sin A \sin B$$
and
$$\cos(A - B) = \cos A \cos B + \sin A \sin B.$$
Subtracting the first from the second gives the identity
$$\sin A \sin B = \frac{1}{2}[\cos(A - B) - \cos(A + B)].$$
If we set $A = k\tau$ and $B = k(t - \tau)$, we can carry out the integration in (7):
$$\begin{aligned}\mathscr{L}^{-1}\left\{\frac{1}{(s^2 + k^2)^2}\right\} &= \frac{1}{2k^2}\int_0^t [\cos k(2\tau - t) - \cos kt]\,d\tau \\ &= \frac{1}{2k^2}\left[\frac{1}{2k}\sin k(2\tau - t) - \tau\cos kt\right]_0^t \\ &= \frac{\sin kt - kt\cos kt}{2k^3}.\end{aligned}$$

Transform of a Periodic Function If a periodic function has period T, $T > 0$, then $f(t + T) = f(t)$. The Laplace transform of a periodic function can be obtained by an integration over one period.

THEOREM 7.10 Transform of a Periodic Function

Let $f(t)$ be piecewise continuous on $[0, \infty)$ and of exponential order. If $f(t)$ is periodic with period T, then
$$\mathscr{L}\{f(t)\} = \frac{1}{1 - e^{-sT}}\int_0^T e^{-st}f(t)\,dt. \tag{8}$$

Proof Write the Laplace transform as two integrals:
$$\mathscr{L}\{f(t)\} = \int_0^T e^{-st}f(t)\,dt + \int_T^\infty e^{-st}f(t)\,dt. \tag{9}$$
When we let $t = u + T$, the last integral in (9) becomes
$$\int_T^\infty e^{-st}f(t)\,dt = \int_0^\infty e^{-s(u+T)}f(u + T)\,du = e^{-sT}\int_0^\infty e^{-su}f(u)\,du = e^{-sT}\mathscr{L}\{f(t)\}.$$

Hence (9) is $$\mathscr{L}\{f(t)\} = \int_0^T e^{-st} f(t)\,dt + e^{-sT}\mathscr{L}\{f(t)\}.$$

Solving for $\mathscr{L}\{f(t)\}$ yields the result given in (8). ■

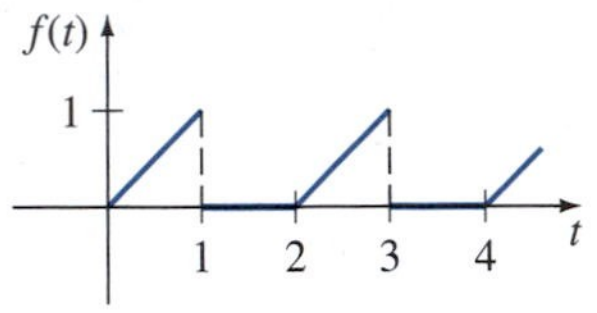

Figure 7.28

EXAMPLE 6 Transform of a Periodic Function

Find the Laplace transform of the periodic function shown in Figure 7.28.

Solution On the interval $0 \le t < 2$ the function can be defined by

$$f(t) = \begin{cases} t, & 0 \le t < 1 \\ 0, & 1 \le t < 2 \end{cases}$$

and outside the interval by $f(t + 2) = f(t)$. Identifying $T = 2$, we use (8) and integration by parts to obtain

$$\mathscr{L}\{f(t)\} = \frac{1}{1 - e^{-2s}} \int_0^2 e^{-st} f(t)\,dt \tag{10}$$

$$= \frac{1}{1 - e^{-2s}} \left[\int_0^1 e^{-st} t\,dt + \int_1^2 e^{-st}\, 0\,dt \right]$$

$$= \frac{1}{1 - e^{-2s}} \left[-\frac{e^{-s}}{s} + \frac{1 - e^{-s}}{s^2} \right] \tag{11}$$

$$= \frac{1 - (s + 1)e^{-s}}{s^2(1 - e^{-2s})}.$$

The result in (11) of Example 6 can be obtained without actually integrating by making use of the second translation theorem. If we define

$$g(t) = \begin{cases} t, & 0 \le t < 1 \\ 0, & t \ge 1 \end{cases}$$

then $f(t) = g(t)$ on the interval $[0, T]$, where $T = 2$. But we can express g in terms of unit step functions as

$$g(t) = t - t\mathscr{U}(t - 1) = t - (t - 1)\mathscr{U}(t - 1) - \mathscr{U}(t - 1).$$

Thus (10) can be written as

$$\mathscr{L}\{f(t)\} = \frac{1}{1 - e^{-2s}} \mathscr{L}\{g(t)\}$$

$$= \frac{1}{1 - e^{-2s}} \mathscr{L}\{t - (t - 1)\mathscr{U}(t - 1) - \mathscr{U}(t - 1)\}$$

$$= \frac{1}{1 - e^{-2s}} \left[\frac{1}{s^2} - \frac{1}{s^2} e^{-s} - \frac{1}{s} e^{-s} \right].$$

Inspection of the expression inside the brackets reveals that it is identical to (11).

EXERCISES 7.4

Answers to odd-numbered problems begin on page A-16.

1. Use the result $(d/dt)e^t = e^t$ and (1) of this section to evaluate $\mathscr{L}\{e^t\}$.
2. Use the result $(d/dt)\cos^2 t = -\sin 2t$ and (1) of this section to evaluate $\mathscr{L}\{\cos^2 t\}$.

In Problems 3 and 4 suppose a function $y(t)$ has the properties that $y(0) = 1$ and $y'(0) = -1$. Find the Laplace transform of the given expression.

3. $y'' + 3y'$
4. $y'' - 4y' + 5y$

In Problems 5 and 6 suppose a function $y(t)$ has the properties that $y(0) = 2$ and $y'(0) = 3$. Solve for the Laplace transform $\mathscr{L}\{y(t)\} = Y(s)$.

5. $y'' - 2y' + y = 0$
6. $y'' + y = 1$

In Problems 7–20 evaluate the given Laplace transform without evaluating the integral.

7. $\mathscr{L}\left\{\int_0^t e^\tau \, d\tau\right\}$
8. $\mathscr{L}\left\{\int_0^t \cos\tau \, d\tau\right\}$
9. $\mathscr{L}\left\{\int_0^t e^{-\tau}\cos\tau \, d\tau\right\}$
10. $\mathscr{L}\left\{\int_0^t \tau\sin\tau \, d\tau\right\}$
11. $\mathscr{L}\left\{\int_0^t \tau e^{t-\tau} \, d\tau\right\}$
12. $\mathscr{L}\left\{\int_0^t \sin\tau\cos(t-\tau) \, d\tau\right\}$
13. $\mathscr{L}\left\{t\int_0^t \sin\tau \, d\tau\right\}$
14. $\mathscr{L}\left\{t\int_0^t \tau e^{-\tau} \, d\tau\right\}$
15. $\mathscr{L}\{1 * t^3\}$
16. $\mathscr{L}\{1 * e^{-2t}\}$
17. $\mathscr{L}\{t^2 * t^4\}$
18. $\mathscr{L}\{t^2 * te^t\}$
19. $\mathscr{L}\{e^{-t} * e^t \cos t\}$
20. $\mathscr{L}\{e^{2t} * \sin t\}$

In Problems 21 and 22 suppose $\mathscr{L}^{-1}\{F(s)\} = f(t)$. Find the inverse Laplace transform of the given function.

21. $\dfrac{1}{s+5}F(s)$
22. $\dfrac{s}{s^2+4}F(s)$

In Problems 23–28 use (6) to find $f(t)$.

23. $\mathscr{L}^{-1}\left\{\dfrac{1}{s(s+1)}\right\}$
24. $\mathscr{L}^{-1}\left\{\dfrac{1}{s(s^2+1)}\right\}$
25. $\mathscr{L}^{-1}\left\{\dfrac{1}{(s+1)(s-2)}\right\}$
26. $\mathscr{L}^{-1}\left\{\dfrac{1}{(s+1)^2}\right\}$
27. $\mathscr{L}^{-1}\left\{\dfrac{s}{(s^2+4)^2}\right\}$
28. $\mathscr{L}^{-1}\left\{\dfrac{1}{(s^2+4s+5)^2}\right\}$
29. Prove the commutative property of the convolution integral $f * g = g * f$.

30. Prove the distributive property of the convolution integral $f * (g + h) = f * g + f * h$.

In Problems 31–38 use Theorem 7.10 to find the Laplace transform of the given periodic function.

31.

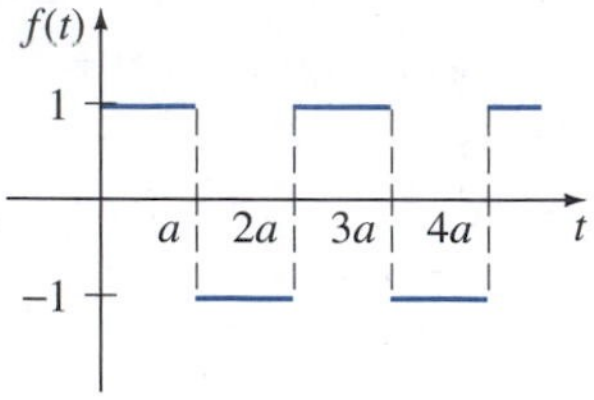

meander function

Figure 7.29

32.

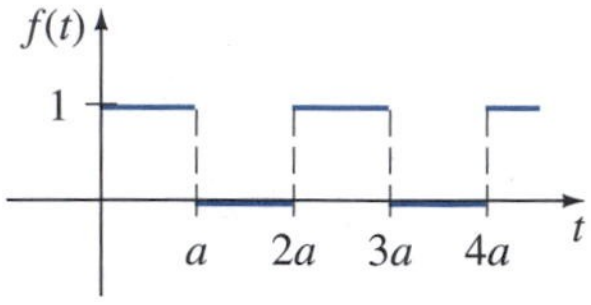

square wave

Figure 7.30

33.

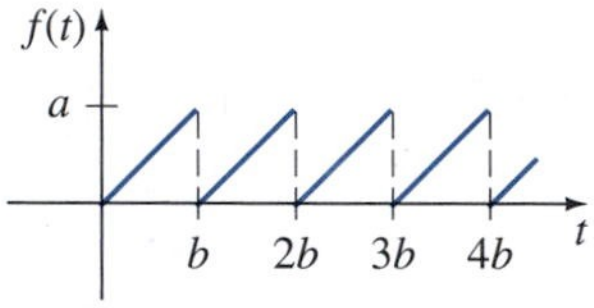

sawtooth function

Figure 7.31

34.

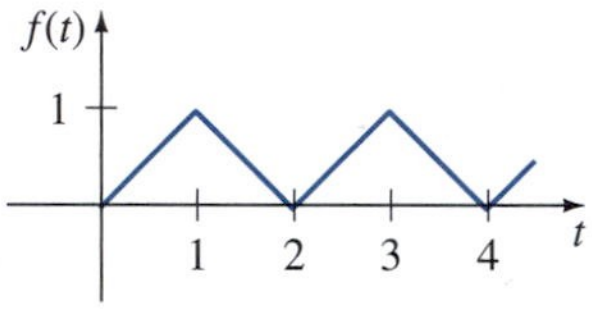

triangular wave

Figure 7.32

35.

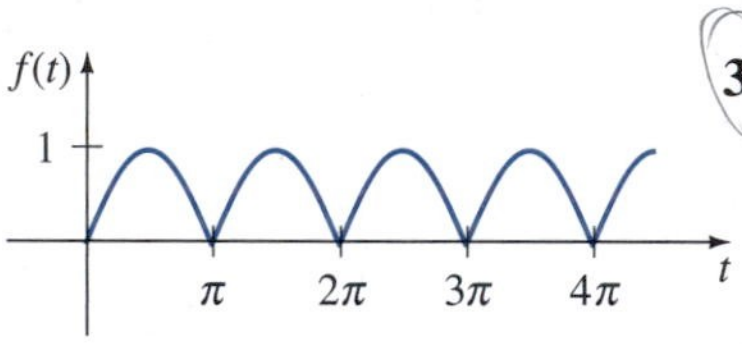

full-wave rectification of sin t

Figure 7.33

36.

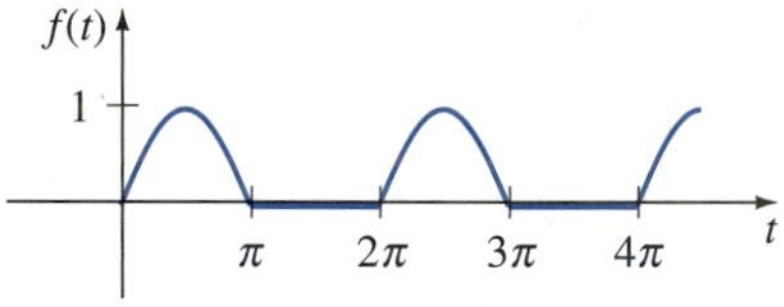

half-wave rectification of sin t

Figure 7.34

37. $f(t) = \sin t$
$f(t + 2\pi) = f(t)$

38. $f(t) = \cos t$
$f(t + 2\pi) = f(t)$

7.5 APPLICATIONS

• Transform of a DE • Volterra integral equation • Integrodifferential equation

Since $\mathscr{L}\{y^{(n)}(t)\}$, $n > 1$ depends on $y(t)$ and its $n - 1$ derivatives evaluated at $t = 0$, the Laplace transform is ideally suited to initial-value problems for linear differential equations with constant coefficients. This kind of differential equation can be reduced to an *algebraic equation* in the transformed function $Y(s)$. To see this, consider the initial-value problem

$$a_n \frac{d^n y}{dt^n} + a_{n-1} \frac{d^{n-1} y}{dt^{n-1}} + \cdots + a_1 \frac{dy}{dt} + a_0 y = g(t)$$

$$y(0) = y_0, \quad y'(0) = y'_0, \quad \ldots, \quad y^{(n-1)}(0) = y_0^{(n-1)},$$

where a_i, $i = 0, 1, \ldots, n$ and $y_0, y'_0, \ldots, y_0^{(n-1)}$ are constants. By the linearity property of the Laplace transform, we can write

$$a_n \mathscr{L}\left\{\frac{d^n y}{dt^n}\right\} + a_{n-1}\mathscr{L}\left\{\frac{d^{n-1}y}{dt^{n-1}}\right\} + \cdots + a_0\mathscr{L}\{y\} = \mathscr{L}\{g(t)\}. \tag{1}$$

From Theorem 7.8, (1) becomes

$$a_n[s^n Y(s) - s^{n-1}y(0) - \cdots - y^{(n-1)}(0)]$$
$$+ a_{n-1}[s^{n-1}Y(s) - s^{n-2}y(0) - \cdots - y^{(n-2)}(0)] + \cdots + a_0 Y(s) = G(s)$$

or

$$[a_n s^n + a_{n-1}s^{n-1} + \cdots + a_0]Y(s) = a_n[s^{n-1}y_0 + \cdots + y_0^{(n-1)}]$$
$$+ a_{n-1}[s^{n-2}y_0 + \cdots + y_0^{(n-2)}] + \cdots + G(s), \tag{2}$$

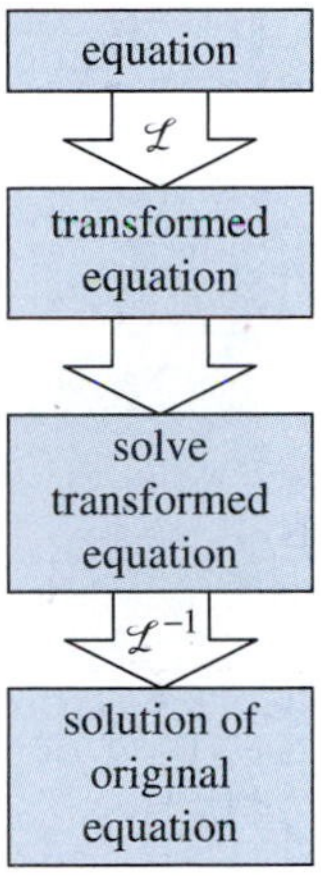

Figure 7.35

where $Y(s) = \mathscr{L}\{y(t)\}$ and $G(s) = \mathscr{L}\{g(t)\}$. By solving (2) for $Y(s)$, we find $y(t)$ by determining the inverse transform

$$y(t) = \mathscr{L}^{-1}\{Y(s)\}.$$

The procedure is outlined in Figure 7.35. Note that this method incorporates the prescribed initial conditions directly into the solution. Hence there is no need for the separate operations of determining constants in the general solution of the differential equation.

EXAMPLE 1 DE Transformed into an Algebraic Equation

Solve $\dfrac{dy}{dt} - 3y = e^{2t}$, $y(0) = 1$.

Solution We first take the transform of each member of the given differential equation:

$$\mathscr{L}\left\{\frac{dy}{dt}\right\} - 3\mathscr{L}\{y\} = \mathscr{L}\{e^{2t}\}.$$

We then use $\mathscr{L}\{dy/dt\} = sY(s) - y(0) = sY(s) - 1$ and $\mathscr{L}\{e^{2t}\} = 1/(s-2)$.

Solving $\qquad sY(s) - 1 - 3Y(s) = \dfrac{1}{s-2}$

for $Y(s)$ gives $\qquad Y(s) = \dfrac{s-1}{(s-2)(s-3)}.$

By partial fractions,

$$\frac{s-1}{(s-2)(s-3)} = \frac{A}{s-2} + \frac{B}{s-3},$$

which yields $\qquad s - 1 = A(s-3) + B(s-2).$

Setting $s = 2$ and $s = 3$ in the last equation, we obtain $A = -1$ and $B = 2$, respectively. Consequently

$$Y(s) = \frac{-1}{s-2} + \frac{2}{s-3}$$

and
$$y(t) = -\mathscr{L}^{-1}\left\{\frac{1}{s-2}\right\} + 2\mathscr{L}^{-1}\left\{\frac{1}{s-3}\right\}.$$

From part (c) of Theorem 7.3 it follows that

$$y(t) = -e^{2t} + 2e^{3t}.$$

EXAMPLE 2 An Initial-Value Problem

Solve $y'' - 6y' + 9y = t^2e^{3t}$, $y(0) = 2$, $y'(0) = 6$.

Solution
$$\mathscr{L}\{y''\} - 6\mathscr{L}\{y'\} + 9\mathscr{L}\{y\} = \mathscr{L}\{t^2e^{3t}\}$$

$$\underbrace{s^2Y(s) - sy(0) - y'(0)}_{\mathscr{L}\{y''\}} - 6\underbrace{[sY(s) - y(0)]}_{\mathscr{L}\{y'\}} + 9\underbrace{Y(s)}_{\mathscr{L}\{y\}} = \underbrace{\frac{2}{(s-3)^3}}_{\mathscr{L}\{t^2e^{3t}\}}.$$

Using the initial conditions and simplifying gives

$$(s^2 - 6s + 9)Y(s) = 2s - 6 + \frac{2}{(s-3)^3}$$

$$(s-3)^2Y(s) = 2(s-3) + \frac{2}{(s-3)^3}$$

$$Y(s) = \frac{2}{s-3} + \frac{2}{(s-3)^5},$$

and so
$$y(t) = 2\mathscr{L}^{-1}\left\{\frac{1}{s-3}\right\} + \frac{2}{4!}\mathscr{L}^{-1}\left\{\frac{4!}{(s-3)^5}\right\}.$$

Recall from the first translation theorem that

$$\mathscr{L}^{-1}\left\{\left.\frac{4!}{s^5}\right|_{s\to s-3}\right\} = t^4e^{3t}.$$

Hence we have
$$y(t) = 2e^{3t} + \frac{1}{12}t^4e^{3t}.$$

EXAMPLE 3 Using the First Translation Theorem

Solve $y'' + 4y' + 6y = 1 + e^{-t}$, $y(0) = 0$, $y'(0) = 0$.

Solution $$\mathscr{L}\{y''\} + 4\mathscr{L}\{y'\} + 6\mathscr{L}\{y\} = \mathscr{L}\{1\} + \mathscr{L}\{e^{-t}\}$$

$$s^2Y(s) - sy(0) - y'(0) + 4[sY(s) - y(0)] + 6Y(s) = \frac{1}{s} + \frac{1}{s+1}$$
$$(s^2 + 4s + 6)Y(s) = \frac{2s+1}{s(s+1)}$$
$$Y(s) = \frac{2s+1}{s(s+1)(s^2+4s+6)}.$$

By partial fractions,

$$\frac{2s+1}{s(s+1)(s^2+4s+6)} = \frac{A}{s} + \frac{B}{s+1} + \frac{Cs+D}{s^2+4s+6},$$

which implies

$$2s + 1 = A(s+1)(s^2+4s+6) + Bs(s^2+4s+6) + (Cs+D)s(s+1).$$

Setting $s = 0$ and $s = -1$ gives, respectively, $A = \frac{1}{6}$ and $B = \frac{1}{3}$. Equating the coefficients of s^3 and s gives

$$A + B + C = 0$$
$$10A + 6B + D = 2,$$

so it follows that $C = -\frac{1}{2}$ and $D = -\frac{5}{3}$. Thus

$$\begin{aligned} Y(s) &= \frac{1/6}{s} + \frac{1/3}{s+1} + \frac{-s/2 - 5/3}{s^2+4s+6} \\ &= \frac{1/6}{s} + \frac{1/3}{s+1} + \frac{(-1/2)(s+2) - 2/3}{(s+2)^2+2} \\ &= \frac{1/6}{s} + \frac{1/3}{s+1} - \frac{1}{2}\frac{s+2}{(s+2)^2+2} - \frac{2}{3}\frac{1}{(s+2)^2+2}. \end{aligned}$$

Finally, from parts (a) and (c) of Theorem 7.3 and the first translation theorem we obtain

$$\begin{aligned} y(t) &= \frac{1}{6}\mathscr{L}^{-1}\left\{\frac{1}{s}\right\} + \frac{1}{3}\mathscr{L}^{-1}\left\{\frac{1}{s+1}\right\} - \frac{1}{2}\mathscr{L}^{-1}\left\{\frac{s+2}{(s+2)^2+2}\right\} \\ &\quad - \frac{2}{3\sqrt{2}}\mathscr{L}^{-1}\left\{\frac{\sqrt{2}}{(s+2)^2+2}\right\} \\ &= \frac{1}{6} + \frac{1}{3}e^{-t} - \frac{1}{2}e^{-2t}\cos\sqrt{2}t - \frac{\sqrt{2}}{3}e^{-2t}\sin\sqrt{2}t. \end{aligned}$$

EXAMPLE 4 Using Theorems 7.3 and 7.7

Solve $x'' + 16x = \cos 4t$, $x(0) = 0$, $x'(0) = 1$.

Solution Recall that this initial-value problem could describe the forced, undamped, and resonant motion of a mass on a spring. The mass starts with an initial velocity of 1 foot per second in the downward direction from the equilibrium position.

Transforming the equation gives

$$(s^2 + 16)X(s) = 1 + \frac{s}{s^2 + 16}$$

$$X(s) = \frac{1}{s^2 + 16} + \frac{s}{(s^2 + 16)^2}.$$

With the aid of part (d) of Theorem 7.3 and Theorem 7.7, we find

$$x(t) = \frac{1}{4}\mathscr{L}^{-1}\left\{\frac{4}{s^2 + 16}\right\} + \frac{1}{8}\mathscr{L}^{-1}\left\{\frac{8s}{(s^2 + 16)^2}\right\}$$

$$= \frac{1}{4}\sin 4t + \frac{1}{8}t\sin 4t.$$

EXAMPLE 5 Using a Unit Step Function

Solve $x'' + 16x = f(t)$, $x(0) = 0$, $x'(0) = 1$,

where
$$f(t) = \begin{cases} \cos 4t, & 0 \le t < \pi \\ 0, & t \ge \pi \end{cases}$$

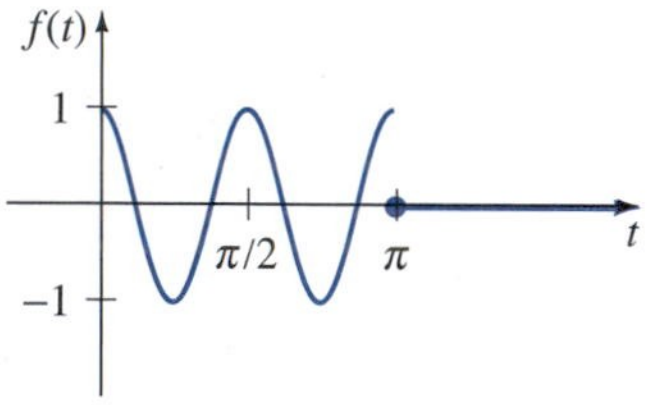

Figure 7.36

Solution The function $f(t)$ can be interpreted as an external force that is acting on a mechanical system for only a short period of time and then is removed. See Figure 7.36. Although this problem could be solved by conventional means, the procedure is not at all convenient when $f(t)$ is defined in a piecewise manner. With the aid of (2) and (3) of Section 7.3 and the periodicity of the cosine, we can rewrite f in terms of the unit step function as

$$f(t) = \cos 4t - \cos 4t\,\mathscr{U}(t - \pi) = \cos 4t - \cos 4(t - \pi)\mathscr{U}(t - \pi).$$

The second translation theorem then yields

$$\mathscr{L}\{x''\} + 16\mathscr{L}\{x\} = \mathscr{L}\{f(t)\}$$

$$s^2X(s) - sx(0) - x'(0) + 16X(s) = \frac{s}{s^2 + 16} - \frac{s}{s^2 + 16}e^{-\pi s}$$

$$(s^2 + 16)X(s) = 1 + \frac{s}{s^2 + 16} - \frac{s}{s^2 + 16}e^{-\pi s}$$

$$X(s) = \frac{1}{s^2 + 16} + \frac{s}{(s^2 + 16)^2} - \frac{s}{(s^2 + 16)^2}e^{-\pi s}.$$

From part (b) of Example 12 in Section 7.3 (with $k = 4$) along with (8) of that section, we find

$$x(t) = \frac{1}{4}\mathscr{L}^{-1}\left\{\frac{4}{s^2 + 16}\right\} + \frac{1}{8}\mathscr{L}^{-1}\left\{\frac{8s}{(s^2 + 16)^2}\right\}$$

$$- \frac{1}{8}\mathscr{L}^{-1}\left\{\frac{8s}{(s^2 + 16)^2}e^{-\pi s}\right\}$$

$$= \frac{1}{4}\sin 4t + \frac{1}{8}t\sin 4t - \frac{1}{8}(t - \pi)\sin 4(t - \pi)\mathscr{U}(t - \pi).$$

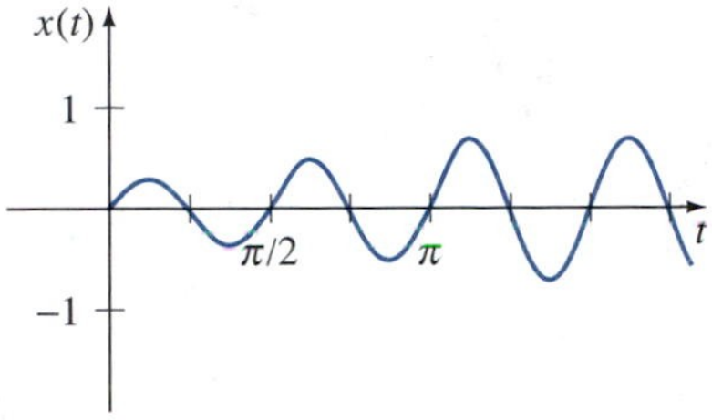

Figure 7.37

The foregoing solution is the same as

$$x(t) = \begin{cases} \dfrac{1}{4}\sin 4t + \dfrac{1}{8}t\sin 4t, & 0 \le t < \pi \\ \dfrac{2+\pi}{8}\sin 4t, & t \ge \pi. \end{cases}$$

Observe from the graph of $x(t)$ in Figure 7.37 that the amplitudes of vibration become steady as soon as the external force is turned off. ■

EXAMPLE 6 Using the Second Translation Theorem

Solve $y'' + 2y' + y = f(t)$, $y(0) = 0$, $y'(0) = 0$,

where $$f(t) = \mathcal{U}(t-1) - 2\mathcal{U}(t-2) + \mathcal{U}(t-3).$$

Solution By the second translation theorem and simplification, the transform of the differential equation is

$$(s+1)^2Y(s) = \frac{e^{-s}}{s} - 2\frac{e^{-2s}}{s} + \frac{e^{-3s}}{s}$$

or $$Y(s) = \frac{e^{-s}}{s(s+1)^2} - 2\frac{e^{-2s}}{s(s+1)^2} + \frac{e^{-3s}}{s(s+1)^2}.$$

With the aid of partial fractions, the last equation becomes

$$Y(s) = \left[\frac{1}{s} - \frac{1}{s+1} - \frac{1}{(s+1)^2}\right]e^{-s} - 2\left[\frac{1}{s} - \frac{1}{s+1} - \frac{1}{(s+1)^2}\right]e^{-2s} + \left[\frac{1}{s} - \frac{1}{s+1} - \frac{1}{(s+1)^2}\right]e^{-3s}.$$

Again using the inverse form of the second translation theorem, we find

$$y(t) = [1 - e^{-(t-1)} - (t-1)e^{-(t-1)}]\mathcal{U}(t-1) - 2[1 - e^{-(t-2)} - (t-2)e^{-(t-2)}]\mathcal{U}(t-2) + [1 - e^{-(t-3)} - (t-3)e^{-(t-3)}]\mathcal{U}(t-3).$$ ■

Volterra Integral Equation The convolution theorem is useful in solving other types of equations in which an unknown function appears under an integral sign. In the next example we solve a **Volterra integral equation**

$$f(t) = g(t) + \int_0^t f(\tau)h(t-\tau)\,d\tau$$

for $f(t)$. The functions $g(t)$ and $h(t)$ are known.

EXAMPLE 7 An Integral Equation

Solve $f(t) = 3t^2 - e^{-t} - \int_0^t f(\tau)e^{t-\tau}\,d\tau$ for $f(t)$.

Solution It follows from Theorem 7.9 that

$$\mathscr{L}\{f(t)\} = 3\mathscr{L}\{t^2\} - \mathscr{L}\{e^{-t}\} - \mathscr{L}\{f(t)\}\mathscr{L}\{e^t\}$$

$$F(s) = 3 \cdot \frac{2}{s^3} - \frac{1}{s+1} - F(s) \cdot \frac{1}{s-1}.$$

Solving the last equation for $F(s)$ gives

$$F(s) = \frac{6(s-1)}{s^4} - \frac{s-1}{s(s+1)}$$

$$= \frac{6}{s^3} - \frac{6}{s^4} + \frac{1}{s} - \frac{2}{s+1}. \qquad \leftarrow \text{termwise division and partial fractions}$$

The inverse transform is

$$f(t) = 3\mathscr{L}^{-1}\left\{\frac{2!}{s^3}\right\} - \mathscr{L}^{-1}\left\{\frac{3!}{s^4}\right\} + \mathscr{L}^{-1}\left\{\frac{1}{s}\right\} - 2\mathscr{L}^{-1}\left\{\frac{1}{s+1}\right\}$$

$$= 3t^2 - t^3 + 1 - 2e^{-t}.$$

Series Circuits In a single loop or series circuit, Kirchhoff's second law states that the sum of the voltage drops across an inductor, resistor, and capacitor is equal to the impressed voltage $E(t)$. Now it is known that

$$\text{the voltage drop across the inductor} = L\frac{di}{dt},$$

$$\text{the voltage drop across the resistor} = Ri(t),$$

and

$$\text{the voltage drop across the capacitor} = \frac{1}{C}\int_0^t i(\tau)\,d\tau,$$

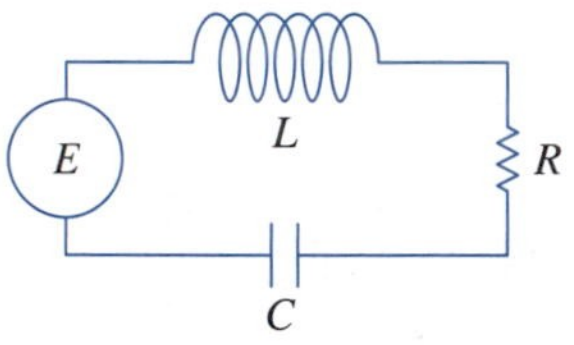

Figure 7.38

where $i(t)$ is the current and L, R, and C are constants. See Section 1.2. It follows that the current in a circuit, such as that shown in Figure 7.38, is governed by the **integrodifferential equation**

$$L\frac{di}{dt} + Ri + \frac{1}{C}\int_0^t i(\tau)\,d\tau = E(t). \tag{3}$$

EXAMPLE 8 An Integrodifferential Equation

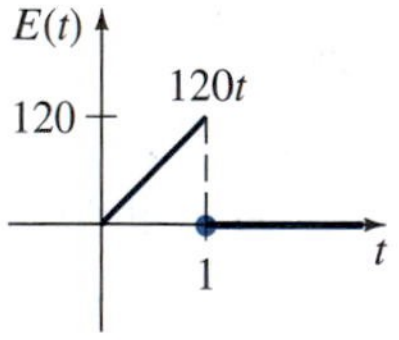

Figure 7.39

Determine the current $i(t)$ in a single loop L-R-C circuit when $L = 0.1$ henry, $R = 20$ ohms, $C = 10^{-3}$ farad, $i(0) = 0$, and the impressed voltage $E(t)$ is as given in Figure 7.39.

Solution Since the voltage is off for $t \geq 1$, we can write

$$E(t) = 120t - 120t\mathscr{U}(t-1). \tag{4}$$

But in order to use the second translation theorem we must rewrite (4) as

$$E(t) = 120t - 120(t-1)\mathscr{U}(t-1) - 120\mathscr{U}(t-1).$$

Equation (3) then becomes

$$0.1\frac{di}{dt} + 20i + 10^3\int_0^t i(\tau)\,d\tau = 120t - 120(t-1)\mathscr{U}(t-1) - 120\mathscr{U}(t-1). \tag{5}$$

Now recall from (5) of Section 7.4 that

$$\mathscr{L}\left\{\int_0^t i(\tau)\,d\tau\right\} = \frac{I(s)}{s},$$

where $I(s) = \mathscr{L}\{i(t)\}$. Thus the transform of equation (5) is

$$0.1sI(s) + 20I(s) + 10^3\frac{I(s)}{s} = 120\left[\frac{1}{s^2} - \frac{1}{s^2}e^{-s} - \frac{1}{s}e^{-s}\right]$$

or, after multiplying by $10s$,

$$(s+100)^2 I(s) = 1200\left[\frac{1}{s} - \frac{1}{s}e^{-s} - e^{-s}\right]$$

$$I(s) = 1200\left[\frac{1}{s(s+100)^2} - \frac{1}{s(s+100)^2}e^{-s} - \frac{1}{(s+100)^2}e^{-s}\right].$$

By partial fractions we can write

$$I(s) = 1200\left[\frac{1/10{,}000}{s} - \frac{1/10{,}000}{s+100} - \frac{1/100}{(s+100)^2} - \frac{1/10{,}000}{s}e^{-s} + \frac{1/10{,}000}{s+100}e^{-s} + \frac{1/100}{(s+100)^2}e^{-s} - \frac{1}{(s+100)^2}e^{-s}\right].$$

Employing the inverse form of the second translation theorem, we obtain

$$i(t) = \frac{3}{25}[1 - \mathscr{U}(t-1)] - \frac{3}{25}[e^{-100t} - e^{-100(t-1)}\mathscr{U}(t-1)] - 12te^{-100t} - 1188(t-1)e^{-100(t-1)}\mathscr{U}(t-1).$$

EXAMPLE 9 A Periodic Impressed Voltage

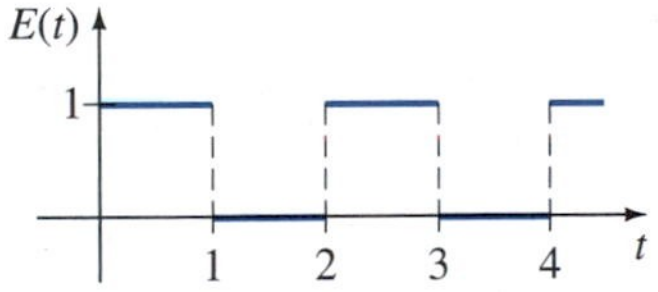

Figure 7.40

The differential equation for the current $i(t)$ in a single loop L-R series circuit is

$$L\frac{di}{dt} + Ri = E(t). \tag{6}$$

Determine the current $i(t)$ when $i(0) = 0$ and $E(t)$ is the square wave function shown in Figure 7.40.

Solution The Laplace transform of the equation is

$$LsI(s) + RI(s) = \mathscr{L}\{E(t)\}. \tag{7}$$

Since $E(t)$ is periodic with period $T = 2$, we use (8) of Section 7.4:

$$\begin{aligned}\mathscr{L}\{E(t)\} &= \frac{1}{1-e^{-2s}}\int_0^2 e^{-st}f(t)\,dt\\ &= \frac{1}{1-e^{-2s}}\left(\int_0^1 1\cdot e^{-st}\,dt + \int_1^2 0\cdot e^{-st}\,dt\right)\\ &= \frac{1}{1-e^{-2s}}\,\frac{1-e^{-s}}{s} \quad \leftarrow 1-e^{-2s} = (1+e^{-s})(1-e^{-s})\\ &= \frac{1}{s(1+e^{-s})}.\end{aligned}$$

Hence from (7) we find

$$I(s) = \frac{1/L}{s(s + R/L)(1 + e^{-s})}. \tag{8}$$

To find the inverse Laplace transform of this function we first make use of a geometric series. Recall, for $|x| < 1$, that

$$\frac{1}{1+x} = 1 - x + x^2 - x^3 + \cdots.$$

With the identification $x = e^{-s}$ we then have for $s > 0$

$$\frac{1}{1+e^{-s}} = 1 - e^{-s} + e^{-2s} - e^{-3s} + \cdots.$$

If we write $$\frac{1}{s(s + R/L)} = \frac{L/R}{s} - \frac{L/R}{s + R/L},$$

(8) becomes

$$I(s) = \frac{1}{R}\left(\frac{1}{s} - \frac{1}{s + R/L}\right)(1 - e^{-s} + e^{-2s} - e^{-3s} + \cdots)$$

$$= \frac{1}{R}\left(\frac{1}{s} - \frac{e^{-s}}{s} + \frac{e^{-2s}}{s} - \frac{e^{-3s}}{s} + \cdots\right) - \frac{1}{R}\left(\frac{1}{s + R/L} - \frac{e^{-s}}{s + R/L} + \frac{e^{-2s}}{s + R/L} - \frac{e^{-3s}}{s + R/L} + \cdots\right).$$

By applying the inverse form of the second translation theorem to each term of both series, we obtain

$$i(t) = \frac{1}{R}(1 - \mathcal{U}(t-1) + \mathcal{U}(t-2) - \mathcal{U}(t-3) + \cdots)$$

$$- \frac{1}{R}(e^{-Rt/L} - e^{-R(t-1)/L}\mathcal{U}(t-1) + e^{-R(t-2)/L}\mathcal{U}(t-2) - e^{-R(t-3)/L}\mathcal{U}(t-3) + \cdots)$$

or, equivalently,

$$i(t) = \frac{1}{R}(1 - e^{-Rt/L}) + \frac{1}{R}\sum_{n=1}^{\infty}(-1)^n(1 - e^{-R(t-n)/L})\mathcal{U}(t-n).$$ ■

To interpret the solution in Example 9 let us assume for the sake of illustration that $R = 1$, $L = 1$, and $0 \le t < 4$. In this case

$$i(t) = 1 - e^{-t} - (1 - e^{t-1})\mathcal{U}(t-1) + (1 - e^{-(t-2)})\mathcal{U}(t-2) - (1 - e^{-(t-3)})\mathcal{U}(t-3);$$

in other words,

$$i(t) = \begin{cases} 1 - e^{-t}, & 0 \le t < 1 \\ -e^{-t} + e^{-(t-1)}, & 1 \le t < 2 \\ 1 - e^{-t} + e^{-(t-1)} - e^{-(t-2)}, & 2 \le t < 3 \\ -e^{-t} + e^{-(t-1)} - e^{-(t-2)} + e^{-(t-3)}, & 3 \le t < 4. \end{cases}$$

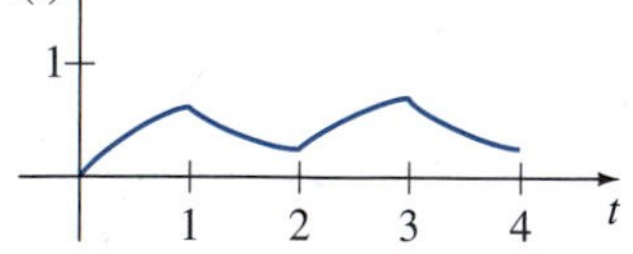

Figure 7.41

The graph of $i(t)$ on the interval $0 \le t < 4$ is given in Figure 7.41.

Beams In Example 9 of Section 1.2 we saw that the static deflection $y(x)$ of a uniform beam of length L carrying a load $w(x)$ per unit length is found from the fourth-order differential equation

$$EI\frac{d^4y}{dx^4} = w(x), \tag{9}$$

where E is Young's modulus of elasticity and I is a moment of inertia of a cross section of the beam. To apply the Laplace transform to (9) we tacitly assume that $w(x)$ and $y(x)$ are defined on $(0, \infty)$ rather than on $(0, L)$. Note too that the next example is a boundary-value problem rather than an initial-value problem.

EXAMPLE 10 A Boundary-Value Problem

A beam of length L is embedded at both ends. See Figure 7.42. In this case the deflection $y(x)$ must satisfy (9) and the conditions

$$y(0) = 0, \quad y(L) = 0, \quad y'(0) = 0, \quad \text{and} \quad y'(L) = 0.$$

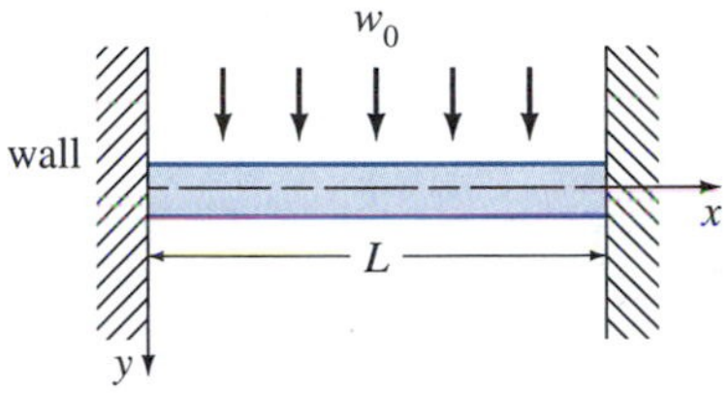

Figure 7.42

The first two conditions indicate that there is no vertical deflection at the ends; the last two conditions mean that the line of deflection is horizontal (zero slope) at the ends. Find the deflection of the beam when a constant load w_0 is uniformly distributed along its length—that is, when $w(x) = w_0$, $0 < x < L$.

Solution Transforming (9) with respect to the variable x gives

$$EI(s^4Y(s) - s^3y(0) - s^2y'(0) - sy''(0) - y'''(0)) = \frac{w_0}{s}$$

$$s^4Y(s) - sy''(0) - y'''(0) = \frac{w_0}{EIs}.$$

If we let $c_1 = y''(0)$ and $c_2 = y'''(0)$, then

$$Y(s) = \frac{c_1}{s^3} + \frac{c_2}{s^4} + \frac{w_0}{EIs^5}$$

and consequently

$$\begin{aligned} y(x) &= \frac{c_1}{2!}\mathscr{L}^{-1}\left\{\frac{2!}{s^3}\right\} + \frac{c_2}{3!}\mathscr{L}^{-1}\left\{\frac{3!}{s^4}\right\} + \frac{w_0}{4!EI}\mathscr{L}^{-1}\left\{\frac{4!}{s^5}\right\} \\ &= \frac{c_1}{2}x^2 + \frac{c_2}{6}x^3 + \frac{w_0}{24EI}x^4. \end{aligned}$$

Applying the given conditions $y(L) = 0$ and $y'(L) = 0$ to the last equation yields the system

$$\frac{c_1}{2}L^2 + \frac{c_2}{6}L^3 + \frac{w_0}{24EI}L^4 = 0$$

$$c_1L + \frac{c_2}{2}L^2 + \frac{w_0}{6EI}L^3 = 0.$$

Solving, we find $c_1 = w_0L^2/12EI$ and $c_2 = -w_0L/2EI$. Thus the deflection is given by

$$y(x) = \frac{w_0L^2}{24EI}x^2 - \frac{w_0L}{12EI}x^3 + \frac{w_0}{24EI}x^4 = \frac{w_0}{24EI}x^2(x - L)^2.$$

EXERCISES 7.5

Answers to odd-numbered problems begin on page A-16.

A table of the transforms of some basic functions is given in Appendix II.

In Problems 1–26 use the Laplace transform to solve the given differential equation subject to the indicated initial conditions. Where appropriate, write f in terms of unit step functions.

1. $\dfrac{dy}{dt} - y = 1, \quad y(0) = 0$

2. $\dfrac{dy}{dt} + 2y = t, \quad y(0) = -1$

3. $y' + 4y = e^{-4t}, \quad y(0) = 2$

4. $y' - y = \sin t, \quad y(0) = 0$

5. $y'' + 5y' + 4y = 0, \quad y(0) = 1, y'(0) = 0$

6. $y'' - 6y' + 13y = 0, \quad y(0) = 0, y'(0) = -3$

7. $y'' - 6y' + 9y = t, \quad y(0) = 0, y'(0) = 1$

8. $y'' - 4y' + 4y = t^3, \quad y(0) = 1, y'(0) = 0$

9. $y'' - 4y' + 4y = t^3e^{2t}, \quad y(0) = 0, y'(0) = 0$

10. $y'' - 2y' + 5y = 1 + t, \quad y(0) = 0, y'(0) = 4$

11. $y'' + y = \sin t, \quad y(0) = 1, y'(0) = -1$

12. $y'' + 16y = 1, \quad y(0) = 1, y'(0) = 2$

13. $y'' - y' = e^t \cos t, \quad y(0) = 0, y'(0) = 0$

14. $y'' - 2y' = e^t \sinh t, \quad y(0) = 0, y'(0) = 0$

15. $2y''' + 3y'' - 3y' - 2y = e^{-t}, \quad y(0) = 0, y'(0) = 0, y''(0) = 1$

16. $y''' + 2y'' - y' - 2y = \sin 3t, \quad y(0) = 0, y'(0) = 0, y''(0) = 1$

17. $y^{(4)} - y = 0, \quad y(0) = 1, y'(0) = 0, y''(0) = -1, y'''(0) = 0$

18. $y^{(4)} - y = t, \quad y(0) = 0, y'(0) = 0, y''(0) = 0, y'''(0) = 0$

19. $y' + y = f(t)$, where $f(t) = \begin{cases} 0, & 0 \le t > 1 \\ 5, & t \ge 1 \end{cases}, y(0) = 0$

20. $y' + y = f(t)$, where $f(t) = \begin{cases} 1, & 0 \le t < 1 \\ -1, & t \ge 1 \end{cases}, y(0) = 0$

21. $y' + 2y = f(t)$, where $f(t) = \begin{cases} t, & 0 \le t < 1 \\ 0, & t \ge 1 \end{cases}, y(0) = 0$

22. $y'' + 4y = f(t)$, where $f(t) = \begin{cases} 1, & 0 \le t < 1 \\ 0, & t \ge 1 \end{cases}, y(0) = 0, y'(0) = -1$

23. $y'' + 4y = \sin t\, \mathcal{U}(t - 2\pi), \quad y(0) = 1, y'(0) = 0$

24. $y'' - 5y' + 6y = \mathcal{U}(t - 1), \quad y(0) = 0, y'(0) = 1$

25. $y'' + y = f(t)$, where $f(t) = \begin{cases} 0, & 0 \le t < \pi \\ 1, & \pi \le t < 2\pi \\ 0, & t \ge 2\pi \end{cases}, y(0) = 0, y'(0) = 1$

26. $y'' + 4y' + 3y = 1 - \mathcal{U}(t - 2) - \mathcal{U}(t - 4) + \mathcal{U}(t - 6),$
$y(0) = 0, y'(0) = 0$

In Problems 27 and 28 use the Laplace transform to solve the given differential equation subject to the indicated boundary conditions.

27. $y'' + 2y' + y = 0, \quad y'(0) = 2, y(1) = 2$

28. $y'' - 9y' + 20y = 1, \quad y(0) = 0, y'(1) = 0$

In Problems 29–38 use the Laplace transform to solve the given integral equation or integrodifferential equation.

29. $f(t) + \int_0^t (t - \tau) f(\tau)\, d\tau = t$

30. $f(t) = 2t - 4\int_0^t \sin \tau f(t - \tau)\, d\tau$

31. $f(t) = te^t + \int_0^t \tau f(t - \tau)\, d\tau$

32. $f(t) + 2\int_0^t f(\tau) \cos(t - \tau)\, d\tau = 4e^{-t} + \sin t$

33. $f(t) + \int_0^t f(\tau)\, d\tau = 1$

34. $f(t) = \cos t + \int_0^t e^{-\tau} f(t - \tau)\, d\tau$

35. $f(t) = 1 + t - \frac{8}{3}\int_0^t (\tau - t)^3 f(\tau)\, d\tau$

36. $t - 2f(t) = \int_0^t (e^{\tau} - e^{-\tau}) f(t - \tau)\, d\tau$

37. $y'(t) = 1 - \sin t - \int_0^t y(\tau)\, d\tau, \quad y(0) = 0$

38. $\frac{dy}{dt} + 6y(t) + 9\int_0^t y(\tau)\, d\tau = 1, \quad y(0) = 0$

39. Use equation (3) to determine the current $i(t)$ in a single loop L-R-C circuit when $L = 0.005$ henry, $R = 1$ ohm, $C = 0.02$ farad, $E(t) = 100[1 - \mathcal{U}(t - 1)]$ volts, and $i(0) = 0$.

40. Solve Problem 39 when $E(t) = 100[t - (t - 1)\mathcal{U}(t - 1)]$.

41. Recall that the differential equation for the charge $q(t)$ on the capacitor in an R-C series circuit is

$$R\frac{dq}{dt} + \frac{1}{C}q = E(t),$$

where $E(t)$ is the impressed voltage. See Section 3.2. Use the Laplace transform to determine the charge $q(t)$ when $q(0) = 0$ and $E(t) = E_0 e^{-kt}$, $k > 0$. Consider two cases: $k \neq 1/RC$ and $k = 1/RC$.

42. Use the Laplace transform to determine the charge on the capacitor in an R-C series circuit if $q(0) = q_0$, $R = 10$ ohms, $C = 0.1$ farad, and $E(t)$ is as given in Figure 7.43.

43. Use the Laplace transform to determine the charge on the capacitor in an R-C series circuit if $q(0) = 0$, $R = 2.5$ ohms, $C = 0.08$ farad, and $E(t)$ is as given in Figure 7.44.

44. Use the Laplace transform to determine the charge $q(t)$ on the capacitor in an R-C series circuit when $q(0) = 0$, $R = 50$ ohms, $C = 0.01$ farad, and $E(t)$ is as given in Figure 7.45.

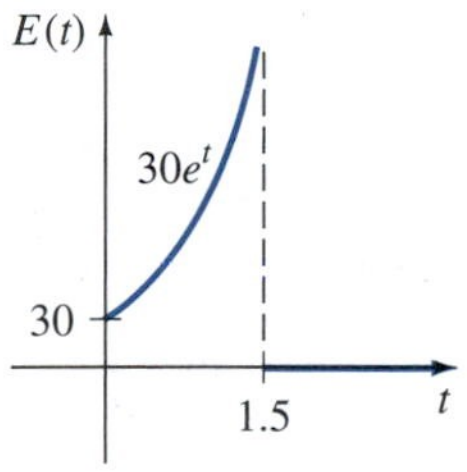

Figure 7.43

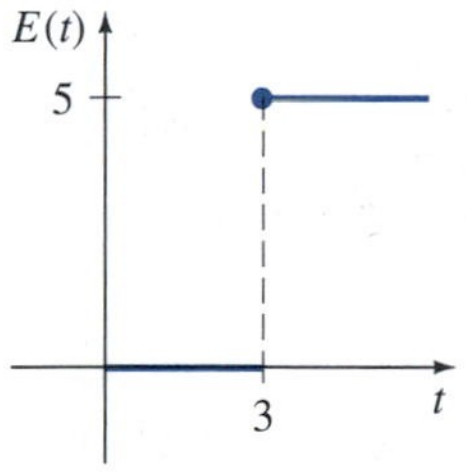

Figure 7.44

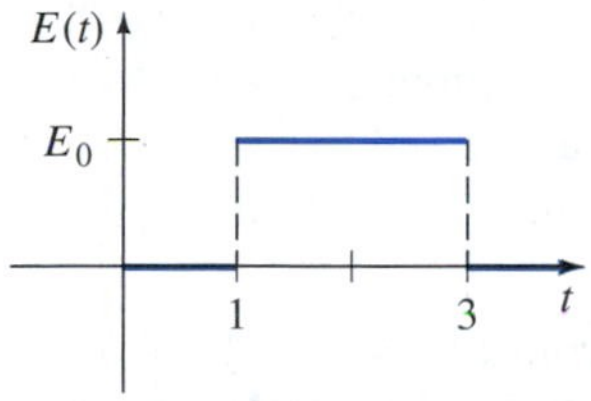

Figure 7.45

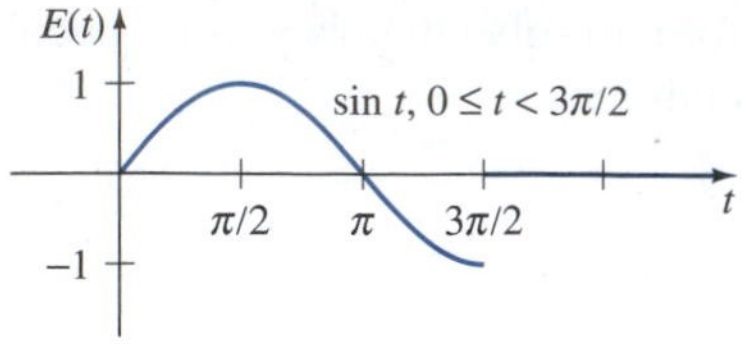

Figure 7.46

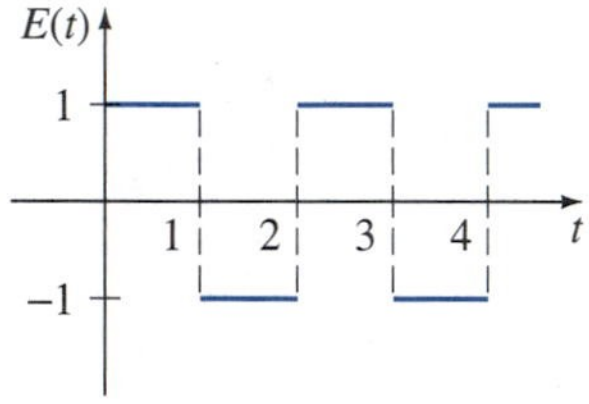

Figure 7.47

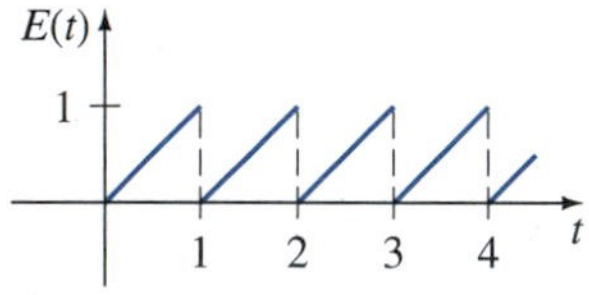

Figure 7.48

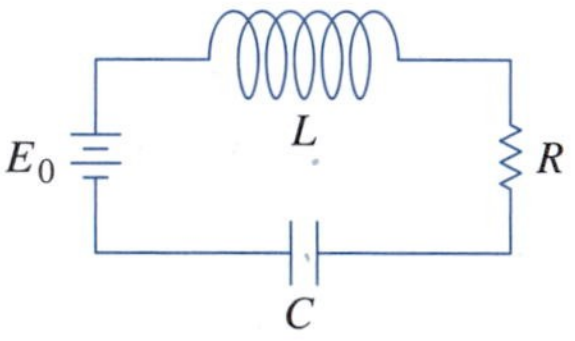

Figure 7.49

45. Use the Laplace transform to determine the current $i(t)$ in a single loop L-R series circuit when $i(0) = 0$, $L = 1$ henry, $R = 10$ ohms, and $E(t)$ is as given in Figure 7.46.

46. Solve equation (6) subject to $i(0) = 0$, where $E(t)$ is as given in Figure 7.47. [*Hint*: See Problem 31, Exercises 7.4.]

47. Solve equation (6) subject to $i(0) = 0$, where $E(t)$ is as given in Figure 7.48. Specify the solution for $0 \le t < 2$. [*Hint*: See Problem 33, Exercises 7.4.]

48. Recall that the differential equation for the instantaneous charge $q(t)$ on the capacitor in an L-R-C series circuit is given by

$$L\frac{d^2q}{dt^2} + R\frac{dq}{dt} + \frac{1}{C}q = E(t). \tag{10}$$

See Section 5.4. Use the Laplace transform to determine $q(t)$ when $L = 1$ henry, $R = 20$ ohms, $C = 0.005$ farad, $E(t) = 150$ volts, $t > 0$, and $q(0) = 0$, $i(0) = 0$. What is the current $i(t)$? What is the charge $q(t)$ if the same constant voltage is turned off for $t \ge 2$?

49. Determine the charge $q(t)$ and current $i(t)$ for a series circuit in which $L = 1$ henry, $R = 20$ ohms, $C = 0.01$ farad, $E(t) = 120 \sin 10t$ volts, $q(0) = 0$, and $i(0) = 0$. What is the steady-state current?

50. Consider the battery of constant voltage E_0 that charges the capacitor shown in Figure 7.49. If we divide by L and define $\lambda = R/2L$ and $\omega^2 = 1/LC$, then (10) becomes

$$\frac{d^2q}{dt^2} + 2\lambda\frac{dq}{dt} + \omega^2 q = \frac{E_0}{L}.$$

Use the Laplace transform to show that the solution of this equation, subject to $q(0) = 0$ and $i(0)$, is

$$q(t) = \begin{cases} E_0C\left[1 - e^{-\lambda t}\left(\cosh\sqrt{\lambda^2 - \omega^2}\,t + \dfrac{\lambda}{\sqrt{\lambda^2 - \omega^2}}\sinh\sqrt{\lambda^2 - \omega^2}\,t\right)\right], & \lambda > \omega \\ E_0C\left[1 - e^{-\lambda t}(1 + \lambda t)\right], & \lambda = \omega \\ E_0C\left[1 - e^{-\lambda t}\left(\cos\sqrt{\omega^2 - \lambda^2}\,t + \dfrac{\lambda}{\sqrt{\omega^2 - \lambda^2}}\sin\sqrt{\omega^2 - \lambda^2}\,t\right)\right], & \lambda < \omega \end{cases}$$

51. Use the Laplace transform to determine the charge $q(t)$ on the capacitor in an L-C series circuit when $q(0) = 0$, $i(0) = 0$, and $E(t) = E_0e^{-kt}$, $k > 0$.

52. Suppose a 32-lb weight stretches a spring 2 ft. If the weight is released from rest at the equilibrium position, determine the equation of motion if an impressed force $f(t) = \sin t$ acts on the system for $0 \le t < 2\pi$ and is then removed. Ignore any damping forces. [*Hint*: Write the impressed force in terms of the unit step function.]

53. A 4-lb weight stretches a spring 2 ft. The weight is released from rest 18 in. above the equilibrium position, and the resulting motion takes place in a medium offering a damping force numerically equal to $\frac{7}{8}$ times the instantaneous velocity. Use the Laplace transform to determine the equation of motion.

54. A 16-lb weight is attached to a spring whose constant is $k = 4.5$ lb/ft. Beginning at $t = 0$, a force equal to $f(t) = 4\sin 3t + 2\cos 3t$ acts on the system. Assuming that no damping forces are present, use the

Laplace transform to find the equation of motion if the weight is released from rest from the equilibrium position.

55. For a cantilever beam embedded at its left end ($x = 0$) and free at its right end ($x = L$), the deflection $y(x)$ must satisfy (9) and

$$y(0) = 0, \quad y'(0) = 0, \quad y''(L) = 0, \quad y'''(L) = 0. \qquad \textbf{(11)}$$

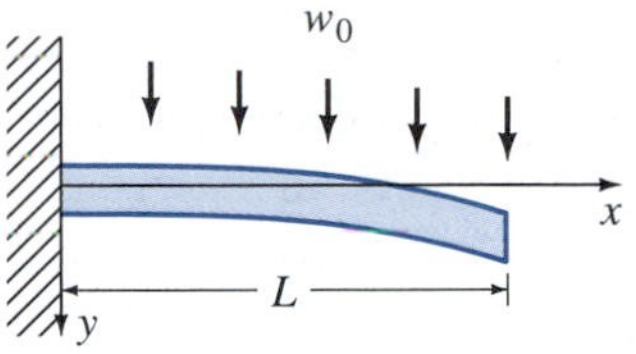

Figure 7.50

The first two conditions state that the deflection and slope are zero at $x = 0$, and the last two conditions state that the bending moment and shear force are zero at $x = L$. Use the Laplace transform to solve equation (9) subject to (11) when a constant load w_0 is uniformly distributed along the length of the beam. See Figure 7.50. Find the deflection at the midpoint of the beam. Find the maximum deflection of the beam.

56. Solve Problem 55 when the load is given by

$$w(x) = \begin{cases} 0, & 0 < x < L/3 \\ w_0, & L/3 < x < 2L/3 \\ 0, & 2L/3 < x < L \end{cases}$$

Write $w(x)$ in terms of unit step functions.

57. Solve Problem 55 when the load is given by

$$w(x) = \begin{cases} w_0, & 0 < x < L/2 \\ 0, & L/2 < x < L \end{cases}$$

58. The static deflection $y(x)$ of a beam that is hinged at both ends must satisfy the differential equation (9) and the conditions

$$y(0) = 0, \quad y''(0) = 0, \quad y(L) = 0, \quad y''(L) = 0. \qquad \textbf{(12)}$$

Use the Laplace transform to solve (9) subject to (12) when $w(x) = w_0$, $0 < x < L$. See Figure 7.51.

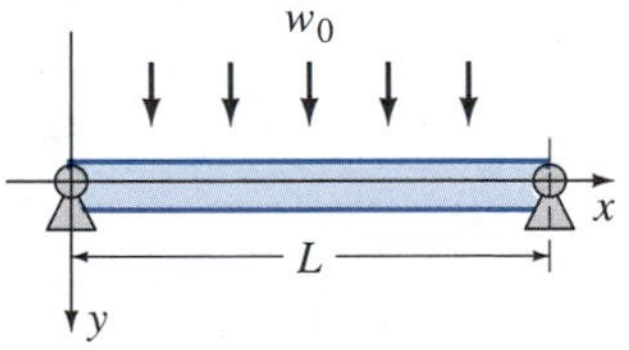

Figure 7.51

In Problems 59 and 60 use the Laplace transform and Theorem 7.7 to find a solution of the given equation.

59. $ty'' - y' = t^2, \quad y(0) = 0$ **60.** $ty'' + 2ty' + 2y = 0, \quad y(0) = 0$

61. In this problem we show how the convolution integral can be used to find a solution of an initial-value problem of the type

$$ay'' + by' + cy = g(t), \quad y(0) = 0, \quad y'(0) = 0. \qquad \textbf{(13)}$$

(a) Show that the solution $y_1(t)$ of the initial-value problem

$$ay'' + by' + cy = 0, \quad y(0) = 0, \quad y'(0) = 1,$$

is
$$y_1(t) = \mathscr{L}^{-1}\left\{\frac{a}{as^2 + bs + c}\right\}.$$

(b) Use the result in part (a) to show that the solution $y_2(t)$ of the initial-value problem in (13) is given by

$$y_2(t) = \frac{1}{a} g * y_1.$$

62. Use the procedure outlined in Problem 61 to find a solution of the initial-value problem

$$y'' + y = \sec t, \quad y(0) = 0, \quad y'(0) = 0.$$

7.6 DIRAC DELTA FUNCTION

• Unit impulse • Dirac delta function • Laplace transform of delta function

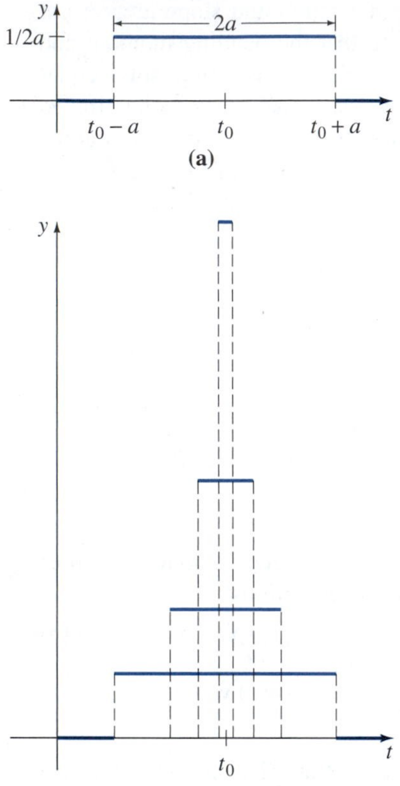

(b) behavior of δ_a as $a \to 0$

Figure 7.52

Unit Impulse Mechanical systems are often acted upon by an external force (or emf in an electrical circuit) of large magnitude that acts only for a very short period of time. For example, a vibrating airplane wing could be struck by lightning, a mass on a spring could be given a sharp blow by a ball peen hammer, a ball (baseball, golf ball, tennis ball) could be sent soaring when struck violently by some kind of club (baseball bat, golf club, tennis racket). The function

$$\delta_a(t - t_0) = \begin{cases} 0, & 0 \le t < t_0 - a \\ \dfrac{1}{2a}, & t_0 - a \le t < t_0 + a \\ 0, & t \ge t_0 + a \end{cases} \tag{1}$$

$a > 0$, $t_0 > 0$, shown in Figure 7.52(a), could serve as a mathematical model for such a force. For a small value of a, $\delta_a(t - t_0)$ is essentially a constant function of large magnitude that is "on" for just a very short period of time around t_0. The behavior of $\delta_a(t - t_0)$ as $a \to 0$ is illustrated in Figure 7.52(b). The function $\delta_a(t - t_0)$ is called a **unit impulse** since it possesses the integration property

$$\int_0^\infty \delta_a(t - t_0)\, dt = 1.$$

Dirac Delta Function In practice it is convenient to work with another type of unit impulse, a "function" that approximates $\delta_a(t - t_0)$ and is defined by the limit

$$\delta(t - t_0) = \lim_{a \to 0} \delta_a(t - t_0). \tag{2}$$

The latter expression, which is not a function at all, can be characterized by the two properties

$$(i)\ \ \delta(t - t_0) = \begin{cases} \infty, & t = t_0 \\ 0, & t \ne t_0 \end{cases}, \quad \text{and} \quad (ii)\ \ \int_0^\infty \delta(t - t_0)\, dt = 1.$$

The expression $\delta(t - t_0)$ is called the **Dirac delta function.***

It is possible to obtain the Laplace transform of the Dirac delta function by the formal assumption that

$$\mathscr{L}\{\delta(t - t_0)\} = \lim_{a \to 0} \mathscr{L}\{\delta_a(t - t_0)\}.$$

---

* PAUL ADRIAN MAURICE DIRAC (1902–1984) The delta function was the invention of the contemporary British physicist P. A. M. Dirac. Along with Max Planck, Werner Heisenberg, Erwin Schrödinger, and Albert Einstein, Dirac was one of the founding fathers, in the era 1900–1930, of a new way of describing the behavior of atoms, molecules, and elementary particles called *quantum mechanics.* For their pioneering work in this field, Dirac and Schrödinger shared the 1933 Nobel Prize in physics. The Dirac delta function was used extensively throughout his 1932 classic treatise, *The Principles of Quantum Mechanics.*

THEOREM 7.11 Transform of the Dirac Delta Function

For $t_0 > 0$,

$$\mathscr{L}\{\delta(t - t_0)\} = e^{-st_0}. \tag{3}$$

Proof To begin, we can write $\delta_a(t - t_0)$ in terms of the unit step function by virtue of (4) and (5) of Section 7.3:

$$\delta_a(t - t_0) = \frac{1}{2a}[\mathscr{U}(t - (t_0 - a)) - \mathscr{U}(t - (t_0 + a))].$$

By linearity and (7) of Section 7.3 the Laplace transform of this last expression is

$$\mathscr{L}\{\delta_a(t - t_0)\} = \frac{1}{2a}\left[\frac{e^{-s(t_0-a)}}{s} - \frac{e^{-s(t_0+a)}}{s}\right]$$

or

$$\mathscr{L}\{\delta_a(t - t_0)\} = e^{-st_0}\left(\frac{e^{sa} - e^{-sa}}{2sa}\right). \tag{4}$$

Since (4) has the indeterminate form 0/0 as $a \to 0$, we apply L'Hôpital's rule:

$$\mathscr{L}\{\delta(t - t_0)\} = \lim_{a\to 0} \mathscr{L}\{\delta_a(t - t_0)\} = e^{-st_0} \lim_{a\to 0}\left(\frac{e^{sa} - e^{-sa}}{2sa}\right) = e^{-st_0}. \quad \blacksquare$$

Now when $t_0 = 0$, it seems plausible to conclude from (3) that

$$\mathscr{L}\{\delta(t)\} = 1.$$

The last result emphasizes the fact that $\delta(t)$ is not the usual type of function that we have been considering, since we expect from Theorem 7.4 that $\mathscr{L}\{f(t)\} \to 0$ as $s \to \infty$.

EXAMPLE 1 Two Initial-Value Problems

Solve $y'' + y = 4\delta(t - 2\pi)$ subject to

(a) $y(0) = 1, y'(0) = 0$ and **(b)** $y(0) = 0, y'(0) = 0$.

The two initial-value problems could serve as models for describing the motion of a mass on a spring moving in a medium in which damping is negligible. At $t = 2\pi$ seconds the mass is given a sharp blow. In (a) the mass is released from rest 1 unit below the equilibrium position. In (b) the mass is at rest in the equilibrium position.

Solution

(a) From (3) the Laplace transform of the differential equation is

$$s^2Y(s) - s + Y(s) = 4e^{-2\pi s} \quad \text{or} \quad Y(s) = \frac{s}{s^2 + 1} + \frac{4e^{-2\pi s}}{s^2 + 1}.$$

Utilizing the inverse form of the second translation theorem, we find

$$y(t) = \cos t + 4\sin(t - 2\pi)\mathscr{U}(t - 2\pi).$$

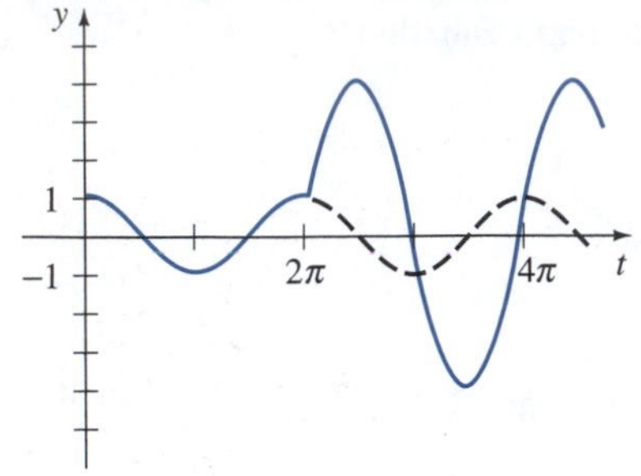

Figure 7.53

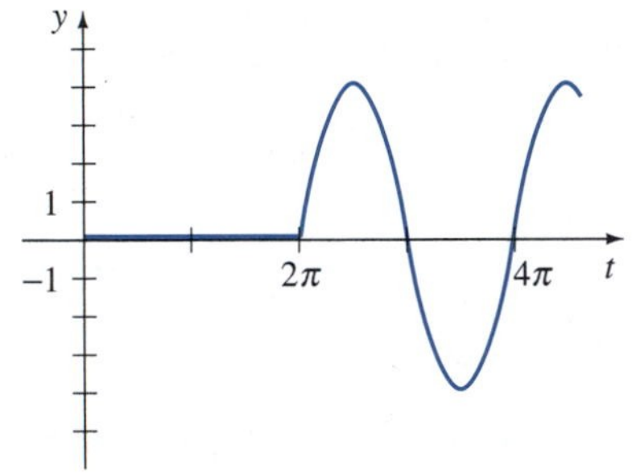

Figure 7.54

Since $\sin(t - 2\pi) = \sin t$, the foregoing solution can be written as

$$y(t) = \begin{cases} \cos t, & 0 \le t < 2\pi \\ \cos t + 4 \sin t, & t \ge 2\pi. \end{cases} \tag{5}$$

In Figure 7.53 we see from the graph of (5) that the mass is exhibiting simple harmonic motion until it is struck at $t = 2\pi$. The influence of the unit impulse is to increase the amplitude of vibration to $\sqrt{17}$ for $t > 2\pi$.

(b) In this case the transform of the equation is simply

$$Y(s) = \frac{4e^{-2\pi s}}{s^2 + 1}$$

and so

$$\begin{aligned} y(t) &= 4 \sin(t - 2\pi)\mathcal{U}(t - 2\pi) \\ &= \begin{cases} 0, & 0 \le t < 2\pi \\ 4 \sin t, & t \ge 2\pi. \end{cases} \end{aligned} \tag{6}$$

The graph of (6) in Figure 7.54 shows, as we would expect from the initial conditions, that the mass exhibits no motion until it is struck at $t = 2\pi$. ■

Remarks If $\delta(t - t_0)$ were a function in the usual sense, then property (*i*) on page 328 would imply $\int_0^\infty \delta(t - t_0)\, dt = 0$ rather than $\int_0^\infty \delta(t - t_0)\, dt = 1$. Since the Dirac delta function did not "behave" like an ordinary function, even though its users produced correct results, it was met initially with great scorn by mathematicians. However, in the 1940s Dirac's controversial function was put on a rigorous footing by the French mathematician Laurent Schwartz in his book *La Théorie de distribution,* and this, in turn, led to an entirely new branch of mathematics known as the *theory of distributions,* or *generalized functions.* In the modern theory of generalized functions, (2) is not an accepted definition of $\delta(t - t_0)$, nor does one speak of a function whose values are either ∞ or 0. Although this theory is much beyond the level of this text, it suffices for our purposes to say that the Dirac delta function is defined in terms of its effect or action on other functions. To see this let us suppose that f is a continuous function on $[0, \infty)$. Then, by the mean value theorem for integrals, it follows that

$$\int_0^\infty f(t)\delta_a(t - t_0)\, dt = \int_{t_0 - a}^{t_0 + a} f(t)\left(\frac{1}{2a}\right) dt = \frac{1}{2a}(2af(c)) = f(c),$$

where c is some number in the interval $t_0 - a < t < t_0 + a$. As $a \to 0$ we must have $c \to t_0$, so

$$\int_0^\infty f(t)\delta(t - t_0)\, dt = \lim_{a \to 0} \int_0^\infty f(t)\delta_a(t - t_0)\, dt = \lim_{a \to 0} f(c)$$

implies

$$\int_0^\infty f(t)\delta(t - t_0)\, dt = f(t_0). \tag{7}$$

Although we have used the intuitive definition (2) to arrive at (7), the result is nevertheless valid and can be obtained in a rigorous fashion. The result in (7) can be taken as the *definition* of $\delta(t - t_0)$. It is known as the *sifting property* since $\delta(t - t_0)$ has the effect of sifting the value $f(t_0)$ from the values of f. Note that property (*ii*) on page 328 is consistent with (7) when $f(t) = 1$, $0 \leq t < \infty$. The integral operation (7) that corresponds a number $f(t_0)$ with a function f leads to the notion of a *linear functional.* We stop at this point and urge the curious reader to consult an advanced text.*

EXERCISES 7.6

Answers to odd-numbered problems begin on page A-17.

In Problems 1–12 use the Laplace transform to solve the given differential equation subject to the indicated initial conditions.

1. $y' - 3y = \delta(t - 2), \quad y(0) = 0$
2. $y' + y = \delta(t - 1), \quad y(0) = 2$
3. $y'' + y = \delta(t - 2\pi), \quad y(0) = 0, y'(0) = 1$
4. $y'' + 16y = \delta(t - 2\pi), \quad y(0) = 0, y'(0) = 0$
5. $y'' + y = \delta\left(t - \dfrac{\pi}{2}\right) + \delta\left(t - \dfrac{3\pi}{2}\right), \quad y(0) = 0, y'(0) = 0$
6. $y'' + y = \delta(t - 2\pi) + \delta(t - 4\pi), \quad y(0) = 1, y'(0) = 0$
7. $y'' + 2y' = \delta(t - 1), \quad y(0) = 0, y'(0) = 1$
8. $y'' - 2y' = 1 + \delta(t - 2), \quad y(0) = 0, y'(0) = 1$
9. $y'' + 4y' + 5y = \delta(t - 2\pi), \quad y(0) = 0, y'(0) = 0$
10. $y'' + 2y' + y = \delta(t - 1), \quad y(0) = 0, y'(0) = 0$
11. $y'' + 4y' + 13y = \delta(t - \pi) + \delta(t - 3\pi), \quad y(0) = 1, y'(0) = 0$
12. $y'' - 7y' + 6y = e^t + \delta(t - 2) + \delta(t - 4), \quad y(0) = 0, y'(0) = 0$
13. A uniform beam of length L carries a concentrated load w_0 at $x = L/2$. The beam is embedded at its left end and is free at its right end. Use the Laplace transform to determine the deflection $y(x)$ from

$$EI\frac{d^4y}{dx^4} = w_0\delta\left(x - \frac{L}{2}\right),$$

where $y(0) = 0$, $y'(0) = 0$, $y''(L) = 0$, and $y'''(L) = 0$.

14. Solve the differential equation in Problem 13 subject to $y(0) = 0$, $y'(0) = 0$, $y(L) = 0$, $y'(L) = 0$. In this case the beam is embedded at both ends. See Figure 7.55.

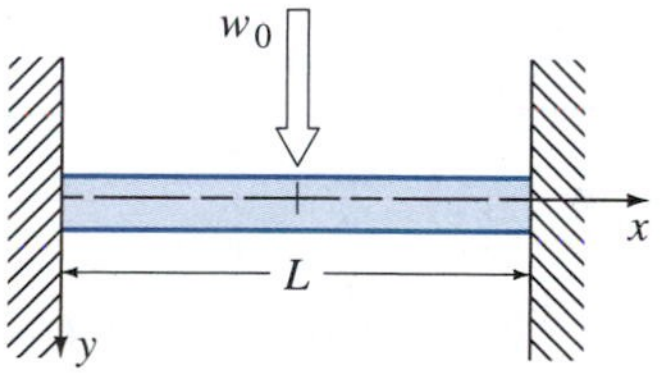

Figure 7.55

15. Use (7) to obtain (3).
16. Use (7) to evaluate $\int_0^\infty t^2 e^{-3t}\delta(t - 4)\,dt$.

* See M. J. Lighthill, *Introduction to Fourier Analysis and Generalized Functions* (New York: Cambridge University Press, 1958).

17. Use the Laplace transform and (7) to solve

$$y'' + 2y' + 2y = \cos t\,\delta(t - 3\pi)$$

subject to $y(0) = 1$ and $y'(0) = -1$.

18. To emphasize the unusual nature of the Dirac delta function, show that the "solution" of the initial-value problem $y'' + \omega^2 y = \delta(t)$, $y(0) = 0$, $y'(0) = 0$ does *not* satisfy the initial condition $y'(0) = 0$.

19. Solve the initial-value problem

$$L\frac{di}{dt} + Ri = \delta(t), \quad i(0) = 0,$$

where L and R are constants. Does the solution satisfy the condition at $t = 0$?

20. If $\delta'(t - t_0)$ denotes the derivative of the Dirac delta function, then it is known that $\mathscr{L}\{\delta'(t - t_0)\} = se^{-st_0}$, $t_0 \geq 0$. Use this result to solve $y' + 5y = \delta'(t)$ subject to $y(0) = 0$.

CHAPTER 7 REVIEW

The **Laplace transform** of a function $f(t)$, $t \geq 0$ is defined by the integral

$$\mathscr{L}\{f(t)\} = \int_0^\infty e^{-st} f(t)\,dt = F(s).$$

The parameter s is usually restricted in such a manner that convergence of the integral is guaranteed. When it is applied to a linear differential equation with constant coefficients such as $ay'' + by' + cy = g(t)$, there results an algebraic equation

$$a[s^2Y(s) - sy(0) - y'(0)] + b[sY(s) - y(0)] + cY(s) = G(s),$$

which depends on the initial conditions $y(0)$ and $y'(0)$. When these values are known, we determine $y(t)$ by evaluating $y(t) = \mathscr{L}^{-1}\{Y(s)\}$.

CHAPTER 7 REVIEW EXERCISES

Answers to odd-numbered problems begin on page A-17.

In Problems 1 and 2 use the definition of the Laplace transform to find $\mathscr{L}\{f(t)\}$.

1. $f(t) = \begin{cases} t, & 0 \leq t < 1 \\ 2 - t, & t \geq 1 \end{cases}$

2. $f(t) = \begin{cases} 0, & 0 \leq t < 2 \\ 1, & 2 \leq t < 4 \\ 0, & t \geq 4 \end{cases}$

In Problems 3–24 fill in the blanks or answer true or false.

3. If f is not piecewise continuous on $[0, \infty)$, then $\mathscr{L}\{f(t)\}$ will not exist. ______

4. The function $f(t) = (e^t)^{10}$ is not of exponential order. ______

5. $F(s) = s^2/(s^2 + 4)$ is not the Laplace transform of a function that is piecewise continuous and of exponential order. ________

6. If $\mathscr{L}\{f(t)\} = F(s)$ and $\mathscr{L}\{g(t)\} = G(s)$, then $\mathscr{L}^{-1}\{F(s)G(s)\} = f(t)g(t)$. ________

7. $\mathscr{L}\{e^{-7t}\} =$ ________

8. $\mathscr{L}\{te^{-7t}\} =$ ________

9. $\mathscr{L}\{\sin 2t\} =$ ________

10. $\mathscr{L}\{e^{-3t}\sin 2t\} =$ ________

11. $\mathscr{L}\{t\sin 2t\} =$ ________

12. $\mathscr{L}\{\sin 2t\,\mathscr{U}(t - \pi)\} =$ ________

13. $\mathscr{L}^{-1}\left\{\dfrac{20}{s^6}\right\} =$ ________

14. $\mathscr{L}^{-1}\left\{\dfrac{1}{3s - 1}\right\} =$ ________

15. $\mathscr{L}^{-1}\left\{\dfrac{1}{(s - 5)^3}\right\} =$ ________

16. $\mathscr{L}^{-1}\left\{\dfrac{1}{s^2 - 5}\right\} =$ ________

17. $\mathscr{L}^{-1}\left\{\dfrac{s}{s^2 - 10s + 29}\right\} =$ ________

18. $\mathscr{L}^{-1}\left\{\dfrac{e^{-5s}}{s^2}\right\} =$ ________

19. $\mathscr{L}^{-1}\left\{\dfrac{s + \pi}{s^2 + \pi^2}e^{-s}\right\} =$ ________

20. $\mathscr{L}^{-1}\left\{\dfrac{1}{L^2s^2 + n^2\pi^2}\right\} =$ ________

21. $\mathscr{L}\{e^{-5t}\}$ exists for $s >$ ________.

22. If $\mathscr{L}\{f(t)\} = F(s)$, then $\mathscr{L}\{te^{8t}f(t)\} =$ ________.

23. If $\mathscr{L}\{f(t)\} = F(s)$ and $k > 0$, then $\mathscr{L}\{e^{a(t-k)}f(t-k)\mathscr{U}(t-k)\} =$ ________.

24. $1 * 1 =$ ________

In Problems 25–28, (**a**) express f in terms of unit step functions, (**b**) find $\mathscr{L}\{f(t)\}$, and (**c**) find $\mathscr{L}\{e^t f(t)\}$.

25.
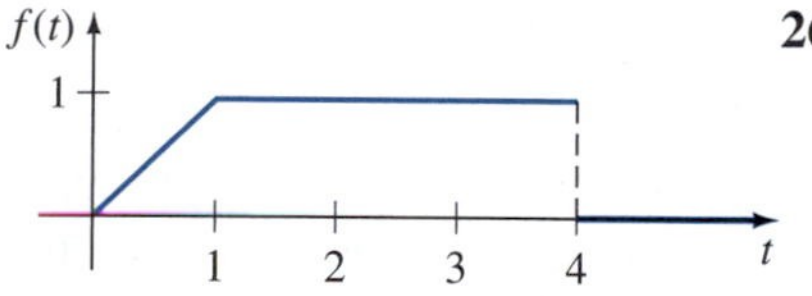

Figure 7.56

26.
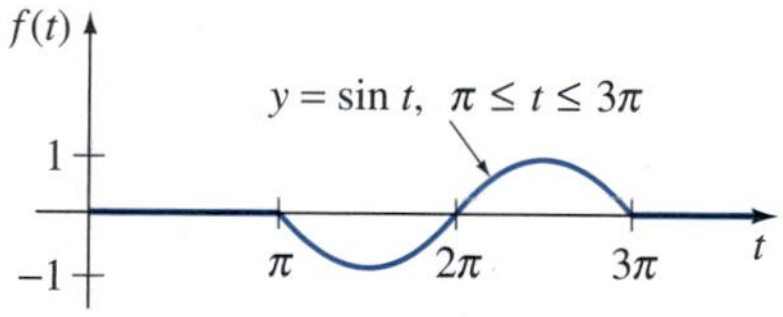

Figure 7.57

27.
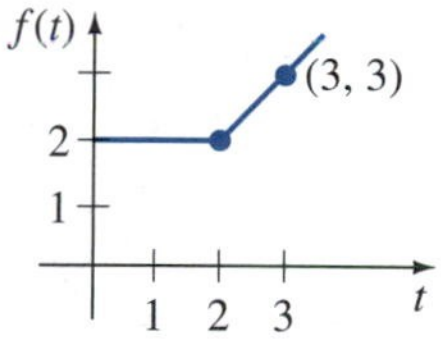

Figure 7.58

28.
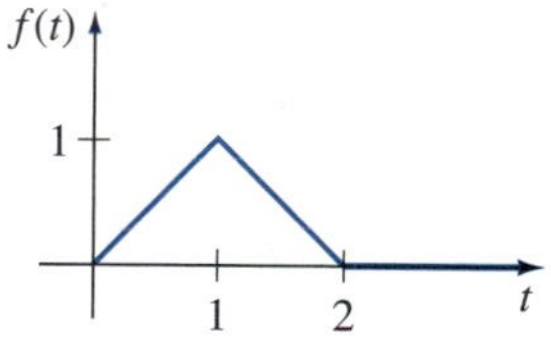

Figure 7.59

In Problems 29–36 use the Laplace transform to solve the given equation.

29. $y'' - 2y' + y = e^t, \quad y(0) = 0, y'(0) = 5$

30. $y'' - 8y' + 20y = te^t, \quad y(0) = 0, y'(0) = 0$

31. $y'' - 4y' + 6y = 30\mathscr{U}(t - \pi), \quad y(0) = 0, y'(0) = 0$

32. $y'' + 6y' + 5y = t - t\,\mathscr{U}(t - 2), \quad y(0) = 1, y'(0) = 0$

33. $y' - 5y = f(t)$, where $f(t) = \begin{cases} t^2, & 0 \le t < 1 \\ 0, & t \ge 1 \end{cases}, \quad y(0) = 1$

34. $f(t) = 1 - 2\int_0^t e^{-3\tau} f(t - \tau)\, d\tau$

35. $y'(t) = \cos t + \int_0^t y(\tau)\cos(t - \tau)\, d\tau, \quad y(0) = 1$

36. $\int_0^t f(\tau) f(t - \tau)\, d\tau = 6t^3$

37. The current $i(t)$ in an R-C series circuit can be determined from the integral equation

$$Ri + \frac{1}{C}\int_0^t i(\tau)\, d\tau = E(t),$$

where $E(t)$ is the impressed voltage. Determine $i(t)$ when $R = 10$ ohms, $C = 0.5$ farad, and $E(t) = 2(t^2 + t)$.

38. A series circuit contains an inductor, a resistor, and a capacitor for which $L = \frac{1}{2}$ henry, $R = 10$ ohms, and $C = 0.01$ farad, respectively. The voltage

$$E(t) = \begin{cases} 10, & 0 \le t < 5 \\ 0, & t \ge 5 \end{cases}$$

is applied to the circuit. Determine the instantaneous charge $q(t)$ on the capacitor for $t > 0$ if $q(0) = 0$ and $q'(0) = 0$.

39. A uniform cantilever beam of length L is embedded at its left end ($x = 0$) and free at its right end. Find the deflection $y(x)$ if the load per unit length is given by

$$w(x) = \frac{2w_0}{L}\left[\frac{L}{2} - x + \left(x - \frac{L}{2}\right)\mathcal{U}\left(x - \frac{L}{2}\right)\right].$$

40. When a uniform beam is supported by an elastic foundation, the differential equation for its deflection $y(x)$ is

$$\frac{d^4y}{dx^4} + 4a^4y = \frac{w(x)}{EI},$$

where a is a constant. In the case when $a = 1$, find the deflection $y(x)$ of an elastically supported beam of length π that is embedded in concrete at both ends when a concentrated load w_0 is applied at $x = \pi/2$. [*Hint*: Use the table of Laplace transforms in Appendix II.]

8 SYSTEMS OF LINEAR DIFFERENTIAL EQUATIONS

INTRODUCTION

In applications considered in previous chapters, a physical system could be described by a single differential equation. For example, in Chapter 5 we saw that the mathematical model for the motion of a mass attached to a spring or for the response of a series electrical circuit was a relatively simple differential equation of the form $ay'' + by' + cy = f(t)$. However, if we attached two (or more) springs together or formed a parallel circuit or network, as shown in the accompanying figures, we would need two (or more) coupled, or simultaneous, differential equations to describe the motion of the masses or the response of the network.

In this chapter we confine our attention to the theory and solution of systems of simultaneous linear differential equations or, simply, linear systems in which all the coefficients are constants.

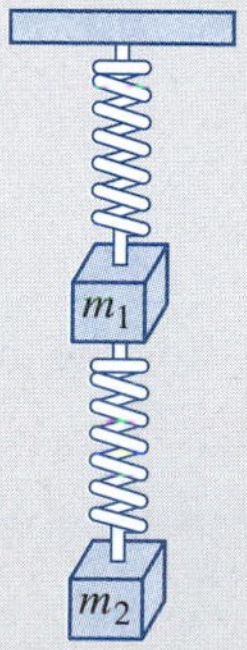

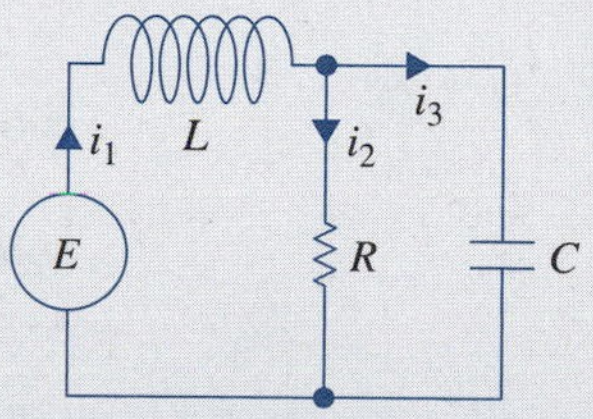

8.1 OPERATOR METHOD

• Systematic elimination • Operator notation • Use of determinants

Simultaneous ordinary differential equations involve two or more equations that contain derivatives of two or more unknown functions of a single independent variable. If x, y, and z are functions of the variable t, then

$$\begin{aligned} 4\frac{d^2x}{dt^2} &= -5x + y \\ 2\frac{d^2y}{dt^2} &= 3x - y \end{aligned} \qquad \text{and} \qquad \begin{aligned} x' - 3x + y' + z' &= 5 \\ x' - y' + 2z' &= t^2 \\ x + y' - 6z' &= t - 1 \end{aligned}$$

are two examples of systems of simultaneous differential equations.

Solution of a System A **solution** of a system of differential equations is a set of differentiable functions $x = f(t)$, $y = g(t)$, $z = h(t)$, and so on, that satisfies each equation of the system on some interval I.

Systematic Elimination The first technique we shall consider for solving systems is based on the fundamental principle of **systematic algebraic elimination** of variables. We shall see that the analogue of multiplying an algebraic equation by a constant is operating on a differential equation with some combination of derivatives. Recall that a linear differential equation

$$a_n y^{(n)} + a_{n-1} y^{(n-1)} + \cdots + a_1 y' + a_0 y = g(t),$$

where the a_i, $i = 0, 1, \ldots, n$ are constants, can be written in terms of differential operators as

$$(a_n D^n + a_{n-1} D^{n-1} + \cdots + a_1 D + a_0)y = g(t).$$

EXAMPLE 1 A System in Operator Notation

Write the system of differential equations

$$\begin{aligned} x'' + 2x' + y'' &= x + 3y + \sin t \\ x' + y' &= -4x + 2y + e^{-t} \end{aligned}$$

in operator notation.

Solution Rewrite the given system as

$$\begin{aligned} x'' + 2x' - x + y'' - 3y &= \sin t \\ x' + 4x + y' - 2y &= e^{-t} \end{aligned} \qquad \text{so that} \qquad \begin{aligned} (D^2 + 2D - 1)x + (D^2 - 3)y &= \sin t \\ (D + 4)x + (D - 2)y &= e^{-t}. \end{aligned}$$

Method of Solution Consider the simple system of linear first-order equations

$$
\begin{aligned}
Dy &= 2x \\
Dx &= 3y
\end{aligned}
\tag{1}
$$

or, equivalently,

$$
\begin{aligned}
2x - Dy &= 0 \\
Dx - 3y &= 0.
\end{aligned}
\tag{2}
$$

Operating on the first equation in (2) by D while multiplying the second by 2 and then subtracting eliminates x from the system. It follows that

$$-D^2y + 6y = 0 \quad \text{or} \quad D^2y - 6y = 0.$$

Since the roots of the auxiliary equation are $m_1 = \sqrt{6}$ and $m_2 = -\sqrt{6}$, we obtain

$$y(t) = c_1e^{\sqrt{6}t} + c_2e^{-\sqrt{6}t}. \tag{3}$$

Multiplying the first equation by -3 while operating on the second by D and then adding gives the differential equation for x, $D^2x - 6x = 0$. It follows immediately that

$$x(t) = c_3e^{\sqrt{6}t} + c_4e^{-\sqrt{6}t}. \tag{4}$$

Now (3) and (4) do not satisfy the system (1) for every choice of c_1, c_2, c_3, and c_4. Substituting $x(t)$ and $y(t)$ into the first equation of the original system (1) gives, after we simplify,

$$(\sqrt{6}c_1 - 2c_3)e^{\sqrt{6}t} + (-\sqrt{6}c_2 - 2c_4)e^{-\sqrt{6}t} = 0.$$

Since the latter expression is to be zero for all values of t, we must have

$$\sqrt{6}c_1 - 2c_3 = 0 \quad \text{and} \quad -\sqrt{6}c_2 - 2c_4 = 0$$

or

$$c_3 = \frac{\sqrt{6}}{2}c_1, \quad c_4 = -\frac{\sqrt{6}}{2}c_2. \tag{5}$$

Hence we conclude that a solution of the system must be

$$x(t) = \frac{\sqrt{6}}{2}c_1e^{\sqrt{6}t} - \frac{\sqrt{6}}{2}c_2e^{-\sqrt{6}t}, \quad y(t) = c_1e^{\sqrt{6}t} + c_2e^{-\sqrt{6}t}.$$

You are urged to substitute (3) and (4) into the second equation of (1) and verify that the same relationship (5) holds between the constants.

EXAMPLE 2 Solving a System by Elimination

Solve

$$
\begin{aligned}
Dx + (D + 2)y &= 0 \\
(D - 3)x - \qquad 2y &= 0.
\end{aligned}
\tag{6}
$$

Solution Operating on the first equation by $D - 3$ and on the second by D and subtracting eliminates x from the system. It follows that the differential equation for y is

$$[(D-3)(D+2) + 2D]y = 0 \quad \text{or} \quad (D^2 + D - 6)y = 0.$$

Since the characteristic equation of this last differential equation is $m^2 + m - 6 = (m-2)(m+3) = 0$, we obtain the solution

$$y(t) = c_1e^{2t} + c_2e^{-3t}. \tag{7}$$

Eliminating y in a similar manner yields $(D^2 + D - 6)x = 0$, from which we find

$$x(t) = c_3e^{2t} + c_4e^{-3t}. \tag{8}$$

As we noted in the foregoing discussion, a solution of (6) does not contain four independent constants since the system itself puts a constraint on the actual number that can be chosen arbitrarily. Substituting (7) and (8) into the first equation of (6) gives

$$(4c_1 + 2c_3)e^{2t} + (-c_2 - 3c_4)e^{-3t} = 0$$

and so

$$4c_1 + 2c_3 = 0 \quad \text{and} \quad -c_2 - 3c_4 = 0.$$

Therefore $\quad c_3 = -2c_1 \quad$ and $\quad c_4 = -\tfrac{1}{3}c_2.$

Accordingly, a solution of the system is

$$x(t) = -2c_1e^{2t} - \tfrac{1}{3}c_2e^{-3t}, \quad y(t) = c_1e^{2t} + c_2e^{-3t}.$$

Since we could just as easily solve for c_3 and c_4 in terms of c_1 and c_2, the solution in Example 2 can be written in the alternative form

$$x(t) = c_3e^{2t} + c_4e^{-3t}, \quad y(t) = -\tfrac{1}{2}c_3e^{2t} - 3c_4e^{-3t}.$$

Also, it sometimes pays to keep one's eyes open when solving systems. Had we solved for x first, then y could be found, along with the relationship between the constants, by simply using the last equation of (6):

$$y = \frac{1}{2}(Dx - 3x) = \frac{1}{2}[2c_3e^{2t} - 3c_4e^{-3t} - 3c_3e^{2t} - 3c_4e^{-3t}]$$

or

$$y = -\frac{1}{2}c_3e^{2t} - 3c_4e^{-3t}.$$

EXAMPLE 3 Solving a System by Elimination

Solve

$$\begin{aligned} x' - 4x + y'' &= t^2 \\ x' + x + y' &= 0. \end{aligned} \tag{9}$$

Solution First we write the system in differential operator notation:

$$\begin{aligned} (D-4)x + D^2y &= t^2 \\ (D+1)x + Dy &= 0. \end{aligned} \tag{10}$$

Then, by eliminating x, we obtain

$$[(D+1)D^2 - (D-4)D]y = (D+1)t^2 - (D-4)0$$

or

$$(D^3 + 4D)y = t^2 + 2t.$$

Since the roots of the auxiliary equation $m(m^2+4) = 0$ are $m_1 = 0,\ m_2 = 2i$, and $m_3 = -2i$, the complementary function is

$$y_c = c_1 + c_2 \cos 2t + c_3 \sin 2t.$$

To determine the particular solution y_p we use undetermined coefficients by assuming $y_p = At^3 + Bt^2 + Ct$. Therefore

$$y_p' = 3At^2 + 2Bt + C, \quad y_p'' = 6At + 2B, \quad y_p''' = 6A,$$

$$y_p''' + 4y_p' = 12At^2 + 8Bt + 6A + 4C = t^2 + 2t.$$

The last equality implies

$$12A = 1, \quad 8B = 2, \quad 6A + 4C = 0,$$

and hence $A = \frac{1}{12},\ B = \frac{1}{4},\ C = -\frac{1}{8}$. Thus

$$y = y_c + y_p = c_1 + c_2 \cos 2t + c_3 \sin 2t + \frac{1}{12}t^3 + \frac{1}{4}t^2 - \frac{1}{8}t. \tag{11}$$

Eliminating y from the system (10) leads to

$$[(D-4) - D(D+1)]x = t^2 \quad \text{or} \quad (D^2+4)x = -t^2.$$

It should be obvious that $x_c = c_4 \cos 2t + c_5 \sin 2t$ and that undetermined coefficients can be applied to obtain a particular solution of the form $x_p = At^2 + Bt + C$. In this case the usual differentiations and algebra yield $x_p = -\frac{1}{4}t^2 + \frac{1}{8}$ and so

$$x = x_c + x_p = c_4 \cos 2t + c_5 \sin 2t - \frac{1}{4}t^2 + \frac{1}{8}. \tag{12}$$

Now c_4 and c_5 can be expressed in terms of c_2 and c_3 by substituting (11) and (12) into either equation of (9). By using the second equation, we find, after combining terms,

$$(c_5 - 2c_4 - 2c_2)\sin 2t + (2c_5 + c_4 + 2c_3)\cos 2t = 0$$

so

$$c_5 - 2c_4 - 2c_2 = 0 \quad \text{and} \quad 2c_5 + c_4 + 2c_3 = 0.$$

Solving for c_4 and c_5 in terms of c_2 and c_3 gives

$$c_4 = -\frac{1}{5}(4c_2 + 2c_3) \quad \text{and} \quad c_5 = \frac{1}{5}(2c_2 - 4c_3).$$

Finally, a solution of (9) is found to be

$$x(t) = -\frac{1}{5}(4c_2 + 2c_3)\cos 2t + \frac{1}{5}(2c_2 - 4c_3)\sin 2t - \frac{1}{4}t^2 + \frac{1}{8}$$

$$y(t) = c_1 + c_2 \cos 2t + c_3 \sin 2t + \frac{1}{12}t^3 + \frac{1}{4}t^2 - \frac{1}{8}t.$$

■

Use of Determinants Symbolically, if L_1, L_2, L_3, and L_4 denote linear differential operators with constant coefficients, then a system of linear differential equations in two variables x and y can be written as

$$\begin{aligned} L_1x + L_2y &= g_1(t) \\ L_3x + L_4y &= g_2(t). \end{aligned} \tag{13}$$

Eliminating variables, as we would for algebraic equations, leads to

$$(L_1L_4 - L_2L_3)x = f_1(t) \quad \text{and} \quad (L_1L_4 - L_2L_3)y = f_2(t), \tag{14}$$

where

$$f_1(t) = L_4g_1(t) - L_2g_2(t) \quad \text{and} \quad f_2(t) = L_1g_2(t) - L_3g_1(t).$$

Formally the results in (14) can be written in terms of determinants similar to those used in Cramer's rule:

$$\begin{vmatrix} L_1 & L_2 \\ L_3 & L_4 \end{vmatrix} x = \begin{vmatrix} g_1 & L_2 \\ g_2 & L_4 \end{vmatrix} \quad \text{and} \quad \begin{vmatrix} L_1 & L_2 \\ L_3 & L_4 \end{vmatrix} y = \begin{vmatrix} L_1 & g_1 \\ L_3 & g_2 \end{vmatrix}. \tag{15}$$

The left-hand determinant in each equation in (15) can be expanded in the usual algebraic sense, with the result then operating on the functions $x(t)$ and $y(t)$. However, some care should be exercised in the expansion of the right-hand determinants in (15). We must expand these determinants in the sense of the internal differential operators actually operating upon the functions $g_1(t)$ and $g_2(t)$.

If
$$\begin{vmatrix} L_1 & L_2 \\ L_3 & L_4 \end{vmatrix} \neq 0$$

in (15) and is a differential operator of order n, then

- The system (13) can be uncoupled into two nth-order differential equations in x and y.
- The characteristic equation and hence the complementary function of each of these differential equations are the same.
- Since x and y both contain n constants, there are a total of $2n$ constants appearing.
- The total number of *independent* constants in the solution of the system is n.

If
$$\begin{vmatrix} L_1 & L_2 \\ L_3 & L_4 \end{vmatrix} = 0$$

in (13), then the system may have a solution containing any number of independent constants or may have no solution at all. Similar remarks hold for systems larger than indicated in (13).

EXAMPLE 4 Solving a System Using Determinants

Solve
$$\begin{aligned} x' &= 3x - y - 1 \\ y' &= x + y + 4e^t. \end{aligned} \tag{16}$$

Solution Write the system in terms of differential operators,

$$\begin{aligned} (D-3)x + \qquad y &= -1 \\ -x + (D-1)y &= 4e^t, \end{aligned}$$

and then use determinants:

$$\begin{vmatrix} D-3 & 1 \\ -1 & D-1 \end{vmatrix} x = \begin{vmatrix} -1 & 1 \\ 4e^t & D-1 \end{vmatrix}$$

$$\begin{vmatrix} D-3 & 1 \\ -1 & D-1 \end{vmatrix} y = \begin{vmatrix} D-3 & -1 \\ -1 & 4e^t \end{vmatrix}.$$

After expanding, we find that

$$(D-2)^2x = 1 - 4e^t$$
$$(D-2)^2y = -1 - 8e^t.$$

By the usual methods it follows that

$$x = x_c + x_p = c_1e^{2t} + c_2te^{2t} + \frac{1}{4} - 4e^t \tag{17}$$

$$y = y_c + y_p = c_3e^{2t} + c_4te^{2t} - \frac{1}{4} - 8e^t. \tag{18}$$

Substituting (17) and (18) into the second equation of (16) gives

$$(c_3 - c_1 + c_4)e^{2t} + (c_4 - c_2)te^{2t} = 0,$$

which then implies

$$c_4 = c_2 \quad \text{and} \quad c_3 = c_1 - c_4 = c_1 - c_2.$$

Thus a solution of (16) is

$$x(t) = c_1e^{2t} + c_2te^{2t} + \frac{1}{4} - 4e^t, \quad y(t) = (c_1 - c_2)e^{2t} + c_2te^{2t} - \frac{1}{4} - 8e^t.$$ ■

EXAMPLE 5 Using Determinants

Given the system

$$\begin{aligned} Dx + \qquad\qquad\quad Dz &= t^2 \\ 2x + D^2y \qquad\qquad &= e^t \\ -2Dx - \quad 2y + (D+1)z &= 0, \end{aligned}$$

find the differential equation for the variable y.

Solution With determinants we can write

$$\begin{vmatrix} D & 0 & D \\ 2 & D^2 & 0 \\ -2D & -2 & D+1 \end{vmatrix} y = \begin{vmatrix} D & t^2 & D \\ 2 & e^t & 0 \\ -2D & 0 & D+1 \end{vmatrix}.$$

In turn, expanding each determinant by cofactors of the first row gives

$$\left(D\begin{vmatrix} D^2 & 0 \\ -2 & D+1 \end{vmatrix} + D\begin{vmatrix} 2 & D^2 \\ -2D & -2 \end{vmatrix}\right)y = D\begin{vmatrix} e^t & 0 \\ 0 & D+1 \end{vmatrix} - \begin{vmatrix} 2 & 0 \\ -2D & D+1 \end{vmatrix}t^2 + D\begin{vmatrix} 2 & e^t \\ -2D & 0 \end{vmatrix}$$

or $$D(3D^3 + D^2 - 4)y = 4e^t - 2t^2 - 4t.$$

Again we remind you that the D symbol on the left-hand side is to be treated as an algebraic quantity, but this is not the case on the right-hand side. ■

EXERCISES 8.1

Answers to odd-numbered problems begin on page A-18.

In Problems 1–22 solve, if possible, the given system of differential equations by either systematic elimination or determinants.

1. $\dfrac{dx}{dt} = 2x - y$, $\quad \dfrac{dy}{dt} = x$

2. $\dfrac{dx}{dt} = 4x + 7y$, $\quad \dfrac{dy}{dt} = x - 2y$

3. $\dfrac{dx}{dt} = -y + t$, $\quad \dfrac{dy}{dt} = x - t$

4. $\dfrac{dx}{dt} - 4y = 1$, $\quad x + \dfrac{dy}{dt} = 2$

5. $(D^2 + 5)x - 2y = 0$, $\quad -2x + (D^2 + 2)y = 0$

6. $(D + 1)x + (D - 1)y = 2$, $\quad 3x + (D + 2)y = -1$

7. $\dfrac{d^2x}{dt^2} = 4y + e^t$, $\quad \dfrac{d^2y}{dt^2} = 4x - e^t$

8. $\dfrac{d^2x}{dt^2} + \dfrac{dy}{dt} = -5x$, $\quad \dfrac{dx}{dt} + \dfrac{dy}{dt} = -x + 4y$

9. $Dx + D^2y = e^{3t}$, $\quad (D + 1)x + (D - 1)y = 4e^{3t}$

10. $D^2x - Dy = t$, $\quad (D + 3)x + (D + 3)y = 2$

11. $(D^2 - 1)x - y = 0$, $\quad (D - 1)x + Dy = 0$

12. $(2D^2 - D - 1)x - (2D + 1)y = 1$, $\quad (D - 1)x + Dy = -1$

13. $2\dfrac{dx}{dt} - 5x + \dfrac{dy}{dt} = e^t$, $\quad \dfrac{dx}{dt} - x + \dfrac{dy}{dt} = 5e^t$

14. $\dfrac{dx}{dt} + \dfrac{dy}{dt} = e^t$, $\quad -\dfrac{d^2x}{dt^2} + \dfrac{dx}{dt} + x + y = 0$

15. $(D - 1)x + (D^2 + 1)y = 1$, $\quad (D^2 - 1)x + (D + 1)y = 2$

16. $D^2x - 2(D^2 + D)y = \sin t$, $\quad x + Dy = 0$

17. $Dx = y$, $\quad Dy = z$, $\quad Dz = x$

18. $Dx + z = e^t$, $\quad (D - 1)x + Dy + Dz = 0$, $\quad x + 2y + Dz = e^t$

19. $\dfrac{dx}{dt} - 6y = 0$, $\quad x - \dfrac{dy}{dt} + z = 0$, $\quad x + y - \dfrac{dz}{dt} = 0$

20. $\dfrac{dx}{dt} = -x + z$, $\quad \dfrac{dy}{dt} = -y + z$, $\quad \dfrac{dz}{dt} = -x + y$

21. $2Dx + (D - 1)y = t$, $\quad Dx + Dy = t^2$

22. $Dx - 2Dy = t^2$, $\quad (D + 1)x - 2(D + 1)y = 1$

In Problems 23 and 24 solve the given system subject to the indicated initial conditions.

23. $\dfrac{dx}{dt} = -5x - y$

$\dfrac{dy}{dt} = 4x - y$

$x(1) = 0,\ y(1) = 1$

24. $\dfrac{dx}{dt} = y - 1$

$\dfrac{dy}{dt} = -3x + 2y$

$x(0) = 0,\ y(0) = 0$

25. Determine, if possible, a system of differential equations having

$$x(t) = c_1 + c_2e^{2t}, \quad y(t) = -c_1 + c_2e^{2t}$$

as its solution.

8.2 LAPLACE TRANSFORM METHOD

• *Reduction to algebraic system* • *Application*

When initial conditions are specified, the Laplace transform reduces a system of linear differential equations with constant coefficients to a set of simultaneous algebraic equations in the transformed functions.

EXAMPLE 1 System of DEs Transformed into an Algebraic System

Solve

$$\begin{aligned} 2x' + y' - y &= t \\ x' + y' &= t^2 \end{aligned} \tag{1}$$

subject to $x(0) = 1$, $y(0) = 0$.

Solution If $X(s) = \mathscr{L}\{x(t)\}$ and $Y(s) = \mathscr{L}\{y(t)\}$, then after transforming each equation we obtain

$$\begin{aligned} 2[sX(s) - x(0)] + sY(s) - y(0) - Y(s) &= \frac{1}{s^2} \\ sX(s) - x(0) + sY(s) - y(0) &= \frac{2}{s^3} \end{aligned}$$

or

$$\begin{aligned} 2sX(s) + (s - 1)Y(s) &= 2 + \frac{1}{s^2} \\ sX(s) + sY(s) &= 1 + \frac{2}{s^3}. \end{aligned} \tag{2}$$

Multiplying the second equation of (2) by 2 and subtracting yields

$$(-s - 1)Y(s) = \frac{1}{s^2} - \frac{4}{s^3} \quad \text{or} \quad Y(s) = \frac{4 - s}{s^3(s + 1)}. \tag{3}$$

Now by partial fractions

$$\frac{4 - s}{s^3(s + 1)} = \frac{A}{s} + \frac{B}{s^2} + \frac{C}{s^3} + \frac{D}{s + 1},$$

so

$$4 - s = As^2(s + 1) + Bs(s + 1) + C(s + 1) + Ds^3.$$

Setting $s = 0$ and $s = -1$ in the last line gives $C = 4$ and $D = -5$, respectively, whereas equating the coefficients of s^3 and s^2 on each side of the equality yields

$$A + D = 0 \quad \text{and} \quad A + B = 0.$$

It follows that $A = 5$, $B = -5$. Thus (3) becomes

$$Y(s) = \frac{5}{s} - \frac{5}{s^2} + \frac{4}{s^3} - \frac{5}{s + 1}$$

and so

$$\begin{aligned} y(t) &= 5\mathscr{L}^{-1}\left\{\frac{1}{s}\right\} - 5\mathscr{L}^{-1}\left\{\frac{1}{s^2}\right\} + 2\mathscr{L}^{-1}\left\{\frac{2!}{s^3}\right\} - 5\mathscr{L}^{-1}\left\{\frac{1}{s + 1}\right\} \\ &= 5 - 5t + 2t^2 - 5e^{-t}. \end{aligned}$$

By the second equation of (2),

$$X(s) = -Y(s) + \frac{1}{s} + \frac{2}{s^4},$$

from which it follows that

$$\begin{aligned} x(t) &= -\mathscr{L}^{-1}\{Y(s)\} + \mathscr{L}^{-1}\left\{\frac{1}{s}\right\} + \frac{2}{3!}\mathscr{L}^{-1}\left\{\frac{3!}{s^4}\right\} \\ &= -4 + 5t - 2t^2 + \frac{1}{3}t^3 + 5e^{-t}. \end{aligned}$$

Hence we conclude that the solution of the given system (1) is

$$x(t) = -4 + 5t - 2t^2 + \frac{1}{3}t^3 + 5e^{-t}, \quad y(t) = 5 - 5t + 2t^2 - 5e^{-t}. \qquad \textbf{(4)}$$

Applications Let us turn now to some elementary applications involving systems of differential equations. The solutions of the problems that we shall consider can be obtained either by the method of the preceding section or through the use of the Laplace transform.

Coupled Springs Two masses m_1 and m_2 are connected to two springs A and B of negligible mass having spring constants k_1 and k_2, respectively. In turn, the two springs are attached as shown in Figure 8.1. Let $x_1(t)$ and $x_2(t)$ denote the vertical displacements of the masses from their equilibrium positions. When the system is in motion, spring B is subject to both an elongation and a compression; hence its net elongation is $x_2 - x_1$. Therefore it follows from Hooke's law that springs A and B exert forces

$$-k_1x_1 \quad \text{and} \quad k_2(x_2 - x_1),$$

respectively, on m_1. If no external force is impressed on the system and if no damping force is present, then the net force on m_1 is $-k_1x_1 + k_2(x_2 - x_1)$. By Newton's second law we can write

$$m_1\frac{d^2x_1}{dt^2} = -k_1x_1 + k_2(x_2 - x_1).$$

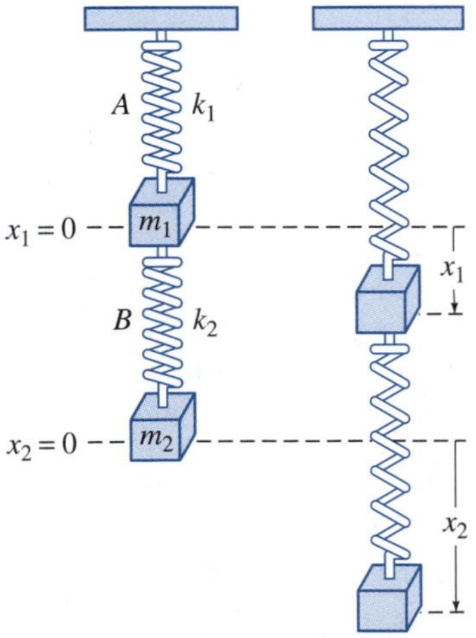

Figure 8.1

Similarly, the net force exerted on mass m_2 is due solely to the net elongation of B; that is, $-k_2(x_2 - x_1)$. Thus it follows that

$$m_2 \frac{d^2x_2}{dt^2} = -k_2(x_2 - x_1).$$

In other words, the motion of the coupled system is represented by the system of simultaneous second-order differential equations

$$\begin{aligned} m_1 x_1'' &= -k_1 x_1 + k_2(x_2 - x_1) \\ m_2 x_2'' &= -k_2(x_2 - x_1). \end{aligned} \tag{5}$$

In the next example we shall solve (5) under the assumptions that

$$k_1 = 6, \quad k_2 = 4, \quad m_1 = 1, \quad m_2 = 1$$

and that the masses start from their equilibrium positions with opposite unit velocities.

EXAMPLE 2 Coupled Springs

Solve

$$\begin{aligned} x_1'' + 10x_1 - 4x_2 &= 0 \\ -4x_1 + x_2'' + 4x_2 &= 0 \end{aligned} \tag{6}$$

subject to $x_1(0) = 0, x_1'(0) = 1, x_2(0) = 0, x_2'(0) = -1$.

Solution The Laplace transform of each equation is

$$\begin{aligned} s^2X_1(s) - sx_1(0) - x_1'(0) + 10X_1(s) - 4X_2(s) &= 0 \\ -4X_1(s) + s^2X_2(s) - sx_2(0) - x_2'(0) + 4X_2(s) &= 0, \end{aligned}$$

where $X_1(s) = \mathscr{L}\{x_1(t)\}$ and $X_2(s) = \mathscr{L}\{x_2(t)\}$. The preceding system is the same as

$$\begin{aligned} (s^2 + 10)X_1(s) - \qquad 4X_2(s) &= 1 \\ -4X_1(s) + (s^2 + 4)X_2(s) &= -1. \end{aligned} \tag{7}$$

Eliminating X_2 gives

$$X_1(s) = \frac{s^2}{(s^2 + 2)(s^2 + 12)}.$$

By partial fractions we can write

$$\frac{s^2}{(s^2 + 2)(s^2 + 12)} = \frac{As + B}{s^2 + 2} + \frac{Cs + D}{s^2 + 12}$$

and $$s^2 = (As + B)(s^2 + 12) + (Cs + D)(s^2 + 2).$$

Comparing the coefficients of s on each side of the last equality gives

$$A + C = 0, \quad B + D = 1, \quad 12A + 2C = 0, \quad 12B + 2D = 0,$$

so $A = 0, C = 0, B = -\frac{1}{5}$, and $D = \frac{6}{5}$. Hence

$$X_1(s) = -\frac{1/5}{s^2 + 2} + \frac{6/5}{s^2 + 12}$$

and therefore

$$x_1(t) = -\frac{1}{5\sqrt{2}}\mathscr{L}^{-1}\left\{\frac{\sqrt{2}}{s^2+2}\right\} + \frac{6}{5\sqrt{12}}\mathscr{L}^{-1}\left\{\frac{\sqrt{12}}{s^2+12}\right\}$$
$$= -\frac{\sqrt{2}}{10}\sin\sqrt{2}t + \frac{\sqrt{3}}{5}\sin 2\sqrt{3}t.$$

From the first equation of (7) it follows that

$$X_2(s) = -\frac{s^2+6}{(s^2+2)(s^2+12)}.$$

Proceeding as before with partial fractions, we obtain

$$X_2(s) = -\frac{2/5}{s^2+2} - \frac{3/5}{s^2+12}$$

and $$x_2(t) = -\frac{2}{5\sqrt{2}}\mathscr{L}^{-1}\left\{\frac{\sqrt{2}}{s^2+2}\right\} - \frac{3}{5\sqrt{12}}\mathscr{L}^{-1}\left\{\frac{\sqrt{12}}{s^2+12}\right\}$$

$$= -\frac{\sqrt{2}}{5}\sin\sqrt{2}t - \frac{\sqrt{3}}{10}\sin 2\sqrt{3}t.$$

Finally, the solution to the given system (6) is

$$\begin{aligned} x_1(t) &= -\frac{\sqrt{2}}{10}\sin\sqrt{2}t + \frac{\sqrt{3}}{5}\sin 2\sqrt{3}t \\ x_2(t) &= -\frac{\sqrt{2}}{5}\sin\sqrt{2}t - \frac{\sqrt{3}}{10}\sin 2\sqrt{3}t. \end{aligned} \tag{8}$$

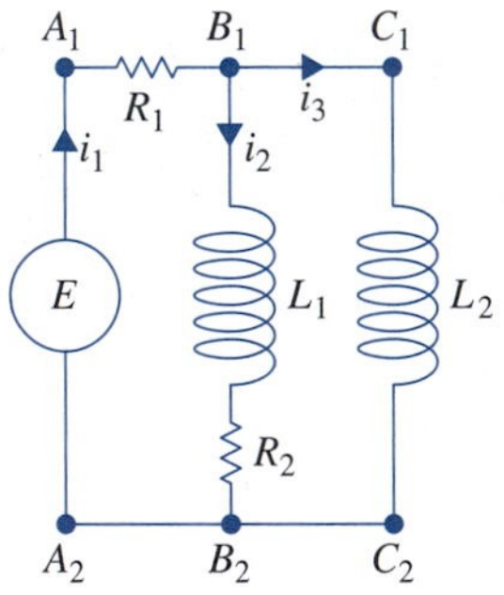

Figure 8.2

Networks An electrical network having more than one loop also gives rise to simultaneous differential equations. As shown in Figure 8.2, the current $i_1(t)$ splits in the directions shown at point B_1, called a branch point of the network. By Kirchhoff's first law we can write

$$i_1(t) = i_2(t) + i_3(t). \tag{9}$$

In addition we can also apply **Kirchhoff's second law** to each loop. For loop $A_1B_1B_2A_2A_1$, summing the voltage drops across each part of the loop gives

$$E(t) = i_1R_1 + L_1\frac{di_2}{dt} + i_2R_2. \tag{10}$$

Similarly, for loop $A_1B_1C_1C_2B_2A_2A_1$ we find

$$E(t) = i_1R_1 + L_2\frac{di_3}{dt}. \tag{11}$$

Using (9) to eliminate i_1 in (10) and (11) yields two first-order equations for the currents $i_2(t)$ and $i_3(t)$:

$$\begin{aligned} L_1\frac{di_2}{dt} + (R_1+R_2)i_2 + R_1i_3 &= E(t) \\ L_2\frac{di_3}{dt} + \qquad\qquad R_1i_2 + R_1i_3 &= E(t). \end{aligned} \tag{12}$$

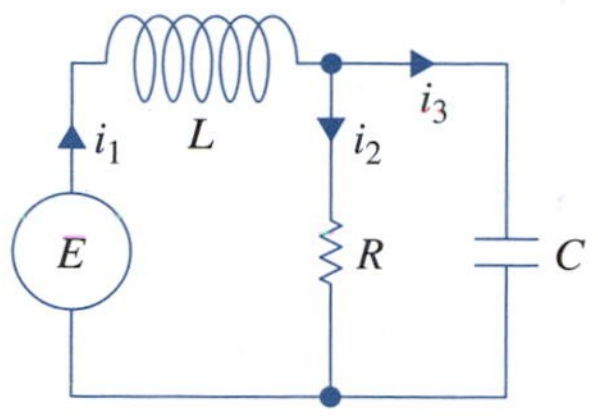

Figure 8.3

Given the natural initial conditions $i_2(0) = 0$, $i_3(0) = 0$, the system (12) is amenable to solution by the Laplace transform.

We leave it as an exercise (see Problem 18) to show that the system of differential equations describing the currents $i_1(t)$ and $i_2(t)$ in the network shown in Figure 8.3 containing a resistor, an inductor, and a capacitor is

$$\begin{aligned} L\frac{di_1}{dt} + Ri_2 \quad &= E(t) \\ RC\frac{di_2}{dt} + i_2 - i_1 &= 0. \end{aligned} \tag{13}$$

EXAMPLE 3 An Electrical Network

Solve the system (13) under the conditions $E = 60$ volts, $L = 1$ henry, $R = 50$ ohms, $C = 10^{-4}$ farad, and i_1 and i_2 are initially zero.

Solution We must solve

$$\begin{aligned} \frac{di_1}{dt} + 50i_2 \quad &= 60 \\ 50(10^{-4})\frac{di_2}{dt} + i_2 - i_1 &= 0 \end{aligned}$$

subject to $i_1(0) = 0$, $i_2(0) = 0$.

Applying the Laplace transform to each equation of the system and simplifying gives

$$\begin{aligned} sI_1(s) + \qquad 50I_2(s) &= \frac{60}{s} \\ -200I_1(s) + (s + 200)I_2(s) &= 0, \end{aligned}$$

where $I_1(s) = \mathscr{L}\{i_1(t)\}$ and $I_2(s) = \mathscr{L}\{i_2(t)\}$. Solving the system for I_1 and I_2 yields

$$I_1(s) = \frac{60s + 12{,}000}{s(s + 100)^2} \quad \text{and} \quad I_2(s) = \frac{12{,}000}{s(s + 100)^2}.$$

By partial fractions we can write

$$\begin{aligned} I_1(s) &= \frac{6/5}{s} - \frac{6/5}{s + 100} - \frac{60}{(s + 100)^2} \\ I_2(s) &= \frac{6/5}{s} - \frac{6/5}{s + 100} - \frac{120}{(s + 100)^2}, \end{aligned}$$

from which it follows that

$$i_1(t) = \frac{6}{5} - \frac{6}{5}e^{-100t} - 60te^{-100t}, \quad i_2(t) = \frac{6}{5} - \frac{6}{5}e^{-100t} - 120te^{-100t}.$$

Note that both $i_1(t)$ and $i_2(t)$ in Example 3 tend toward the value $E/R = \frac{6}{5}$ as $t \to \infty$. Furthermore since the current through the capacitor is $i_3(t) = i_1(t) - i_2(t) = 60te^{-100t}$, we observe that $i_3(t) \to 0$ as $t \to \infty$.

EXERCISES 8.2

Answers to odd-numbered problems begin on page A-18.

In Problems 1–12 use the Laplace transform to solve the given system of differential equations.

1. $\dfrac{dx}{dt} = -x + y$

$\dfrac{dy}{dt} = 2x$

$x(0) = 0, y(0) = 1$

2. $\dfrac{dx}{dt} = 2y + e^t$

$\dfrac{dy}{dt} = 8x - t$

$x(0) = 1, y(0) = 1$

3. $\dfrac{dx}{dt} = x - 2y$

$\dfrac{dy}{dt} = 5x - y$

$x(0) = -1, y(0) = 2$

4. $\dfrac{dx}{dt} + 3x + \dfrac{dy}{dt} = 1$

$\dfrac{dx}{dt} - x + \dfrac{dy}{dt} - y = e^t$

$x(0) = 0, y(0) = 0$

5. $2\dfrac{dx}{dt} + \dfrac{dy}{dt} - 2x = 1$

$\dfrac{dx}{dt} + \dfrac{dy}{dt} - 3x - 3y = 2$

$x(0) = 0, y(0) = 0$

6. $\dfrac{dx}{dt} + x - \dfrac{dy}{dt} + y = 0$

$\dfrac{dx}{dt} + \dfrac{dy}{dt} + 2y = 0$

$x(0) = 0, y(0) = 1$

7. $\dfrac{d^2x}{dt^2} + x - y = 0$

$\dfrac{d^2y}{dt^2} + y - x = 0$

$x(0) = 0, x'(0) = -2,$

$y(0) = 0, y'(0) = 1$

8. $\dfrac{d^2x}{dt^2} + \dfrac{dx}{dt} + \dfrac{dy}{dt} = 0$

$\dfrac{d^2y}{dt^2} + \dfrac{dy}{dt} - 4\dfrac{dx}{dt} = 0$

$x(0) = 1, x'(0) = 0,$

$y(0) = -1, y'(0) = 5$

9. $\dfrac{d^2x}{dt^2} + \dfrac{d^2y}{dt^2} = t^2$

$\dfrac{d^2x}{dt^2} - \dfrac{d^2y}{dt^2} = 4t$

$x(0) = 8, x'(0) = 0,$

$y(0) = 0, y'(0) = 0$

10. $\dfrac{dx}{dt} - 4x + \dfrac{d^3y}{dt^3} = 6 \sin t$

$\dfrac{dx}{dt} + 2x - 2\dfrac{d^3y}{dt^3} = 0$

$x(0) = 0, y(0) = 0,$

$y'(0) = 0, y''(0) = 0$

11. $\dfrac{d^2x}{dt^2} + 3\dfrac{dy}{dt} + 3y = 0$

$\dfrac{d^2x}{dt^2} + 3y = te^{-t}$

$x(0) = 0, x'(0) = 2, y(0) = 0$

12. $\dfrac{dx}{dt} = 4x - 2y + 2\mathcal{U}(t - 1)$

$\dfrac{dy}{dt} = 3x - y + \mathcal{U}(t - 1)$

$x(0) = 0, y(0) = \frac{1}{2}$

13. Solve system (5) when $k_1 = 3, k_2 = 2, m_1 = 1, m_2 = 1, x_1(0) = 0,$ $x_1'(0) = 1, x_2(0) = 1,$ and $x_2'(0) = 0.$

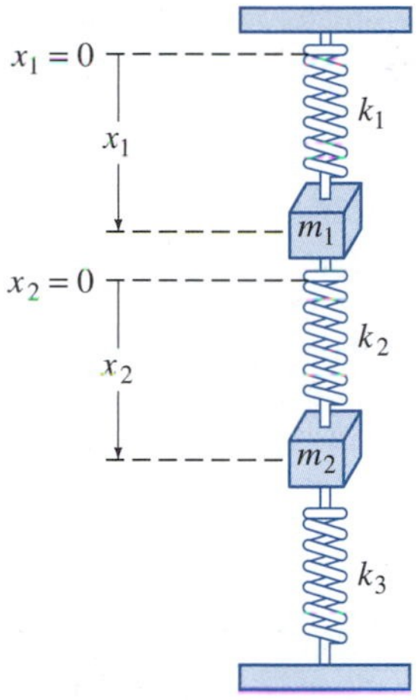

Figure 8.4

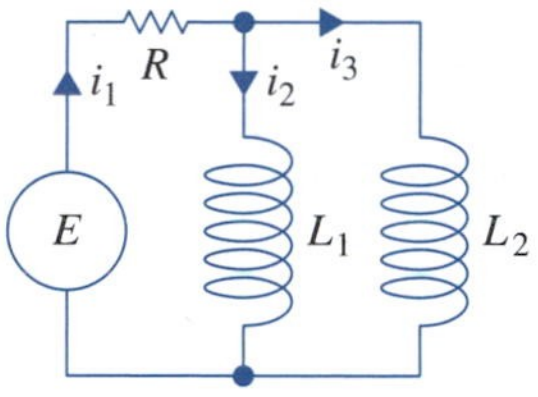

Figure 8.5

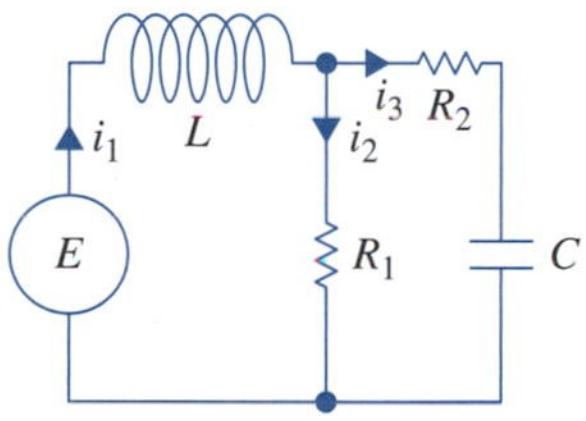

Figure 8.6

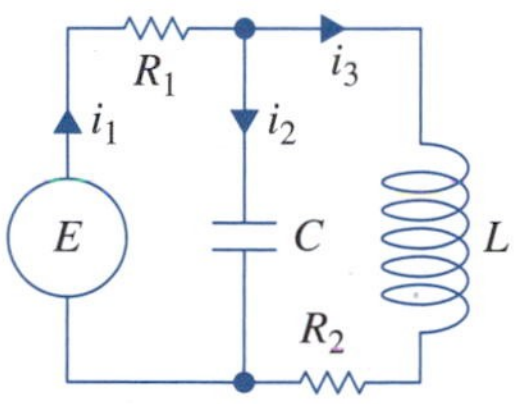

Figure 8.7

14. Derive the system of differential equations describing the straight-line vertical motion of the coupled springs shown in Figure 8.4. Use the Laplace transform to solve the system when $k_1 = 1$, $k_2 = 1$, $k_3 = 1$, $m_1 = 1$, $m_2 = 1$, $x_1(0) = 0$, $x_1'(0) = -1$, $x_2(0) = 0$, and $x_2'(0) = 1$.

15. **(a)** Show that the system of differential equations for the currents $i_2(t)$ and $i_3(t)$ in the electrical network shown in Figure 8.5 is

$$L_1 \frac{di_2}{dt} + Ri_2 + Ri_3 = E(t)$$
$$L_2 \frac{di_3}{dt} + Ri_2 + Ri_3 = E(t).$$

(b) Solve the system in part (a) if $R = 5$ ohms, $L_1 = 0.01$ henry, $L_2 = 0.0125$ henry, $E = 100$ volts, $i_2(0) = 0$, and $i_3(0) = 0$.

(c) Determine the current $i_1(t)$.

16. **(a)** Show that the system of differential equations for the currents $i_2(t)$ and $i_3(t)$ in the electrical network shown in Figure 8.6 is

$$L \frac{di_2}{dt} + L \frac{di_3}{dt} + R_1 i_2 = E(t)$$
$$-R_1 \frac{di_2}{dt} + R_2 \frac{di_3}{dt} + \frac{1}{C} i_3 = 0.$$

(b) Solve the system in part (a) if $R_1 = 10$ ohms, $R_2 = 5$ ohms, $L = 1$ henry, $C = 0.2$ farad,

$$E(t) = \begin{cases} 120, & 0 \le t < 2 \\ 0, & t \ge 2 \end{cases}$$

$i_2(0) = 0$, and $i_3(0) = 0$.

(c) Determine the current $i_1(t)$.

17. Solve the system given in (12) when $R_1 = 6$ ohms, $R_2 = 5$ ohms, $L_1 = 1$ henry, $L_2 = 1$ henry, and $E(t) = 50 \sin t$ volts.

18. Derive the system of equations (13).

19. Solve (13) when $E = 60$ volts, $L = \frac{1}{2}$ henry, $R = 50$ ohms, $C = 10^{-4}$ farad, $i_1(0) = 0$, and $i_2(0) = 0$.

20. Solve (13) when $E = 60$ volts, $L = 2$ henrys, $R = 50$ ohms, $C = 10^{-4}$ farad, $i_1(0) = 0$, and $i_2(0) = 0$.

21. **(a)** Show that the system of differential equations for the charge on the capacitor $q(t)$ and the current $i_3(t)$ in the electrical network shown in Figure 8.7 is

$$R_1 \frac{dq}{dt} + \frac{1}{C} q + R_1 i_3 = E(t)$$
$$L \frac{di_3}{dt} + R_2 i_3 - \frac{1}{C} q = 0.$$

(b) Find the charge on the capacitor when $L = 1$ henry, $R_1 = 1$ ohm, $R_2 = 1$ ohm, $C = 1$ farad,

$$E(t) = \begin{cases} 0, & 0 < t < 1 \\ 50e^{-t}, & t \ge 1 \end{cases}$$

$i_3(0) = 0$, and $q(0) = 0$.

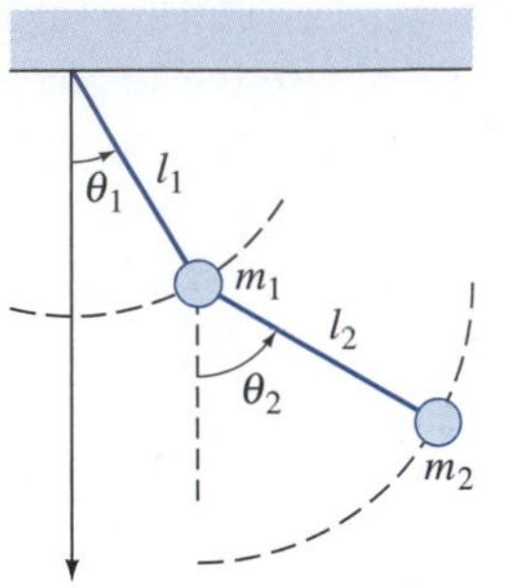

Figure 8.8

22. A double pendulum oscillates in a vertical plane under the influence of gravity (see Figure 8.8). For small displacements $\theta_1(t)$ and $\theta_2(t)$, it can be shown that the differential equations of motion are

$$\begin{aligned}(m_1 + m_2)l_1^2\theta_1'' + m_2l_1l_2\theta_2'' + (m_1 + m_2)l_1g\theta_1 &= 0\\ m_2l_2^2\theta_2'' + m_2l_1l_2\theta_1'' + m_2l_2g\theta_2 &= 0.\end{aligned}$$

Use the Laplace transform to solve the system when $m_1 = 3, m_2 = 1$, $l_1 = l_2 = 16$, $\theta_1(0) = 1$, $\theta_2(0) = -1$, $\theta_1'(0) = 0$, and $\theta_2'(0) = 0$.

8.3 SYSTEMS OF LINEAR FIRST-ORDER EQUATIONS

• Normal form • Homogeneous system • Nonhomogeneous system

In the preceding two sections we dealt with linear systems that were of the form

$$\begin{aligned}P_{11}(D)x_1 + P_{12}(D)x_2 + \cdots + P_{1n}(D)x_n &= b_1(t)\\ P_{21}(D)x_1 + P_{22}(D)x_2 + \cdots + P_{2n}(D)x_n &= b_2(t)\\ &\vdots\\ P_{n1}(D)x_1 + P_{n2}(D)x_2 + \cdots + P_{nn}(D)x_n &= b_n(t),\end{aligned} \tag{1}$$

where the P_{ij} were polynomials in the differential operator D. However, the study of systems of *first-order* differential equations

$$\begin{aligned}\frac{dx_1}{dt} &= g_1(t, x_1, x_2, \ldots, x_n)\\ \frac{dx_2}{dt} &= g_2(t, x_1, x_2, \ldots, x_n)\\ &\vdots\\ \frac{dx_n}{dt} &= g_n(t, x_1, x_2, \ldots, x_n)\end{aligned} \tag{2}$$

is particularly important in advanced mathematics since every nth-order differential equation

$$y^{(n)} = F(t, y, y', \ldots, y^{(n-1)}),$$

as well as most systems of differential equations, can be reduced to form (2). System (2) of n first-order equations is called an ***n*th-order system.**

Linear Normal Form Of course, a system such as (2) need not be linear and need not have constant coefficients. Consequently, the system may not be readily solvable, if at all. In the remaining sections of this chapter we shall be interested only in a particular, but important, case of (2)—namely, those systems having the linear **normal,** or **canonical,** form

$$\begin{aligned}\frac{dx_1}{dt} &= a_{11}(t)x_1 + a_{12}(t)x_2 + \cdots + a_{1n}(t)x_n + f_1(t)\\ \frac{dx_2}{dt} &= a_{21}(t)x_1 + a_{22}(t)x_2 + \cdots + a_{2n}(t)x_n + f_2(t)\\ &\vdots\\ \frac{dx_n}{dt} &= a_{n1}(t)x_1 + a_{n2}(t)x_2 + \cdots + a_{nn}(t)x_n + f_n(t),\end{aligned} \tag{3}$$

where the coefficients a_{ij} and the functions f_i are continuous on a common interval I. When $f_i(t) = 0$, $i = 1, 2, \ldots, n$, the system (3) is said to be **homogeneous;** otherwise it is called **nonhomogeneous.**

We shall now show that every linear nth-order differential equation can be reduced to a linear system having the normal form (3).

Equation to a System Suppose a linear nth-order differential equation is first written as

$$\frac{d^n y}{dt^n} = -\frac{a_0}{a_n}y - \frac{a_1}{a_n}y' - \cdots - \frac{a_{n-1}}{a_n}y^{(n-1)} + f(t). \tag{4}$$

If we then introduce the variables

$$y = x_1, \quad y' = x_2, \quad y'' = x_3, \quad \ldots, \quad y^{(n-1)} = x_n, \tag{5}$$

it follows that $y' = x_1' = x_2$, $y'' = x_2' = x_3$, $\ldots$, $y^{(n-1)} = x_{n-1}' = x_n$, and $y^{(n)} = x_n'$. Hence from (4) and (5) we find that a linear nth-order differential equation can be expressed as an nth-order system:

$$\begin{aligned}x_1' &= x_2\\ x_2' &= x_3\\ x_3' &= x_4\\ &\vdots\\ x_{n-1}' &= x_n\\ x_n' &= -\frac{a_0}{a_n}x_1 - \frac{a_1}{a_n}x_2 - \cdots - \frac{a_{n-1}}{a_n}x_n + f(t).\end{aligned} \tag{6}$$

Inspection of (6) reveals that it has the same form as (3).

EXAMPLE 1 Writing a DE as a System

Reduce the third-order equation $2y''' - 6y'' + 4y' + y = \sin t$ to the normal form (3).

Solution Write the differential equation as

$$y''' = -\frac{1}{2}y - 2y' + 3y'' + \frac{1}{2}\sin t$$

and then let $y = x_1$, $y' = x_2$, $y'' = x_3$. Since

$$x_1' = y' = x_2, \quad x_2' = y'' = x_3, \quad \text{and} \quad x_3' = y''',$$

we find

$$\begin{aligned} x_1' &= x_2 \\ x_2' &= x_3 \\ x_3' &= -\frac{1}{2}x_1 - 2x_2 + 3x_3 + \frac{1}{2}\sin t. \end{aligned}$$

Systems Reduced to Normal Form Using a procedure similar to that just outlined, we can reduce most systems of the linear form (1) to the linear normal form (3). To accomplish this it is necessary to first solve the system for the highest-order derivative of each dependent variable. As we shall see, this may not always be possible.

EXAMPLE 2 Writing a System as a System of First-Order DEs

Reduce

$$\begin{aligned} (D^2 - D + 5)x + \qquad 2D^2y &= e^t \\ -2x + (D^2 + 2)y &= 3t^2 \end{aligned}$$

to the normal form (3).

Solution Write the system as

$$\begin{aligned} D^2x + 2D^2y &= e^t - 5x + Dx \\ D^2y &= 3t^2 + 2x - 2y \end{aligned}$$

and then eliminate D^2y by multiplying the second equation by 2 and subtracting. We have

$$D^2x = e^t - 6t^2 - 9x + 4y + Dx.$$

Since the second equation of the system already expresses the highest-order derivative of y in terms of the remaining functions, we are now in a position to introduce new variables. If we let

$$Dx = u \quad \text{and} \quad Dy = v,$$

the expressions for D^2x and D^2y become, respectively,

$$\begin{aligned} Du &= e^t - 6t^2 - 9x + 4y + u \\ Dv &= 3t^2 + 2x - 2y. \end{aligned}$$

Thus the original system can be written in the normal form

$$\begin{aligned} Dx &= u \\ Dy &= v \\ Du &= -9x + 4y + u + e^t - 6t^2 \\ Dv &= 2x - 2y + 3t^2. \end{aligned}$$

Degenerate Systems Those systems of differential equations of form (1) that cannot be reduced to a linear system in normal form are said to be **degenerate.**

For example, it is a straightforward matter to show that it is impossible to solve the system

$$\begin{aligned} (D+1)x + (D+1)y &= 0 \\ 2Dx + (2D+1)y &= 0 \end{aligned} \qquad \textbf{(7)}$$

for the highest derivative of each variable, and hence the system is degenerate.*

You may be wondering why anyone would want to convert a single differential equation to a system of equations, or for that matter a system of differential equations to an even larger system. While we are not in a position to completely justify their importance, suffice it to say that these procedures are more than a theoretical exercise. There are times when it is actually desirable to work with a system rather than with one equation. In the numerical analysis of differential equations, almost all computational algorithms are established for first-order equations. Since these algorithms can be generalized directly to systems, to compute numerically, say, a second-order equation, we could reduce it to a system of two first-order equations (see Chapter 9).

A linear system such as (3) also arises naturally in some physical applications. The following example illustrates a homogeneous system in two dependent variables.

EXAMPLE 3 Two-Container Mixture Model

Tank A contains 50 gallons of water in which 25 pounds of salt are dissolved. A second tank, B, contains 50 gallons of pure water. Liquid is pumped in and out of the tanks at the rates shown in Figure 8.9. Derive the differential equations that describe the number of pounds $x_1(t)$ and $x_2(t)$ of salt at any time in tanks A and B, respectively.

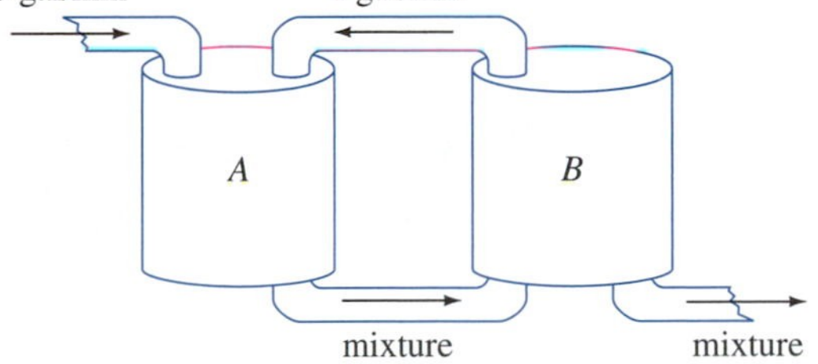

Figure 8.9

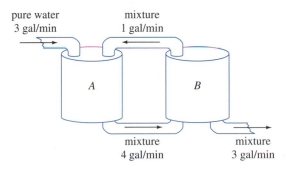

Solution By an analysis similar to that used in Section 3.2, we see that the net rate of change in $x_1(t)$ in lb/min is

$$\frac{dx_1}{dt} = \overbrace{(3 \text{ gal/min}) \cdot (0 \text{ lb/gal}) + (1 \text{ gal/min}) \cdot \left(\frac{x_2}{50} \text{ lb/gal}\right)}^{\text{input}} - \overbrace{(4 \text{ gal/min}) \cdot \left(\frac{x_1}{50} \text{ lb/gal}\right)}^{\text{output}}$$

$$= -\frac{2}{25}x_1 + \frac{1}{50}x_2.$$

* This does not mean that the system does not have a solution (see Problem 21).

In addition we find that the net rate of change in $x_2(t)$ is

$$\begin{aligned}\frac{dx_2}{dt} &= 4 \cdot \frac{x_1}{50} - 3 \cdot \frac{x_2}{50} - 1 \cdot \frac{x_2}{50} \\ &= \frac{2}{25}x_1 - \frac{2}{25}x_2.\end{aligned}$$

Thus we obtain the first-order system

$$\begin{aligned}\frac{dx_1}{dt} &= -\frac{2}{25}x_1 + \frac{1}{50}x_2 \\ \frac{dx_2}{dt} &= \frac{2}{25}x_1 - \frac{2}{25}x_2.\end{aligned} \qquad (8)$$

Observe that the foregoing system is accompanied by the initial conditions $x_1(0) = 25, x_2(0) = 0$. ■

It is left as an exercise to solve (8) by the Laplace transform. See Problem 17.

EXERCISES 8.3

Answers to odd-numbered problems begin on page A-18.

In Problems 1–8 rewrite the given differential equation as a system in normal form (3).

1. $y'' - 3y' + 4y = \sin 3t$

2. $2\dfrac{d^2y}{dt^2} + 4\dfrac{dy}{dt} - 5y = 0$

3. $y''' - 3y'' + 6y' - 10y = t^2 + 1$

4. $4y''' + y = e^t$

5. $\dfrac{d^4y}{dt^4} - 2\dfrac{d^2y}{dt^2} + 4\dfrac{dy}{dt} + y = t$

6. $2\dfrac{d^4y}{dt^4} + \dfrac{d^3y}{dt^3} - 8y = 10$

7. $(t + 1)y'' = ty$

8. $t^2y'' + ty' + (t^2 - 4)y = 0$

In Problems 9–16 rewrite, if possible, the given system in the normal form (3).

9. $\begin{aligned} x' + 4x - y' &= 7t \\ x' + y' - 2y &= 3t \end{aligned}$

10. $\begin{aligned} x'' + y' &= 1 \\ x'' + y' &= -1 \end{aligned}$

11. $\begin{aligned} (D - 1)x - Dy &= t^2 \\ x + Dy &= 5t - 2 \end{aligned}$

12. $\begin{aligned} x'' - 2y'' &= \sin t \\ x'' + y'' &= \cos t \end{aligned}$

13. $\begin{aligned} (2D + 1)x - 2Dy &= 4 \\ Dx - Dy &= e^t \end{aligned}$

14. $\begin{aligned} m_1x_1'' &= -k_1x_1 + k_2(x_2 - x_1) \\ m_2x_2'' &= -k_2(x_2 - x_1) \end{aligned}$

15. $\begin{aligned} \frac{d^3x}{dt^3} &= 4x - 3\frac{d^2x}{dt^2} + 4\frac{dy}{dt} \\ \frac{d^2y}{dt^2} &= 10t^2 - 4\frac{dx}{dt} + 3\frac{dy}{dt} \end{aligned}$

16. $\begin{aligned} D^2x + Dy &= 4t \\ -D^2x + (D + 1)y &= 6t^2 + 10 \end{aligned}$

17. Use the Laplace transform to solve system (8) subject to $x_1(0) = 25$ and $x_2(0) = 0$.

18. Consider two tanks A and B with liquid being pumped in and out at the same rates as given in Example 3. What is the system of differential equations if, instead of pure water, a brine solution containing 2 lb of salt per gallon is pumped into tank A?

19. Using the information given in Figure 8.10, derive the system of differential equations describing the numbers of pounds of salt x_1, x_2, and x_3 at any time in tanks A, B, and C, respectively.

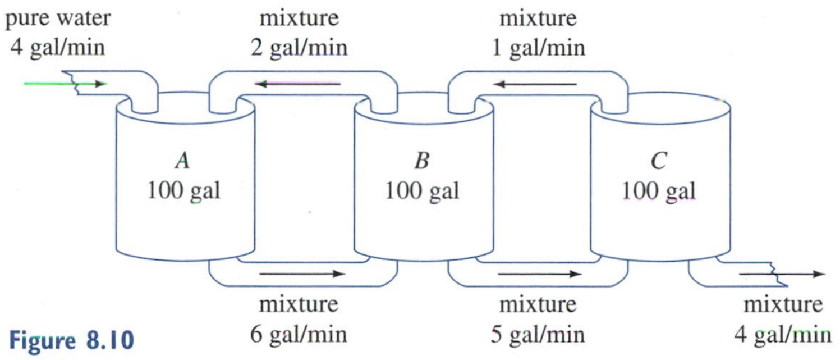

Figure 8.10

20. Consider the first-order system

$$(a_1D - b_1)x + (a_2D - b_2)y = 0$$
$$(a_3D - b_3)x + (a_4D - b_4)y = 0,$$

where the a_i are nonzero constants. Determine a condition on the a_i such that the system is degenerate.

21. Verify that the degenerate system (7) possesses the solution $x(t) = c_1e^{-t}$, $y(t) = -2c_1e^{-t}$.

8.4 INTRODUCTION TO MATRICES

- *Matrix* • *Multiplicative inverse* • *Nonsingular matrix* • *Singular matrix*
- *Augmented matrix* • *Elementary row operations* • *Gaussian elimination*
- *Gauss-Jordan elimination* • *Reduced row-echelon form* • *Eigenvalue*
- *Eigenvector* • *Characteristic equation*

8.4.1 BASIC DEFINITIONS AND THEORY

Before examining a systematic procedure for solving linear first-order systems in normal form, we need the new and useful concept of a **matrix.**

DEFINITION 8.1 Matrix

A **matrix A** is any rectangular array of numbers or functions:

$$\mathbf{A} = \begin{pmatrix} a_{11} & a_{12} & \cdots & a_{1n} \\ a_{21} & a_{22} & \cdots & a_{2n} \\ \vdots & & & \vdots \\ a_{m1} & a_{m2} & \cdots & a_{mn} \end{pmatrix}. \qquad (1)$$

If a matrix has m rows and n columns, we say that its **size** is m by n (written $m \times n$). An $n \times n$ matrix is called a **square matrix** of order n.

The element, or entry, in the ith row and jth column of an $m \times n$ matrix **A** is written a_{ij}. An $m \times n$ matrix **A** is then abbreviated as $\mathbf{A} = (a_{ij})_{m \times n}$, or simply $\mathbf{A} = (a_{ij})$. A 1×1 matrix is one constant or function.

DEFINITION 8.2 **Equality of Matrices**

Two $m \times n$ matrices **A** and **B** are **equal** if $a_{ij} = b_{ij}$ for each i and j.

DEFINITION 8.3 **Column Matrix**

A column matrix X is any matrix having n rows and one column:

$$\mathbf{X} = \begin{pmatrix} b_{11} \\ b_{21} \\ \vdots \\ b_{n1} \end{pmatrix} = (b_{i1})_{n\times 1}.$$

A column matrix is also called a **column vector** or simply a **vector.**

DEFINITION 8.4 **Multiples of Matrices**

A **multiple** of a matrix **A** is defined to be

$$k\mathbf{A} = \begin{pmatrix} ka_{11} & ka_{12} & \cdots & ka_{1n} \\ ka_{21} & ka_{22} & \cdots & ka_{2n} \\ \vdots & & & \vdots \\ ka_{m1} & ka_{m2} & \cdots & ka_{mn} \end{pmatrix} = (ka_{ij})_{m\times n},$$

where k is a constant or a function.

EXAMPLE 1 **Multiples of Matrices**

(a) $5\begin{pmatrix} 2 & -3 \\ 4 & -1 \\ \frac{1}{5} & 6 \end{pmatrix} = \begin{pmatrix} 10 & -15 \\ 20 & -5 \\ 1 & 30 \end{pmatrix}$ **(b)** $e^t\begin{pmatrix} 1 \\ -2 \\ 4 \end{pmatrix} = \begin{pmatrix} e^t \\ -2e^t \\ 4e^t \end{pmatrix}$ ■

We note in passing that for any matrix **A**, the product $k\mathbf{A}$ is the same as $\mathbf{A}k$. For example,

$$e^{-3t}\begin{pmatrix} 2 \\ 5 \end{pmatrix} = \begin{pmatrix} 2e^{-3t} \\ 5e^{-3t} \end{pmatrix} = \begin{pmatrix} 2 \\ 5 \end{pmatrix}e^{-3t}.$$

DEFINITION 8.5 **Addition of Matrices**

The **sum** of two $m \times n$ matrices **A** and **B** is defined to be the matrix

$$\mathbf{A} + \mathbf{B} = (a_{ij} + b_{ij})_{m\times n}.$$

In other words, when adding two matrices of the same size, we add the corresponding elements.

EXAMPLE 2 Matrix Addition

The sum of $\mathbf{A} = \begin{pmatrix} 2 & -1 & 3 \\ 0 & 4 & 6 \\ -6 & 10 & -5 \end{pmatrix}$ and $\mathbf{B} = \begin{pmatrix} 4 & 7 & -8 \\ 9 & 3 & 5 \\ 1 & -1 & 2 \end{pmatrix}$ is

$$\mathbf{A} + \mathbf{B} = \begin{pmatrix} 2+4 & -1+7 & 3+(-8) \\ 0+9 & 4+3 & 6+5 \\ -6+1 & 10+(-1) & -5+2 \end{pmatrix} = \begin{pmatrix} 6 & 6 & -5 \\ 9 & 7 & 11 \\ -5 & 9 & -3 \end{pmatrix}.$$

EXAMPLE 3 Matrix Written as a Sum of Column Matrices

The single matrix $\begin{pmatrix} 3t^2 - 2e^t \\ t^2 + 7t \\ 5t \end{pmatrix}$ can be written as the sum of three column vectors:

$$\begin{pmatrix} 3t^2 - 2e^t \\ t^2 + 7t \\ 5t \end{pmatrix} = \begin{pmatrix} 3t^2 \\ t^2 \\ 0 \end{pmatrix} + \begin{pmatrix} 0 \\ 7t \\ 5t \end{pmatrix} + \begin{pmatrix} -2e^t \\ 0 \\ 0 \end{pmatrix} = \begin{pmatrix} 3 \\ 1 \\ 0 \end{pmatrix} t^2 + \begin{pmatrix} 0 \\ 7 \\ 5 \end{pmatrix} t + \begin{pmatrix} -2 \\ 0 \\ 0 \end{pmatrix} e^t.$$

The **difference** of two $m \times n$ matrices is defined in the usual manner: $\mathbf{A} - \mathbf{B} = \mathbf{A} + (-\mathbf{B})$, where $-\mathbf{B} = (-1)\mathbf{B}$.

DEFINITION 8.6 Multiplication of Matrices

Let **A** be a matrix having m rows and n columns and **B** be a matrix having n rows and p columns. We define the **product AB** to be the $m \times p$ matrix

$$\mathbf{AB} = \begin{pmatrix} a_{11} & a_{12} & \cdots & a_{1n} \\ a_{21} & a_{22} & \cdots & a_{2n} \\ \vdots & & & \vdots \\ a_{m1} & a_{m2} & \cdots & a_{mn} \end{pmatrix} \begin{pmatrix} b_{11} & b_{12} & \cdots & b_{1p} \\ b_{21} & b_{22} & \cdots & b_{2p} \\ \vdots & & & \vdots \\ b_{n1} & b_{n2} & \cdots & b_{np} \end{pmatrix}$$

$$= \begin{pmatrix} a_{11}b_{11} + a_{12}b_{21} + \cdots + a_{1n}b_{n1} & \cdots & a_{11}b_{1p} + a_{12}b_{2p} + \cdots + a_{1n}b_{np} \\ a_{21}b_{11} + a_{22}b_{21} + \cdots + a_{2n}b_{n1} & \cdots & a_{21}b_{1p} + a_{22}b_{2p} + \cdots + a_{2n}b_{np} \\ \vdots & & \vdots \\ a_{m1}b_{11} + a_{m2}b_{21} + \cdots + a_{mn}b_{n1} & \cdots & a_{m1}b_{1p} + a_{m2}b_{2p} + \cdots + a_{mn}b_{np} \end{pmatrix}$$

$$= \left(\sum_{k=1}^{n} a_{ik}b_{kj} \right)_{m \times p}.$$

Note carefully in Definition 8.6 that the product **AB** = **C** is defined only when the number of columns in the matrix **A** is the same as the number of rows in **B.** The size of the product can be determined from

$$\mathbf{A}_{m\times n}\mathbf{B}_{n\times p} = \mathbf{C}_{m\times p}.$$

Also, you might recognize that the entries in, say, the ith row of the final matrix **AB** are formed by using the component definition of the inner or dot product of the ith row of **A** with each of the columns of **B.**

EXAMPLE 4 Multiplication of Matrices

(a) For $\mathbf{A} = \begin{pmatrix} 4 & 7 \\ 3 & 5 \end{pmatrix}$ and $\mathbf{B} = \begin{pmatrix} 9 & -2 \\ 6 & 8 \end{pmatrix}$,

$$\mathbf{AB} = \begin{pmatrix} 4\cdot 9 + 7\cdot 6 & 4\cdot(-2) + 7\cdot 8 \\ 3\cdot 9 + 5\cdot 6 & 3\cdot(-2) + 5\cdot 8 \end{pmatrix} = \begin{pmatrix} 78 & 48 \\ 57 & 34 \end{pmatrix}.$$

(b) For $\mathbf{A} = \begin{pmatrix} 5 & 8 \\ 1 & 0 \\ 2 & 7 \end{pmatrix}$ and $\mathbf{B} = \begin{pmatrix} -4 & -3 \\ 2 & 0 \end{pmatrix}$,

$$\mathbf{AB} = \begin{pmatrix} 5\cdot(-4) + 8\cdot 2 & 5\cdot(-3) + 8\cdot 0 \\ 1\cdot(-4) + 0\cdot 2 & 1\cdot(-3) + 0\cdot 0 \\ 2\cdot(-4) + 7\cdot 2 & 2\cdot(-3) + 7\cdot 0 \end{pmatrix} = \begin{pmatrix} -4 & -15 \\ -4 & -3 \\ 6 & -6 \end{pmatrix}.$$

In general, matrix multiplication is not commutative; that is, **AB** ≠ **BA.** Observe in part (a) of Example 4 that $\mathbf{BA} = \begin{pmatrix} 30 & 53 \\ 48 & 82 \end{pmatrix}$, whereas in part (b) the product **BA** is *not defined* since Definition 8.6 requires that the first matrix (in this case **B**) have the same number of columns as the second matrix has rows.

We are particularly interested in the product of a square matrix and a column vector.

EXAMPLE 5 Multiplication of Matrices

(a) $$\begin{pmatrix} 2 & -1 & 3 \\ 0 & 4 & 5 \\ 1 & -7 & 9 \end{pmatrix}\begin{pmatrix} -3 \\ 6 \\ 4 \end{pmatrix} = \begin{pmatrix} 2\cdot(-3) + (-1)\cdot 6 + 3\cdot 4 \\ 0\cdot(-3) + 4\cdot 6 + 5\cdot 4 \\ 1\cdot(-3) + (-7)\cdot 6 + 9\cdot 4 \end{pmatrix} = \begin{pmatrix} 0 \\ 44 \\ -9 \end{pmatrix}$$

(b) $$\begin{pmatrix} -4 & 2 \\ 3 & 8 \end{pmatrix}\begin{pmatrix} x \\ y \end{pmatrix} = \begin{pmatrix} -4x + 2y \\ 3x + 8y \end{pmatrix}$$

Multiplicative Identity For a given positive integer n, the $n \times n$ matrix

$$\mathbf{I} = \begin{pmatrix} 1 & 0 & 0 & \cdots & 0 \\ 0 & 1 & 0 & \cdots & 0 \\ \vdots & & & & \vdots \\ 0 & 0 & 0 & \cdots & 1 \end{pmatrix}$$

is called the **multiplicative identity matrix.** It follows from Definition 8.6 that for any $n \times n$ matrix **A,**

$$\mathbf{AI} = \mathbf{IA} = \mathbf{A}.$$

Also, it is readily verified that if **X** is an $n \times 1$ column matrix, then $\mathbf{IX} = \mathbf{X}$.

Zero Matrix A matrix consisting of all zero entries is called a **zero matrix** and is denoted by **0.** For example,

$$\mathbf{0} = \begin{pmatrix} 0 \\ 0 \end{pmatrix}, \quad \mathbf{0} = \begin{pmatrix} 0 & 0 \\ 0 & 0 \end{pmatrix}, \quad \mathbf{0} = \begin{pmatrix} 0 & 0 \\ 0 & 0 \\ 0 & 0 \end{pmatrix},$$

and so on. If **A** and **0** are $m \times n$ matrices, then

$$\mathbf{A} + \mathbf{0} = \mathbf{0} + \mathbf{A} = \mathbf{A}.$$

Associative Law Although we shall not prove it, matrix multiplication is **associative.** If **A** is an $m \times p$ matrix, **B** a $p \times r$ matrix, and **C** an $r \times n$ matrix, then

$$\mathbf{A}(\mathbf{BC}) = (\mathbf{AB})\mathbf{C}$$

is an $m \times n$ matrix.

Distributive Law If **B** and **C** are $r \times n$ matrices and **A** is an $m \times r$ matrix, then the **distributive law** is

$$\mathbf{A}(\mathbf{B} + \mathbf{C}) = \mathbf{AB} + \mathbf{AC}.$$

Furthermore, if the product $(\mathbf{B} + \mathbf{C})\mathbf{A}$ is defined, then

$$(\mathbf{B} + \mathbf{C})\mathbf{A} = \mathbf{BA} + \mathbf{CA}.$$

Determinant of a Matrix Associated with every *square* matrix **A** of constants, there is a number called the **determinant of the matrix,** which is denoted by det **A** or $|\mathbf{A}|$.

EXAMPLE 6 Determinant of a Square Matrix

For $\mathbf{A} = \begin{pmatrix} 3 & 6 & 2 \\ 2 & 5 & 1 \\ -1 & 2 & 4 \end{pmatrix}$ we expand det **A** by cofactors of the first row:

$$\det \mathbf{A} = \begin{vmatrix} 3 & 6 & 2 \\ 2 & 5 & 1 \\ -1 & 2 & 4 \end{vmatrix} = 3\begin{vmatrix} 5 & 1 \\ 2 & 4 \end{vmatrix} - 6\begin{vmatrix} 2 & 1 \\ -1 & 4 \end{vmatrix} + 2\begin{vmatrix} 2 & 5 \\ -1 & 2 \end{vmatrix}$$

$$= 3(20 - 2) - 6(8 + 1) + 2(4 + 5) = 18.$$

See Appendix III for a brief review of the properties of determinants.

DEFINITION 8.7 **Transpose of a Matrix**

The **transpose** of the $m \times n$ matrix (1) is the $n \times m$ matrix $\mathbf{A}^T$ given by

$$\mathbf{A}^T = \begin{pmatrix} a_{11} & a_{21} & \cdots & a_{m1} \\ a_{12} & a_{22} & \cdots & a_{m2} \\ \vdots & & & \vdots \\ a_{1n} & a_{2n} & \cdots & a_{mn} \end{pmatrix}.$$

In other words, the rows of a matrix **A** become the columns of its transpose $\mathbf{A}^T$.

EXAMPLE 7 **Transpose of a Matrix**

(a) The transpose of matrix **A** in Example 6 is

$$\mathbf{A}^T = \begin{pmatrix} 3 & 2 & -1 \\ 6 & 5 & 2 \\ 2 & 1 & 4 \end{pmatrix}.$$

(b) If $\mathbf{X} = \begin{pmatrix} 5 \\ 0 \\ 3 \end{pmatrix}$, then $\mathbf{X}^T = (5 \quad 0 \quad 3)$.

DEFINITION 8.8 **Multiplicative Inverse of a Matrix**

Let **A** be an $n \times n$ matrix. If there exists an $n \times n$ matrix **B** such that

$$\mathbf{AB} = \mathbf{BA} = \mathbf{I},$$

where **I** is the multiplicative identity, then **B** is said to be the **multiplicative inverse of A** and is denoted by $\mathbf{B} = \mathbf{A}^{-1}$.

DEFINITION 8.9 **Nonsingular/Singular Matrices**

Let **A** be an $n \times n$ matrix. If det $\mathbf{A} \neq 0$, then **A** is said to be **nonsingular.** If det $\mathbf{A} = 0$, then **A** is said to be **singular.**

The following gives a necessary and sufficient condition for a square matrix to have a multiplicative inverse.

THEOREM 8.1 Nonsingularity Implies A Has an Inverse

An $n \times n$ matrix $\mathbf{A}$ has a multiplicative inverse $\mathbf{A}^{-1}$ if and only if $\mathbf{A}$ is nonsingular.

The following theorem gives one way of finding the multiplicative inverse for a nonsingular matrix.

THEOREM 8.2 A Formula for the Inverse of a Matrix

Let $\mathbf{A}$ be an $n \times n$ nonsingular matrix and let $C_{ij} = (-1)^{i+j}M_{ij}$, where M_{ij} is the determinant of the $(n-1) \times (n-1)$ matrix obtained by deleting the ith row and jth column from $\mathbf{A}$. Then

$$\mathbf{A}^{-1} = \frac{1}{\det \mathbf{A}}(C_{ij})^T. \tag{2}$$

Each C_{ij} in Theorem 8.2 is simply the **cofactor** (signed minor) of the corresponding entry a_{ij} in $\mathbf{A}$. Note that the transpose is utilized in formula (2).

For future reference we observe in the case of a 2×2 nonsingular matrix

$$\mathbf{A} = \begin{pmatrix} a_{11} & a_{12} \\ a_{21} & a_{22} \end{pmatrix}$$

that $C_{11} = a_{22}$, $C_{12} = -a_{21}$, $C_{21} = -a_{12}$, and $C_{22} = a_{11}$. Thus

$$\mathbf{A}^{-1} = \frac{1}{\det \mathbf{A}}\begin{pmatrix} a_{22} & -a_{21} \\ -a_{12} & a_{11} \end{pmatrix}^T = \frac{1}{\det \mathbf{A}}\begin{pmatrix} a_{22} & -a_{12} \\ -a_{21} & a_{11} \end{pmatrix}. \tag{3}$$

For a 3×3 nonsingular matrix

$$\mathbf{A} = \begin{pmatrix} a_{11} & a_{12} & a_{13} \\ a_{21} & a_{22} & a_{23} \\ a_{31} & a_{32} & a_{33} \end{pmatrix},$$

$$C_{11} = \begin{vmatrix} a_{22} & a_{23} \\ a_{32} & a_{33} \end{vmatrix}, \qquad C_{12} = -\begin{vmatrix} a_{21} & a_{23} \\ a_{31} & a_{33} \end{vmatrix}, \qquad C_{13} = \begin{vmatrix} a_{21} & a_{22} \\ a_{31} & a_{32} \end{vmatrix},$$

and so on. Carrying out the transposition gives

$$\mathbf{A}^{-1} = \frac{1}{\det \mathbf{A}}\begin{pmatrix} C_{11} & C_{21} & C_{31} \\ C_{12} & C_{22} & C_{32} \\ C_{13} & C_{23} & C_{33} \end{pmatrix}. \tag{4}$$

EXAMPLE 8 **Inverse of a 2 × 2 Matrix**

Find the multiplicative inverse for $\mathbf{A} = \begin{pmatrix} 1 & 4 \\ 2 & 10 \end{pmatrix}$.

Solution Since det $\mathbf{A} = 10 - 8 = 2 \neq 0$, $\mathbf{A}$ is nonsingular. It follows from Theorem 8.1 that $\mathbf{A}^{-1}$ exists. From (3) we find

$$\mathbf{A}^{-1} = \frac{1}{2}\begin{pmatrix} 10 & -4 \\ -2 & 1 \end{pmatrix} = \begin{pmatrix} 5 & -2 \\ -1 & \frac{1}{2} \end{pmatrix}.$$

EXAMPLE 9 **Matrix with No Inverse**

The matrix $\mathbf{A} = \begin{pmatrix} 2 & 2 \\ 3 & 3 \end{pmatrix}$ is singular since det $\mathbf{A} = 2(3) - 2(3) = 0$. We conclude that $\mathbf{A}^{-1}$ does not exist.

EXAMPLE 10 **Inverse of a 3 × 3 Matrix**

Find the multiplicative inverse for $\mathbf{A} = \begin{pmatrix} 2 & 2 & 0 \\ -2 & 1 & 1 \\ 3 & 0 & 1 \end{pmatrix}$.

Solution Since det $\mathbf{A} = 12 \neq 0$, the given matrix is nonsingular. The cofactors corresponding to the entries in each row of det $\mathbf{A}$ are

$$C_{11} = \begin{vmatrix} 1 & 1 \\ 0 & 1 \end{vmatrix} = 1 \qquad C_{12} = -\begin{vmatrix} -2 & 1 \\ 3 & 1 \end{vmatrix} = 5 \qquad C_{13} = \begin{vmatrix} -2 & 1 \\ 3 & 0 \end{vmatrix} = -3$$

$$C_{21} = -\begin{vmatrix} 2 & 0 \\ 0 & 1 \end{vmatrix} = -2 \qquad C_{22} = \begin{vmatrix} 2 & 0 \\ 3 & 1 \end{vmatrix} = 2 \qquad C_{23} = -\begin{vmatrix} 2 & 2 \\ 3 & 0 \end{vmatrix} = 6$$

$$C_{31} = \begin{vmatrix} 2 & 0 \\ 1 & 1 \end{vmatrix} = 2 \qquad C_{32} = -\begin{vmatrix} 2 & 0 \\ -2 & 1 \end{vmatrix} = -2 \qquad C_{33} = \begin{vmatrix} 2 & 2 \\ -2 & 1 \end{vmatrix} = 6$$

It follows from (4) that

$$\mathbf{A}^{-1} = \frac{1}{12}\begin{pmatrix} 1 & -2 & 2 \\ 5 & 2 & -2 \\ -3 & 6 & 6 \end{pmatrix} = \begin{pmatrix} \frac{1}{12} & -\frac{1}{6} & \frac{1}{6} \\ \frac{5}{12} & \frac{1}{6} & -\frac{1}{6} \\ -\frac{1}{4} & \frac{1}{2} & \frac{1}{2} \end{pmatrix}.$$

You are urged to verify that $\mathbf{A}^{-1}\mathbf{A} = \mathbf{A}\mathbf{A}^{-1} = \mathbf{I}$.

Formula (2) presents obvious difficulties for nonsingular matrices larger than 3 × 3. For example, to apply (2) to a 4 × 4 matrix we would have to calculate *sixteen* 3×3 determinants.* In the case of a large matrix, there are more efficient ways of finding $\mathbf{A}^{-1}$. The curious reader is referred to any text in linear algebra.

Since our goal is to apply the concept of a matrix to systems of linear differential equations in normal form, we need the following definitions.

DEFINITION 8.10 Derivative of a Matrix of Functions

If $\mathbf{A}(t) = (a_{ij}(t))_{m\times n}$ is a matrix whose entries are functions differentiable on a common interval, then

$$\frac{d\mathbf{A}}{dt} = \left(\frac{d}{dt}a_{ij}\right)_{m\times n}.$$

DEFINITION 8.11 Integral of a Matrix of Functions

If $\mathbf{A}(t) = (a_{ij}(t))_{m\times n}$ is a matrix whose entries are functions continuous on a common interval containing t and t_0, then

$$\int_{t_0}^{t}\mathbf{A}(s)\,ds = \left(\int_{t_0}^{t}a_{ij}(s)\,ds\right)_{m\times n}.$$

To differentiate (integrate) a matrix of functions we simply differentiate (integrate) each entry. The derivative of a matrix is also denoted by $\mathbf{A}'(t)$.

EXAMPLE 11 Derivative/Integral of a Matrix

If $\mathbf{X}(t) = \begin{pmatrix} \sin 2t \\ e^{3t} \\ 8t-1 \end{pmatrix}$, then $\mathbf{X}'(t) = \begin{pmatrix} \dfrac{d}{dt}\sin 2t \\ \dfrac{d}{dt}e^{3t} \\ \dfrac{d}{dt}(8t-1) \end{pmatrix} = \begin{pmatrix} 2\cos 2t \\ 3e^{3t} \\ 8 \end{pmatrix}$

and

$$\int_0^t \mathbf{X}(s)\,ds = \begin{pmatrix} \int_0^t \sin 2s\,ds \\ \int_0^t e^{3s}\,ds \\ \int_0^t (8s-1)\,ds \end{pmatrix} = \begin{pmatrix} -\frac{1}{2}\cos 2t + \frac{1}{2} \\ \frac{1}{3}e^{3t} - \frac{1}{3} \\ 4t^2 - t \end{pmatrix}$$

* Strictly speaking, a determinant is a number, but it is sometimes convenient to refer to a determinant as if it were an array.

8.4.2 GAUSSIAN AND GAUSS-JORDAN ELIMINATION METHODS

In preparation for Section 8.6, we need to know more about solving algebraic systems of n linear equations in n unknowns

$$
\begin{aligned}
a_{11}x_1 + a_{12}x_2 + \cdots + a_{1n}x_n &= b_1 \\
a_{21}x_1 + a_{22}x_2 + \cdots + a_{2n}x_n &= b_2 \\
&\vdots \\
a_{n1}x_1 + a_{n}x_2 + \cdots + a_{nn}x_n &= b_n.
\end{aligned}
\qquad (5)
$$

If **A** denotes the matrix of coefficients in (5), we know that Cramer's rule (see Appendix III) could be used to solve the system whenever det $\mathbf{A} \neq 0$. However, that rule requires a herculean effort if **A** is larger than 3×3. The procedure that we shall now consider has the distinct advantage of being not only an efficient way of handling large systems but also a means of solving consistent systems (5) in which det $\mathbf{A} = 0$ and a means of solving m linear equations in n unknowns.

DEFINITION 8.12 Augmented Matrix

The **augmented matrix** of the system (5) is the $n \times (n + 1)$ matrix

$$
\left(\begin{array}{cccc|c}
a_{11} & a_{12} & \cdots & a_{1n} & b_1 \\
a_{21} & a_{22} & \cdots & a_{2n} & b_2 \\
\vdots & & & & \vdots \\
a_{n1} & a_{n2} & \cdots & a_{nn} & b_n
\end{array}\right).
$$

If **B** is the column matrix of the b_i, $i = 1, 2, \ldots, n$, the augmented matrix of (5) is denoted by $(\mathbf{A}|\mathbf{B})$.

Elementary Row Operations Recall from algebra that we can transform an algebraic system of equations into an equivalent system (that is, one having the same solution) by multiplying an equation by a nonzero constant, interchanging the positions of any two equations in the system, and adding a nonzero constant multiple of an equation to another equation. These operations on equations in a system are, in turn, equivalent to **elementary row operations** on an augmented matrix:

(*i*) Multiply a row by a nonzero constant.
(*ii*) Interchange any two rows.
(*iii*) Add a nonzero constant multiple of one row to any other row.

Elimination Methods To solve a system such as (5) using an augmented matrix we use either **Gaussian elimination** or the **Gauss-Jordan elimination**

method.* In the former method we carry out a succession of elementary row operations until we arrive at an augmented matrix in **row-echelon form:**

(*i*) The first nonzero entry in a nonzero row is 1.
(*ii*) In consecutive nonzero rows, the first entry 1 in the lower row appears to the right of the first 1 in the higher row.
(*iii*) Rows consisting of all 0's are at the bottom of the matrix.

In the Gauss-Jordan method the row operations are continued until we obtain an augmented matrix that is in **reduced row-echelon form.** A reduced row-echelon matrix has the same three properties listed above in addition to

(*iv*) A column containing a first entry 1 has 0's everywhere else.

EXAMPLE 12 Row Echelon/Reduced Row-Echelon Form

(a) The augmented matrices

$$\left(\begin{array}{ccc|c} 1 & 5 & 0 & 2 \\ 0 & 1 & 0 & -1 \\ 0 & 0 & 0 & 0 \end{array}\right) \quad \text{and} \quad \left(\begin{array}{ccccc|c} 0 & 0 & 1 & -6 & 2 & 2 \\ 0 & 0 & 0 & 0 & 1 & 4 \end{array}\right)$$

are in row-echelon form. You should verify that the three criteria are satisfied.

---

* **KARL FRIEDRICH GAUSS** (1777–1855) Gauss was the first of a new breed of precise and demanding mathematicians–the "rigorists." As a child, Gauss was a prodigy in mathematics. As an adult he often remarked that he could calculate or "reckon" before he could talk. However, as a college student, Gauss was torn between two loves: philology and mathematics. But he was inspired by some original mathematical achievements as a teenager and encouraged by the mathematician Wolfgang Bolyai, so the choice was not too difficult. At the age of twenty Gauss settled on a career in mathematics. At the age of twenty-two he completed a book on number theory, *Disquisitiones Arithmeticae*. Published in 1801, this text was recognized as a masterpiece, and even today remains a classic in its field. Gauss's doctoral dissertation of 1799 also remains a memorable document. Using the theory of functions of a complex variable, he was the first to prove the so-called *fundamental theorem of algebra:* Every polynomial equation has at least one root.

Although Gauss was certainly recognized and respected as an outstanding mathematician during his lifetime, the full extent of his genius was not realized until the publication of his scientific diary in 1898, forty-four years after his death. Much to the chagrin of some nineteenth-century mathematicians, the diary revealed that Gauss had foreseen, sometimes by decades, many of their discoveries or, perhaps more accurately, rediscoveries. He was oblivious to fame; his mathematical researches were often pursued, like a child playing on a beach, simply for pleasure and self-satisfaction and not for the instruction that could be given to others through publication.

On any list of "Greatest Mathematicians Who Ever Lived," Karl Friedrich Gauss must surely rank near or at the top. For his profound impact on so many branches of mathematics, Gauss is sometimes referred to as "the prince of mathematicians."

WILHELM JORDAN (1842–1899) Jordan, a German engineer, used this method to solve linear systems in his 1888 text, *Handbook of Geodesy.*

(b) The augmented matrices

$$\left(\begin{array}{ccc|c} 1 & 0 & 0 & 7 \\ 0 & 1 & 0 & -1 \\ 0 & 0 & 0 & 0 \end{array}\right) \quad \text{and} \quad \left(\begin{array}{ccccc|c} 0 & 0 & 1 & -6 & 0 & -6 \\ 0 & 0 & 0 & 0 & 1 & 4 \end{array}\right)$$

are in reduced row-echelon form. Note that the remaining entries in the columns containing a leading entry 1 are all 0's. ■

It should be noted that in Gaussian elimination we stop once we have obtained *an* augmented matrix in row-echelon form. In other words, by using different sequences of row operations we may arrive at different row-echelon forms. This method then requires the use of back-substitution. In Gauss-Jordan elimination we stop when we have obtained *the* augmented matrix in reduced row-echelon form. Any sequence of row operations will lead to the same augmented matrix in reduced row-echelon form. This method does not require back-substitution; the solution of the system will be apparent by inspection of the final matrix. In terms of the equations of the original system, our goal in both methods is simply to make the coefficient of x_1 in the first equation* equal to 1 and then use multiples of that equation to eliminate x_1 from other equations. The process is repeated on the other variables.

To keep track of the row operations on an augmented matrix, we utilize the following notation:

Symbol	**Meaning**
R_{ij}	Interchange rows i and j
cR_i	Multiply the ith row by the nonzero constant c
$cR_i + R_j$	Multiply the ith row by c and add to the jth row

EXAMPLE 13 Solution by Elimination

Solve

$$\begin{aligned} 2x_1 + 6x_2 + x_3 &= 7 \\ x_1 + 2x_2 - x_3 &= -1 \\ 5x_1 + 7x_2 - 4x_3 &= 9 \end{aligned}$$

using **(a)** Gaussian elimination and **(b)** Gauss-Jordan elimination.

Solution

(a) Using row operations on the augmented matrix of the system, we obtain

$$\left(\begin{array}{ccc|c} 2 & 6 & 1 & 7 \\ 1 & 2 & -1 & -1 \\ 5 & 7 & -4 & 9 \end{array}\right) \xrightarrow{R_{12}} \left(\begin{array}{ccc|c} 1 & 2 & -1 & -1 \\ 2 & 6 & 1 & 7 \\ 5 & 7 & -4 & 9 \end{array}\right) \xrightarrow[-5R_1 + R_3]{-2R_1 + R_2} \left(\begin{array}{ccc|c} 1 & 2 & -1 & -1 \\ 0 & 2 & 3 & 9 \\ 0 & -3 & 1 & 14 \end{array}\right)$$

$$\xrightarrow{\frac{1}{2}R_{12}} \left(\begin{array}{ccc|c} 1 & 2 & -1 & -1 \\ 0 & 1 & \frac{3}{2} & \frac{9}{2} \\ 0 & -3 & 1 & 14 \end{array}\right) \xrightarrow{3R_2 + R_3} \left(\begin{array}{ccc|c} 1 & 2 & -1 & -1 \\ 0 & 1 & \frac{3}{2} & \frac{9}{2} \\ 0 & 0 & \frac{11}{2} & \frac{55}{2} \end{array}\right) \xrightarrow{\frac{2}{11}R_3} \left(\begin{array}{ccc|c} 1 & 2 & -1 & -1 \\ 0 & 1 & \frac{3}{2} & \frac{9}{2} \\ 0 & 0 & 1 & 5 \end{array}\right).$$

* We can always interchange equations so that the first equation contains the variable x_1.

The last matrix is in row-echelon form and represents the system

$$\begin{aligned} x_1 + 2x_2 - \quad x_3 &= -1 \\ x_2 + \frac{3}{2}x_3 &= \frac{9}{2} \\ x_3 &= 5. \end{aligned}$$

Substituting $x_3 = 5$ into the second equation then gives $x_2 = -3$. Substituting both these values back into the first equation finally yields $x_1 = 10$.

(b) We start with the last matrix above. Since the first entries in the second and third rows are 1's, we must, in turn, make the remaining entries in the second and third columns 0's:

$$\left(\begin{array}{ccc|c} 1 & 2 & -1 & -1 \\ 0 & 1 & \frac{3}{2} & \frac{9}{2} \\ 0 & 0 & 1 & 5 \end{array}\right) \xrightarrow{-2R_2 + R_1} \left(\begin{array}{ccc|c} 1 & 0 & -4 & -10 \\ 0 & 1 & \frac{3}{2} & \frac{9}{2} \\ 0 & 0 & 1 & 5 \end{array}\right) \xrightarrow[-\frac{3}{2}R_3 + R_2]{4R_3 + R_1} \left(\begin{array}{ccc|c} 1 & 0 & 0 & 10 \\ 0 & 1 & 0 & -3 \\ 0 & 0 & 1 & 5 \end{array}\right).$$

The last matrix is now in reduced row-echelon form. Because of what the matrix means in terms of equations, it evident that the solution of the system is $x_1 = 10, x_2 = -3, x_3 = 5$.

EXAMPLE 14 Gauss-Jordan Elimination

Use Gauss-Jordan elimination to solve

$$\begin{aligned} x + 3y - 2z &= -7 \\ 4x + \quad y + 3z &= 5 \\ 2x - 5y + 7z &= 19. \end{aligned}$$

Solution We solve the system using Gauss-Jordan elimination:

$$\left(\begin{array}{ccc|c} 1 & 3 & -2 & -7 \\ 4 & 1 & 3 & 5 \\ 2 & -5 & 7 & 19 \end{array}\right) \xrightarrow[-2R_1 + R_3]{-4R_1 + R_2} \left(\begin{array}{ccc|c} 1 & 3 & -2 & -7 \\ 0 & -11 & 11 & 33 \\ 0 & -11 & 11 & 33 \end{array}\right)$$

$$\xrightarrow[-\frac{1}{11}R_3]{-\frac{1}{11}R_2} \left(\begin{array}{ccc|c} 1 & 3 & -2 & -7 \\ 0 & 1 & -1 & -3 \\ 0 & 1 & -1 & -3 \end{array}\right) \xrightarrow[-R_2 + R_3]{-3R_2 + R_1} \left(\begin{array}{ccc|c} 1 & 0 & 1 & 2 \\ 0 & 1 & -1 & -3 \\ 0 & 0 & 0 & 0 \end{array}\right).$$

In this case the last matrix in reduced row-echelon form implies that the original system of three equations in three unknowns is really equivalent to two equations in three unknowns. Since only z is common to both equations (the nonzero rows), we can assign its values arbitrarily. If we let $z = t$, where t represents any real number, then we see that the system has infinitely many solutions: $x = 2 - t, y = -3 + t, z = t$. Geometrically, these equations are the parametric equations for the line of intersection of the planes $x + 0y + z = 2$ and $0x + y - z = -3$.

8.4.3 The Eigenvalue Problem

Eigenvalues and Eigenvectors Gauss-Jordan elimination can be used to find the **eigenvectors** for a square matrix.

DEFINITION 8.13 Eigenvalues and Eigenvectors

> Let **A** be an $n \times n$ matrix. A number λ is said to be an **eigenvalue** of **A** if there exists a *nonzero* solution vector **K** of the linear system
>
> $$\mathbf{AK} = \lambda\mathbf{K}. \tag{6}$$
>
> The solution vector **K** is said to be an **eigenvector** corresponding to the eigenvalue λ.

The word *eigenvalue* is a combination of German and English terms adapted from the German word *eigenwert,* which, translated literally, is "proper value." Eigenvalues and eigenvectors are also called **characteristic values** and **characteristic vectors,** respectively.

EXAMPLE 15 Eigenvector of a Matrix

Verify that $\mathbf{K} = \begin{pmatrix} 1 \\ -1 \\ 1 \end{pmatrix}$ is an eigenvector of the matrix $\mathbf{A} = \begin{pmatrix} 0 & -1 & -3 \\ 2 & 3 & 3 \\ -2 & 1 & 1 \end{pmatrix}$.

Solution By carrying out the multiplication **AK,** we see that

$$\mathbf{AK} = \begin{pmatrix} 0 & -1 & -3 \\ 2 & 3 & 3 \\ -2 & 1 & 1 \end{pmatrix}\begin{pmatrix} 1 \\ -1 \\ 1 \end{pmatrix} = \begin{pmatrix} -2 \\ 2 \\ -2 \end{pmatrix} = (-2)\begin{pmatrix} 1 \\ -1 \\ 1 \end{pmatrix} = (\underset{\text{eigenvalue}}{-2})\mathbf{K}.$$

We see from the preceding line and Definition 8.13 that $\lambda = -2$ is an eigenvalue of **A.** ■

Using properties of matrix algebra, we can write (6) in the alternative form

$$(\mathbf{A} - \lambda\mathbf{I})\mathbf{K} = \mathbf{0}, \tag{7}$$

where **I** is the multiplicative identity. If we let

$$\mathbf{K} = \begin{pmatrix} k_1 \\ k_2 \\ \vdots \\ k_n \end{pmatrix},$$

then (7) is the same as

$$\begin{aligned}(a_{11}-\lambda)k_1 + \quad a_{12}k_2 + \cdots + \quad a_{1n}k_n &= 0\\ a_{21}k_1 + (a_{22}-\lambda)k_2 + \cdots + \quad a_{2n}k_n &= 0\\ &\vdots\\ a_{n1}k_1 + \quad a_{n2}k_2 + \cdots + (a_{nn}-\lambda)k_n &= 0.\end{aligned} \tag{8}$$

Although an obvious solution of (8) is $k_1 = 0, k_2 = 0, \ldots, k_n = 0$, we are seeking only nontrivial solutions. Now it is known that a homogeneous system of n linear equations in n unknowns (that is, $b_i = 0, i = 1, 2, \ldots, n$ in (5)) has a nontrivial solution if and only if the determinant of the coefficient matrix is equal to zero. Thus to find a nonzero solution $\mathbf{K}$ for (7), we must have

$$\det(\mathbf{A} - \lambda\mathbf{I}) = 0. \tag{9}$$

Inspection of (8) shows that the expansion of $\det(\mathbf{A} - \lambda\mathbf{I})$ by cofactors results in an nth-degree polynomial in λ. The equation (9) is called the **characteristic equation** of **A**. Thus *the eigenvalues of* **A** *are the roots of the characteristic equation.* To find an eigenvector corresponding to an eigenvalue λ we simply solve the system of equations $(\mathbf{A} - \lambda\mathbf{I})\mathbf{K} = \mathbf{0}$ by applying Gauss-Jordan elimination to the augmented matrix $(\mathbf{A} - \lambda\mathbf{I} \mid \mathbf{0})$.

EXAMPLE 16 Eigenvalues/Eigenvectors

Find the eigenvalues and eigenvectors of $\mathbf{A} = \begin{pmatrix} 1 & 2 & 1 \\ 6 & -1 & 0 \\ -1 & -2 & -1 \end{pmatrix}$.

Solution To expand the determinant in the characteristic equation we use the cofactors of the second row:

$$\det(\mathbf{A}-\lambda\mathbf{I}) = \begin{vmatrix} 1-\lambda & 2 & 1 \\ 6 & -1-\lambda & 0 \\ -1 & -2 & -1-\lambda \end{vmatrix} = -\lambda^3 - \lambda^2 + 12\lambda = 0.$$

From $-\lambda^3 - \lambda^2 + 12\lambda = -\lambda(\lambda + 4)(\lambda - 3) = 0$ we see that the eigenvalues are $\lambda_1 = 0$, $\lambda_2 = -4$, and $\lambda_3 = 3$. To find the eigenvectors we must now reduce $(\mathbf{A} - \lambda\mathbf{I} \mid \mathbf{0})$ three times corresponding to the three distinct eigenvalues.

For $\lambda_1 = 0$ we have

$$(\mathbf{A} - 0\mathbf{I} \mid \mathbf{0}) = \left(\begin{array}{ccc|c} 1 & 2 & 1 & 0 \\ 6 & -1 & 0 & 0 \\ -1 & -2 & -1 & 0 \end{array}\right) \xrightarrow[R_1 + R_3]{-6R_1 + R_2} \left(\begin{array}{ccc|c} 1 & 2 & 1 & 0 \\ 0 & -13 & -6 & 0 \\ 0 & 0 & 0 & 0 \end{array}\right)$$

$$\xrightarrow{-\frac{1}{13}R_2} \left(\begin{array}{ccc|c} 1 & 2 & 1 & 0 \\ 0 & 1 & \frac{6}{13} & 0 \\ 0 & 0 & 0 & 0 \end{array}\right) \xrightarrow{-2R_2 + R_1} \left(\begin{array}{ccc|c} 1 & 0 & \frac{1}{13} & 0 \\ 0 & 1 & \frac{6}{13} & 0 \\ 0 & 0 & 0 & 0 \end{array}\right).$$

Thus we see that $k_1 = -\frac{1}{13}k_3$ and $k_2 = -\frac{6}{13}k_3$. Choosing $k_3 = -13$, we get the eigenvector*

$$\mathbf{K}_1 = \begin{pmatrix} 1 \\ 6 \\ -13 \end{pmatrix}.$$

For $\lambda_2 = -4$,

$$(\mathbf{A} + 4\mathbf{I}|0) = \left(\begin{array}{ccc|c} 5 & 2 & 1 & 0 \\ 6 & 3 & 0 & 0 \\ -1 & -2 & 3 & 0 \end{array}\right) \xrightarrow[R_{31}]{-R_3} \left(\begin{array}{ccc|c} 1 & 2 & -3 & 0 \\ 6 & 3 & 0 & 0 \\ 5 & 2 & 1 & 0 \end{array}\right)$$

$$\xrightarrow[-5R_1 + R_3]{-6R_1 + R_2} \left(\begin{array}{ccc|c} 1 & 2 & -3 & 0 \\ 0 & -9 & 18 & 0 \\ 0 & -8 & 16 & 0 \end{array}\right) \xrightarrow[-\frac{1}{8}R_3]{-\frac{1}{9}R_2} \left(\begin{array}{ccc|c} 1 & 2 & -3 & 0 \\ 0 & 1 & -2 & 0 \\ 0 & 1 & -2 & 0 \end{array}\right) \xrightarrow[-R_2 + R_3]{-2R_2 + R_1} \left(\begin{array}{ccc|c} 1 & 0 & 1 & 0 \\ 0 & 1 & -2 & 0 \\ 0 & 0 & 0 & 0 \end{array}\right)$$

implies $k_1 = -k_3$ and $k_2 = 2k_3$. Choosing $k_3 = 1$ then yields the second eigenvector

$$\mathbf{K}_2 = \begin{pmatrix} -1 \\ 2 \\ 1 \end{pmatrix}.$$

Finally, for $\lambda_3 = 3$ Gauss-Jordan elimination gives

$$(\mathbf{A} - 3\mathbf{I}|\mathbf{0}) = \left(\begin{array}{ccc|c} -2 & 2 & 1 & 0 \\ 6 & -4 & 0 & 0 \\ -1 & -2 & -4 & 0 \end{array}\right) \xrightarrow{\text{row operations}} \left(\begin{array}{ccc|c} 1 & 0 & 1 & 0 \\ 0 & 1 & \frac{3}{2} & 0 \\ 0 & 0 & 0 & 0 \end{array}\right),$$

and so $k_1 = -k_3$ and $k_2 = -\frac{3}{2}k_3$. The choice of $k_3 = -2$ leads to the third eigenvector:

$$\mathbf{K}_3 = \begin{pmatrix} 2 \\ 3 \\ -2 \end{pmatrix}.$$

When an $n \times n$ matrix $\mathbf{A}$ possesses n distinct eigenvalues $\lambda_1, \lambda_2, \ldots, \lambda_n$, it can be proved that a set of n linearly independent† eigenvectors $\mathbf{K}_1, \mathbf{K}_2, \ldots, \mathbf{K}_n$ can be found. However, when the characteristic equation has repeated roots, it may not be possible to find n linearly independent eigenvectors for $\mathbf{A}$.

EXAMPLE 17 Eigenvalues/Eigenvectors

Find the eigenvalues and eigenvectors of $\mathbf{A} = \begin{pmatrix} 3 & 4 \\ -1 & 7 \end{pmatrix}$.

* Of course k_3 could be chosen as any nonzero number. In other words, a nonzero constant multiple of an eigenvector is also an eigenvector.

† Linear independence of column vectors is defined in exactly the same manner as for functions.

Solution From the characteristic equation

$$\det(\mathbf{A} - \lambda\mathbf{I}) = \begin{vmatrix} 3-\lambda & 4 \\ -1 & 7-\lambda \end{vmatrix} = (\lambda - 5)^2 = 0,$$

we see that $\lambda_1 = \lambda_2 = 5$ is an eigenvalue of multiplicity two. In the case of a 2×2 matrix there is no need to use Gauss-Jordan elimination. To find the eigenvector(s) corresponding to $\lambda_1 = 5$ we resort to the system $(\mathbf{A} - 5\mathbf{I}\,|\,\mathbf{0})$ in its equivalent form

$$\begin{aligned} -2k_1 + 4k_2 &= 0 \\ -k_1 + 2k_2 &= 0. \end{aligned}$$

It is apparent from this system that $k_1 = 2k_2$. Thus if we choose $k_2 = 1$, we find the single eigenvector

$$\mathbf{K}_1 = \begin{pmatrix} 2 \\ 1 \end{pmatrix}.$$

EXAMPLE 18 Eigenvalues/Eigenvectors

Find the eigenvalues and eigenvectors of $\mathbf{A} = \begin{pmatrix} 9 & 1 & 1 \\ 1 & 9 & 1 \\ 1 & 1 & 9 \end{pmatrix}$.

Solution The characteristic equation

$$\det(\mathbf{A} - \lambda\mathbf{I}) = \begin{vmatrix} 9-\lambda & 1 & 1 \\ 1 & 9-\lambda & 1 \\ 1 & 1 & 9-\lambda \end{vmatrix} = -(\lambda - 11)(\lambda - 8)^2 = 0$$

shows that $\lambda_1 = 11$ and that $\lambda_2 = \lambda_3 = 8$ is an eigenvalue of multiplicity two.

For $\lambda_1 = 11$ Gauss-Jordan elimination gives

$$(\mathbf{A} - 11\mathbf{I}\,|\,\mathbf{0}) = \left(\begin{array}{ccc|c} -2 & 1 & 1 & 0 \\ 1 & -2 & 1 & 0 \\ 1 & 1 & -2 & 0 \end{array}\right) \xrightarrow{\text{row operations}} \left(\begin{array}{ccc|c} 1 & 0 & -1 & 0 \\ 0 & 1 & -1 & 0 \\ 0 & 0 & 0 & 0 \end{array}\right).$$

Hence $k_1 = k_3$ and $k_2 = k_3$. If $k_3 = 1$, then

$$\mathbf{K}_1 = \begin{pmatrix} 1 \\ 1 \\ 1 \end{pmatrix}.$$

Now for $\lambda_2 = 8$ we have

$$(\mathbf{A} - 8\mathbf{I}\,|\,\mathbf{0}) = \left(\begin{array}{ccc|c} 1 & 1 & 1 & 0 \\ 1 & 1 & 1 & 0 \\ 1 & 1 & 1 & 0 \end{array}\right) \xrightarrow{\text{row operations}} \left(\begin{array}{ccc|c} 1 & 1 & 1 & 0 \\ 0 & 0 & 0 & 0 \\ 0 & 0 & 0 & 0 \end{array}\right).$$

In the equation $k_1 + k_2 + k_3 = 0$, we are free to select two of the variables arbitrarily. Choosing, on the one hand, $k_2 = 1, k_3 = 0$ and, on the other, $k_2 = 0, k_3 = 1$, we obtain two linearly independent eigenvectors

$$\mathbf{K}_2 = \begin{pmatrix} -1 \\ 1 \\ 0 \end{pmatrix} \quad \text{and} \quad \mathbf{K}_3 = \begin{pmatrix} -1 \\ 0 \\ 1 \end{pmatrix}$$

corresponding to a single eigenvalue.

EXERCISES 8.4

Answers to odd-numbered problems begin on page A-19.

8.4.1 Basic Definitions and Theory

1. If $\mathbf{A} = \begin{pmatrix} 4 & 5 \\ -6 & 9 \end{pmatrix}$ and $\mathbf{B} = \begin{pmatrix} -2 & 6 \\ 8 & -10 \end{pmatrix}$,

find **(a)** $\mathbf{A} + \mathbf{B}$, **(b)** $\mathbf{B} - \mathbf{A}$, **(c)** $2\mathbf{A} + 3\mathbf{B}$.

2. If $\mathbf{A} = \begin{pmatrix} -2 & 0 \\ 4 & 1 \\ 7 & 3 \end{pmatrix}$ and $\mathbf{B} = \begin{pmatrix} 3 & -1 \\ 0 & 2 \\ -4 & -2 \end{pmatrix}$,

find **(a)** $\mathbf{A} - \mathbf{B}$, **(b)** $\mathbf{B} - \mathbf{A}$, **(c)** $2(\mathbf{A} + \mathbf{B})$.

3. If $\mathbf{A} = \begin{pmatrix} 2 & -3 \\ -5 & 4 \end{pmatrix}$ and $\mathbf{B} = \begin{pmatrix} -1 & 6 \\ 3 & 2 \end{pmatrix}$,

find **(a)** $\mathbf{AB}$, **(b)** $\mathbf{BA}$, **(c)** $\mathbf{A}^2 = \mathbf{AA}$, **(d)** $\mathbf{B}^2 = \mathbf{BB}$.

4. If $\mathbf{A} = \begin{pmatrix} 1 & 4 \\ 5 & 10 \\ 8 & 12 \end{pmatrix}$ and $\mathbf{B} = \begin{pmatrix} -4 & 6 & -3 \\ 1 & -3 & 2 \end{pmatrix}$,

find **(a)** $\mathbf{AB}$, **(b)** $\mathbf{BA}$.

5. If $\mathbf{A} = \begin{pmatrix} 1 & -2 \\ -2 & 4 \end{pmatrix}$, $\mathbf{B} = \begin{pmatrix} 6 & 3 \\ 2 & 1 \end{pmatrix}$, and $\mathbf{C} = \begin{pmatrix} 0 & 2 \\ 3 & 4 \end{pmatrix}$,

find **(a)** $\mathbf{BC}$, **(b)** $\mathbf{A}(\mathbf{BC})$, **(c)** $\mathbf{C}(\mathbf{BA})$, **(d)** $\mathbf{A}(\mathbf{B} + \mathbf{C})$.

6. If $\mathbf{A} = (5 \quad -6 \quad 7)$, $\mathbf{B} = \begin{pmatrix} 3 \\ 4 \\ -1 \end{pmatrix}$, and $\mathbf{C} = \begin{pmatrix} 1 & 2 & 4 \\ 0 & 1 & -1 \\ 3 & 2 & 1 \end{pmatrix}$,

find **(a)** $\mathbf{AB}$, **(b)** $\mathbf{BA}$, **(c)** $(\mathbf{BA})\mathbf{C}$, **(d)** $(\mathbf{AB})\mathbf{C}$.

7. If $\mathbf{A} = \begin{pmatrix} 4 \\ 8 \\ -10 \end{pmatrix}$ and $\mathbf{B} = (2 \quad 4 \quad 5)$,

find **(a)** $\mathbf{A}^T\mathbf{A}$, **(b)** $\mathbf{B}^T\mathbf{B}$, **(c)** $\mathbf{A} + \mathbf{B}^T$.

8. If $\mathbf{A} = \begin{pmatrix} 1 & 2 \\ 2 & 4 \end{pmatrix}$ and $\mathbf{B} = \begin{pmatrix} -2 & 3 \\ 5 & 7 \end{pmatrix}$,
find **(a)** $\mathbf{A} + \mathbf{B}^T$, **(b)** $2\mathbf{A}^T - \mathbf{B}^T$, **(c)** $\mathbf{A}^T(\mathbf{A} - \mathbf{B})$.

9. If $\mathbf{A} = \begin{pmatrix} 3 & 4 \\ 8 & 1 \end{pmatrix}$ and $\mathbf{B} = \begin{pmatrix} 5 & 10 \\ -2 & -5 \end{pmatrix}$,
find **(a)** $(\mathbf{AB})^T$, **(b)** $\mathbf{B}^T\mathbf{A}^T$.

10. If $\mathbf{A} = \begin{pmatrix} 5 & 9 \\ -4 & 6 \end{pmatrix}$ and $\mathbf{B} = \begin{pmatrix} -3 & 11 \\ -7 & 2 \end{pmatrix}$,
find **(a)** $\mathbf{A}^T + \mathbf{B}^T$, **(b)** $(\mathbf{A} + \mathbf{B})^T$.

In Problems 11–14 write the given sum as a single column matrix.

11. $4\begin{pmatrix} -1 \\ 2 \end{pmatrix} - 2\begin{pmatrix} 2 \\ 8 \end{pmatrix} + 3\begin{pmatrix} -2 \\ 3 \end{pmatrix}$

12. $3t\begin{pmatrix} 2 \\ t \\ -1 \end{pmatrix} + (t-1)\begin{pmatrix} -1 \\ -t \\ 3 \end{pmatrix} - 2\begin{pmatrix} 3t \\ 4 \\ -5t \end{pmatrix}$

13. $\begin{pmatrix} 2 & -3 \\ 1 & 4 \end{pmatrix}\begin{pmatrix} -2 \\ 5 \end{pmatrix} - \begin{pmatrix} -1 & 6 \\ -2 & 3 \end{pmatrix}\begin{pmatrix} -7 \\ 2 \end{pmatrix}$

14. $\begin{pmatrix} 1 & -3 & 4 \\ 2 & 5 & -1 \\ 0 & -4 & -2 \end{pmatrix}\begin{pmatrix} t \\ 2t-1 \\ -t \end{pmatrix} + \begin{pmatrix} -t \\ 1 \\ 4 \end{pmatrix} - \begin{pmatrix} 2 \\ 8 \\ -6 \end{pmatrix}$

In Problems 15–22 determine whether the given matrix is singular or nonsingular. If it is nonsingular, find $\mathbf{A}^{-1}$.

15. $\mathbf{A} = \begin{pmatrix} -3 & 6 \\ -2 & 4 \end{pmatrix}$

16. $\mathbf{A} = \begin{pmatrix} 2 & 5 \\ 1 & 4 \end{pmatrix}$

17. $\mathbf{A} = \begin{pmatrix} 4 & 8 \\ -3 & -5 \end{pmatrix}$

18. $\mathbf{A} = \begin{pmatrix} 7 & 10 \\ 2 & 2 \end{pmatrix}$

19. $\mathbf{A} = \begin{pmatrix} 2 & 1 & 0 \\ -1 & 2 & 1 \\ 1 & 2 & 1 \end{pmatrix}$

20. $\mathbf{A} = \begin{pmatrix} 3 & 2 & 1 \\ 4 & 1 & 0 \\ -2 & 5 & -1 \end{pmatrix}$

21. $\mathbf{A} = \begin{pmatrix} 2 & 1 & 1 \\ 1 & -2 & -3 \\ 3 & 2 & 4 \end{pmatrix}$

22. $\mathbf{A} = \begin{pmatrix} 4 & 1 & -1 \\ 6 & 2 & -3 \\ -2 & -1 & 2 \end{pmatrix}$

In Problems 23 and 24 show that the given matrix is nonsingular for every real value of t. Find $\mathbf{A}^{-1}(t)$.

23. $\mathbf{A}(t) = \begin{pmatrix} 2e^{-t} & e^{4t} \\ 4e^{-t} & 3e^{4t} \end{pmatrix}$

24. $\mathbf{A}(t) = \begin{pmatrix} 2e^t \sin t & -2e^t \cos t \\ e^t \cos t & e^t \sin t \end{pmatrix}$

In Problems 25–28 find $d\mathbf{X}/dt$.

25. $\mathbf{X} = \begin{pmatrix} 5e^{-t} \\ 2e^{-t} \\ -7e^{-t} \end{pmatrix}$

26. $\mathbf{X} = \begin{pmatrix} \frac{1}{2}\sin 2t - 4\cos 2t \\ -3\sin 2t + 5\cos 2t \end{pmatrix}$

27. $\mathbf{X} = 2\begin{pmatrix} 1 \\ -1 \end{pmatrix}e^{2t} + 4\begin{pmatrix} 2 \\ 1 \end{pmatrix}e^{-3t}$

28. $\mathbf{X} = \begin{pmatrix} 5te^{2t} \\ t\sin 3t \end{pmatrix}$

29. Let $\mathbf{A}(t) = \begin{pmatrix} e^{4t} & \cos \pi t \\ 2t & 3t^2 - 1 \end{pmatrix}$.

Find **(a)** $\dfrac{d\mathbf{A}}{dt}$, **(b)** $\int_0^2 \mathbf{A}(t)\,dt$, **(c)** $\int_0^t \mathbf{A}(s)\,ds$.

30. Let $\mathbf{A}(t) = \begin{pmatrix} \dfrac{1}{t^2+1} & 3t \\ t^2 & t \end{pmatrix}$ and $\mathbf{B}(t) = \begin{pmatrix} 6t & 2 \\ 1/t & 4t \end{pmatrix}$.

Find **(a)** $\dfrac{d\mathbf{A}}{dt}$, **(b)** $\dfrac{d\mathbf{B}}{dt}$, **(c)** $\int_0^1 \mathbf{A}(t)\,dt$, **(d)** $\int_1^2 \mathbf{B}(t)\,dt$, **(e)** $\mathbf{A}(t)\mathbf{B}(t)$,

(f) $\dfrac{d}{dt}\mathbf{A}(t)\mathbf{B}(t)$, **(g)** $\int_1^t \mathbf{A}(s)\mathbf{B}(s)\,ds$.

8.4.2 Gaussian and Gauss-Jordan Elimination Methods

In Problems 31–38 solve the given system of equations by either Gaussian elimination or Gauss-Jordan elimination.

31. $\begin{aligned} x + y - 2z &= 14 \\ 2x - y + z &= 0 \\ 6x + 3y + 4z &= 1 \end{aligned}$

32. $\begin{aligned} 5x - 2y + 4z &= 10 \\ x + y + z &= 9 \\ 4x - 3y + 3z &= 1 \end{aligned}$

33. $\begin{aligned} y + z &= -5 \\ 5x + 4y - 16z &= -10 \\ x - y - 5z &= 7 \end{aligned}$

34. $\begin{aligned} 3x + y + z &= 4 \\ 4x + 2y - z &= 7 \\ x + y - 3z &= 6 \end{aligned}$

35. $\begin{aligned} 2x + y + z &= 4 \\ 10x - 2y + 2z &= -1 \\ 6x - 2y + 4z &= 8 \end{aligned}$

36. $\begin{aligned} x + 2z &= 8 \\ x + 2y - 2z &= 4 \\ 2x + 5y - 6z &= 6 \end{aligned}$

37. $\begin{aligned} x_1 + x_2 - x_3 - x_4 &= -1 \\ x_1 + x_2 + x_3 + x_4 &= 3 \\ x_1 - x_2 + x_3 - x_4 &= 3 \\ 4x_1 + x_2 - 2x_3 + x_4 &= 0 \end{aligned}$

38. $\begin{aligned} 2x_1 + x_2 + x_3 &= 0 \\ x_1 + 3x_2 + x_3 &= 0 \\ 7x_1 + x_2 + 3x_3 &= 0 \end{aligned}$

In Problems 39 and 40 use Gauss-Jordan elimination to demonstrate that the given system of equations has no solution.

39. $\begin{aligned} x + 2y + 4z &= 2 \\ 2x + 4y + 3z &= 1 \\ x + 2y - z &= 7 \end{aligned}$

40. $\begin{aligned} x_1 + x_2 - x_3 + 3x_4 &= 1 \\ x_2 - x_3 - 4x_4 &= 0 \\ x_1 + 2x_2 - 2x_3 - x_4 &= 6 \\ 4x_1 + 7x_2 - 7x_3 &= 9 \end{aligned}$

8.4.3 The Eigenvalue Problem

In Problems 41–48 find the eigenvalues and eigenvectors of the given matrix.

41. $\begin{pmatrix} -1 & 2 \\ -7 & 8 \end{pmatrix}$

42. $\begin{pmatrix} 2 & 1 \\ 2 & 1 \end{pmatrix}$

43. $\begin{pmatrix} -8 & -1 \\ 16 & 0 \end{pmatrix}$

44. $\begin{pmatrix} 1 & 1 \\ \frac{1}{4} & 1 \end{pmatrix}$

45. $\begin{pmatrix} 5 & -1 & 0 \\ 0 & -5 & 9 \\ 5 & -1 & 0 \end{pmatrix}$

46. $\begin{pmatrix} 3 & 0 & 0 \\ 0 & 2 & 0 \\ 4 & 0 & 1 \end{pmatrix}$

47. $\begin{pmatrix} 0 & 4 & 0 \\ -1 & -4 & 0 \\ 0 & 0 & -2 \end{pmatrix}$ **48.** $\begin{pmatrix} 1 & 6 & 0 \\ 0 & 2 & 1 \\ 0 & 1 & 2 \end{pmatrix}$

In Problems 49 and 50 show that the given matrix has complex eigenvalues. Find the eigenvectors of the matrix.

49. $\begin{pmatrix} -1 & 2 \\ -5 & 1 \end{pmatrix}$ **50.** $\begin{pmatrix} 2 & -1 & 0 \\ 5 & 2 & 4 \\ 0 & 1 & 2 \end{pmatrix}$

51. If $\mathbf{A}(t)$ is a 2×2 matrix of differentiable functions and $\mathbf{X}(t)$ is a 2×1 column matrix of differentiable functions, prove the product rule

$$\frac{d}{dt}[\mathbf{A}(t)\mathbf{X}(t)] = \mathbf{A}(t)\mathbf{X}'(t) + \mathbf{A}'(t)\mathbf{X}(t).$$

52. Derive formula (3). [*Hint*: Find a matrix $\mathbf{B} = \begin{pmatrix} b_{11} & b_{12} \\ b_{21} & b_{22} \end{pmatrix}$ for which $\mathbf{AB} = \mathbf{I}$. Solve for b_{11}, b_{12}, b_{21}, and b_{22}. Then show that $\mathbf{BA} = \mathbf{I}$.]

53. If $\mathbf{A}$ is nonsingular and $\mathbf{AB} = \mathbf{AC}$, show that $\mathbf{B} = \mathbf{C}$.

54. If $\mathbf{A}$ and $\mathbf{B}$ are nonsingular, show that $(\mathbf{AB})^{-1} = \mathbf{B}^{-1}\mathbf{A}^{-1}$.

55. Let $\mathbf{A}$ and $\mathbf{B}$ be $n \times n$ matrices. In general, is

$$(\mathbf{A} + \mathbf{B})^2 = \mathbf{A}^2 + 2\mathbf{AB} + \mathbf{B}^2?$$

8.5 MATRICES AND SYSTEMS OF LINEAR FIRST-ORDER EQUATIONS

• *Solution vector* • *Superposition principle* • *Linear dependence*
• *Linear independence* • *Wronskian* • *Fundamental set* • *General solution*
• *Fundamental matrix*

8.5.1 PRELIMINARY THEORY

Matrix Form of a System If $\mathbf{X}$, $\mathbf{A}(t)$, and $\mathbf{F}(t)$ denote the respective matrices

$$\mathbf{X} = \begin{pmatrix} x_1(t) \\ x_2(t) \\ \vdots \\ x_n(t) \end{pmatrix}, \quad \mathbf{A}(t) = \begin{pmatrix} a_{11}(t) & a_{12}(t) & \cdots & a_{1n}(t) \\ a_{21}(t) & a_{22}(t) & \cdots & a_{2n}(t) \\ \vdots & & & \vdots \\ a_{n1}(t) & a_{n2}(t) & \cdots & a_{nn}(t) \end{pmatrix}, \quad \mathbf{F}(t) = \begin{pmatrix} f_1(t) \\ f_2(t) \\ \vdots \\ f_n(t) \end{pmatrix},$$

then the system of linear first-order differential equations

$$\begin{aligned} \frac{dx_1}{dt} &= a_{11}(t)x_1 + a_{12}(t)x_2 + \cdots + a_{1n}(t)x_n + f_1(t) \\ \frac{dx_2}{dt} &= a_{21}(t)x_1 + a_{22}(t)x_2 + \cdots + a_{2n}(t)x_n + f_2(t) \\ &\vdots \\ \frac{dx_n}{dt} &= a_{n1}(t)x_1 + a_{n2}(t)x_2 + \cdots + a_{nn}(t)x_n + f_n(t) \end{aligned} \tag{1}$$

can be written as

$$\frac{d}{dt}\begin{pmatrix} x_1 \\ x_2 \\ \vdots \\ x_n \end{pmatrix} = \begin{pmatrix} a_{11}(t) & a_{12}(t) & \dots & a_{1n}(t) \\ a_{21}(t) & a_{22}(t) & \dots & a_{2n}(t) \\ \vdots & & & \vdots \\ a_{n1}(t) & a_{n2}(t) & \dots & a_{nn}(t) \end{pmatrix}\begin{pmatrix} x_1 \\ x_2 \\ \vdots \\ x_n \end{pmatrix} + \begin{pmatrix} f_1(t) \\ f_2(t) \\ \vdots \\ f_n(t) \end{pmatrix},$$

or simply

$$\frac{d\mathbf{X}}{dt} = \mathbf{A}(t)\mathbf{X} + \mathbf{F}(t). \tag{2}$$

If the system is homogeneous, (2) becomes

$$\frac{d\mathbf{X}}{dt} = \mathbf{A}(t)\mathbf{X}. \tag{3}$$

Equations (2) and (3) are also written as $\mathbf{X}' = \mathbf{AX} + \mathbf{F}$ and $\mathbf{X}' = \mathbf{AX}$, respectively.

EXAMPLE 1 Systems Written in Matrix Notation

In matrix terms the nonhomogeneous system

$$\frac{dx}{dt} = -2x + 5y + e^t - 2t$$
$$\frac{dy}{dt} = 4x - 3y + 10t$$

can be written as

$$\frac{d\mathbf{X}}{dt} = \begin{pmatrix} -2 & 5 \\ 4 & -3 \end{pmatrix}\mathbf{X} + \begin{pmatrix} e^t - 2t \\ 10t \end{pmatrix}$$

or

$$\mathbf{X}' = \begin{pmatrix} -2 & 5 \\ 4 & -3 \end{pmatrix}\mathbf{X} + \begin{pmatrix} 1 \\ 0 \end{pmatrix}e^t + \begin{pmatrix} -2 \\ 10 \end{pmatrix}t,$$

where $\mathbf{X} = \begin{pmatrix} x \\ y \end{pmatrix}$. ■

EXAMPLE 2 Systems Written in Matrix Notation

The matrix form of the homogeneous system

$$\frac{dx}{dt} = 2x - 3y \qquad \text{is} \qquad \frac{d\mathbf{X}}{dt} = \begin{pmatrix} 2 & -3 \\ 6 & 5 \end{pmatrix}\mathbf{X},$$
$$\frac{dy}{dt} = 6x + 5y$$

where $\mathbf{X} = \begin{pmatrix} x \\ y \end{pmatrix}$. ■

DEFINITION 8.14 Solution Vector

A solution vector on an interval I is any column matrix

$$\mathbf{X} = \begin{pmatrix} x_1(t) \\ x_2(t) \\ \vdots \\ x_n(t) \end{pmatrix}$$

whose entries are differentiable functions satisfying the system (2) on the interval.

EXAMPLE 3 Verification of Solutions

Verify that

$$\mathbf{X}_1 = \begin{pmatrix} 1 \\ -1 \end{pmatrix} e^{-2t} = \begin{pmatrix} e^{-2t} \\ -e^{-2t} \end{pmatrix} \quad \text{and} \quad \mathbf{X}_2 = \begin{pmatrix} 3 \\ 5 \end{pmatrix} e^{6t} = \begin{pmatrix} 3e^{6t} \\ 5e^{6t} \end{pmatrix}$$

are solutions of

$$\mathbf{X}' = \begin{pmatrix} 1 & 3 \\ 5 & 3 \end{pmatrix} \mathbf{X} \tag{4}$$

on the interval $(-\infty, \infty)$.

Solution We have

$$\mathbf{X}_1' = \begin{pmatrix} -2e^{-2t} \\ 2e^{-2t} \end{pmatrix}$$

and

$$\mathbf{AX}_1 = \begin{pmatrix} 1 & 3 \\ 5 & 3 \end{pmatrix} \begin{pmatrix} e^{-2t} \\ -e^{-2t} \end{pmatrix} = \begin{pmatrix} e^{-2t} - 3e^{-2t} \\ 5e^{-2t} - 3e^{-2t} \end{pmatrix} = \begin{pmatrix} -2e^{-2t} \\ 2e^{-2t} \end{pmatrix} = \mathbf{X}_1'.$$

Now

$$\mathbf{X}_2' = \begin{pmatrix} 18e^{6t} \\ 30e^{5t} \end{pmatrix}$$

and

$$\mathbf{AX}_2 = \begin{pmatrix} 1 & 3 \\ 5 & 3 \end{pmatrix} \begin{pmatrix} 3e^{6t} \\ 5e^{6t} \end{pmatrix} = \begin{pmatrix} 3e^{6t} + 15e^{6t} \\ 15e^{6t} + 15e^{6t} \end{pmatrix} = \begin{pmatrix} 18e^{6t} \\ 30e^{6t} \end{pmatrix} = \mathbf{X}_2'.$$

Much of the theory of systems of n linear first-order differential equations is similar to that of linear nth-order differential equations.

Initial-Value Problem Let t_0 denote a point on an interval I and

$$\mathbf{X}(t_0) = \begin{pmatrix} x_1(t_0) \\ x_2(t_0) \\ \vdots \\ x_n(t_0) \end{pmatrix} \quad \text{and} \quad \mathbf{X}_0 = \begin{pmatrix} \gamma_1 \\ \gamma_2 \\ \vdots \\ \gamma_n \end{pmatrix},$$

where the γ_i, $i = 1, 2, \ldots, n$ are given constants. Then the problem

$$\begin{aligned} &\textit{Solve:} && \frac{d\mathbf{X}}{dt} = \mathbf{A}(t)\mathbf{X} + \mathbf{F}(t) \\ &\textit{Subject to:} && \mathbf{X}(t_0) = \mathbf{X}_0 \end{aligned} \tag{5}$$

is an **initial-value problem** on the interval.

THEOREM 8.3 Existence of a Unique Solution

Let the entries of the matrices $\mathbf{A}(t)$ and $\mathbf{F}(t)$ be functions continuous on a common interval I that contains the point t_0. Then there exists a unique solution of the initial-value problem (5) on the interval.

Homogeneous Systems In the next several definitions and theorems, we are concerned only with homogeneous systems. Without stating it, we shall always assume that the a_{ij} and the f_i are continuous functions of t on some common interval I.

Superposition Principle The following result is a **superposition principle** for solutions of linear systems.

THEOREM 8.4 Superposition Principle

Let $\mathbf{X}_1, \mathbf{X}_2, \ldots, \mathbf{X}_k$ be a set of solution vectors of the homogeneous system (3) on an interval I. Then the linear combination

$$\mathbf{X} = c_1\mathbf{X}_1 + c_2\mathbf{X}_2 + \cdots + c_k\mathbf{X}_k,$$

where the c_i, $i = 1, 2, \ldots, k$ are arbitrary constants, is also a solution on the interval.

It follows from Theorem 8.4 that a constant multiple of any solution vector of a homogeneous system of linear first-order differential equations is also a solution.

EXAMPLE 4 Constant Multiple of a Solution Is a Solution

One solution of the system

$$\mathbf{X}' = \begin{pmatrix} 1 & 0 & 1 \\ 1 & 1 & 0 \\ -2 & 0 & -1 \end{pmatrix}\mathbf{X} \tag{6}$$

is

$$\mathbf{X}_1 = \begin{pmatrix} \cos t \\ -\frac{1}{2}\cos t + \frac{1}{2}\sin t \\ -\cos t - \sin t \end{pmatrix}.$$

For any constant c_1 the vector $\mathbf{X} = c_1\mathbf{X}_1$ is also a solution since

$$\frac{d\mathbf{X}}{dt} = \begin{pmatrix} -c_1\sin t \\ \frac{1}{2}c_1\sin t + \frac{1}{2}c_1\cos t \\ c_1\sin t - c_1\cos t \end{pmatrix}$$

and

$$\mathbf{AX} = \begin{pmatrix} 1 & 0 & 1 \\ 1 & 1 & 0 \\ -2 & 0 & -1 \end{pmatrix} \begin{pmatrix} -c_1 \cos t \\ -\frac{1}{2}c_1 \cos t + \frac{1}{2}c_1 \sin t \\ -c_1 \cos t - c_1 \sin t \end{pmatrix} = \begin{pmatrix} -c_1 \sin t \\ \frac{1}{2}c_1 \cos t + \frac{1}{2}c_1 \sin t \\ -c_1 \cos t + c_1 \sin t \end{pmatrix}.$$

Inspection of the resulting matrices shows that $\mathbf{X}' = \mathbf{AX}$.

EXAMPLE 5 Using the Superposition Principle

Consider the system (6) of Example 4. If

$$\mathbf{X}_2 = \begin{pmatrix} 0 \\ e^t \\ 0 \end{pmatrix}, \quad \text{then} \quad \mathbf{X}_2' = \begin{pmatrix} 0 \\ e^t \\ 0 \end{pmatrix}$$

and

$$\mathbf{AX}_2 = \begin{pmatrix} 1 & 0 & 1 \\ 1 & 1 & 0 \\ -2 & 0 & -1 \end{pmatrix} \begin{pmatrix} 0 \\ e^t \\ 0 \end{pmatrix} = \begin{pmatrix} 0 \\ e^t \\ 0 \end{pmatrix} = \mathbf{X}_2'.$$

Thus we see that $\mathbf{X}_2$ is also a solution vector of (6). By the superposition principle the linear combination

$$\mathbf{X} = c_1\mathbf{X}_1 + c_2\mathbf{X}_2 = c_1 \begin{pmatrix} \cos t \\ -\frac{1}{2}\cos t + \frac{1}{2}\sin t \\ -\cos t - \sin t \end{pmatrix} + c_2 \begin{pmatrix} 0 \\ e^t \\ 0 \end{pmatrix}$$

is yet another solution of the system.

Linear Independence We are primarily interested in linearly independent solutions of the homogeneous system (3).

DEFINITION 8.15 Linear Dependence/Independence

Let $\mathbf{X}_1, \mathbf{X}_2, \ldots, \mathbf{X}_k$ be a set of solution vectors of the homogeneous system (3) on an interval I. We say that the set is **linearly dependent** on the interval if there exist constants $c_1, c_2, \ldots, c_k$, not all zero, such that

$$c_1\mathbf{X}_1 + c_2\mathbf{X}_2 + \cdots + c_k\mathbf{X}_k = \mathbf{0}$$

for every t in the interval. If the set of vectors is not linearly dependent on the interval, it is said to be **linearly independent.**

The case when $k = 2$ should be clear; two solution vectors $\mathbf{X}_1$ and $\mathbf{X}_2$ are linearly dependent if one is a constant multiple of the other, and conversely. For

$k > 2$ a set of solution vectors is linearly dependent if we can express at least one solution vector as a linear combination of the remaining vectors.

EXAMPLE 6 Linearly Independent Solutions

It can be verified that $\mathbf{X}_1 = \begin{pmatrix} 3 \\ 1 \end{pmatrix} e^t$ and $\mathbf{X}_2 = \begin{pmatrix} 1 \\ 1 \end{pmatrix} e^{-t}$ are solution vectors of the system

$$\mathbf{X}' = \begin{pmatrix} 2 & -3 \\ 1 & -2 \end{pmatrix} \mathbf{X}. \tag{7}$$

Now $\mathbf{X}_1$ and $\mathbf{X}_2$ are linearly independent on the interval $(-\infty, \infty)$ since

$$c_1\mathbf{X}_1 + c_2\mathbf{X}_2 = \mathbf{0} \quad \text{or} \quad c_1\begin{pmatrix} 3 \\ 1 \end{pmatrix} e^t + c_2\begin{pmatrix} 1 \\ 1 \end{pmatrix} e^{-t} = \begin{pmatrix} 0 \\ 0 \end{pmatrix}$$

is equivalent to

$$\begin{aligned} 3c_1e^t + c_2e^{-t} &= 0 \\ c_1e^t + c_2e^{-t} &= 0. \end{aligned}$$

Solving this system for c_1 and c_2 immediately yields $c_1 = 0$ and $c_2 = 0$. ■

EXAMPLE 7 Linearly Dependent Solutions

The vector $\mathbf{X}_3 = \begin{pmatrix} e^t + \cosh t \\ \cosh t \end{pmatrix}$ is also a solution of the system (7) given in Example 6. However, $\mathbf{X}_1$, $\mathbf{X}_2$, and $\mathbf{X}_3$ are linearly dependent since

$$\mathbf{X}_3 = \frac{1}{2}\mathbf{X}_1 + \frac{1}{2}\mathbf{X}_2.$$ ■

Wronskian As in our earlier consideration of the theory of a single ordinary differential equation, we can introduce the concept of the **Wronskian** determinant as a test for linear independence. We state the following theorem without proof.

THEOREM 8.5 Criterion for Linearly Independent Solutions

Let
$$\mathbf{X}_1 = \begin{pmatrix} x_{11} \\ x_{21} \\ \vdots \\ x_{n1} \end{pmatrix}, \quad \mathbf{X}_2 = \begin{pmatrix} x_{12} \\ x_{22} \\ \vdots \\ x_{n2} \end{pmatrix}, \quad \ldots, \quad \mathbf{X}_n = \begin{pmatrix} x_{1n} \\ x_{2n} \\ \vdots \\ x_{nn} \end{pmatrix}$$

be n solution vectors of the homogeneous system (3) on an interval I. A necessary and sufficient condition that the set of solutions be linearly independent is that the Wronskian

$$W(\mathbf{X}_1, \mathbf{X}_2, \ldots, \mathbf{X}_n) = \begin{vmatrix} x_{11} & x_{12} & \cdots & x_{1n} \\ x_{21} & x_{22} & \cdots & x_{2n} \\ \vdots & & & \vdots \\ x_{n1} & x_{n2} & \cdots & x_{nn} \end{vmatrix} \neq 0 \tag{8}$$

for every t in I.

In fact, it can be shown that if $\mathbf{X}_1, \mathbf{X}_2, \ldots, \mathbf{X}_n$ are solution vectors of (3), then either $W(\mathbf{X}_1, \mathbf{X}_2, \ldots, \mathbf{X}_n) \neq 0$ for every t in I or $W(\mathbf{X}_1, \mathbf{X}_2, \ldots, \mathbf{X}_n) = 0$ for every t in the interval. Thus if we can show that $W \neq 0$ for some t_0 in I, then $W \neq 0$ for every t and hence the solutions are linearly independent on the interval.

Notice that, unlike our previous definition of the Wronskian, the determinant (8) does not involve differentiation.

EXAMPLE 8 Using the Wronskian

In Example 3 we saw that

$$\mathbf{X}_1 = \begin{pmatrix} 1 \\ -1 \end{pmatrix} e^{-2t} \quad \text{and} \quad \mathbf{X}_2 = \begin{pmatrix} 3 \\ 5 \end{pmatrix} e^{6t}$$

are solutions of the system (4). Clearly, $\mathbf{X}_1$ and $\mathbf{X}_2$ are linearly independent on $(-\infty, \infty)$ since neither vector is a constant multiple of the other. In addition, we have

$$W(\mathbf{X}_1, \mathbf{X}_2) = \begin{vmatrix} e^{-2t} & 3e^{6t} \\ -e^{-2t} & 5e^{6t} \end{vmatrix} = 8e^{4t} \neq 0$$

for all real values of t.

Fundamental Set of Solutions

DEFINITION 8.16 Fundamental Set of Solutions

Any set $\mathbf{X}_1, \mathbf{X}_2, \ldots, \mathbf{X}_n$ of n linearly independent solution vectors of the homogeneous system (3) on an interval I is said to be a **fundamental set of solutions** on the interval.

THEOREM 8.6 Existence of a Fundamental Set

There exists a fundamental set of solutions for the homogeneous system (3) on an interval I.

DEFINITION 8.17 General Solution—Homogeneous Systems

Let $\mathbf{X}_1, \mathbf{X}_2, \ldots, \mathbf{X}_n$ be a fundamental set of solutions of the homogeneous system (3) on an interval I. The **general solution** of the system on the interval is defined to be

$$\mathbf{X} = c_1\mathbf{X}_1 + c_2\mathbf{X}_2 + \cdots + c_n\mathbf{X}_n,$$

where the c_i, $i = 1, 2, \ldots, n$ are arbitrary constants.

Although we shall not give the proof, it can be shown that, for appropriate choices of the constants $c_1, c_2, \ldots, c_n$, *any* solution of (3) on the interval I can be obtained from the general solution.

EXAMPLE 9 General Solution

From Example 8 we know that

$$\mathbf{X}_1 = \begin{pmatrix} 1 \\ -1 \end{pmatrix} e^{-2t} \quad \text{and} \quad \mathbf{X}_2 = \begin{pmatrix} 3 \\ 5 \end{pmatrix} e^{6t}$$

are linearly independent solutions of (4) on $(-\infty, \infty)$. Hence $\mathbf{X}_1$ and $\mathbf{X}_2$ form a fundamental set of solutions on the interval. The general solution of the system on the interval is then

$$\mathbf{X} = c_1\mathbf{X}_1 + c_2\mathbf{X}_2 = c_1\begin{pmatrix} 1 \\ -1 \end{pmatrix} e^{-2t} + c^2\begin{pmatrix} 3 \\ 5 \end{pmatrix} e^{6t}. \tag{9}$$

EXAMPLE 10 General Solution

The vectors

$$\mathbf{X}_1 = \begin{pmatrix} \cos t \\ -\frac{1}{2}\cos t + \frac{1}{2}\sin t \\ -\cos t - \sin t \end{pmatrix}, \quad \mathbf{X}_2 = \begin{pmatrix} 0 \\ 1 \\ 0 \end{pmatrix} e^t, \quad \mathbf{X}_3 = \begin{pmatrix} \sin t \\ -\frac{1}{2}\sin t - \frac{1}{2}\cos t \\ -\sin t + \cos t \end{pmatrix}$$

are solutions of the system (6)* in Example 4. Now

$$W(\mathbf{X}_1, \mathbf{X}_2, \mathbf{X}_3) = \begin{vmatrix} \cos t & 0 & \sin t \\ -\frac{1}{2}\cos t + \frac{1}{2}\sin t & e^t & -\frac{1}{2}\sin t - \frac{1}{2}\cos t \\ -\cos t - \sin t & 0 & -\sin t + \cos t \end{vmatrix}$$

$$= e^t \begin{vmatrix} \cos t & \sin t \\ -\cos t - \sin t & -\sin t + \cos t \end{vmatrix} = e^t \neq 0$$

for all real values of t. We conclude that $\mathbf{X}_1$, $\mathbf{X}_2$, and $\mathbf{X}_3$ form a fundamental set of solutions on $(-\infty, \infty)$. Thus the general solution of the system on the interval is

$$\mathbf{X} = c_1\mathbf{X}_1 + c_2\mathbf{X}_2 + c_3\mathbf{X}_3$$

$$= c_1\begin{pmatrix} \cos t \\ -\frac{1}{2}\cos t + \frac{1}{2}\sin t \\ -\cos t - \sin t \end{pmatrix} + c_2\begin{pmatrix} 0 \\ 1 \\ 0 \end{pmatrix} e^t + c_3\begin{pmatrix} \sin t \\ -\frac{1}{2}\sin t - \frac{1}{2}\cos t \\ -\sin t + \cos t \end{pmatrix}.$$

Nonhomogeneous Systems For nonhomogeneous systems a **particular solution $\mathbf{X}_p$** on an interval I is any vector, free of arbitrary parameters, whose entries are functions that satisfy the system (2).

---

* On pages 378 and 379 it was verified that $\mathbf{X}_1$ and $\mathbf{X}_2$ are solutions; it is left as an exercise to demonstrate that $\mathbf{X}_3$ is also a solution. See Problem 16.

EXAMPLE 11 **Particular Solution**

Verify that the vector $\mathbf{X}_p = \begin{pmatrix} 3t - 4 \\ -5t + 6 \end{pmatrix}$ is a particular solution of the nonhomogeneous system

$$\mathbf{X}' = \begin{pmatrix} 1 & 3 \\ 5 & 3 \end{pmatrix}\mathbf{X} + \begin{pmatrix} 12t - 11 \\ -3 \end{pmatrix} \tag{10}$$

on the interval $(-\infty, \infty)$.

Solution We have $\mathbf{X}_p' = \begin{pmatrix} 3 \\ -5 \end{pmatrix}$ and

$$\begin{aligned}
\begin{pmatrix} 1 & 3 \\ 5 & 3 \end{pmatrix}\mathbf{X}_p + \begin{pmatrix} 12t - 11 \\ -3 \end{pmatrix} &= \begin{pmatrix} 1 & 3 \\ 5 & 3 \end{pmatrix}\begin{pmatrix} 3t - 4 \\ -5t + 6 \end{pmatrix} + \begin{pmatrix} 12t - 11 \\ -3 \end{pmatrix} \\
&= \begin{pmatrix} (3t - 4) + 3(-5t + 6) \\ 5(3t - 4) + 3(-5t + 6) \end{pmatrix} + \begin{pmatrix} 12t - 11 \\ -3 \end{pmatrix} \\
&= \begin{pmatrix} -12t + 14 \\ -2 \end{pmatrix} + \begin{pmatrix} 12t - 11 \\ -3 \end{pmatrix} = \begin{pmatrix} 3 \\ -5 \end{pmatrix} = \mathbf{X}_p'.
\end{aligned}$$

THEOREM 8.7

Let $\mathbf{X}_1, \mathbf{X}_2, \ldots, \mathbf{X}_k$ be a set of solution vectors of the homogeneous system (3) on an interval I and let $\mathbf{X}_p$ be any solution vector of the nonhomogeneous system (2) on the same interval. Then

$$\mathbf{X} = c_1\mathbf{X}_1 + c_2\mathbf{X}_2 + \cdots + c_k\mathbf{X}_k + \mathbf{X}_p$$

is also a solution of the nonhomogeneous system on the interval for any constants $c_1, c_2, \ldots, c_k$.

DEFINITION 8.18 **General Solution—Nonhomogeneous Systems**

Let $\mathbf{X}_p$ be a given solution of the nonhomogeneous system (2) on an interval I, and let

$$\mathbf{X}_c = c_1\mathbf{X}_1 + c_2\mathbf{X}_2 + \cdots + c_n\mathbf{X}_n$$

denote the general solution on the same interval of the associated homogeneous system (3). The **general solution** of the nonhomogeneous system on the interval is defined to be

$$\mathbf{X} = \mathbf{X}_c + \mathbf{X}_p.$$

The general solution $\mathbf{X}_c$ of the homogeneous system (3) is called the **complementary function** of the nonhomogeneous system (2).

EXAMPLE 12 **General Solution**

In Example 11 it was verified that a particular solution of the nonhomogeneous system (10) on $(-\infty, \infty)$ is

$$\mathbf{X}_p = \begin{pmatrix} 3t - 5 \\ -5t + 6 \end{pmatrix}.$$

The complementary function of (10) on the same interval, or the general solution of

$$\mathbf{X}' = \begin{pmatrix} 1 & 3 \\ 5 & 3 \end{pmatrix}\mathbf{X},$$

was seen in Example 9 to be

$$\mathbf{X}_c = c_1\begin{pmatrix} 1 \\ -1 \end{pmatrix}e^{-2t} + c_2\begin{pmatrix} 3 \\ 5 \end{pmatrix}e^{6t}.$$

Hence by Definition 8.18,

$$\mathbf{X} = \mathbf{X}_c + \mathbf{X}_p = c_1\begin{pmatrix} 1 \\ -1 \end{pmatrix}e^{-2t} + c_2\begin{pmatrix} 3 \\ 5 \end{pmatrix}e^{6t} + \begin{pmatrix} 3t - 4 \\ -5t + 6 \end{pmatrix}$$

is the general solution of (10) on $(-\infty, \infty)$. ■

As one might expect, if $\mathbf{X}$ is *any* solution of the nonhomogeneous system (2) on an interval I, then it is always possible to find appropriate constants $c_1, c_2, \ldots, c_n$ so that $\mathbf{X}$ can be obtained from the general solution.

8.5.2 A FUNDAMENTAL MATRIX

If $\mathbf{X}_1, \mathbf{X}_2, \ldots, \mathbf{X}_n$ is a fundamental set of solutions of the homogeneous system (3) on an interval I, then its general solution on the interval is

$$\mathbf{X} = c_1\mathbf{X}_1 + c_2\mathbf{X}_2 + \cdots + c_n\mathbf{X}_n$$
$$= c_1\begin{pmatrix} x_{11} \\ x_{21} \\ \vdots \\ x_{n1} \end{pmatrix} + c_2\begin{pmatrix} x_{12} \\ x_{22} \\ \vdots \\ x_{n2} \end{pmatrix} + \cdots + c_n\begin{pmatrix} x_{1n} \\ x_{2n} \\ \vdots \\ x_{nn} \end{pmatrix} = \begin{pmatrix} c_1x_{11} + c_2x_{12} + \cdots + c_nx_{1n} \\ c_1x_{21} + c_2x_{22} + \cdots + c_nx_{2n} \\ \vdots \\ c_1x_{n1} + c_2x_{n2} + \cdots + c_nx_{nn} \end{pmatrix}. \quad \textbf{(11)}$$

Observe that (11) can be written as the matrix product

$$\mathbf{X} = \begin{pmatrix} x_{11} & x_{12} & \cdots & x_{1n} \\ x_{21} & x_{22} & \cdots & x_{2n} \\ \vdots & & & \vdots \\ x_{n1} & x_{n2} & \cdots & x_{nn} \end{pmatrix}\begin{pmatrix} c_1 \\ c_2 \\ \vdots \\ c_n \end{pmatrix}. \quad \textbf{(12)}$$

We are led to the following definition.

DEFINITION 8.19 Fundamental Matrix

Let $$\mathbf{X}_1 = \begin{pmatrix} x_{11} \\ x_{21} \\ \vdots \\ x_{n1} \end{pmatrix}, \quad \mathbf{X}_2 = \begin{pmatrix} x_{12} \\ x_{22} \\ \vdots \\ x_{n2} \end{pmatrix}, \quad \ldots, \quad \mathbf{X}_n = \begin{pmatrix} x_{1n} \\ x_{2n} \\ \vdots \\ x_{nn} \end{pmatrix}$$

be a fundamental set of n solution vectors of the homogeneous system (3) on an interval I. The matrix

$$\mathbf{\Phi}(t) = \begin{pmatrix} x_{11} & x_{12} & \cdots & x_{1n} \\ x_{21} & x_{22} & \cdots & x_{2n} \\ \vdots & & & \vdots \\ x_{n1} & x_{n2} & \cdots & x_{nn} \end{pmatrix}$$

is said to be a **fundamental matrix** of the system on the interval.

EXAMPLE 13 A Fundamental Matrix

The vectors

$$\mathbf{X}_1 = \begin{pmatrix} 1 \\ -1 \end{pmatrix} e^{-2t} = \begin{pmatrix} e^{-2t} \\ -e^{-2t} \end{pmatrix} \quad \text{and} \quad \mathbf{X}_2 = \begin{pmatrix} 3 \\ 5 \end{pmatrix} e^{6t} = \begin{pmatrix} 3e^{6t} \\ 5e^{6t} \end{pmatrix}$$

have been shown to form a fundamental set of solutions of the system (4) on $(-\infty, \infty)$. A fundamental matrix of the system on the interval is then

$$\mathbf{\Phi}(t) = \begin{pmatrix} e^{-2t} & 3e^{6t} \\ -e^{-2t} & 5e^{6t} \end{pmatrix}. \qquad \textbf{(13)}$$

The result given in (12) states that the general solution of any homogeneous system $\mathbf{X}' = \mathbf{A}(t)\mathbf{X}$ can always be written in terms of a fundamental matrix of the system: $\mathbf{X} = \mathbf{\Phi}(t)\mathbf{C}$, where $\mathbf{C}$ is an $n \times 1$ column vector of arbitrary constants.

EXAMPLE 14 General Solution Using a Fundamental Matrix

The general solution given in (9) can be written as

$$\mathbf{X} = \begin{pmatrix} e^{-2t} & 3e^{6t} \\ -e^{-2t} & 5e^{6t} \end{pmatrix} \begin{pmatrix} c_1 \\ c_2 \end{pmatrix}.$$

Furthermore, to say that $\mathbf{X} = \mathbf{\Phi}(t)\mathbf{C}$ is a solution of $\mathbf{X}' = \mathbf{A}(t)\mathbf{X}$ means

$$\mathbf{\Phi}'(t)\mathbf{C} = \mathbf{A}(t)\mathbf{\Phi}(t)\mathbf{C}$$

or $(\mathbf{\Phi}'(t) - \mathbf{A}(t)\mathbf{\Phi}(t))\mathbf{C} = \mathbf{0}$. Since the last equation is to hold for every t in the interval I and for every possible column matrix of constants $\mathbf{C}$, we must have $\mathbf{\Phi}'(t) - \mathbf{A}(t)\mathbf{\Phi}(t) = \mathbf{0}$ or

$$\mathbf{\Phi}'(t) = \mathbf{A}(t)\mathbf{\Phi}(t). \tag{14}$$

This result will be useful in Section 8.8.

Fundamental Matrix Is Nonsingular Comparison of Theorem 8.5 and Definition 8.19 shows that $\det \mathbf{\Phi}(t)$ is the same as the Wronskian $W(\mathbf{X}_1, \mathbf{X}_2, \ldots, \mathbf{X}_n)$.* Hence the linear independence of the columns of $\mathbf{\Phi}(t)$ on an interval I guarantees that $\det \mathbf{\Phi}(t) \neq 0$ for every t in the interval; that is, $\mathbf{\Phi}(t)$ is nonsingular on the interval.

THEOREM 8.8 A Fundamental Matrix Has an Inverse

Let $\mathbf{\Phi}(t)$ be a fundamental matrix of the homogeneous system (3) on an interval I. Then $\mathbf{\Phi}^{-1}(t)$ exists for every value of t in the interval.

EXAMPLE 15 Inverse of a Fundamental Matrix

For the fundamental matrix given in (13) we see that $\det \mathbf{\Phi}(t) = 8e^{4t}$. It then follows from (3) of Section 8.4 that

$$\mathbf{\Phi}^{-1}(t) = \frac{1}{8e^{4t}}\begin{pmatrix} 5e^{6t} & -3e^{6t} \\ e^{-2t} & e^{-2t} \end{pmatrix} = \begin{pmatrix} \frac{5}{8}e^{2t} & -\frac{3}{8}e^{2t} \\ \frac{1}{8}e^{-6t} & \frac{1}{8}e^{-6t} \end{pmatrix}.$$

Special Matrix In some instances it is convenient to form another special $n \times n$ matrix, a matrix in which the column vectors $\mathbf{V}_i$ are solutions of $\mathbf{X}' = \mathbf{A}(t)\mathbf{X}$ that satisfy the conditions

$$\mathbf{V}_1(t_0) = \begin{pmatrix} 1 \\ 0 \\ \vdots \\ 0 \end{pmatrix}, \quad \mathbf{V}_2(t_0) = \begin{pmatrix} 0 \\ 1 \\ \vdots \\ 0 \end{pmatrix}, \quad \ldots, \quad \mathbf{V}_n(t_0) = \begin{pmatrix} 0 \\ 0 \\ \vdots \\ 1 \end{pmatrix}. \tag{15}$$

Here t_0 is an arbitrarily chosen point in the interval on which the general solution of the system is defined. We denote this special matrix by the symbol $\mathbf{\Psi}(t)$. Observe that $\mathbf{\Psi}(t)$ has the property

$$\mathbf{\Psi}(t_0) = \begin{pmatrix} 1 & 0 & 0 & \cdots & 0 \\ 0 & 1 & 0 & \cdots & 0 \\ \vdots & & & & \vdots \\ 0 & 0 & 0 & \cdots & 1 \end{pmatrix} = \mathbf{I}, \tag{16}$$

where $\mathbf{I}$ is the $n \times n$ multiplicative identity.

* For this reason some texts call $\mathbf{\Phi}(t)$ a *Wronski matrix.*

EXAMPLE 16 Finding $\Psi(t)$

Find the matrix $\Psi(t)$ satisfying $\Psi(0) = \mathbf{I}$ for the system given in (4).

Solution From (9) we know that the general solution of (4) is given by

$$\mathbf{X} = c_1\begin{pmatrix} 1 \\ -1 \end{pmatrix}e^{-2t} + c_2\begin{pmatrix} 3 \\ 5 \end{pmatrix}e^{6t}.$$

When $t = 0$, we first solve for constants c_1 and c_2 such that

$$c_1\begin{pmatrix} 1 \\ -1 \end{pmatrix} + c_2\begin{pmatrix} 3 \\ 5 \end{pmatrix} = \begin{pmatrix} 1 \\ 0 \end{pmatrix} \quad \text{or} \quad \begin{aligned} c_1 + 3c_2 &= 1 \\ -c_1 + 5c_2 &= 0. \end{aligned}$$

We find that $c_1 = \frac{5}{8}$ and $c_2 = \frac{1}{8}$. Hence we define the vector $\mathbf{V}_1$ to be the linear combination

$$\mathbf{V}_1 = \frac{5}{8}\begin{pmatrix} 1 \\ -1 \end{pmatrix}e^{-2t} + \frac{1}{8}\begin{pmatrix} 3 \\ 5 \end{pmatrix}e^{6t}.$$

Again when $t = 0$ we wish to find another pair of constants c_1 and c_2 for which

$$c_1\begin{pmatrix} 1 \\ -1 \end{pmatrix} + c_2\begin{pmatrix} 3 \\ 5 \end{pmatrix} = \begin{pmatrix} 0 \\ 1 \end{pmatrix} \quad \text{or} \quad \begin{aligned} c_1 + 3c_2 &= 0 \\ -c_1 + 5c_2 &= 1. \end{aligned}$$

In this case we find $c_1 = -\frac{3}{8}$ and $c_2 = \frac{1}{8}$. We then define

$$\mathbf{V}_2 = -\frac{3}{8}\begin{pmatrix} 1 \\ -1 \end{pmatrix}e^{-2t} + \frac{1}{8}\begin{pmatrix} 3 \\ 5 \end{pmatrix}e^{6t}.$$

Hence

$$\Psi(t) = \begin{pmatrix} \frac{5}{8}e^{-2t} + \frac{3}{8}e^{6t} & -\frac{3}{8}e^{-2t} + \frac{3}{8}e^{6t} \\ -\frac{5}{8}e^{-2t} + \frac{5}{8}e^{6t} & \frac{3}{8}e^{-2t} + \frac{5}{8}e^{6t} \end{pmatrix}. \tag{17}$$

Observe that $\Psi(0) = \begin{pmatrix} 1 & 0 \\ 0 & 1 \end{pmatrix} = \mathbf{I}.$ ■

Note in Example 16 that since the columns of $\Psi(t)$ are linear combinations of the solutions Ψ of $\mathbf{X}' = \mathbf{A}(t)\mathbf{X}$, we know from the superposition principle that each column is a solution of the system.

$\Psi(t)$ Is a Fundamental Matrix From (16) it is seen that det $\Psi(t_0) \neq 0$, and hence we conclude from Theorem 8.5 that the columns of $\Psi(t)$ are linearly independent on the interval under consideration. Therefore $\Psi(t)$ is a fundamental matrix. Also, it follows from Theorem 8.3 that $\Psi(t)$ is the unique matrix that satisfies the condition $\Psi(t_0) = \mathbf{I}$. Last, the fundamental matrices $\Phi(t)$ and $\Psi(t)$ are related by

$$\Psi(t) = \Phi(t)\Phi^{-1}(t_0). \tag{18}$$

Equation (18) provides an alternative method for determining $\Psi(t)$ (see Problem 37).

The answer to why anyone would want to form an obviously complicated looking fundamental matrix such as (17) will be answered in Sections 8.8 and 8.9.

EXERCISES 8.5

Answers to odd-numbered problems begin on page A-19.

8.5.1 Preliminary Theory

In Problems 1–6 write the given system in matrix form.

1. $\dfrac{dx}{dt} = 3x - 5y$

$\dfrac{dy}{dt} = 4x + 8y$

2. $\dfrac{dx}{dt} = 4x - 7y$

$\dfrac{dy}{dt} = 5x$

3. $\dfrac{dx}{dt} = -3x + 4y - 9z$

$\dfrac{dy}{dt} = 6x - y$

$\dfrac{dz}{dt} = 10x + 4y + 3z$

4. $\dfrac{dx}{dt} = x - y$

$\dfrac{dy}{dt} = x + 2z$

$\dfrac{dz}{dt} = -x + z$

5. $\dfrac{dx}{dt} = x - y + z + t - 1$

$\dfrac{dy}{dt} = 2x + y - z - 3t^2$

$\dfrac{dz}{dt} = x + y + z + t^2 - t + 2$

6. $\dfrac{dx}{dt} = -3x + 4y + e^{-t}\sin 2t$

$\dfrac{dy}{dt} = 5x + 9y + 4e^{-t}\cos 2t$

In Problems 7–10 write the given system without the use of matrices.

7. $\mathbf{X}' = \begin{pmatrix} 4 & 2 \\ -1 & 3 \end{pmatrix}\mathbf{X} + \begin{pmatrix} 1 \\ -1 \end{pmatrix}e^{t}$

8. $\mathbf{X}' = \begin{pmatrix} 7 & 5 & -9 \\ 4 & 1 & 1 \\ 0 & -2 & 3 \end{pmatrix}\mathbf{X} + \begin{pmatrix} 0 \\ 2 \\ 1 \end{pmatrix}e^{5t} - \begin{pmatrix} 8 \\ 0 \\ 3 \end{pmatrix}e^{-2t}$

9. $\dfrac{d}{dt}\begin{pmatrix} x \\ y \\ z \end{pmatrix} = \begin{pmatrix} 1 & -1 & 2 \\ 3 & -4 & 1 \\ -2 & 5 & 6 \end{pmatrix}\begin{pmatrix} x \\ y \\ z \end{pmatrix} + \begin{pmatrix} 1 \\ 2 \\ 2 \end{pmatrix}e^{-t} - \begin{pmatrix} 3 \\ -1 \\ 1 \end{pmatrix}t$

10. $\dfrac{d}{dt}\begin{pmatrix} x \\ y \end{pmatrix} = \begin{pmatrix} 3 & -7 \\ 1 & 1 \end{pmatrix}\begin{pmatrix} x \\ y \end{pmatrix} + \begin{pmatrix} 4 \\ 8 \end{pmatrix}\sin t + \begin{pmatrix} t - 4 \\ 2t + 1 \end{pmatrix}e^{4t}$

In Problems 11–16 verify that the vector **X** is a solution of the given system.

11. $\dfrac{dx}{dt} = 3x - 4y$

$\dfrac{dy}{dt} = 4x - 7y; \quad \mathbf{X} = \begin{pmatrix} 1 \\ 2 \end{pmatrix}e^{-5t}$

12. $\dfrac{dx}{dt} = -2x + 5y$

$\dfrac{dy}{dt} = -2x + 4y;\quad \mathbf{X} = \begin{pmatrix} 5\cos t \\ 3\cos t - \sin t \end{pmatrix} e^{t}$

13. $\mathbf{X}' = \begin{pmatrix} -1 & \frac{1}{4} \\ 1 & -1 \end{pmatrix}\mathbf{X};\quad \mathbf{X} = \begin{pmatrix} -1 \\ 2 \end{pmatrix} e^{-3t/2}$

14. $\mathbf{X}' = \begin{pmatrix} 2 & 1 \\ -1 & 0 \end{pmatrix}\mathbf{X};\quad \mathbf{X} = \begin{pmatrix} 1 \\ 3 \end{pmatrix} e^{t} + \begin{pmatrix} 4 \\ -4 \end{pmatrix} te^{t}$

15. $\dfrac{d\mathbf{X}}{dt} = \begin{pmatrix} 1 & 2 & 1 \\ 6 & -1 & 0 \\ -1 & -2 & -1 \end{pmatrix}\mathbf{X};\quad \mathbf{X} = \begin{pmatrix} 1 \\ 6 \\ -13 \end{pmatrix}$

16. $\mathbf{X}' = \begin{pmatrix} 1 & 0 & 1 \\ 1 & 1 & 0 \\ -2 & 0 & -1 \end{pmatrix}\mathbf{X};\quad \mathbf{X} = \begin{pmatrix} \sin t \\ -\frac{1}{2}\sin t - \frac{1}{2}\cos t \\ -\sin t + \cos t \end{pmatrix}$

In Problems 17–20 the given vectors are solutions of a system $\mathbf{X}' = \mathbf{AX}$. Determine whether the vectors form a fundamental set on $(-\infty, \infty)$.

17. $\mathbf{X}_1 = \begin{pmatrix} 1 \\ 1 \end{pmatrix} e^{-2t},\ \mathbf{X}_2 = \begin{pmatrix} 1 \\ -1 \end{pmatrix} e^{-6t}$

18. $\mathbf{X}_1 = \begin{pmatrix} 1 \\ -1 \end{pmatrix} e^{t},\ \mathbf{X}_2 = \begin{pmatrix} 2 \\ 6 \end{pmatrix} e^{t} + \begin{pmatrix} 8 \\ -8 \end{pmatrix} te^{t}$

19. $\mathbf{X}_1 = \begin{pmatrix} 1 \\ -2 \\ 4 \end{pmatrix} + t\begin{pmatrix} 1 \\ 2 \\ 2 \end{pmatrix},\ \mathbf{X}_2 = \begin{pmatrix} 1 \\ -2 \\ 4 \end{pmatrix},\ \mathbf{X}_3 = \begin{pmatrix} 3 \\ -6 \\ 12 \end{pmatrix} + t\begin{pmatrix} 2 \\ 4 \\ 4 \end{pmatrix}$

20. $\mathbf{X}_1 = \begin{pmatrix} 1 \\ 6 \\ -13 \end{pmatrix},\ \mathbf{X}_2 = \begin{pmatrix} 1 \\ -2 \\ -1 \end{pmatrix} e^{-4t},\ \mathbf{X}_3 = \begin{pmatrix} 2 \\ 3 \\ -2 \end{pmatrix} e^{3t}$

In Problems 21–24 verify that the vector $\mathbf{X}_p$ is a particular solution of the given system.

21. $\dfrac{dx}{dt} = x + 4y + 2t - 7$

$\dfrac{dy}{dt} = 3x + 2y - 4t - 18;\quad \mathbf{X}_p = \begin{pmatrix} 2 \\ -1 \end{pmatrix} t + \begin{pmatrix} 5 \\ 1 \end{pmatrix}$

22. $\mathbf{X}' = \begin{pmatrix} 2 & 1 \\ 1 & -1 \end{pmatrix}\mathbf{X} + \begin{pmatrix} -5 \\ 2 \end{pmatrix};\quad \mathbf{X}_p = \begin{pmatrix} 1 \\ 3 \end{pmatrix}$

23. $\mathbf{X}' = \begin{pmatrix} 2 & 1 \\ 3 & 4 \end{pmatrix}\mathbf{X} - \begin{pmatrix} 1 \\ 7 \end{pmatrix} e^{t};\quad \mathbf{X}_p = \begin{pmatrix} 1 \\ 1 \end{pmatrix} e^{t} + \begin{pmatrix} 1 \\ -1 \end{pmatrix} te^{t}$

24. $\mathbf{X}' = \begin{pmatrix} 1 & 2 & 3 \\ -4 & 2 & 0 \\ -6 & 1 & 0 \end{pmatrix}\mathbf{X} + \begin{pmatrix} -1 \\ 4 \\ 3 \end{pmatrix}\sin 3t;\quad \mathbf{X}_p = \begin{pmatrix} \sin 3t \\ 0 \\ \cos 3t \end{pmatrix}$

25. Prove that the general solution of

$$\mathbf{X}' = \begin{pmatrix} 0 & 6 & 0 \\ 1 & 0 & 1 \\ 1 & 1 & 0 \end{pmatrix}\mathbf{X}$$

on the interval $(-\infty, \infty)$ is

$$\mathbf{X} = c_1\begin{pmatrix} 6 \\ -1 \\ -5 \end{pmatrix}e^{-t} + c_2\begin{pmatrix} -3 \\ 1 \\ 1 \end{pmatrix}e^{-2t} + c_3\begin{pmatrix} 2 \\ 1 \\ 1 \end{pmatrix}e^{3t}.$$

26. Prove that the general solution of

$$\mathbf{X}' = \begin{pmatrix} -1 & -1 \\ -1 & 1 \end{pmatrix}\mathbf{X} + \begin{pmatrix} 1 \\ 1 \end{pmatrix}t^2 + \begin{pmatrix} 4 \\ -6 \end{pmatrix}t + \begin{pmatrix} -1 \\ 5 \end{pmatrix}$$

on the interval $(-\infty, \infty)$ is

$$\mathbf{X} = c_1\begin{pmatrix} 1 \\ -1 - \sqrt{2} \end{pmatrix}e^{\sqrt{2}t} + c_2\begin{pmatrix} 1 \\ -1 + \sqrt{2} \end{pmatrix}e^{-\sqrt{2}t} + \begin{pmatrix} 1 \\ 0 \end{pmatrix}t^2 + \begin{pmatrix} -2 \\ 4 \end{pmatrix}t + \begin{pmatrix} 1 \\ 0 \end{pmatrix}.$$

8.5.2 A Fundamental Matrix

In Problems 27–30 the indicated column vectors form a fundamental set of solutions for the given system on $(-\infty, \infty)$. Form a fundamental matrix $\mathbf{\Phi}(t)$ and compute $\mathbf{\Phi}^{-1}(t)$.

27. $\mathbf{X}' = \begin{pmatrix} 4 & 1 \\ 6 & 5 \end{pmatrix}\mathbf{X}$; $\mathbf{X}_1 = \begin{pmatrix} 1 \\ -2 \end{pmatrix}e^{2t}$, $\mathbf{X}_2 = \begin{pmatrix} 1 \\ 3 \end{pmatrix}e^{7t}$

28. $\mathbf{X}' = \begin{pmatrix} 2 & 3 \\ 3 & 2 \end{pmatrix}\mathbf{X}$; $\mathbf{X}_1 = \begin{pmatrix} -1 \\ 1 \end{pmatrix}e^{-t}$, $\mathbf{X}_2 = \begin{pmatrix} 1 \\ 1 \end{pmatrix}e^{5t}$

29. $\mathbf{X}' = \begin{pmatrix} 4 & 1 \\ -9 & -2 \end{pmatrix}\mathbf{X}$; $\mathbf{X}_1 = \begin{pmatrix} -1 \\ 3 \end{pmatrix}e^{t}$, $\mathbf{X}_2 = \begin{pmatrix} -1 \\ 3 \end{pmatrix}te^{t} + \begin{pmatrix} 0 \\ -1 \end{pmatrix}e^{t}$

30. $\mathbf{X}' = \begin{pmatrix} 3 & -2 \\ 5 & -3 \end{pmatrix}\mathbf{X}$; $\mathbf{X}_1 = \begin{pmatrix} 2\cos t \\ 3\cos t + \sin t \end{pmatrix}$, $\mathbf{X}_2 = \begin{pmatrix} -2\sin t \\ \cos t - 3\sin t \end{pmatrix}$

31. Find the fundamental matrix $\mathbf{\Psi}(t)$ satisfying $\mathbf{\Psi}(0) = \mathbf{I}$ for the system given in Problem 27.

32. Find the fundamental matrix $\mathbf{\Psi}(t)$ satisfying $\mathbf{\Psi}(0) = \mathbf{I}$ for the system given in Problem 28.

33. Find the fundamental matrix $\mathbf{\Psi}(t)$ satisfying $\mathbf{\Psi}(0) = \mathbf{I}$ for the system given in Problem 29.

34. Find the fundamental matrix $\mathbf{\Psi}(t)$ satisfying $\mathbf{\Psi}(\pi/2) = \mathbf{I}$ for the system given in Problem 30.

35. If $\mathbf{X} = \mathbf{\Phi}(t)\mathbf{C}$ is the general solution of $\mathbf{X}' = \mathbf{AX}$, show that the solution of the initial-value problem $\mathbf{X}' = \mathbf{AX}$, $\mathbf{X}(t_0) = \mathbf{X}_0$ is $\mathbf{X} = \mathbf{\Phi}(t)\mathbf{\Phi}^{-1}(t_0)\mathbf{X}_0$.

36. Show that the solution of the initial-value problem given in Problem 35 is also given by $\mathbf{X} = \mathbf{\Psi}(t)\mathbf{X}_0$.

37. Show that $\mathbf{\Psi}(t) = \mathbf{\Phi}(t)\mathbf{\Phi}^{-1}(t_0)$. [*Hint*: Compare Problems 35 and 36.]

8.6 HOMOGENEOUS LINEAR SYSTEMS

• *Eigenvalues* • *Eigenvectors* • *General solutions*

8.6.1 DISTINCT REAL EIGENVALUES

For the remainder of this chapter we shall be concerned only with linear systems with real constant coefficients.

We saw in Example 9 of Section 8.5 that the general solution of the homogeneous system

$$\frac{dx}{dt} = x + 3y$$

$$\frac{dy}{dt} = 5x + 3y$$

is
$$\mathbf{X} = c_1\begin{pmatrix} 1 \\ -1 \end{pmatrix}e^{-2t} + c_2\begin{pmatrix} 3 \\ 5 \end{pmatrix}e^{6t}.$$

Since both solution vectors have the basic form

$$\mathbf{X}_i = \begin{pmatrix} k_1 \\ k_2 \end{pmatrix}e^{\lambda_i t}, \quad i = 1, 2,$$

k_1 and k_2 constants, we are prompted to ask whether we can always find a solution of the form

$$\mathbf{X} = \begin{pmatrix} k_1 \\ k_2 \\ \vdots \\ k_n \end{pmatrix}e^{\lambda t} = \mathbf{K}e^{\lambda t} \tag{1}$$

for the general homogeneous linear first-order system

$$\mathbf{X}' = \mathbf{AX}, \tag{2}$$

where $\mathbf{A}$ is an $n \times n$ matrix of constants.

Eigenvalues and Eigenvectors If (1) is to be a solution vector of (2), then $\mathbf{X}' = \mathbf{K}\lambda e^{\lambda t}$ so the system becomes

$$\mathbf{K}\lambda e^{\lambda t} = \mathbf{AK}e^{\lambda t}.$$

After dividing out $e^{\lambda t}$ and rearranging, we obtain $\mathbf{AK} = \lambda\mathbf{K}$ or

$$(\mathbf{A} - \lambda\mathbf{I})\mathbf{K} = \mathbf{0}. \tag{3}$$

Equation (3) is equivalent to the simultaneous algebraic equations (8) of Section 8.4. To find a nontrivial solution $\mathbf{X}$ of (2) we must find a nontrivial vector $\mathbf{K}$ satisfying (3). But in order for (3) to have nontrivial solutions we must have

$$\det(\mathbf{A} - \lambda\mathbf{I}) = 0.$$

The latter equation is recognized as the characteristic equation of the matrix $\mathbf{A}$. In other words, $\mathbf{X} = \mathbf{K}e^{\lambda t}$ will be a solution of the system of differential equations (2) if and only if λ is an **eigenvalue** of $\mathbf{A}$ and $\mathbf{K}$ is an **eigenvector** corresponding to λ.

When the $n \times n$ matrix $\mathbf{A}$ possesses n distinct real eigenvalues $\lambda_1, \lambda_2, \ldots, \lambda_n$, then a set of n linearly independent eigenvectors $\mathbf{K}_1, \mathbf{K}_2, \ldots, \mathbf{K}_n$ can always be found and

$$\mathbf{X}_1 = \mathbf{K}_1 e^{\lambda_1 t}, \quad \mathbf{X}_2 = \mathbf{K}_2 e^{\lambda_2 t}, \quad \ldots, \quad \mathbf{X}_n = \mathbf{K}_n e^{\lambda_n t}$$

is a fundamental set of solutions of (2) on $(-\infty, \infty)$.

THEOREM 8.9 General Solution—Homogeneous Systems

Let $\lambda_1, \lambda_2, \ldots, \lambda_n$ be n distinct real eigenvalues of the coefficient matrix $\mathbf{A}$ of the homogeneous system (2) and let $\mathbf{K}_1, \mathbf{K}_2, \ldots, \mathbf{K}_n$ be the corresponding eigenvectors. Then the **general solution** of (2) on the interval $(-\infty, \infty)$ is given by

$$\mathbf{X} = c_1 \mathbf{K}_1 e^{\lambda_1 t} + c_2 \mathbf{K}_2 e^{\lambda_2 t} + \cdots + c_n \mathbf{K}_n e^{\lambda_n t}.$$

EXAMPLE 1 Distinct Eigenvalues

Solve

$$\begin{aligned} \frac{dx}{dt} &= 2x + 3y \\ \frac{dy}{dt} &= 2x + y. \end{aligned} \tag{4}$$

Solution We first find the eigenvalues and eigenvectors of the matrix of coefficients.

The characteristic equation is

$$\det(\mathbf{A} - \lambda \mathbf{I}) = \begin{vmatrix} 2-\lambda & 3 \\ 2 & 1-\lambda \end{vmatrix} = \lambda^2 - 3\lambda - 4 = 0.$$

Since $\lambda^2 - 3\lambda - 4 = (\lambda + 1)(\lambda - 4)$, we see that the eigenvalues are $\lambda_1 = -1$ and $\lambda_2 = 4$.

Now for $\lambda_1 = -1$, (3) is equivalent to

$$\begin{aligned} 3k_1 + 3k_2 &= 0 \\ 2k_1 + 2k_2 &= 0. \end{aligned}$$

Thus $k_1 = -k_2$. When $k_2 = -1$, the related eigenvector is

$$\mathbf{K}_1 = \begin{pmatrix} 1 \\ -1 \end{pmatrix}.$$

For $\lambda_2 = 4$ we have

$$\begin{aligned} -2k_1 + 3k_2 &= 0 \\ 2k_1 - 3k_2 &= 0 \end{aligned}$$

so $k_1 = 3k_2/2$, and therefore with $k_2 = 2$ the corresponding eigenvector is

$$\mathbf{K}_2 = \begin{pmatrix} 3 \\ 2 \end{pmatrix}.$$

Since the matrix of coefficients **A** is a 2×2 matrix and since we have found two linearly independent solutions of (4),

$$\mathbf{X}_1 = \begin{pmatrix} 1 \\ -1 \end{pmatrix} e^{-t} \quad \text{and} \quad \mathbf{X}_2 = \begin{pmatrix} 3 \\ 2 \end{pmatrix} e^{4t},$$

we conclude that the general solution of the system is

$$\mathbf{X} = c_1\mathbf{X}_1 + c_2\mathbf{X}_2 = c_1\begin{pmatrix} 1 \\ -1 \end{pmatrix} e^{-t} + c_2\begin{pmatrix} 3 \\ 2 \end{pmatrix} e^{4t}. \tag{5}$$

For the sake of review, you should keep firmly in mind that a solution of a system of first-order differential equations, when written in terms of matrices, is simply an alternative to the method that we employed in Section 8.1—namely, listing the individual functions and the relationships between the constants. By adding the vectors given in (5), we obtain

$$\begin{pmatrix} x(t) \\ y(t) \end{pmatrix} = \begin{pmatrix} c_1e^{-t} + 3c_2e^{4t} \\ -c_1e^{-t} + 2c_2e^{4t} \end{pmatrix},$$

and this in turn yields the more familiar statement

$$x(t) = c_1e^{-t} + 3c_2e^{4t}, \quad y(t) = -c_1e^{-t} + 2c_2e^{4t}.$$

EXAMPLE 2 Distinct Eigenvalues

Solve

$$\begin{aligned} \frac{dx}{dt} &= -4x + y + z \\ \frac{dy}{dy} &= x + 5y - z \\ \frac{dz}{dt} &= y - 3z. \end{aligned} \tag{6}$$

Solution Using the cofactors of the third row, we find

$$\det(\mathbf{A} - \lambda\mathbf{I}) = \begin{vmatrix} -4-\lambda & 1 & 1 \\ 1 & 5-\lambda & -1 \\ 0 & 1 & -3-\lambda \end{vmatrix} = -(\lambda + 3)(\lambda + 4)(\lambda - 5) = 0,$$

and so the eigenvalues are $\lambda_1 = -3$, $\lambda_2 = -4$, $\lambda_3 = 5$.

Now for $\lambda_1 = -3$ Gauss-Jordan elimination gives

$$(\mathbf{A} + 3\mathbf{I}|\mathbf{0}) = \left(\begin{array}{ccc|c} -1 & 1 & 1 & 0 \\ 1 & 8 & -1 & 0 \\ 0 & 1 & 0 & 0 \end{array}\right) \xrightarrow{\text{row operations}} \left(\begin{array}{ccc|c} 1 & 0 & -1 & 0 \\ 0 & 1 & 0 & 0 \\ 0 & 0 & 0 & 0 \end{array}\right).$$

Therefore $k_1 = k_3$, $k_2 = 0$. The choice $k_3 = 1$ gives the eigenvector

$$\mathbf{K}_1 = \begin{pmatrix} 1 \\ 0 \\ 1 \end{pmatrix}. \tag{7}$$

Similarly, for $\lambda_2 = -4$

$$(\mathbf{A} + 4\mathbf{I}|\mathbf{0}) = \left(\begin{array}{ccc|c} 0 & 1 & 1 & 0 \\ 1 & 9 & -1 & 0 \\ 0 & 1 & 1 & 0 \end{array}\right) \xrightarrow{\text{row operations}} \left(\begin{array}{ccc|c} 1 & 0 & -10 & 0 \\ 0 & 1 & 1 & 0 \\ 0 & 0 & 0 & 0 \end{array}\right)$$

implies $k_1 = 10k_3$, $k_2 = -k_3$. Choosing $k_3 = 1$, we get the second eigenvector

$$\mathbf{K}_2 = \begin{pmatrix} 10 \\ -1 \\ 1 \end{pmatrix}. \qquad \mathbf{(8)}$$

Finally, when $\lambda_3 = 5$, the augmented matrices

$$(\mathbf{A} - 5\mathbf{I}|\mathbf{0}) = \left(\begin{array}{ccc|c} -9 & 1 & 1 & 0 \\ 1 & 0 & -1 & 0 \\ 0 & 1 & -8 & 0 \end{array}\right) \xrightarrow{\text{row operations}} \left(\begin{array}{ccc|c} 1 & 0 & -1 & 0 \\ 0 & 1 & -8 & 0 \\ 0 & 0 & 0 & 0 \end{array}\right)$$

yield

$$\mathbf{K}_3 = \begin{pmatrix} 1 \\ 8 \\ 1 \end{pmatrix}. \qquad \mathbf{(9)}$$

Multiplying the vectors (7), (8), and (9) by e^{-3t}, e^{-4t}, and e^{5t}, respectively, gives three solutions of (6):

$$\mathbf{X}_1 = \begin{pmatrix} 1 \\ 0 \\ 1 \end{pmatrix} e^{-3t}, \quad \mathbf{X}_2 = \begin{pmatrix} 10 \\ -1 \\ 1 \end{pmatrix} e^{-4t}, \quad \mathbf{X}_3 = \begin{pmatrix} 1 \\ 8 \\ 1 \end{pmatrix} e^{5t}.$$

The general solution of the system is then

$$\mathbf{X} = c_1 \begin{pmatrix} 1 \\ 0 \\ 1 \end{pmatrix} e^{-3t} + c_2 \begin{pmatrix} 10 \\ -1 \\ 1 \end{pmatrix} e^{-4t} + c_3 \begin{pmatrix} 1 \\ 8 \\ 1 \end{pmatrix} e^{5t}.$$

8.6.2 COMPLEX EIGENVALUES

If $\qquad \lambda_1 = \alpha + i\beta \quad \text{and} \quad \lambda_2 = \alpha - i\beta, \quad i^2 = -1, \beta > 0$

are complex eigenvalues of the coefficient matrix **A**, we can then certainly expect their corresponding eigenvectors to also have complex entries.*

For example, the characteristic equation of the system

$$\begin{aligned} \frac{dx}{dt} &= 6x - y \\ \frac{dy}{dt} &= 5x + 4y \end{aligned} \qquad \mathbf{(10)}$$

* When the characteristic equation has real coefficients, complex eigenvalues always appear in conjugate pairs.

is $$\det(\mathbf{A} - \lambda\mathbf{I}) = \begin{vmatrix} 6-\lambda & -1 \\ 5 & 4-\lambda \end{vmatrix} = \lambda^2 - 10\lambda + 29 = 0.$$

From the quadratic formula we find $\lambda_1 = 5 + 2i$, $\lambda_2 = 5 - 2i$.

Now for $\lambda_1 = 5 + 2i$ we must solve

$$\begin{aligned} (1-2i)k_1 - k_2 &= 0 \\ 5k_1 - (1+2i)k_2 &= 0. \end{aligned}$$

Since $k_2 = (1 - 2i)k_1$,* it follows, after we choose $k_1 = 1$, that one eigenvector is

$$\mathbf{K}_1 = \begin{pmatrix} 1 \\ 1-2i \end{pmatrix}.$$

Similarly, for $\lambda_2 = 5 - 2i$ we find the other eigenvector to be

$$\mathbf{K}_2 = \begin{pmatrix} 1 \\ 1+2i \end{pmatrix}.$$

Consequently two solutions of (10) are

$$\mathbf{X}_1 = \begin{pmatrix} 1 \\ 1-2i \end{pmatrix} e^{(5+2i)t} \quad \text{and} \quad \mathbf{X}_2 = \begin{pmatrix} 1 \\ 1+2i \end{pmatrix} e^{(5-2i)t}.$$

By the superposition principle another solution is

$$\mathbf{X} = c_1\begin{pmatrix} 1 \\ 1-2i \end{pmatrix} e^{(5+2i)t} + c_2\begin{pmatrix} 1 \\ 1+2i \end{pmatrix} e^{(5-2i)t}. \tag{11}$$

Note that the entries in $\mathbf{K}_2$ corresponding to λ_2 are the conjugates of the entries in $\mathbf{K}_1$ corresponding to λ_1. The conjugate of λ_1 is, of course, λ_2. We write this as $\lambda_2 = \overline{\lambda}_1$ and $\mathbf{K}_2 = \overline{\mathbf{K}}_1$. We have illustrated the following general result.

THEOREM 8.10 Solutions Corresponding to a Complex Eigenvalue

Let $\mathbf{A}$ be the coefficient matrix having real entries of the homogeneous system (2), and let $\mathbf{K}$ be an eigenvector corresponding to the complex eigenvalue $\lambda_1 = \alpha + i\beta$, α and β real. Then

$$\mathbf{K}_1 e^{\lambda_1 t} \quad \text{and} \quad \overline{\mathbf{K}}_1 e^{\overline{\lambda}_1 t}$$

are solutions of (2).

It is desirable and relatively easy to rewrite a solution such as (11) in terms of real functions. Since

$$\begin{aligned} x &= c_1 e^{(5+2i)t} + c_2 e^{(5-2i)t} \\ y &= c_1(1-2i)e^{(5+2i)t} + c_2(1+2i)e^{(5-2i)t}, \end{aligned}$$

* Note that the second equation is simply $1 + 2i$ times the first.

it follows from Euler's formula that

$$\begin{aligned} x &= e^{5t}[c_1e^{2it} + c_2e^{-2it}] \\ &= e^{5t}[(c_1 + c_2)\cos 2t + (c_1i - c_2i)\sin 2t] \\ y &= e^{5t}[(c_1(1 - 2i) + c_2(1 + 2i))\cos 2t + (c_1i(1 - 2i) - c_2i(1 + 2i))\sin 2t] \\ &= e^{5t}[(c_1 + c_2) - 2(c_1i - c_2i)]\cos 2t + e^{5t}[2(c_1 + c_2) + (c_1i - c_2i)]\sin 2t. \end{aligned}$$

If we replace $c_1 + c_2$ by C_1 and $c_1i - c_2i$ by C_2, then

$$\begin{aligned} x &= e^{5t}[C_1\cos 2t + C_2\sin 2t] \\ y &= e^{5t}[C_1 - 2C_2]\cos 2t + e^{5t}[2C_1 + C_2]\sin 2t \end{aligned}$$

or, in terms of vectors,

$$\mathbf{X} = \begin{pmatrix} x \\ y \end{pmatrix} = C_1\begin{pmatrix} \cos 2t \\ \cos 2t + 2\sin 2t \end{pmatrix}e^{5t} + C_2\begin{pmatrix} \sin 2t \\ -2\cos 2t + \sin 2t \end{pmatrix}e^{5t}. \quad \textbf{(12)}$$

Here, of course, it can be verified that each vector in (12) is a solution of (10). In addition, the solutions are linearly independent on the interval $(-\infty, \infty)$. We may further assume that C_1 and C_2 are completely arbitrary and real. Thus (12) is the general solution of (10).

The foregoing process can be generalized. Let $\mathbf{K}_1$ be an eigenvector of the matrix $\mathbf{A}$ corresponding to the complex eigenvalue $\lambda_1 = \alpha + i\beta$. Then $\mathbf{X}_1$ and $\mathbf{X}_2$ in Theorem 8.10 can be written as

$$\begin{aligned} \mathbf{K}_1e^{\lambda_1 t} &= \mathbf{K}_1e^{\alpha t}e^{i\beta t} = \mathbf{K}_1e^{\alpha t}(\cos\beta t + i\sin\beta t) \\ \overline{\mathbf{K}}_1e^{\overline{\lambda}_1 t} &= \overline{\mathbf{K}}_1e^{\alpha t}e^{-i\beta t} = \overline{\mathbf{K}}_1e^{\alpha t}(\cos\beta t - i\sin\beta t). \end{aligned}$$

The foregoing equations then yield

$$\frac{1}{2}(\mathbf{K}_1e^{\lambda_1 t} + \overline{\mathbf{K}}_1e^{\overline{\lambda}_1 t}) = \frac{1}{2}(\mathbf{K}_1 + \overline{\mathbf{K}}_1)e^{\alpha t}\cos\beta t - \frac{i}{2}(-\mathbf{K}_1 + \overline{\mathbf{K}}_1)e^{\alpha t}\sin\beta t$$

$$\frac{i}{2}(-\mathbf{K}_1e^{\lambda_1 t} + \overline{\mathbf{K}}_1e^{\overline{\lambda}_1 t}) = \frac{i}{2}(-\mathbf{K}_1 + \overline{\mathbf{K}}_1)e^{\alpha t}\cos\beta t + \frac{1}{2}(\mathbf{K}_1 + \overline{\mathbf{K}}_1)e^{\alpha t}\sin\beta t.$$

For *any* complex number $z = a + ib$, we note that $\frac{1}{2}(z + \overline{z}) = a$ and $(i/2)(-z + \overline{z}) = b$ are *real* numbers. Therefore, the entries in the column vectors $\frac{1}{2}(\mathbf{K}_1 + \overline{\mathbf{K}}_1)$ and $(i/2)(-\mathbf{K}_1 + \overline{\mathbf{K}}_1)$ are real numbers. By defining

$$\mathbf{B}_1 = \frac{1}{2}[\mathbf{K}_1 + \overline{\mathbf{K}}_1] \quad \text{and} \quad \mathbf{B}_2 = \frac{i}{2}[-\mathbf{K}_1 + \overline{\mathbf{K}}_1], \quad \textbf{(13)}$$

we are led to the following theorem.

THEOREM 8.11 Real Solutions Corresponding to a Complex Eigenvalue

Let $\lambda_1 = \alpha + i\beta$ be a complex eigenvalue of the coefficient matrix $\mathbf{A}$ in the homogeneous system (2) and let $\mathbf{B}_1$ and $\mathbf{B}_2$ denote the column vectors defined in (13). Then

$$\begin{aligned} \mathbf{X}_1 &= (\mathbf{B}_1\cos\beta t - \mathbf{B}_2\sin\beta t)e^{\alpha t} \\ \mathbf{X}_2 &= (\mathbf{B}_2\cos\beta t + \mathbf{B}_1\sin\beta t)e^{\alpha t} \end{aligned} \quad \textbf{(14)}$$

are linearly independent solutions of (2) on $(-\infty, \infty)$.

The matrices $\mathbf{B}_1$ and $\mathbf{B}_2$ in (13) are often denoted by

$$\mathbf{B}_1 = \text{Re}(\mathbf{K}_1) \quad \text{and} \quad \mathbf{B}_2 = \text{Im}(\mathbf{K}_1) \tag{15}$$

since these vectors are, in turn, the *real* and *imaginary* parts of the eigenvector $\mathbf{K}_1$. For example, (12) follows from (14) with

$$\mathbf{K}_1 = \begin{pmatrix} 1 \\ 1 - 2i \end{pmatrix} = \begin{pmatrix} 1 \\ 1 \end{pmatrix} + i\begin{pmatrix} 0 \\ -2 \end{pmatrix}$$

$$\mathbf{B}_1 = \text{Re}(\mathbf{K}_1) = \begin{pmatrix} 1 \\ 1 \end{pmatrix} \quad \text{and} \quad \mathbf{B}_2 = \text{Im}(\mathbf{K}_1) = \begin{pmatrix} 0 \\ -2 \end{pmatrix}.$$

EXAMPLE 3 Complex Eigenvalues

Solve $\mathbf{X}' = \begin{pmatrix} 2 & 8 \\ -1 & -2 \end{pmatrix}\mathbf{X}$.

Solution First we obtain the eigenvalues from

$$\det(\mathbf{A} - \lambda\mathbf{I}) = \begin{vmatrix} 2-\lambda & 8 \\ -1 & -2-\lambda \end{vmatrix} = \lambda^2 + 4 = 0.$$

Thus the eigenvalues are $\lambda_1 = 2i$ and $\lambda_2 = \overline{\lambda_1} = -2i$. For λ_1 we see that the system

$$\begin{aligned} (2 - 2i)k_1 + \qquad\quad 8k_2 &= 0 \\ -k_1 + (-2 - 2i)k_2 &= 0 \end{aligned}$$

gives $k_1 = -(2 + 2i)k_2$. By choosing $k_2 = -1$, we get

$$\mathbf{K}_1 = \begin{pmatrix} 2 + 2i \\ -1 \end{pmatrix} = \begin{pmatrix} 2 \\ -1 \end{pmatrix} + i\begin{pmatrix} 2 \\ 0 \end{pmatrix}.$$

Now from (15) we form

$$\mathbf{B}_1 = \text{Re}(\mathbf{K}_1) = \begin{pmatrix} 2 \\ -1 \end{pmatrix} \quad \text{and} \quad \mathbf{B}_2 = \text{Im}(\mathbf{K}_1) = \begin{pmatrix} 2 \\ 0 \end{pmatrix}.$$

Since $\alpha = 0$, it follows from (14) that the general solution of the system is

$$\begin{aligned} \mathbf{X} &= c_1\left[\begin{pmatrix} 2 \\ -1 \end{pmatrix}\cos 2t - \begin{pmatrix} 2 \\ 0 \end{pmatrix}\sin 2t\right] + c_2\left[\begin{pmatrix} 2 \\ 0 \end{pmatrix}\cos 2t + \begin{pmatrix} 2 \\ -1 \end{pmatrix}\sin 2t\right] \\ &= c_1\begin{pmatrix} 2\cos 2t - 2\sin 2t \\ -\cos 2t \end{pmatrix} + c_2\begin{pmatrix} 2\cos 2t + 2\sin 2t \\ -\sin 2t \end{pmatrix}. \end{aligned}$$

EXAMPLE 4 Complex Eigenvalues

Solve $\mathbf{X}' = \begin{pmatrix} 1 & 2 \\ -\frac{1}{2} & 1 \end{pmatrix}\mathbf{X}$.

Solution The solutions of the characteristic equation

$$\det(\mathbf{A} - \lambda\mathbf{I}) = \begin{vmatrix} 1-\lambda & 2 \\ -\frac{1}{2} & 1-\lambda \end{vmatrix} = \lambda^2 - 2\lambda + 2 = 0$$

are $\lambda_1 = 1 + i$ and $\lambda_2 = \overline{\lambda}_1 = 1 - i$. Now an eigenvector associated with λ_1 is

$$\mathbf{K}_1 = \begin{pmatrix} 2 \\ i \end{pmatrix} = \begin{pmatrix} 2 \\ 0 \end{pmatrix} + i\begin{pmatrix} 0 \\ 1 \end{pmatrix}.$$

From (15) we find $\mathbf{B}_1 = \begin{pmatrix} 2 \\ 0 \end{pmatrix}$ and $\mathbf{B}_2 = \begin{pmatrix} 0 \\ 1 \end{pmatrix}$. Thus (14) gives

$$\begin{aligned} \mathbf{X} &= c_1\left[\begin{pmatrix} 2 \\ 0 \end{pmatrix}\cos t - \begin{pmatrix} 0 \\ 1 \end{pmatrix}\sin t\right]e^t + c_2\left[\begin{pmatrix} 0 \\ 1 \end{pmatrix}\cos t + \begin{pmatrix} 2 \\ 0 \end{pmatrix}\sin t\right]e^t \\ &= c_1\begin{pmatrix} 2\cos t \\ -\sin t \end{pmatrix}e^t + c_2\begin{pmatrix} 2\sin t \\ \cos t \end{pmatrix}e^t. \end{aligned}$$

Alternative Method When $\mathbf{A}$ is a 2×2 matrix having a complex eigenvalue $\lambda = \alpha + i\beta$, the general solution of the system can also be obtained from the assumption

$$\mathbf{X} = \begin{pmatrix} c_1 \\ c_2 \end{pmatrix}e^{\alpha t}\sin\beta t + \begin{pmatrix} c_3 \\ c_4 \end{pmatrix}e^{\alpha t}\cos\beta t$$

and then the substitution of $x(t)$ and $y(t)$ into one of the equations of the original system. This procedure is basically that of Section 8.1.

8.6.3 REPEATED EIGENVALUES

Up to this point we have not considered the case in which some of the n eigenvalues $\lambda_1, \lambda_2, \ldots, \lambda_n$ of an $n \times n$ matrix are repeated. For example, the characteristic equation of the coefficient matrix in

$$\mathbf{X}' = \begin{pmatrix} 3 & -18 \\ 2 & -9 \end{pmatrix}\mathbf{X} \tag{16}$$

is readily shown to be $(\lambda + 3)^2 = 0$, and therefore $\lambda_1 = \lambda_2 = -3$ is a root of *multiplicity two.* Now for this value we find the single eigenvector

$$\mathbf{K}_1 = \begin{pmatrix} 3 \\ 1 \end{pmatrix},$$

and so one solution of (16) is

$$\mathbf{X}_1 = \begin{pmatrix} 3 \\ 1 \end{pmatrix}e^{-3t}. \tag{17}$$

But since we are obviously interested in forming the general solution of the system, we need to pursue the question of finding a second solution.

In general, if m is a positive integer and $(\lambda - \lambda_1)^m$ is a factor of the characteristic equation, while $(\lambda - \lambda_1)^{m+1}$ is not a factor, then λ_1 is said to be an **eigenvalue of multiplicity m.** The next three examples illustrate the following cases:

(*i*) For some $n \times n$ matrices $\mathbf{A}$ it may be possible to find m linearly independent eigenvectors $\mathbf{K}_1, \mathbf{K}_2, \ldots, \mathbf{K}_m$ corresponding to an eigenvalue

λ_1 of multiplicity $m \le n$. In this case the general solution of the system contains the linear combination

$$c_1\mathbf{K}_1e^{\lambda_1 t} + c_2\mathbf{K}_2e^{\lambda_1 t} + \cdots + c_m\mathbf{K}_me^{\lambda_1 t}.$$

(*ii*) If there is only one eigenvector corresponding to the eigenvalue λ_1 of multiplicity m, then m linearly independent solutions of the form

$$\begin{aligned}\mathbf{X}_1 &= \mathbf{K}_{11}e^{\lambda_1 t}\\ \mathbf{X}_2 &= \mathbf{K}_{21}te^{\lambda_1 t} + \mathbf{K}_{22}e^{\lambda_1 t}\\ &\vdots\\ \mathbf{X}_m &= \mathbf{K}_{m1}\frac{t^{m-1}}{(m-1)!}e^{\lambda_1 t} + \mathbf{K}_{m2}\frac{t^{m-2}}{(m-2)!}e^{\lambda_1 t} + \cdots + \mathbf{K}_{mm}e^{\lambda_1 t},\end{aligned}$$

where $\mathbf{K}_{ij}$ are column vectors, can always be found.

Eigenvalue of Multiplicity Two We begin by considering eigenvalues of multiplicity two. In the first example we illustrate a matrix for which we can find two distinct eigenvectors corresponding to a double eigenvalue.

EXAMPLE 5 Repeated Eigenvalues

Solve $\mathbf{X}' = \begin{pmatrix} 1 & -2 & 2\\ -2 & 1 & -2\\ 2 & -2 & 1\end{pmatrix}\mathbf{X}$.

Solution Expanding the determinant in the characteristic equation

$$\det(\mathbf{A} - \lambda\mathbf{I}) = \begin{vmatrix} 1-\lambda & -2 & 2\\ -2 & 1-\lambda & -2\\ 2 & -2 & 1-\lambda\end{vmatrix} = 0$$

yields $-(\lambda + 1)^2(\lambda - 5) = 0$. We see that $\lambda_1 = \lambda_2 = -1$ and $\lambda_3 = 5$.

For $\lambda_1 = -1$ Gauss-Jordan elimination gives immediately

$$(\mathbf{A} + \mathbf{I}|\mathbf{0}) = \left(\begin{array}{rrr|r} 2 & -2 & 2 & 0\\ -2 & 2 & -2 & 0\\ 2 & -2 & 2 & 0\end{array}\right) \xrightarrow{\text{row operations}} \left(\begin{array}{rrr|r} 1 & -1 & 1 & 0\\ 0 & 0 & 0 & 0\\ 0 & 0 & 0 & 0\end{array}\right).$$

The first row of the last matrix means $k_1 - k_2 + k_3 = 0$ or $k_1 = k_2 - k_3$. The choices $k_2 = 1, k_3 = 0$ and $k_2 = 1, k_3 = 1$ yield, in turn, $k_1 = 1$ and $k_1 = 0$. Thus two eigenvectors corresponding to $\lambda_1 = -1$ are

$$\mathbf{K}_1 = \begin{pmatrix}1\\1\\0\end{pmatrix} \quad\text{and}\quad \mathbf{K}_2 = \begin{pmatrix}0\\1\\1\end{pmatrix}.$$

Since neither eigenvector is a constant multiple of the other, we have found, corresponding to the same eigenvalue, two linearly independent solutions

$$\mathbf{X}_1 = \begin{pmatrix}1\\1\\0\end{pmatrix}e^{-t} \quad\text{and}\quad \mathbf{X}_2 = \begin{pmatrix}0\\1\\1\end{pmatrix}e^{-t}.$$

Last, for $\lambda_3 = 5$ the reduction

$$(\mathbf{A} + 5\mathbf{I}|\mathbf{0}) = \left(\begin{array}{ccc|c} -4 & -2 & 2 & 0 \\ -2 & -4 & -2 & 0 \\ 2 & -2 & -4 & 0 \end{array}\right) \xrightarrow{\text{row operations}} \left(\begin{array}{ccc|c} 1 & 0 & -1 & 0 \\ 0 & 1 & 1 & 0 \\ 0 & 0 & 0 & 0 \end{array}\right)$$

implies $k_1 = k_3$ and $k_2 = -k_3$. Picking $k_3 = 1$ gives $k_1 = 1$, $k_2 = -1$, and thus a third eigenvector is

$$\mathbf{K}_3 = \begin{pmatrix} 1 \\ -1 \\ 1 \end{pmatrix}.$$

We conclude that the general solution of the system is

$$\mathbf{X} = c_1\begin{pmatrix} 1 \\ 1 \\ 0 \end{pmatrix}e^{-t} + c_2\begin{pmatrix} 0 \\ 1 \\ 1 \end{pmatrix}e^{-t} + c_3\begin{pmatrix} 1 \\ -1 \\ 1 \end{pmatrix}e^{5t}.$$

Second Solution Now suppose λ_1 is an eigenvalue of multiplicity two and there is only one eigenvector associated with this value. A second solution can be found of the form

$$\mathbf{X}_2 = \mathbf{K}te^{\lambda_1 t} + \mathbf{P}e^{\lambda_1 t}, \tag{18}$$

where $\mathbf{K} = \begin{pmatrix} k_1 \\ k_2 \\ \vdots \\ k_n \end{pmatrix}$ and $\mathbf{P} = \begin{pmatrix} p_1 \\ p_2 \\ \vdots \\ p_n \end{pmatrix}$.

To see this we substitute (18) into the system $\mathbf{X}' = \mathbf{AX}$ and simplify:

$$(\mathbf{AK} - \lambda_1\mathbf{K})te^{\lambda_1 t} + (\mathbf{AP} - \lambda_1\mathbf{P} - \mathbf{K})e^{\lambda_1 t} = \mathbf{0}.$$

Since this last equation is to hold for all values of t, we must have

$$(\mathbf{A} - \lambda_1\mathbf{I})\mathbf{K} = \mathbf{0} \tag{19}$$

and

$$(\mathbf{A} - \lambda_1\mathbf{I})\mathbf{P} = \mathbf{K}. \tag{20}$$

Equation (19) simply states that $\mathbf{K}$ must be an eigenvector of $\mathbf{A}$ associated with λ_1. By solving (19), we find one solution, $\mathbf{X}_1 = \mathbf{K}e^{\lambda_1 t}$. To find the second solution $\mathbf{X}_2$ we need only solve the additional system (20) for the vector $\mathbf{P}$.

EXAMPLE 6 Repeated Eigenvalues

Find the general solution of the system given in (16).

Solution From (17) we know that $\lambda_1 = -3$ and that one solution is $\mathbf{X}_1 = \begin{pmatrix} 3 \\ 1 \end{pmatrix} e^{-3t}$. Identifying $\mathbf{K} = \begin{pmatrix} 3 \\ 1 \end{pmatrix}$ and $\mathbf{P} = \begin{pmatrix} p_1 \\ p_2 \end{pmatrix}$, we find from (20) that we must now solve

$$(\mathbf{A} + 3\mathbf{I})\mathbf{P} = \mathbf{K} \quad \text{or} \quad \begin{aligned} 6p_1 - 18p_2 &= 3 \\ 2p_1 - 6p_2 &= 1. \end{aligned}$$

Since this system is obviously equivalent to one equation, we have an infinite number of choices for p_1 and p_2. For example, by choosing $p_1 = 1$, we find $p_2 = \frac{1}{6}$. However, for simplicity, we shall choose $p_1 = \frac{1}{2}$ so that $p_2 = 0$. Hence $\mathbf{P} = \begin{pmatrix} \frac{1}{2} \\ 0 \end{pmatrix}$. Thus from (18) we find

$$\mathbf{X}_2 = \begin{pmatrix} 3 \\ 1 \end{pmatrix} te^{-3t} + \begin{pmatrix} \frac{1}{2} \\ 0 \end{pmatrix} e^{-3t}.$$

The general solution of (16) is then

$$\mathbf{X} = c_1 \begin{pmatrix} 3 \\ 1 \end{pmatrix} e^{-3t} + c_2 \left[\begin{pmatrix} 3 \\ 1 \end{pmatrix} te^{-3t} + \begin{pmatrix} \frac{1}{2} \\ 0 \end{pmatrix} e^{-3t} \right].$$

Eigenvalues of Multiplicity Three When a matrix **A** has only one eigenvector associated with an eigenvalue λ_1 of multiplicity three, we can find a second solution of form (18) and a third solution of the form

$$\mathbf{X}_3 = \mathbf{K}\frac{t^2}{2} e^{\lambda_1 t} + \mathbf{P}te^{\lambda_1 t} + \mathbf{Q}e^{\lambda_1 t}, \tag{21}$$

where $\quad \mathbf{K} = \begin{pmatrix} k_1 \\ k_2 \\ \vdots \\ k_n \end{pmatrix}, \quad \mathbf{P} = \begin{pmatrix} p_1 \\ p_2 \\ \vdots \\ p_n \end{pmatrix}, \quad$ and $\quad \mathbf{Q} = \begin{pmatrix} q_1 \\ q_2 \\ \vdots \\ q_n \end{pmatrix}.$

By substituting (21) into the system $\mathbf{X}' = \mathbf{AX}$, we find that the column vectors **K, P,** and **Q** must satisfy

$$(\mathbf{A} - \lambda_1\mathbf{I})\mathbf{K} = \mathbf{0} \tag{22}$$

$$(\mathbf{A} - \lambda_1\mathbf{I})\mathbf{P} = \mathbf{K} \tag{23}$$

and

$$(\mathbf{A} - \lambda_1\mathbf{I})\mathbf{Q} = \mathbf{P}. \tag{24}$$

Of course, the solutions of (22) and (23) can be utilized in the formulation of the solutions $\mathbf{X}_1$ and $\mathbf{X}_2$.

EXAMPLE 7 Repeated Eigenvalues

Solve $\mathbf{X}' = \begin{pmatrix} 2 & 1 & 6 \\ 0 & 2 & 5 \\ 0 & 0 & 2 \end{pmatrix} \mathbf{X}.$

Solution The characteristic equation $(\lambda - 2)^3 = 0$ shows that $\lambda_1 = 2$ is an eigenvalue of multiplicity three. In succession we find that a solution of

$$(\mathbf{A} - 2\mathbf{I})\mathbf{K} = \mathbf{0} \quad \text{is} \quad \mathbf{K} = \begin{pmatrix} 1 \\ 0 \\ 0 \end{pmatrix};$$

a solution of

$$(\mathbf{A} - 2\mathbf{I})\mathbf{P} = \mathbf{K} \quad \text{is} \quad \mathbf{P} = \begin{pmatrix} 0 \\ 1 \\ 0 \end{pmatrix};$$

and finally a solution of

$$(\mathbf{A} - 2\mathbf{I})\mathbf{Q} = \mathbf{P} \quad \text{is} \quad \mathbf{Q} = \begin{pmatrix} 0 \\ -\frac{6}{5} \\ \frac{1}{5} \end{pmatrix}.$$

We see from (18) and (21) that the general solution of the system is

$$\mathbf{X} = c_1\begin{pmatrix} 1 \\ 0 \\ 0 \end{pmatrix}e^{2t} + c^2\left[\begin{pmatrix} 1 \\ 0 \\ 0 \end{pmatrix}te^{2t} + \begin{pmatrix} 0 \\ 1 \\ 0 \end{pmatrix}e^{2t}\right] + c^3\left[\begin{pmatrix} 1 \\ 0 \\ 0 \end{pmatrix}\frac{t^2}{2}e^{2t} + \begin{pmatrix} 0 \\ 1 \\ 0 \end{pmatrix}te^{2t} + \begin{pmatrix} 0 \\ -\frac{6}{5} \\ \frac{1}{5} \end{pmatrix}e^{2t}\right].$$

Remarks When an eigenvalue λ_1 has multiplicity m, either we can find m linearly independent eigenvectors or the number of corresponding eigenvectors is less than m. Hence the two cases listed on pages 398–399 are not all the possibilities under which a repeated eigenvalue can occur. It could happen, say, that a 5×5 matrix has an eigenvalue of multiplicity five and there exist three linearly independent eigenvectors.

EXERCISES 8.6

Answers to odd-numbered problems begin on page A-21.

8.6.1 Distinct Real Eigenvalues

In Problems 1–12 find the general solution of the given system.

1. $\dfrac{dx}{dt} = x + 2y$

$\dfrac{dy}{dt} = 4x + 3y$

2. $\dfrac{dx}{dt} = 2y$

$\dfrac{dy}{dt} = 8x$

3. $\dfrac{dx}{dt} = -4x + 2y$

$\dfrac{dy}{dt} = -\dfrac{5}{2}x + 2y$

4. $\dfrac{dx}{dt} = \dfrac{1}{2}x + 9y$

$\dfrac{dy}{dt} = \dfrac{1}{2}x + 2y$

5. $\mathbf{X}' = \begin{pmatrix} 10 & -5 \\ 8 & -12 \end{pmatrix}\mathbf{X}$

6. $\mathbf{X}' = \begin{pmatrix} -6 & 2 \\ -3 & 1 \end{pmatrix}\mathbf{X}$

7. $\dfrac{dx}{dt} = x + y - z$

$\dfrac{dy}{dt} = 2y$

$\dfrac{dz}{dt} = y - z$

8. $\dfrac{dx}{dt} = 2x - 7y$

$\dfrac{dy}{dt} = 5x + 10y + 4z$

$\dfrac{dz}{dt} = 5y + 2z$

9. $\mathbf{X}' = \begin{pmatrix} -1 & 1 & 0 \\ 1 & 2 & 1 \\ 0 & 3 & -1 \end{pmatrix}\mathbf{X}$

10. $\mathbf{X}' = \begin{pmatrix} 1 & 0 & 1 \\ 0 & 1 & 0 \\ 1 & 0 & 1 \end{pmatrix}\mathbf{X}$

11. $\mathbf{X}' = \begin{pmatrix} -1 & -1 & 0 \\ \frac{3}{4} & -\frac{3}{2} & 3 \\ \frac{1}{8} & \frac{1}{4} & -\frac{1}{2} \end{pmatrix}\mathbf{X}$

12. $\mathbf{X}' = \begin{pmatrix} -1 & 4 & 2 \\ 4 & -1 & -2 \\ 0 & 0 & 6 \end{pmatrix}\mathbf{X}$

In Problems 13 and 14 solve the given system subject to the indicated initial condition.

13. $\mathbf{X}' = \begin{pmatrix} \frac{1}{2} & 0 \\ 1 & -\frac{1}{2} \end{pmatrix}\mathbf{X}, \quad \mathbf{X}(0) = \begin{pmatrix} 3 \\ 5 \end{pmatrix}$

14. $\mathbf{X}' = \begin{pmatrix} 1 & 1 & 4 \\ 0 & 2 & 0 \\ 1 & 1 & 1 \end{pmatrix}\mathbf{X}, \quad \mathbf{X}(0) = \begin{pmatrix} 1 \\ 3 \\ 0 \end{pmatrix}$

8.6.2 Complex Eigenvalues

In Problems 15–26 find the general solution of the given system.

15. $\dfrac{dx}{dt} = 6x - y$

$\dfrac{dy}{dt} = 5x + 2y$

16. $\dfrac{dx}{dt} = x + y$

$\dfrac{dy}{dt} = -2x - y$

17. $\dfrac{dx}{dt} = 5x + y$

$\dfrac{dy}{dt} = -2x + 3y$

18. $\dfrac{dx}{dt} = 4x + 5y$

$\dfrac{dy}{dt} = -2x + 6y$

19. $\mathbf{X}' = \begin{pmatrix} 4 & -5 \\ 5 & -4 \end{pmatrix}\mathbf{X}$

20. $\mathbf{X}' = \begin{pmatrix} 1 & -8 \\ 1 & -3 \end{pmatrix}\mathbf{X}$

21. $\dfrac{dx}{dt} = z$

$\dfrac{dy}{dt} = -z$

$\dfrac{dz}{dt} = y$

22. $\dfrac{dx}{dt} = 2x + y + 2z$

$\dfrac{dy}{dt} = 3x + 6z$

$\dfrac{dz}{dt} = -4x - 3z$

23. $\mathbf{X}' = \begin{pmatrix} 1 & -1 & 2 \\ -1 & 1 & 0 \\ -1 & 0 & 1 \end{pmatrix}\mathbf{X}$

24. $\mathbf{X}' = \begin{pmatrix} 4 & 0 & 1 \\ 0 & 6 & 0 \\ -4 & 0 & 4 \end{pmatrix}\mathbf{X}$

25. $\mathbf{X}' = \begin{pmatrix} 2 & 5 & 1 \\ -5 & -6 & 4 \\ 0 & 0 & 2 \end{pmatrix}\mathbf{X}$

26. $\mathbf{X}' = \begin{pmatrix} 2 & 4 & 4 \\ -1 & -2 & 0 \\ -1 & 0 & -2 \end{pmatrix}\mathbf{X}$

In Problems 27 and 28 solve the given system subject to the indicated initial condition.

27. $\mathbf{X}' = \begin{pmatrix} 1 & -12 & -14 \\ 1 & 2 & -3 \\ 1 & 1 & -2 \end{pmatrix}\mathbf{X}, \quad \mathbf{X}(0) = \begin{pmatrix} 4 \\ 6 \\ -7 \end{pmatrix}$

28. $\mathbf{X}' = \begin{pmatrix} 6 & -1 \\ 5 & 4 \end{pmatrix}\mathbf{X}, \quad \mathbf{X}(0) = \begin{pmatrix} -2 \\ 8 \end{pmatrix}$

8.6.3 Repeated Eigenvalues

In Problems 29–38 find the general solution of the given system.

29. $\dfrac{dx}{dt} = 3x - y$

$\dfrac{dy}{dt} = 9x - 3y$

30. $\dfrac{dx}{dt} = -6x + 5y$

$\dfrac{dy}{dt} = -5x + 4y$

31. $\dfrac{dx}{dt} = -x + 3y$

$\dfrac{dy}{dt} = -3x + 5y$

32. $\dfrac{dx}{dt} = 12x - 9y$

$\dfrac{dy}{dt} = 4x$

33. $\dfrac{dx}{dt} = 3x - y - z$

$\dfrac{dy}{dt} = x + y - z$

$\dfrac{dz}{dt} = x - y + z$

34. $\dfrac{dx}{dt} = 3x + 2y + 4z$

$\dfrac{dy}{dt} = 2x + 2z$

$\dfrac{dz}{dt} = 4x + 2y + 3z$

35. $\mathbf{X}' = \begin{pmatrix} 5 & -4 & 0 \\ 1 & 0 & 2 \\ 0 & 2 & 5 \end{pmatrix}\mathbf{X}$

36. $\mathbf{X}' = \begin{pmatrix} 1 & 0 & 0 \\ 0 & 3 & 1 \\ 0 & -1 & 1 \end{pmatrix}\mathbf{X}$

37. $\mathbf{X}' = \begin{pmatrix} 1 & 0 & 0 \\ 2 & 2 & -1 \\ 0 & 1 & 0 \end{pmatrix}\mathbf{X}$

38. $\mathbf{X}' = \begin{pmatrix} 4 & 1 & 0 \\ 0 & 4 & 1 \\ 0 & 0 & 4 \end{pmatrix}\mathbf{X}$

In Problems 39 and 40 solve the given system subject to the indicated initial condition.

39. $\mathbf{X}' = \begin{pmatrix} 2 & 4 \\ -1 & 6 \end{pmatrix}\mathbf{X}, \quad \mathbf{X}(0) = \begin{pmatrix} -1 \\ 6 \end{pmatrix}$

40. $\mathbf{X}' = \begin{pmatrix} 0 & 0 & 1 \\ 0 & 1 & 0 \\ 1 & 0 & 0 \end{pmatrix}\mathbf{X}, \quad \mathbf{X}(0) = \begin{pmatrix} 1 \\ 2 \\ 5 \end{pmatrix}$

Miscellaneous Problems

If $\Phi(t)$ is a fundamental matrix of the system, the initial-value problem $\mathbf{X}' = \mathbf{AX}$, $\mathbf{X}(t_0) = \mathbf{X}_0$ has the solution $\mathbf{X} = \Phi(t)\Phi^{-1}(t_0)\mathbf{X}_0$ (see Problem 35, Exercises 8.5). In Problems 41 and 42 use this result to solve the given system subject to the indicated initial condition.

41. $\mathbf{X}' = \begin{pmatrix} 4 & 3 \\ 3 & -4 \end{pmatrix}\mathbf{X}, \quad \mathbf{X}(0) = \begin{pmatrix} 1 \\ 1 \end{pmatrix}$

42. $\mathbf{X}' = \begin{pmatrix} -\frac{2}{25} & \frac{1}{50} \\ \frac{2}{25} & -\frac{2}{25} \end{pmatrix}\mathbf{X}, \quad \mathbf{X}(0) = \begin{pmatrix} 25 \\ 0 \end{pmatrix}$

In Problems 43 and 44 find a solution of the given system of the form $\mathbf{X} = t^{\lambda}\mathbf{K}$, $t > 0$, where $\mathbf{K}$ is a column vector of constants.

43. $t\mathbf{X}' = \begin{pmatrix} 1 & 3 \\ -1 & 5 \end{pmatrix}\mathbf{X}$

44. $t\mathbf{X}' = \begin{pmatrix} 2 & -2 \\ 2 & 7 \end{pmatrix}\mathbf{X}$

8.7* UNDETERMINED COEFFICIENTS

• *Undetermined coefficients* • *Particular solution*

The methods of **undetermined coefficients** and **variation of parameters** can both be adapted to the solution of a nonhomogeneous linear system $\mathbf{X}' = \mathbf{AX} + \mathbf{F}(t)$. Of these two methods, variation of parameters is the more powerful technique. However, there are a few instances when the method of undetermined coefficients gives a quick means of finding a particular solution $\mathbf{X}_p$.

EXAMPLE 1 **Undetermined Coefficients**

Solve the system $\mathbf{X}' = \begin{pmatrix} -1 & 2 \\ -1 & 1 \end{pmatrix}\mathbf{X} + \begin{pmatrix} -8 \\ 3 \end{pmatrix}$ on $(-\infty, \infty)$.

* This section is an optional section.

Solution We first solve the homogeneous system

$$\mathbf{X}' = \begin{pmatrix} -1 & 2 \\ -1 & 1 \end{pmatrix}\mathbf{X}.$$

The characteristic equation

$$\det(\mathbf{A} - \lambda\mathbf{I}) = \begin{vmatrix} -1-\lambda & 2 \\ -1 & 1-\lambda \end{vmatrix} = \lambda^2 + 1 = 0$$

yields the complex eigenvalues $\lambda_1 = i$ and $\lambda_2 = \overline{\lambda_1} = -i$. By the procedures of the last section, we find

$$\mathbf{X}_c = c_1\begin{pmatrix} \cos t + \sin t \\ \cos t \end{pmatrix} + c_2\begin{pmatrix} \cos t - \sin t \\ -\sin t \end{pmatrix}.$$

Now since $\mathbf{F}(t)$ is a constant vector, we assume a constant particular solution vector $\mathbf{X}_p = \begin{pmatrix} a_1 \\ b_1 \end{pmatrix}$. Substituting this latter assumption into the original system leads to

$$\begin{aligned} 0 &= -a_1 + 2b_1 - 8 \\ 0 &= -a_1 + b_1 + 3. \end{aligned}$$

Solving this system of algebraic equations gives $a_1 = 14$ and $b_1 = 11$, and so $\mathbf{X}_p = \begin{pmatrix} 14 \\ 11 \end{pmatrix}$. The general solution of the system is

$$\mathbf{X} = c_1\begin{pmatrix} \cos t + \sin t \\ \cos t \end{pmatrix} + c_2\begin{pmatrix} \cos t - \sin t \\ -\sin t \end{pmatrix} + \begin{pmatrix} 14 \\ 11 \end{pmatrix}.$$

EXAMPLE 2 Undetermined Coefficients

Solve the system

$$\begin{aligned} \frac{dx}{dt} &= 6x + y + 6t \\ \frac{dy}{dt} &= 4x + 3y - 10t + 4 \end{aligned}$$

on $(-\infty, \infty)$.

Solution We first solve the homogeneous system

$$\begin{aligned} \frac{dx}{dt} &= 6x + y \\ \frac{dy}{dt} &= 4x + 3y \end{aligned}$$

by the method of Section 8.6. The eigenvalues are determined from

$$\det(\mathbf{A} - \lambda\mathbf{I}) = \begin{vmatrix} 6-\lambda & 1 \\ 4 & 3-\lambda \end{vmatrix} = \lambda^2 - 9\lambda + 14 = 0.$$

Since $\lambda^2 - 9\lambda + 14 = (\lambda - 2)(\lambda - 7)$, we have $\lambda_1 = 2$ and $\lambda_2 = 7$. It is then easily verified that the respective eigenvectors of the coefficient matrix are

$$\mathbf{K}_1 = \begin{pmatrix} 1 \\ -4 \end{pmatrix} \quad \text{and} \quad \mathbf{K}_2 = \begin{pmatrix} 1 \\ 1 \end{pmatrix}.$$

Consequently the complementary function is

$$\mathbf{X}_c = c_1\begin{pmatrix} 1 \\ -4 \end{pmatrix}e^{2t} + c_2\begin{pmatrix} 1 \\ 1 \end{pmatrix}e^{7t}.$$

Because $\mathbf{F}(t)$ can be written as

$$\mathbf{F}(t) = \begin{pmatrix} 6 \\ -10 \end{pmatrix}t + \begin{pmatrix} 0 \\ 4 \end{pmatrix},$$

we shall try to find a particular solution of the system possessing the *same* form:

$$\mathbf{X}_p = \begin{pmatrix} a_2 \\ b_2 \end{pmatrix}t + \begin{pmatrix} a_1 \\ b_1 \end{pmatrix}.$$

In matrix terms we must have

$$\mathbf{X}_p' = \begin{pmatrix} 6 & 1 \\ 4 & 3 \end{pmatrix}\mathbf{X}_p + \begin{pmatrix} 6 \\ -10 \end{pmatrix}t + \begin{pmatrix} 0 \\ 4 \end{pmatrix}$$

or

$$\begin{pmatrix} a_2 \\ b_2 \end{pmatrix} = \begin{pmatrix} 6 & 1 \\ 4 & 3 \end{pmatrix}\left[\begin{pmatrix} a_2 \\ b_2 \end{pmatrix}t + \begin{pmatrix} a_1 \\ b_1 \end{pmatrix}\right] + \begin{pmatrix} 6 \\ -10 \end{pmatrix}t + \begin{pmatrix} 0 \\ 4 \end{pmatrix}$$

$$\begin{pmatrix} 0 \\ 0 \end{pmatrix} = \begin{pmatrix} (6a_2 + b_2 + 6)t + 6a_1 + b_1 - a_2 \\ (4a_2 + 3b_2 - 10)t + 4a_1 + 3b_1 - b_2 + 4 \end{pmatrix}.$$

From this last identity we conclude that

$$\begin{aligned} 6a_2 + b_2 + 6 &= 0 \\ 4a_2 + 3b_2 - 10 &= 0 \end{aligned} \quad \text{and} \quad \begin{aligned} 6a_1 + b_1 - a_2 &= 0 \\ 4a_1 + 3b_1 - b_2 + 4 &= 0. \end{aligned}$$

Solving the first two equations simultaneously yields $a_2 = -2$ and $b_2 = 6$. Substituting these values into the last two equations and solving for a_1 and b_1 gives $a_1 = -\frac{4}{7}$, $b_1 = \frac{10}{7}$. It follows, therefore, that a particular solution vector is

$$\mathbf{X}_p = \begin{pmatrix} -2 \\ 6 \end{pmatrix}t + \begin{pmatrix} -\frac{4}{7} \\ \frac{10}{7} \end{pmatrix},$$

and so the general solution of the system on $(-\infty, \infty)$ is

$$\begin{aligned} \mathbf{X} &= \mathbf{X}_c + \mathbf{X}_p \\ &= c_1\begin{pmatrix} 1 \\ -4 \end{pmatrix}e^{2t} + c_2\begin{pmatrix} 1 \\ 1 \end{pmatrix}e^{7t} + \begin{pmatrix} -2 \\ 6 \end{pmatrix}t + \begin{pmatrix} -\frac{4}{7} \\ \frac{10}{7} \end{pmatrix}. \end{aligned}$$

EXAMPLE 3 Form of a Particular Solution

Determine the form of the particular solution vector $\mathbf{X}_p$ for

$$\frac{dx}{dt} = 5x + 3y - 2e^{-t} + 1$$
$$\frac{dy}{dt} = -x + y + e^{-t} - 5t + 7.$$

Solution Proceeding in the usual manner, we find

$$\mathbf{X}_c = c_1\begin{pmatrix} 1 \\ -1 \end{pmatrix}e^{2t} + c_2\begin{pmatrix} 3 \\ -1 \end{pmatrix}e^{4t}.$$

Now since
$$\mathbf{F}(t) = \begin{pmatrix} -2 \\ 1 \end{pmatrix}e^{-t} + \begin{pmatrix} 0 \\ -5 \end{pmatrix}t + \begin{pmatrix} 1 \\ 7 \end{pmatrix},$$

we assume a particular solution of the form

$$\mathbf{X}_p = \begin{pmatrix} a_3 \\ b_3 \end{pmatrix}e^{-t} + \begin{pmatrix} a_2 \\ b_2 \end{pmatrix}t + \begin{pmatrix} a_1 \\ b_1 \end{pmatrix}.$$

Remarks The method of undetermined coefficients is not as simple as the last three examples seem to indicate. As in Section 4.4, the method can be applied only when the entries in the matrix $\mathbf{F}(t)$ are constants, polynomials, exponential functions, sines and cosines, or finite sums and products of these functions. There are further difficulties. The assumption for $\mathbf{X}_p$ is actually predicated on a prior knowledge of the complementary function $\mathbf{X}_c$. For example, if $\mathbf{F}(t)$ is a constant vector and $\lambda = 0$ is an eigenvalue, then $\mathbf{X}_c$ contains a constant vector. In this case $\mathbf{X}_p$ is not a constant vector as in Example 1 but rather

$$\mathbf{X}_p = \begin{pmatrix} a_2 \\ b_2 \end{pmatrix}t + \begin{pmatrix} a_1 \\ b_1 \end{pmatrix}.$$

See Problem 11.

Similarly, in Example 3, if we replace e^{-t} in $\mathbf{F}(t)$ by e^{2t} ($\lambda = 2$ is an eigenvalue), then the correct form of the particular solution is

$$\mathbf{X}_p = \begin{pmatrix} a_4 \\ b_4 \end{pmatrix}te^{2t} + \begin{pmatrix} a_3 \\ b_3 \end{pmatrix}e^{2t} + \begin{pmatrix} a_2 \\ b_2 \end{pmatrix}t + \begin{pmatrix} a_1 \\ b_1 \end{pmatrix}.$$

Rather than pursue these difficulties, we turn our attention now to the method of variation of parameters.

EXERCISES 8.7

Answers to odd-numbered problems begin on page A-21.

In Problems 1–8 use the method of undetermined coefficients to solve the given system on $(-\infty, \infty)$.

1. $$\frac{dx}{dt} = 2x + 3y - 7$$
$$\frac{dy}{dt} = -x - 2y + 5$$

2. $$\frac{dx}{dt} = 5x + 9y + 2$$
$$\frac{dy}{dt} = -x + 11y + 6$$

3. $\dfrac{dx}{dt} = x + 3y - 2t^2$

$\dfrac{dy}{dt} = 3x + y + t + 5$

4. $\dfrac{dx}{dt} = x - 4y + 4t + 9e^{6t}$

$\dfrac{dy}{dt} = 4x + y - t + e^{6t}$

5. $\mathbf{X}' = \begin{pmatrix} 4 & \frac{1}{3} \\ 9 & 6 \end{pmatrix}\mathbf{X} + \begin{pmatrix} -3 \\ 10 \end{pmatrix}e^t$

6. $\mathbf{X}' = \begin{pmatrix} -1 & 5 \\ -1 & 1 \end{pmatrix}\mathbf{X} + \begin{pmatrix} \sin t \\ -2\cos t \end{pmatrix}$

7. $\mathbf{X}' = \begin{pmatrix} 1 & 1 & 1 \\ 0 & 2 & 3 \\ 0 & 0 & 5 \end{pmatrix}\mathbf{X} + \begin{pmatrix} 1 \\ -1 \\ 2 \end{pmatrix}e^{4t}$

8. $\mathbf{X}' = \begin{pmatrix} 0 & 0 & 5 \\ 0 & 5 & 0 \\ 5 & 0 & 0 \end{pmatrix}\mathbf{X} + \begin{pmatrix} 5 \\ -10 \\ 40 \end{pmatrix}$

9. Solve $\mathbf{X}' = \begin{pmatrix} -1 & -2 \\ 3 & 4 \end{pmatrix}\mathbf{X} + \begin{pmatrix} 3 \\ 3 \end{pmatrix}$ subject to $\mathbf{X}(0) = \begin{pmatrix} -4 \\ 5 \end{pmatrix}$.

10. **(a)** Show that the system of differential equations for the currents $i_2(t)$ and $i_3(t)$ in the electrical network shown in Figure 8.11 is

$$\frac{d}{dt}\begin{pmatrix} i_2 \\ i_3 \end{pmatrix} = \begin{pmatrix} -R_1/L_1 & -R_1/L_1 \\ -R_1/L_2 & -(R_1 + R_2)/L_2 \end{pmatrix}\begin{pmatrix} i_2 \\ i_3 \end{pmatrix} + \begin{pmatrix} E/L_1 \\ E/L_2 \end{pmatrix}.$$

(b) Solve the system in part (a) if $R_1 = 2$ ohms, $R_2 = 3$ ohms, $L_1 = 1$ henry, $L_2 = 1$ henry, $E = 60$ volts, $i_2(0) = 0$, and $i_3(0) = 0$.

(c) Determine the current $i_1(t)$.

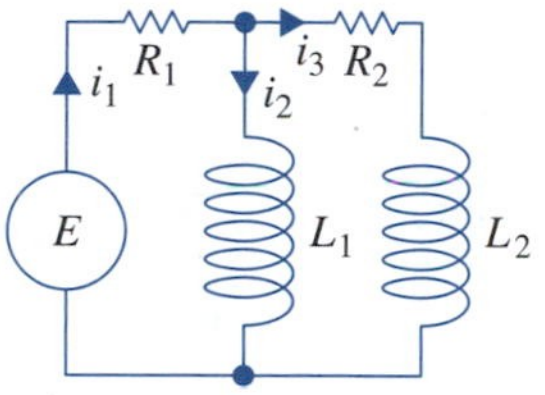

Figure 8.11

11. Solve the system $\mathbf{X}' = \begin{pmatrix} 1 & -1 \\ -1 & 1 \end{pmatrix}\mathbf{X} + \begin{pmatrix} 3 \\ -5 \end{pmatrix}$ on $(-\infty, \infty)$.

[*Hint*: A particular solution may not be unique.]

8.8 VARIATION OF PARAMETERS

• *Variation of parameters* • *Particular solution*

In Section 8.5 we saw that the general solution of a homogeneous system $\mathbf{X}' = \mathbf{AX}$ can be written as the product

$$\mathbf{X} = \mathbf{\Phi}(t)\mathbf{C},$$

where $\mathbf{\Phi}(t)$ is a fundamental matrix of the system and $\mathbf{C}$ is an $n \times 1$ column vector of constants. As in the procedure of Section 4.5, we ask whether it is possible to replace $\mathbf{C}$ by a column matrix of functions

$$\mathbf{U}(t) = \begin{pmatrix} u_1(t) \\ u_2(t) \\ \vdots \\ u_n(t) \end{pmatrix} \quad \text{so that} \quad \mathbf{X}_p = \mathbf{\Phi}(t)\mathbf{U}(t) \qquad \textbf{(1)}$$

is a particular solution of the nonhomogeneous system

$$\mathbf{X}' = \mathbf{A}\mathbf{X} + \mathbf{F}(t). \tag{2}$$

By the product rule* the derivative of (1) is

$$\mathbf{X}_p' = \mathbf{\Phi}(t)\mathbf{U}'(t) + \mathbf{\Phi}'(t)\mathbf{U}(t). \tag{3}$$

Substituting (3) and (1) into (2) gives

$$\mathbf{\Phi}(t)\mathbf{U}'(t) + \mathbf{\Phi}'(t)\mathbf{U}(t) = \mathbf{A}\mathbf{\Phi}(t)\mathbf{U}(t) + \mathbf{F}(t). \tag{4}$$

Now recall from (14) of Section 8.5 that $\mathbf{\Phi}'(t) = \mathbf{A}\mathbf{\Phi}(t)$. Thus (4) becomes

$$\mathbf{\Phi}(t)\mathbf{U}'(t) + \mathbf{A}\mathbf{\Phi}(t)\mathbf{U}(t) = \mathbf{A}\mathbf{\Phi}(t)\mathbf{U}(t) + \mathbf{F}(t)$$

or

$$\mathbf{\Phi}(t)\mathbf{U}'(t) = \mathbf{F}(t). \tag{5}$$

Multiplying both sides of equation (5) by $\mathbf{\Phi}^{-1}(t)$ gives

$$\mathbf{U}'(t) = \mathbf{\Phi}^{-1}(t)\mathbf{F}(t) \quad \text{or} \quad \mathbf{U}(t) = \int \mathbf{\Phi}^{-1}(t)\mathbf{F}(t)\,dt.$$

Hence by assumption (1) we conclude that a particular solution of (2) is given by

$$\mathbf{X}_p = \mathbf{\Phi}(t)\int \mathbf{\Phi}^{-1}(t)\mathbf{F}(t)\,dt. \tag{6}$$

To calculate the indefinite integral of the column matrix $\mathbf{\Phi}^{-1}(t)\mathbf{F}(t)$ in (6) we integrate each entry. Thus the general solution of the system (2) is $\mathbf{X} = \mathbf{X}_c + \mathbf{X}_p$ or

$$\mathbf{X} = \mathbf{\Phi}(t)\mathbf{C} + \mathbf{\Phi}(t)\int \mathbf{\Phi}^{-1}(t)\mathbf{F}(t)\,dt. \tag{7}$$

EXAMPLE 1 Variation of Parameters

Find the general solution of the nonhomogeneous system

$$\mathbf{X}' = \begin{pmatrix} -3 & 1 \\ 2 & -4 \end{pmatrix}\mathbf{X} + \begin{pmatrix} 3t \\ e^{-t} \end{pmatrix} \tag{8}$$

on the interval $(-\infty, \infty)$.

Solution We first solve the homogeneous system

$$\mathbf{X}' = \begin{pmatrix} -3 & 2 \\ 1 & -4 \end{pmatrix}\mathbf{X}. \tag{9}$$

The characteristic equation of the coefficient matrix is

$$\det(\mathbf{A} - \lambda\mathbf{I}) = \begin{vmatrix} -3-\lambda & 1 \\ 2 & -4-\lambda \end{vmatrix} = (\lambda + 2)(\lambda + 5) = 0$$

* See Problem 51, Exercises 8.4. Note that the order of the products is very important. Since $\mathbf{U}(t)$ is a column matrix, the products $\mathbf{U}'(t)\mathbf{\Phi}(t)$ and $\mathbf{U}(t)\mathbf{\Phi}'(t)$ are not defined.

and so the eigenvalues are $\lambda_1 = -2$ and $\lambda_2 = -5$. By the usual method we find that the eigenvectors corresponding to λ_1 and λ_2 are, respectively,

$$\begin{pmatrix} 1 \\ 1 \end{pmatrix} \quad \text{and} \quad \begin{pmatrix} 1 \\ -2 \end{pmatrix}.$$

The solution vectors of the system (9) are then

$$\mathbf{X}_1 = \begin{pmatrix} 1 \\ 1 \end{pmatrix} e^{-2t} \quad \text{and} \quad \mathbf{X}_2 = \begin{pmatrix} 1 \\ -2 \end{pmatrix} e^{-5t}.$$

Next we form

$$\mathbf{\Phi}(t) = \begin{pmatrix} e^{-2t} & e^{-2t} \\ e^{-5t} & -2e^{-5t} \end{pmatrix} \quad \text{and} \quad \mathbf{\Phi}^{-1}(t) = \begin{pmatrix} \frac{2}{3}e^{2t} & \frac{1}{3}e^{2t} \\ \frac{1}{3}e^{5t} & -\frac{1}{3}e^{5t} \end{pmatrix}.$$

From (6) we then obtain

$$\begin{aligned}
\mathbf{X}_p = \mathbf{\Phi}(t) \int \mathbf{\Phi}^{-1}(t)\mathbf{F}(t)\, dt &= \begin{pmatrix} e^{-2t} & e^{-5t} \\ e^{-2t} & -2e^{-5t} \end{pmatrix} \int \begin{pmatrix} \frac{2}{3}e^{2t} & \frac{1}{3}e^{2t} \\ \frac{1}{3}e^{5t} & -\frac{1}{3}e^{5t} \end{pmatrix} \begin{pmatrix} 3t \\ e^{-t} \end{pmatrix} dt \\
&= \begin{pmatrix} e^{-2t} & e^{-5t} \\ e^{-2t} & -2e^{-5t} \end{pmatrix} \int \begin{pmatrix} 2te^{2t} + \frac{1}{3}e^{t} \\ te^{5t} - \frac{1}{3}e^{4t} \end{pmatrix} dt \\
&= \begin{pmatrix} e^{-2t} & e^{-5t} \\ e^{-2t} & -2e^{-5t} \end{pmatrix} \begin{pmatrix} te^{2t} - \frac{1}{2}e^{2t} + \frac{1}{3}e^{t} \\ \frac{1}{5}te^{5t} - \frac{1}{25}e^{5t} - \frac{1}{12}e^{4t} \end{pmatrix} \\
&= \begin{pmatrix} \frac{6}{5}t - \frac{27}{50} + \frac{1}{4}e^{-t} \\ \frac{3}{5}t - \frac{21}{50} + \frac{1}{2}e^{-t} \end{pmatrix}.
\end{aligned}$$

Hence from (7) the general solution of (8) on the interval is

$$\begin{aligned}
\mathbf{X} &= \begin{pmatrix} e^{-2t} & e^{-5t} \\ e^{-2t} & -2e^{-5t} \end{pmatrix} \begin{pmatrix} c_1 \\ c_2 \end{pmatrix} + \begin{pmatrix} \frac{6}{5}t - \frac{27}{50} + \frac{1}{4}e^{-t} \\ \frac{3}{5}t - \frac{21}{50} + \frac{1}{2}e^{-t} \end{pmatrix} \\
&= c_1 \begin{pmatrix} 1 \\ 1 \end{pmatrix} e^{-2t} + c_2 \begin{pmatrix} 1 \\ -2 \end{pmatrix} e^{-5t} + \begin{pmatrix} \frac{6}{5} \\ \frac{3}{5} \end{pmatrix} t - \begin{pmatrix} \frac{27}{50} \\ \frac{21}{50} \end{pmatrix} + \begin{pmatrix} \frac{1}{4} \\ \frac{1}{2} \end{pmatrix} e^{-t}.
\end{aligned}$$

The general solution of (2) on an interval can be written in the alternative manner

$$\mathbf{X} = \mathbf{\Phi}(t)\mathbf{C} + \mathbf{\Phi}(t) \int_{t_0}^{t} \mathbf{\Phi}^{-1}(s)\mathbf{F}(s)\, ds, \tag{10}$$

where t and t_0 are points in the interval. This last form is useful in solving (2) subject to an initial condition $\mathbf{X}(t_0) = \mathbf{X}_0$. Substituting $t = t_0$ in (10) yields

$$\mathbf{X}_0 = \mathbf{\Phi}(t_0)\mathbf{C},$$

from which we see immediately that

$$\mathbf{C} = \mathbf{\Phi}^{-1}(t_0)\mathbf{X}_0.$$

We conclude that the solution of the initial-value problem is given by

$$\mathbf{X} = \mathbf{\Phi}(t)\mathbf{\Phi}^{-1}(t_0)\mathbf{X}_0 + \mathbf{\Phi}(t) \int_{t_0}^{t} \mathbf{\Phi}^{-1}(s)\mathbf{F}(s)\, ds. \tag{11}$$

Recall from Section 8.5 that an alternative way of forming a fundamental matrix is to choose its column vectors $\mathbf{V}_i$ in such a manner that

$$\mathbf{V}_1(t_0) = \begin{pmatrix} 1 \\ 0 \\ \vdots \\ 0 \end{pmatrix}, \quad \mathbf{V}_2(t_0) = \begin{pmatrix} 0 \\ 1 \\ \vdots \\ 0 \end{pmatrix}, \quad \ldots, \quad \mathbf{V}_n(t_0) = \begin{pmatrix} 0 \\ 0 \\ \vdots \\ 1 \end{pmatrix}. \tag{12}$$

This fundamental matrix is denoted by $\mathbf{\Psi}(t)$. As a consequence of (12) we know that $\mathbf{\Psi}(t)$ has the property

$$\mathbf{\Psi}(t_0) = \mathbf{I}. \tag{13}$$

But since $\mathbf{\Psi}(t)$ is nonsingular for all values of t in an interval, (13) implies

$$\mathbf{\Psi}^{-1}(t_0) = \mathbf{I}. \tag{14}$$

Thus when $\mathbf{\Psi}(t)$ is used rather than $\mathbf{\Phi}(t)$, it follows from (14) that (11) can be written

$$\mathbf{X} = \mathbf{\Psi}(t)\mathbf{X}_0 + \mathbf{\Psi}(t)\int_{t_0}^{t} \mathbf{\Psi}^{-1}(s)\mathbf{F}(s)\, ds. \tag{15}$$

EXERCISES 8.8

Answers to odd-numbered problems begin on page A-22.

In Problems 1–20 use variation of parameters to solve the given system.

1. $\dfrac{dx}{dt} = 3x - 3y + 4$

$\dfrac{dy}{dt} = 2x - 2y - 1$

2. $\dfrac{dx}{dt} = 2x - y$

$\dfrac{dy}{dt} = 3x - 2y + 4t$

3. $\mathbf{X}' = \begin{pmatrix} 3 & -5 \\ \frac{3}{4} & -1 \end{pmatrix}\mathbf{X} + \begin{pmatrix} 1 \\ -1 \end{pmatrix} e^{t/2}$

4. $\mathbf{X}' = \begin{pmatrix} 2 & -1 \\ 4 & 2 \end{pmatrix}\mathbf{X} + \begin{pmatrix} \sin 2t \\ 2\cos 2t \end{pmatrix} e^{2t}$

5. $\mathbf{X}' = \begin{pmatrix} 0 & 2 \\ -1 & 3 \end{pmatrix}\mathbf{X} + \begin{pmatrix} 1 \\ -1 \end{pmatrix} e^{t}$

6. $\mathbf{X}' = \begin{pmatrix} 0 & 2 \\ -1 & 3 \end{pmatrix}\mathbf{X} + \begin{pmatrix} 2 \\ e^{-3t} \end{pmatrix}$

7. $\mathbf{X}' = \begin{pmatrix} 1 & 8 \\ 1 & -1 \end{pmatrix}\mathbf{X} + \begin{pmatrix} 12 \\ 12 \end{pmatrix} t$

8. $\mathbf{X}' = \begin{pmatrix} 1 & 8 \\ 1 & -1 \end{pmatrix}\mathbf{X} + \begin{pmatrix} e^{-t} \\ te^{t} \end{pmatrix}$

9. $\mathbf{X}' = \begin{pmatrix} 3 & 2 \\ -2 & -1 \end{pmatrix}\mathbf{X} + \begin{pmatrix} 2e^{-t} \\ e^{-t} \end{pmatrix}$

10. $\mathbf{X}' = \begin{pmatrix} 3 & 2 \\ -2 & -1 \end{pmatrix}\mathbf{X} + \begin{pmatrix} 1 \\ 1 \end{pmatrix}$

11. $\mathbf{X}' = \begin{pmatrix} 0 & -1 \\ 1 & 0 \end{pmatrix}\mathbf{X} + \begin{pmatrix} \sec t \\ 0 \end{pmatrix}$

12. $\mathbf{X}' = \begin{pmatrix} 1 & -1 \\ 1 & 1 \end{pmatrix}\mathbf{X} + \begin{pmatrix} 3 \\ 3 \end{pmatrix} e^{t}$

13. $\mathbf{X}' = \begin{pmatrix} 1 & -1 \\ 1 & 1 \end{pmatrix}\mathbf{X} + \begin{pmatrix} \cos t \\ \sin t \end{pmatrix} e^{t}$

14. $\mathbf{X}' = \begin{pmatrix} 2 & -2 \\ 8 & -6 \end{pmatrix}\mathbf{X} + \begin{pmatrix} 1 \\ 3 \end{pmatrix} \dfrac{e^{-2t}}{t}$

15. $\mathbf{X}' = \begin{pmatrix} 0 & 1 \\ -1 & 0 \end{pmatrix}\mathbf{X} + \begin{pmatrix} 0 \\ \sec t \tan t \end{pmatrix}$

16. $\mathbf{X}' = \begin{pmatrix} 0 & 1 \\ -1 & 0 \end{pmatrix}\mathbf{X} + \begin{pmatrix} 1 \\ \cot t \end{pmatrix}$

17. $\mathbf{X}' = \begin{pmatrix} 1 & 2 \\ -\frac{1}{2} & 1 \end{pmatrix}\mathbf{X} + \begin{pmatrix} \csc t \\ \sec t \end{pmatrix} e^t$

18. $\mathbf{X}' = \begin{pmatrix} 1 & -2 \\ 1 & -1 \end{pmatrix}\mathbf{X} + \begin{pmatrix} \tan t \\ 1 \end{pmatrix}$

19. $\mathbf{X}' = \begin{pmatrix} 1 & 1 & 0 \\ 1 & 1 & 0 \\ 0 & 0 & 3 \end{pmatrix}\mathbf{X} + \begin{pmatrix} e^t \\ e^{2t} \\ te^{3t} \end{pmatrix}$

20. $\mathbf{X}' = \begin{pmatrix} 3 & -1 & -1 \\ 1 & 1 & -1 \\ 1 & -1 & 1 \end{pmatrix}\mathbf{X} + \begin{pmatrix} 0 \\ t \\ 2e^t \end{pmatrix}$

In Problems 21 and 22 use (11) to solve the given system subject to the indicated initial condition.

21. $\mathbf{X}' = \begin{pmatrix} 3 & -1 \\ -1 & 3 \end{pmatrix}\mathbf{X} + \begin{pmatrix} 4e^{2t} \\ 4e^{4t} \end{pmatrix}, \quad \mathbf{X}(0) = \begin{pmatrix} 1 \\ 1 \end{pmatrix}$

22. $\mathbf{X}' = \begin{pmatrix} 1 & -1 \\ 1 & -1 \end{pmatrix}\mathbf{X} + \begin{pmatrix} 1/t \\ 1/t \end{pmatrix}, \quad \mathbf{X}(1) = \begin{pmatrix} 2 \\ -1 \end{pmatrix}$

In Problems 23 and 24 use (15) to solve the given system subject to the indicated initial condition. Use the results of Problems 31 and 34 in Exercises 8.5.

23. $\mathbf{X}' = \begin{pmatrix} 4 & 1 \\ 6 & 5 \end{pmatrix}\mathbf{X} + \begin{pmatrix} 50e^{7t} \\ 0 \end{pmatrix}, \quad \mathbf{X}(0) = \begin{pmatrix} 5 \\ -5 \end{pmatrix}$

24. $\mathbf{X}' = \begin{pmatrix} 3 & -2 \\ 5 & -3 \end{pmatrix}\mathbf{X} + \begin{pmatrix} 2 \\ 3 \end{pmatrix}, \quad \mathbf{X}(\pi/2) = \begin{pmatrix} 0 \\ 0 \end{pmatrix}$

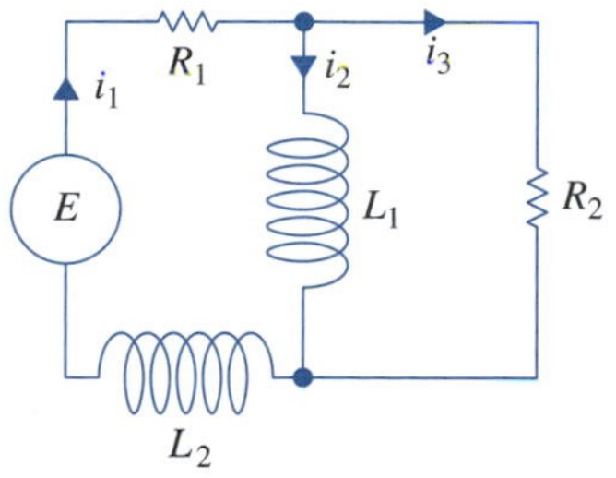

Figure 8.12

25. **(a)** Show that the system of differential equations for the currents $i_1(t)$ and $i_2(t)$ in the electrical network shown in Figure 8.12 is

$$\frac{d}{dt}\begin{pmatrix} i_1 \\ i_2 \end{pmatrix} = \begin{pmatrix} -(R_1 + R_2)/L_2 & R_2/L_2 \\ R_2/L_1 & -R_2/L_1 \end{pmatrix}\begin{pmatrix} i_1 \\ i_2 \end{pmatrix} + \begin{pmatrix} E/L_2 \\ 0 \end{pmatrix}.$$

(b) Solve the system in part (a) if $R_1 = 8$ ohms, $R_2 = 3$ ohms, $L_1 = 1$ henry, $L_2 = 1$ henry, $E(t) = 100 \sin t$ volts, $i_1(0) = 0$, and $i_2(0) = 0$.

8.9* MATRIX EXPONENTIAL

• *Matrix exponential* • *Solution of linear systems*

Matrices can be utilized in an entirely different manner to solve a homogeneous system of linear first-order differential equations.

Recall that the simple linear first-order differential equation $x' = ax$, where a is a constant, has the general solution $x = ce^{at}$. It seems natural then

* This section is an optional section.

to ask whether we can define a matrix exponential $e^{t\mathbf{A}}$ so that the homogeneous system $\mathbf{X}' = \mathbf{AX}$, where $\mathbf{A}$ is an $n \times n$ matrix of constants, has a solution

$$\mathbf{X} = e^{t\mathbf{A}}\mathbf{C}. \tag{1}$$

Since $\mathbf{C}$ is to be an $n \times 1$ column vector of arbitrary constants, we want $e^{t\mathbf{A}}$ to be an $n \times n$ matrix. While the complete development of the meaning of the **matrix exponential** would necessitate a more thorough investigation of matrix algebra, one means of computing $e^{t\mathbf{A}}$ is given in the following definition.

DEFINITION 8.20 Matrix Exponential

For any $n \times n$ matrix $\mathbf{A}$,

$$e^{t\mathbf{A}} = \sum_{n=0}^{\infty} \frac{(t\mathbf{A})^n}{n!} = \mathbf{I} + t\mathbf{A} + \frac{t^2}{2!}\mathbf{A}^2 + \frac{t^2}{3!}\mathbf{A}^3 + \cdots. \tag{2}$$

It can be shown that the series given in (2) converges to an $n \times n$ matrix for every value of t. Also, $\mathbf{A}^2 = \mathbf{AA}$, $\mathbf{A}^3 = \mathbf{A}(\mathbf{A}^2)$, and so on.

Now the general solution of the single differential equation

$$x' = ax + f(t),$$

where a is a constant, can be expressed as

$$x = x_c + x_p = ce^{at} + e^{at}\int_{t_0}^{t} e^{-as}f(s)\,ds.$$

For systems of linear first-order differential equations, it can be shown that the general solution of

$$\mathbf{X}' = \mathbf{AX} + \mathbf{F}(t),$$

where $\mathbf{A}$ is an $n \times n$ matrix of constants, is

$$\mathbf{X} = \mathbf{X}_c + \mathbf{X}_p = e^{t\mathbf{A}}\mathbf{C} + e^{t\mathbf{A}}\int_{t_0}^{t} e^{-s\mathbf{A}}\mathbf{F}(s)\,ds. \tag{3}$$

The matrix exponential $e^{t\mathbf{A}}$ is always nonsingular and $e^{-s\mathbf{A}} = (e^{s\mathbf{A}})^{-1}$. In practice, $e^{-s\mathbf{A}}$ can be obtained from $e^{t\mathbf{A}}$ by replacing t by $-s$.

Additional Properties From (2) it is seen that

$$e^{\mathbf{0}} = \mathbf{I}. \tag{4}$$

Also, formal termwise differentiation of (2) shows that

$$\frac{d}{dt}e^{t\mathbf{A}} = \mathbf{A}e^{t\mathbf{A}}. \tag{5}$$

If we denote the matrix exponential by $\mathbf{\Psi}(t)$, then (5) and (4) are equivalent to

$$\mathbf{\Psi}'(t) = \mathbf{A}\mathbf{\Psi}(t) \tag{6}$$

and
$$\Psi(0) = \mathbf{I}, \tag{7}$$

respectively. The notation here is chosen deliberately. Comparing (6) with (14) of Section 8.5 reveals that $e^{t\mathbf{A}}$ is a fundamental matrix of the system $\mathbf{X}' = \mathbf{AX}$. It is precisely this formulation of the fundamental matrix that was discussed on page 386 of Section 8.5.

By multiplying the series defining $e^{t\mathbf{A}}$ and $e^{-s\mathbf{A}}$, we can prove that

$$e^{t\mathbf{A}}e^{-s\mathbf{A}} = e^{(t-s)\mathbf{A}} \quad \text{or, equivalently,} \quad \Psi(t)\Psi^{-1}(s) = \Psi(t-s).*$$

This last result enables us to relate (3) to (10) of the preceding section:

$$\mathbf{X} = e^{t\mathbf{A}}\mathbf{C} + \int_{t_0}^{t} e^{t\mathbf{A}}e^{-s\mathbf{A}}\mathbf{F}(s)\,ds$$

$$= e^{t\mathbf{A}}\mathbf{C} + \int_{t_0}^{t} e^{(t-s)\mathbf{A}}\mathbf{F}(s)\,ds \tag{8}$$

$$\mathbf{X} = \Psi(t)\mathbf{C} + \int_{t_0}^{t} \Psi(t-s)\mathbf{F}(s)\,ds. \tag{9}$$

Equation (9) possesses a form simpler than (10) of Section 8.8. In other words, there is no need to compute Ψ^{-1}; we need only replace t by $t - s$ in $\Psi(t)$.

EXERCISES 8.9

Answers to odd-numbered problems begin on page A-22.

In Problems 1 and 2 use (2) to compute $e^{t\mathbf{A}}$ and $e^{-t\mathbf{A}}$.

1. $\mathbf{A} = \begin{pmatrix} 0 & 1 \\ 1 & 0 \end{pmatrix}$

2. $\mathbf{A} = \begin{pmatrix} 1 & 0 \\ 0 & 2 \end{pmatrix}$

In Problems 3 and 4 use (1) to find the general solution of each system.

3. $\mathbf{X}' = \begin{pmatrix} 0 & 1 \\ 1 & 0 \end{pmatrix}\mathbf{X}$

4. $\mathbf{X}' = \begin{pmatrix} 1 & 0 \\ 0 & 2 \end{pmatrix}\mathbf{X}$

In Problems 5–8 use (3) to find the general solution of each system.

5. $\mathbf{X}' = \begin{pmatrix} 0 & 1 \\ 1 & 0 \end{pmatrix}\mathbf{X} + \begin{pmatrix} 1 \\ 1 \end{pmatrix}$

6. $\mathbf{X}' = \begin{pmatrix} 0 & 1 \\ 1 & 0 \end{pmatrix}\mathbf{X} + \begin{pmatrix} \cosh t \\ \sinh t \end{pmatrix}$

7. $\mathbf{X}' = \begin{pmatrix} 1 & 0 \\ 0 & 2 \end{pmatrix}\mathbf{X} + \begin{pmatrix} t \\ e^{4t} \end{pmatrix}$

8. $\mathbf{X}' = \begin{pmatrix} 1 & 0 \\ 0 & 2 \end{pmatrix}\mathbf{X} + \begin{pmatrix} 3 \\ -1 \end{pmatrix}$

*Although $e^{t\mathbf{A}}e^{-s\mathbf{A}} = e^{(t-s)\mathbf{A}}$, it is interesting to note that $e^{\mathbf{A}}e^{\mathbf{B}}$ is, in general, not the same as $e^{\mathbf{A}+\mathbf{B}}$ for $n \times n$ matrices $\mathbf{A}$ and $\mathbf{B}$.

Miscellaneous Problems

Let $\mathbf{P}$ denote a matrix whose columns are eigenvectors $\mathbf{K}_1, \mathbf{K}_2, \ldots, \mathbf{K}_n$ responding to *distinct* eigenvalues $\lambda_1, \lambda_2, \ldots, \lambda_n$ of an $n \times n$ matrix $\mathbf{A}$. Then it can be shown that $\mathbf{A} = \mathbf{PDP}^{-1}$, where $\mathbf{D}$ is defined by

$$\mathbf{D} = \begin{pmatrix} \lambda_1 & 0 & \cdots & 0 \\ 0 & \lambda_2 & \cdots & 0 \\ \vdots & & & \vdots \\ 0 & 0 & \cdots & \lambda_n \end{pmatrix}. \tag{10}$$

In Problems 9 and 10 verify the above result for the given matrix.

9. $\mathbf{A} = \begin{pmatrix} 2 & 1 \\ -3 & 6 \end{pmatrix}$

10. $\mathbf{A} = \begin{pmatrix} 2 & 1 \\ 1 & 2 \end{pmatrix}$

11. Suppose $\mathbf{A} = \mathbf{PDP}^{-1}$, where $\mathbf{D}$ is defined in (10). Use (2) to show that $e^{t\mathbf{A}} = \mathbf{P}e^{t\mathbf{D}}\mathbf{P}^{-1}$.

12. Use (2) to show that

$$e^{t\mathbf{D}} = \begin{pmatrix} e^{\lambda_1 t} & 0 & \cdots & 0 \\ 0 & e^{\lambda_2 t} & \cdots & 0 \\ \vdots & & & \vdots \\ 0 & 0 & \cdots & e^{\lambda_n t} \end{pmatrix},$$

where $\mathbf{D}$ is defined in (10).

In Problems 13 and 14 use the results of Problems 9–12 to solve the given system.

13. $\mathbf{X}' = \begin{pmatrix} 2 & 1 \\ -3 & 6 \end{pmatrix}\mathbf{X}$

14. $\mathbf{X}' = \begin{pmatrix} 2 & 1 \\ 1 & 2 \end{pmatrix}\mathbf{X}$

CHAPTER 8 REVIEW

Throughout this chapter we have considered **systems of linear differential equations.**

For linear systems probably the most basic technique of solution consists of rewriting the entire system in **operator notation** and then using **systematic elimination** to obtain single differential equations in one dependent variable that can be solved by the usual procedures. Also, **determinants** can usually be utilized to accomplish the same result. Once all dependent variables have been determined, it is necessary to use the system itself to find various relationships between the parameters.

When initial conditions are specified, the **Laplace transform** can be used to reduce the system to simultaneous algebraic equations in the transformed functions.

A first-order system in **normal form** in *two* dependent variables is any system

$$\begin{aligned} \frac{dx}{dt} &= a_{11}(t)x + a_{12}(t)y + f_1(t) \\ \frac{dy}{dt} &= a_{21}(t)x + a_{22}(t)y + f_2(t), \end{aligned} \tag{1}$$

where the coefficients $a_{ij}(t)$, $f_1(t)$, and $f_2(t)$ are continuous on some common interval I. When $f_1(t) = 0, f_2(t) = 0$, the system is said to be **homogeneous;**

otherwise it is **nonhomogeneous.** Any linear second-order differential equation can be expressed in this form.

Using matrices, we can write the system (1) compactly as

$$\frac{d\mathbf{X}}{dt} = \mathbf{A}(t)\mathbf{X} + \mathbf{F}(t), \tag{2}$$

where

$$\mathbf{X} = \begin{pmatrix} x \\ y \end{pmatrix}, \quad \mathbf{A}(t) = \begin{pmatrix} a_{11}(t) & a_{12}(t) \\ a_{21}(t) & a_{22}(t) \end{pmatrix}, \quad \text{and} \quad \mathbf{F}(t) = \begin{pmatrix} f_1(t) \\ f_2(t) \end{pmatrix}.$$

The **general solution of the homogeneous system** in two dependent variables

$$\frac{d\mathbf{X}}{dt} = \mathbf{A}(t)\mathbf{X} \tag{3}$$

is defined to be the linear combination

$$\mathbf{X} = c_1\mathbf{X}_1 + c_2\mathbf{X}_2, \tag{4}$$

where $\mathbf{X}_1$ and $\mathbf{X}_2$ form a **fundamental set of solutions** of (3) on I. The **general solution of the nonhomogeneous system** (2) is defined to be

$$\mathbf{X} = \mathbf{X}_c + \mathbf{X}_p,$$

where $\mathbf{X}_c$ is defined by (4) and $\mathbf{X}_p$ is *any* solution vector of (2).

To solve a homogeneous system (3) we determine the **eigenvalues** of the coefficient matrix **A** and then find the corresponding **eigenvectors.**

To solve a nonhomogeneous system we first solve the associated homogeneous system. A particular solution vector $\mathbf{X}_p$ of the nonhomogeneous system is found by either **undetermined coefficients** or **variation of parameters.**

A fundamental matrix of a homogeneous system (3) in two dependent variables is defined to be

$$\mathbf{\Phi}(t) = \begin{pmatrix} x_1 & x_2 \\ y_1 & y_2 \end{pmatrix}. \tag{5}$$

The columns in (5) are obtained from two linearly independent solution vectors $\mathbf{X}_1$ and $\mathbf{X}_2$ of (3). In terms of matrices, the method of variation of parameters leads to a particular solution given by

$$\mathbf{X}_p = \mathbf{\Phi}(t)\int \mathbf{\Phi}^{-1}(t)\mathbf{F}(t)\,dt.$$

The general solution of (2) on an interval is

$$\mathbf{X} = \mathbf{\Phi}(t)\mathbf{C} + \mathbf{\Phi}(t)\int \mathbf{\Phi}^{-1}(t)\mathbf{F}(t)\,dt,$$

where **C** is a column matrix containing two arbitrary constants. The matrix $\mathbf{\Phi}^{-1}(t)$ is called the **multiplicative inverse** of $\mathbf{\Phi}(t)$; the multiplicative inverse satisfies $\mathbf{\Phi}(t)\mathbf{\Phi}^{-1}(t) = \mathbf{\Phi}^{-1}(t)\mathbf{\Phi}(t) = \mathbf{I}$, where **I** is the 2×2 **multiplicative identity.**

CHAPTER 8 REVIEW EXERCISES

Answers to odd-numbered problems begin on page A-22.

Answer Problems 1–12 without referring back to the text. Fill in the blank or answer true or false.

1. Every second-order linear differential equation can be expressed as a system of two linear first-order differential equations.________

2. If $\mathbf{A} = \begin{pmatrix} 1 \\ 2 \end{pmatrix}$ and $\mathbf{B} = (3 \quad 4)$, then $\mathbf{AB} =$ ________ and $\mathbf{BA} =$ ________.

3. If $\mathbf{A} = \begin{pmatrix} 1 & 2 \\ 3 & 4 \end{pmatrix}$, then $\mathbf{A}^{-1} =$ ________.

4. If $\mathbf{A}$ is a nonsingular matrix for which $\mathbf{AB} = \mathbf{AC}$, then $\mathbf{B} = \mathbf{C}$.________

5. If $\mathbf{X}_1$ is a solution of $\mathbf{X}' = \mathbf{AX}$ and $\mathbf{X}_2$ is a solution of $\mathbf{X}' = \mathbf{AX} + \mathbf{F}$, then $\mathbf{X} = \mathbf{X}_1 + \mathbf{X}_2$ is a solution of $\mathbf{X}' = \mathbf{AX} + \mathbf{F}$.________

6. A fundamental matrix $\mathbf{\Phi}$ of a system $\mathbf{X}' = \mathbf{AX}$ is always nonsingular. ________

7. Let $\mathbf{A}$ be an $n \times n$ matrix. The eigenvalues of $\mathbf{A}$ are the nonzero solutions of $\det(\mathbf{A} - \lambda\mathbf{I}) = 0$.________

8. A nonzero constant multiple of an eigenvector is an eigenvector corresponding to the same eigenvalue.________

9. An $n \times 1$ column vector $\mathbf{K}$ with all zero entries is never an eigenvector of an $n \times n$ matrix $\mathbf{A}$.________

10. Let $\mathbf{A}$ be an $n \times n$ matrix with real entries. If λ is a complex eigenvalue, then $\bar{\lambda}$ is also an eigenvalue of $\mathbf{A}$.________

11. An $n \times n$ matrix $\mathbf{A}$ always possesses n linearly independent eigenvectors. ________

12. The augmented matrix

$$\left(\begin{array}{ccc|c} 1 & 1 & 1 & 2 \\ 0 & 1 & 0 & 3 \\ 0 & 0 & 0 & 0 \end{array}\right)$$

is in reduced row-echelon form.________

In Problems 13–16 use systematic elimination or determinants to solve the given system.

13. $$\begin{aligned} x' + y' &= 2x + 2y + 1 \\ x' + 2y' &= y + 3 \end{aligned}$$

14. $$\begin{aligned} \frac{dx}{dt} &= 2x + y + t - 2 \\ \frac{dy}{dt} &= 3x + 4y - 4t \end{aligned}$$

15. $$\begin{aligned} (D - 2)x - y &= -e^t \\ -3x + (D - 4)y &= -7e^t \end{aligned}$$

16. $$\begin{aligned} (D + 2)x + (D + 1)y &= \sin 2t \\ 5x + (D + 3)y &= \cos 2t \end{aligned}$$

In Problems 17 and 18 use the Laplace transform to solve each system.

17. $x' + y = t$
$4x + y' = 0$
$x(0) = 1, y(0) = 2$

18. $x'' + y'' = e^{2t}$
$2x' + y'' = -e^{2t}$
$x(0) = 0, y(0) = 0$
$x'(0) = 0, y'(0) = 0$

19. (**a**) Write as one column matrix **X**:

$$\begin{pmatrix} 3 & 1 & 1 \\ -1 & 2 & -1 \\ 0 & -2 & 4 \end{pmatrix}\begin{pmatrix} t \\ t^2 \\ t^3 \end{pmatrix} - \begin{pmatrix} 2 \\ -2 \\ -1 \end{pmatrix} + 2t\begin{pmatrix} 1 \\ 0 \\ 4 \end{pmatrix} + 2t^2\begin{pmatrix} 1 \\ -1 \\ 7 \end{pmatrix}.$$

(**b**) Find $d\mathbf{X}/dt$.

20. Write the differential equation

$$3y^{(4)} - 5y'' + 9y = 6e^t - 2t$$

as a system of first-order equations in linear normal form.

21. Write the system

$$\begin{aligned} (2D^2 + D)y - \qquad D^2x &= \ln t \\ D^2y + (D + 1)x &= 5t - 2 \end{aligned}$$

as a system of first-order equations in linear normal form.

22. Verify that the general solution of the system

$$\frac{dx}{dt} = y$$

$$\frac{dy}{dt} = -x + 2y - 2\cos t$$

on the interval $(-\infty, \infty)$ is

$$\mathbf{X} = c_1\begin{pmatrix} 1 \\ 1 \end{pmatrix}e^t + c_2\left[\begin{pmatrix} 1 \\ 1 \end{pmatrix}te^t + \begin{pmatrix} 0 \\ 1 \end{pmatrix}e^t\right] + \begin{pmatrix} \sin t \\ \cos t \end{pmatrix}.$$

In Problems 23–28 use the concept of eigenvalues and eigenvectors to solve each system.

23. $\dfrac{dx}{dt} = 2x + y$

$\dfrac{dy}{dt} = -x$

24. $\dfrac{dx}{dt} = -4x + 2y$

$\dfrac{dy}{dt} = 2x - 4y$

25. $\mathbf{X}' = \begin{pmatrix} 1 & 2 \\ -2 & 1 \end{pmatrix}\mathbf{X}$

26. $\mathbf{X}' = \begin{pmatrix} -2 & 5 \\ -2 & 4 \end{pmatrix}\mathbf{X}$

27. $\mathbf{X}' = \begin{pmatrix} 1 & 1 & 1 \\ 1 & 1 & 1 \\ 1 & 1 & 1 \end{pmatrix}\mathbf{X}$

28. $\mathbf{X}' = \begin{pmatrix} 1 & -1 & 1 \\ 0 & 1 & 3 \\ 4 & 3 & 1 \end{pmatrix}\mathbf{X}$

In Problems 29–32 use either undetermined coefficients or variation of parameters to solve the given system.

29. $\mathbf{X}' = \begin{pmatrix} 2 & 8 \\ 0 & 4 \end{pmatrix}\mathbf{X} + \begin{pmatrix} 2 \\ 16t \end{pmatrix}$

30. $\dfrac{dx}{dt} = x + 2y$

$\dfrac{dy}{dt} = -\dfrac{1}{2}x + y + e^t \tan t$

31. $\mathbf{X}' = \begin{pmatrix} -1 & 1 \\ -2 & 1 \end{pmatrix}\mathbf{X} + \begin{pmatrix} 1 \\ \cot t \end{pmatrix}$

32. $\mathbf{X}' = \begin{pmatrix} 3 & 1 \\ -1 & 1 \end{pmatrix}\mathbf{X} + \begin{pmatrix} -2 \\ 1 \end{pmatrix}e^{2t}$

9 NUMERICAL METHODS FOR ORDINARY DIFFERENTIAL EQUATIONS

Introduction

A differential equation does not have to have a solution, and even if a solution exists, we may not always be able to find (that is, exhibit) an explicit or implicit solution of the equation. In many instances, particularly in the study of nonlinear equations, we may have to be content with an approximation to the solution.

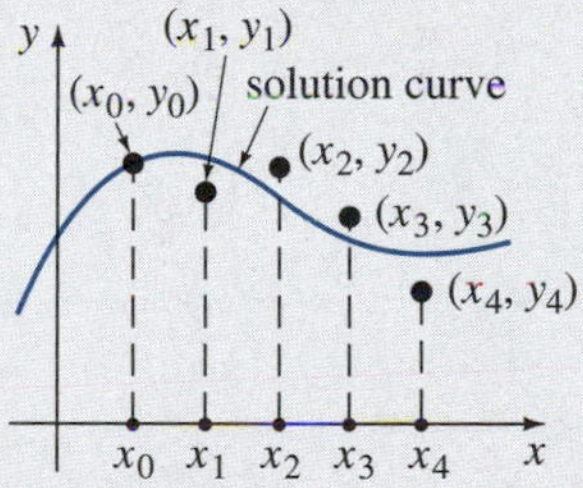

If a solution of a differential equation exists, it represents a locus of points (points connected by a smooth curve) in the Cartesian plane. Beginning in Section 9.2 we shall consider numerical procedures that utilize the differential equation to obtain a sequence of distinct points whose coordinates, as shown in the accompanying figure, approximate the coordinates of the points on the actual solution curve.

Our primary focus in this chapter is on first-order initial-value problems: $dy/dx = f(x, y)$, $y(x_0) = y_0$. We shall see that the numerical procedures developed for first-order equations may be adapted to systems of first-order equations in a very natural manner. As a consequence, we can approximate solutions of higher-order initial-value problems by simply reducing the differential equation to a system of first-order equations (see Section 8.3). In Section 9.8 we shall turn our attention to a numerical procedure for linear second-order boundary-value problems.

We begin with the study of direction fields. Although it is not a numerical method, the concept of a direction field enables us to obtain a rough sketch of a solution of a first-order differential equation without actually solving it.

9.1 DIRECTION FIELDS

• *Lineal elements* • *Isocline* • *Slope field*

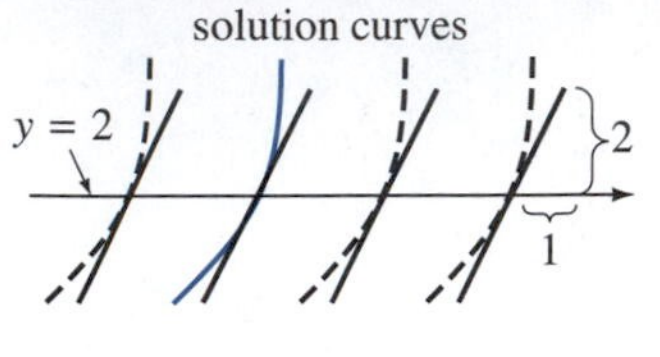

Figure 9.1

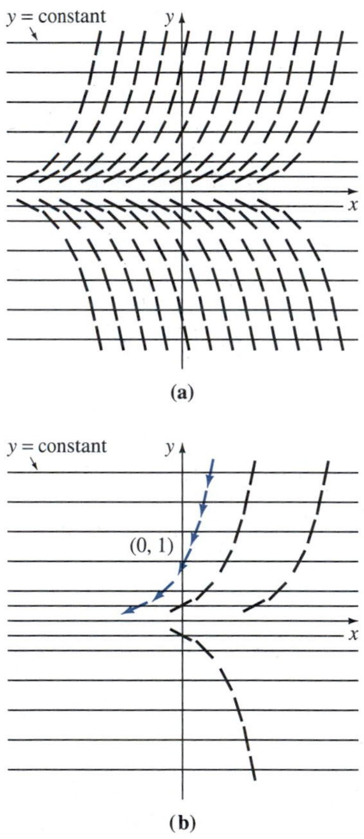

Figure 9.2

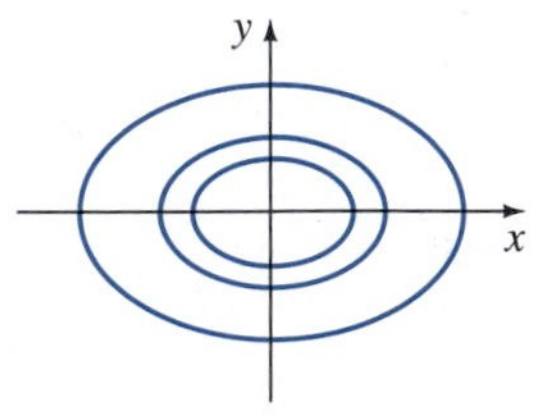

Figure 9.3

Lineal Elements Suppose for the moment that we do not know the general solution of the simple equation $y' = y$. Specifically, the differential equation implies that the slopes of tangent lines to a solution curve are given by the function $f(x, y) = y$. When $f(x, y)$ is held constant—that is, when

$$y = c, \tag{1}$$

where c is any constant—we are in effect stating that the slope of the tangents to the solution curves is the same constant value along a horizontal line. For example, for $y = 2$ let us draw a sequence of short line segments, or **lineal elements,** each having slope 2 and its midpoint on the line. As shown in Figure 9.1, the solution curves pass through this horizontal line at every point tangent to the lineal elements.

Isoclines and Direction Fields Equation (1) represents a one-parameter family of horizontal lines. In general, any member of the family $f(x, y) = c$ is called an **isocline,** which literally means a curve along which the inclination (of the tangents) is the same. As the parameter c is varied, we obtain a collection of isoclines on which the lineal elements are judiciously constructed. The totality of these lineal elements is called a **direction field, slope field,** or **lineal element field** of the differential equation $y' = f(x, y)$. As we see in Figure 9.2(a), the direction field suggests the "flow pattern" for the family of solution curves of the differential equation $y' = y$. In particular, if we want the one solution that passes through the point (0, 1), then, as indicated in Figure 9.2(b), we construct a curve through this point and passing through the isoclines with the appropriate slopes.

EXAMPLE 1 Isoclines

Determine the isoclines for the differential equation

$$\frac{dy}{dx} = 4x^2 + 9y^2.$$

Solution For $c > 0$ the isoclines are the curves

$$4x^2 + 9y^2 = c.$$

As Figure 9.3 shows, the curves are a concentric family of ellipses with major axis along the x-axis. ■

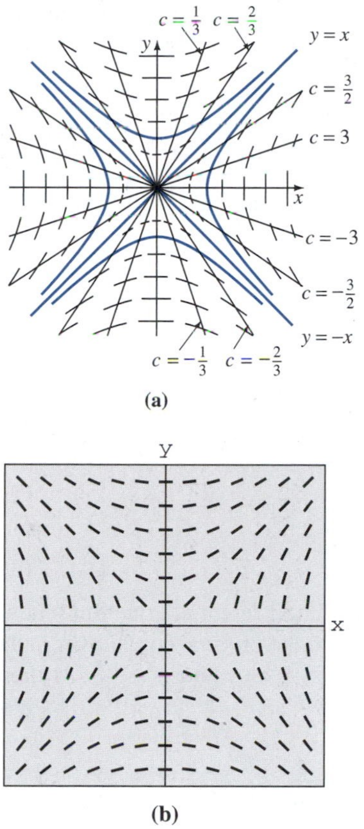

Figure 9.4

EXAMPLE 2 Direction Field

Sketch the direction field and indicate several possible members of the family of solution curves for

$$\frac{dy}{dx} = \frac{x}{y}.$$

Solution Before sketching the direction field corresponding to the isoclines $x/y = c$ or $y = x/c$, we note that the differential equation gives the following information:

(a) If a solution curve crosses the x-axis ($y = 0$), it does so tangent to a vertical lineal element at every point except possibly (0, 0).

(b) If a solution curve crosses the y-axis ($x = 0$), it does so tangent to a horizontal lineal element at every point except possibly (0, 0).

(c) The lineal elements corresponding to the isoclines $c = 1$ and $c = -1$ are collinear with the lines $y = x$ and $y = -x$, respectively. Indeed, it is easily verified that these isoclines are both particular solutions of the given differential equation. However, it should be noted that *in general* isoclines are themselves not solutions to a differential equation. See Example 4.

Figure 9.4(a) shows the direction field and several possible solution curves. Remember that on any particular isocline all the lineal elements are parallel. Also, the lineal elements may be drawn in such a manner as to suggest the flow of a particular curve. In other words, imagine the isoclines so close together that if the lineal elements were connected, we would have a polygonal curve suggestive of the shape of a smooth curve. Alternatively, the lineal elements can be drawn uniformly spaced on their isoclines as shown in the computer-generated version of the same direction field given in Figure 9.4(b). Note, however, that the isoclines themselves are not drawn in part (b). ■

EXAMPLE 3 Approximate Solution Curve

In Section 2.1 we indicated that the differential equation

$$\frac{dy}{dx} = x^2 + y^2$$

cannot be solved in terms of elementary functions. Use a direction field to locate an approximate solution satisfying $y(0) = 1$.

Solution The isoclines are concentric circles defined by

$$x^2 + y^2 = c, \quad c > 0.$$

By choosing $c = \frac{1}{4}$, $c = 1$, $c = \frac{9}{4}$, and $c = 4$, we obtain the circles with radii $\frac{1}{2}$, 1, $\frac{3}{2}$, and 2 shown in Figure 9.5(a). The lineal elements superimposed on each circle have slope corresponding to the particular value of c. It seems plausible from inspection of Figure 9.5(a) that a solution curve of the given initial-value problem might have the shape given in Figure 9.5(b). ■

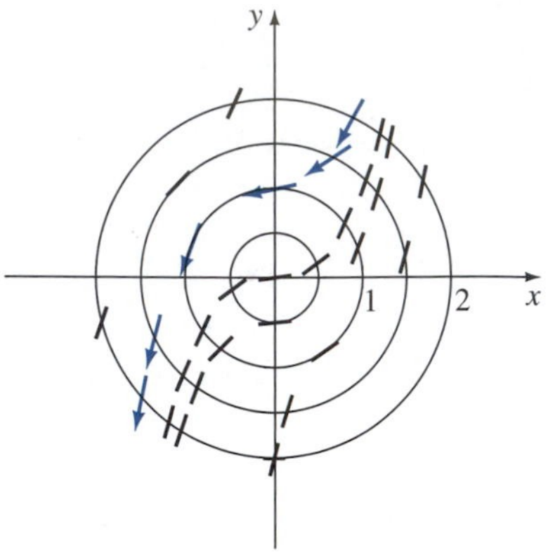

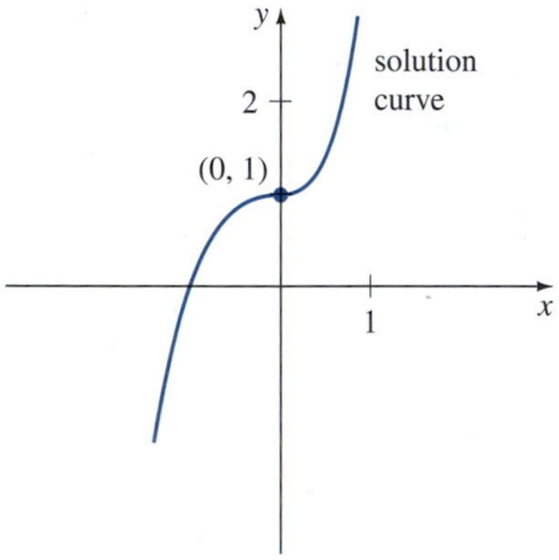

Figure 9.5

The concept of the direction field is used primarily to establish the existence of, and possibly to locate an approximate solution curve for, a first-order differential equation that cannot be solved by the usual standard techniques. However, the preceding discussion is of little value in determining specific values of a solution $y(x)$ at given points. For example, if we want to know the approximate value of $y(0.5)$ for the solution of

$$\frac{dy}{dx} = x^2 + y^2, \quad y(0) = 1,$$

then Figure 9.5(b) can do nothing more than indicate that $y(0.5)$ may be in the same "ball park" as $y = 2$.

When the isoclines are straight lines, it is easy to determine which, if any, of these isoclines are also particular solutions of the differential equation.

EXAMPLE 4 Solution of a DE

The isoclines of the differential equation

$$y' = 2x + y \tag{2}$$

are the straight lines

$$2x + y = c. \tag{3}$$

A line in this latter family will be a solution of the differential equation whenever its slope is the same as c. In other words, both the original equation and the line satisfy $y' = c$. Since the slope of (3) is –2 if we choose $c = -2$, then $2x + y = -2$ is a solution of (2). ■

Remarks Sketching a direction field by hand is a straightforward but time-consuming task. If available to you, computer software is recommended.

EXERCISES 9.1

Answers to odd-numbered problems begin on page A-23.

In Problems 1–4 use the given computer-generated direction field to sketch several possible solution curves for the indicated differential equation.

1. $y' = xy$

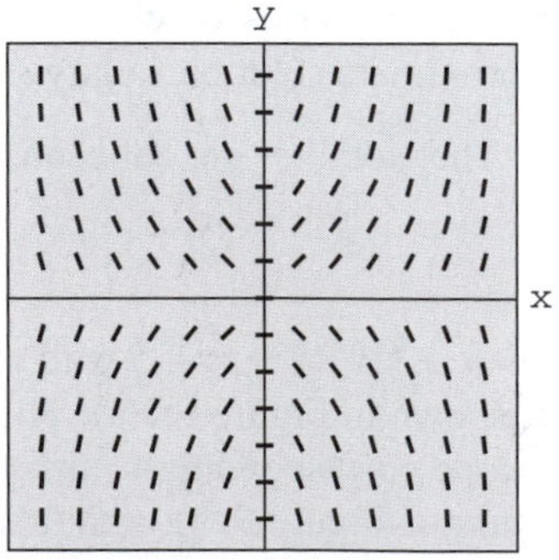

Figure 9.6

2. $y' = 1 - xy$

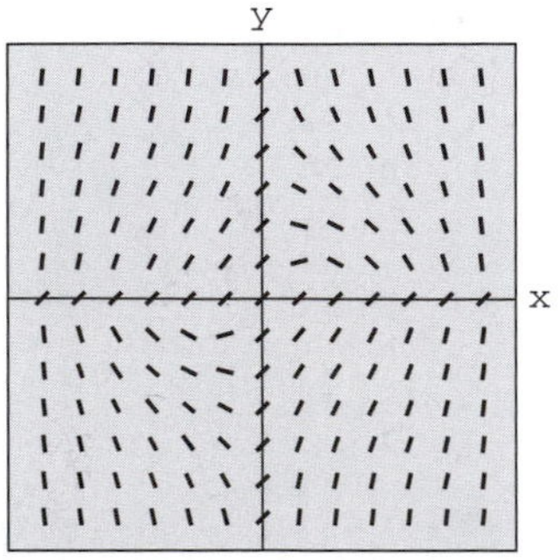

Figure 9.7

3. $y' = y - x$

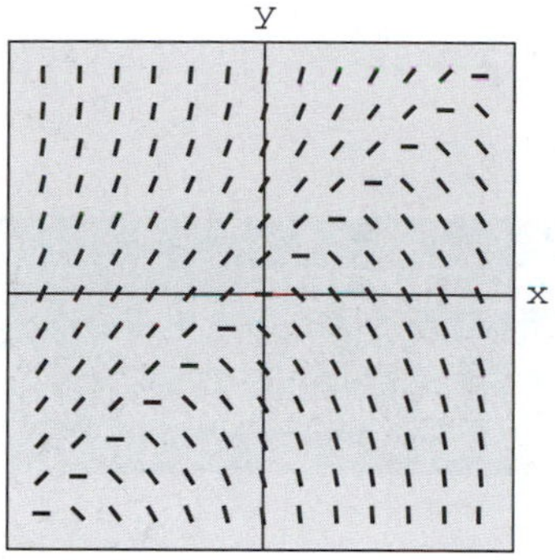

Figure 9.8

4. $y' = \cos x/\sin y$

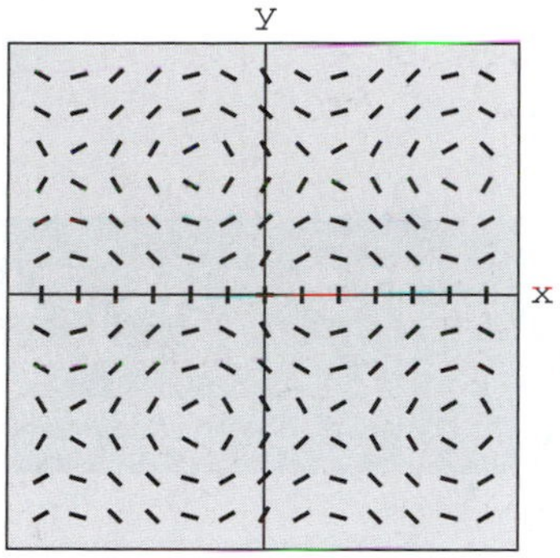

Figure 9.9

In Problems 5–14 identify the isoclines for the given differential equation.

5. $\dfrac{dy}{dx} = x + 4$

6. $\dfrac{dy}{dx} = 2x + y$

7. $\dfrac{dy}{dx} = x^2 - y^2$

8. $\dfrac{dy}{dx} = y - x^2$

9. $y' = \sqrt{x^2 + y^2 + 2y + 1}$

10. $y' = (x^2 + y^2)^{-1}$

11. $\dfrac{dy}{dx} = y(x + y)$

12. $\dfrac{dy}{dx} = y + e^x$

13. $\dfrac{dy}{dx} = \dfrac{y - 1}{x - 2}$

14. $\dfrac{dy}{dx} = \dfrac{x - y}{x + y}$

In Problems 15–22 sketch—or use a computer to obtain—the direction field for the given differential equation. Indicate several possible solution curves.

15. $y' = x$

16. $y' = x + y$

17. $y\dfrac{dy}{dx} = -x$

18. $\dfrac{dy}{dx} = \dfrac{1}{y}$

19. $\dfrac{dy}{dx} = 0.2x^2 + y$

20. $\dfrac{dy}{dx} = xe^y$

21. $y' = y - \cos \dfrac{\pi}{2} x$

22. $y' = 1 - \dfrac{y}{x}$

23. Formally show that the isoclines for the differential equation

$$\frac{dy}{dx} = \frac{\alpha x + \beta y}{\gamma x + \delta y}$$

are straight lines through the origin.

24. Show that $y = cx$ is a solution of the differential equation in Problem 23 if and only if $(\beta - \gamma)^2 + 4\alpha\delta \geq 0$.

In Problems 25–30 find those isoclines that are also solutions of the given differential equation. See Problems 23 and 24 and Example 4.

25. $y' = 3x + 2y$

26. $y' = 6x - 2y$

27. $y' = \dfrac{2x}{y}$

28. $y' = \dfrac{2y}{x + y}$

29. $\dfrac{dy}{dx} = \dfrac{4x + 3y}{y}$

30. $\dfrac{dy}{dx} = \dfrac{5x + 10y}{-4x + 3y}$

9.2 THE EULER METHODS

• *Euler's method* • *Step size* • *Absolute error* • *Relative error*
• *Percentage relative error* • *Improved Euler's method*

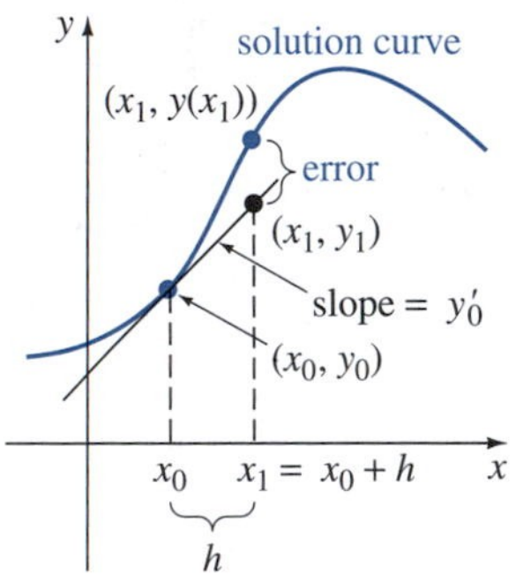

Figure 9.10

Euler's Method One of the simplest techniques for approximating solutions of differential equations is known as **Euler's method,** or the method of **tangent lines.** Suppose we wish to approximate the solution of the initial-value problem

$$y' = f(x, y), \quad y(x_0) = y_0.$$

If h is a positive increment on the x-axis, then as Figure 9.10 shows, we can find a point $(x_1, y_1) = (x_0 + h, y_1)$ on the line tangent to the unknown solution curve at (x_0, y_0).

By the point-slope form of the equation of a line, we have

$$\frac{y_1 - y_0}{(x_0 + h) - x_0} = y_0' \quad \text{or} \quad y_1 = y_0 + hy_0',$$

where $y_0' = f(x_0, y_0)$. If we label $x_0 + h$ by x_1, then the point (x_1, y_1) on the tangent line is an approximation to the point $(x_1, y(x_1))$ on the solution curve; that is, $y_1 \approx y(x_1)$. Of course the accuracy of the approximation depends heavily on the size of the increment h. Usually we must choose this **step size** to be "reasonably small."

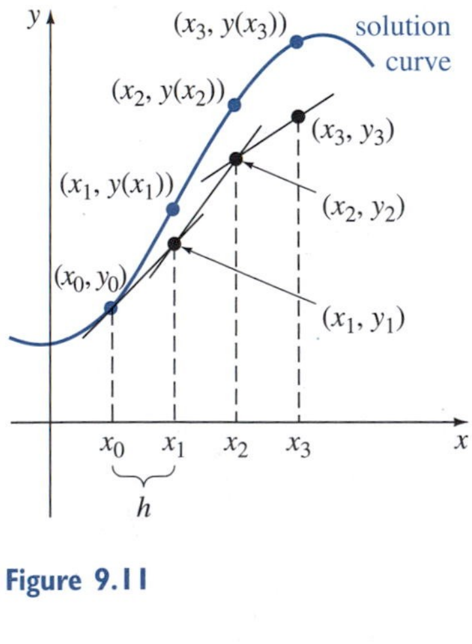

Figure 9.11

Assuming a uniform (constant) value of h, we can obtain a succession of points $(x_1, y_1), (x_2, y_2), \ldots, (x_n, y_n)$, which we hope are close to the points $(x_1, y(x_1)), (x_2, y(x_2)), \ldots, (x_n, y(x_n))$. See Figure 9.11. Now using (x_1, y_1), we can obtain the value of y_2, which is the ordinate of a point on a new "tangent" line. We have

$$\frac{y_2 - y_1}{h} = y_1' \quad \text{or} \quad y_2 = y_1 + hy_1' = y_1 + hf(x_1, y_1).$$

In general it follows that

$$y_{n+1} = y_n + hf(x_n, y_n), \tag{1}$$

where $x_n = x_0 + nh$.

As an example, suppose we try the iteration scheme (1) on a differential equation for which we know the explicit solution; in this way we can compare the estimated values y_n and the true values $y(x_n)$.

EXAMPLE 1 **Euler's Method**

Consider the initial-value problem $y' = 0.2xy$, $y(1) = 1$.

Use Euler's method to obtain an approximation to $y(1.5)$ using first $h = 0.1$ and then $h = 0.05$.

Solution We first identify $f(x, y) = 0.2xy$ so that (1) becomes

$$y_{n+1} = y_n + h(0.2x_n y_n).$$

Then for $h = 0.1$ we find

$$y_1 = y_0 + (0.1)(0.2x_0 y_0) = 1 + (0.1)[0.2(1)(1)] = 1.02,$$

which is an estimate to the value of $y(1.1)$. However, if we use $h = 0.05$, it takes *two* iterations to reach $x = 1.1$. We have

$$y_1 = 1 + (0.05)[0.2(1)(1)] = 1.01$$
$$y_2 = 1.01 + (0.05)[0.2(1.05)(1.01)] = 1.020605.$$

Here we note that $y_1 \approx y(1.05)$ and $y_2 \approx y(1.1)$. The remainder of the calculations are summarized in Tables 9.1 and 9.2. Each entry is rounded to four decimal places.

Table 9.1 Euler's Method with $h = 0.1$

x_n	y_n	True value	Abs. error	% Rel. error
1.00	1.0000	1.0000	0.0000	0.00
1.10	1.0200	1.0212	0.0012	0.12
1.20	1.0424	1.0450	0.0025	0.24
1.30	1.0675	1.0714	0.0040	0.37
1.40	1.0952	1.1008	0.0055	0.50
1.50	1.1259	1.1331	0.0073	0.64

Table 9.2 Euler's Method with $h = 0.05$

x_n	y_n	True value	Abs. error	% Rel. error
1.00	1.0000	1.0000	0.0000	0.00
1.05	1.0100	1.0103	0.0003	0.03
1.10	1.0206	1.0212	0.0006	0.06
1.15	1.0318	1.0328	0.0009	0.09
1.20	1.0437	1.0450	0.0013	0.12
1.25	1.0562	1.0579	0.0016	0.16
1.30	1.0694	1.0714	0.0020	0.19
1.35	1.0833	1.0857	0.0024	0.22
1.40	1.0980	1.1008	0.0028	0.25
1.45	1.1133	1.1166	0.0032	0.29
1.50	1.1295	1.1331	0.0037	0.32

In Example 1 the true values were calculated from the known solution $y = e^{0.1(x^2-1)}$. Also, the **absolute error** is defined to be

$$|true\ value - approximation|.$$

The **relative error** and the **percentage relative error** are, in turn,

$$\frac{|true\ value - approximation|}{|true\ value|}$$

and $$\frac{|true\ value - approximation|}{|true\ value|} \times 100 = \frac{absolute\ error}{|true\ value|} \times 100.$$

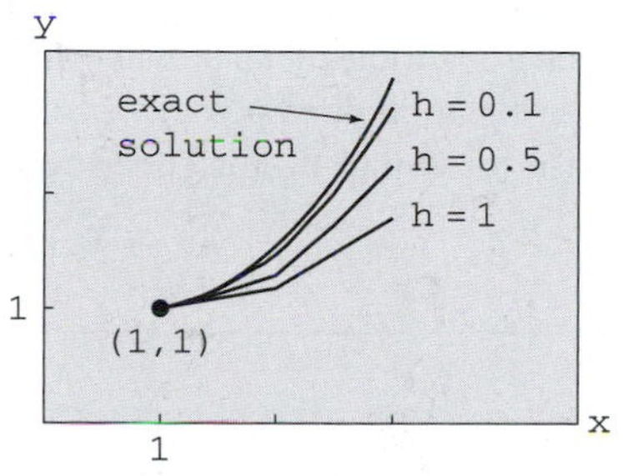

Figure 9.12

Computer software enables us to examine approximations to the graph of the solution $y(x)$ of an initial-value problem by plotting straight lines through the points (x_n, y_n) generated by Euler's method. In Figure 9.12 we have compared, on the interval $[1, 3]$, the graph of the exact solution of the initial-value problem in Example 1 with the graphs obtained from Euler's method using the step sizes $h = 1$, $h = 0.5$, and $h = 0.1$. It is apparent from the figure that the approximation improves as the step size decreases.

Although we see that the percentage relative error in Tables 9.1 and 9.2 is growing, it does not appear to be that bad. But you should not be deceived by Example 1 and Figure 9.12. Watch what happens in the next example when we simply change the coefficient 0.2 of the differential equation in Example 1 to the number 2.

EXAMPLE 2 Euler's Method

Use Euler's method to obtain the approximate value of $y(1.5)$ for the solution of $y' = 2xy$, $y(1) = 1$.

Solution You should verify that the exact or analytic solution is now $y = e^{x^2-1}$. Proceeding as in Example 1, we obtain the following results.

Table 9.3 Euler's Method with $h = 0.1$

x_n	y_n	True value	Abs. error	% Rel. error
1.00	1.0000	1.0000	0.0000	0.00
1.10	1.2000	1.2337	0.0337	2.73
1.20	1.4640	1.5527	0.0887	5.71
1.30	1.8154	1.9937	0.1784	8.95
1.40	2.2874	2.6117	0.3244	12.42
1.50	2.9278	3.4904	0.5625	16.12

Table 9.4 Euler's Method with $h = 0.05$

x_n	y_n	True value	Abs. error	% Rel. error
1.00	1.0000	1.0000	0.0000	0.00
1.05	1.1000	1.1079	0.0079	0.72
1.10	1.2155	1.2337	0.0182	1.47
1.15	1.3492	1.3806	0.0314	2.27
1.20	1.5044	1.5527	0.0483	3.11
1.25	1.6849	1.7551	0.0702	4.00
1.30	1.8955	1.9937	0.0982	4.93
1.35	2.1419	2.2762	0.1343	5.90
1.40	2.4311	2.6117	0.1806	6.92
1.45	2.7714	3.0117	0.2403	7.98
1.50	3.1733	3.4904	0.3171	9.08

In this case, with a step size $h = 0.1$, a 16% relative error in the calculation of the approximation to $y(1.5)$ is totally unacceptable. At the expense of doubling the number of calculations, a slight improvement in accuracy is obtained by halving the step size to $h = 0.05$.

We will discuss the errors for Euler's method in detail in Section 9.6.

Of course in many instances we may not know the solution of a particular differential equation, or for that matter whether a solution of an initial-value problem actually exists. The following nonlinear equation does possess a solution in closed form, but we leave it as an exercise for you to find it. (See Problem 1.)

EXAMPLE 3 Euler's Method

Use Euler's method to obtain the approximate value of $y(0.5)$ for the solution of $y' = (x + y - 1)^2$, $y(0) = 2$.

Table 9.5 Euler's Method with $h = 0.1$

x_n	y_n
0.00	2.0000
0.10	2.1000
0.20	2.2440
0.30	2.4525
0.40	2.7596
0.50	3.2261

Table 9.6 Euler's Method with $h = 0.05$

x_n	y_n
0.00	2.0000
0.05	2.0500
0.10	2.1105
0.15	2.1838
0.20	2.2727
0.25	2.3812
0.30	2.5142
0.35	2.6788
0.40	2.8845
0.45	3.1455
0.50	3.4823

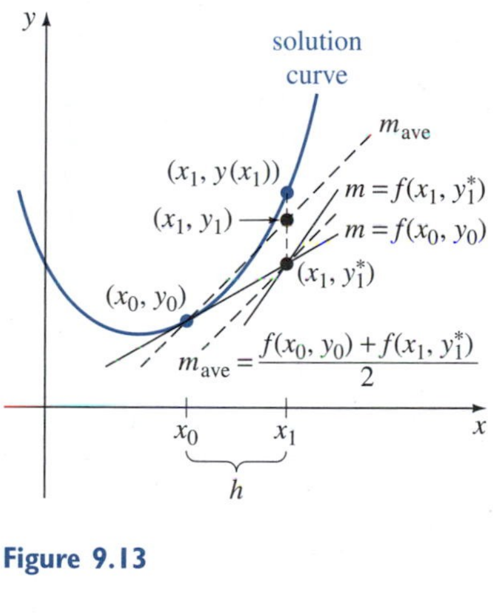

Figure 9.13

Solution For $n = 0$ and $h = 0.1$ we have

$$y_1 = y_0 + (0.1)(x_0 + y_0 - 1)^2 = 2 + (0.1)(1)^2 = 2.1.$$

The remaining calculations are summarized in Tables 9.5 and 9.6 for $h = 0.1$ and $h = 0.05$, respectively. ■

We may want greater accuracy than that displayed, say, in Table 9.4, and so we could try a step size even smaller than $h = 0.05$. However, rather than resorting to this extra labor, it probably would be more advantageous to employ an alternative numerical procedure. Euler's formula by itself, though attractive in its simplicity, is seldom used in serious calculations.

Improved Euler's Method The formula

$$y_{n+1} = y_n + h\frac{f(x_n, y_n) + f(x_{n+1}, y^*_{n+1})}{2},$$

where
$$y^*_{n+1} = y_n + hf(x_n, y_n), \tag{2}$$

is known as the **improved Euler's formula,** or **Heun's formula.** The values $f(x_n, y_n)$ and $f(x_{n+1}, y^*_{n+1})$ are approximations to the slope of the curve at $(x_n, y(x_n))$ and $(x_{n+1}, y(x_{n+1}))$, and consequently the quotient

$$\frac{f(x_n, y_n) + f(x_{n+1}, y^*_{n+1})}{2}$$

can be interpreted as an average slope on the interval between x_n and x_{n+1}.

The equations in (2) can be readily visualized. In Figure 9.13 we show the case in which $n = 0$. Note that

$$f(x_0, y_0) \text{ and } f(x_1, y_1^*)$$

are slopes of the indicated straight lines passing through the points (x_0, y_0) and (x_1, y_1^*), respectively. By taking an average of these slopes, we obtain the slope of the dashed skew lines. Rather than advancing along the line with slope $m = f(x_0, y_0)$ to the point with ordinate y_1^* obtained by the usual Euler's method, we advance along the line through (x_0, y_0) with slope m_{ave} until we reach x_1. It seems plausible from inspection of the figure that y_1 is an improvement over y_1^*. In fact, in Section 9.6 we will see that the improved Euler's method is more accurate than Euler's method.

We might also say that the value of

$$y_1^* = y_0 + hf(x_0, y_0)$$

predicts a value of $y(x_1)$, whereas

$$y_1 = y_0 + h\frac{f(x_0, y_0) + f(x_1, y_1^*)}{2}$$

corrects this estimate.

EXAMPLE 4 Improved Euler's Method

Use the improved Euler's formula to obtain the approximate value of $y(1.5)$ for the solution of the initial-value problem in Example 2. Compare the results for $h = 0.1$ and $h = 0.05$.

Solution For $n = 0$ and $h = 0.1$ we first compute

$$y_1^* = y_0 + (0.1)(2x_0 y_0) = 1.2.$$

Then from (2)

$$y_1 = y_0 + (0.1)\frac{2x_0y_0 + 2x_1y_1^*}{2} = 1 + (0.1)\frac{2(1)(1) + 2(1.1)(1.2)}{2} = 1.232.$$

The comparative values of the calculations for $h = 0.1$ and $h = 0.5$ are given in Tables 9.7 and 9.8, respectively.

Table 9.7 Improved Euler's Method with $h = 0.1$

x_n	y_n	True value	Abs. error	% Rel. error
1.00	1.0000	1.0000	0.0000	0.00
1.10	1.2320	1.2337	0.0017	0.14
1.20	1.5479	1.5527	0.0048	0.31
1.30	1.9832	1.9937	0.0106	0.53
1.40	2.5908	2.6117	0.0209	0.80
1.50	3.4509	3.4904	0.0394	1.13

Table 9.8 Improved Euler's Method with $h = 0.05$

x_n	y_n	True value	Abs. error	% Rel. error
1.00	1.0000	1.0000	0.0000	0.00
1.05	1.1077	1.1079	0.0002	0.02
1.10	1.2332	1.2337	0.0004	0.04
1.15	1.3798	1.3806	0.0008	0.06
1.20	1.5514	1.5527	0.0013	0.08
1.25	1.7531	1.7551	0.0020	0.11
1.30	1.9909	1.9937	0.0029	0.14
1.35	2.2721	2.2762	0.0041	0.18
1.40	2.6060	2.6117	0.0057	0.22
1.45	3.0038	3.0117	0.0079	0.26
1.50	3.4795	3.4904	0.0108	0.31

A brief word of caution is in order here. We cannot compute all the values of y_n^* first and then substitute these values in the first formula of (2). In other words, we cannot use the data in Table 9.3 to help construct the values in Table 9.7. Why not?

EXAMPLE 5 Improved Euler's Method

Use the improved Euler's formula to obtain the approximate value of $y(0.5)$ for the solution of the initial-value problem in Example 3.

Table 9.9 Improved Euler's Method with $h = 0.1$

x_n	y_n
0.00	2.0000
0.10	2.1220
0.20	2.3049
0.30	2.5858
0.40	3.0378
0.50	3.8254

Table 9.10 Improved Euler's Method with $h = 0.05$

x_n	y_n
0.00	2.0000
0.05	2.0553
0.10	2.1228
0.15	2.2056
0.20	2.3075
0.25	2.4342
0.30	2.5931
0.35	2.7953
0.40	3.0574
0.45	3.4057
0.50	3.8840

Solution For $n = 0$ and $h = 0.1$ we have

$$y_1^* = y_0 + (0.1)(x_0 + y_0 - 1)^2 = 2.1,$$

and so

$$y_1 = y_0 + (0.1)\frac{(x_0 + y_0 - 1)^2 + (x_1 + y_1^* - 1)^2}{2}$$

$$= 2 + (0.1)\frac{1 + 1.44}{2} = 2.122.$$

The remaining calculations are summarized in Tables 9.9 and 9.10 for $h = 0.1$ and $h = 0.05$, respectively.

EXERCISES 9.2

Answers to odd-numbered problems begin on page A-24.

1. Solve the initial-value problem

$$y' = (x + y - 1)^2, \quad y(0) = 2$$

in terms of elementary functions.

2. Let $y(x)$ be the solution of the initial-value problem given in Problem 1. Round to four decimal places as you compute the exact values of $y(0.1)$, $y(0.2)$, $y(0.3)$, $y(0.4)$, and $y(0.5)$. Compare these values with the entries in Tables 9.5, 9.6, 9.9, and 9.10.

Given the initial-value problems in Problems 3–12, use Euler's formula to obtain a four-decimal approximation to the indicated value. First use **(a)** $h = 0.1$ and then **(b)** $h = 0.05$.

3. $y' = 2x - 3y + 1, y(1) = 5; \quad y(1.5)$
4. $y' = 4x - 2y, y(0) = 2; \quad y(0.5)$
5. $y' = 1 + y^2, y(0) = 0; \quad y(0.5)$
6. $y' = x^2 + y^2, y(0) = 1; \quad y(0.5)$
7. $y' = e^{-y}, y(0) = 0; \quad y(0.5)$
8. $y' = x + y^2, y(0) = 0; \quad y(0.5)$
9. $y' = (x - y)^2, y(0) = 0.5; \quad y(0.5)$
10. $y' = xy + \sqrt{y}, y(0) = 1; \quad y(0.5)$
11. $y' = xy^2 - \dfrac{y}{x}, y(1) = 1; \quad y(1.5)$
12. $y' = y - y^2, y(0) = 0.5; \quad y(0.5)$
13. As parts **(a)**–**(e)** of this problem, repeat the calculations of Problems 3, 5, 7, 9, and 11 using the improved Euler's formula.
14. As parts **(a)**–**(e)** of this problem, repeat the calculations of Problems 4, 6, 8, 10, and 12 using the improved Euler's formula.

15. Although it may not be obvious from the differential equation, its solution could "behave badly" near a point x at which we wish to approximate $y(x)$. Numerical procedures may then give widely differing results near this point. Let $y(x)$ be the solution of the initial-value problem

$$y' = x^2 + y^3, \quad y(1) = 1.$$

Using the step size $h = 0.1$, compare the results obtained from Euler's formula with the results from the improved Euler's formula in the approximation of $y(1.4)$.

16. Derive the basic Euler's formula by integrating both sides of the equation $y' = f(x, y)$ on the interval $x_n \le x \le x_{n+1}$. Approximate the integral of the right side by replacing the function $f(x, y)$ by its value at the left endpoint of the interval of integration.

17. By following the procedure outlined in Problem 16, derive the improved Euler's formula. [*Hint*: Replace the integrand of the right side by the average of its values at the endpoints of the interval of integration.]

9.3 THE THREE-TERM TAYLOR METHOD

• *Taylor series expansion* • *Taylor's method*

The numerical method considered in this section, the **three-term Taylor method,** is more of theoretical interest than of practical importance since the results obtained using the following formula (5) do not differ substantially from those obtained using the improved Euler's method.

In this study of numerical solutions of differential equations, many computational algorithms can be derived from a Taylor series expansion. Recall from calculus that the form of this expansion centered at a point $x = a$ is

$$y(x) = y(a) + y'(a)\frac{(x-a)}{1!} + y''(a)\frac{(x-a)^2}{2!} + \cdots. \tag{1}$$

It is understood that the function $y(x)$ possesses derivatives of all orders and that the series (1) converges in some interval defined by $|x - a| < R$. Notice, in particular, that if we set $a = x_n$ and $x = x_n + h$, then (1) becomes

$$y(x_n + h) = y(x_n) + y'(x_n)h + y''(x_n)\frac{h^2}{2} + \cdots. \tag{2}$$

Euler's Method Revisited Furthermore, let us now assume that the function $y(x)$ is a solution of the first-order differential equation $y' = f(x, y)$. If we then truncate the series (2) after, say, two terms, we obtain the approximation

$$y(x_n + h) \approx y(x_n) + y'(x_n)h \quad \text{or} \quad y(x_n + h) \approx y(x_n) + f(x_n, y(x_n))h. \tag{3}$$

Observe that we can obtain Euler's formula

$$y_{n+1} = y_n + hf(x_n, y_n) \tag{4}$$

of the preceding section by replacing $y(x_n + h)$ and $y(x_n)$ in (3) by their approximations y_{n+1} and y_n, respectively. The approximation symbol $\approx$ is replaced by an equality since we are defining the left side of (4) by the numbers obtained from the right-hand member.

Taylor's Method By retaining three terms in the series (2), we can write

$$y(x_n + h) \approx y(x_n) + y'(x_n)h + y''(x_n)\frac{h^2}{2}.$$

After the replacements noted in the preceding material, it follows that

$$y_{n+1} = y_n + y'_n h + y''_n \frac{h^2}{2}. \tag{5}$$

The second derivative y'' can be obtained by differentiating $y' = f(x, y)$.

At this point let us reexamine two initial-value problems from Section 9.2.

EXAMPLE 1 Taylor's Method

Use the three-term Taylor formula to obtain the approximate value of $y(1.5)$ for the solution of $y' = 2xy$, $y(1) = 1$. Compare the results for $h = 0.1$ and $h = 0.05$.

Solution Since $y' = 2xy$, it follows by the product rule that $y'' = 2xy' + 2y$. Thus, for example, when $h = 0.1$, $n = 0$, we can first calculate

$$y'_0 = 2x_0 y_0 = 2(1)(1) = 2$$

and then $$y''_0 = 2x_0 y'_0 + 2y_0 = 2(1)(2) + 2(1) = 6.$$

Hence (5) becomes

$$y_1 = y_0 + y'_0(0.1) + y''_0\frac{(0.1)^2}{2} = 1 + 2(0.1) + 6(0.005) = 1.23.$$

The results of the iteration, along with the comparative exact values, are summarized in Tables 9.11 and 9.12.

Table 9.11 Three-Term Taylor Method with $h = 0.1$

x_n	y_n	True value	Abs. error	% Rel. error
1.00	1.0000	1.0000	0.0000	0.00
1.10	1.2300	1.2337	0.0037	0.30
1.20	1.5427	1.5527	0.0100	0.65
1.30	1.9728	1.9937	0.0210	1.05
1.40	2.5721	2.6117	0.0396	1.52
1.50	3.4188	3.4904	0.0715	2.05

Table 9.12 Three-Term Taylor Method with $h = 0.05$

x_n	y_n	True value	Abs. error	% Rel. error
1.00	1.0000	1.0000	0.0000	0.00
1.05	1.1075	1.1079	0.0004	0.04
1.10	1.2327	1.2337	0.0010	0.08
1.15	1.3788	1.3806	0.0018	0.13
1.20	1.5499	1.5527	0.0028	0.18
1.25	1.7509	1.7551	0.0041	0.23
1.30	1.9879	1.9937	0.0059	0.29
1.35	2.2681	2.2762	0.0081	0.36
1.40	2.6006	2.6117	0.0111	0.43
1.45	2.9967	3.0117	0.0150	0.50
1.50	3.4702	3.4904	0.0202	0.58

EXAMPLE 2 Taylor's Method

Use the three-term Taylor formula to obtain the approximate value of $y(0.5)$ for the solution of $y' = (x + y - 1)^2, y(0) = 2$.

Solution In this case we compute y'' by the power rule. We have

$$y'' = 2(x + y - 1)(1 + y').$$

The results are summarized in Tables 9.13 and 9.14 for $h = 0.1$ and $h = 0.05$, respectively.

Table 9.13 Three-Term Taylor Method with $h = 0.1$

x_n	y_n
0.00	2.0000
0.10	2.1200
0.20	2.2992
0.30	2.5726
0.40	3.0077
0.50	3.7511

Table 9.14 Three-Term Taylor Method with $h = 0.05$

x_n	y_n
0.00	2.0000
0.05	2.0550
0.10	2.1222
0.15	2.2045
0.20	2.3058
0.25	2.4315
0.30	2.5890
0.35	2.7889
0.40	3.0475
0.45	3.3898
0.50	3.8574

A comparison of the last two examples with the corresponding results obtained from the improved Euler's method shows no startling dissimilarities. In fact, when $f(x, y)$ is linear in both variables x and y, Taylor's method gives the *same* values of y_n as the improved Euler's method for a given value of h. See Problems 12 and 13. In the error analysis given in Section 9.6 we will see that these two methods have the same accuracy.

EXERCISES 9.3

Answers to odd-numbered problems begin on page A-25.

Given the initial-value problems in Problems 1–10, use the three-term Taylor formula to obtain a four-decimal approximation to the indicated value. First use **(a)** $h = 0.1$ and then **(b)** $h = 0.05$.

1. $y' = 2x - 3y + 1, y(1) = 5;\quad y(1.5)$
2. $y' = 4x - 2y, y(0) = 2;\quad y(0.5)$
3. $y' = 1 + y^2, y(0) = 0;\quad y(0.5)$
4. $y' = x^2 + y^2, y(0) = 1;\quad y(0.5)$
5. $y' = e^{-y}, y(0) = 0;\quad y(0.5)$
6. $y' = x + y^2, y(0) = 0;\quad y(0.5)$

7. $y' = (x - y)^2, y(0) = 0.5;\quad y(0.5)$

8. $y' = xy + \sqrt{y}, y(0) = 1;\quad y(0.5)$

9. $y' = xy^2 - \dfrac{y}{x}, y(1) = 1;\quad y(1.5)$

10. $y' = y - y^2, y(0) = 0.5;\quad y(0.5)$

11. Let $y(x)$ be the solution of the initial-value problem

$$y' = x^2 + y^3, \quad y(1) = 1.$$

Use $h = 0.1$ and the three-term Taylor formula to obtain an approximation to $y(1.4)$. Compare your answer with the results obtained in Problem 15 of Exercises 9.2.

12. Consider the differential equation $y' = f(x, y)$, where f is linear in x and y. In this case prove that the improved Euler's formula is the same as the three-term Taylor formula. [*Hint*: Recall from calculus that a Taylor series for a function g of two variables is

$$g(a + h, b + k) = g(a, b) + g_x(a, b)h + g_y(a, b)k + \frac{1}{2}(h^2 g_{xx} + 2hk g_{xy} + k^2 g_{yy})\Big|_{(a,b)}$$

$$+ \textit{ terms involving higher-order derivatives.}$$

Apply this result to $f(x_n + h, y_n + hf(x_n, y_n))$ in the improved Euler's formula. Also use the fact that $y''(x) = (d/dx)y'(x) = f_x + f_y y'$.]

13. Compare the approximate values of $y(1.5)$ for

$$y' = x + y - 1, \quad y(1) = 5,$$

using the three-term Taylor method and the improved Euler's method with $h = 0.1$. Solve the initial-value problem and compute the true values $y(x_n), n = 0, 1, \ldots, 5$.

9.4 THE RUNGE-KUTTA METHOD

• *First- and second-order methods* • *Fourth-order Runge-Kutta method*

Probably one of the most popular as well as most accurate numerical procedures used in obtaining approximate solutions to differential equations is the **fourth-order Runge-Kutta method.*** As the name suggests, there are Runge-Kutta methods of different orders.

* CARL D. T. RUNGE (1856–1927) A German professor of applied mathematics at the University of Göttingen, Runge devised this numerical method around 1895. (M. W. Kutta expanded on this work in 1901.) In mathematics Runge also did notable work in the field of Diophantine equations. In physics Runge is remembered for his work on the Zeeman effect (the spectral lines in the emission spectrum of an element are affected by the presence of a magnetic field).

MARTIN W. KUTTA (1867–1944) Also a German applied mathematician, Kutta made significant contributions to the field of aerodynamics.

For the moment let us consider a **second-order** procedure. This consists of finding constants a, b, α, and β such that the formula

$$y_{n+1} = y_n + ak_1 + bk_2, \tag{1}$$

where

$$\begin{aligned} k_1 &= hf(x_n, y_n) \\ k_2 &= hf(x_n + \alpha h, y_n + \beta k_1), \end{aligned} \tag{2}$$

agrees with a Taylor series expansion to as many terms as possible. The obvious purpose is to achieve the accuracy of the Taylor method without having to compute higher-order derivatives. Now it can be shown that whenever the constants satisfy $a + b = 1, b\alpha = \frac{1}{2}, b\beta = \frac{1}{2}$, (1) agrees with a Taylor expansion out to the h^2, or third, term. It should be of interest to observe that when $a = \frac{1}{2}, b = \frac{1}{2}, \alpha = 1, \beta = 1$, then (1) reduces to the improved Euler's method. Thus we can conclude that the three-term Taylor formula is essentially equivalent to the improved Euler's formula. Also, the basic Euler's method is a **first-order** Runge-Kutta procedure.

Notice too that the sum $ak_1 + bk_2, a + b = 1$ in equation (1) is simply a *weighted average* of k_1 and k_2. The numbers k_1 and k_2 are multiples of approximations to the slope at two different points.

Fourth-Order Runge-Kutta Formula The **fourth-order** Runge-Kutta method consists of determining appropriate constants so that a formula such as

$$y_{n+1} = y_n + ak_1 + bk_2 + ck_3 + dk_4$$

agrees with a Taylor expansion out to the h^4, or fifth, term. As in (2), the k_i are constant multiples of $f(x, y)$ evaluated at select points. The derivation of the actual method is tedious, to say the least, so we state the results:

$$\begin{aligned} y_{n+1} &= y_n + \tfrac{1}{6}(k_1 + 2k_2 + 2k_3 + k_4) \\ k_1 &= hf(x_n, y_n) \\ k_2 &= hf(x_n + \tfrac{1}{2}h, y_n + \tfrac{1}{2}k_1) \\ k_3 &= hf(x_n + \tfrac{1}{2}h, y_n + \tfrac{1}{2}k_2) \\ k_4 &= hf(x_n + h, y_n + k_3). \end{aligned} \tag{3}$$

You are advised to look carefully at the formulas in (3); note that k_2 depends on k_1, k_3 depends on k_2, and so on. Also, k_2 and k_3 involve approximations to the slope at the midpoint of the interval between x_n and $x_{n+1} = x_n + h$.

EXAMPLE 1 Fourth-Order Runge-Kutta Method

Use the Runge-Kutta method with $h = 0.1$ to obtain an approximation to $y(1.5)$ for the solution of $y' = 2xy, y(1) = 1$.

Solution For the sake of illustration let us compute the case when $n = 0$. From (3) we find

$$\begin{aligned} k_1 &= (0.1)f(x_0, y_0) = (0.1)(2x_0y_0) = 0.2 \\ k_2 &= (0.1)f(x_0 + \tfrac{1}{2}(0.1), y_0 + \tfrac{1}{2}(0.2)) \\ &= (0.1)2(x_0 + \tfrac{1}{2}(0.1))(y_0 + \tfrac{1}{2}(0.2)) = 0.231 \end{aligned}$$

$$\begin{aligned} k_3 &= (0.1)f(x_0 + \tfrac{1}{2}(0.1), y_0 + \tfrac{1}{2}(0.231)) \\ &= (0.1)2(x_0 + \tfrac{1}{2}(0.1))(y_0 + \tfrac{1}{2}(0.231)) = 0.234255 \\ k_4 &= (0.1)f(x_0 + 0.1, y_0 + 0.234255) \\ &= (0.1)2(x_0 + 0.1)(y_0 + 0.234255) = 0.2715361 \end{aligned}$$

and therefore

$$\begin{aligned} y_1 &= y_0 + \tfrac{1}{6}(k_1 + 2k_2 + 2k_3 + k_4) \\ &= 1 + \tfrac{1}{6}(0.2 + 2(0.231) + 2(0.234255) + 0.2715361) = 1.23367435. \end{aligned}$$

Table 9.15, whose entries are rounded to four decimal places, should convince you why the Runge-Kutta method is so popular. Of course there is no need to use any smaller step size. The reason for the high accuracy of the Runge-Kutta method will be seen in Section 9.6.

Table 9.15 Runge-Kutta Method with $h = 0.1$

x_n	y_n	True value	Abs. error	% Rel. error
1.00	1.0000	1.0000	0.0000	0.00
1.10	1.2337	1.2337	0.0000	0.00
1.20	1.5527	1.5527	0.0000	0.00
1.30	1.9937	1.9937	0.0000	0.00
1.40	2.6116	2.6117	0.0001	0.00
1.50	3.4902	3.4904	0.0001	0.00

You might be interested in inspecting Tables 9.16 and 9.17 at this point. These tables compare the results obtained from applying the various formulas that we have examined to the two specific problems

$$y' = 2xy, \quad y(1) = 1 \quad \text{and} \quad y' = (x + y - 1)^2, \quad y(0) = 2$$

that we have considered throughout the last three sections.

Table 9.16 $y' = 2xy, y(1) = 1$

Comparison of Numerical Methods with $h = 0.1$						Comparison of Numerical Methods with $h = 0.05$					
x_n	Euler's	Improved Euler's	Three-Term Taylor	Runge-Kutta	True value	x_n	Euler's	Improved Euler's	Three-Term Taylor	Runge-Kutta	True value
1.00	1.0000	1.0000	1.0000	1.0000	1.0000	1.00	1.0000	1.0000	1.0000	1.0000	1.0000
1.10	1.2000	1.2320	1.2300	1.2337	1.2337	1.05	1.1000	1.1077	1.1075	1.1079	1.1079
1.20	1.4640	1.5479	1.5427	1.5527	1.5527	1.10	1.2155	1.2332	1.2327	1.2337	1.2337
1.30	1.8154	1.9832	1.9728	1.9937	1.9937	1.15	1.3492	1.3798	1.3788	1.3806	1.3806
1.40	2.2874	2.5908	2.5721	2.6116	2.6117	1.20	1.5044	1.5514	1.5499	1.5527	1.5527
1.50	2.9278	3.4509	3.4188	3.4902	3.4904	1.25	1.6849	1.7531	1.7509	1.7551	1.7551
						1.30	1.8955	1.9909	1.9879	1.9937	1.9937
						1.35	2.1419	2.2721	2.2681	2.2762	2.2762
						1.40	2.4311	2.6060	2.6006	2.6117	2.6117
						1.45	2.7714	3.0038	2.9967	3.0117	3.0117
						1.50	3.1733	3.4795	3.4702	3.4903	3.4904

Table 9.17 $y' = (x + y - 1)^2$, $y(0) = 2$

Comparison of Numerical Methods with $h = 0.1$						Comparison of Numerical Methods with $h = 0.05$					
x_n	**Euler's**	**Improved Euler's**	**Three-Term Taylor**	**Runge-Kutta**	**True value**	x_n	**Euler's**	**Improved Euler's**	**Three-Term Taylor**	**Runge-Kutta**	**True value**
0.00	2.0000	2.0000	2.0000	2.0000	2.0000	0.00	2.0000	2.0000	2.0000	2.0000	2.0000
0.10	2.1000	2.1220	2.1200	2.1230	2.1230	0.05	2.0500	2.0553	2.0550	2.0554	2.0554
0.20	2.2440	2.3049	2.2992	2.3085	2.3085	0.10	2.1105	2.1228	2.1222	2.1230	2.1230
0.30	2.4525	2.5858	2.5726	2.5958	2.5958	0.15	2.1838	2.2056	2.2045	2.2061	2.2061
0.40	2.7596	3.0378	3.0077	3.0649	3.0650	0.20	2.2727	2.3075	2.3058	2.3085	2.3085
0.50	3.2261	3.8254	3.7511	3.9078	3.9082	0.25	2.3812	2.4342	2.4315	2.4358	2.4358
						0.30	2.5142	2.5931	2.5890	2.5958	2.5958
						0.35	2.6788	2.7953	2.7889	2.7998	2.7997
						0.40	2.8845	3.0574	3.0475	3.0650	3.0650
						0.45	3.1455	3.4057	3.3898	3.4189	3.4189
						0.50	3.4823	3.8840	3.8574	3.9082	3.9082

EXERCISES 9.4

Answers to odd-numbered problems begin on page A-26.

Given the initial-value problems in Problems 1–10, use the Runge-Kutta method with $h = 0.1$ to obtain a four-decimal approximation to the indicated value.

1. $y' = 2x - 3y + 1, y(1) = 5, \quad y(1.5)$
2. $y' = 4x - 2y, y(0) = 2, \quad y(0.5)$
3. $y' = 1 + y^2, y(0) = 0, \quad y(0.5)$
4. $y' = x^2 + y^2, y(0) = 1, \quad y(0.5)$
5. $y' = e^{-y}, y(0) = 0, \quad y(0.5)$
6. $y' = x + y^2, y(0) = 0, \quad y(0.5)$
7. $y' = (x - y)^2, y(0) = 0.5, \quad y(0.5)$
8. $y' = xy + \sqrt{y}, y(0) = 1, \quad y(0.5)$
9. $y' = xy^2 - \dfrac{y}{x}, y(1) = 1, \quad y(1.5)$
10. $y' = y - y^2, y(0) = 0.5, \quad y(0.5)$
11. If air resistance is proportional to the square of the instantaneous velocity, then the velocity v of a mass m dropped from a given height is determined from

$$m\frac{dv}{dt} = mg - kv^2, \quad k > 0$$

(see Problem 8, Chapter 3 Review Exercises). If $v(0) = 0$, $k = 0.125$, $m = 5$ slugs, and $g = 32$ ft/s^2, use the Runge-Kutta method to find an approximation to the velocity of the falling mass at $t = 5$ seconds. Use $h = 1$.

12. Solve the initial-value problem in Problem 11 by one of the methods of Chapter 2. Find the true value of $v(5)$.

13. A mathematical model for the area A (in cm^2) that a colony of bacteria (*B. dendroides*) occupies is given by*

$$\frac{dA}{dt} = A(2.128 - 0.0432A).$$

If $A(0) = 0.24$ cm^2, use the Runge-Kutta method to complete the following table. Use $h = 0.5$.

t (days)	1	2	3	4	5
A (observed)	2.78	13.53	36.30	47.50	49.40
A (approximated)					

14. Solve the initial-value problem in Problem 13. Compute the values $A(1)$, $A(2)$, $A(3)$, $A(4)$, and $A(5)$. [*Hint*: See Section 3.3.]

15. Let $y(x)$ be the solution of the initial-value problem

$$y' = x^2 + y^3, \quad y(1) = 1.$$

Determine whether the Runge-Kutta formula can be used to obtain an approximation for $y(1.4)$. Use $h = 0.1$.

16. Consider the differential equation $y' = f(x)$. In this case show that the fourth-order Runge-Kutta method reduces to Simpson's rule for the integral of $f(x)$ on the interval $x_n \leq x \leq x_{n+1}$.

9.5 MULTISTEP METHODS

- *Single-step methods*
- *Multistep methods*
- *Adams-Bashforth/Adams-Moulton method*
- *Milne's method*

Adams-Bashforth/Adams-Moulton Method There are many additional formulas that can be applied to obtain approximations to solutions of differential equations. Although it is not our intention to survey the vast field of numerical methods, several additional formulas deserve mention. The **Adams-Bashforth/Adams-Moulton method,** like the improved Euler formula, is a predictor-corrector method. By using the Adams-Bashforth formula

$$y^*_{n+1} = y_n + \frac{h}{24}(55y'_n - 59y'_{n-1} + 37y'_{n-2} - 9y'_{n-3}), \tag{1}$$

where

$$\begin{aligned} y'_n &= f(x_n, y_n) \\ y'_{n-1} &= f(x_{n-1}, y_{n-1}) \\ y'_{n-2} &= f(x_{n-2}, y_{n-2}) \\ y'_{n-3} &= f(x_{n-3}, y_{n-3}) \end{aligned}$$

---

* See V. A. Kostitzin, *Mathematical Biology* (London: Harrap, 1939).

for $n \geq 3$, as a predictor, we are able to substitute the value of y^*_{n+1} into the Adams-Moulton corrector

$$y_{n+1} = y_n + \frac{h}{24}(9y'_{n+1} + 19y'_n - 5y'_{n-1} + y'_{n-2}), \qquad (2)$$
$$y'_{n+1} = f(x_{n+1}, y^*_{n+1}).$$

Notice formula (1) requires that we know the values of y_0, y_1, y_2, and y_3 in order to obtain y_4. The value of y_0 is, of course, the given initial condition; the values of y_1, y_2, and y_3 are computed by an accurate method such as the Runge-Kutta formula.

Since the Adams-Bashforth/Adams-Moulton formulas demand that we know more than just y_n to compute y_{n+1}, the procedure is called a **multistep,** or **continuing,** method. The Euler formulas, the three-term Taylor formula, and the Runge-Kutta formulas are examples of **single-step,** or **starting,** methods.

EXAMPLE 1 Adams-Bashforth/Adams-Moulton Method

Use the Adams-Bashforth/Adams-Moulton method with $h = 0.2$ to obtain an approximation to $y(0.8)$ for the solution of $y' = x + y - 1$, $y(0) = 1$.

Solution With a step size of $h = 0.2$, $y(0.8)$ will be approximated by y_4. To get started we use the Runge-Kutta method with $x_0 = 0$, $y_0 = 1$, and $h = 0.2$ to obtain

$$y_1 = 1.02140000, \quad y_2 = 1.09181796, \quad y_3 = 1.22210646.$$

Now with the identifications $x_0 = 0$, $x_1 = 0.2$, $x_2 = 0.4$, $x_3 = 0.6$, and $f(x, y) = x + y - 1$, we find

$$\begin{aligned} y'_0 &= f(x_0, y_0) = (0) + (1) - 1 = 0 \\ y'_1 &= f(x_1, y_1) = (0.2) + (1.02140000) - 1 = 0.22140000 \\ y'_2 &= f(x_2, y_2) = (0.4) + (1.09181796) - 1 = 0.49181796 \\ y'_3 &= f(x_3, y_3) = (0.6) + (1.22210646) - 1 = 0.82210646. \end{aligned}$$

With the foregoing values the predictor (1) then gives

$$y^*_4 = y_3 + \frac{0.2}{24}(55y'_3 - 59y'_2 + 37y'_1 - 9y'_0) = 1.42535975.$$

To use the corrector (2) we first need

$$y'_4 = f(x_4, y^*_4) = (0.8) + (1.42535975) - 1 = 1.22535975.$$

Finally, (2) yields

$$y_4 = y_3 + \frac{0.2}{24}(9y'_4 + 19y'_3 - 5y'_2 + y'_1) = 1.42552788.$$

You should verify that the exact value of $y(0.8)$ in Example 1 is $y(0.8) = 1.42554093$.

Milne's Method Another multistep method, which admittedly has limited use (for a reason explained in Section 9.6), is called **Milne's method.** In this method the predictor is

$$y^*_{n+1} = y_{n-3} + \frac{4h}{3}(2y'_n - y'_{n-1} + 2y'_{n-2}), \tag{3}$$

where

$$\begin{aligned} y'_n &= f(x_n, y_n) \\ y'_{n-1} &= f(x_{n-1}, y_{n-1}) \\ y'_{n-2} &= f(x_{n-2}, y_{n-2}) \end{aligned}$$

for $n \geq 3$. The corrector for this formula, based on Simpson's rule of integration, is given by

$$y_{n+1} = y_{n-1} + \frac{h}{3}(y'_{n+1} + 4y'_n + y'_{n-1}), \tag{4}$$

where

$$y'_{n+1} = f(x_{n+1}, y^*_{n+1}).$$

As in the Adams-Bashforth/Adams-Moulton method, we generally use the Runge-Kutta method to compute y_1, y_2, and y_3.

Choice of Starting Methods As stated earlier in this section, a multistep method requires the use of a single-step method, or starting method. Some care should be taken in the selection of this latter method. In practice, starting methods are usually chosen to have the same accuracy as the multistep method with which they are used. By *accuracy* we mean the same order of error (see Section 9.6). Using a less accurate starting method with a highly accurate multistep method is not advisable because then the values used in the multistep method will be unreliable. On the other hand, a highly accurate starting method is complicated and expensive to calculate and this effort will be wasted if the multistep method is of low accuracy.

Advantages and Disadvantages of Multistep Methods Many considerations enter into the choice of a method to solve a differential equation numerically. Single-step methods, particularly the Runge-Kutta method, are often chosen because of their accuracy and the fact that they are easy to program. However, a major drawback to them is that the right-hand side of the differential equation must be evaluated many times at each step. For instance, the fourth-order Runge-Kutta method requires four function evaluations for each step. On the other hand, if the function evaluations in the previous step have been calculated and stored, a multistep method requires only one new function evaluation for each step. This can lead to a great savings in time and expense.

As an example, to solve $y' = f(x, y)$, $y(x_0) = y_0$ numerically using n steps by the fourth-order Runge-Kutta method requires $4n$ function evaluations. The Adams-Bashforth multistep method requires 16 function evaluations for the Runge-Kutta fourth-order starter and $n - 4$ for the n Adams-Bashforth steps, giving a total of $n + 12$ function evaluations for this method. In general the Adams-Bashforth multistep method requires slightly more than a quarter of the number of function evaluations required for the fourth-order Runge-Kutta method. If the evaluation of $f(x, y)$ is complicated, the multistep method will be more efficient.

Another issue involved with multistep methods is how many times the corrector formula should be repeated in each step. Each time the corrector is used, another function evaluation is done, and so the accuracy is increased at the expense of losing an advantage of the multistep method. In practice, the corrector is calculated once, and if the value of y_{n+1} is changed by a large amount, the entire problem is restarted using a smaller step size. This is often the basis of the variable step size methods, whose discussion is beyond the scope of this text.

EXERCISES 9.5

Answers to odd-numbered problems begin on page A-27.

1. Find the exact solution of the initial-value problem in Example 1. Compare the exact values of $y(0.2)$, $y(0.4)$, $y(0.6)$, and $y(0.8)$ with the approximations y_1, y_2, y_3, and y_4.
2. Write a computer program for the Adams-Bashforth/Adams-Moulton method.

In Problems 3 and 4 use the Adams-Bashforth/Adams-Moulton method to approximate $y(0.8)$, where $y(x)$ is the solution of the given initial-value problem. Use $h = 0.2$ and the Runge-Kutta method to compute y_1, y_2, and y_3.

3. $y' = 2x - 3y + 1,\ y(0) = 1$
4. $y' = 4x - 2y,\ y(0) = 2$

In Problems 5–8 use the Adams-Bashforth/Adams-Moulton method to approximate $y(1.0)$, where $y(x)$ is the solution of the given initial-value problem. Use $h = 0.2$ and $h = 0.1$ and the Runge-Kutta method to compute y_1, y_2, and y_3.

5. $y' = 1 + y^2,\ y(0) = 0$
6. $y' = y + \cos x,\ y(0) = 1$
7. $y' = (x - y)^2,\ y(0) = 0$
8. $y' = xy + \sqrt{y},\ y(0) = 1$
9. Use Milne's method to approximate the value of $y(0.4)$, where $y(x)$ is the solution of the initial-value problem in Example 1. Use $h = 0.1$ and the Runge-Kutta method to compute y_1, y_2, and y_3.

9.6 ERRORS AND STABILITY

• *Stable method* • *Round-off error* • *Local truncation error*
• *Global truncation error*

Errors in Numerical Methods In choosing and using a numerical method for the solution of an initial-value problem, we must be aware of the various sources of errors. For some kinds of computation, the accumulation of errors might reduce the accuracy of an approximation to the point of being useless. On the other hand, depending upon the use to which a numerical solution may be put, extreme accuracy may not be worth the added expense and complication.

One source of error always present in calculations is **round-off error.** This is caused by the fact that any calculator or computer can compute to only, at most, a finite number of decimal places. Suppose for the sake of illustration that we have a calculator that can display six digits while carrying eight digits internally. If we multiply two numbers, each having six decimals, then the product actually contains 12 decimal places. But the number that we see is rounded to six decimal places, while the machine has stored a number rounded to eight decimal places. In one calculation such as this, the round-off error may not be significant, but a problem could arise if many calculations are performed with rounded numbers. One way to reduce the effect of round-off error is to minimize the number of calculations. Another technique on a computer is to use double precision capabilities. Since round-off error is unpredictable and difficult to analyze, we will neglect it in the error analysis that follows. We will concentrate on investigating the error made by using a formula or algorithm to approximate the values of the solution.

Truncation Errors for Euler's Method When iterating Euler's formula

$$y_{n+1} = y_n + hf(x_n, y_n)$$

we obtain a sequence of values $y_1, y_2, y_3, \ldots$. The value y_1 will not agree with $y(x_1)$, the actual solution evaluated at x_1, because the algorithm gives only a straight-line approximation to the solution. See Figure 9.10 in Section 9.2. This error is called the **local truncation error, formula error,** or **discretization error.** It occurs at each step; that is, if we assume that y_n is accurate, then y_{n+1} will contain local truncation error.

To derive a formula for the local truncation error for Euler's method we will use Taylor's formula with remainder. This is very similar to the Taylor series expansion (1) in Section 9.3. If a function $y(x)$ possesses $k + 1$ continuous derivatives on an open interval containing x and a, then

$$y(x) = y(a) + y'(a)\frac{(x-a)}{1!} + \cdots + y^{(k)}(a)\frac{(x-a)^k}{k!} + y^{(k+1)}(c)\frac{(x-a)^{k+1}}{(k+1)!}, \tag{1}$$

where c is some point between a and x. Setting $k = 1$, $a = x_n$, and $x = x_{n+1} = x_n + h$, we get

$$y(x_{n+1}) = y(x_n) + y'(x_n)\frac{h}{1!} + y''(c)\frac{h^2}{2!} \quad \text{or} \quad y_{n+1} = y_n + hf(x_n, y_n) + y''(c)\frac{h^2}{2!}.$$

Euler's method is this formula without the last term; hence, the local truncation error in y_{n+1} is

$$y''(c)\frac{h^2}{2!}, \quad \text{where} \quad x_n < c < x_{n+1}.$$

Unfortunately, the value of c is usually unknown (it exists theoretically) and so the exact error cannot be calculated, but an upper bound on the absolute value of the error is

$$M\frac{h^2}{2}, \quad \text{where} \quad M = \max_{x_n < x < x_{n+1}} |y''(x)|.$$

The following terminology is used when discussing errors of numerical methods. A calculation is said to be of **order** α, denoted $O(h^\alpha)$, if the error behaves like Ch^α, where C is a constant. From this we see that if the step size is halved, then the new error should be approximately $C(h^\alpha/2^\alpha)$; that is, the error is reduced by a factor of $1/2^\alpha$.

Using this definition, we see that the local truncation error for Euler's method is of order two, or $O(h^2)$.

EXAMPLE 1 Bound for Local Truncation Errors

Find a bound for the local truncation errors for Euler's method applied to $y' = 2xy$, $y(1) = 1$.

Solution This differential equation was studied in Example 2 of Section 9.2, and its analytic solution is $y(x) = e^{x^2-1}$.

The local truncation error is

$$y''(c)\frac{h^2}{2} = (2 + 4c^2)e^{(c^2-1)}\frac{h^2}{2},$$

where c is between x_n and $x_n + h$. In particular, for $h = 0.1$ we can get an upper bound on the local truncation error for y_1 by replacing c by 1.1:

$$[2 + (4)(1.1)^2]e^{((1.1)^2-1)}\frac{(0.1)^2}{2} = 0.0422.$$

From Table 9.3 we see that the error after the first step is 0.0337, less than the value given by the bound.

Similarly, we can get a bound for the local truncation error for any of the five steps given in Table 9.3 by replacing c by 1.5 (this value of c gives the largest value of $y''(c)$ for any of the steps and may be too generous for the first few steps). Doing this gives

$$[2 + (4)(1.5)^2]e^{((1.5)^2-1)}\frac{(0.1)^2}{2} = 0.1920 \quad \textbf{(2)}$$

as an upper bound for the local truncation error in each step. ■

Note in Example 1 that if h was halved to 0.05, then the error bound (2) would be 0.0480, about one fourth as much. This is expected since the local truncation error for Euler's method is $O(h^2)$.

In the above analysis we assumed in the calculation of y_{n+1} that the value of y_n was correct, but it is not because it contains local truncation errors from previous steps. The total error in y_{n+1} is an accumulation of the errors in each of the previous steps. See Figure 9.11 in Section 9.2. This total error is called the **global truncation error.** A complete analysis of the global truncation error is beyond the scope of this text, but it can be shown that the global truncation error for Euler's method is $O(h)$.

We expect that for Euler's method if the step size was halved, then the error would be approximately halved as well. This is borne out in Example 2 of Section 9.2, where the absolute error at $x = 1.50$ with $h = 0.1$ is 0.5625 and the absolute error with $h = 0.05$ is 0.3171, approximately half as large. See Tables 9.3 and 9.4 in Section 9.2.

In general it can be shown that if a method for the numerical solution of a differential equation has local truncation error $O(h^{\alpha+1})$, then the global truncation error is $O(h^{\alpha})$. Hence we can say that the order of global truncation error for any method is one less than the order of the local truncation error.

Truncation Errors for Other Methods The local and global truncation errors for the other methods we have studied can be analyzed similarly. For instance, it was pointed out in Section 9.4 that the improved Euler's method is a second-order Runge-Kutta method and thus one condition it satisfies is that it agrees with the Taylor polynomial through $k = 2$. Thus the local truncation error is

$$y^{(3)}(c)\,\frac{h^3}{3!}.$$

The derivation of this result is left for you. (See Problem 1.)

Since the local truncation error is $O(h^3)$, we know that the global truncation error is $O(h^2)$. This can be seen in Tables 9.7 and 9.8 of Section 9.2; when the step size is halved from $h = 0.1$ to $h = 0.05$, the global truncation error at $x = 1.50$ is reduced from 0.0394 to 0.0108. This is a reduction of approximately one fourth.

Similarly, a condition in the derivation of the fourth-order Runge-Kutta method is that it should agree with the Taylor polynomial through $k = 4$. Hence the local truncation error is

$$y^{(5)}(c)\,\frac{h^5}{5!} \qquad \text{or} \qquad O(h^5)$$

and the global truncation error is $O(h^4)$. It is now obvious why this is called the fourth-order Runge-Kutta method.

EXAMPLE 2 Bound for Local/Global Truncation Errors

Analyze the local and global truncation errors for the fourth-order Runge-Kutta method applied to $y' = 2xy$, $y(1) = 1$.

Solution By differentiating the known solution $y(x) = e^{x^2-1}$, we get

$$y^{(5)}(c)\,\frac{h^5}{5!} = (120c + 160c^3 + 32c^5)e^{c^2-1}\,\frac{h^5}{5!}. \tag{3}$$

Thus with $c = 1.5$, (3) yields a bound of 0.00028 on the local discretization error for each of the five steps when $h = 0.1$. Note that in Table 9.15 the error in y_1 is much less than this bound.

In Table 9.18 we summarize the approximations to the solution of the initial-value problem at $x = 1.5$ given by the fourth-order Runge-Kutta method with $h = 0.1$ and $h = 0.05$. Because the method is so exact, many decimal places must be kept in the numerical solution to see the effect of halving the step size. Note that when we halve h, the error is divided by a factor of about $2^4 = 16$, as expected. ■

Table 9.18 Runge-Kutta Method

h	Approximation	Error
0.1	3.49021064	$1.323210889 \times 10^{-4}$
0.05	3.49033382	$9.137760898 \times 10^{-6}$

A similar analysis can be done for the errors in each of the methods we have studied. Table 9.19 gives the results.

Table 9.19 Orders of Error for Different Methods

Method	Order of global truncation error
Euler	1
Improved Euler	2
Three-term Taylor	2
Fourth-order Runge-Kutta	4
Adams-Bashforth	4
Adams-Moulton	4
Milne	4

Stability of Numerical Methods In accepting the results of a numerical method, a user should always be concerned with the stability of the method. Put simply, a method is **stable** if small changes in the data make only small changes in the results. This is important since the numbers used in a calculator or computer are not exact. Also, in physical applications the data are often obtained by imprecise measurements. If a method is not stable, it is possible to have an error greater than would be expected from global truncation or round-off error. All single-step methods are stable if the step size taken is small enough; however, some multistep methods, particularly Milne's method, are not stable.

EXAMPLE 3 Milne's Method

Consider the initial-value problem $y' = -10y + 5$, $y(1) = 1$. Use Milne's method to obtain an approximation to $y(2.0)$ using $h = 0.1$.

Solution The instability of the method for this relatively simple problem is apparent from inspection of the results given in Table 9.20. The starting method is the fourth-order Runge-Kutta method, which is very accurate and stable, but nevertheless the very large errors at the end of the calculations make the approximation for $y(2.0)$ quite useless.

Table 9.20 Instability of Milne's Method

x	Actual	Milne	Error
1.0	1.00000	1.00000	
1.1	0.68393	0.68750	0.00357
1.2	0.56767	0.57031	0.00264
1.3	0.52489	0.52637	0.00148
1.4	0.50917	0.52344	0.01427
1.5	0.50337	0.47266	−0.03071
1.6	0.50125	0.60417	0.10292
1.7	0.50045	0.14963	−0.35082
1.8	0.50017	1.66956	1.16939
1.9	0.50005	−3.39111	−3.89116
2.0	0.50002	13.47418	12.97416

Milne's method is the only numerical procedure we have studied that has serious stability problems, but stability must be considered when any numerical method is used.

EXERCISES 9.6

Answers to odd-numbered problems begin on page A-27.

1. Use the Taylor polynomial with remainder (1) to find the formula for the local truncation error for the improved Euler's method (recall that it is a second-order Runge-Kutta method).
2. Repeat Problem 1 for the three-term Taylor method.
3. Repeat Problem 1 for the fourth-order Runge-Kutta method.

4. Consider the initial-value problem $y' = 2y$, $y(0) = 1$. The analytic solution is $y(x) = e^{2x}$.

(**a**) Approximate $y(0.1)$ using one step and Euler's method.

(**b**) Find a bound for the local truncation error in y_1.

(**c**) Compare the actual error in y_1 with your error bound.

(**d**) Approximate $y(0.1)$ using two steps and Euler's method.

(**e**) Verify that the global truncation error for Euler's method is $O(h)$ by comparing the errors in parts (a) and (d).

5. Repeat Problem 4 using the improved Euler's method. Its global truncation error is $O(h^2)$.

6. Repeat Problem 4 using the three-term Taylor method. Its global truncation error is $O(h^2)$.

7. Repeat Problem 4 using the fourth-order Runge-Kutta method. Its global truncation error is $O(h^4)$. Keep eight or nine decimal places in your calculations.

8.–11. Repeat Problems 4–7 using the initial-value problem $y' = -2y + x$, $y(0) = 1$. The analytic solution is $y(x) = \frac{1}{2}x - \frac{1}{4} + \frac{5}{4}e^{-2x}$.

12. Consider the initial-value problem $y' = 2x - 3y + 1$, $y(1) = 5$. The analytic solution is $y(x) = \frac{1}{9} + \frac{2}{3}x + \frac{38}{9}e^{-3(x-1)}$.

(**a**) Find a formula involving c and h for the local truncation error in the nth step if Euler's method is used.

(**b**) Find a bound for the local truncation error in each step if $h = 0.1$ is used to approximate $y(1.5)$.

(**c**) Approximate $y(1.5)$ using $h = 0.1$ and $h = 0.05$ with Euler's method. See Problem 3, Exercises 9.2.

(**d**) Calculate the errors in part (c) and verify that the global truncation error of Euler's method is $O(h)$.

13. Repeat Problem 12 using the improved Euler's method, which has global truncation error $O(h^2)$. See Problem 13(a), Exercises 9.2. You may need to keep more than four decimal places to see the effect of reducing the order of error.

14. Repeat Problem 12 using the three-term Taylor series method, which has global truncation error $O(h^2)$. See Problem 1, Exercises 9.3. You may need to keep more than four decimal places to see the effect of reducing the order of error.

15. Repeat Problem 12 using the fourth-order Runge-Kutta method, which has global truncation error $O(h^4)$. See Problem 1, Exercise 9.4. You will need to keep more than six decimal places to see the effect of reducing the order of error.

16.–19. Repeat Problems 12–15 for the initial-value problem $y' = e^{-y}$, $y(0) = 0$. The analytic solution is $y(x) = \ln(x + 1)$. Approximate $y(0.5)$. See Problems 7 and 13(c) in Exercises 9.2, Problem 5 in Exercises 9.3, and Problem 5 in Exercises 9.4.

9.7 HIGHER-ORDER EQUATIONS AND SYSTEMS

- *Numerical methods applied to systems*
- *Euler's method*
- *Runge-Kutta method*

Second-Order Initial-Value Problem The numerical procedures we have discussed so far in this chapter applied to only a first-order equation $dy/dx = f(x, y)$ subject to an initial condition $y(x_0) = y_0$. To approximate the solution to, say, a second-order initial-value problem

$$\frac{d^2y}{dx^2} = f(x, y, y'), \quad y(x_0) = y_0, \quad y'(x_0) = y'_0, \tag{1}$$

we reduce the second-order differential equation to a second-order system. (See Section 8.3.) If we let $y' = u$, then the equation in (1) becomes

$$\begin{aligned} y' &= u \\ u' &= f(x, y, u). \end{aligned} \tag{2}$$

We now apply a particular method to each equation in the resulting system. For example, the basic **Euler's method** for the system (2) would be

$$\begin{aligned} y_{n+1} &= y_n + hu_n \\ u_{n+1} &= u_n + hf(x_n, y_n, u_n). \end{aligned} \tag{3}$$

EXAMPLE 1 Second-Order DE

Use Euler's method to obtain the approximate value of $y(0.2)$, where $y(x)$ is the solution of the initial-value problem

$$y'' + xy' + y = 0, \quad y(0) = 1, \quad y'(0) = 2.$$

Solution In terms of the substitution $y' = u$, the equation is equivalent to the system

$$\begin{aligned} y' &= u \\ u' &= -xu - y. \end{aligned}$$

Thus from (3) we obtain

$$\begin{aligned} y_{n+1} &= y_n + hu_n \\ u_{n+1} &= u_n + h[-x_n u_n - y_n]. \end{aligned}$$

Using the step size $h = 0.1$ and $y_0 = 1$, $u_0 = 2$, we find

$$\begin{aligned} y_1 &= y_0 + (0.1)u_0 = 1 + (0.1)2 = 1.2 \\ u_1 &= u_0 + (0.1)[-x_0u_0 - y_0] = 2 + (0.1)[-(0)(2) - 1] = 1.9 \\ y_2 &= y_1 + (0.1)u_1 = 1.2 + (0.1)(1.9) = 1.39 \\ u_2 &= u_1 + (0.1)[-x_1u_1 - y_1] = 1.9 + (0.1)[-(0.1)(1.9) - 1.2] = 1.761. \end{aligned}$$

In other words, $y(0.2) \approx 1.39$ and $y'(0.2) \approx 1.761$. ■

Systems An initial-value problem for a second-order system—that is, a system of two first-order differential equations—is

$$\begin{aligned} x' &= f(t, x, y) \\ y' &= g(t, x, y) \\ x(t_0) &= x_0, \quad y(t_0) = y_0. \end{aligned} \tag{4}$$

As we did in (3), to approximate a solution of this problem we apply a numerical method to each equation in (4). The **fourth-order Runge-Kutta method** looks like this:

$$\begin{aligned} x_{n+1} &= x_n + \frac{1}{6}(m_1 + 2m_2 + 2m_3 + m_4) \\ y_{n+1} &= y_n + \frac{1}{6}(k_1 + 2k_2 + 2k_3 + k_4), \end{aligned} \tag{5}$$

where

$$\begin{aligned} m_1 &= hf(t_n, x_n, y_n) & k_1 &= hg(t_n, x_n, y_n) \\ m_2 &= hf(t_n + \tfrac{1}{2}h, x_n + \tfrac{1}{2}m_1, y_n + \tfrac{1}{2}k_1) & k_2 &= hg(t_n + \tfrac{1}{2}h, x_n + \tfrac{1}{2}m_1, y_n + \tfrac{1}{2}k_1) \\ m_3 &= hf(t_n + \tfrac{1}{2}h, x_n + \tfrac{1}{2}m_2, y_n + \tfrac{1}{2}k_2) & k_3 &= hg(t_n + \tfrac{1}{2}h, x_n + \tfrac{1}{2}m_2, y_n + \tfrac{1}{2}k_2) \\ m_4 &= hf(t_n + h, x_n + m_3, y_n + k_3) & k_4 &= hg(t_n + h, x_n + m_3, y_n + k_3). \end{aligned} \tag{6}$$

EXAMPLE 2 Runge-Kutta Method Applied to a System

Consider the initial-value problem

$$\begin{aligned} x' &= 2x + 4y \\ y' &= -x + 6y \\ x(0) &= -1, \quad y(0) = 6. \end{aligned}$$

Use the fourth-order Runge-Kutta method to approximate $x(0.6)$ and $y(0.6)$. Compare the results for $h = 0.2$ and $h = 0.1$.

Solution We illustrate the computations of x_1 and y_1 with the step size $h = 0.2$. With the identifications $f(t, x, y) = 2x + 4y$, $g(t, x, y) = -x + 6y$, $t_0 = 0$, $x_0 = -1$, and $y_0 = 6$, we see from (6) that

$$\begin{aligned} m_1 &= hf(t_0, x_0, y_0) = 0.2f(0, -1, 6) = 0.2[2(-1) + 4(6)] = 4.4000 \\ k_1 &= hg(t_0, x_0, y_0) = 0.2g(0, -1, 6) = 0.2[-1(-1) + 6(6)] = 7.4000 \\ m_2 &= hf(t_0 + \tfrac{1}{2}h, x_0 + \tfrac{1}{2}m_1, y_0 + \tfrac{1}{2}k_1) = 0.2f(0.1, 1.2, 9.7) = 8.2400 \\ k_2 &= hg(t_0 + \tfrac{1}{2}h, x_0 + \tfrac{1}{2}m_1, y_0 + \tfrac{1}{2}k_1) = 0.2g(0.1, 1.2, 9.7) = 11.4000 \\ m_3 &= hf(t_0 + \tfrac{1}{2}h, x_0 + \tfrac{1}{2}m_2, y_0 + \tfrac{1}{2}k_2) = 0.2f(0.1, 3.12, 11.7) = 10.6080 \\ k_3 &= hg(t_0 + \tfrac{1}{2}h, x_0 + \tfrac{1}{2}m_2, y_0 + \tfrac{1}{2}k_2) = 0.2g(0.1, 3.12, 11.7) = 13.4160 \\ m_4 &= hf(t_0 + h, x_0 + m_3, y_0 + k_3) = 0.2f(0.2, 8, 20.216) = 19.3760 \\ k_4 &= hg(t_0 + h, x_0 + m_3, y_0 + k_3) = 0.2g(0.2, 8, 20.216) = 21.3776. \end{aligned}$$

Therefore, from (5) we get

$$x_1 = x_0 + \frac{1}{6}(m_1 + 2m_2 + 2m_3 + m_4)$$

$$= -1 + \frac{1}{6}(4.4 + 2(8.24) + 2(10.608) + 19.3760) = 9.2453$$

$$y_1 = y_0 + \frac{1}{6}(k_1 + 2k_2 + 2k_3 + k_4)$$

$$= 6 + \frac{1}{6}(7.4 + 2(11.4) + 2(13.416) + 21.3776) = 19.0683,$$

where, as usual, the computed values are rounded to four decimal places. These numbers give us the approximations $x_1 \approx x(0.2)$ and $y_1 \approx y(0.2)$. The subsequent values, obtained with the aid of a computer, are summarized in Tables 9.21 and 9.22.

Table 9.21 Runge-Kutta Method with $h = 0.2$

m_1	m_2	m_3	m_4	k_1	k_2	k_3	k_4	t_n	x_n	y_n
								0.00	−1.0000	6.0000
4.4000	8.2400	10.6080	19.3760	7.4000	11.4000	13.4160	21.3776	0.20	9.2453	19.0683
18.9527	31.1564	37.8870	63.6848	21.0329	31.7573	36.9716	57.8214	0.40	46.0327	55.1203
62.5093	97.7863	116.0063	187.3669	56.9378	84.8495	98.0688	151.4191	0.60	158.9430	150.8192

Table 9.22 Runge-Kutta Method with $h = 0.1$

m_1	m_2	m_3	m_4	k_1	k_2	k_3	k_4	t_n	x_n	y_n
								0.00	−1.0000	6.0000
2.2000	3.1600	3.4560	4.8720	3.7000	4.7000	4.9520	6.3256	0.10	2.3840	10.8883
4.8321	6.5742	7.0778	9.5870	6.2946	7.9413	8.3482	10.5957	0.20	9.3379	19.1332
9.5208	12.5821	13.4258	17.7609	10.5461	13.2339	13.8872	17.5358	0.30	22.5541	32.8539
17.6524	22.9090	24.3055	31.6554	17.4569	21.8114	22.8549	28.7393	0.40	46.5103	55.4420
31.4788	40.3496	42.6387	54.9202	28.6141	35.6245	37.2840	46.7207	0.50	88.5729	93.3006
54.6348	69.4029	73.1247	93.4107	46.5231	57.7482	60.3774	75.4370	0.60	160.7563	152.0025

You should verify that the solution of the initial-value problem in Example 2 (see Problem 39, Exercises 8.6) is given by $x(t) = (26t - 1)e^{4t}$, $y(t) = (13t + 6)e^{4t}$. From these equations we see that the exact values are $x(0.6) = 160.9384$ and $y(0.6) = 152.1198$.

In conclusion, we state **Euler's method** for the general system (4):

$$x_{n+1} = x_n + hf(t_n, x_n, y_n)$$
$$y_{n+1} = y_n + hg(t_n, x_n, y_n).$$

EXERCISES 9.7

Answers to odd-numbered problems begin on page A-28.

1. Use Euler's method to approximate $y(0.2)$, where $y(x)$ is the solution of the initial-value problem

$$y'' - 4y' + 4y = 0, \quad y(0) = -2, \quad y'(0) = 1.$$

Use $h = 0.1$. Find the exact solution of the problem and compare the exact value of $y(0.2)$ with y_2.

2. Use Euler's method to approximate $y(1.2)$, where $y(x)$ is the solution of the initial-value problem

$$x^2y'' - 2xy' + 2y = 0, \quad y(1) = 4, \quad y'(1) = 9,$$

where $x > 0$. Use $h = 0.1$. Find the exact solution of the problem and compare the exact value of $y(1.2)$ with y_2.

3. Repeat Problem 1 using the Runge-Kutta method with $h = 0.2$ and $h = 0.1$.

4. Repeat Problem 2 using the Runge-Kutta method with $h = 0.2$ and $h = 0.1$.

5. Use the Runge-Kutta method to obtain the approximate value of $y(0.2)$, where $y(x)$ is a solution of the initial-value problem

$$y'' - 2y' + 2y = e^t \cos t, \quad y(0) = 1, \quad y'(0) = 2.$$

Use $h = 0.2$ and $h = 0.1$.

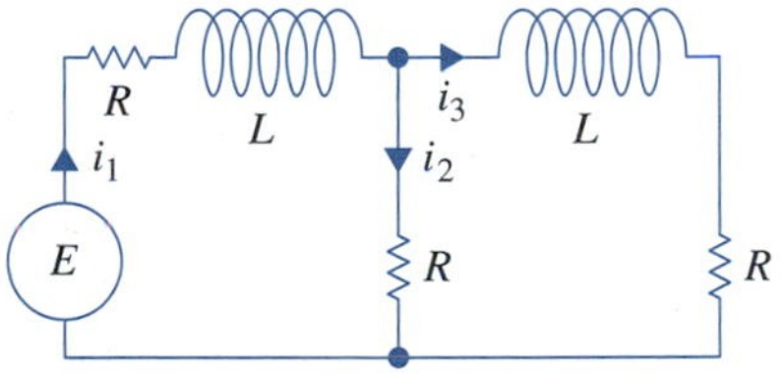

Figure 9.14

6. When $E = 100$ volts, $R = 10$ ohms, and $L = 1$ henry, the system of differential equations for the currents $i_1(t)$ and $i_3(t)$ in the electrical network given in Figure 9.14 is

$$\frac{di_1}{dt} = -20i_1 + 10i_3 + 100$$

$$\frac{di_3}{dt} = 10i_1 - 20i_3,$$

where $i_1(0) = 0$ and $i_3(0) = 0$. Use the Runge-Kutta method to approximate $i_1(t)$ and $i_3(t)$ at $t = 0.1, 0.2, 0.3, 0.4$, and 0.5. Use $h = 0.1$.

In Problems 7–10 use the Runge-Kutta method to approximate $x(0.2)$ and $y(0.2)$. Compare the results for $h = 0.2$ and $h = 0.1$.

7. $x' = 2x - y$
$y' = x$
$x(0) = 6, y(0) = 2$

8. $x' = x + 2y$
$y' = 4x + 3y$
$x(0) = 1, y(0) = 1$

9. $x' = -y + t$
$y' = x - t$
$x(0) = -3, y(0) = 5$

10. $x' = 6x + y + 6t$
$y' = 4x + 3y - 10t + 4$
$x(0) = 0.5, y(0) = 0.2$

9.8 SECOND-ORDER BOUNDARY-VALUE PROBLEMS

• *Difference quotient* • *Finite difference* • *Finite difference equation*

In Sections 9.2–9.5 we focused on techniques that yield an approximation to a solution of a first-order initial-value problem $y' = f(x, y), y(x_0) = y_0$. In addition, we saw in Section 9.7 that we can adapt the approximation techniques to a second-order *initial-value problem* $y'' = f(x, y, y'), y(x_0) = y_0, y'(x_0) = y'_0$ by reducing the second-order differential equation to a system of first-order equations. In this section we are going to examine a method for approximating a solution to a second-order *boundary-value problem* $y'' = f(x, y, y'), y(a) = \alpha$, $y(b) = \beta$. We note at the outset that this method does not require reducing the second-order differential equation to a system of equations.

Finite Difference Approximations Recall from (2) of Section 9.3 that a Taylor series expansion of a function $y(x)$ can be written as

$$y(x + h) = y(x) + y'(x)h + y''(x)\frac{h^2}{2} + y'''(x)\frac{h^3}{6} + \cdots \tag{1}$$

and so $$y(x - h) = y(x) - y'(x)h + y''(x)\frac{h^2}{2} - y'''(x)\frac{h^3}{6} + \cdots. \tag{2}$$

If h is small, we can ignore terms involving $h^4, h^5, \ldots$ since these values are negligible. Indeed, if we ignore all terms involving h^2 and higher, then (1) and (2) yield, in turn, the following approximations for the first derivative $y'(x)$:

$$y'(x) \approx \frac{1}{h}[y(x + h) - y(x)] \tag{3}$$

$$y'(x) \approx \frac{1}{h}[y(x) - y(x - h)]. \tag{4}$$

Subtracting (1) and (2) also gives

$$y'(x) \approx \frac{1}{2h}[y(x + h) - y(x - h)]. \tag{5}$$

On the other hand, if we ignore terms involving h^3 and higher, then by adding (1) and (2) we obtain an approximation for the second derivative $y''(x)$:

$$y''(x) \approx \frac{1}{h^2}[y(x + h) - 2y(x) + y(x - h)]. \tag{6}$$

The right sides of (3), (4), (5), and (6) are called **difference quotients.** The expressions

$$y(x + h) - y(x) \qquad y(x) - y(x - h)$$
$$y(x + h) - y(x - h) \quad \text{and} \quad y(x + h) - 2y(x) + y(x - h)$$

are called **finite differences.** Specifically, $y(x + h) - y(x)$ is called a **forward difference,** $y(x) - y(x - h)$ is a **backward difference,** and both $y(x + h) - y(x - h)$ and $y(x + h) - 2y(x) + y(x - h)$ are called **central dif-**

ferences. The results given in (5) and (6) are referred to as **central difference approximations** for the derivatives y' and y''.

Finite Difference Equation Consider now a linear second-order boundary-value problem

$$y'' + P(x)y' + Q(x)y = f(x), \quad y(a) = \alpha, \quad y(b) = \beta. \tag{7}$$

Suppose $a = x_0 < x_1 < x_2 < \cdots < x_{n-1} < x_n = b$ represents a regular partition of the interval $[a, b]$; that is, $x_i = a + ih$, where $i = 0, 1, 2, \ldots, n$ and $h = (b - a)/n$. The points

$$x_1 = a + h, \quad x_2 = a + 2h, \quad \ldots, \quad x_{n-1} = a + (n - 1)h$$

are called **interior mesh points** of the interval $[a, b]$. If we let

$$y_i = y(x_i), \quad P_i = P(x_i), \quad Q_i = Q(x_i), \quad \text{and} \quad f_i = f(x_i)$$

and if y'' and y' in (7) are replaced by the central difference approximations (5) and (6), we get

$$\frac{y_{i+1} - 2y_i + y_{i-1}}{h^2} + P_i\frac{y_{i+1} - y_{i-1}}{2h} + Q_iy_i = f_i$$

or, after simplifying,

$$\left(1 + \frac{h}{2}P_i\right)y_{i+1} + (-2 + h^2Q_i)y_i + \left(1 - \frac{h}{2}P_i\right)y_{i-1} = h^2f_i. \tag{8}$$

The last equation, known as a **finite difference equation,** is an approximation to the differential equation. It enables us to approximate the solution $y(x)$ of (7) at the interior mesh points $x_1, x_2, \ldots, x_{n-1}$ of the interval $[a, b]$. By letting i take on the values $1, 2, \ldots, n - 1$ in (8), we obtain $n - 1$ equations in the $n - 1$ unknowns $y_1, y_2, \ldots, y_{n-1}$. Bear in mind that we know y_0 and y_n since these are the prescribed boundary conditions $y_0 = y(x_0) = y(a) = \alpha$ and $y_n = y(x_n) = y(b) = \beta$.

In Example 1 we consider a boundary-value problem for which we can compare the approximate values found with the exact values of an explicit solution.

EXAMPLE 1 Finite Difference Method

Use the difference equation (8) with $n = 4$ to approximate the solution of the boundary-value problem $y'' - 4y = 0$, $y(0) = 0$, $y(1) = 5$.

Solution To use (8) we identify $P(x) = 0$, $Q(x) = -4$, $f(x) = 0$, and $h = (1 - 0)/4 = \frac{1}{4}$. Hence the difference equation is

$$y_{i+1} - 2.25y_i + y_{i-1} = 0. \tag{9}$$

Now the interior points are $x_1 = 0 + \frac{1}{4}$, $x_2 = 0 + \frac{2}{4}$, $x_3 = 0 + \frac{3}{4}$, and so for $i = 1, 2$, and 3, (9) yields the following system for the corresponding y_1, y_2, and y_3:

$$\begin{aligned} y_2 - 2.25y_1 + y_0 &= 0 \\ y_3 - 2.25y_2 + y_1 &= 0 \\ y_4 - 2.25y_3 + y_2 &= 0. \end{aligned}$$

With the boundary conditions $y_0 = 0$ and $y_4 = 5$, the foregoing system becomes

$$\begin{aligned} -2.25y_1 + y_2 &= 0 \\ y_1 - 2.25y_2 + y_3 &= 0 \\ y_2 - 2.25y_3 &= -5. \end{aligned}$$

Solving the system gives $y_1 = 0.7256$, $y_2 = 1.6327$, and $y_3 = 2.9479$.

Now the general solution of the given differential equation is $y = c_1 \cosh 2x + c_2 \sinh 2x$. The condition $y(0) = 0$ implies $c_1 = 0$. The other boundary condition gives c_2. In this way we see that an explicit solution of the boundary-value problem is $y(x) = (5 \sinh 2x)/\sinh 2$. Thus the exact values (rounded to four decimal places) of this solution at the interior points are $y(0.25) = 0.7184$, $y(0.5) = 1.6201$, and $y(0.75) = 2.9354$. ■

The accuracy of the approximations in Example 1 can be improved by using a smaller value of h. Of course, the trade-off here is that a smaller value of h necessitates solving a larger system of equations. It is left as an exercise to show that with $h = \frac{1}{8}$, approximations to $y(0.25)$, $y(0.5)$, and $y(0.75)$ are 0.7202, 1.6233, and 2.9386, respectively. See Problem 11.

EXAMPLE 2 Finite Difference Method

Use the difference equation (8) with $n = 10$ to approximate the solution of

$$y'' + 3y' + 2y = 4x^2, \quad y(1) = 1, \quad y(2) = 6.$$

Solution In this case we identify $P(x) = 3$, $Q(x) = 2$, $f(x) = 4x^2$, and $h = (2 - 1)/10 = 0.1$, and so (8) becomes

$$1.15y_{i+1} - 1.98y_i + 0.85y_{i-1} = 0.04x_i^2. \tag{10}$$

Now the interior points are $x_1 = 1.1, x_2 = 1.2, x_3 = 1.3, x_4 = 1.4, x_5 = 1.5, x_6 = 1.6, x_7 = 1.7, x_8 = 1.8$, and $x_9 = 1.9$. Thus, for $i = 1, 2, \ldots, 9$ and $y_0 = 1$, $y_{10} = 6$, (10) gives a system of nine equations and nine unknowns:

$$\begin{aligned} 1.15y_2 - 1.98y_1 &= -0.8016 \\ 1.15y_3 - 1.98y_2 + 0.85y_1 &= 0.0576 \\ 1.15y_4 - 1.98y_3 + 0.85y_2 &= 0.0676 \\ 1.15y_5 - 1.98y_4 + 0.85y_3 &= 0.0784 \\ 1.15y_6 - 1.98y_5 + 0.85y_4 &= 0.0900 \\ 1.15y_7 - 1.98y_6 + 0.85y_5 &= 0.1024 \\ 1.15y_8 - 1.98y_7 + 0.85y_6 &= 0.1156 \\ 1.15y_9 - 1.98y_8 + 0.85y_7 &= 0.1296 \\ - 1.98y_9 + 0.85y_8 &= -6.7556. \end{aligned}$$

We can solve this large system using Gaussian elimination or, with relative ease, by means of a computer algebra system such as *Mathematica*. The result is found to be $y_1 = 2.4047$, $y_2 = 3.4432$, $y_3 = 4.2010$, $y_4 = 4.7469$, $y_5 = 5.1359$, $y_6 = 5.4124$, $y_7 = 5.6117$, $y_8 = 5.7620$, and $y_9 = 5.8855$. ■

Remarks The approximation method that we have just considered can be extended to boundary-value problems in which the first derivative is specified at a boundary—for example, a problem such as $y'' = f(x, y, y')$, $y'(a) = \alpha$, $y(b) = \beta$. See Problem 13.

EXERCISES 9.8

Answers to odd-numbered problems begin on page A-28.

In Problems 1–10 use the finite difference method and the indicated value of n to approximate the solution of the given boundary-value problem.

1. $y'' + 9y = 0$, $y(0) = 4$, $y(2) = 1$; $n = 4$

2. $y'' - y = x^2$, $y(0) = 0$, $y(1) = 0$; $n = 4$

3. $y'' + 2y' + y = 5x$, $y(0) = 0$, $y(1) = 0$; $n = 5$

4. $y'' - 10y' + 25y = 1$, $y(0) = 1$, $y(1) = 0$; $n = 5$

5. $y'' - 4y' + 4y = (x + 1)e^{2x}$, $y(0) = 3$, $y(1) = 0$; $n = 6$

6. $y'' + 5y' = 4\sqrt{x}$, $y(1) = 1$, $y(2) = -1$; $n = 6$

7. $x^2y'' + 3xy' + 3y = 0$, $y(1) = 5$, $y(2) = 0$; $n = 8$

8. $x^2y'' - xy' + y = \ln x$, $y(1) = 0$, $y(2) = -2$; $n = 8$

9. $y'' + (1 - x)y' + xy = x$, $y(0) = 0$, $y(1) = 2$; $n = 10$

10. $y'' + xy' + y = x$, $y(0) = 1$, $y(1) = 0$; $n = 10$

11. Rework Example 1 using $n = 8$.

12. The electrostatic potential u between two concentric spheres of radius $r = 1$ and $r = 4$ is determined from

$$\frac{d^2u}{dr^2} + \frac{2}{r}\frac{du}{dr} = 0, \quad u(1) = 50, \quad u(4) = 100.$$

Use the method of this section with $n = 6$ to approximate the solution of this boundary-value problem.

13. Consider the boundary-value problem $y'' + xy = 0$, $y'(0) = 1$, $y(1) = -1$.

(a) Find the difference equation corresponding to the differential equation. Show that for $i = 0, 1, 2, \ldots, n - 1$ the difference equation yields n equations in $n + 1$ unknowns $y_{-1}, y_0, y_1, y_2, \ldots, y_{n-1}$. Here y_{-1} and y_0 are unknowns since y_{-1} represents an approximation to y at the exterior point $x = -h$ and y_0 is not specified at $x = 0$.

(b) Use the central difference approximation (5) to show that $y_1 - y_{-1} = 2h$. Use this equation to eliminate y_{-1} from the system in part (a).

(c) Use $n = 5$ and the system of equations found in parts (a) and (b) to approximate the solution of the original boundary-value problem.

CHAPTER 9 REVIEW

A solution of a differential equation may exist and yet we may not be able to determine it in terms of the familiar elementary functions. A way of convincing oneself that a first-order equation $y' = f(x, y)$ possesses a solution passing through a specific point (x_0, y_0) is to sketch the **direction field** associated with the equation. The equation $f(x, y) = c$ determines the **isoclines,** or curves of constant inclination. This means that every solution curve that passes through a particular isocline does so with the same slope. The direction field is the totality of short line segments throughout two-dimensional space that have midpoints on the isoclines and that possess slope equal to the value of the parameter c. A carefully plotted sequence of these **lineal elements** can suggest the shape of a solution curve passing through the given point (x_0, y_0).

At best, a direction field can give only the crudest form of an approximation to a numerical value of the solution $y(x)$ of the initial-value problem when x is close to x_0.

To obtain the approximate values of $y(x)$, we used **Euler's formula,**

$$y_{n+1} = y_n + hf(x_n, y_n);$$

the **improved Euler's formula,**

$$y_{n+1} = y_n + h\frac{f(x_n, y_n) + f(x_{n+1}, y^*_{n+1})}{2},$$

where $$y^*_{n+1} = y_n + hf(x_n, y_n);$$

the **three-term Taylor formula,**

$$y_{n+1} = y_n + y'_n h + y''_n \frac{h^2}{2};$$

the **Runge-Kutta formula,**

$$y_{n+1} = y_n + \frac{1}{6}(k_1 + 2k_2 + 2k_3 + k_4),$$

where
$$k_1 = hf(x_n, y_n)$$
$$k_2 = hf(x_n + \tfrac{1}{2}h, y_n + \tfrac{1}{2}k_1)$$
$$k_3 = hf(x_n + \tfrac{1}{2}h, y_n + \tfrac{1}{2}k_2)$$
$$k_4 = hf(x_n + h, y_n + k_3);$$

and the **Adams-Bashforth/Adams-Moulton formulas,**

$$y^*_{n+1} = y_n + \frac{h}{24}(55y'_n - 59y'_{n-1} + 37y'_{n-2} - 9y'_{n-3})$$
$$y_{n+1} = y_n + \frac{h}{24}(9y'_{n+1} + 19y'_n - 5y'_{n-1} + y'_{n-2}),$$

where $$y'_{n+1} = f(x_{n+1}, y^*_{n+1}).$$

In each of the foregoing formulas, the number h is the length of a uniform step. In other words, $x_1 = x_0 + h, x_2 = x_1 + h = x_0 + 2h, \ldots, x_n = x_{n-1} + h = x_0 + nh$.

The first four methods are known as **single-step,** or **starting, methods.** The **Adams-Bashforth/Adams-Moulton** and **Milne methods** are examples of **multistep,** or **continuing, methods.** To use these multistep methods we need to first compute y_1, y_2, and y_3 by some starting method. The **Adams-Bashforth/Adams-Moulton** and **Milne methods** are examples of a class of approximating formulas known as **predictor-corrector formulas.**

An important question in using numerical techniques is the size of error. The order of the global truncation error is a good way to measure the accuracy of a method.

To obtain numerical approximations to higher-order differential equations, we can reduce the differential equation to a system of first-order equations. We then apply a particular numerical technique to each equation of the system.

We can approximate a solution of a second-order boundary-value problem by replacing the differential equation by a **finite difference equation.**

CHAPTER 9 REVIEW EXERCISES

Answers to odd-numbered problems begin on page A-29.

In Problems 1 and 2 sketch the direction field for the given differential equation. Indicate several possible solution curves.

1. $y\,dx - x\,dy = 0$

2. $y' = 2x - y$

In Problems 3–6 construct a table comparing the indicated values of $y(x)$ using Euler's, improved Euler's, three-term Taylor, and Runge-Kutta methods. Compute to four rounded decimal places. Use $h = 0.1$ and $h = 0.05$.

3. $y' = 2 \ln xy$, $y(1) = 2$
$y(1.1), y(1.2), y(1.3), y(1.4), y(1.5)$

4. $y' = \sin x^2 + \cos y^2$, $y(0) = 0$
$y(0.1), y(0.2), y(0.3), y(0.4), y(0.5)$

5. $y' = \sqrt{x + y}$, $y(0.5) = 0.5$
$y(0.6), y(0.7), y(0.8), y(0.9), y(1.0)$

6. $y' = xy + y^2$, $y(1) = 1$
$y(1.1), y(1.2), y(1.3), y(1.4), y(1.5)$

7. Use Euler's method to obtain the approximate value of $y(0.2)$, where $y(x)$ is the solution of the initial-value problem

$$y'' - (2x + 1)y = 1, \quad y(0) = 3, \quad y'(0) = 1.$$

First use one step with $h = 0.2$ and then repeat the calculations using $h = 0.1$.

8. Use the Adams-Bashforth/Adams-Moulton method to approximate the value of $y(0.4)$, where $y(x)$ is the solution of

$$y' = 4x - 2y, \quad y(0) = 2.$$

Use the Runge-Kutta formula and $h = 0.1$ to obtain the values of y_1, y_2, and y_3.

9. Use Euler's method and $h = 0.1$ to approximate the values of $x(0.2)$, $y(0.2)$, where $x(t)$ and $y(t)$ are solutions of

$$x' = x + y$$
$$y' = x - y,$$
$$x(0) = 1, \quad y(0) = 2.$$

10. Use the finite difference method with $n = 10$ to approximate the solution of the boundary-value problem

$$y'' + 6.55(1 + x)y = 1, \quad y(0) = 0, \quad y(1) = 0.$$

Essay

by
C. J. Knickerbocker
Department of Mathematics, St. Lawrence University

NERVE IMPULSE MODELS

A biologically interesting and mathematically rich problem is the transmission of an electrical impulse down a nerve axon. The impulse, or **action potential,** is represented pictorially in Figure 9.15 as a voltage change over time at some point on the nerve axon.

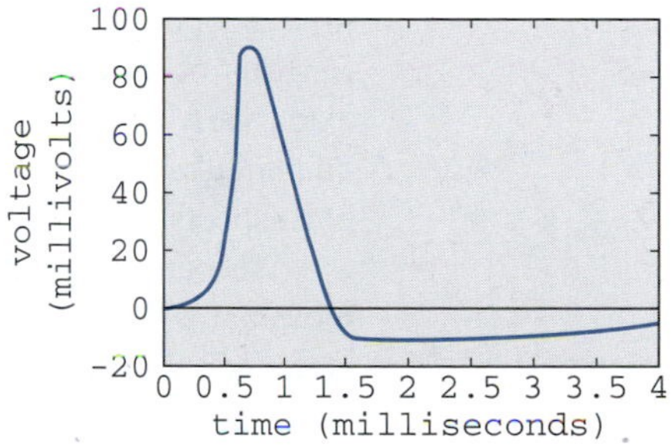

Figure 9.15

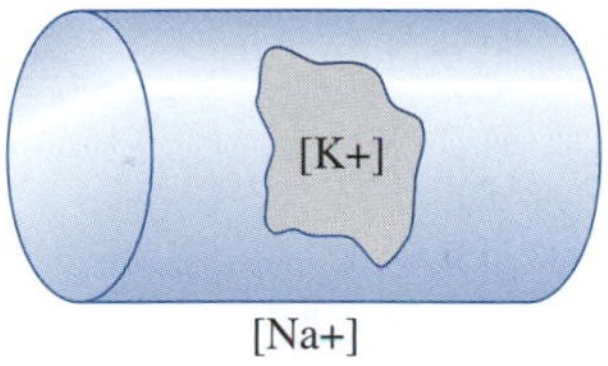

Figure 9.16 Nerve axon

Physically, a nerve axon can be thought of as a cylinder surrounded by a thin permeable membrane with sodium ions outside the axon membrane and potassium ions inside the membrane. See Figure 9.16.

Pretend we are standing at a point on the axon as a nerve impulse approaches. This "disturbance" induces a change in the voltage near us by having the sodium ions rush through the membrane, making the inside of the axon more positively charged. In Figure 9.15 this corresponds to the voltage rise at the left of the pulse. At this stage the potassium ions inside react to the voltage change by rushing out through the membrane, trying to re-equalize the potential across the membrane. This is seen as the voltage decline in Figure 9.15. But the potassium ions overcompensate and the charge becomes negative relative to the undisturbed case. A slow recovery period, the long tail in Figure 9.15, now occurs as the ions move back to their original sides of the membrane.

In about 1950, A. L. Hodgkin and A. F. Huxley developed a system of nonlinear partial differential equations that model the action potential in a nerve axon. The Hodgkin-Huxley system is similar to a nonlinear heat equation in four dependent variables and two independent variables. In general, the solution to this type of equation is difficult, if not impossible, to find, and in particular the Hodgkin-Huxley equation cannot be solved analytically. Therefore, we need some alternatives to finding an analytical solution. The first possibility is to find an analytical solution to a simplified version of the Hodgkin-Huxley equation. Much research has been done on this idea, but in order to find an exact solution to the simplified equations the Hodgkin-Huxley equation must be reduced to the point where most of the interesting physical properties are lost. A second possibility is to find a numerical solution to a simplified Hodgkin-Huxley equation; that is, we approximate the Hodgkin-Huxley equation and then find a numerical solution to this approximate version. The simplified model we will consider is called the **Fitzhugh-Nagumo equation,** given by

$$\frac{\partial^2 u}{\partial x^2} = \frac{\partial u}{\partial t} + f(u) + z$$

$$\frac{\partial z}{\partial t} = \varepsilon u$$

$$f(u) = u(1 - u)(a - u),$$

where ε is very small and $0 \leq a \leq 1$. The Hodgkin-Huxley equation is simplified by scaling variables, noting that two of the dependent variables are roughly time independent and that a certain nonlinear term can be simulated by a cubic function of the voltage. This equation was first described by Fitzhugh and was extensively studied by Nagumo. This simplified equation does not yield any real quantitative information about the action potential, but it is rich in the qualitative aspects of the physical problem.

We can make another simplification by noting that the action potential moves at a constant velocity θ and does not change its shape as it moves; that is, we can transfer the equation from an x, t-coordinate system to a traveling coordinate system in ξ by assuming a solution of the form $u = u(\xi) = u(x - \theta t)$. This yields the following ordinary differential equation in ξ:

$$\frac{d^2 u}{d\xi^2} - \theta \frac{du}{d\xi} - f(u) - z = 0$$

$$\theta \frac{dz}{d\xi} = \varepsilon u$$

$$f(u) = u(1 - u)(a - u).$$

Some analytical results are known for this equation when specific values are chosen for the parameters. For example, when $\theta = 0$, $\varepsilon = 0$, and u is assumed to be small (so that $f(u) \approx au$), an analytical solution is easily determined.

Before any attempt is made to solve this equation numerically, we should look for any analytic "clues" that may help guide us in the choice and implementation of a numerical scheme. An analysis of the Fitzhugh-Nagumo equation produces two important pieces of information. First, using a technique called perturbation analysis, we anticipate a solution whose shape is similar to that of the action potential (see Figure 9.15). Second, using phase-plane analysis, we find that the ordinary differential equation is very "sensitive" to the choices of the parameters θ, ε, and a. That is to say that the numerical solutions might vary dramatically with small changes in any of these parameters.

The sensitive nature of the equation can be seen in the numerical results. Using the fourth-order Runge-Kutta method, we begin by choosing $\varepsilon = 0.005$, $a = 0.1$, $u(0) = 0.001$, and $\theta = 0.5184627034239536284$. This yields the solution given in Figure 9.17.

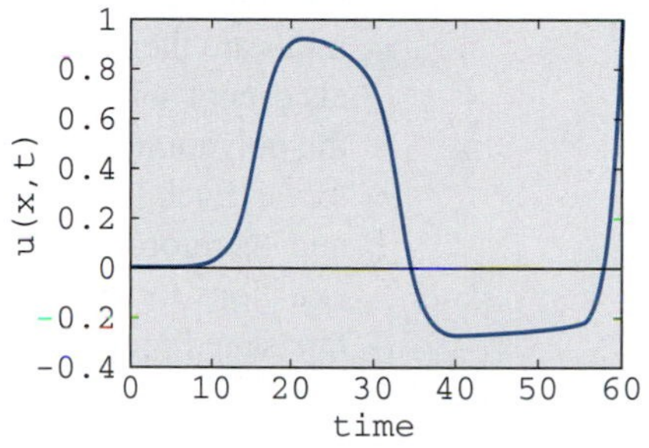

Figure 9.17 Runge-Kutta solution 1

Note that all the qualitative features of the action potential are represented: a sharp rise in voltage followed by a sharp decline followed by a long slow recovery period. But, note that the numerical solution "blows up" to positive infinity on the right-hand side. Now by changing the last digit of θ, yielding $\theta = 0.5184627034239536283$, we get the solution shown in Figure 9.18.

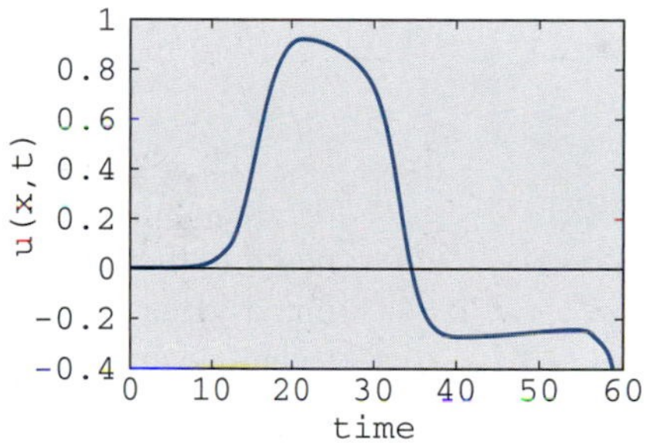

Figure 9.18 Runge-Kutta solution 2

This solution appears identical to the first solution except on the right-hand side, where it goes to negative infinity. This clearly shows how sensitive the numerical solution is to the choice of parameters.

A careful application of the fourth-order Runge-Kutta method also generates other possible solutions. For example, a numerical solution can be found in the form of a small slow pulse followed by a slightly larger pulse, both of which are considerably smaller than the "normal" solution.

A more advanced numerical method that can be applied to this ordinary differential equation is a technique called the **finite elements method.** In contrast to the finite difference approach, where an exact solution is found to an approximation of the differential equation, the finite element approach is to find an approximate solution to the exact equation. This is accomplished by subdividing the ξ-axis with mesh width $\Delta\xi$. Over each interval a set of polynomials is defined. The higher the order of the set of polynomials, the more accurate the numerical solution will be. The polynomials are selected so that the values of the functions and the corresponding derivatives have certain properties at the boundaries of the interval. For the problem at hand we have selected over each interval six fifth-degree polynomials, called the Hermite quintics. The polynomials are then "plugged into" the differential equation. Clearly these polynomials cannot solve the differential equation on the complete interval, so we "force" the polynomials to work at a few points (called the Gaussian points) on each interval. Each Gaussian point now yields a nonlinear algebraic equation to solve, and therefore the method generates a system of coupled nonlinear algebraic equations to solve. The results of this type of approach used with the parameters stated previously are given in Figure 9.19.

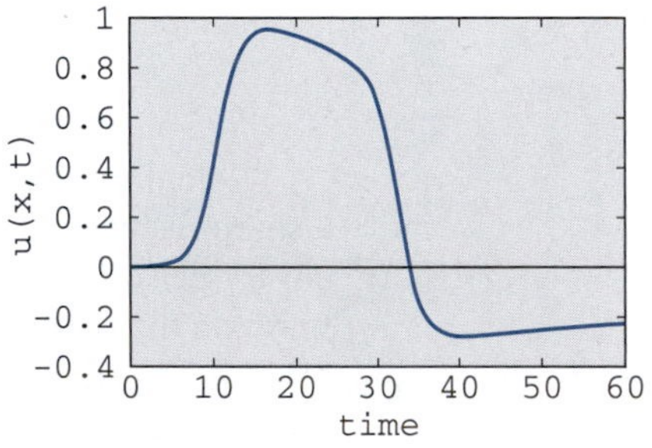

Figure 9.19 Finite element solution

The advantage to this method is that it invokes conditions at both ends of the domain, thereby treating the problem as a boundary-value problem, as opposed to the fourth-order Runge-Kutta method, which treats the problem as an initial-value problem. The disadvantage to the method is the complexity of the technique and difficulties in programming.

The interested reader can find a survey on research that has been done on the various aspects of the Hodgkin-Huxley equation in an article entitled "The Electrophysics of a Nerve Fiber" by A. C. Scott in *Reviews of Modern Physics* (Volume 47, No. 2, April 1975). This article contains a very extensive list of references.

APPENDIX

I GAMMA FUNCTION

Euler's integral definition of the **gamma function*** is

$$\Gamma(x) = \int_0^{\infty} t^{x-1}e^{-t}\,dt. \quad \textbf{(1)}$$

Convergence of the integral requires that $x - 1 > -1$, or $x > 0$. The recurrence relation

$$\Gamma(x + 1) = x\Gamma(x), \quad \textbf{(2)}$$

which we saw in Section 6.5, can be obtained from (1) with integration by parts. Now when $x = 1$,

$$\Gamma(1) = \int_0^{\infty} e^{-t}dt = 1$$

and thus (2) gives

$$\begin{aligned}\Gamma(2) &= 1\Gamma(1) = 1\\ \Gamma(3) &= 2\Gamma(2) = 2\cdot 1\\ \Gamma(4) &= 3\Gamma(3) = 3\cdot 2\cdot 1\end{aligned}$$

and so on. In this manner it is seen that when n is a positive integer,

$$\Gamma(n + 1) = n!$$

For this reason the gamma function is often called the **generalized factorial function.**

Although the integral form (1) does not converge for $x < 0$, it can be shown by means of alternative definitions that the gamma function is defined for all real and complex numbers *except* $x = -n$, $n = 0, 1, 2, \ldots$. As a consequence, (2) is actually valid for $x \neq -n$. Considered as a function of a real variable x, the graph of $\Gamma(x)$ is as given in Figure A.1. Observe that the nonpositive integers correspond to vertical asymptotes of the graph.

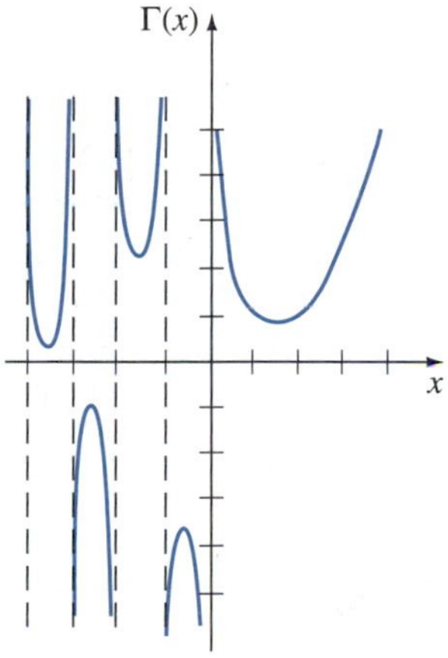

Figure A.1

In Problems 27–33 of Exercises 6.5 we utilized the fact that $\Gamma(\frac{1}{2}) = \sqrt{\pi}$. This result can be derived from (1) by setting $x = \frac{1}{2}$:

$$\Gamma(\tfrac{1}{2}) = \int_0^{\infty} t^{-1/2}e^{-t}\,dt. \quad \textbf{(3)}$$

---

* This function was first defined by Leonhard Euler in his text *Institutiones calculi integralis,* published in 1768.

When we let $t = u^2$, (3) can be written as

$$\Gamma(\tfrac{1}{2}) = 2\int_0^\infty e^{-u^2}\,du.$$

But
$$\int_0^\infty e^{-u^2}\,du = \int_0^\infty e^{-v^2}\,dv,$$

and so
$$[\Gamma(\tfrac{1}{2})]^2 = \left(2\int_0^\infty e^{-u^2}\,du\right)\left(2\int_0^\infty e^{-v^2}\,dv\right)$$
$$= 4\int_0^\infty\int_0^\infty e^{-(u^2+v^2)}\,du\,dv.$$

Switching to polar coordinates $u = r\cos\theta$, $v = \sin\theta$ enables us to evaluate the double integral:

$$4\int_0^\infty\int_0^\infty e^{-(u^2+v^2)}\,du\,dv = 4\int_0^{\pi/2}\int_0^\infty e^{-r^2}r\,dr\,d\theta = \pi.$$

Hence
$$[\Gamma(\tfrac{1}{2})]^2 = \pi \quad \text{or} \quad \Gamma(\tfrac{1}{2}) = \sqrt{\pi}.$$

EXAMPLE 1 Value of $\Gamma(-1/2)$

Evaluate $\Gamma(-\tfrac{1}{2})$.

Solution In view of (2) it follows that, with $x = -\tfrac{1}{2}$,

$$\Gamma(\tfrac{1}{2}) = -\tfrac{1}{2}\,\Gamma(-\tfrac{1}{2}).$$

Therefore
$$\Gamma(-\tfrac{1}{2}) = -2\Gamma(\tfrac{1}{2}) = -2\sqrt{\pi}.$$

EXERCISES FOR APPENDIX I

Answers to odd-numbered problems begin on page A-29.

1. Evaluate.
   (a) $\Gamma(5)$ (b) $\Gamma(7)$ (c) $\Gamma(-\tfrac{3}{2})$ (d) $\Gamma(-\tfrac{5}{2})$

2. Use (1) and the fact that $\Gamma(\tfrac{6}{5}) = 0.92$ to evaluate $\int_0^\infty x^5 e^{-x^5}\,dx$. [*Hint*: Let $t = x^5$.]

3. Use (1) and the fact that $\Gamma(\tfrac{5}{3}) = 0.89$ to evaluate $\int_0^\infty x^4 e^{-x^3}\,dx$.

4. Evaluate $\int_0^1 x^3\left(\ln\frac{1}{x}\right)^3 dx$. [*Hint*: Let $t = -\ln x$.]

5. Use the fact that $\Gamma(x) > \int_0^1 t^{x-1}e^{-t}\,dt$ to show that $\Gamma(x)$ is unbounded as $x \to 0^+$.

6. Use (1) to derive (2) for $x > 0$.

APPENDIX

II

LAPLACE TRANSFORMS

$f(t)$	$\mathscr{L}\{f(t)\} = F(s)$
1. 1	$\dfrac{1}{s}$
2. t	$\dfrac{1}{s^2}$
3. t^n	$\dfrac{n!}{s^{n+1}}$, n a positive integer
4. $t^{-1/2}$	$\sqrt{\dfrac{\pi}{s}}$
5. $t^{1/2}$	$\dfrac{\sqrt{\pi}}{2s^{3/2}}$
6. t^{α}	$\dfrac{\Gamma(\alpha+1)}{s^{\alpha+1}}, \alpha > -1$
7. $\sin kt$	$\dfrac{k}{s^2+k^2}$
8. $\cos kt$	$\dfrac{s}{s^2+k^2}$
9. $\sin^2 kt$	$\dfrac{2k^2}{s(s^2+4k^2)}$
10. $\cos^2 kt$	$\dfrac{s^2+2k^2}{s(s^2+4k^2)}$
11. e^{at}	$\dfrac{1}{s-a}$
12. $\sinh kt$	$\dfrac{k}{s^2-k^2}$
13. $\cosh kt$	$\dfrac{s}{s^2-k^2}$
14. $\sinh^2 kt$	$\dfrac{2k^2}{s(s^2-4k^2)}$
15. $\cosh^2 kt$	$\dfrac{s^2-2k^2}{s(s^2-4k^2)}$
16. te^{at}	$\dfrac{1}{(s-a)^2}$
17. $t^n e^{at}$	$\dfrac{n!}{(s-a)^{n+1}}$, n a positive integer

$f(t)$	$\mathscr{L}\{f(t)\} = F(s)$
18. $e^{at} \sin kt$	$\dfrac{k}{(s-a)^2 + k^2}$
19. $e^{at} \cos kt$	$\dfrac{s-a}{(s-a)^2 + k^2}$
20. $e^{at} \sinh kt$	$\dfrac{k}{(s-a)^2 - k^2}$
21. $e^{at} \cosh kt$	$\dfrac{s-a}{(s-a)^2 - k^2}$
22. $t \sin kt$	$\dfrac{2ks}{(s^2 + k^2)^2}$
23. $t \cos kt$	$\dfrac{s^2 - k^2}{(s^2 + k^2)^2}$
24. $\sin kt + kt \cos kt$	$\dfrac{2ks^2}{(s^2 + k^2)^2}$
25. $\sin kt - kt \cos kt$	$\dfrac{2k^3}{(s^2 + k^2)^2}$
26. $t \sinh kt$	$\dfrac{2ks}{(s^2 - k^2)^2}$
27. $t \cosh kt$	$\dfrac{s^2 + k^2}{(s^2 - k^2)^2}$
28. $\dfrac{e^{at} - e^{bt}}{a - b}$	$\dfrac{1}{(s-a)(s-b)}$
29. $\dfrac{ae^{at} - be^{bt}}{a - b}$	$\dfrac{s}{(s-a)(s-b)}$
30. $1 - \cos kt$	$\dfrac{k^2}{s(s^2 + k^2)}$
31. $kt - \sin kt$	$\dfrac{k^3}{s^2(s^2 + k^2)}$
32. $\dfrac{a \sin bt - b \sin at}{ab(a^2 - b^2)}$	$\dfrac{1}{(s^2 + a^2)(s^2 + b^2)}$
33. $\dfrac{\cos bt - \cos at}{a^2 - b^2}$	$\dfrac{s}{(s^2 + a^2)(s^2 + b^2)}$
34. $\sin kt \sinh kt$	$\dfrac{2k^2 s}{s^4 + 4k^4}$
35. $\sin kt \cosh kt$	$\dfrac{k(s^2 + 2k^2)}{s^4 + 4k^4}$
36. $\cos kt \sinh kt$	$\dfrac{k(s^2 - 2k^2)}{s^4 + 4k^4}$
37. $\cos kt \cosh kt$	$\dfrac{s^3}{s^4 + 4k^4}$
38. $J_0(kt)$	$\dfrac{1}{\sqrt{s^2 + k^2}}$

	$f(t)$	$\mathscr{L}\{f(t)\} = F(s)$
39.	$\dfrac{e^{bt} - e^{at}}{t}$	$\ln \dfrac{s - a}{s - b}$
40.	$\dfrac{2(1 - \cos kt)}{t}$	$\ln \dfrac{s^2 + k^2}{s^2}$
41.	$\dfrac{2(1 - \cosh kt)}{t}$	$\ln \dfrac{s^2 - k^2}{s^2}$
42.	$\dfrac{\sin at}{t}$	$\arctan\left(\dfrac{a}{s}\right)$
43.	$\dfrac{\sin at \cos bt}{t}$	$\dfrac{1}{2} \arctan \dfrac{a + b}{s} + \dfrac{1}{2} \arctan \dfrac{a - b}{s}$
44.	$\dfrac{1}{\sqrt{\pi t}} e^{-a^2/4t}$	$\dfrac{e^{-a\sqrt{s}}}{\sqrt{s}}$
45.	$\dfrac{a}{2\sqrt{\pi t^3}} e^{-a^2/4t}$	$e^{-a\sqrt{s}}$
46.	$\operatorname{erfc}\left(\dfrac{a}{2\sqrt{t}}\right)$	$\dfrac{e^{-a\sqrt{s}}}{s}$
47.	$2\sqrt{\dfrac{t}{\pi}}\, e^{-a^2/4t} - a \operatorname{erfc}\left(\dfrac{a}{2\sqrt{t}}\right)$	$\dfrac{e^{-a\sqrt{s}}}{s\sqrt{s}}$
48.	$e^{ab} e^{b^2 t} \operatorname{erfc}\left(b\sqrt{t} + \dfrac{a}{2\sqrt{t}}\right)$	$\dfrac{e^{-a\sqrt{s}}}{\sqrt{s}(\sqrt{s} + b)}$
49.	$-e^{ab} e^{b^2 t} \operatorname{erfc}\left(b\sqrt{t} + \dfrac{a}{2\sqrt{t}}\right) + \operatorname{erfc}\left(\dfrac{a}{2\sqrt{t}}\right)$	$\dfrac{be^{-a\sqrt{s}}}{s(\sqrt{s} + b)}$
50.	$\delta(t)$	1
51.	$\delta(t - t_0)$	e^{-st_0}
52.	$e^{at} f(t)$	$F(s - a)$
53.	$f(t - a)\mathscr{U}(t - a)$	$e^{-as} F(s)$
54.	$\mathscr{U}(t - a)$	$\dfrac{e^{-as}}{s}$
55.	$f^{(n)}(t)$	$s^n F(s) - s^{(n-1)} f(0) - \cdots - f^{(n-1)}(0)$
56.	$t^n f(t)$	$(-1)^n \dfrac{d^n}{ds^n} F(s)$
57.	$\int_0^t f(\tau) g(t - \tau)\, d\tau$	$F(s)G(s)$

APPENDIX

III REVIEW OF DETERMINANTS

The determinant of a 2×2 matrix $\mathbf{A}$ is defined by

$$\begin{vmatrix} a_{11} & a_{12} \\ a_{21} & a_{22} \end{vmatrix} = a_{11}a_{22} - a_{12}a_{21}.$$

Minors and Cofactors For an $n \times n$ matrix $\mathbf{A}$, let a_{ij} be the entry in the ith row and jth column. The **minor** M_{ij} associated with a_{ij} is the determinant of the $(n-1) \times (n-1)$ matrix obtained by deleting the ith row and the jth column of the matrix. The **cofactor** C_{ij} associated with a_{ij} is a *signed minor,* specifically

$$C_{ij} = (-1)^{i+j}M_{ij}.$$

EXAMPLE 1 Cofactors

The cofactors of the entries in the 3×3 matrix

$$\begin{pmatrix} 2 & 4 & 7 \\ 1 & 2 & 3 \\ 1 & 5 & 3 \end{pmatrix} \tag{1}$$

are

$$C_{11} = (-1)^{1+1}M_{11} = \begin{vmatrix} 2 & 3 \\ 5 & 3 \end{vmatrix} = -9 \qquad C_{12} = (-1)^{1+2}M_{12} = (-1)\begin{vmatrix} 1 & 3 \\ 1 & 3 \end{vmatrix} = 0 \qquad C_{13} = (-1)^{1+3}M_{13} = \begin{vmatrix} 1 & 2 \\ 1 & 5 \end{vmatrix} = 3$$

$$C_{21} = (-1)^{2+1}M_{21} = (-1)\begin{vmatrix} 4 & 7 \\ 5 & 3 \end{vmatrix} = 23 \qquad C_{22} = (-1)^{2+2}M_{22} = \begin{vmatrix} 2 & 7 \\ 1 & 3 \end{vmatrix} = -1 \qquad C_{23} = (-1)^{2+3}M_{23} = (-1)\begin{vmatrix} 2 & 4 \\ 1 & 5 \end{vmatrix} = -6$$

$$C_{31} = (-1)^{3+1}M_{31} = \begin{vmatrix} 4 & 7 \\ 2 & 3 \end{vmatrix} = -2 \qquad C_{32} = (-1)^{3+2}M_{32} = (-1)\begin{vmatrix} 2 & 7 \\ 1 & 3 \end{vmatrix} = 1 \qquad C_{33} = (-1)^{3+3}M_{33} = \begin{vmatrix} 2 & 4 \\ 1 & 2 \end{vmatrix} = 0.$$

Expansion by Cofactors It can be proved that a determinant can be **expanded** in terms of cofactors:

Multiply the entries a_{ij} in any row (or column) by their corresponding cofactors C_{ij} and add the n products.

Thus a 3×3 determinant* can be expanded into three 2×2 determinants, a 4×4 determinant can be expanded into four 3×3 determinants, and so on.

EXAMPLE 2 Determinant of a 3×3 Matrix

Evaluate the determinant of the matrix in (1).

Solution Expanding by the first row gives

$$\begin{vmatrix} 2 & 4 & 7 \\ 1 & 2 & 3 \\ 1 & 5 & 3 \end{vmatrix} = 2\begin{vmatrix} 2 & 3 \\ 5 & 3 \end{vmatrix} + 4(-1)\begin{vmatrix} 1 & 3 \\ 1 & 3 \end{vmatrix} + 7\begin{vmatrix} 1 & 2 \\ 1 & 5 \end{vmatrix} = 3.$$

Alternatively, we can expand the determinant by, say, the second column:

$$\begin{vmatrix} 2 & 4 & 7 \\ 1 & 2 & 3 \\ 1 & 5 & 3 \end{vmatrix} = 4(-1)\begin{vmatrix} 1 & 3 \\ 1 & 3 \end{vmatrix} + 2\begin{vmatrix} 2 & 7 \\ 1 & 3 \end{vmatrix} + 5(-1)\begin{vmatrix} 2 & 7 \\ 1 & 3 \end{vmatrix} = 3.$$

We note that if a determinant has a row (or a column) containing many zero entries, then wisdom dictates that we expand the determinant by that row (or column).

Cramer's Rule Determinants are sometimes useful in solving algebraic systems of n linear equations in n unknowns:

$$\begin{aligned} a_{11}x_1 + a_{12}x_2 + \cdots + a_{1n}x_n &= b_1 \\ a_{21}x_1 + a_{22}x_2 + \cdots + a_{2n}x_n &= b_2 \\ &\vdots \\ a_{n1}x_1 + a_{n2}x_2 + \cdots + a_{nn}x_n &= b_n. \end{aligned} \tag{2}$$

Let $\mathbf{A}$ be the matrix of coefficients of (2) and let

$$\det \mathbf{A} = \begin{vmatrix} a_{11} & a_{12} & \cdots & a_{1n} \\ a_{21} & a_{22} & \cdots & a_{2n} \\ \vdots & & & \vdots \\ a_{n1} & a_{n2} & \cdots & a_{nn} \end{vmatrix}$$

If

$$\det \mathbf{A}_k = \begin{vmatrix} a_{11} & a_{12} & \cdots & a_{1,k-1} & \overset{\overset{k\text{th column}}{\downarrow}}{b_1} & a_{1,k+1} & \cdots & a_{1n} \\ a_{21} & a_{22} & \cdots & a_{2,k-1} & b_2 & a_{2,k+1} & \cdots & a_{2n} \\ \vdots & & & \vdots & \vdots & \vdots & & \vdots \\ a_{n1} & a_{n2} & \cdots & a_{n,k-1} & b_n & a_{n,k+1} & \cdots & a_n \end{vmatrix}$$

is the same as $\det \mathbf{A}$ except that its kth column has been replaced by the column

$$\begin{matrix} b_1 \\ b_2 \\ \vdots \\ b_n, \end{matrix}$$

---

* Even though a determinant of a matrix of numbers is a number, it is sometimes convenient to refer to it as if it were an array.

then (2) has the unique solution

$$x_1 = \frac{\det \mathbf{A}_1}{\det \mathbf{A}}, \quad x_2 = \frac{\det \mathbf{A}_2}{\det \mathbf{A}}, \quad \ldots, \quad x_n = \frac{\det \mathbf{A}_n}{\det \mathbf{A}} \tag{3}$$

whenever $\det \mathbf{A} \neq 0$. This method of solving (2) by determinants is known as **Cramer's rule.***

EXAMPLE 3 Solution Using Cramer's Rule

Solve the system

$$\begin{aligned} 3x + 2y + z &= 7 \\ x - y + 3z &= 3 \\ 5x + 4y - 2z &= 1 \end{aligned}$$

by Cramer's rule.

Solution The solution requires the calculation of four determinants:

$$\det \mathbf{A} = \begin{vmatrix} 3 & 2 & 1 \\ 1 & -1 & 3 \\ 5 & 4 & -2 \end{vmatrix} = 13, \quad \det \mathbf{A}_1 = \begin{vmatrix} 7 & 2 & 1 \\ 3 & -1 & 3 \\ 1 & 4 & -2 \end{vmatrix} = -39$$

$$\det \mathbf{A}_2 = \begin{vmatrix} 3 & 7 & 1 \\ 1 & 3 & 3 \\ 5 & 1 & -2 \end{vmatrix} = 78, \quad \det \mathbf{A}_3 = \begin{vmatrix} 3 & 2 & 7 \\ 1 & -1 & 3 \\ 5 & 4 & 1 \end{vmatrix} = 52.$$

Hence (3) gives

$$x = \frac{\det \mathbf{A}_1}{\det \mathbf{A}} = -3, \quad y = \frac{\det \mathbf{A}_2}{\det \mathbf{A}} = 6, \quad z = \frac{\det \mathbf{A}_3}{\det \mathbf{A}} = 4.$$

Homogeneous Systems If $b_i = 0, i = 1, 2, \ldots, n$, then the system of equations (2) is said to be **homogeneous.** If at least one of the b_i is not zero, the system is **nonhomogeneous.** Now if $\det \mathbf{A} \neq 0$, (3) implies that the only solution of a homogeneous system is $x_1 = 0, x_2 = 0, \ldots, x_n = 0$. If $\det \mathbf{A} = 0$, then a homogeneous system of n linear equations in n unknowns has infinitely many solutions. These solutions can be found by solving the system through elimination. If $\det \mathbf{A} = 0$, then a nonhomogeneous system may have either infinitely many solutions or no solution at all.

* This rule was named after **Gabriel Cramer** (1704–1752), a Swiss mathematician who was the first to publish this result in 1750.

EXERCISES FOR APPENDIX III

Answers to odd-numbered problems begin on page A-29.

In Problems 1–8 evaluate the given determinant.

1. $\begin{vmatrix} 2 & 4 & 6 \\ -1 & 5 & 1 \\ 0 & 2 & -3 \end{vmatrix}$

2. $\begin{vmatrix} 1 & 4 & 2 \\ -2 & 6 & 3 \\ 9 & 8 & 4 \end{vmatrix}$

3. $\begin{vmatrix} 2 & 0 & 5 \\ 0 & 7 & 9 \\ -6 & 1 & 4 \end{vmatrix}$

4. $\begin{vmatrix} 79 & 81 & 40 \\ 22 & 16 & 59 \\ 0 & 0 & 0 \end{vmatrix}$

5. $\begin{vmatrix} 1 & 2 & 3 & 4 \\ 1 & 1 & 0 & 0 \\ 8 & 7 & 0 & 0 \\ 9 & 5 & 3 & 0 \end{vmatrix}$

6. $\begin{vmatrix} 1 & 0 & 9 & 0 & 3 \\ 2 & 1 & 7 & 0 & 0 \\ 0 & 0 & 2 & 0 & 0 \\ -1 & 1 & 5 & 2 & 2 \\ 2 & 2 & 8 & 1 & 1 \end{vmatrix}$

7. $\begin{vmatrix} e^t & e^{3t} & e^{-t} \\ e^t & 3e^{3t} & -e^{-t} \\ e^t & 9e^{3t} & e^{-t} \end{vmatrix}$

8. $\begin{vmatrix} e^{2t} & \sin t & \cos t \\ 2e^{2t} & \cos t & -\sin t \\ 4e^{2t} & -\sin t & -\cos t \end{vmatrix}$

In Problems 9–12 use Cramer's rule to solve the given system of equations.

9. $\begin{aligned} 2x + y &= 1 \\ 3x + 2y &= -2 \end{aligned}$

10. $\begin{aligned} 5x + 4y &= -1 \\ 10x - 6y &= 5 \end{aligned}$

11. $\begin{aligned} x + 2y + z &= 8 \\ 2x - 2y + 2z &= 7 \\ x - 4y + 3z &= 1 \end{aligned}$

12. $\begin{aligned} 4x + 3y + 2z &= 8 \\ -x \qquad + 2z &= 12 \\ 3x + 2y + z &= 3 \end{aligned}$

13. For the system

$$\begin{aligned} x - y + 2z &= 0 \\ 2x + y - z &= 0 \\ 4x - y + 3z &= 0, \end{aligned}$$

let **A** denote the matrix of coefficients.

(a) Show that det $\mathbf{A} = 0$.

(b) Show that the system has infinitely many solutions.

(c) Explain the geometric significance of the system.

14. For the system

$$\begin{aligned} a_1x + b_1y &= c_1 \\ a_2x + b_2y &= c_2, \end{aligned}$$

let **A** denote the matrix of coefficients.

(a) Explain the geometric significance of det $\mathbf{A} \neq 0$.

(b) Explain the geometric significance of det $\mathbf{A} = 0$.

APPENDIX IV

COMPLEX NUMBERS

A **complex number** is any expression of the form

$$z = a + bi, \quad \text{where} \quad i^2 = -1.$$

The real numbers a and b are called the **real** and **imaginary parts** of z, respectively. In practice, the symbol i is written $i = \sqrt{-1}$. The number $\bar{z} = a - bi$ is called the **conjugate** of z.

EXAMPLE 1 A Complex Number

From the properties of radicals, we have

$$\sqrt{-25} = \sqrt{25}\sqrt{-1} = 5i.$$ ■

EXAMPLE 2 Conjugates of Complex Numbers

The conjugates of the complex numbers $z_1 = 4 + 5i$ and $z_2 = 3 - 2i$ are, in turn, $\bar{z}_1 = 4 - 5i$ and $\bar{z}_2 = 3 + 2i$. ■

Sum, Difference, and Product The **sum, difference,** and **product** of two complex numbers $z_1 = a_1 + b_1 i$ and $z_2 = a_2 + b_2 i$ are defined as follows:

$$(i)\ z_1 + z_2 = (a_1 + a_2) + (b_1 + b_2)i$$
$$(ii)\ z_1 - z_2 = (a_1 - a_2) + (b_1 - b_2)i$$
$$(iii)\ z_1 z_2 = (a_1 a_2 - b_1 b_2) + (a_1 b_2 + b_1 a_2)i.$$

In other words, to add or subtract two complex numbers we simply add or subtract the corresponding real and imaginary parts. To multiply two complex numbers we use the distributive law and the fact that $i^2 = -1$.

EXAMPLE 3 Addition/Subtraction/Multiplication

If $z_1 = 4 + 5i$ and $z_2 = 3 - 2i$, then

$$z_1 + z_2 = (4 + 3) + (5 + (-2))i = 7 + 3i$$
$$z_1 - z_2 = (4 - 3) + (5 - (-2))i = 1 + 7i$$
$$z_1 z_2 = (4 + 5i)(3 - 2i)$$

$$= (4 + 5i)3 + (4 + 5i)(-2i)$$
$$= 12 + 15i - 8i - 10i^2$$
$$= (12 + 10) + (15 - 8)i = 22 + 7i.$$

The product of a complex number $z = a + bi$ and its conjugate $\bar{z} = a - bi$ is the real number.

$$z\bar{z} = a^2 + b^2. \tag{1}$$

Quotient The quotient of two complex numbers z_1 and z_2 is found by multiplying the numerator and denominator of z_1/z_2 by the conjugate of the denominator z_2 and using (1). The next example illustrates the procedure.

EXAMPLE 4 Quotient of Complex Numbers

$$\frac{z_1}{z_2} = \frac{4 + 5i}{3 - 2i}$$
$$= \frac{4 + 5i}{3 - 2i}\frac{3 + 2i}{3 + 2i}$$
$$= \frac{12 + 15i + 8i + 10i^2}{9 + 4} = \frac{2}{13} + \frac{23}{13}i$$

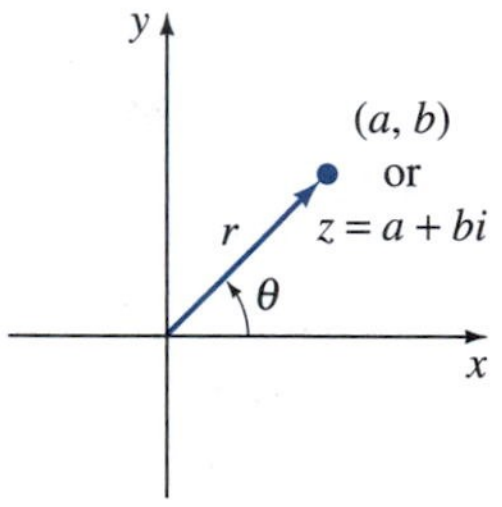

Figure A.2

Geometric Interpretation By taking a and b to be the x- and y-coordinates of a point in the plane, we can interpret a complex number $z = a + bi$ as a **vector** from the origin terminating at (a, b) (see Figure A.2). The length of the vector is called the **modulus** of z and is written as r, or $|z|$. From the Pythagorean theorem it follows that

$$r = |z| = \sqrt{a^2 + b^2}.$$

If θ is an angle the vector makes with the positive x-axis, then from Figure A.2 we see that

$$a = r\cos\theta \quad \text{and} \quad b = r\sin\theta.$$

Thus

$$z = r\cos\theta + ir\sin\theta \quad \text{or} \quad z = r(\cos\theta + i\sin\theta). \tag{2}$$

This latter form is called the **polar form** of the complex number z. The angle θ is called an **argument** of z.

Euler's Formula The power series $e^X = \sum_{n=0}^{\infty} \frac{X^n}{n!}$ is known to converge for all real and complex numbers. If we let $X = i\theta, \theta$ a real number, then

$$e^{i\theta} = \sum_{n=0}^{\infty} \frac{(i\theta)^n}{n!} = 1 + i\theta + \frac{i^2\theta^2}{2!} + \frac{i^3\theta^3}{3!} + \frac{i^4\theta^4}{4!} + \frac{i^5\theta^5}{5!} + \frac{i^6\theta^6}{6!} + \frac{i^7\theta^7}{7!} + \cdots. \tag{3}$$

Now $i^2 = -1$, $i^3 = -i$, $i^4 = 1$, $i^5 = i$, and so on. Thus (3) can be separated into real and imaginary parts:

$$e^{i\theta} = \left(1 - \frac{\theta^2}{2!} + \frac{\theta^4}{4!} - \frac{\theta^6}{6!} + \cdots\right) + i\left(\theta - \frac{\theta^3}{3!} + \frac{\theta^5}{5!} - \frac{\theta^7}{7!} + \cdots\right). \quad (4)$$

But from calculus we recall that

$$\cos\theta = \sum_{n=0}^{\infty} \frac{(-1)^n}{(2n)!}\theta^{2n} \quad \text{and} \quad \sin\theta = \sum_{n=0}^{\infty} \frac{(-1)^n}{(2n+1)!}\theta^{2n+1},$$

where each series converges for every real number θ. Hence (4) can be written as

$$e^{i\theta} = \cos\theta + i\sin\theta. \quad (5)$$

This last result is known as **Euler's formula.** Note that in view of (5), the polar form (2) of a complex number can be expressed in the compact form

$$z = re^{i\theta}. \quad (6)$$

EXAMPLE 5 Polar Form

Find the polar form (6) of $z = 1 - i$.

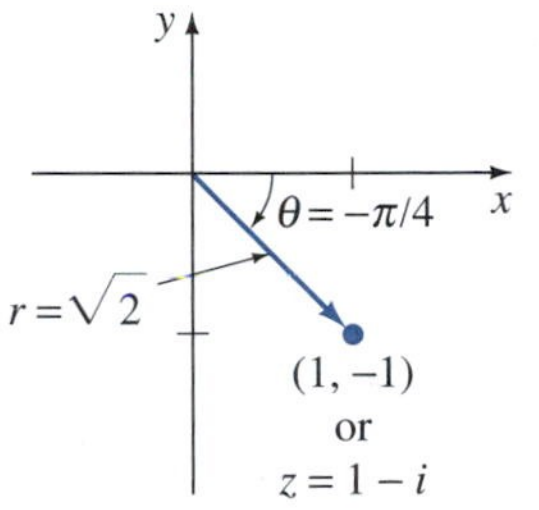

Figure A.3

Solution The graph of the complex number is given in Figure A.3. Since $a = 1$ and $b = -1$, the modulus of z is

$$r = \sqrt{1^2 + (-1)^2} = \sqrt{2}.$$

As seen in Figure A.3, $\tan\theta = -1$ and so we can take an argument of z to be $\theta = -\pi/4$. Therefore, the polar form of the number is

$$z = \sqrt{2}\, e^{-i\pi/4}.$$

EXERCISES FOR APPENDIX IV

Answers to odd-numbered problems begin on page A-29.

In Problems 1–10 let $z_1 = 2 - i$ and $z_2 = 5 + 3i$. Perform the indicated operation.

1. $z_1 + \bar{z}_2$
2. $4z_1 + z_2$
3. $2z_1 - 3z_2$
4. $z_1 z_2$
5. $(z_1)^2$
6. $\bar{z}_1(i + z_2)$
7. z_1/z_2
8. z_2/z_1
9. $1/z_2$
10. z_1/i

In Problems 11–20 write the given complex number in the polar form (6).

11. $z = i$
12. $z = -4i$
13. $z = i^2$
14. $z = 6i^5$

15. $z = 2 + 2i$

16. $z = -\sqrt{5} - \sqrt{5}i$

17. $z = 6 + 6\sqrt{3}i$

18. $z = -10\sqrt{3} + 10i$

19. $z = i(1 - \sqrt{3}i)$

20. $z = -7 + 7i$

In Problems 21 and 22 express the given complex number in the polar form in the form $z = a + bi$.

21. $z = 8e^{-i\pi}$

22. $z = 2e^{i7\pi/4}$

23. Prove **DeMoivre's theorem:*** For any positive integer n,

$$[r(\cos\theta + i\sin\theta)]^n = r^n[\cos n\theta + i\sin n\theta].$$

24. Use DeMoivre's theorem of Problem 23 to evaluate $(1 + i)^{10}$.

25. Use Euler's formula to show that

$$\cos\theta = \frac{e^{i\theta} + e^{-i\theta}}{2} \quad \text{and} \quad \sin\theta = \frac{e^{i\theta} - e^{-i\theta}}{2i}.$$

* This theorem was named after the French mathematician **Abraham DeMoivre** (1667–1754).

ANSWERS TO ODD-NUMBERED PROBLEMS

EXERCISES 1.1 (PAGE 9)

1. second-order linear **3.** first-order nonlinear
5. fourth-order linear **7.** second-order nonlinear
9. third-order linear **11.** $2y' + y = 2(-\frac{1}{2})e^{-x/2} + e^{-x/2} = 0$

13. $\dfrac{dy}{dx} - 2y - e^{3x}$
$= (3e^{3x} + 20e^{2x}) - 2(e^{3x} + 10e^{2x}) - e^{3x} = 0$

15. $y' - 25 - y^2 = 25\sec^2 5x - 25(1 + \tan^2 5x)$
$= 25\sec^2 5x - 25\sec^2 5x = 0$

17. $y' + y - \sin x = \frac{1}{2}\cos x + \frac{1}{2}\sin x - 10e^{-x} + \frac{1}{2}\sin x$
$- \frac{1}{2}\cos x + 10e^{-x} - \sin x = 0$

19. $yx^2 = -1$ implies $d(yx^2) = 0$
or $2yx\,dx + x^2\,dy = 0$

21. $y - 2xy' - y(y')^2 = y - 2x\dfrac{c_1}{2y} - y\dfrac{c_1^2}{4y^2}$
$= \dfrac{y^2 - (c_1x + (c_1^2/4))}{y}$
$= \dfrac{y^2 - y^2}{y} = 0$

23. $y' - \dfrac{1}{x}y - 1 = 1 + \ln x - \ln x - 1 = 0$

25. $\dfrac{d}{dt}\ln\dfrac{2 - X}{1 - X} = 1, \quad \left[\dfrac{-1}{2 - X} + \dfrac{1}{1 - X}\right]\dfrac{dX}{dt} = 1$
simplifies to $\dfrac{dX}{dt} = (2 - X)(1 - X)$

27. The differential of $c_1 = xe^{y/x}/(x + y)^2$ is
$\{(x + y)^2[xe^{y/x}(x\,dy - y\,dx/x^2) + e^{y/x}\,dx]$
$- xe^{y/x}2(x + y)(dx + dy)\}/(x + y)^4 = 0.$
Multiplying by $-x^2(x + y)^3e^{-y/x}$ and simplifying yields
$(x^2 + y^2)\,dx + (x^2 - xy)\,dy = 0.$

29. $y'' - 6y' + 13y = 5e^{3x}\cos 2x - 12e^{3x}\sin 2x$
$+ 12e^{3x}\sin 2x - 18e^{3x}\cos 2x$
$+ 13e^{3x}\cos 2x = 0$

31. $y'' = \cosh x + \sinh x = y$

33. $y'' + (y')^2 = -\dfrac{1}{(x + c_1)^2} + \dfrac{1}{(x + c_1)^2} = 0$

35. $x\dfrac{d^2y}{dx^2} + 2\dfrac{dy}{dx} = x(2c_2x^{-3}) + 2(-c_2x^{-2}) = 0$

37. $x^2y'' - 3xy' + 4y$
$= x^2(5 + 2\ln x) - 3x(3x + 2x\ln x)$
$+ 4(x^2 + x^2\ln x)$
$= 9x^2 - 9x^2 + 6x^2\ln x - 6x^2\ln x = 0$

39. $y''' - 3y'' + 3y' - y$
$= x^2e^x + 6xe^x + 6e^x - 3x^2e^x - 12xe^x$
$- 6e^x + 3x^2e^x + 6xe^x - x^2e^x = 0$

41. For $x < 0$, $xy' - 2y = x(-2x) - 2(-x^2) = 0$;
for $x \ge 0$, $xy' - 2y = x(2x) - 2(x^2) = 0$.

43. $y - xy' - (y')^2 = cx + c^2 - x(c) - c^2 = 0$;
$k = -\frac{1}{4}$

45. $y = -1$ **47.** $m = 2$ and $m = 3$

49. $m = \dfrac{1 \pm \sqrt{5}}{2}$

51. For $y = x^2$,
$x^2y'' - 4xy' + 6y = x^2(2) - 4x(2x) + 6x^2$
$= 8x^2 - 8x^2 = 0$;
for $y = x^3$,
$x^2y'' - 4xy' + 6y = x^2(6x) - 4x(3x^2) + 6x^3$
$= 12x^3 - 12x^3 = 0$;
yes; yes

53. **(a)** $y = 0$ **(b)** no real solution **(c)** $y = 1$ or $y = -1$

EXERCISES 1.2 (PAGE 22)

1. $\dfrac{dv}{dt} + \dfrac{k}{m}v = g$

3. **(a)** $k = gR^2$ **(b)** $\dfrac{d^2r}{dt^2} - \dfrac{gR^2}{r^2} = 0$

(c) $v\dfrac{dv}{dr} - \dfrac{gR^2}{r^2} = 0$

5. $L\dfrac{di}{dt} + Ri = E(t)$ **7.** $\dfrac{dh}{dt} = -\dfrac{\pi}{750}\sqrt{h}$

9. $\dfrac{dh}{dt} = -\dfrac{1}{30\sqrt{h}(10 - h)}$ **11.** $\dfrac{dx}{dt} + kx = r, \quad k > 0$

13. $mx'' = -k\cos\theta \qquad my'' = -mg - k\sin\theta$
$= -k\cdot\dfrac{1}{v}\dfrac{dx}{dt} \qquad = -mg - k\cdot\dfrac{1}{v}\dfrac{dy}{dt}$
$= -|c|\dfrac{dx}{dt} \qquad = -mg - |c|\dfrac{dy}{dt}$

15. Using $\tan\phi = \dfrac{x}{y}$, $\tan\left(\dfrac{\pi}{2} - \theta\right) = \dfrac{dy}{dx}$, $\tan\theta = \dfrac{dx}{dy}$,
and $\tan\phi = \tan 2\theta = \dfrac{2\tan\theta}{1 - \tan^2\theta}$,
we obtain $x\left(\dfrac{dx}{dy}\right)^2 + 2y\dfrac{dx}{dy} = x$.

17. By combining Newton's second law of motion with his law of gravitation, we obtain

$$m\frac{d^2y}{dt^2} = -k_1\frac{mM}{y^2},$$

where M is the mass of the earth and k_1 is a constant of proportionality. Dividing by m gives

$$\frac{d^2y}{dt^2} = -\frac{k}{y^2},$$

where $k = k_1M$. The constant k is gR^2, where R is the radius of the earth. This follows from the fact that on the surface of the earth $y = R$ so

$$k_1\frac{mM}{R^2} = mg$$

$$k_1M = gR^2 \quad \text{or} \quad k = gR^2.$$

If $t = 0$ is the time at which burnout occurs, then

$$y(0) = R + y_B,$$

where y_B is the distance from the earth's surface to the rocket at the time of burnout and

$$y'(0) = V_B$$

is the corresponding velocity at that time.

19. $\frac{dy}{dx} = -\frac{y}{\sqrt{s^2 - y^2}}$ 21. $\frac{dA}{dt} = k(M - A), \quad k > 0$

CHAPTER 1 REVIEW EXERCISES (PAGE 26)

1. ordinary, first-order, nonlinear 3. partial, second-order

9. $y = x^2$ 11. $y = \frac{x^2}{2}$ 13. $y = 0,\ y = e^x$

15. $y = 0,\ y = \cos x,\ y = \sin x$

17. $x < 0$ or $x > 1$ 19. $\frac{dh}{dt} = -\frac{25\sqrt{2g}}{16\pi}h^{-3/2}$

EXERCISES 2.1 (PAGE 32)

1. half-planes defined by either $y > 0$ or $y < 0$
3. half-planes defined by either $x > 0$ or $x < 0$
5. the regions defined by either $y > 2$, $y < -2$, or $-2 < y < 2$
7. any region not containing (0, 0) 9. the entire xy-plane
11. $y = 0,\ y = x^3$
13. There is some interval around $x = 0$ on which the unique solution is $y = 0$.
15. $y = 0, y = x$. No, the given function is nondifferentiable at $x = 0$.
17. yes 19. no

EXERCISES 2.2 (PAGE 38)

1. $y = -\frac{1}{5}\cos 5x + c$ 3. $y = \frac{1}{3}e^{-3x} + c$

5. $y = x + 5\ln|x + 1| + c$ 7. $y = cx^4$

9. $y^{-2} = 2x^{-1} + c$ 11. $-3 + 3x\ln|x| = xy^3 + cx$

13. $-3e^{-2y} = 2e^{3x} + c$ 15. $2 + y^2 = c(4 + x^2)$

17. $y^2 = x - \ln|x + 1| + c$

19. $\frac{x^3}{3}\ln x - \frac{1}{9}x^3 = \frac{y^2}{2} + 2y + \ln|y| + c$

21. $S = ce^{kr}$ 23. $\frac{P}{1 - P} = ce^t$ or $P = \frac{ce^t}{1 + ce^t}$

25. $4\cos y = 2x + \sin 2x + c$

27. $-2\cos x + e^y + ye^{-y} + e^{-y} = c$

29. $(e^x + 1)^{-2} + 2(e^y + 1)^{-1} = c$

31. $(y + 1)^{-1} + \ln|y + 1| = \frac{1}{2}\ln\left|\frac{x + 1}{x - 1}\right| + c$

33. $y - 5\ln|y + 3| = x - 5\ln|x + 4| + c$ or $\left(\frac{y + 3}{x + 4}\right)^5 = c_1e^{y-x}$

35. $-\cot y = \cos x + c$ 37. $y = \sin\left(\frac{x^2}{2} + c\right)$

39. $-y^{-1} = \tan^{-1}(e^x) + c$

41. $(1 + \cos x)(1 + e^y) = 4$

43. $\sqrt{y^2 + 1} = 2x^2 + \sqrt{2}$ 45. $x = \tan(4y - 3\pi/4)$

47. $xy = e^{-(1+1/x)}$

49. (a) $y = 3\frac{1 - e^{6x}}{1 + e^{6x}}$ (b) $y = 3$

(c) $y = 3\frac{2 - e^{6x-2}}{2 + e^{6x-2}}$

51. $y = 1$ 53. $y = 1$ 55. $y = 1 + \frac{1}{10}\tan\frac{x}{10}$

57. $y = -x - 1 + \tan(x + c)$

59. $2y - 2x + \sin 2(x + y) = c$

61. $4(y - 2x + 3) = (x + c)^2$

EXERCISES 2.3 (PAGE 45)

1. homogeneous of degree 3 3. homogeneous of degree 2
5. not homogeneous 7. homogeneous of degree 0
9. homogeneous of degree −2 11. $x\ln|x| + y = cx$
13. $(x - y)\ln|x - y| = y + c(x - y)$ 15. $x + y\ln|x| = cy$
17. $\ln(x^2 + y^2) + 2\tan^{-1}(y/x) = c$ 19. $4x = y(\ln|y| - c)^2$
21. $y^9 = c(x^3 + y^3)^2$ 23. $(y/x)^2 = 2\ln|x| + c$
25. $e^{2x/y} = 8\ln|y| + c$ 27. $x\cos(y/x) = c$
29. $y + x = cx^2e^{y/x}$ 31. $y^3 + 3x^3\ln|x| = 8x^3$
33. $y^2 = 4x(x + y)^2$ 35. $\ln|x| = e^{y/x} - 1$
37. $4x\ln|y/x| + x\ln x + y - x = 0$
39. $3x^{3/2}\ln x + 3x^{1/2}y + 2y^{3/2} = 5x^{3/2}$
41. $(x + y)\ln|y| + x = 0$
43. $\ln|y| = -2(1 - x/y)^{1/2} + \sqrt{2}$
45. By homogeneity the equation can be written as

$$M(x/y, 1)\,dx + N(x/y, 1)\,dy = 0.$$

With $v = x/y$, it follows that

$$M(v,1)(v\,dy + y\,dv) + N(v,1)\,dy = 0$$

$$[vM(v,1) + N(v,1)]\,dy + yM(v,1)\,dv = 0$$

or

$$\frac{dy}{y} + \frac{M(v,1)\,dv}{vM(v,1) + N(v,1)} = 0.$$

47. $\dfrac{dy}{dx} = -\dfrac{M(x,y)}{N(x,y)} = -\dfrac{y^n M(x/y,1)}{y^n N(x/y,1)}$
$= -\dfrac{M(x/y,1)}{N(x/y,1)} = G(x/y)$

EXERCISES 2.4 (PAGE 52)

1. $x^2 - x + \frac{3}{2}y^2 + 7y = c$ **3.** $\frac{5}{2}x^2 + 4xy - 2y^4 = c$
5. $x^2y^2 - 3x + 4y = c$ **7.** not exact, but is homogeneous
9. $xy^3 + y^2 \cos x - \frac{1}{2}x^2 = c$ **11.** not exact
13. $xy - 2xe^x + 2e^x - 2x^3 = c$
15. $x + y + xy - 3\ln|xy| = c$ **17.** $x^3y^3 - \tan^{-1}3x = c$
19. $-\ln|\cos x| + \cos x \sin y = c$
21. $y - 2x^2y - y^2 - x^4 = c$ **23.** $x^4y - 5x^3 - xy + y^3 = c$
25. $\frac{1}{3}x^3 + x^2y + xy^2 - y = \frac{4}{3}$
27. $4xy + x^2 - 5x + 3y^2 - y = 8$
29. $y^2 \sin x - x^3y - x^2 + y\ln y - y = 0$ **31.** $k = 10$
33. $k = 1$ **35.** $M(x,y) = ye^{xy} + y^2 - (y/x^2) + h(x)$
37. $M(x,y) = 6xy^3$
$N(x,y) = 4y^3 + 9x^2y^2$
$\partial M/\partial y = 18xy^2 = \partial N/\partial x$
Solution is $3x^2y^3 + y^4 = c$.
39. $M(x,y) = -x^2y^2 \sin x + 2xy^2 \cos x$
$N(x,y) = 2x^2y \cos x$
$\partial M/\partial y = -2x^2y \sin x + 4xy \cos x = \partial N/\partial x$
Solution is $x^2y^2 \cos x = c$.
41. $M(x,y) = 2xy^2 + 3x^2$
$N(x,y) = 2x^2y$
$\partial M/\partial y = 4xy = \partial N/\partial x$
Solution is $x^2y^2 + x^3 = c$.
43. A separable first-order differential equation can be written $h(y)\,dy - g(x)\,dx = 0$. Identifying $M(x,y) = -g(x)$ and $N(x,y) = h(y)$, we find that $\partial M/\partial y = 0 = \partial N/\partial x$.

EXERCISES 2.5 (PAGE 60)

1. $y = ce^{5x}, \quad -\infty < x < \infty$
3. $y = \frac{1}{3} + ce^{-4x}, \quad -\infty < x < \infty$
5. $y = \frac{1}{4}e^{3x} + ce^{-x}, \quad -\infty < x < \infty$
7. $y = \frac{1}{3} + ce^{-x^3}, \quad -\infty < x < \infty$
9. $y = x^{-1}\ln x + cx^{-1}, \quad 0 < x < \infty$
11. $x = -\frac{4}{5}y^2 + cy^{-1/2}, \quad 0 < y < \infty$
13. $y = -\cos x + \dfrac{\sin x}{x} + \dfrac{c}{x}, \quad 0 < x < \infty$
15. $y = \dfrac{c}{e^x + 1}, \quad -\infty < x < \infty$
17. $y = \sin x + c\cos x, \quad -\pi/2 < x < \pi/2$
19. $y = \frac{1}{7}x^3 - \frac{1}{5}x + cx^{-4}, \quad 0 < x < \infty$
21. $y = \dfrac{1}{2x^2}e^x + \dfrac{c}{x^2}e^{-x}, \quad 0 < x < \infty$
23. $y = \sec x + c\csc x, \quad 0 < x < \pi/2$
25. $x = \dfrac{1}{2}e^y - \dfrac{1}{2y}e^y + \dfrac{1}{4y^2}e^y + \dfrac{c}{y^2}e^{-y}, \quad 0 < y < \infty$
27. $y = e^{-3x} + \dfrac{c}{x}e^{-3x}, \quad 0 < x < \infty$
29. $x = 2y^6 + cy^4, \quad 0 < y < \infty$
31. $y = e^{-x}\ln(e^x + e^{-x}) + ce^{-x}, \quad -\infty < x < \infty$
33. $x = \dfrac{1}{y} + \dfrac{c}{y}e^{-y^2}, \quad 0 < y < \infty$
35. $(\sec\theta + \tan\theta)r = \theta - \cos\theta + c, \quad -\pi/2 < \theta < \pi/2$
37. $y = \frac{5}{3}(x+2)^{-1} + c(x+2)^{-4}, \quad -2 < x < \infty$
39. $y = 10 + ce^{-\sinh x}, \quad -\infty < x < \infty$
41. $y = 4 - 2e^{-5x}, \quad -\infty < x < \infty$
43. $i(t) = E/R + (i_0 - E/R)e^{-Rt/L}, \quad -\infty < t < \infty$
45. $y = \sin x \cos x - \cos x, \quad -\pi/2 < x < \pi/2$
47. $T(t) = 50 + 150e^{kt}, \quad -\infty < t < \infty$
49. $(x+1)y = x\ln x - x + 21, \quad 0 < x < \infty$
51. $y = \dfrac{2x}{x-2}, \quad 2 < x < \infty$
53. $x = \frac{1}{2}y + 8/y, \quad 0 < y < \infty$
55. $y = \begin{cases} \frac{1}{2}(1 - e^{-2x}), & 0 \le x \le 3 \\ \frac{1}{2}(e^6 - 1)e^{-2x}, & x > 3 \end{cases}$
57. $y = \begin{cases} \frac{1}{2} + \frac{3}{2}e^{-x^2}, & 0 \le x < 1 \\ (\frac{1}{2}e + \frac{3}{2})e^{-x^2}, & x \ge 1 \end{cases}$

EXERCISES 2.6 (PAGE 65)

1. $y^3 = 1 + cx^{-3}$ **3.** $y^{-3} = x + \frac{1}{3} + ce^{3x}$
5. $e^{x/y} = cx$ **7.** $y^{-3} = -\frac{9}{5}x^{-1} + \frac{49}{5}x^{-6}$
9. $x^{-1} = 2 - y^2 - e^{-y^2/2}$, the equation is Bernoulli in the variable x.
11. $y = 2 + \dfrac{1}{ce^{-3x} - 1/3}$ **13.** $y = \dfrac{2}{x} + \dfrac{1}{cx^{-3} - x/4}$
15. $y = -e^x + \dfrac{1}{ce^{-x} - 1}$ **17.** $y = -2 + \dfrac{1}{ce^{-x} - 1}$
19. $y = cx + 1 - \ln c$; $y = 2 + \ln x$
21. $y = cx - c^3$; $27y^2 = 4x^3$
23. $y = cx - e^c$; $y = x\ln x - x$

EXERCISES 2.7 (PAGE 69)

1. $x^2e^{2y} = 2x\ln x - 2x + c$ **3.** $e^{-x} = y\ln|y| + cy$
5. $-e^{-y/x^4} = x^2 + c$ **7.** $x^2 + y^2 = x - 1 + ce^{-x}$
9. $\ln(\tan y) = x + cx^{-1}$ **11.** $x^3y^3 = 2x^3 - 9\ln|x| + c$
13. $e^y = -e^{-x}\cos x + ce^{-x}$ **15.** $y^2\ln x = ye^y - e^y + c$
17. $y = \ln|\cos(c_1 - x)| + c_2$
19. $y = -\dfrac{1}{c_1}(1 - c_1^2x^2)^{1/2} + c_2$
21. The given equation is a Clairaut equation in $u = y'$. The solution is $y = c_1x^2/2 + x + c_1^3x + c_2$.
23. $y = c_1 + c_2x^2$ **25.** $\frac{1}{3}y^3 - c_1y = x + c_2$
27. $y = -\sqrt{1 - x^2}$

EXERCISES 2.8 (PAGE 72)

1. $y_1(x) = 1 - x$

$$y_2(x) = 1 - \frac{x}{1!} + \frac{x^2}{2!}$$

$$y_3(x) = 1 - \frac{x}{1!} + \frac{x^2}{2!} - \frac{x^3}{3!}$$

$$y_4(x) = 1 - \frac{x}{1!} + \frac{x^2}{2!} - \frac{x^3}{3!} + \frac{x^4}{4!}$$

$y_n(x) \to e^{-x}$ as $n \to \infty$

3. $y_1(x) = 1 + x^2$

$$y_2(x) = 1 + \frac{x^2}{1!} + \frac{x^4}{2!}$$

$$y_3(x) = 1 + \frac{x^2}{1!} + \frac{x^4}{2!} + \frac{x^6}{3!}$$

$$y_4(x) = 1 + \frac{x^2}{1!} + \frac{x^4}{2!} + \frac{x^6}{3!} + \frac{x^8}{4!}$$

$y_n(x) \to e^{x^2}$ as $n \to \infty$

5. $y_1(x) = y_2(x) = y_3(x) = y_4(x) = 0$

$y_n(x) \to 0$ as $n \to \infty$

7. **(a)** $y_1(x) = x$

$y_2(x) = x + \frac{1}{3}x^3$

$y_3(x) = x + \frac{1}{3}x^3 + \frac{2}{15}x^5 + \frac{1}{63}x^7$

(b) $y = \tan x$

(c) The Maclaurin series expansion of $\tan x$ is $x + \frac{1}{3}x^3 + \frac{2}{15}x^5 + \frac{17}{315}x^7 + \cdots,\ |x| < \pi/2$.

CHAPTER 2 REVIEW EXERCISES (PAGE 73)

1. the regions defined by $x^2 + y^2 > 25$ and $x^2 + y^2 < 25$

3. false

5. **(a)** linear in x **(b)** homogeneous, exact, linear in y **(c)** Clairaut **(d)** Bernoulli in x **(e)** separable **(f)** separable, Ricatti **(g)** linear in x **(h)** homogeneous **(i)** Bernoulli **(j)** homogeneous, exact, Bernoulli **(k)** separable, homogeneous, exact, linear in x and in y **(l)** exact, linear in y **(m)** homogeneous **(n)** separable **(o)** Clairaut **(p)** Ricatti

7. $2y^2 \ln y - y^2 = 4xe^x - 4e^x - 1$ **9.** $2y^2 + x^2 = 9x^6$

11. $e^{xy} - 4y^3 = 5$ **13.** $y = \frac{1}{4} - 320(x^2 + 4)^{-4}$

15. $y = \dfrac{1}{x^4 - x^4 \ln|x|}$ **17.** $x^2 - \sin\dfrac{1}{y^2} = c$

19. $y_1(x) = 1 + x + \frac{1}{3}x^3$

$y_2(x) = 1 + x + x^2 + \frac{2}{3}x^3 + \frac{1}{6}x^4 + \frac{2}{15}x^5 + \frac{1}{63}x^7$

EXERCISES 3.1 (PAGE 80)

1. $x^2 + y^2 = c_2^2$ **3.** $2y^2 + x^2 = c_2$

5. $2\ln|y| = x^2 + y^2 + c_2$ **7.** $y^2 = 2x + c_2$

9. $2x^2 + 3y^2 = c_2$ **11.** $x^3 + y^3 = c_2$

13. $y^2 \ln|y| + x^2 = c_2 y^2$ **15.** $y^2 - x^2 = c_2 x$

17. $2y^2 = 2\ln|x| + x^2 + c_2$ **19.** $y = \frac{1}{4} - \frac{1}{6}x^2 + c_2 x^{-4}$

21. $2y^3 = 3x^2 + c_2$ **23.** $2\ln(\cosh y) + x^2 = c_2$

25. $y^{5/3} = x^{5/3} + c_2$ **27.** $y = 2 - x + 3e^{-x}$

29. $r = c_2 \sin\theta$ **31.** $r^2 = c_2 \cos 2\theta$ **33.** $r = c_2 \csc\theta$

35. Let β be the angle of inclination, measured from the positive x-axis, of the tangent line to a member of the given family, and ϕ the angle of inclination of the tangent to a trajectory. At the point where the curves intersect, the angle between the tangents is α. From the accompanying figures we conclude that there exist two possible cases and that $\phi = \beta \pm \alpha$. Thus the slope of the tangent line to a trajectory is

$$\frac{dy}{dx} = \tan\phi = \tan(\beta \pm \alpha) = \frac{\tan\beta \pm \tan\alpha}{1 \mp \tan\beta\tan\alpha} = \frac{f(x, y) \pm \tan\alpha}{1 \mp f(x, y)\tan\alpha}.$$

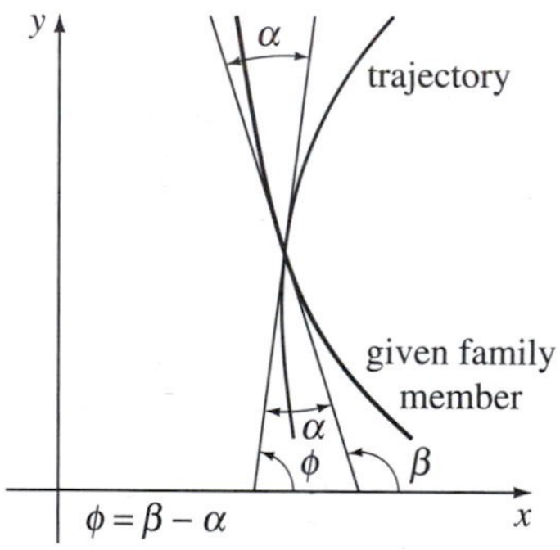

(a)

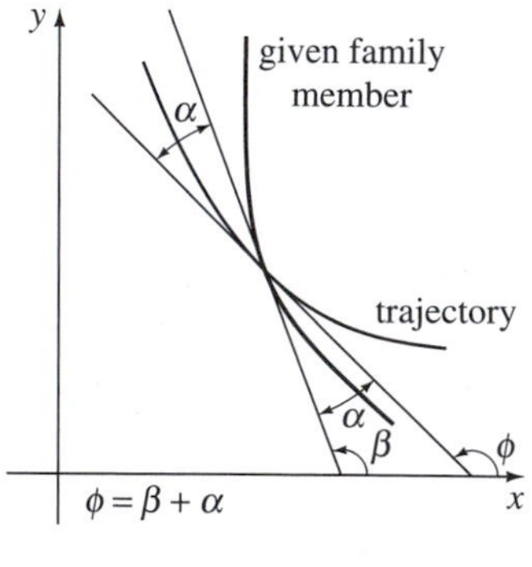

(b)

37. $\mp \dfrac{2}{\sqrt{3}} \tan^{-1}\left(\dfrac{y}{x}\right) + \ln c_2(x^2 + y^2) = 0$

39. Since the given equation is quadratic in c_1, it follows from the quadratic formula that

$$c_1 = -x \pm \sqrt{x^2 + y^2}.$$

Differentiating this last expression and solving for dy/dx gives

$$\frac{dy}{dx} = \frac{-x + \sqrt{x^2 + y^2}}{y}$$

and

$$\frac{dy}{dx} = \frac{-x - \sqrt{x^2 + y^2}}{y}.$$

These two equations correspond to choosing $c_1 > 0$ and $c_1 < 0$ in the given family, respectively. Forming the product of these derivatives yields

$$\left(\frac{dy}{dx}\right)_{(1)} \cdot \left(\frac{dy}{dx}\right)_{(2)} = \frac{x^2 - x^2 - y^2}{y^2} = -1.$$

This shows that the family is self-orthogonal.

41. The differential equation of the orthogonal family is $(x - y)\,dx + (x + y)\,dy = 0$. The verification follows by substituting $x = c_2e^{-t}\cos t$ and $y = c_2e^{-t}\sin t$ into the equation.

EXERCISES 3.2 (PAGE 89)

1. 7.9 yr; 10 yr **3.** 760 **5.** 11 h **7.** 136.5 h
9. $I(15) = 0.00098I_0$, or $I(15)$ is approximately 0.1% of I_0.
11. 15,600 yr
13. $T(1) = 36.67°$; approximately 3.06 min
15. $i(t) = \frac{3}{5} - \frac{3}{5}e^{-500t}$; $i \to \frac{3}{5}$ as $t \to \infty$
17. $q(t) = \frac{1}{100} - \frac{1}{100}e^{-50t}$; $i(t) = \frac{1}{2}e^{-50t}$
19. $i(t) = \begin{cases} 60 - 60e^{-t/10}, & 0 \le t \le 20 \\ 60(e^2 - 1)e^{-t/10}, & t > 20 \end{cases}$
21. $A(t) = 200 - 170e^{-t/50}$
23. $A(t) = 1000 - 1000e^{-t/100}$ **25.** 64.38 lb
27. **(a)** $v(t) = \dfrac{mg}{k} + \left(v_0 - \dfrac{mg}{k}\right)e^{-kt/m}$

(b) $v \to \dfrac{mg}{k}$ as $t \to \infty$

(c) $s(t) = \dfrac{mg}{k}t - \dfrac{m}{k}\left(v_0 - \dfrac{mg}{k}\right)e^{-kt/m} + \dfrac{m}{k}\left(v_0 - \dfrac{mg}{k}\right) + s_0$

29. $E(t) = E_0e^{-(t-t_1)/RC}$
31. **(a)** $P(t) = P_0e^{(k_1 - k_2)t}$

(b) $k_1 > k_2$, births surpass deaths so population increases. $k_1 = k_2$, a constant population since the number of births equals the number of deaths. $k_1 < k_2$, deaths surpass births so population decreases.

33. From $r^2\,d\theta = \dfrac{L}{M}\,dt$ we get

$$A = \frac{1}{2}\int_{\theta_1}^{\theta_2} r^2\,d\theta = \frac{1}{2}\frac{L}{M}\int_a^b dt = \frac{1}{2}\frac{L}{M}(b - a).$$

EXERCISES 3.3 (PAGE 99)

1. 1834; 2000 **3.** 1,000,000; 52.9 mo
5. **(a)** Separating variables gives

$$\frac{dP}{P(a - b\ln P)} = dt,$$

so

$$-(1/b)\ln|a - b\ln P| = t + c_1$$
$$a - b\ln P = c_2e^{-bt} \qquad (e^{-bc_1} = c_2)$$
$$\ln P = (a/b) - ce^{-bt} \qquad (c_2/b = c)$$
$$P(t) = e^{a/b} \cdot e^{-ce^{-bt}}.$$

(b) If $P(0) = P_0$, then

$$P_0 = e^{a/b}e^{-c} = e^{a/b - c}$$

and so

$$\ln P_0 = (a/b) - c$$
$$c = (a/b) - \ln P_0.$$

7. 29.3 g; $X \to 60$ as $t \to \infty$; 0 g of A and 30 g of B
9. For $\alpha \neq \beta$ the differential equation separates as

$$\frac{1}{\alpha - \beta}\left[-\frac{1}{\alpha - X} + \frac{1}{\beta - X}\right]dx = k\,dt.$$

It follows immediately that

$$\frac{1}{\alpha - \beta}[\ln|\alpha - X| - \ln|\beta - X|] = kt + c$$

or

$$\frac{1}{\alpha - \beta}\ln\left|\frac{\alpha - X}{\beta - X}\right| = kt + c.$$

For $\alpha = \beta$ the differential equation can be written as

$$(\alpha - X)^{-2}\,dX = k\,dt.$$

It follows that $(\alpha - X)^{-1} = kt + c$ or

$$X = \alpha - \frac{1}{kt + c}.$$

11. **(a)** $v^2 = (2gR^2/y) + v_0^2 - 2gR$. We note that as y increases, v decreases. In particular, if $v_0^2 - 2gR < 0$, then there must be some value of y for which $v = 0$; the rocket stops and returns to earth under the influence of gravity. However, if $v_0^2 - 2gR \ge 0$, then $v > 0$ for all values of y. Hence we should have $v_0 \ge \sqrt{2gR}$.

(b) With the values $R = 4000$ mi, $g = 32$ ft/s^2, 1 ft = 1/5280 mi, and 1 s = 1/3600 h, it follows that $v_0 \ge 25{,}067$ mi/h.

13. Using the condition $y'(1) = 0$, we find

$$\frac{dy}{dx} = \frac{1}{2}[x^{v_1/v_2} - x^{-v_1/v_2}].$$

Now if $v_1 = v_2$, then $y = \frac{1}{4}x^2 - \frac{1}{2}\ln x - \frac{1}{4}$; if $v_1 \neq v_2$, then

$$y = \frac{1}{2}\left[\frac{x^{1+(v_1/v_2)}}{1 + \frac{v_1}{v_2}} - \frac{x^{1-(v_1/v_2)}}{1 - \frac{v_1}{v_2}}\right] + \frac{v_1 v_2}{v_2^2 - v_1^2}.$$

15. $2h^{1/2} = -\frac{1}{25}t + 2\sqrt{20}$; $t = 50\sqrt{20}$ s

17. To evaluate the indefinite integral of the left side of

$$\frac{\sqrt{100 - y^2}}{y}\,dy = -dx$$

we use the substitution $y = 10\cos\theta$. It follows that

$$x = 10\ln\left(\frac{10 + \sqrt{100 - y^2}}{y}\right) - \sqrt{100 - y^2}.$$

19. Under the substitution $w = x^2$, the differential equation becomes

$$w = y\frac{dw}{dy} + \frac{1}{4}\left(\frac{dw}{dy}\right)^2,$$

which is Clairaut's equation. The solution is

$$x^2 = cy + \frac{c^2}{4}.$$

If $2c_1 = c$, then we recognize

$$x^2 = 2c_1 y + c_1^2$$

as describing a family of parabolas.

21. $-\gamma \ln y + \delta y = \alpha \ln x - \beta x + c$

23. **(a)** The equation $2\dfrac{d^2\theta}{dt^2}\dfrac{d\theta}{dt} + 2\dfrac{g}{l}\sin\theta\dfrac{d\theta}{dt} = 0$ is the same as

$$\frac{d}{dt}\left(\frac{d\theta}{dt}\right)^2 + 2\frac{g}{l}\sin\theta\frac{d\theta}{dt} = 0.$$

Integrating this last equation with respect to t and using the initial conditions gives the result.

(b) From (a),

$$dt = \sqrt{\frac{l}{2g}}\frac{d\theta}{\sqrt{\cos\theta - \cos\theta_0}}.$$

Integrating this last equation gives the time for the pendulum to move from $\theta = \theta_0$ to $\theta = 0$:

$$t = \sqrt{\frac{l}{2g}}\int_0^{\theta_0}\frac{d\theta}{\sqrt{\cos\theta - \cos\theta_0}}.$$

The period is the total time T to go from $\theta = \theta_0$ to $\theta = -\theta_0$ and back again to $\theta = \theta_0$. This is

$$T = 4\sqrt{\frac{l}{2g}}\int_0^{\theta_0}\frac{d\theta}{\sqrt{\cos\theta - \cos\theta_0}}$$

$$= 2\sqrt{\frac{2l}{g}}\int_0^{\theta_0}\frac{d\theta}{\sqrt{\cos\theta - \cos\theta_0}}.$$

CHAPTER 3 REVIEW EXERCISES (PAGE 104)

1. $y^3 + 3/x = c_2$ **3.** $2(y - 2)^2 + (x - 1)^2 = c_2^2$

5. $P(45) = 8.99$ billion

7. $x(t) = \dfrac{\alpha c_1 e^{\alpha k_1 t}}{1 + c_1 e^{\alpha k_1 t}}$, $y(t) = c_2(1 + c_1 e^{\alpha k_1 t})^{k_2/k_1}$

9. **(a)** $T(t) = \dfrac{T_2 + BT_1}{1 + B} + \dfrac{T_1 - T_2}{1 + B}e^{k(1+B)t}$

(b) $\dfrac{T_2 + BT_1}{1 + B}$ **(c)** $\dfrac{T_2 + BT_1}{1 + B}$

EXERCISES 4.1 (PAGE 129)

1. $y = \frac{1}{2}e^x - \frac{1}{2}e^{-x}$ **3.** $y = \frac{3}{5}e^{4x} + \frac{2}{5}e^{-x}$

5. $y = 3x - 4x\ln x$ **7.** $y = 0, y = x^2$

9. **(a)** $y = e^x\cos x - e^x\sin x$ **(b)** no solution

(c) $y = e^x\cos x + e^{-\pi/2}e^x\sin x$

(d) $y = c_2 e^x \sin x$, where c_2 is arbitrary

11. $(-\infty, 2)$ **13.** $\lambda = n, n = 1, 2, 3, \ldots$ **15.** dependent

17. dependent **19.** dependent **21.** independent

23. $W(x^{1/2}, x^2) = \frac{3}{2}x^{3/2} \neq 0$ on $(0, \infty)$

25. $W(\sin x, \csc x) = -2\cot x$.
$W = 0$ only at $x = \pi/2$ in the interval.

27. $W(e^x, e^{-x}, e^{4x}) = -30e^{4x} \neq 0$ on $(-\infty, \infty)$ **29.** no

31. **(a)** $y'' - 2y^3 = \dfrac{2}{x^3} - 2\left(\dfrac{1}{x}\right)^3 = 0$

(b) $y'' - 2y^3 = \dfrac{2c}{x^2} - 2\dfrac{c^3}{x^3} = \dfrac{2}{x^3}c(1 - c^2) \neq 0$ for $c \neq 0, \pm 1$

33. The functions satisfy the differential equation and are linearly independent on the interval since

$$W(e^{-3x}, e^{4x}) = 7e^x \neq 0; \quad y = c_1 e^{-3x} + c_2 e^{4x}.$$

35. The functions satisfy the differential equation and are linearly independent on the interval since

$$W(e^x\cos 2x, e^x\sin 2x) = 2e^{2x} \neq 0;$$
$$y = c_1 e^x\cos 2x + c_2 e^x\sin 2x.$$

37. The functions satisfy the differential equation and are linearly independent on the interval since

$$W(x^3, x^4) = x^6 \neq 0; \quad y = c_1x^3 + c_2x^4.$$

39. The functions satisfy the differential equation and are linearly independent on the interval since

$$W(x, x^{-2}, x^{-2}\ln x) = 9x^{-6} \neq 0;$$
$$y = c_1x + c_2x^{-2} + c_3x^{-2}\ln x.$$

41. e^{2x} and e^{5x} form a fundamental set of solutions of the homogeneous equation; $6e^x$ is a particular solution of the nonhomogeneous equation.

43. e^{2x} and xe^{2x} form a fundamental set of solutions of the homogeneous equation; $x^2e^{2x} + x - 2$ is a particular solution of the nonhomogeneous equation.

45. **(a)** The accompanying graphs show that y_1 and y_2 are not multiples of each other. Also,

$$x^2y_1'' - 4xy_1' + 6y_1 = x^2(6x) - 4x(3x^2) + 6x^3$$
$$= 12x^3 - 12x^3 = 0.$$

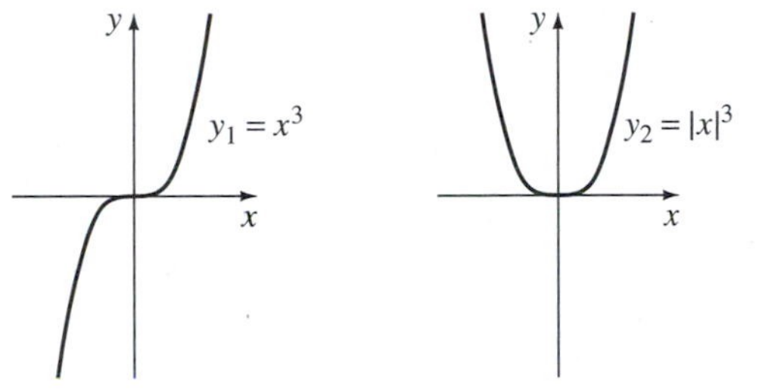

(a) **(b)**

For $x \geq 0$ the demonstration that y_2 is a solution of the equation is exactly as given above for y_1.

For $x < 0$, $y_2 = -x^3$ and so

$$x^2y_2'' - 4xy_2' + 6y_2 = x^2(-6x) - 4x(-3x^2) + 6(-x^3)$$
$$= -12x^3 + 12x^3 = 0.$$

(b) For $x \geq 0$,

$$W(y_1, y_2) = \begin{vmatrix} x^3 & x^3 \\ 3x^2 & 3x^2 \end{vmatrix} = 3x^5 - 3x^5 = 0.$$

For $x < 0$,

$$W(y_1, y_2) = \begin{vmatrix} x^3 & -x^3 \\ 3x^2 & -3x^2 \end{vmatrix} = -3x^5 + 3x^5 = 0.$$

Thus $W(y_1, y_2) = 0$ for every real value of x.

(c) No, $a_2(x) = x^2$ is zero at $x = 0$.

(d) Since $Y_1 = y_1$, we need only show that

$$x^2Y_2'' - 4xY_2' + 6Y_2 = x^2(2) - 4x(2x) + 6x^2$$
$$= 8x^2 - 8x^2 = 0$$

and $W(x^3, x^2) = -x^4$. Thus Y_1 and Y_2 are linearly independent solutions on the interval.

(e) $Y_1 = x^3$, $Y_2 = x^2$, or $y_2 = |x|^3$

(f) Neither; we form a general solution on an interval for which $a_2(x) \neq 0$ for every x in the interval. The linear combination

$$y = c_1Y_1 + c_2Y_2$$

would be a general solution of the equation on, say, the interval $(0, \infty)$.

47. **(a)** Since y_1 and y_2 are solutions of the given differential equation, we have

$$a_2(x)y_1'' + a_1(x)y_1' + a_0(x)y_1 = 0$$

and $$a_2(x)y_2'' + a_1(x)y_2' + a_0(x)y_2 = 0.$$

We multiply the first equation by y_2 and the second by y_1 and subtract the first from the second:

$$a_2(x)[y_1y_2'' - y_2y_1''] + a_1(x)[y_1y_2' - y_2y_1'] = 0.$$

Now it is easily verified that

$$\frac{dW}{dx} = \frac{d}{dx}(y_1y_2' - y_2y_1') = y_1y_2'' - y_2y_1'',$$

and so it follows that

$$a_2(x)\frac{dW}{dx} + a_1(x)W = 0.$$

(b) Since this last equation is a linear first-order differential equation, the integrating factor is

$$e^{\int[a_1(x)/a_2(x)]dx}.$$

Therefore, from

$$\frac{d}{dx}[e^{\int[a_1(x)/a_2(x)]dx}W] = 0$$

we obtain $W = ce^{-\int[a_1(x)/a_2(x)]dx}$.

(c) Substituting $x = x_0$ in the given result, we find $c = W(x_0)$.

(d) Since an exponential function is never zero when $W(x_0) \neq 0$, it follows from part (c) that $W \neq 0$. On the other hand, if $W(x_0) = 0$, we have immediately that $W = 0$.

49. From part (c) of Problem 47 we have

$$W(y_1, y_2) = W(y_1(x_0), y_2(x_0))e^{-\int_{x_0}^{x} dt/t}$$
$$= \begin{vmatrix} k_1 & k_3 \\ k_2 & k_4 \end{vmatrix} e^{-\ln(x/x_0)}$$
$$= (k_1k_4 - k_3k_2)\left(\frac{x_0}{x}\right).$$

EXERCISES 4.2 (PAGE 137)

1. $y_2 = e^{-5x}$ **3.** $y_2 = xe^{2x}$ **5.** $y_2 = \sin 4x$
7. $y_2 = \sinh x$ **9.** $y_2 = xe^{2x/3}$ **11.** $y_2 = x^4\ln|x|$

13. $y_2 = 1$ **15.** $y_2 = x^2 + x + 2$ **17.** $y_2 = x\cos(\ln x)$
19. $y_2 = x$ **21.** $y_2 = x\ln x$ **23.** $y_2 = x^3$ **25.** $y_2 = x^2$
27. $y_2 = 3x + 2$
29. $y_2 = \frac{1}{2}[\tan x \sec x + \ln|\sec x + \tan x|]$
31. $y_2 = e^{2x}, y_p = -\frac{1}{2}$ **33.** $y_2 = e^{2x}, y_p = \frac{5}{2}e^{3x}$

EXERCISES 4.3 (PAGE 144)

1. $y = c_1 + c_2e^{-x/4}$ **3.** $y = c_1e^{-6x} + c_2e^{6x}$
5. $y = c_1\cos 3x + c_2\sin 3x$ **7.** $y = c_1e^{3x} + c_2e^{-2x}$
9. $y = c_1e^{-4x} + c_2xe^{-4x}$
11. $y = c_1e^{(-3+\sqrt{29})x/2} + c_2e^{(-3-\sqrt{29})x/2}$
13. $y = c_1e^{2x/3} + c_2e^{-x/4}$ **15.** $y = e^{2x}(c_1\cos x + c_2\sin x)$
17. $y = e^{-x/3}\left(c_1\cos\frac{\sqrt{2}}{3}x + c_2\sin\frac{\sqrt{2}}{3}x\right)$
19. $y = c_1 + c_2e^{-x} + c_3e^{5x}$
21. $y = c_1e^{x} + e^{-x/2}\left(c_2\cos\frac{\sqrt{3}}{2}x + c_3\sin\frac{\sqrt{3}}{2}x\right)$
23. $y = c_1e^{-x} + c_2e^{3x} + c_3xe^{3x}$
25. $y = c_1e^{x} + e^{-x}(c_2\cos x + c_3\sin x)$
27. $y = c_1e^{-x} + c_2xe^{-x} + c_3x^2e^{-x}$
29. $y = c_1 + c_2x + e^{-x/2}\left(c_3\cos\frac{\sqrt{3}}{2}x + c_4\sin\frac{\sqrt{3}}{2}x\right)$
31. $y = c_1\cos\frac{\sqrt{3}}{2}x + c_2\sin\frac{\sqrt{3}}{2}x$
$\quad + c_3x\cos\frac{\sqrt{3}}{2}x + c_4x\sin\frac{\sqrt{3}}{2}x$
33. $y = c_1 + c_2e^{-2x} + c_3e^{2x} + c_4\cos 2x + c_5\sin 2x$
35. $y = c_1e^{x} + c_2xe^{x} + c_3e^{-x} + c_4xe^{-x} + c_5e^{-5x}$
37. $y = 2\cos 4x - \frac{1}{2}\sin 4x$ **39.** $y = -\frac{3}{4}e^{-5x} + \frac{3}{4}e^{-x}$
41. $y = -e^{x/2}\cos(x/2) + e^{x/2}\sin(x/2)$
43. $y = 0$ **45.** $y = e^{2(x-1)} - e^{x-1}$
47. $y = \frac{5}{36} - \frac{5}{36}e^{-6x} + \frac{1}{6}xe^{-6x}$
49. $y = -\frac{1}{6}e^{2x} + \frac{1}{6}e^{-x}\cos\sqrt{3}x - \frac{\sqrt{3}}{6}e^{-x}\sin\sqrt{3}x$
51. $y = 2 - 2e^{x} + 2xe^{x} - \frac{1}{2}x^2e^{x}$
53. $y = e^{5x} - xe^{5x}$ **55.** $y = -2\cos x$
57. $\frac{d^3y}{dx^3} + 6\frac{d^2y}{dx^2} - 15\frac{dy}{dx} - 100y = 0$
59. $y = c_1e^{x} + e^{4x}(c_2\cos x + c_3\sin x)$
61. $y'' - 3y' - 18y = 0$ **63.** $y''' - 7y'' = 0$
65. $y = e^{-\sqrt{2}x/2}\left(c_1\cos\frac{\sqrt{2}}{2}x + c_2\sin\frac{\sqrt{2}}{2}x\right)$
$\quad + e^{\sqrt{2}x/2}\left(c_3\cos\frac{\sqrt{2}}{2}x + c_4\sin\frac{\sqrt{2}}{2}x\right)$

EXERCISES 4.4 (PAGE 155)

1. $y = c_1e^{-x} + c_2e^{-2x} + 3$
3. $y = c_1e^{5x} + c_2xe^{5x} + \frac{6}{5}x + \frac{3}{5}$
5. $y = c_1e^{-2x} + c_2xe^{-2x} + x^2 - 4x + \frac{7}{2}$
7. $y = c_1\cos\sqrt{3}x + c_2\sin\sqrt{3}x + (-4x^2 + 4x - \frac{4}{3})e^{3x}$
9. $y = c_1 + c_2e^{x} + 3x$
11. $y = c_1e^{x/2} + c_2xe^{x/2} + 12 + \frac{1}{2}x^2e^{x/2}$
13. $y = c_1\cos 2x + c_2\sin 2x - \frac{3}{4}x\cos 2x$
15. $y = c_1\cos x + c_2\sin x - \frac{1}{2}x^2\cos x + \frac{1}{2}x\sin x$
17. $y = c_1e^{x}\cos 2x + c_2e^{x}\sin 2x + \frac{1}{4}xe^{x}\sin 2x$
19. $y = c_1e^{-x} + c_2xe^{-x} - \frac{1}{2}\cos x + \frac{12}{25}\sin 2x - \frac{9}{25}\cos 2x$
21. $y = c_1 + c_2x + c_3e^{6x} - \frac{1}{4}x^2 - \frac{6}{37}\cos x + \frac{1}{37}\sin x$
23. $y = c_1e^{x} + c_2xe^{x} + c_3x^2e^{x} - x - 3 - \frac{2}{3}x^3e^{x}$
25. $y = c_1\cos x + c_2\sin x + c_3x\cos x$
$\quad + c_4x\sin x + x^2 - 2x - 3$
27. $y_p = 4 + \frac{4}{3}\cos 2x$ **29.** $y = \sqrt{2}\sin 2x - \frac{1}{2}$
31. $y = -200 + 200e^{-x/5} - 3x^2 + 30x$
33. $y = -10e^{-2x}\cos x + 9e^{-2x}\sin x + 7e^{-4x}$
35. $x = \frac{F_0}{2\omega^2}\sin\omega t - \frac{F_0}{2\omega}t\cos\omega t$
37. $y = -\frac{1}{6}\cos x - \frac{\pi}{4}\sin x + \frac{1}{2}x\sin x + \frac{1}{3}\sin 2x$
39. $y = 11 - 11e^{x} + 9xe^{x} + 2x - 12x^2e^{x} + \frac{1}{2}e^{5x}$
41. $y = 6\cos x - 6(\cot 1)\sin x + x^2 - 1$
43. $y = \begin{cases} \cos 2x + \frac{5}{6}\sin 2x + \frac{1}{3}\sin x, & 0 \le x \le \pi/2 \\ \frac{2}{3}\cos 2x + \frac{5}{6}\sin 2x, & x > \pi/2 \end{cases}$

EXERCISES 4.5 (PAGE 161)

1. $(D + 5)y = 9\sin x$ **3.** $(3D^2 - 5D + 1)y = e^{x}$
5. $(D^3 - 4D^2 + 5D)y = 4x$ **7.** $(3D - 2)(3D + 2)$
9. $(D - 6)(D + 2)$ **11.** $D(D + 5)^2$
13. $(D - 1)(D - 2)(D + 5)$ **15.** $D(D + 2)(D^2 - 2D + 4)$
21. D^4 **23.** $D(D - 2)$ **25.** $D^2 + 4$ **27.** $D^3(D^2 + 16)$
29. $(D + 1)(D - 1)^3$ **31.** $D(D^2 - 2D + 5)$
33. $1, x, x^2, x^3, x^4$ **35.** $e^{6x}, e^{-3x/2}$ **37.** $\cos\sqrt{5}x, \sin\sqrt{5}x$
39. $1, e^{5x}, xe^{5x}$

EXERCISES 4.6 (PAGE 167)

1. $y = c_1e^{-3x} + c_2e^{3x} - 6$ **3.** $y = c_1 + c_2e^{-x} + 3x$

5. $y = c_1e^{-2x} + c_2xe^{-2x} + \frac{1}{2}x + 1$

7. $y = c_1 + c_2x + c_3e^{-x} + \frac{2}{3}x^4 - \frac{8}{3}x^3 + 8x^2$

9. $y = c_1e^{-3x} + c_2e^{4x} + \frac{1}{7}xe^{4x}$

11. $y = c_1e^{-x} + c_2e^{3x} - e^x + 3$

13. $y = c_1 \cos 5x + c_2 \sin 5x + \frac{1}{4} \sin x$

15. $y = c_1e^{-3x} + c_2xe^{-3x} - \frac{1}{49}xe^{4x} + \frac{2}{343}e^{4x}$

17. $y = c_1e^{-x} + c_2e^x + \frac{1}{6}x^3e^x - \frac{1}{4}x^2e^x + \frac{1}{4}xe^x - 5$

19. $y = e^x(c_1 \cos 2x + c_2 \sin 2x) + \frac{1}{3}e^x \sin x$

21. $y = c_1 \cos 5x + c_2 \sin 5x - 2x \cos 5x$

23. $y = e^{-x/2}\left(c_1 \cos \frac{\sqrt{3}}{2}x + c_2 \sin \frac{\sqrt{3}}{2}x\right)$
$+ \sin x + 2 \cos x - x \cos x$

25. $y = c_1 + c_2x + c_3e^{-8x} + \frac{11}{256}x^2 + \frac{7}{32}x^3 - \frac{1}{16}x^4$

27. $y = c_1e^x + c_2xe^x + c_3x^2e^x + \frac{1}{6}x^3e^x + x - 13$

29. $y = c_1 + c_2x + c_3e^x + c_4xe^x + \frac{1}{2}x^2e^x + \frac{1}{2}x^2$

31. $y = c_1e^{x/2} + c_2e^{-x/2} + c_3 \cos \frac{x}{2} + c_4 \sin \frac{x}{2} + \frac{1}{8}xe^{x/2}$

33. $y = \frac{5}{8}e^{-8x} + \frac{5}{8}e^{8x} - \frac{1}{4}$

35. $y = -\frac{41}{125} + \frac{41}{125}e^{5x} - \frac{1}{10}x^2 + \frac{9}{25}x$

37. $y = -\pi \cos x - \frac{11}{3} \sin x - \frac{8}{3} \cos 2x + 2x \cos x$

39. $y = 2e^{2x} \cos 2x - \frac{3}{64}e^{2x} \sin 2x + \frac{1}{8}x^3 + \frac{3}{16}x^2 + \frac{3}{32}x$

41. $y_p = Axe^x + Be^x \cos 2x + Ce^x \sin 2x$
$+ Exe^x \cos 2x + Fxe^x \sin 2x$

EXERCISES 4.7 (PAGE 174)

1. $y = c_1 \cos x + c_2 \sin x + x \sin x + \cos x \ln|\cos x|$; $(-\pi/2, \pi/2)$

3. $y = c_1 \cos x + c_2 \sin x + \frac{1}{2} \sin x - \frac{1}{2}x \cos x$
$= c_1 \cos x + c_3 \sin x - \frac{1}{2}x \cos x$; $(-\infty, \infty)$

5. $y = c_1 \cos x + c_2 \sin x + \frac{1}{2} - \frac{1}{6} \cos 2x$; $(-\infty, \infty)$

7. $y = c_1e^x + c_2e^{-x} + \frac{1}{4}xe^x - \frac{1}{4}xe^{-x}$
$= c_1e^x + c_2e^{-x} + \frac{1}{2}x \sinh x$; $(-\infty, \infty)$

9. $y = c_1e^{2x} + c_2e^{-2x}$
$+ \frac{1}{4}\left(e^{2x} \ln|x| - e^{-2x}\int_{x_0}^{x} \frac{e^{4t}}{t}\,dt\right)$,
$x_0 > 0$; $(0, \infty)$

11. $y = c_1e^{-x} + c_2e^{-2x} + (e^{-x} + e^{-2x}) \ln(1 + e^x)$; $(-\infty, \infty)$

13. $y = c_1e^{-2x} + c_2e^{-x} - e^{-2x} \sin e^x$; $(-\infty, \infty)$

15. $y = c_1e^x + c_2xe^x - \frac{1}{2}e^x \ln(1 + x^2) + xe^x \tan^{-1}x$; $(-\infty, \infty)$

17. $y = c_1e^{-x} + c_2xe^{-x} + \frac{1}{2}x^2e^{-x} \ln x - \frac{3}{4}x^2e^{-x}$; $(0, \infty)$

19. $y = c_1e^x \cos 3x + c_2e^x \sin x$
$- \frac{1}{27}e^x \cos 3x \ln|\sec 3x + \tan 3x|$;
$(-\pi/6, \pi/6)$

21. $y = c_1 + c_2 \cos x + c_3 \sin x - \ln|\cos x|$
$- \sin x \ln|\sec x + \tan x|$; $(-\pi/2, \pi/2)$

23. $y = c_1e^x + c_2e^{2x} + c_3e^{-x} + \frac{1}{8}e^{3x}$; $(-\infty, \infty)$

25. $y = \frac{1}{4}e^{-x/2} + \frac{3}{4}e^{x/2} + \frac{1}{8}x^2e^{x/2} - \frac{1}{4}xe^{x/2}$

27. $y = \frac{4}{9}e^{-4x} + \frac{25}{36}e^{2x} - \frac{1}{4}e^{-2x} + \frac{1}{9}e^{-x}$

29. $y = c_1x + c_2x \ln x + \frac{2}{3}x(\ln x)^3$

31. $y = c_1x^{-1/2} \cos x + c_2x^{-1/2} \sin x + x^{-1/2}$

33. **(a)** $y_{p_1} = 4x^2 - 16x + 21$ **(b)** $y_{p_2} = xe^{-x} \ln x$
(c) $y_p = 4x^2 - 16x + 21 + xe^{-x} \ln x$

CHAPTER 4 REVIEW EXERCISES (PAGE 177)

1. $y = 0$

3. False; the functions $f_1(x) = 0$ and $f_2(x) = e^x$ are linearly dependent on $(-\infty, \infty)$ but f_2 is not a constant multiple of f_1.

5. $(-\infty, 0)$; $(0, \infty)$ 7. false 9. $y_p = A + Bxe^x$

11. $y_2 = \sin 2x$ 13. $y = c_1e^{(1+\sqrt{3})x} + c_2e^{(1-\sqrt{3})x}$

15. $y = c_1 + c_2e^{-5x} + c_3xe^{-5x}$

17. $y = c_1e^{-x/3} + e^{-3x/2}\left(c_2 \cos \frac{\sqrt{7}}{2}x + c_3 \sin \frac{\sqrt{7}}{2}x\right)$

19. $y = e^{3x/2}\left(c_1 \cos \frac{\sqrt{11}}{2}x + c_2 \sin \frac{\sqrt{11}}{2}x\right)$
$+ \frac{4}{5}x^3 + \frac{36}{25}x^2 + \frac{46}{125}x - \frac{222}{625}$

21. $y = c_1 + c_2e^{2x} + c_3e^{3x} + \frac{1}{5} \sin x - \frac{1}{5} \cos x + \frac{4}{3}x$

23. $y = e^{x-\pi} \cos x$

25. $y = e^x(c_1 \cos x + c_2 \sin x)$
$- e^x \cos x \ln|\sec x + \tan x|$

27. $y = \frac{2}{5}e^{x/2} - \frac{2}{5}e^{3x} + xe^{3x} - 4$

EXERCISES 5.1 (PAGE 189)

1. A weight of 4 lb ($\frac{1}{8}$ slug), attached to a spring, is released from a point 3 units above the equilibrium position with an initial upward velocity of 2 ft/s. The spring costant is 3 lb/ft.

3. $x(t) = 2\sqrt{2} \sin\left(5t - \frac{\pi}{4}\right)$

5. $x(t) = \sqrt{5} \sin(\sqrt{2}t + 3.6052)$

7. $x(t) = \frac{\sqrt{101}}{10} \sin(10t + 1.4711)$ 9. 8 lb 11. $\sqrt{2}\pi/8$

13. $x(t) = -\frac{1}{4} \cos 4\sqrt{6}t$

15. (a) $x(\pi/12) = -1/4$; $x(\pi/8) = -1/2$;
$x(\pi/6) = -1/4$; $x(\pi/4) = 1/2$;
$x(9\pi/32) = \sqrt{2}/4$

(b) 4 ft/s; downward

(c) $t = (2n + 1)\pi/16$, $n = 0, 1, 2, \dots$

17. (a) the 20-kg mass (b) the 20-kg mass; the 50-kg mass

(c) $t = n\pi$, $n = 0, 1, 2, \dots$; at the equilibrium position; the 50-kg mass is moving upward whereas the 20-kg mass is moving upward when n is even and downward when n is odd.

19. $x(t) = \frac{1}{2}\cos 2t + \frac{3}{4}\sin 2t$
$= \dfrac{\sqrt{13}}{4}\sin(2t + 0.5880)$

21. (a) $x(t) = -\frac{2}{3}\cos 10t + \frac{1}{2}\sin 10t$
$= \frac{5}{6}\sin(10t - 0.927)$

(b) 5/6 ft; $\pi/5$ (c) 15 cycles (d) 0.721 s
(e) $(2n + 1)\pi/20 + 0.0927$, $n = 0, 1, 2, \dots$
(f) $x(3) = -0.597$ ft
(g) $x'(3) = -5.814$ ft/s (h) $x''(3) = 59.702$ ft/s^2
(i) $\pm 8\frac{1}{3}$ ft/s
(j) $0.1451 + n\pi/5$; $0.3545 + n\pi/5$, $n = 0, 1, 2, \dots$
(k) $0.3545 + n\pi/5$, $n = 0, 1, 2, \dots$

23. 120 lb/ft; $x(t) = \dfrac{\sqrt{3}}{12}\sin 8\sqrt{3}t$

25. Using $x(t) = c_1\cos\omega t + c_2\sin\omega t$, $x(0) = x_0$, and $x'(0) = v_0$, we find $c_1 = x_0$ and $c_2 = v_0/\omega$. The result follows from $A = \sqrt{c_1^2 + c_2^2}$.

27. $x(t) = 2\sqrt{2}\cos\left(5t + \dfrac{5\pi}{4}\right)$

29. When $\omega t + \phi = (2m + 1)\pi/2$, $|x''| = A\omega^2$. But $T = 2\pi/\omega$ implies $\omega = 2\pi/T$ and $\omega^2 = 4\pi^2/T^2$. Therefore, the magnitude of the acceleration is $|x''| = 4\pi^2 A/T^2$.

EXERCISES 5.2 (PAGE 197)

1. A 2-lb weight is attached to a spring whose constant is 1 lb/ft. The system is damped with a resisting force numerically equal to 2 times the instantaneous velocity. The weight starts from the equilibrium position with an upward velocity of 1.5 ft/s.

3. (a) above (b) heading upward

5. (a) below (b) heading upward

7. $\frac{1}{4}$s; $\frac{1}{2}$s, $x(\frac{1}{2}) = e^{-2}$; that is, the weight is approximately 0.14 ft below the equilibrium position.

9. (a) $x(t) = \frac{4}{3}e^{-2t} - \frac{1}{3}e^{-8t}$ (b) $x(t) = -\frac{2}{3}e^{-2t} + \frac{5}{3}e^{-8t}$

11. (a) $x(t) = e^{-2t}[-\cos 4t - \frac{1}{2}\sin 4t]$

(b) $x(t) = \dfrac{\sqrt{5}}{2}e^{-2t}\sin(4t + 4.249)$ (c) $t = 1.249$ s

13. (a) $\beta > \frac{5}{2}$ (b) $\beta = \frac{5}{2}$ (c) $0 < \beta < \frac{5}{2}$

15. $x(t) = \frac{2}{7}e^{-7t}\sin 7t$ 17. $v_0 > 2$ ft/s

19. Suppose $\gamma = \sqrt{\omega^2 - \lambda^2}$. Then the derivative of $x(t) = Ae^{-\lambda t}\sin(\gamma t + \phi)$ is

$$x'(t) = Ae^{-\lambda t}[\gamma\cos(\gamma t + \phi) - \lambda\sin(\gamma t + \phi)].$$

So $x'(t) = 0$ implies

$$\tan(\gamma t + \phi) = \frac{\gamma}{\lambda},$$

from which it follows that

$$t = \frac{1}{\gamma}\left[\tan^{-1}\frac{\gamma}{\lambda} + k\pi - \phi\right].$$

The difference between the t values between two successive maxima (or minima) is then

$$t_{k+2} - t_k = (k + 2)(\pi/\gamma) - k(\pi/\gamma) = 2\pi/\gamma.$$

21. $$t^*_{k+1} - t^*_k = \frac{(2k + 3)\pi/2 - \phi}{\sqrt{\omega^2 - \lambda^2}} - \frac{(2k + 1)\pi/2 - \phi}{\sqrt{\omega^2 - \lambda^2}} = \frac{\pi}{\sqrt{\omega^2 - \lambda^2}}$$

23. Let the quasi-period $2\pi/\sqrt{\omega^2 - \lambda^2}$ be denoted by T_q. From equation (15) we find

$$\frac{x_n}{x_{n+2}} = \frac{x(t)}{x(t + T_q)} = \frac{e^{-\lambda t}\sin(\sqrt{\omega^2 - \lambda^2}\,t + \phi)}{e^{-\lambda(t + T_q)}\sin(\sqrt{\omega^2 - \lambda^2}\,(t + T_q) + \phi)} = e^{\lambda T_q}$$

since

$$\sin(\sqrt{\omega^2 - \lambda^2}\,t + \phi) = \sin(\sqrt{\omega^2 - \lambda^2}\,(t + T_q) + \phi).$$

Therefore

$$\ln\left(\frac{x_n}{x_{n+2}}\right) = \lambda T_q = \frac{2\pi\lambda}{\sqrt{\omega^2 - \lambda^2}}.$$

EXERCISES 5.3 (PAGE 206)

1. $x(t) = e^{-t/2}\left(-\dfrac{4}{3}\cos\dfrac{\sqrt{47}}{2}t - \dfrac{64}{3\sqrt{47}}\sin\dfrac{\sqrt{47}}{2}t\right) + \frac{10}{3}(\cos 3t + \sin 3t)$

3. $x(t) = \frac{1}{4}e^{-4t} + te^{-4t} - \frac{1}{4}\cos 4t$

5. $x(t) = -\frac{1}{2}\cos 4t + \frac{9}{4}\sin 4t + \frac{1}{2}e^{-2t}\cos 4t - 2e^{-2t}\sin 4t$

7. $m\dfrac{d^2x}{dt^2} = -k(x - h) - \beta\dfrac{dx}{dt}$ or
$\dfrac{d^2x}{dt^2} + 2\lambda\dfrac{dx}{dt} + \omega^2 x = \omega^2 h(t)$, where
$2\lambda = \beta/m$ and $\omega^2 = k/m$

9. (a) $x(t) = \frac{2}{3}\sin 4t - \frac{1}{3}\sin 8t$
(b) $t = n\pi/4$, $n = 0, 1, 2, \dots$

(c) $t = \pi/6 + n\pi/2, \quad n = 0, 1, 2, \ldots$ and $t = \pi/3 + n\pi/2, \quad n = 0, 1, 2, \ldots$

(d) $\sqrt{3}/2$ cm, $-\sqrt{3}/2$ cm

(e)

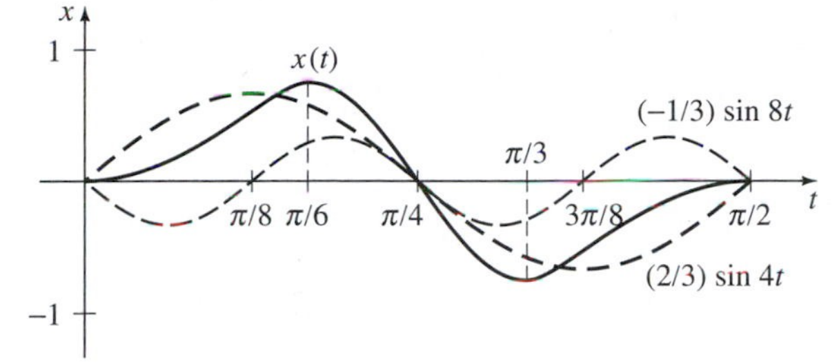

11. (a) $g'(\gamma) = 0$ implies $\gamma(\gamma^2 - \omega^2 + 2\lambda^2) = 0$, so either $\gamma = 0$ or $\gamma = \sqrt{\omega^2 - 2\lambda^2}$. The first derivative test can be used to verify that $g(\gamma)$ is a maximum at the latter value.

(b) $g(\sqrt{\omega^2 - 2\lambda^2}) = F_0/2\lambda\sqrt{\omega^2 - \lambda^2}$

13. $x_p = -5 \cos 2t + 5 \sin 2t$

$$= 5\sqrt{2} \sin\left(2t - \frac{\pi}{4}\right)$$

15. (a) $x(t) = x_c + x_p$

$$= c_1 \cos \omega t + c_2 \sin \omega t + \frac{F_0}{\omega^2 - \gamma^2} \cos \gamma t,$$

where the initial conditions imply that

$$c_1 = -F_0/(\omega^2 - \gamma^2) \text{ and } c_2 = 0.$$

(b) By L'Hôpital's rule the given limit is the same as

$$\lim_{\gamma \to \omega} \frac{F_0(-t \sin \gamma t)}{-2\gamma} = \frac{F_0}{2\omega} t \sin \omega t.$$

17. $x(t) = -\cos 2t - \frac{1}{8} \sin 2t + \frac{3}{4} t \sin 2t + \frac{5}{4} t \cos 2t$

19. (a) Recall that

$$\cos(u - v) = \cos u \cos v + \sin u \sin v$$
$$\cos(u + v) = \cos u \cos v - \sin u \sin v.$$

Subtracting gives

$$\sin u \sin v = \tfrac{1}{2}[\cos(u - v) - \cos(u + v)].$$

Setting $u = \frac{1}{2}(\gamma - \omega)t$ and $v = \frac{1}{2}(\gamma + \omega)t$ then gives

$$\sin \tfrac{1}{2}(\gamma - \omega)t \sin \tfrac{1}{2}(\gamma + \omega)t = \tfrac{1}{2}[\cos \omega t - \cos \gamma t],$$

from which the result follows.

(b) For small ε, $\gamma \approx \omega$ so $\gamma + \omega \approx 2\gamma$ and therefore

$$\frac{-2F_0}{(\omega + \gamma)(\omega - \gamma)} \sin \frac{1}{2}(\gamma - \omega)t \sin \frac{1}{2}(\gamma + \omega)t$$
$$\approx \frac{F_0}{2\gamma\varepsilon} \sin \varepsilon t \sin \frac{1}{2}(2\gamma)t.$$

(c) By L'Hôpital's rule the given limit is the same as

$$\lim_{\varepsilon \to 0} \frac{F_0 t \cos \varepsilon t \sin \gamma t}{2\gamma} = \frac{F_0}{2\gamma} t \sin \gamma t = \frac{F_0}{2\omega} t \sin \omega t.$$

EXERCISES 5.4 (PAGE 212)

1. $q(t) = -\frac{15}{4} \cos 4t + \frac{15}{4}$; $i(t) = 15 \sin 4t$

3. underdamped 5. 4.568 C; 0.0509 s

7. $q(t) = 10 - 10e^{-3t}(\cos 3t + \sin 3t)$; $i(t) = 60e^{-3t} \sin 3t$; 10.432 C

9. $q_p = \frac{100}{13} \sin t + \frac{150}{13} \cos t$; $i_p = \frac{100}{13} \cos t - \frac{150}{13} \sin t$

13. $q(t) = -\frac{1}{2}e^{-10t}(\cos 10t + \sin 10t) + \frac{3}{2}$; $\frac{3}{2}$ C

15. Show that $dZ/dC = 0$ when $C = 1/L\gamma^2$. At this value, Z is a minimum and, correspondingly, the amplitude E_0/Z is a maximum.

17. $$q(t) = \left(q_0 - \frac{E_0 C}{1 - \gamma^2 LC}\right) \cos \frac{t}{\sqrt{LC}} + \sqrt{LC}\, i_0 \sin \frac{t}{\sqrt{LC}} + \frac{E_0 C}{1 - \gamma^2 LC} \cos \gamma t;$$

$$i(t) = i_0 \cos \frac{t}{\sqrt{LC}} - \frac{1}{\sqrt{LC}}\left(q_0 - \frac{E_0 C}{1 - \gamma^2 LC}\right) \sin \frac{t}{\sqrt{LC}} - \frac{E_0 C\gamma}{1 - \gamma^2 LC} \sin \gamma t$$

19. $\theta(t) = \dfrac{1}{2} \cos 4t + \dfrac{\sqrt{3}}{2} \sin 4t$; 1; $\pi/2$; $2/\pi$

CHAPTER 5 REVIEW EXERCISES (PAGE 215)

1. 8 ft 3. $\frac{5}{4}$ m

5. False; there could be an impressed force driving the system. 7. overdamped 9. $\frac{9}{2}$ lb/ft

11. $x(t) = -\frac{2}{3}e^{-2t} + \frac{1}{3}e^{-4t}$ 13. $0 < m \le 2$

15. $\gamma = 8\sqrt{3}/3$

17. $x(t) = e^{-4t}\left(\dfrac{26}{17} \cos 2\sqrt{2}t + \dfrac{28\sqrt{2}}{17} \sin 2\sqrt{2}t\right) + \dfrac{8}{17}e^{-t}$

19. (a) $q(t) = -\frac{1}{150} \sin 100t + \frac{1}{75} \sin 50t$

(b) $i(t) = -\frac{2}{3} \cos 100t + \frac{2}{3} \cos 50t$

(c) $t = n\pi/50, \quad n = 0, 1, 2, \ldots$

EXERCISES 6.1 (PAGE 229)

1. $y = c_1 x^{-1} + c_2 x^2$ 3. $y = c_1 + c_2 \ln x$

5. $y = c_1 \cos(2 \ln x) + c_2 \sin(2 \ln x)$

7. $y = c_1 x^{(2-\sqrt{6})} + c_2 x^{(2+\sqrt{6})}$

9. $y_1 = c_1 \cos(\frac{1}{5}\ln x) + c_2 \sin(\frac{1}{5}\ln x)$

11. $y = c_1 x^{-2} + c_2 x^{-2}\ln x$

13. $y = x[c_1 \cos(\ln x) + c_2 \sin(\ln x)]$

15. $y = x^{-1/2}\left[c_1 \cos\left(\frac{\sqrt{3}}{6}\ln x\right) + c_2 \sin\left(\frac{\sqrt{3}}{6}\ln x\right)\right]$

17. $y = c_1 x^3 + c_2 \cos(\sqrt{2}\ln x) + c_3 \sin(\sqrt{2}\ln x)$

19. $y = c_1 x^{-1} + c_2 x^2 + c_3 x^4$

21. $y = c_1 + c_2 x + c_3 x^2 + c_4 x^{-3}$ **23.** $y = 2 - 2x^{-2}$

25. $y = \cos(\ln x) + 2\sin(\ln x)$

27. $y = 2(-x)^{1/2} - 5(-x)^{1/2}\ln(-x)$

29. $y = c_1 + c_2 \ln x + \dfrac{x^2}{4}$

31. $y = c_1 x^{-1/2} + c_2 x^{-1} + \frac{1}{15}x^2 - \frac{1}{6}x$

33. $y = c_1 x + c_2 x\ln x + x(\ln x)^2$

35. $y = c_1 x^{-1} + c_2 x^{-8} + \frac{1}{30}x^2$

37. $y = x^2[c_1 \cos(3\ln x) + c_2 \sin(3\ln x)] + \frac{4}{13} + \frac{3}{10}x$

39. $y = c_1 x^2 + c_2 x^{-10} - \frac{1}{7}x^{-3}$

41. $u(r) = \left(\dfrac{u_0 - u_1}{b - a}\right)\dfrac{ab}{r} + \dfrac{u_1 b - u_0 a}{b - a}$

43. $y = c_1(x-1)^{-1} + c_2(x-1)^4$

45. $y = c_1 \cos(\ln(x+2)) + c_2 \sin(\ln(x+2))$

EXERCISES 6.2 (PAGE 239)

1. $(-1, 1]$ **3.** $[-\frac{1}{2}, \frac{1}{2})$ **5.** $[2, 4]$ **7.** $(-5, 15)$ **9.** $\{0\}$

11. $x + x^2 + \frac{1}{3}x^3 - \frac{1}{30}x^5 + \cdots$

13. $x - \frac{2}{3}x^3 + \frac{2}{15}x^5 - \frac{4}{315}x^7 + \cdots$

15. $x^2 - \frac{2}{3}x^4 + \frac{23}{45}x^6 - \frac{44}{105}x^8 + \cdots$

17. $x + \frac{1}{3}x^3 + \frac{2}{15}x^5 + \frac{17}{315}x^7 + \cdots$

19. $1 + \frac{1}{2}x^2 - \frac{1}{12}x^4 + \frac{1}{24}x^6 - \cdots$

21. $y = ce^{-x}$; $\quad y = c_0 \displaystyle\sum_{n=0}^{\infty} \frac{(-1)^n}{n!}x^n$

23. $y = ce^{x^3/3}$; $\quad y = c_0 \displaystyle\sum_{n=0}^{\infty} \frac{1}{n!}\left(\frac{x^3}{3}\right)^n$

25. $y = c/(1-x)$; $\quad y = c_0 \displaystyle\sum_{n=0}^{\infty} x^n$

27. $y = C_1 \cos x + C_2 \sin x$;

$$y = c_0 \sum_{n=0}^{\infty} \frac{(-1)^n}{(2n)!}x^{2n} + c_1 \sum_{n=0}^{\infty} \frac{(-1)^n}{(2n+1)!}x^{2n+1}$$

29. $y = C_1 + C_2 e^x$;

$$y = c_0 + c_1 \sum_{n=0}^{\infty} \frac{x^n}{n!} = c_0 - c_1 + c_1 \sum_{n=0}^{\infty} \frac{x^n}{n!}$$

$$= c_0 - c_1 + c_1 e^x$$

EXERCISES 6.3 (PAGE 247)

1. $$y_1(x) = c_0\left[1 + \frac{1}{3\cdot 2}x^3 + \frac{1}{6\cdot 5\cdot 3\cdot 2}x^6 + \frac{1}{9\cdot 8\cdot 6\cdot 5\cdot 3\cdot 2}x^9 + \cdots\right]$$

$$y_2(x) = c_1\left[x + \frac{1}{4\cdot 3}x^4 + \frac{1}{7\cdot 6\cdot 4\cdot 3}x^7 + \frac{1}{10\cdot 9\cdot 7\cdot 6\cdot 4\cdot 3}x^{10} + \cdots\right]$$

3. $$y_1(x) = c_0\left[1 - \frac{1}{2!}x^2 - \frac{3}{4!}x^4 - \frac{21}{6!}x^6 - \cdots\right]$$

$$y_2(x) = c_1\left[x + \frac{1}{3!}x^3 + \frac{5}{5!}x^5 + \frac{45}{7!}x^7 + \cdots\right]$$

5. $$y_1(x) = c_0\left[1 - \frac{1}{3!}x^3 + \frac{4^2}{6!}x^6 - \frac{7^2\cdot 4^2}{9!}x^9 + \cdots\right]$$

$$y_2(x) = c_1\left[x - \frac{2^2}{4!}x^4 + \frac{5^2\cdot 2^2}{7!}x^7 - \frac{8^2\cdot 5^2\cdot 2^2}{10!}x^{10} + \cdots\right]$$

7. $y_1(x) = c_0$; $\quad y_2(x) = c_1 \displaystyle\sum_{n=1}^{\infty} \frac{1}{n}x^n$

9. $y_1(x) = c_0 \displaystyle\sum_{n=0}^{\infty} x^{2n}$; $\quad y_2(x) = c_1 \displaystyle\sum_{n=0}^{\infty} x^{2n+1}$

11. $$y_1(x) = c_0\left[1 + \frac{1}{4}x^2 - \frac{7}{4\cdot 4!}x^4 + \frac{23\cdot 7}{8\cdot 6!}x^6 - \cdots\right]$$

$$y_2(x) = c_1\left[x - \frac{1}{6}x^3 + \frac{14}{2\cdot 5!}x^5 - \frac{34\cdot 14}{4\cdot 7!}x^7 - \cdots\right]$$

13. $y_1(x) = c_0[1 + \frac{1}{2}x^2 + \frac{1}{6}x^3 + \frac{1}{6}x^4 + \cdots]$

$y_2(x) = c_1[x + \frac{1}{2}x^2 + \frac{1}{2}x^3 + \frac{1}{4}x^4 + \cdots]$

15. $$y(x) = -2\left[1 + \frac{1}{2!}x^2 + \frac{1}{3!}x^3 + \frac{1}{4!}x^4 + \cdots\right] + 6x$$

$$= 8x - 2e^x$$

17. $y(x) = 3 - 12x^2 + 4x^4$

19. $y_1(x) = c_0[1 - \frac{1}{6}x^3 + \frac{1}{120}x^5 + \cdots]$

$y_2(x) = c_1[x - \frac{1}{12}x^4 + \frac{1}{180}x^6 + \cdots]$

21. $y_1(x) = c_0[1 - \frac{1}{2}x^2 + \frac{1}{6}x^3 - \frac{1}{40}x^5 + \cdots]$

$y_2(x) = c_1[x - \frac{1}{6}x^3 + \frac{1}{12}x^4 - \frac{1}{60}x^5 + \cdots]$

23. $$y_1(x) = c_0\left[1 + \frac{1}{3!}x^3 + \frac{4}{6!}x^6 + \frac{7\cdot 4}{9!}x^9 + \cdots\right] + c_1\left[x + \frac{2}{4!}x^4 + \frac{5\cdot 2}{7!}x^7 + \frac{8\cdot 5\cdot 2}{10!}x^{10} + \cdots\right]$$

$$+ \frac{1}{2!}x^2 + \frac{3}{5!}x^5 + \frac{6\cdot 3}{8!}x^8 + \frac{9\cdot 6\cdot 3}{11!}x^{11} + \cdots$$

25. For $n = 1$: $y = x$; for $n = 2$: $y = 1 - 2x^2$.

EXERCISES 6.4 (PAGE 264)

1. $x = 0$, irregular singular point
3. $x = -3$, regular singular point; $x = 3$, irregular singular point
5. $x = 0, 2i, -2i$, regular singular points
7. $x = -3, 2$, regular singular points
9. $x = 0$, irregular singular point; $x = -5, 5, 2$, regular singular points

11. $r_1 = \frac{3}{2}, r_2 = 0$;
$$y(x) = C_1x^{3/2}\left[1 - \frac{2}{5}x + \frac{2^2}{7 \cdot 5 \cdot 2}x^2 - \frac{2^3}{9 \cdot 7 \cdot 5 \cdot 3!}x^3 + \cdots\right] + C_2\left[1 + 2x - 2x^2 + \frac{2^3}{3 \cdot 3!}x^3 - \cdots\right]$$

13. $r_1 = \frac{7}{8}, r_2 = 0$;
$$y(x) = C_1x^{7/8}\left[1 - \frac{2}{15}x + \frac{2^2}{23 \cdot 15 \cdot 2}x^2 - \frac{2^3}{31 \cdot 23 \cdot 15 \cdot 3!}x^3 + \cdots\right] + C_2\left[1 - 2x + \frac{2^2}{9 \cdot 2}x^2 - \frac{2^3}{17 \cdot 9 \cdot 3!}x^3 + \cdots\right]$$

15. $r_1 = \frac{1}{3}, r_2 = 0$;
$$y(x) = C_1x^{1/3}\left[1 + \frac{1}{3}x + \frac{1}{3^2 \cdot 2}x^2 + \frac{1}{3^3 \cdot 3!}x^3 + \cdots\right] + C_2\left[1 + \frac{1}{2}x + \frac{1}{5 \cdot 2}x^2 + \frac{1}{8 \cdot 5 \cdot 2}x^3 + \cdots\right]$$

17. $r_1 = \frac{5}{2}, r_2 = 0$;
$$y(x) = C_1x^{5/2}\left[1 + \frac{2 \cdot 2}{7}x + \frac{2^2 \cdot 3}{9 \cdot 7}x^2 + \frac{2^3 \cdot 4}{11 \cdot 9 \cdot 7}x^3 + \cdots\right] + C_2\left[1 + \frac{1}{3}x - \frac{1}{6}x^2 - \frac{1}{6}x^3 - \cdots\right]$$

19. $r_1 = \frac{2}{3}, r_2 = \frac{1}{3}$;
$$y(x) = C_1x^{2/3}[1 - \tfrac{1}{2}x + \tfrac{5}{28}x^2 - \tfrac{1}{21}x^3 + \cdots] + C_2x^{1/3}[1 - \tfrac{1}{2}x + \tfrac{1}{5}x^2 - \tfrac{7}{120}x^3 + \cdots]$$

21. $r_1 = 1, r_2 = -\frac{1}{2}$;
$$y(x) = C_1x\left[1 + \frac{1}{5}x + \frac{1}{5 \cdot 7}x^2 + \frac{1}{5 \cdot 7 \cdot 9}x^3 + \cdots\right] + C_2x^{-1/2}\left[1 + \frac{1}{2}x + \frac{1}{2 \cdot 4}x^2 + \frac{1}{2 \cdot 4 \cdot 6}x^3 + \cdots\right]$$

23. $r_1 = 0, r_2 = -1$;
$$y(x) = C_1x^{-1}\sum_{n=0}^{\infty}\frac{1}{(2n)!}x^{2n} + C_2x^{-1}\sum_{n=0}^{\infty}\frac{1}{(2n+1)!}x^{2n+1} = \frac{1}{x}[C_1\cosh x + C_2\sinh x]$$

25. $r_1 = 4, r_2 = 0$;
$$y(x) = C_1\left[1 + \frac{2}{3}x + \frac{1}{3}x^2\right] + C_2\sum_{n=0}^{\infty}(n+1)x^{n+4}$$

27. $r_1 = r_2 = 0$;
$$y(x) = C_1y_1(x) + C_2\left[y_1(x)\ln x + y_1(x)\times\left(-x + \frac{1}{4}x^2 - \frac{1}{3 \cdot 3!}x^3 + \frac{1}{4 \cdot 4!}x^4 - \cdots\right)\right],$$
where $y_1(x) = \sum_{n=0}^{\infty}\frac{1}{n!}x^n = e^x$

29. $r_1 = r_2 = 0$;
$$y(x) = C_1y_1(x) + C_2[y_1(x)\ln x + y_1(x) \times (2x + \tfrac{5}{4}x^2 + \tfrac{23}{27}x^3 + \cdots)],$$
where $y_1(x) = \sum_{n=0}^{\infty}\frac{(-1)^n}{(n!)^{2n}}x^n$

31. $r_1 = r_2 = 1$;
$$y(x) = C_1xe^{-x} + C_2xe^{-x} \times \left[\ln x + x + \frac{1}{4}x^2 + \frac{1}{3 \cdot 3!}x^3 + \cdots\right]$$

33. $r_1 = 2, r_2 = 0$;
$$y(x) = C_1x^2 + C_2\left[\frac{1}{2}x^2\ln x - \frac{1}{2} + x - \frac{1}{3!}x^3 + \cdots\right]$$

35. The method of Frobenius yields only the trivial solution $y(x) = 0$.

37. The assumption $y = \sum_{n=0}^{\infty}c_nx^{n+r}$ leads to
$$c_n[(n + r)(n + r + 2) - 8] = 0$$
for $n \geq 0$. For $n = 0$ and $c_0 \neq 0$ we have $r^2 + 2r - 8 = 0$ and so $r_1 = 2, r_2 = -4$. For these values we are forced to take $c_n = 0$ for $n > 0$. Hence a solution exists of the form $y = c_0x^r$. It follows that the general solution on $0 < x < \infty$ is $y = C_1x^2 + C_2x^{-4}$.

39. $r(r - 1) + \frac{5}{3}r - \frac{1}{3} = 0$; $r_1 = \frac{1}{3}, r_2 = -1$

EXERCISES 6.5 (PAGE 275)

1. $y = c_1J_{1/3}(x) + c_2J_{-1/3}(x)$ 3. $y = c_1J_{5/2}(x) + c_2J_{-5/2}(x)$
5. $y = c_1J_0(x) + c_2Y_0(x)$ 7. $y = c_1J_2(3x) + c_2Y_2(3x)$
9. After we use the change of variables, the differential equation becomes
$$x^2v'' + xv' + (\lambda^2x^2 - \tfrac{1}{4})v = 0.$$

Since the solution of the last equation is

$$v = c_1 J_{1/2}(\lambda x) + c_2 J_{-1/2}(\lambda x),$$

we find

$$y = c_1 x^{-1/2} J_{1/2}(\lambda x) + c_2 x^{-1/2} J_{-1/2}(\lambda x).$$

11. After substituting into the differential equation, we find

$$\begin{aligned} xy'' + (1 + 2n)y' + xy &\\ &= x^{-n-1}[x^2 J_n'' + xJ_n' + (x^2 - n^2)J_n] \\ &= x^{-n-1} \cdot 0 = 0. \end{aligned}$$

13. From Problem 10 with $n = \frac{1}{2}$ we find $y = x^{1/2}J_{1/2}(x)$; from Problem 11 with $n = -\frac{1}{2}$ we find $y = x^{1/2}J_{-1/2}(x)$.

15. From Problem 10 with $n = -1$ we find $y = x^{-1}J_{-1}(x)$; from Problem 11 with $n = 1$ we find $y = x^{-1}J_1(x)$, but since $J_{-1}(x) = -J_1(x)$, no new solution results.

17. From Problem 12 with $\lambda = 1$ and $\nu = \pm\frac{3}{2}$ we find $y = \sqrt{x}J_{3/2}(x)$ and $y = \sqrt{x}J_{-3/2}(x)$.

19. Using the hint, we can write

$$\begin{aligned} xJ_\nu'(x) &= -\nu \sum_{n=0}^{\infty} \frac{(-1)^n}{n!\Gamma(1+\nu+n)} \left(\frac{x}{2}\right)^{2n+\nu} \\ &\quad + 2\sum_{n=0}^{\infty} \frac{(-1)^n (n+\nu)}{n!(n+\nu)\Gamma(n+\nu)} \left(\frac{x}{2}\right)^{2n+\nu} \\ &= -\nu \sum_{n=0}^{\infty} \frac{(-1)^n}{n!\Gamma(1+\nu+n)} \left(\frac{x}{2}\right)^{2n+\nu} \\ &\quad + x\sum_{n=0}^{\infty} \frac{(-1)^n}{n!\Gamma(n+\nu)} \left(\frac{x}{2}\right)^{2n+\nu-1} \\ &= -\nu J_\nu(x) + xJ_{\nu-1}(x). \end{aligned}$$

21. Subtracting the equations

$$xJ_\nu'(x) = \nu J_\nu(x) - xJ_{\nu+1}(x)$$
$$xJ_\nu'(x) = -\nu J_\nu(x) + xJ_{\nu-1}(x)$$

gives $2\nu J_\nu(x) = xJ_{\nu+1}(x) + xJ_{\nu-1}(x)$.

23. From Problem 20, $\frac{d}{dr}[rJ_1(r)] = rJ_0(r)$. Therefore

$$\int_0^x rJ_0(r)dr = \int_0^x \frac{d}{dr}[rJ_1(r)]\,dr = rJ_1(r)\Big|_0^x = xJ_1(x).$$

25. Start with

$$x^n J_0 = x^{n-1} x J_0 = x^{n-1}\frac{d}{dx}(xJ_1)$$

and then integrate by parts.

27. $J_{-1/2}(x) = \sqrt{\frac{2}{\pi x}}\cos x$

29. $J_{-3/2}(x) = \sqrt{\frac{2}{\pi x}}\left[-\sin x - \frac{\cos x}{x}\right]$

31. $J_{-5/2}(x) = \sqrt{\frac{2}{\pi x}}\left[\frac{3}{x}\sin x + \left(\frac{3}{x^2} - 1\right)\cos x\right]$

33. $J_{-7/2}(x) = \sqrt{\frac{2}{\pi x}}\left[\left(1 - \frac{15}{x^2}\right)\sin x + \left(\frac{6}{x} - \frac{15}{x^3}\right)\cos x\right]$

35. $y = c_1 I_\nu(x) + c_2 I_{-\nu}(x), \quad \nu \neq$ integer

37. Since $1/\Gamma(1 - m + n) = 0$ when $n \le m - 1$, m a positive integer,

$$\begin{aligned} J_{-m}(x) &= \sum_{n=0}^{\infty} \frac{(-1)^n}{n!\Gamma(1-m+n)}\left(\frac{x}{2}\right)^{2n-m} \\ &= \sum_{n=m}^{\infty} \frac{(-1)^n}{n!\Gamma(1-m+n)}\left(\frac{x}{2}\right)^{2n-m} \\ &= \sum_{k=0}^{\infty} \frac{(-1)^{k+m}}{(k+m)!\Gamma(1+k)}\left(\frac{x}{2}\right)^{2k+m} \quad (n = k+m) \\ &= (-1)^m \sum_{k=0}^{\infty} \frac{(-1)^k}{\Gamma(1+k+m)k!}\left(\frac{x}{2}\right)^{2k+m} \\ &= (-1)^m J_m(x). \end{aligned}$$

39. **(a)** $P_6(x) = \frac{1}{16}(231x^6 - 315x^4 + 105x^2 - 5)$

$P_7(x) = \frac{1}{16}[429x^7 - 693x^5 + 315x^3 - 35x]$

(b) $y = P_6(x)$ satisfies $(1 - x^2)y'' - 2xy' + 42y = 0$.
$y = P_7(x)$ satisfies $(1 - x^2)y'' - 2xy' + 56y = 0$.

41. If $x = \cos\theta$, then $\frac{dy}{d\theta} = \frac{dy}{dx}\frac{dx}{d\theta} = -\sin\theta\frac{dy}{dx}$ and $\frac{d^2y}{d\theta^2} = \sin^2\theta\frac{d^2y}{dx^2} - \cos\theta\frac{dy}{dx}$. Now the original equation can be written as

$$\frac{d^2y}{d\theta^2} + \frac{\cos\theta}{\sin\theta}\frac{dy}{d\theta} + n(n+1)y = 0,$$

and so

$$\sin^2\theta\frac{d^2y}{dx^2} - 2\cos\theta\frac{dy}{dx} + n(n+1)y = 0.$$

Since $x = \cos\theta$ and $\sin^2\theta = 1 - \cos^2\theta = 1 - x^2$, we obtain

$$(1 - x^2)\frac{d^2y}{dx^2} - 2x\frac{dy}{dx} + n(n+1)y = 0.$$

43. Using binomial series, we have formally

$$(1 - 2xt + t^2)^{-1/2} = 1 + \frac{1}{2}(2xt - t^2) + \frac{1\cdot 3}{2^2 2!} \times (2xt - t^2)^2 + \cdots.$$

Grouping by powers of t, we then find

$$\begin{aligned} (1 - 2xt + t^2)^{-1/2} &= 1\cdot t^0 + x\cdot t + \tfrac{1}{2}(3x^2 - 1)t^2 + \cdots \\ &= P_0(x)t^0 + P_1(x)t + P_2(x)t^2 + \cdots. \end{aligned}$$

45. For $k = 1$, $P_2(x) = \frac{1}{2}[3xP_1(x) - P_0(x)] = \frac{1}{2}(3x^2 - 1)$.

For $k = 2$, $P_3(x) = \frac{1}{3}[5xP_2(x) - 2P_1(x)] = \frac{1}{2}(5x^3 - 3x)$.

For $k = 3$, $P_4(x) = \frac{1}{4}[7xP_3(x) - 3P_2(x)]$
$= \frac{1}{8}(35x^4 - 30x^2 + 3)$.

For $k = 4$, $P_5(x) = \frac{1}{5}[9xP_4(x) - 4P_3(x)]$
$= \frac{1}{8}(63x^5 - 70x^3 + 15x)$.

For $k = 5$, $P_6(x) = \frac{1}{6}[11xP_5(x) - 5P_4(x)]$
$= \frac{1}{16}(231x^6 - 315x^4 + 105x^2 - 5)$.

47. For $n = 0, 1, 2, 3$, the values of the integrals are $2, \frac{2}{3}, \frac{2}{5}$, and $\frac{2}{7}$, respectively. In general,

$$\int_{-1}^{1} P_n^2(x)\,dx = \frac{2}{2n+1}, \quad n = 0, 1, 2, \ldots.$$

49. y_2 is obtained from (4) of Section 4.2.

CHAPTER 6 REVIEW EXERCISES (PAGE 278)

1. $y = c_1 x^{-1/3} + c_2 x^{1/2}$

3. $y(x) = c_1 x^2 + c_2 x^3 + x^4 - x^2 \ln x$

5. The singular points are $x = 0$, $x = -1 + \sqrt{3}i$, $x = -1 - \sqrt{3}i$. All other finite values of x, real or complex, are ordinary points.

7. regular singular point $x = 0$; irregular singular point $x = 5$

9. regular singular points $x = -3, x = 3$; irregular singular point $x = 0$

11. $|x| < \infty$

13. $$y_1(x) = c_0\left[1 - \frac{1}{3 \cdot 2}x^3 + \frac{1}{6 \cdot 5 \cdot 3 \cdot 2}x^6 - \frac{1}{9 \cdot 8 \cdot 6 \cdot 5 \cdot 3 \cdot 2}x^9 + \cdots\right]$$

$$y_2(x) = c_1\left[x - \frac{1}{4 \cdot 3}x^4 + \frac{1}{7 \cdot 6 \cdot 4 \cdot 3}x^7 - \frac{1}{10 \cdot 9 \cdot 7 \cdot 6 \cdot 4 \cdot 3}x^{10} + \cdots\right]$$

15. $y_1(x) = c_0[1 + \frac{3}{2}x^2 + \frac{1}{2}x^3 + \frac{5}{8}x^4 + \cdots]$
$y_2(x) = c_1[x + \frac{1}{2}x^3 + \frac{1}{4}x^4 + \cdots]$

17. $r_1 = 1, r_2 = -\frac{1}{2}$;

$$y(x) = C_1 x\left[1 + \frac{1}{5}x + \frac{1}{7 \cdot 5 \cdot 2}x^2 + \frac{1}{9 \cdot 7 \cdot 5 \cdot 3 \cdot 2}x^3 + \cdots\right] + C_2 x^{-1/2}\left[1 - x - \frac{1}{2}x^2 - \frac{1}{3^2 \cdot 2}x^3 - \cdots\right]$$

19. $r_1 = 3, r_2 = 0$;

$$y_1(x) = C_3\left[x^3 + \frac{5}{4}x^4 + \frac{11}{8}x^5 + \cdots\right]$$

$$y(x) = C_1 y_1(x) + C_2\left[-\frac{1}{36}y_1(x)\ln x + y_1(x) \times \left(-\frac{1}{3}\frac{1}{x^3} + \frac{1}{4}\frac{1}{x^2} + \frac{1}{16}\frac{1}{x} + \cdots\right)\right]$$

21. $r_1 = r_2 = 0$; $y(x) = C_1 e^x + C_2 e^x \ln x$

23. $$y(x) = c_0\left[1 - \frac{1}{2^2}x^2 + \frac{1}{2^4(1.2)^2}x^4 - \frac{1}{2^6(1 \cdot 2 \cdot 3)^2}x^6 + \cdots\right]$$

EXERCISES 7.1 (PAGE 288)

1. $\frac{2}{s}e^{-s} - \frac{1}{s}$ **3.** $\frac{1}{s^2} - \frac{1}{s^2}e^{-s}$ **5.** $\frac{1 + e^{-s\pi}}{s^2 + 1}$ **7.** $\frac{e^{-s}}{s} + \frac{e^{-s}}{s^2}$

9. $\frac{1}{s} - \frac{1}{s^2} + \frac{e^{-s}}{s^2}$ **11.** $\frac{e^7}{s - 1}$ **13.** $\frac{1}{(s-4)^2}$

15. $\frac{1}{s^2 + 2s + 2}$ **17.** $\frac{s^2 - 1}{(s^2 + 1)^2}$ **19.** $\frac{48}{s^5}$ **21.** $\frac{4}{s^2} - \frac{10}{s}$

23. $\frac{2}{s^3} + \frac{6}{s^2} - \frac{3}{s}$ **25.** $\frac{6}{s^4} + \frac{6}{s^3} + \frac{3}{s^2} + \frac{1}{s}$ **27.** $\frac{1}{s} + \frac{1}{s - 4}$

29. $\frac{1}{s} + \frac{2}{s - 2} + \frac{1}{s - 4}$ **31.** $\frac{8}{s^3} - \frac{15}{s^2 + 9}$

33. Use $\sinh kt = \frac{e^{kt} - e^{-kt}}{2}$ to show that

$$\mathscr{L}\{\sinh kt\} = \frac{k}{s^2 - k^2}.$$

35. $\frac{1}{2(s-2)} - \frac{1}{2s}$ **37.** $\frac{2}{s^2 + 16}$

39. $\frac{1}{2}\left(\frac{s}{s^2 + 9} + \frac{s}{s^2 + 1}\right)$ **41.** $\frac{1}{2}\left(\frac{3}{s^2 + 9} - \frac{1}{s^2 + 1}\right)$

43. The result follows by letting $u = st$ in

$$\mathscr{L}\{t^\alpha\} = \int_0^\infty t^\alpha e^{-st}\,dt.$$

45. $\frac{\frac{1}{2}\Gamma(\frac{1}{2})}{s^{3/2}} = \frac{\sqrt{\pi}}{2s^{3/2}}$

47. On $0 \le t \le 1$, $e^{-st} \ge e^{-s}$ $(s > 0)$. Therefore

$$\int_0^1 e^{-st}\frac{1}{t^2}\,dt \ge e^{-s}\int_0^1 \frac{1}{t^2}\,dt.$$

The latter integral diverges.

EXERCISES 7.2 (PAGE 295)

1. $\frac{1}{2}t^2$ **3.** $t - 2t^4$ **5.** $1 + 3t + \frac{3}{2}t^2 + \frac{1}{6}t^3$ **7.** $t - 1 + e^{2t}$

9. $\frac{1}{4}e^{-t/4}$ **11.** $\frac{5}{7}\sin 7t$ **13.** $\cos(t/2)$ **15.** $\frac{1}{4}\sinh 4t$

17. $2\cos 3t - 2\sin 3t$ **19.** $\frac{1}{3} - \frac{1}{3}e^{-3t}$ **21.** $\frac{3}{4}e^{-3t} + \frac{1}{4}e^t$

23. $0.3e^{0.1t} + 0.6e^{-0.2t}$ **25.** $\frac{1}{2}e^{2t} - e^{3t} + \frac{1}{2}e^{6t}$

27. $-\frac{1}{3}e^{-t} + \frac{8}{15}e^{2t} - \frac{1}{5}e^{-3t}$ **29.** $\frac{1}{4}t - \frac{1}{8}\sin 2t$

31. $-\frac{1}{4}e^{-2t} + \frac{1}{4}\cos 2t + \frac{1}{4}\sin 2t$ **33.** $\frac{1}{3}\sin t - \frac{1}{6}\sin 2t$

35. $1/s$

EXERCISES 7.3 (PAGE 304)

1. $\dfrac{1}{(s-10)^2}$ **3.** $\dfrac{6}{(s+2)^4}$ **5.** $\dfrac{3}{(s-1)^2+9}$

7. $\dfrac{3}{(s-5)^2-9}$ **9.** $\dfrac{1}{(s-2)^2}+\dfrac{2}{(s-3)^2}+\dfrac{1}{(s-4)^2}$

11. $\dfrac{1}{2}\left[\dfrac{1}{s+1}-\dfrac{s+1}{(s+1)^2+4}\right]$ **13.** $\frac{1}{2}t^2e^{-2t}$ **15.** $e^{3t}\sin t$

17. $e^{-2t}\cos t-2e^{-2t}\sin t$ **19.** $e^{-t}-te^{-t}$

21. $5-t-5e^{-t}-4te^{-t}-\frac{3}{2}t^2e^{-t}$ **23.** $\dfrac{e^{-s}}{s^2}$

25. $\dfrac{e^{-2s}}{s^2}+2\dfrac{e^{-2s}}{s}$ **27.** $\dfrac{s}{s^2+4}e^{-\pi s}$ **29.** $\dfrac{6e^{-s}}{(s-1)^4}$

31. $\frac{1}{2}(t-2)^2\mathcal{U}(t-2)$ **33.** $-\sin t\,\mathcal{U}(t-\pi)$

35. $\mathcal{U}(t-1)-e^{-(t-1)}\mathcal{U}(t-1)$ **37.** $\dfrac{s^2-4}{(s^2+4)^2}$

39. $\dfrac{6s^2+2}{(s^2-1)^3}$ **41.** $\dfrac{12s-24}{[(s-2)^2+36]^2}$ **43.** $\frac{1}{2}t\sin t$ **45.** (c)

47. (f) **49.** (a)

51. $f(t)=2-4\mathcal{U}(t-3)$; $\mathcal{L}\{f(t)\}=\dfrac{2}{s}-\dfrac{4}{s}e^{-3s}$

53. $f(t)=t^2\,\mathcal{U}(t-1)$
$=(t-1)^2\mathcal{U}(t-1)+2(t-1)\mathcal{U}(t-1)+\mathcal{U}(t-1)$;
$\mathcal{L}\{f(t)\}=2\dfrac{e^{-s}}{s^3}+2\dfrac{e^{-s}}{s^2}+\dfrac{e^{-s}}{s}$

55. $f(t)=t-t\mathcal{U}(t-2)$
$=t-(t-2)\mathcal{U}(t-2)-2\mathcal{U}(t-2)$;
$\mathcal{L}\{f(t)\}=\dfrac{1}{s^2}-\dfrac{e^{-2s}}{s^2}-2\dfrac{e^{-2s}}{s}$

57. $f(t)=\mathcal{U}(t-a)-\mathcal{U}(t-b)$; $\mathcal{L}\{f(t)\}=\dfrac{e^{-as}}{s}-\dfrac{e^{-bs}}{s}$

59.

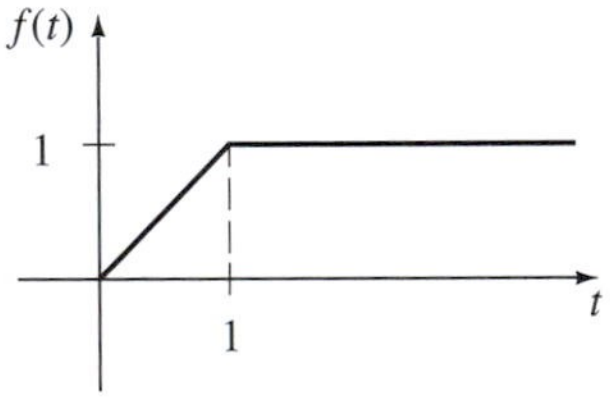

61. $\dfrac{e^{-t}-e^{3t}}{t}$ **63.** $\dfrac{\sin 2t}{t}$

EXERCISES 7.4 (PAGE 313)

1. Since $f'(t)=e^t$, $f(0)=1$, it follows from (1) that $\mathcal{L}\{e^t\}=s\mathcal{L}\{e^t\}-1$. Solving gives $\mathcal{L}\{e^t\}=1/(s-1)$.

3. $(s^2+3s)F(s)-s-2$ **5.** $F(s)=\dfrac{2s-1}{(s-1)^2}$

7. $\dfrac{1}{s(s-1)}$ **9.** $\dfrac{s+1}{s[(s+1)^2+1]}$

11. $\dfrac{1}{s^2(s-1)}$ **13.** $\dfrac{3s^2+1}{s^2(s^2+1)^2}$ **15.** $\dfrac{6}{s^5}$ **17.** $\dfrac{48}{s^8}$

19. $\dfrac{s-1}{(s+1)[(s-1)^2+1]}$ **21.** $\displaystyle\int_0^t f(\tau)e^{-5(t-\tau)}\,d\tau$

23. $1-e^{-t}$ **25.** $-\frac{1}{3}e^{-t}+\frac{1}{3}e^{2t}$ **27.** $\frac{1}{4}t\sin 2t$

29. The result follows from letting $u=t-\tau$ in the first integral.

31. $\dfrac{(1-e^{-as})^2}{s(1-e^{-2as})}=\dfrac{1-e^{-as}}{s(1+e^{-as})}$ **33.** $\dfrac{a}{s}\left(\dfrac{1}{bs}-\dfrac{1}{e^{bs}-1}\right)$

35. $\dfrac{\coth(\pi s/2)}{s^2+1}$ **37.** $\dfrac{1}{s^2+1}$

EXERCISES 7.5 (PAGE 324)

1. $y=-1+e^t$ **3.** $y=te^{-4t}+2e^{-4t}$

5. $y=\frac{4}{3}e^{-t}-\frac{1}{3}e^{-4t}$ **7.** $y=\frac{1}{9}t+\frac{2}{27}-\frac{2}{27}e^{3t}+\frac{10}{9}te^{3t}$

9. $y=\frac{1}{20}t^5e^{2t}$ **11.** $y=\cos t-\frac{1}{2}\sin t-\frac{1}{2}t\cos t$

13. $y=\frac{1}{2}-\frac{1}{2}e^t\cos t+\frac{1}{2}e^t\sin t$

15. $y=-\frac{8}{9}e^{-t/2}+\frac{1}{9}e^{-2t}+\frac{5}{18}e^t+\frac{1}{2}e^{-t}$ **17.** $y=\cos t$

19. $y=[5-5e^{-(t-1)}]\mathcal{U}(t-1)$

21. $y=-\frac{1}{4}+\frac{1}{2}t+\frac{1}{4}e^{-2t}-\frac{1}{4}\mathcal{U}(t-1)-\frac{1}{2}(t-1)\mathcal{U}(t-1)$
$+\frac{1}{4}e^{-2(t-1)}\mathcal{U}(t-1)$

23. $y=\cos 2t-\frac{1}{6}\sin 2(t-2\pi)\mathcal{U}(t-2\pi)$
$+\frac{1}{3}\sin(t-2\pi)\mathcal{U}(t-2\pi)$

25. $y=\sin t+[1-\cos(t-\pi)]\mathcal{U}(t-\pi)$
$-[1-\cos(t-2\pi)]\mathcal{U}(t-2\pi)$

27. $y=(e+1)te^{-t}+(e-1)e^{-t}$ **29.** $f(t)=\sin t$

31. $f(t)=-\frac{1}{8}e^{-t}+\frac{1}{8}e^t+\frac{3}{4}te^t+\frac{1}{4}t^2e^t$ **33.** $f(t)=e^{-t}$

35. $f(t)=\frac{3}{8}e^{2t}+\frac{1}{8}e^{-2t}+\frac{1}{2}\cos 2t+\frac{1}{4}\sin 2t$

37. $y=\sin t-\frac{1}{2}t\sin t$

39. $i(t)=20{,}000[te^{-100t}-(t-1)e^{-100(t-1)}\mathcal{U}(t-1)]$

41. $q(t)=\dfrac{E_0C}{1-kRC}(e^{-kt}-e^{-t/RC})$;

$q(t)=\dfrac{E_0}{R}te^{-t/RC}$

43. $q(t)=\frac{2}{5}\mathcal{U}(t-3)-\frac{2}{5}e^{-5(t-3)}\mathcal{U}(t-3)$

45. $i(t)=\dfrac{1}{101}e^{-10t}-\dfrac{1}{101}\cos t+\dfrac{10}{101}\sin t$
$-\dfrac{10}{101}e^{-10(t-3\pi/2)}\mathcal{U}\left(t-\dfrac{3\pi}{2}\right)$
$+\dfrac{10}{101}\cos\left(t-\dfrac{3\pi}{2}\right)\mathcal{U}\left(t-\dfrac{3\pi}{2}\right)$
$+\dfrac{1}{101}\sin\left(t-\dfrac{3\pi}{2}\right)\mathcal{U}\left(t-\dfrac{3\pi}{2}\right)$

47. $i(t) = \frac{t}{R} + \frac{L}{R^2}(e^{-Rt/L} - 1) + \frac{1}{R}\sum_{n=1}^{\infty}(e^{-R(t-n)/L} - 1)\mathcal{U}(t-n)$

For $0 \le t < 2$,

$$i(t) = \begin{cases} \frac{t}{R} + \frac{L}{R^2}(e^{-Rt/L} - 1), & 0 \le t < 1 \\ \frac{t}{R} + \frac{L}{R^2}(e^{-Rt/L} - 1) + \frac{1}{R}(e^{-R(t-1)/L} - 1), & 1 \le t < 2 \end{cases}$$

49. $q(t) = \frac{3}{5}e^{-10t} + 6te^{-10t} - \frac{3}{5}\cos 10t$;

$i(t) = -60te^{-10t} + 6\sin 10t$;
steady-state current is $6\sin 10t$

51. $q(t) = \frac{E_0}{L\left(k^2 + \frac{1}{LC}\right)}[e^{-kt} - \cos(t/\sqrt{LC})] + \frac{kE_0\sqrt{C/L}}{k^2 + \frac{1}{LC}}\sin(t/\sqrt{LC})$

53. $x(t) = -\frac{3}{2}e^{-7t/2}\cos\frac{\sqrt{15}}{2}t - \frac{7\sqrt{15}}{10}e^{-7t/2}\sin\frac{\sqrt{15}}{2}t$

55. $y(x) = \frac{w_0}{EI}\left(\frac{L^2}{4}x^2 - \frac{L}{6}x^3 + \frac{1}{24}x^4\right)$;

$\frac{17w_0L^4}{384EI}; \frac{w_0L^4}{8EI}$

57. $y(x) = \frac{w_0L^2}{16EI}x^2 - \frac{w_0L}{12EI}x^3 + \frac{w_0}{24EI}x^4 - \frac{w_0}{24EI}\left(x - \frac{L}{2}\right)^4\mathcal{U}\left(x - \frac{L}{2}\right)$

59. $y = \frac{1}{3}t^3 + \frac{1}{2}ct^2$

EXERCISES 7.6 (PAGE 331)

1. $y = e^{3(t-2)}\mathcal{U}(t-2)$

3. $y = \sin t + \sin t\,\mathcal{U}(t - 2\pi)$

5. $y = -\cos t\,\mathcal{U}\left(t - \frac{\pi}{2}\right) + \cos t\,\mathcal{U}\left(t - \frac{3\pi}{2}\right)$

7. $y = \frac{1}{2} - \frac{1}{2}e^{-2t} + [\frac{1}{2} - \frac{1}{2}e^{-2(t-1)}]\mathcal{U}(t-1)$

9. $y = e^{-2(t-2\pi)}\sin t\,\mathcal{U}(t - 2\pi)$

11. $y = e^{-2t}\cos 3t + \frac{2}{3}e^{-2t}\sin 3t + \frac{1}{3}e^{-2(t-\pi)}\sin 3(t-\pi)\mathcal{U}(t-\pi) + \frac{1}{3}e^{-2(t-3\pi)}\sin 3(t - 3\pi)\mathcal{U}(t - 3\pi)$

13. $y(x) = \begin{cases} \frac{P_0}{EI}\left(\frac{L}{4}x^2 - \frac{1}{6}x^3\right), & 0 \le x < L/2 \\ \frac{P_0L^2}{4EI}\left(\frac{1}{2}x - \frac{L}{12}\right), & L/2 \le x \le L \end{cases}$

15. From (7) with $f(t) = e^{-st}$, we have

$$\mathcal{L}\{\delta(t - t_0)\} = \int_0^{\infty} e^{-st}\delta(t - t_0)\,dt = e^{-st_0}.$$

17. $y = e^{-t}\cos t + e^{-(t-3\pi)}\sin t\,\mathcal{U}(t - 3\pi)$

19. $i(t) = \frac{1}{L}e^{-Rt/L}$; no

CHAPTER 7 REVIEW EXERCISES (PAGE 332)

1. $\frac{1}{s^2} - \frac{2}{s^2}e^{-s}$ **3.** false **5.** true **7.** $\frac{1}{s+7}$ **9.** $\frac{2}{s^2+4}$

11. $\frac{4s}{(s^2+4)^2}$ **13.** $\frac{1}{6}t^5$ **15.** $\frac{1}{2}t^2e^{5t}$

17. $e^{5t}\cos 2t + \frac{5}{2}e^{5t}\sin 2t$

19. $\cos\pi(t-1)\mathcal{U}(t-1) + \sin\pi(t-1)\mathcal{U}(t-1)$

21. -5 **23.** $e^{-ks}F(s-a)$

25. **(a)** $f(t) = t - (t-1)\mathcal{U}(t-1) - \mathcal{U}(t-4)$

(b) $\mathcal{L}\{f(t)\} = \frac{1}{s^2} - \frac{1}{s^2}e^{-s} - \frac{1}{s}e^{-4s}$

(c) $\mathcal{L}\{e^tf(t)\} = \frac{1}{(s-1)^2} - \frac{1}{(s-1)^2}e^{-(s-1)} - \frac{1}{s-1}e^{-4(s-1)}$

27. **(a)** $f(t) = 2 + (t-2)\mathcal{U}(t-2)$

(b) $\mathcal{L}\{f(t)\} = \frac{2}{s} + \frac{1}{s^2}e^{-2s}$

(c) $\mathcal{L}\{e^tf(t)\} = \frac{2}{s-1} + \frac{1}{(s-1)^2}e^{-2(s-1)}$

29. $y = 5te^t + \frac{1}{2}t^2e^t$

31. $y = 5\mathcal{U}(t-\pi) - 5e^{2(t-\pi)}\cos\sqrt{2}(t-\pi)\mathcal{U}(t-\pi) + 5\sqrt{2}e^{2(t-\pi)}\sin\sqrt{2}(t-\pi)\mathcal{U}(t-\pi)$

33. $y = -\frac{2}{125} - \frac{2}{25}t - \frac{1}{5}t^2 + \frac{127}{125}e^{5t} - [-\frac{37}{125} - \frac{12}{25}(t-1) - \frac{1}{5}(t-1)^2 + \frac{37}{125}e^{5(t-1)}]\mathcal{U}(t-1)$

35. $y = 1 + t + \frac{1}{2}t^2$

37. $i(t) = -9 + 2t + 9e^{-t/5}$

39. $y(x) = \frac{w_0}{12EIL}\left[-\frac{1}{5}x^5 + \frac{L}{2}x^4 - \frac{L^2}{2}x^3 + \frac{L^3}{4}x^2 + \frac{1}{5}\left(x - \frac{L}{2}\right)^5\mathcal{U}\left(x - \frac{L}{2}\right)\right]$

EXERCISES 8.1 (PAGE 342)

1. $x = c_1e^t + c_2te^t$

$y = (c_1 - c_2)e^t + c_2te^t$

3. $x = c_1 \cos t + c_2 \sin t + t + 1$

$y = c_1 \sin t - c_2 \cos t + t - 1$

5. $x = \frac{1}{2}c_1 \sin t + \frac{1}{2}c_2 \cos t - 2c_3 \sin \sqrt{6}t - 2c_4 \cos \sqrt{6}t$

$y = c_1 \sin t + c_2 \cos t + c_3 \sin \sqrt{6}t + c_4 \cos \sqrt{6}t$

7. $x = c_1e^{2t} + c_2e^{-2t} + c_3 \sin 2t + c_4 \cos 2t + \frac{1}{5}e^t$

$y = c_1e^{2t} + c_2e^{-2t} - c_3 \sin 2t - c_4 \cos 2t - \frac{1}{5}e^t$

9. $x = c_1 - c_2 \cos t + c_3 \sin t + \frac{17}{15}e^{3t}$

$y = c_1 + c_2 \sin t + c_3 \cos t - \frac{4}{15}e^{3t}$

11. $x = c_1e^t + c_2e^{-t/2} \cos \frac{\sqrt{3}}{2}t + c_3e^{-t/2} \sin \frac{\sqrt{3}}{2}t$

$$y = \left(-\frac{3}{2}c_2 - \frac{\sqrt{3}}{2}c_3\right)e^{-t/2} \cos \frac{\sqrt{3}}{2}t + \left(\frac{\sqrt{3}}{2}c_2 - \frac{3}{2}c_3\right)e^{-t/2} \sin \frac{\sqrt{3}}{2}t$$

13. $x = c_1e^{4t} + \frac{4}{3}e^t$

$y = -\frac{3}{4}c_1e^{4t} + c_2 + 5e^t$

15. $x = c_1 + c_2t + c_3e^t + c_4e^{-t} - \frac{1}{2}t^2$

$y = (c_1 - c_2 + 2) + (c_2 + 1)t + c_4e^{-t} - \frac{1}{2}t^2$

17. $x = c_1e^t + c_2e^{-t/2} \sin \frac{\sqrt{3}}{2}t + c_3e^{-t/2} \cos \frac{\sqrt{3}}{2}t$

$$y = c_1e^t + \left(-\frac{1}{2}c_2 - \frac{\sqrt{3}}{2}c_3\right)e^{-t/2} \sin \frac{\sqrt{3}}{2}t + \left(\frac{\sqrt{3}}{2}c_2 - \frac{1}{2}c_3\right)e^{-t/2} \cos \frac{\sqrt{3}}{2}t$$

$$z = c_1e^t + \left(-\frac{1}{2}c_2 + \frac{\sqrt{3}}{2}c_3\right)e^{-t/2} \sin \frac{\sqrt{3}}{2}t + \left(-\frac{\sqrt{3}}{2}c_2 - \frac{1}{2}c_3\right)e^{-t/2} \cos \frac{\sqrt{3}}{2}t$$

19. $x = -6c_1e^{-t} - 3c_2e^{-2t} + 2c_3e^{3t}$

$y = c_1e^{-t} + c_2e^{-2t} + c_3e^{3t}$

$z = 5c_1e^{-t} + c_2e^{-2t} + c_3e^{3t}$

21. $x = -c_1e^{-t} + c_2 + \frac{1}{3}t^3 - 2t^2 + 5t$

$y = c_1e^{-t} + 2t^2 - 5t + 5$

23. $x = e^{-3t+3} - te^{-3t+3}$

$y = -e^{-3t+3} + 2te^{-3t+3}$

25. $Dx - Dy = 0$

$(D - 1)x - y = 0$

EXERCISES 8.2 (PAGE 348)

1. $x = -\frac{1}{3}e^{-2t} + \frac{1}{3}e^t$

$y = \frac{1}{3}e^{-2t} + \frac{2}{3}e^t$

3. $x = -\cos 3t - \frac{5}{3}\sin 3t$

$y = 2 \cos 3t - \frac{7}{3}\sin 3t$

5. $x = -2e^{3t} + \frac{5}{2}e^{2t} - \frac{1}{2}$

$y = \frac{8}{3}e^{3t} - \frac{5}{2}e^{2t} - \frac{1}{6}$

7. $x = -\frac{1}{2}t - \frac{3}{4}\sqrt{2} \sin \sqrt{2}t$

$y = -\frac{1}{2}t + \frac{3}{4}\sqrt{2} \sin \sqrt{2}t$

9. $x = 8 + \frac{2}{3!}t^3 + \frac{1}{4!}t^4$

$y = -\frac{2}{3!}t^3 + \frac{1}{4!}t^4$

11. $x = \frac{1}{2}t^2 + t + 1 - e^{-t}$

$y = -\frac{1}{3} + \frac{1}{3}e^{-t} + \frac{1}{3}te^{-t}$

13. $x_1 = \frac{1}{5}\sin t + \frac{2\sqrt{6}}{15}\sin \sqrt{6}t + \frac{2}{5}\cos t - \frac{2}{5}\cos \sqrt{6}t$

$x_2 = \frac{2}{5}\sin t - \frac{\sqrt{6}}{15}\sin \sqrt{6}t + \frac{4}{5}\cos t + \frac{1}{5}\cos \sqrt{6}t$

15. **(b)** $i_2 = \frac{100}{9} - \frac{100}{9}e^{-900t}$ **(c)** $i_1 = 20 - 20e^{-900t}$

$i_3 = \frac{80}{9} - \frac{80}{9}e^{-900t}$

17. $i_2 = -\frac{20}{13}e^{-2t} + \frac{375}{1469}e^{-15t} + \frac{145}{113}\cos t + \frac{85}{113}\sin t$

$i_3 = \frac{30}{13}e^{-2t} + \frac{250}{1469}e^{-15t} - \frac{280}{113}\cos t + \frac{810}{113}\sin t$

19. $i_1 = \frac{6}{5} - \frac{6}{5}e^{-100t}\cos 100t$

$i_2 = \frac{6}{5} - \frac{6}{5}e^{-100t}\cos 100t - \frac{6}{5}e^{-100t}\sin 100t$

21. **(b)** $q = 50e^{-t}\sin(t - 1)\mathcal{U}(t - 1)$

EXERCISES 8.3 (PAGE 354)

1. $x_1' = x_2$

$x_2' = -4x_1 + 3x_2 + \sin 3t$

3. $x_1' = x_2$

$x_2' = x_3$

$x_3' = 10x_1 - 6x_2 + 3x_3 + t^2 + 1$

5. $x_1' = x_2$

$x_2' = x_3$

$x_3' = x_4$

$x_4' = -x_1 - 4x_2 + 2x_3 + t$

7. $x_1' = x_2$

$x_2' = \frac{t}{t + 1}x_1$

9. $x' = -2x + y + 5t$

$y' = 2x + y - 2t$

11. $Dx = t^2 + 5t - 2$

$Dy = -x + 5t - 2$

13. The system is degenerate.

15. $Dx = u$
$Dy = v$
$Du = w$
$Dv = 10t^2 - 4u + 3v$
$Dw = 4x + 4v - 3w$

17. $x_1 = \frac{25}{2}e^{-t/25} + \frac{25}{2}e^{-3t/25}$

$x_2 = \frac{25}{4}e^{-t/25} - \frac{25}{4}e^{-3t/25}$

19. $x_1' = \frac{1}{50}x_2 - \frac{3}{50}x_1$

$x_2' = \frac{3}{50}x_1 - \frac{7}{100}x_2 + \frac{1}{100}x_3$

$x_3' = \frac{1}{20}x_2 - \frac{1}{20}x_3$

EXERCISES 8.4 (PAGE 372)

1. (a) $\begin{pmatrix} 2 & 11 \\ 2 & -1 \end{pmatrix}$ (b) $\begin{pmatrix} -6 & 1 \\ 14 & -19 \end{pmatrix}$ (c) $\begin{pmatrix} 2 & 28 \\ 12 & -12 \end{pmatrix}$

3. (a) $\begin{pmatrix} -11 & 6 \\ 17 & -22 \end{pmatrix}$ (b) $\begin{pmatrix} -32 & 27 \\ -4 & -1 \end{pmatrix}$

(c) $\begin{pmatrix} 19 & -18 \\ -30 & 31 \end{pmatrix}$ (d) $\begin{pmatrix} 19 & 6 \\ 3 & 22 \end{pmatrix}$

5. (a) $\begin{pmatrix} 9 & 24 \\ 3 & 8 \end{pmatrix}$ (b) $\begin{pmatrix} 3 & 8 \\ -6 & -16 \end{pmatrix}$ (c) $\begin{pmatrix} 0 & 0 \\ 0 & 0 \end{pmatrix}$

(d) $\begin{pmatrix} -4 & -5 \\ 8 & 10 \end{pmatrix}$

7. (a) 180 (b) $\begin{pmatrix} 4 & 8 & 10 \\ 8 & 16 & 20 \\ 10 & 20 & 25 \end{pmatrix}$ (c) $\begin{pmatrix} 6 \\ 12 \\ -5 \end{pmatrix}$

9. (a) $\begin{pmatrix} 7 & 38 \\ 10 & 75 \end{pmatrix}$ (b) $\begin{pmatrix} 7 & 38 \\ 10 & 75 \end{pmatrix}$ 11. $\begin{pmatrix} -14 \\ 1 \end{pmatrix}$

13. $\begin{pmatrix} -38 \\ -2 \end{pmatrix}$ 15. singular

17. nonsingular; $\mathbf{A}^{-1} = \frac{1}{4}\begin{pmatrix} -5 & -8 \\ 3 & 4 \end{pmatrix}$

19. nonsingular; $\mathbf{A}^{-1} = \frac{1}{2}\begin{pmatrix} 0 & -1 & 1 \\ 2 & 2 & -2 \\ -4 & -3 & 5 \end{pmatrix}$

21. nonsingular; $\mathbf{A}^{-1} = -\frac{1}{9}\begin{pmatrix} -2 & -2 & -1 \\ -13 & 5 & 7 \\ 8 & -1 & -5 \end{pmatrix}$

23. $\det \mathbf{A}(t) = 2e^{3t} \neq 0$ for every value of t;

$$\mathbf{A}^{-1}(t) = \frac{1}{2e^{3t}}\begin{pmatrix} 3e^{4t} & -e^{4t} \\ -4e^{-t} & 2e^{-t} \end{pmatrix}$$

25. $\dfrac{d\mathbf{X}}{dt} = \begin{pmatrix} -5e^{-t} \\ -2e^{-t} \\ 7e^{-t} \end{pmatrix}$ 27. $\dfrac{d\mathbf{X}}{dt} = 4\begin{pmatrix} 1 \\ -1 \end{pmatrix}e^{2t} - 12\begin{pmatrix} 2 \\ 1 \end{pmatrix}e^{-3t}$

29. (a) $\begin{pmatrix} 4e^{4t} & -\pi \sin \pi t \\ 2 & 6t \end{pmatrix}$ (b) $\begin{pmatrix} \frac{1}{4}e^{8} - \frac{1}{4} & 0 \\ 4 & 6 \end{pmatrix}$

(c) $\begin{pmatrix} \frac{1}{4}e^{4t} - \frac{1}{4} & (1/\pi) \sin \pi t \\ t^2 & t^3 - t \end{pmatrix}$

31. $x = 3, y = 1, z = -5$

33. $x = 2 + 4t, y = -5 - t, z = t$ 35. $x = -\frac{1}{2}, y = \frac{3}{2}, z = \frac{7}{2}$

37. $x_1 = 1, x_2 = 0, x_3 = 2, x_4 = 0$

41. $\lambda_1 = 6, \lambda_2 = 1, \mathbf{K}_1 = \begin{pmatrix} 2 \\ 7 \end{pmatrix}, \mathbf{K}_2 = \begin{pmatrix} 1 \\ 1 \end{pmatrix}$

43. $\lambda_1 = \lambda_2 = -4, \mathbf{K}_1 = \begin{pmatrix} 1 \\ -4 \end{pmatrix}$

45. $\lambda_1 = 0, \lambda_2 = 4, \lambda_3 = -4,$

$$\mathbf{K}_1 = \begin{pmatrix} 9 \\ 45 \\ 25 \end{pmatrix}, \mathbf{K}_2 = \begin{pmatrix} 1 \\ 1 \\ 1 \end{pmatrix}, \mathbf{K}_3 = \begin{pmatrix} 1 \\ 9 \\ 1 \end{pmatrix}$$

47. $\lambda_1 = \lambda_2 = \lambda_3 = -2,$

$$\mathbf{K}_1 = \begin{pmatrix} 2 \\ -1 \\ 0 \end{pmatrix}, \mathbf{K}_2 = \begin{pmatrix} 0 \\ 0 \\ 1 \end{pmatrix}$$

49. $\lambda_1 = 3i, \lambda_2 = -3i,$

$$\mathbf{K}_1 = \begin{pmatrix} 1 - 3i \\ 5 \end{pmatrix}, \mathbf{K}_2 = \begin{pmatrix} 1 + 3i \\ 5 \end{pmatrix}$$

51. $$\frac{d}{dt}\begin{pmatrix} a_{11}(t) & a_{12}(t) \\ a_{21}(t) & a_{22}(t) \end{pmatrix}\begin{pmatrix} x_1(t) \\ x_2(t) \end{pmatrix}$$

$$= \frac{d}{dt}\begin{pmatrix} a_{11}(t)x_1(t) + a_{12}(t)x_2(t) \\ a_{21}(t)x_1(t) + a_{22}(t)x_2(t) \end{pmatrix}$$

$$= \begin{pmatrix} a_{11}(t)x_1'(t) + a_{11}'(t)x_1(t) + a_{12}(t)x_2'(t) + a_{12}'(t)x_2(t) \\ a_{21}(t)x_1'(t) + a_{21}'(t)x_1(t) + a_{22}(t)x_2'(t) + a_{22}'(t)x_2(t) \end{pmatrix}$$

$$= \begin{pmatrix} a_{11}(t)x_1'(t) + a_{12}(t)x_2'(t) + a_{11}'(t)x_1(t) + a_{12}'(t)x_2(t) \\ a_{21}(t)x_1'(t) + a_{22}(t)x_2'(t) + a_{21}'(t)x_1(t) + a_{22}'(t)x_2(t) \end{pmatrix}$$

$$= \begin{pmatrix} a_{11}(t) & a_{12}(t) \\ a_{21}(t) & a_{22}(t) \end{pmatrix}\begin{pmatrix} x_1'(t) \\ x_2'(t) \end{pmatrix} + \begin{pmatrix} a_{11}'(t) & a_{12}'(t) \\ a_{21}'(t) & a_{22}'(t) \end{pmatrix}\begin{pmatrix} x_1(t) \\ x_2(t) \end{pmatrix}$$

$$= \mathbf{A}(t)\mathbf{X}'(t) + \mathbf{A}'(t)\mathbf{X}(t)$$

53. Since $\mathbf{A}^{-1}$ exists, $\mathbf{AB} = \mathbf{AC}$ implies $\mathbf{A}^{-1}(\mathbf{AB}) = \mathbf{A}^{-1}(\mathbf{AC}), (\mathbf{A}^{-1}\mathbf{A})\mathbf{B} = (\mathbf{A}^{-1}\mathbf{A})\mathbf{C}, \mathbf{IB} = \mathbf{IC}$, or $\mathbf{B} = \mathbf{C}$.

55. No, since in general $\mathbf{AB} \neq \mathbf{BA}$.

EXERCISES 8.5 (PAGE 388)

1. $\mathbf{X}' = \begin{pmatrix} 3 & -5 \\ 4 & 8 \end{pmatrix}\mathbf{X}$, where $\mathbf{X} = \begin{pmatrix} x \\ y \end{pmatrix}$

3. $\mathbf{X}' = \begin{pmatrix} -3 & 4 & -9 \\ 6 & -1 & 0 \\ 10 & 4 & 3 \end{pmatrix}\mathbf{X}$, where $\mathbf{X} = \begin{pmatrix} x \\ y \\ z \end{pmatrix}$

5. $\mathbf{X}' = \begin{pmatrix} 1 & -1 & 1 \\ 2 & 1 & -1 \\ 1 & 1 & 1 \end{pmatrix}\mathbf{X} + \begin{pmatrix} 0 \\ -3t^2 \\ t^2 \end{pmatrix} + \begin{pmatrix} t \\ 0 \\ -t \end{pmatrix}$

$+ \begin{pmatrix} -1 \\ 0 \\ 2 \end{pmatrix}$, where $\mathbf{X} = \begin{pmatrix} x \\ y \\ z \end{pmatrix}$

7. $\dfrac{dx}{dt} = 4x + 2y + e^t$

$\dfrac{dy}{dt} = -x + 3y - e^t$

9. $\dfrac{dx}{dt} = x - y + 2z + e^{-t} - 3t$

$\dfrac{dy}{dt} = 3x - 4y + z + 2e^{-t} + t$

$\dfrac{dz}{dt} = -2x + 5y + 6z + 2e^{-t} - t$

11. $\mathbf{X}' = \begin{pmatrix} -5e^{-5t} \\ -10e^{-5t} \end{pmatrix}$

$$\begin{pmatrix} 3 & -4 \\ 4 & -7 \end{pmatrix}\mathbf{X} = \begin{pmatrix} 3-8 \\ 4-14 \end{pmatrix}e^{-5t} = \begin{pmatrix} -5 \\ -10 \end{pmatrix}e^{-5t} = \mathbf{X}'$$

13. $\mathbf{X}' = \begin{pmatrix} \frac{3}{2} \\ -3 \end{pmatrix}$

$$\begin{pmatrix} -1 & \frac{1}{4} \\ 1 & -1 \end{pmatrix}\mathbf{X} = \begin{pmatrix} 1 + \frac{1}{2} \\ -1 - 2 \end{pmatrix}e^{-3t/2} = \begin{pmatrix} \frac{3}{2} \\ -3 \end{pmatrix}e^{-3t/2} = \mathbf{X}'$$

15. $\dfrac{d\mathbf{X}}{dt} = \begin{pmatrix} 0 \\ 0 \\ 0 \end{pmatrix}$

$$\begin{pmatrix} 1 & 2 & 1 \\ 6 & -1 & 0 \\ -1 & -2 & -1 \end{pmatrix}\mathbf{X} = \begin{pmatrix} 1 + 12 - 13 \\ 6 - 6 \\ -1 - 12 + 13 \end{pmatrix} = \begin{pmatrix} 0 \\ 0 \\ 0 \end{pmatrix} = \frac{d\mathbf{X}}{dt}$$

17. Yes; $W(\mathbf{X}_1, \mathbf{X}_2) = -2e^{-8t} \neq 0$ implies $\mathbf{X}_1$ and $\mathbf{X}_2$ are linearly independent on $(-\infty, \infty)$.

19. No; $W(\mathbf{X}_1, \mathbf{X}_2, \mathbf{X}_3) =$

$$\begin{vmatrix} 1 + t & 1 & 3 + 2t \\ -2 + 2t & -2 & -6 + 4t \\ 4 + 2t & 4 & 12 + 4t \end{vmatrix} = 0 \text{ for every } t.$$

The solution vectors are linearly dependent on $(-\infty, \infty)$. Note that $\mathbf{X}_3 = 2\mathbf{X}_1 + \mathbf{X}_2$.

21. $\dfrac{d\mathbf{X}_p}{dt} = \begin{pmatrix} 2 \\ -1 \end{pmatrix}$

$$\begin{pmatrix} 1 & 4 \\ 3 & 2 \end{pmatrix}\mathbf{X}_p + \begin{pmatrix} 2 \\ -4 \end{pmatrix}t - \begin{pmatrix} 7 \\ 18 \end{pmatrix}$$

$$= \begin{pmatrix} (2-4)t + 9 + 2t - 7 \\ (6-2)t + 17 - 4t - 18 \end{pmatrix} = \begin{pmatrix} 2 \\ -1 \end{pmatrix} = \frac{d\mathbf{X}_p}{dt}$$

23. $\mathbf{X}_p' = \begin{pmatrix} 2e^t + te^t \\ -te^t \end{pmatrix}$

$$\begin{pmatrix} 2 & 1 \\ 3 & 4 \end{pmatrix}\mathbf{X}_p - \begin{pmatrix} 1 \\ 7 \end{pmatrix}e^t = \begin{pmatrix} 3e^t + te^t - e^t \\ 7e^t - te^t - 7e^t \end{pmatrix}$$

$$= \begin{pmatrix} 2e^t + te^t \\ -te^t \end{pmatrix} = \mathbf{X}_p'$$

25. Let $\mathbf{X}_1 = \begin{pmatrix} 6 \\ -1 \\ -5 \end{pmatrix}e^{-t}$, $\mathbf{X}_2 = \begin{pmatrix} -3 \\ 1 \\ 1 \end{pmatrix}e^{-2t}$,

$\mathbf{X}_3 = \begin{pmatrix} 2 \\ 1 \\ 1 \end{pmatrix}e^{3t}$, and $\mathbf{A} = \begin{pmatrix} 0 & 6 & 0 \\ 1 & 0 & 1 \\ 1 & 1 & 0 \end{pmatrix}$. Then

$$\mathbf{A}\mathbf{X}_1 = \begin{pmatrix} -6 \\ 1 \\ 5 \end{pmatrix}e^{-t} = \mathbf{X}_1',$$

$$\mathbf{A}\mathbf{X}_2 = \begin{pmatrix} 6 \\ -2 \\ -2 \end{pmatrix}e^{-2t} = \mathbf{X}_2',$$

$$\mathbf{A}\mathbf{X}_3 = \begin{pmatrix} 6 \\ 3 \\ 3 \end{pmatrix}e^{3t} = \mathbf{X}_3', \text{ and}$$

$$W(\mathbf{X}_1, \mathbf{X}_2, \mathbf{X}_3) = \begin{vmatrix} 6e^{-t} & -3e^{-2t} & 2e^{3t} \\ -e^{-t} & e^{-2t} & e^{3t} \\ -5e^{-t} & e^{-2t} & e^{3t} \end{vmatrix}$$

$$= 20 \neq 0.$$

Therefore, $\mathbf{X}_1, \mathbf{X}_2, \mathbf{X}_3$ form a fundamental set of solutions of $\mathbf{X}' = \mathbf{A}\mathbf{X}$ on $(-\infty, \infty)$. By definition,

$$\mathbf{X} = c_1\mathbf{X}_1 + c_2\mathbf{X}_2 + c_3\mathbf{X}_3$$

is the general solution.

27. $\mathbf{\Phi}(t) = \begin{pmatrix} e^{2t} & e^{7t} \\ -2e^{2t} & 3e^{7t} \end{pmatrix}$

$$\mathbf{\Phi}^{-1}(t) = \frac{1}{5e^{9t}}\begin{pmatrix} 3e^{7t} & -e^{7t} \\ 2e^{2t} & e^{2t} \end{pmatrix}$$

29. $\mathbf{\Phi}(t) = \begin{pmatrix} -e^t & -te^t \\ 3e^t & 3te^t - e^t \end{pmatrix}$

$$\mathbf{\Phi}^{-1}(t) = \frac{1}{e^{2t}}\begin{pmatrix} 3te^t - e^t & te^t \\ -3e^t & -e^t \end{pmatrix}$$

31. $\mathbf{\Psi}(t) = \begin{pmatrix} \frac{3}{5}e^{2t} + \frac{2}{5}e^{7t} & -\frac{1}{5}e^{2t} + \frac{1}{5}e^{7t} \\ -\frac{6}{5}e^{2t} + \frac{6}{5}e^{7t} & \frac{2}{5}e^{2t} + \frac{3}{5}e^{7t} \end{pmatrix}$

33. $\Psi(t) = \begin{pmatrix} 3te^t + e^t & te^t \\ -9te^t & -3te^t + e^t \end{pmatrix}$

35. $\mathbf{X}(t_0) = \mathbf{\Phi}(t_0)\mathbf{C}$ implies $\mathbf{C} = \mathbf{\Phi}^{-1}(t_0)\mathbf{X}(t_0)$. Substituting in $\mathbf{X} = \mathbf{\Phi}(t)\mathbf{C}$ gives $\mathbf{X} = \mathbf{\Phi}(t)\mathbf{\Phi}^{-1}(t_0)\mathbf{X}_0$.

37. Comparing $\mathbf{X} = \mathbf{\Phi}(t)\mathbf{\Phi}^{-1}(t_0)\mathbf{X}_0$ and $\mathbf{X} = \mathbf{\Psi}(t)\mathbf{X}_0$ implies $[\mathbf{\Psi}(t) - \mathbf{\Phi}(t)\mathbf{\Phi}^{-1}(t_0)]\mathbf{X}_0 = \mathbf{0}$. Since this last equation is to hold for any $\mathbf{X}_0$, we conclude that $\mathbf{\Psi}(t) = \mathbf{\Phi}(t)\mathbf{\Phi}^{-1}(t_0)$.

EXERCISES 8.6 (PAGE 402)

1. $\mathbf{X} = c_1\begin{pmatrix} 1 \\ 2 \end{pmatrix}e^{5t} + c_2\begin{pmatrix} 1 \\ -1 \end{pmatrix}e^{-t}$

3. $\mathbf{X} = c_1\begin{pmatrix} 2 \\ 1 \end{pmatrix}e^{-3t} + c_2\begin{pmatrix} 2 \\ 5 \end{pmatrix}e^{t}$

5. $\mathbf{X} = c_1\begin{pmatrix} 5 \\ 2 \end{pmatrix}e^{8t} + c_2\begin{pmatrix} 1 \\ 4 \end{pmatrix}e^{-10t}$

7. $\mathbf{X} = c_1\begin{pmatrix} 1 \\ 0 \\ 0 \end{pmatrix}e^{t} + c_2\begin{pmatrix} 2 \\ 3 \\ 1 \end{pmatrix}e^{2t} + c_3\begin{pmatrix} 1 \\ 0 \\ 2 \end{pmatrix}e^{-t}$

9. $\mathbf{X} = c_1\begin{pmatrix} -1 \\ 0 \\ 1 \end{pmatrix}e^{-t} + c_2\begin{pmatrix} 1 \\ 4 \\ 3 \end{pmatrix}e^{3t} + c_3\begin{pmatrix} 1 \\ -1 \\ 3 \end{pmatrix}e^{-2t}$

11. $\mathbf{X} = c_1\begin{pmatrix} 4 \\ 0 \\ -1 \end{pmatrix}e^{-t} + c_2\begin{pmatrix} -12 \\ 6 \\ 5 \end{pmatrix}e^{-t/2} + c_3\begin{pmatrix} 4 \\ 2 \\ -1 \end{pmatrix}e^{-3t/2}$

13. $\mathbf{X} = 3\begin{pmatrix} 1 \\ 1 \end{pmatrix}e^{t/2} + 2\begin{pmatrix} 0 \\ 1 \end{pmatrix}e^{-t/2}$

15. $\mathbf{X} = c_1\begin{pmatrix} \cos t \\ 2\cos t + \sin t \end{pmatrix}e^{4t} + c_2\begin{pmatrix} \sin t \\ 2\sin t - \cos t \end{pmatrix}e^{4t}$

17. $\mathbf{X} = c_1\begin{pmatrix} \cos t \\ -\cos t - \sin t \end{pmatrix}e^{4t} + c_2\begin{pmatrix} \sin t \\ -\sin t + \cos t \end{pmatrix}e^{4t}$

19. $\mathbf{X} = c_1\begin{pmatrix} 5\cos 3t \\ 4\cos 3t + 3\sin 3t \end{pmatrix} + c_2\begin{pmatrix} 5\sin 3t \\ 4\sin 3t - 3\cos 3t \end{pmatrix}$

21. $\mathbf{X} = c_1\begin{pmatrix} 1 \\ 0 \\ 0 \end{pmatrix} + c_2\begin{pmatrix} -\cos t \\ \cos t \\ \sin t \end{pmatrix} + c_3\begin{pmatrix} \sin t \\ -\sin t \\ \cos t \end{pmatrix}$

23. $\mathbf{X} = c_1\begin{pmatrix} 0 \\ 2 \\ 1 \end{pmatrix}e^{t} + c_2\begin{pmatrix} \sin t \\ \cos t \\ \cos t \end{pmatrix}e^{t} + c_3\begin{pmatrix} \cos t \\ -\sin t \\ -\sin t \end{pmatrix}e^{t}$

25. $\mathbf{X} = \begin{pmatrix} 28 \\ -5 \\ 25 \end{pmatrix}e^{2t} + c_2\begin{pmatrix} 5\cos 3t \\ -4\cos 3t - 3\sin 3t \\ 0 \end{pmatrix}e^{-2t}$

$+ c_3\begin{pmatrix} 5\sin 3t \\ -4\sin 3t + 3\cos 3t \\ 0 \end{pmatrix}e^{-2t}$

27. $\mathbf{X} = -\begin{pmatrix} 25 \\ -7 \\ 6 \end{pmatrix}e^{t} - \begin{pmatrix} \cos 5t - 5\sin 5t \\ \cos 5t \\ \cos 5t \end{pmatrix}$

$+ 6\begin{pmatrix} 5\cos 5t + \sin 5t \\ \sin 5t \\ \sin 5t \end{pmatrix}$

29. $\mathbf{X} = c_1\begin{pmatrix} 1 \\ 3 \end{pmatrix} + c_2\left\{\begin{pmatrix} 1 \\ 3 \end{pmatrix}t + \begin{pmatrix} \frac{1}{4} \\ -\frac{1}{4} \end{pmatrix}\right\}$

31. $\mathbf{X} = c_1\begin{pmatrix} 1 \\ 1 \end{pmatrix}e^{2t} + c_2\left\{\begin{pmatrix} 1 \\ 1 \end{pmatrix}te^{2t} + \begin{pmatrix} -\frac{1}{3} \\ 0 \end{pmatrix}e^{2t}\right\}$

33. $\mathbf{X} = c_1\begin{pmatrix} 1 \\ 1 \\ 1 \end{pmatrix}e^{t} + c_2\begin{pmatrix} 1 \\ 1 \\ 0 \end{pmatrix}e^{2t} + c_3\begin{pmatrix} 1 \\ 0 \\ 1 \end{pmatrix}e^{2t}$

35. $\mathbf{X} = c_1\begin{pmatrix} -4 \\ -5 \\ 2 \end{pmatrix} + c_2\begin{pmatrix} 2 \\ 0 \\ -1 \end{pmatrix}e^{5t}$

$+ c_3\left\{\begin{pmatrix} 2 \\ 0 \\ -1 \end{pmatrix}te^{5t} + \begin{pmatrix} -\frac{1}{2} \\ -\frac{1}{2} \\ -1 \end{pmatrix}e^{5t}\right\}$

37. $\mathbf{X} = c_1\begin{pmatrix} 0 \\ 1 \\ 1 \end{pmatrix}e^{t} + c_2\left\{\begin{pmatrix} 0 \\ 1 \\ 1 \end{pmatrix}te^{t} + \begin{pmatrix} 0 \\ 1 \\ 0 \end{pmatrix}e^{t}\right\}$

$+ c_3\left\{\begin{pmatrix} 0 \\ 1 \\ 1 \end{pmatrix}\frac{t^2}{2}e^{t} + \begin{pmatrix} 0 \\ 1 \\ 0 \end{pmatrix}te^{t} + \begin{pmatrix} \frac{1}{2} \\ 0 \\ 0 \end{pmatrix}e^{t}\right\}$

39. $\mathbf{X} = -7\begin{pmatrix} 2 \\ 1 \end{pmatrix}e^{4t} + 13\begin{pmatrix} 2t + 1 \\ t + 1 \end{pmatrix}e^{4t}$

41. $\mathbf{X} = \begin{pmatrix} \frac{6}{5}e^{5t} - \frac{1}{5}e^{-5t} \\ \frac{2}{5}e^{5t} + \frac{3}{5}e^{-5t} \end{pmatrix}$

43. $\mathbf{X} = c_1t^2\begin{pmatrix} 3 \\ 1 \end{pmatrix} + c_2t^4\begin{pmatrix} 1 \\ 1 \end{pmatrix}$

EXERCISES 8.7 (PAGE 408)

1. $\mathbf{X} = c_1\begin{pmatrix} -1 \\ 1 \end{pmatrix}e^{-t} + c_2\begin{pmatrix} -3 \\ 1 \end{pmatrix}e^{t} + \begin{pmatrix} -1 \\ 3 \end{pmatrix}$

3. $\mathbf{X} = c_1\begin{pmatrix} 1 \\ -1 \end{pmatrix}e^{-2t} + c_2\begin{pmatrix} 1 \\ 1 \end{pmatrix}e^{4t} + \begin{pmatrix} -\frac{1}{4} \\ \frac{3}{4} \end{pmatrix}t^2$

$+ \begin{pmatrix} \frac{1}{4} \\ -\frac{1}{4} \end{pmatrix}t + \begin{pmatrix} -2 \\ \frac{3}{4} \end{pmatrix}$

5. $\mathbf{X} = c_1\begin{pmatrix} 1 \\ -3 \end{pmatrix}e^{3t} + c_2\begin{pmatrix} 1 \\ 9 \end{pmatrix}e^{7t} + \begin{pmatrix} \frac{55}{36} \\ -\frac{19}{4} \end{pmatrix}e^{t}$

7. $\mathbf{X} = c_1\begin{pmatrix}1\\0\\0\end{pmatrix}e^{t} + c_2\begin{pmatrix}1\\1\\0\end{pmatrix}e^{2t} + c_3\begin{pmatrix}1\\2\\2\end{pmatrix}e^{5t} - \begin{pmatrix}\frac{3}{2}\\\frac{7}{2}\\2\end{pmatrix}e^{4t}$

9. $\mathbf{X} = 13\begin{pmatrix}1\\-1\end{pmatrix}e^{t} + 2\begin{pmatrix}-4\\6\end{pmatrix}e^{2t} + \begin{pmatrix}-9\\6\end{pmatrix}$

11. $\mathbf{X} = c_1\begin{pmatrix}1\\1\end{pmatrix} + c_2\begin{pmatrix}1\\-1\end{pmatrix}e^{2t} + \begin{pmatrix}-\frac{3}{2}\\-\frac{3}{2}\end{pmatrix}t + \begin{pmatrix}-\frac{5}{2}\\1\end{pmatrix}$

EXERCISES 8.8 (PAGE 412)

1. $\mathbf{X} = c_1\begin{pmatrix}1\\1\end{pmatrix} + c_2\begin{pmatrix}3\\2\end{pmatrix}e^{t} - \begin{pmatrix}11\\11\end{pmatrix}t - \begin{pmatrix}15\\10\end{pmatrix}$

3. $\mathbf{X} = c_1\begin{pmatrix}2\\1\end{pmatrix}e^{t/2} + c_2\begin{pmatrix}10\\3\end{pmatrix}e^{3t/2} - \begin{pmatrix}\frac{13}{2}\\\frac{13}{4}\end{pmatrix}te^{t/2} - \begin{pmatrix}\frac{15}{2}\\\frac{9}{4}\end{pmatrix}e^{t/2}$

5. $\mathbf{X} = c_1\begin{pmatrix}2\\1\end{pmatrix}e^{t} + c_2\begin{pmatrix}1\\1\end{pmatrix}e^{2t} + \begin{pmatrix}3\\3\end{pmatrix}e^{t} + \begin{pmatrix}4\\2\end{pmatrix}te^{t}$

7. $\mathbf{X} = c_1\begin{pmatrix}4\\1\end{pmatrix}e^{3t} + c_2\begin{pmatrix}-2\\1\end{pmatrix}e^{-3t} + \begin{pmatrix}-12\\0\end{pmatrix}t - \begin{pmatrix}\frac{4}{3}\\\frac{4}{3}\end{pmatrix}$

9. $\mathbf{X} = c_1\begin{pmatrix}1\\-1\end{pmatrix}e^{t} + c_2\begin{pmatrix}-t\\\frac{1}{2}-t\end{pmatrix}e^{t} + \begin{pmatrix}\frac{1}{2}\\-2\end{pmatrix}e^{-t}$

11. $\mathbf{X} = c_1\begin{pmatrix}\cos t\\\sin t\end{pmatrix} + c_2\begin{pmatrix}\sin t\\-\cos t\end{pmatrix}$

$+ \begin{pmatrix}\cos t\\\sin t\end{pmatrix}t + \begin{pmatrix}-\sin t\\\cos t\end{pmatrix}\ln|\cos t|$

13. $\mathbf{X} = c_1\begin{pmatrix}\cos t\\\sin t\end{pmatrix}e^{t} + c_2\begin{pmatrix}\sin t\\-\cos t\end{pmatrix}e^{t} + \begin{pmatrix}\cos t\\\sin t\end{pmatrix}te^{t}$

15. $\mathbf{X} = c_1\begin{pmatrix}\cos t\\-\sin t\end{pmatrix} + c_2\begin{pmatrix}\sin t\\\cos t\end{pmatrix} + \begin{pmatrix}\cos t\\-\sin t\end{pmatrix}t$

$+ \begin{pmatrix}-\sin t\\\sin t\tan t\end{pmatrix} - \begin{pmatrix}\sin t\\\cos t\end{pmatrix}\ln|\cos t|$

17. $\mathbf{X} = c_1\begin{pmatrix}2\sin t\\\cos t\end{pmatrix}e^{t} + c_2\begin{pmatrix}2\cos t\\-\sin t\end{pmatrix}e^{t} + \begin{pmatrix}3\sin t\\\frac{3}{2}\cos t\end{pmatrix}te^{t}$

$+ \begin{pmatrix}\cos t\\-\frac{1}{2}\sin t\end{pmatrix}e^{t}\ln|\sin t| + \begin{pmatrix}2\cos t\\-\sin t\end{pmatrix}e^{t}\ln|\cos t|$

19. $\mathbf{X} = c_1\begin{pmatrix}1\\-1\\0\end{pmatrix} + c_2\begin{pmatrix}1\\1\\0\end{pmatrix}e^{2t} + c_3\begin{pmatrix}0\\0\\1\end{pmatrix}e^{3t}$

$+ \begin{pmatrix}-\frac{1}{4}e^{2t} + \frac{1}{2}te^{2t}\\-e^{t} + \frac{1}{4}e^{2t} + \frac{1}{2}te^{2t}\\\frac{1}{2}t^2e^{3t}\end{pmatrix}$

21. $\mathbf{X} = \begin{pmatrix}2\\2\end{pmatrix}te^{2t} + \begin{pmatrix}-1\\1\end{pmatrix}e^{2t} + \begin{pmatrix}-2\\2\end{pmatrix}te^{4t} + \begin{pmatrix}2\\0\end{pmatrix}e^{4t}$

23. $\mathbf{X} = \begin{pmatrix}-2\\4\end{pmatrix}e^{2t} + \begin{pmatrix}7\\-9\end{pmatrix}e^{7t} + \begin{pmatrix}20\\60\end{pmatrix}te^{7t}$

25. (b) $\begin{pmatrix}i_1\\i_2\end{pmatrix} = 2\begin{pmatrix}1\\3\end{pmatrix}e^{-2t} + \frac{6}{29}\begin{pmatrix}3\\-1\end{pmatrix}e^{-12t}$

$+ \begin{pmatrix}\frac{332}{29}\\\frac{276}{29}\end{pmatrix}\sin t - \begin{pmatrix}\frac{76}{29}\\\frac{168}{29}\end{pmatrix}\cos t$

EXERCISES 8.9 (PAGE 415)

1. $\begin{pmatrix}\cosh t & \sinh t\\\sinh t & \cosh t\end{pmatrix}, \begin{pmatrix}\cosh t - \sinh t\\\sinh t - \cosh t\end{pmatrix}$

3. $\mathbf{X} = \begin{pmatrix}\cosh t & \sinh t\\\sinh t & \cosh t\end{pmatrix}\begin{pmatrix}c_1\\c_2\end{pmatrix} = c_1\begin{pmatrix}\cosh t\\\sinh t\end{pmatrix} + c_2\begin{pmatrix}\sinh t\\\cosh t\end{pmatrix}$

5. $\mathbf{X} = c_1\begin{pmatrix}\cosh t\\\sinh t\end{pmatrix} + c_2\begin{pmatrix}\sinh t\\\cosh t\end{pmatrix} - \begin{pmatrix}1\\1\end{pmatrix}$

7. $\mathbf{X} = c_1\begin{pmatrix}1\\0\end{pmatrix}e^{t} + c_2\begin{pmatrix}0\\1\end{pmatrix}e^{2t} + \begin{pmatrix}-t-1\\\frac{1}{2}e^{4t}\end{pmatrix}$

9. $\mathbf{P} = \begin{pmatrix}1 & 1\\1 & 3\end{pmatrix}, \mathbf{P}^{-1} = \begin{pmatrix}\frac{3}{2} & -\frac{1}{2}\\-\frac{1}{2} & \frac{1}{2}\end{pmatrix},$

$\mathbf{D} = \begin{pmatrix}3 & 0\\0 & 5\end{pmatrix}, \mathbf{PDP}^{-1} = \begin{pmatrix}2 & 1\\-3 & 6\end{pmatrix}$

11. $e^{t\mathbf{A}} = e^{\mathbf{PDP}^{-1}}$

$= \mathbf{I} + t\mathbf{PDP}^{-1} + \frac{t^2}{2!}(\mathbf{PDP}^{-1})^2 + \cdots$

$= \mathbf{PP}^{-1} + t\mathbf{PDP}^{-1} + \frac{t^2}{2!}\mathbf{PD}^2\mathbf{P}^{-1} + \cdots$

$= \mathbf{P}\left[\mathbf{I} + t\mathbf{D} + \frac{t^2}{2!}\mathbf{D}^2 + \cdots\right]\mathbf{P}^{-1}$

$= \mathbf{P}e^{t\mathbf{D}}\mathbf{P}^{-1}$

13. $\mathbf{X} = \begin{pmatrix}\frac{3}{2}e^{3t} - \frac{1}{2}e^{5t} & -\frac{1}{2}e^{3t} + \frac{1}{2}e^{5t}\\\frac{3}{2}e^{3t} - \frac{3}{2}e^{5t} & -\frac{1}{2}e^{3t} + \frac{3}{2}e^{5t}\end{pmatrix}\begin{pmatrix}c_1\\c_2\end{pmatrix}$

CHAPTER 8 REVIEW EXERCISES (PAGE 418)

1. true **3.** $\begin{pmatrix}-2 & 1\\\frac{3}{2} & -\frac{1}{2}\end{pmatrix}$ **5.** true **7.** false **9.** true

11. false

13. $x = -c_1e^{t} - \frac{3}{2}c_2e^{2t} + \frac{5}{2}$

$y = c_1e^{t} + c_2e^{2t} - 3$

15. $x = c_1e^{t} + c_2e^{5t} + te^{t}$

$y = -c_1e^{t} + 3c_2e^{5t} - te^{t} + 2e^{t}$

17. $x = -\frac{1}{4} + \frac{9}{8}e^{-2t} + \frac{1}{8}e^{2t}$

$y = t + \frac{9}{4}e^{-2t} - \frac{1}{4}e^{2t}$

19. (a) $\begin{pmatrix}t^3 + 3t^2 + 5t - 2\\-t^3 - t + 2\\4t^3 + 12t^2 + 8t + 1\end{pmatrix}$ **(b)** $\begin{pmatrix}3t^2 + 6t + 5\\-3t^2 - 1\\12t^2 + 24t + 8\end{pmatrix}$

21. $Dx = u$

$Dy = v$

$Du = -2u + v - 2x - \ln t + 10t - 4$

$Dv = -u - x + 5t - 2$

23. $\mathbf{X} = c_1\begin{pmatrix} 1 \\ -1 \end{pmatrix}e^t + c_2\left\{\begin{pmatrix} 1 \\ -1 \end{pmatrix}te^t + \begin{pmatrix} 0 \\ 1 \end{pmatrix}e^t\right\}$

25. $\mathbf{X} = c_1\begin{pmatrix} \cos 2t \\ -\sin 2t \end{pmatrix}e^t + c_2\begin{pmatrix} \sin 2t \\ \cos 2t \end{pmatrix}e^t$

27. $\mathbf{X} = c_1\begin{pmatrix} -1 \\ 1 \\ 0 \end{pmatrix} + c_2\begin{pmatrix} -1 \\ 0 \\ 1 \end{pmatrix} + c_3\begin{pmatrix} 1 \\ 1 \\ 1 \end{pmatrix}e^{3t}$

29. $\mathbf{X} = c_1\begin{pmatrix} 1 \\ 0 \end{pmatrix}e^{2t} + c_2\begin{pmatrix} 4 \\ 1 \end{pmatrix}e^{4t} + \begin{pmatrix} 16 \\ -4 \end{pmatrix}t + \begin{pmatrix} 11 \\ -1 \end{pmatrix}$

31. $\mathbf{X} = c_1\begin{pmatrix} \cos t \\ \cos t - \sin t \end{pmatrix} + c_2\begin{pmatrix} \sin t \\ \sin t + \cos t \end{pmatrix} - \begin{pmatrix} 1 \\ 1 \end{pmatrix}$

$+ \begin{pmatrix} \sin t \\ \sin t + \cos t \end{pmatrix}\ln|\csc t - \cot t|$

EXERCISES 9.1 (PAGE 424)

1.

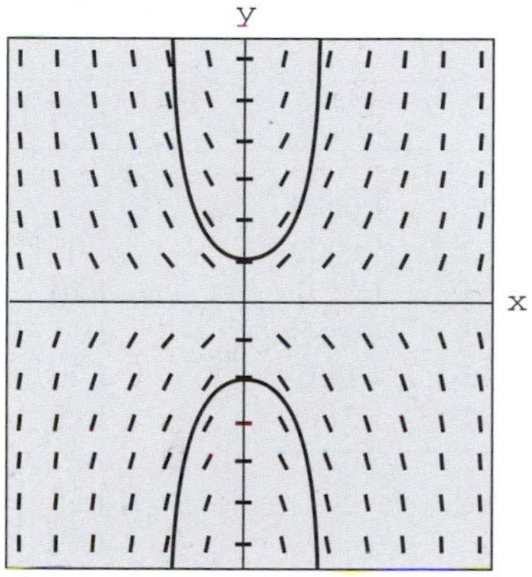

3.

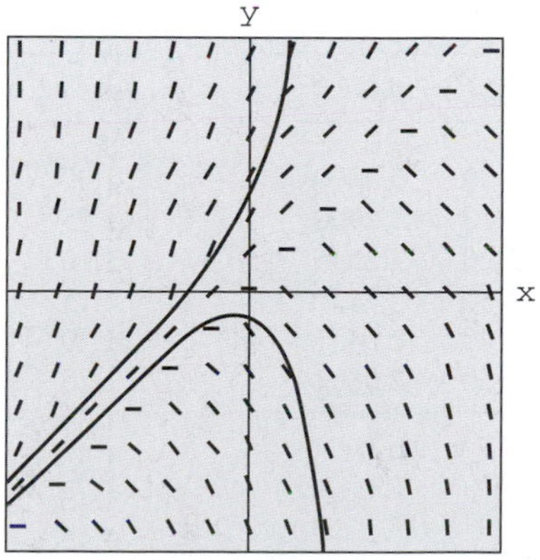

5. a family of vertical lines $x = c - 4$

7. a family of hyperbolas $x^2 - y^2 = c$

9. a family of circles $x^2 + (y + 1)^2 = c^2$ with center at $(0, -1)$

11. a family of hyperbolas $xy + y^2 = c$

13. a family of straight lines $y = c(x - 2) + 1$ passing through $(2, 1)$

15.

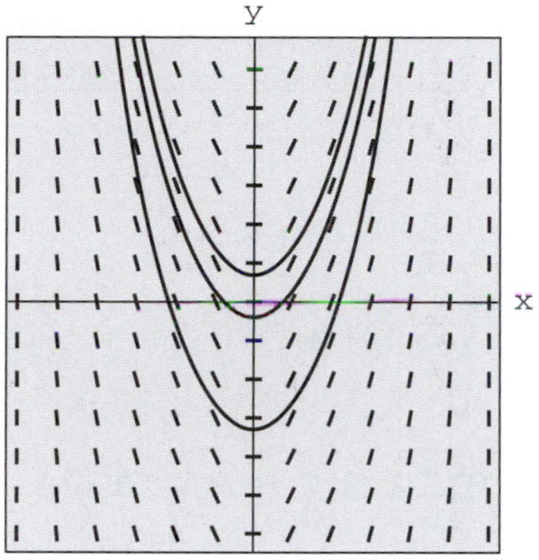

17.

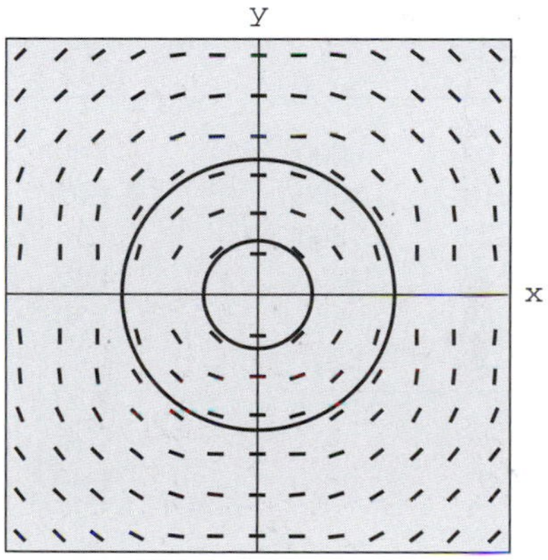

19.

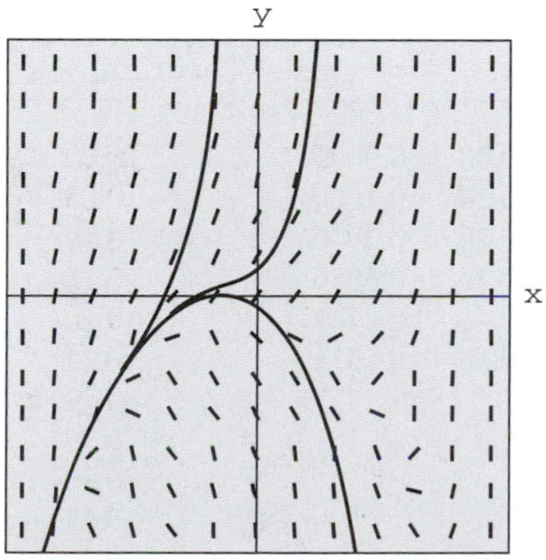

21.

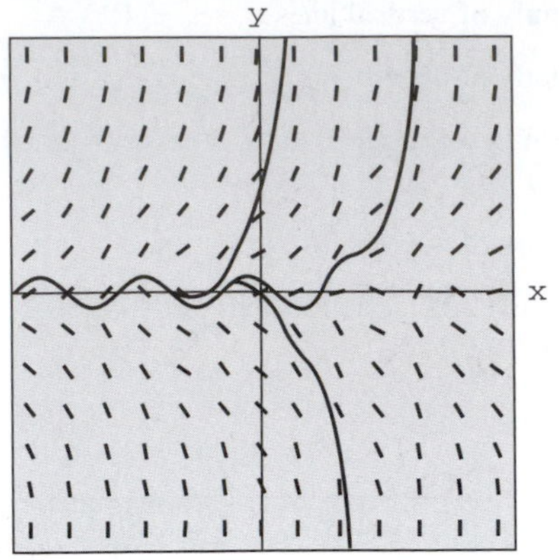

23. $y = \dfrac{\alpha - c\gamma}{c\delta - \beta}x$ **25.** $3x + 2y = -\frac{3}{2}$

27. $y = \pm\sqrt{2}x$ **29.** $y = 4x; y = -x$

EXERCISES 9.2 (PAGE 431)

1. $y = 1 - x + \tan(x + \pi/4)$

3. (a)

x_n	y_n
1.00	5.0000
1.10	3.8000
1.20	2.9800
1.30	2.4260
1.40	2.0582
1.50	1.8207

(b)

x_n	y_n
1.00	5.0000
1.05	4.4000
1.10	3.8950
1.15	3.4707
1.20	3.1151
1.25	2.8179
1.30	2.5702
1.35	2.3647
1.40	2.1950
1.45	2.0557
1.50	1.9424

5. (a)

x_n	y_n
0.00	0.0000
0.10	0.1000
0.20	0.2010
0.30	0.3050
0.40	0.4143
0.50	0.5315

(b)

x_n	y_n
0.00	0.0000
0.05	0.0500
0.10	0.1001
0.15	0.1506
0.20	0.2018
0.25	0.2538
0.30	0.3070
0.35	0.3617
0.40	0.4183
0.45	0.4770
0.50	0.5384

7. (a)

x_n	y_n
0.00	0.0000
0.10	0.1000
0.20	0.1905
0.30	0.2731
0.40	0.3492
0.50	0.4198

(b)

x_n	y_n
0.00	0.0000
0.05	0.0500
0.10	0.0976
0.15	0.1429
0.20	0.1863
0.25	0.2278
0.30	0.2676
0.35	0.3058
0.40	0.3427
0.45	0.3782
0.50	0.4124

9. (a)

x_n	y_n
0.00	0.5000
0.10	0.5250
0.20	0.5431
0.30	0.5548
0.40	0.5613
0.50	0.5639

(b)

x_n	y_n
0.00	0.5000
0.05	0.5125
0.10	0.5232
0.15	0.5322
0.20	0.5395
0.25	0.5452
0.30	0.5496
0.35	0.5527
0.40	0.5547
0.45	0.5559
0.50	0.5565

11. (a)

x_n	y_n
1.00	1.0000
1.10	1.0000
1.20	1.0191
1.30	1.0588
1.40	1.1231
1.50	1.2194

(b)

x_n	y_n
1.00	1.0000
1.05	1.0000
1.10	1.0049
1.15	1.0147
1.20	1.0298
1.25	1.0506
1.30	1.0775
1.35	1.1115
1.40	1.1538
1.45	1.2057
1.50	1.2696

13. (a) $h = 0.1$

x_n	y_n
1.00	5.0000
1.10	3.9900
1.20	3.2545
1.30	2.7236
1.40	2.3451
1.50	2.0801

$h = 0.05$

x_n	y_n
1.00	5.0000
1.05	4.4475
1.10	3.9763
1.15	3.5751
1.20	3.2342
1.25	2.9452
1.30	2.7009
1.35	2.4952
1.40	2.3226
1.45	2.1786
1.50	2.0592

13. **(b)** $h = 0.1$

x_n	y_n
0.00	0.0000
0.10	0.1005
0.20	0.2030
0.30	0.3098
0.40	0.4234
0.50	0.5470

$h = 0.05$

x_n	y_n
0.00	0.0000
0.05	0.0501
0.10	0.1004
0.15	0.1512
0.20	0.2028
0.25	0.2554
0.30	0.3095
0.35	0.3652
0.40	0.4230
0.45	0.4832
0.50	0.5465

(c) $h = 0.1$

x_n	y_n
0.00	0.0000
0.10	0.0952
0.20	0.1822
0.30	0.2622
0.40	0.3363
0.50	0.4053

$h = 0.05$

x_n	y_n
0.00	0.0000
0.05	0.0488
0.10	0.0953
0.15	0.1397
0.20	0.1823
0.25	0.2231
0.30	0.2623
0.35	0.3001
0.40	0.3364
0.45	0.3715
0.50	0.4054

(d) $h = 0.1$

x_n	y_n
0.00	0.5000
0.10	0.5215
0.20	0.5362
0.30	0.5449
0.40	0.5490
0.50	0.5503

$h = 0.05$

x_n	y_n
0.00	0.5000
0.05	0.5116
0.10	0.5214
0.15	0.5294
0.20	0.5359
0.25	0.5408
0.30	0.5444
0.35	0.5469
0.40	0.5484
0.45	0.5492
0.50	0.5495

(e) $h = 0.1$

x_n	y_n
1.00	1.0000
1.10	1.0095
1.20	1.0404
1.30	1.0967
1.40	1.1866
1.50	1.3260

$h = 0.05$

x_n	y_n
1.00	1.0000
1.05	1.0024
1.10	1.0100
1.15	1.0228
1.20	1.0414
1.25	1.0663
1.30	1.0984
1.35	1.1389
1.40	1.1895
1.45	1.2526
1.50	1.3315

15.

x_n	**Euler**	**Improved Euler**
1.0	1.0000	1.0000
1.1	1.2000	1.2469
1.2	1.4938	1.6668
1.3	1.9711	2.6427
1.4	2.9060	8.7989

17. Using $y' = f(x, y)$ gives

$$\int_{x_n}^{x_{n+1}} y'\,dx = \int_{x_n}^{x_{n+1}} f(x, y)\,dx$$

$$y(x_{n+1}) - y(x_n) \approx f(x_n, y_n)(x_{n+1} - x_n)$$

$$= hf(x_n, y_n)$$

$$y(x_{n+1}) \approx y(x_n) + hf(x_n, y_n).$$

We write this as

$$y_{n+1} = y_n + hf(x_n, y_n).$$

EXERCISES 9.3 (PAGE 434)

1. **(a)**

x_n	y_n
1.00	5.0000
1.10	3.9900
1.20	3.2545
1.30	2.7236
1.40	2.3451
1.50	2.0801

(b)

x_n	y_n
1.00	5.0000
1.05	4.4475
1.10	3.9763
1.15	3.5751
1.20	3.2342
1.25	2.9452
1.30	2.7009
1.35	2.4952
1.40	2.3226
1.45	2.1786
1.50	2.0592

3. (a)

x_n	y_n
0.00	0.0000
0.10	0.1000
0.20	0.2020
0.30	0.3082
0.40	0.4211
0.50	0.5438

(b)

x_n	y_n
0.00	0.0000
0.05	0.0500
0.10	0.1003
0.15	0.1510
0.20	0.2025
0.25	0.2551
0.30	0.3090
0.35	0.3647
0.40	0.4223
0.45	0.4825
0.50	0.5456

5. (a)

x_n	y_n
0.00	0.0000
0.10	0.0950
0.20	0.1818
0.30	0.2617
0.40	0.3357
0.50	0.4046

(b)

x_n	y_n
0.00	0.0000
0.05	0.0488
0.10	0.0952
0.15	0.1397
0.20	0.1822
0.25	0.2230
0.30	0.2622
0.35	0.2999
0.40	0.3363
0.45	0.3714
0.50	0.4053

7. (a)

x_n	y_n
0.00	0.5000
0.10	0.5213
0.20	0.5355
0.30	0.5438
0.40	0.5475
0.50	0.5482

(b)

x_n	y_n
0.00	0.5000
0.05	0.5116
0.10	0.5213
0.15	0.5293
0.20	0.5357
0.25	0.5406
0.30	0.5441
0.35	0.5466
0.40	0.5480
0.45	0.5487
0.50	0.5490

9. (a)

x_n	y_n
1.00	1.0000
1.10	1.0100
1.20	1.0410
1.30	1.0969
1.40	1.1857
1.50	1.3226

(b)

x_n	y_n
1.00	1.0000
1.05	1.0025
1.10	1.0101
1.15	1.0229
1.20	1.0415
1.25	1.0663
1.30	1.0983
1.35	1.1387
1.40	1.1891
1.45	1.2518
1.50	1.3301

11.

x_n	Euler	Improved Euler	Three-Term Taylor
1.0	1.0000	1.0000	1.0000
1.1	1.2000	1.2469	1.2400
1.2	1.4938	1.6668	1.6345
1.3	1.9711	2.6427	2.4600
1.4	2.9060	8.7988	5.6353

13.

x_n	Improved Euler	Three-Term Taylor	True Value
1.00	5.0000	5.0000	5.0000
1.10	5.5300	5.5300	5.5310
1.20	6.1262	6.1262	6.1284
1.30	6.7954	6.7954	6.7992
1.40	7.5454	7.5454	7.5510
1.50	8.3847	8.3847	8.3923

EXERCISES 9.4 (PAGE 438)

1.

x_n	y_n
1.00	5.0000
1.10	3.9724
1.20	3.2284
1.30	2.6945
1.40	2.3163
1.50	2.0533

3.

x_n	y_n
0.00	0.0000
0.10	0.1003
0.20	0.2027
0.30	0.3093
0.40	0.4228
0.50	0.5463

5.

x_n	y_n
0.00	0.0000
0.10	0.0953
0.20	0.1823
0.30	0.2624
0.40	0.3365
0.50	0.4055

7.

x_n	y_n
0.00	0.5000
0.10	0.5213
0.20	0.5358
0.30	0.5443
0.40	0.5482
0.50	0.5493

9.

x_n	y_n
1.00	1.0000
1.10	1.0101
1.20	1.0417
1.30	1.0989
1.40	1.1905
1.50	1.3333

11. $v(5) \approx 35.7678$

13. 1.93 12.50 36.46 47.23 49.00

15.

x_n	y_n
1.00	1.0000
1.10	1.2511
1.20	1.6934
1.30	2.9425
1.40	903.0283

EXERCISES 9.5 (PAGE 442)

1. $y(x) = -x + e^x$; $y(0.2) = 1.0214$, $y(0.4) = 1.0918$, $y(0.6) = 1.2221$, $y(0.8) = 1.4255$

3.

x_n	y_n
0.00	1.0000
0.20	0.7328
0.40	0.6461
0.60	0.6585
0.80	0.7232

5.

x_n	y_n
0.00	0.0000
0.20	0.2027
0.40	0.4228
0.60	0.6841
0.80	1.0297
1.00	1.5569

x_n	y_n
0.00	0.0000
0.10	0.1003
0.20	0.2027
0.30	0.3093
0.40	0.4228
0.50	0.5463
0.60	0.6842
0.70	0.8423
0.80	1.0297
0.90	1.2603
1.00	1.5576

7.

x_n	y_n
0.00	0.0000
0.20	0.0026
0.40	0.0201
0.60	0.0630
0.80	0.1360
1.00	0.2385

x_n	y_n
0.00	0.0000
0.10	0.0003
0.20	0.0026
0.30	0.0087
0.40	0.0200
0.50	0.0379
0.60	0.0629
0.70	0.0956
0.80	0.1360
0.90	0.1837
1.00	0.2384

9.

x_n	y_n
0.00	1.0000
0.10	1.0052
0.20	1.0214
0.30	1.0499
0.40	1.0918

EXERCISES 9.6 (PAGE 446)

1. The improved Euler method is a second-order Runge-Kutta method, so letting $k = 3$, $a = x_n$, $x = x_{n+1} = x_n + h$, we get

$$y(x_{n+1}) = y(x_n) + y'(x_n)h + y''(x_n)\frac{h^2}{2} + y^{(3)}(c)\frac{h^3}{3!}.$$

The derivation of the Runge-Kutta method requires that the formula agree with the Taylor series through the h^2 term, so the local truncation error is $y^{(3)}(c)(h^3/3!)$.

3. For the fourth-order Runge-Kutta method, we let $k = 4$, $a = x_n$, $x = x_{n+1} = x_n + h$ and get

$$y(x_{n+1}) = y(x_n) + y'(x_n)h + \cdots + y^{(4)}(x_n)\frac{h^4}{4!} + y^{(5)}(c)\frac{h^5}{5!}.$$

The derivation of this Runge-Kutta method requires that the formula agree with the Taylor series through the h^4 term, so the local truncation error is $y^{(5)}(c)(h^5/5!)$.

5. **(a)** $y_1 = 1.2200$

(b) $y^{(3)}(c)\dfrac{h^3}{3!} = 8e^{2c}\dfrac{h^3}{3!} \le 8e^{2(0.1)}\dfrac{(0.1)^3}{3!} = 0.001629$

(c) Actual value is $y(0.1) = e^{2(0.1)} = 1.221403$.
Error is $1.221403 - 1.22 = 0.001403 \le 0.001629$.

(d) If $h = 0.05$, $y_2 = 1.221025$.

(e) Error with $h = 0.1$ is 0.001403.
Error with $h = 0.05$ is
$1.221403 - 1.221025 = 0.000378$.

7. **(a)** $y_1 = 1.221400$

(b) $y^{(5)}(c)\dfrac{h^5}{5!} = 32e^{2c}\dfrac{h^5}{5!} \le 32e^{2(0.1)}\dfrac{(0.1)^5}{5!}$
$= 3.257 \times 10^{-6}$

(c) Actual value is $y(0.1) = e^{2(0.1)} = 1.221402758$.
Error is $1.221402758 - 1.2214 = 2.758 \times 10^{-6} \le 3.257 \times 10^{-6}$.

(d) If $h = 0.05$, $y_2 = 1.22140257$.

(e) Error with $h = 0.1$ is 2.758×10^{-6}.
Error with $h = 0.05$ is 1.873×10^{-7}.

9. **(a)** $y_1 = 0.8250$

(b) $y^{(3)}(c)\dfrac{h^3}{3!} = 10e^{-2c}\dfrac{h^3}{3!} \le 10e^{-2(0)}\dfrac{(0.1)^3}{3!}$
$= 0.001667$

(c) Actual value is $y(0.1) = 0.8234$.
Error is $0.8250 - 0.8234 = 0.001600 \le 0.001667$.

(d) If $h = 0.05$, $y_2 = 0.8238$.

(e) Error with $h = 0.1$ is 0.001600.
Error with $h = 0.05$ is
$0.008238 - 0.008234 = 0.0004$.

11. (a) $y_1 = 0.82341667$

(b) $y^{(5)}(c)\dfrac{h^5}{5!} = 40e^{-2c}\dfrac{h^5}{5!} \le 40e^{2(0)}\dfrac{(0.1)^5}{5!}$

$= 3.333 \times 10^{-6}$

(c) Actual value is $y(0.1) = 0.8234134413$.
Error is $3.225 \times 10^{-6} \le 3.333 \times 10^{-6}$.

(d) If $h = 0.05$, $y_2 = 0.82341363$.

(e) Error with $h = 0.1$ is 3.225×10^{-6}.
Error with $h = 0.05$ is 1.854×10^{-7}.

13. (a) $\left| y^{(3)}(c)\dfrac{h^3}{3!} \right| = 114e^{-3(c-1)}\dfrac{h^3}{3!}$

(b) $114e^{-3(c-1)}\dfrac{h^3}{3!} \le 114e^{-3(1-1)}\dfrac{(0.1)^3}{3!} = 0.019$

(c) From Problem 3 in Exercises 9.2 with $h = 0.1$, $y_5 = 2.0801$. From Problem 3 in Exercises 9.2 with $h = 0.05$, $y_{10} = 2.0592$.

(d) The error with $h = 0.1$ is 0.0269.
The error with $h = 0.05$ is 0.0060.

15. (a) $\left| y^{(5)}(c)\dfrac{h^5}{5!} \right| = 1026e^{-3(c-1)}\dfrac{h^5}{5!}$

(b) $1026e^{-3(c-1)}\dfrac{h^5}{5!} \le 1026e^{-3(1-1)}\dfrac{(0.1)^5}{5!}$

$= 8.55 \times 10^{-5}$

(c) From calculation with $h = 0.1$, $y_5 = 2.05333883$.
From calculation with $h = 0.05$, $y_{10} = 2.05322299$.

(d) The error with $h = 0.1$ is 0.00012259.
The error with $h = 0.05$ is 0.00006757.

17. (a) $y^{(3)}(c)\dfrac{h^3}{3!} = \dfrac{2}{(c+1)^3}\dfrac{h^3}{3!}$

(b) $\dfrac{2}{(c+1)^3}\dfrac{h^3}{3!} \le 2\dfrac{(0.1)^3}{3!} = 0.0003333$

(c) From Problem 13(c) in Exercises 9.2 with $h = 0.1$, $y_5 = 0.40528104$. From Problem 13(c) in Exercises 9.2 with $h = 0.05$, $y_{10} = 0.40541888$.

(d) The error with $h = 0.1$ is 0.0001840683428.
The error with $h = 0.05$ is 0.00004622850912.

19. (a) $y^{(5)}(c)\dfrac{h^5}{5!} = \dfrac{24}{(c+1)^5}\dfrac{h^5}{5!}$

(b) $\dfrac{24}{(c+1)^5}\dfrac{h^3}{5!} \le 24\dfrac{(0.1)^3}{5!} = 2.0000 \times 10^{-6}$

(c) From calculation with $h = 0.1$, $y_5 = 0.40546517$.
From calculation with $h = 0.05$, $y_{10} = 0.40546511$.

(d) The error with $h = 0.1$ is $5.978114648 \times 10^{-8}$.
The error with $h = 0.05$ is $3.388367642 \times 10^{-9}$.

EXERCISES 9.7 (PAGE 451)

1. $y(x) = -2e^{2x} + 5xe^{2x}$; $y(0.2) = -1.4918$, $y_2 = -1.6800$

3. $y_1 = -1.4928$, $y_2 = -1.4919$

5. $y_1 = 1.4640$, $y_2 = 1.4640$

7. $x_1 = 8.3055$, $y_1 = 3.4199$;
$x_2 = 8.3055$, $y_2 = 3.4199$

9. $x_1 = -3.9123$, $y_1 = 4.2857$;
$x_2 = -3.9123$, $y_2 = 4.2857$

EXERCISES 9.8 (PAGE 455)

1. $y_1 = -5.6774$
$y_2 = -2.5807$
$y_3 = 6.3226$

3. $y_1 = -0.2259$
$y_2 = -0.3356$
$y_3 = -0.3308$
$y_4 = -0.2167$

5. $y_1 = 3.3751$
$y_2 = 3.6306$
$y_3 = 3.6448$
$y_4 = 3.2355$
$y_5 = 2.1411$

7. $y_1 = 3.8842$
$y_2 = 2.9640$
$y_3 = 2.2064$
$y_4 = 1.5826$
$y_5 = 1.0681$
$y_6 = 0.6430$
$y_7 = 0.2913$

9. $y_1 = 0.2660$
$y_2 = 0.5097$
$y_3 = 0.7357$
$y_4 = 0.9471$
$y_5 = 1.1465$
$y_6 = 1.3353$
$y_7 = 1.5149$
$y_8 = 1.6855$
$y_9 = 1.8474$

11. $y_1 = 0.3492$
$y_2 = 0.7202$
$y_3 = 1.1363$
$y_4 = 1.6233$
$y_5 = 2.2118$
$y_6 = 2.9386$
$y_7 = 3.8490$

13. (c) $y_0 = -2.2755$
$y_1 = -2.0755$
$y_2 = -1.8589$
$y_3 = -1.6126$
$y_4 = -1.3275$

CHAPTER 9 REVIEW EXERCISES (PAGE 457)

1. All isoclines $y = cx$ are solutions of the differential equation.

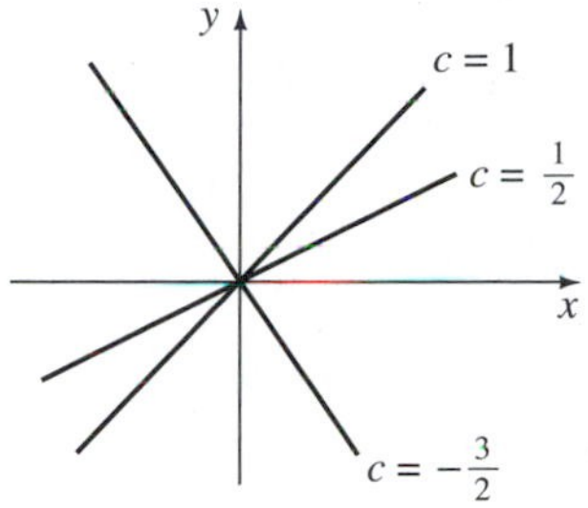

3. Comparison of Numerical Methods with $h = 0.1$

x_n	Euler	Improved Euler	Three-Term Taylor	Runge-Kutta
1.00	2.0000	2.0000	2.0000	2.0000
1.10	2.1386	2.1549	2.1556	2.1556
1.20	2.3097	2.3439	2.3453	2.3454
1.30	2.5136	2.5672	2.5694	2.5695
1.40	2.7504	2.8246	2.8277	2.8278
1.50	3.0201	3.1157	3.1198	3.1197

Comparison of Numerical Methods with $h = 0.05$

x_n	Euler	Improved Euler	Three-Term Taylor	Runge-Kutta
1.00	2.0000	2.0000	2.0000	2.0000
1.05	2.0693	2.0735	2.0735	2.0736
1.10	2.1469	2.1554	2.1556	2.1556
1.15	2.2329	2.2459	2.2462	2.2462
1.20	2.3272	2.3450	2.3454	2.3454
1.25	2.4299	2.4527	2.4532	2.4532
1.30	2.5410	2.5689	2.5695	2.5695
1.35	2.6604	2.6937	2.6944	2.6944
1.40	2.7883	2.8269	2.8278	2.8278
1.45	2.9245	2.9686	2.9696	2.9696
1.50	3.0690	3.1187	3.1198	3.1197

5. Comparison of Numerical Methods with $h = 0.1$

x_n	Euler	Improved Euler	Three-Term Taylor	Runge-Kutta
0.50	0.5000	0.5000	0.5000	0.5000
0.60	0.6000	0.6048	0.6050	0.6049
0.70	0.7095	0.7191	0.7195	0.7194
0.80	0.8283	0.8427	0.8433	0.8431
0.90	0.9559	0.9752	0.9759	0.9757
1.00	1.0921	1.1163	1.1172	1.1169

Comparison of Numerical Methods with $h = 0.05$

x_n	Euler	Improved Euler	Three-Term Taylor	Runge-Kutta
0.50	0.5000	0.5000	0.5000	0.5000
0.55	0.5500	0.5512	0.5512	0.5512
0.60	0.6024	0.6049	0.6049	0.6049
0.65	0.6573	0.6609	0.6610	0.6610
0.70	0.7144	0.7193	0.7194	0.7194
0.75	0.7739	0.7800	0.7802	0.7801
0.80	0.8356	0.8430	0.8431	0.8431
0.85	0.8996	0.9082	0.9083	0.9083
0.90	0.9657	0.9755	0.9757	0.9757
0.95	1.0340	1.0451	1.0453	1.0452
1.00	1.1044	1.1168	1.1170	1.1169

7. $h = 0.2$: $y(0.2) \approx 3.2$
$h = 0.1$: $y(0.2) \approx 3.23$

9. $x(0.2) \approx 1.62$, $y(0.2) \approx 1.84$

EXERCISES FOR APPENDIX I (PAGE APP-2)

1. **(a)** 24 **(b)** 720 **(c)** $4\sqrt{\pi}/3$ **(d)** $-8\sqrt{\pi}/15$

3. 0.297

5. $\Gamma(x) > \int_0^1 t^{x-1}e^{-t}\,dt > e^{-1}\int_0^1 t^{x-1}\,dt = \dfrac{1}{xe}$

for $x > 0$. As $x \to 0^+$, $1/x \to +\infty$.

EXERCISES FOR APPENDIX III (PAGE APP-10)

1. -58 **3.** 248 **5.** 12 **7.** $16e^{3t}$

9. $x = 4$, $y = -7$ **11.** $x = 4$, $y = \frac{3}{2}$, $z = 1$

13. Let $z = t$, t any real number. The solution is $x = -t/3$, $y = 5t/3$, $z = t$; the system represents the line of intersection of two planes.

EXERCISES FOR APPENDIX IV (PAGE APP-13)

1. $7 - 4i$ **3.** $-11 - 11i$ **5.** $3 - 4i$

7. $\frac{7}{34} - \frac{11}{34}i$ **9.** $\frac{5}{34} - \frac{3}{34}i$ **11.** $z = e^{i\pi/2}$

13. $z = e^{i\pi}$ **15.** $z = 2\sqrt{2}\,e^{i\pi/4}$ **17.** $z = 12e^{i\pi/3}$

19. $z = 2e^{i\pi/6}$ **21.** $z = -8$

INDEX

TABLE OF INTEGRALS

1. $\displaystyle\int u\,dv = uv - \int v\,du$

2. $\displaystyle\int u^n\,du = \frac{1}{n+1}u^{n+1} + C,\ n \neq -1$

3. $\displaystyle\int \frac{du}{u} = \ln|u| + C$

4. $\displaystyle\int e^u\,du = e^u + C$

5. $\displaystyle\int a^u\,du = \frac{1}{\ln a}a^u + C$

6. $\displaystyle\int \sin u\,du = -\cos u + C$

7. $\displaystyle\int \cos u\,du = \sin u + C$

8. $\displaystyle\int \sec^2 u\,du = \tan u + C$

9. $\displaystyle\int \csc^2 u\,du = -\cot u + C$

10. $\displaystyle\int \sec u \tan u\,du = \sec u + C$

11. $\displaystyle\int \csc u \cot u\,du = -\csc u + C$

12. $\displaystyle\int \tan u\,du = -\ln|\cos u| + C$

13. $\displaystyle\int \cot u\,du = \ln|\sin u| + C$

14. $\displaystyle\int \sec u\,du = \ln|\sec u + \tan u| + C$

15. $\displaystyle\int \csc u\,du = \ln|\csc u - \cot u| + C$

16. $\displaystyle\int \frac{du}{\sqrt{a^2 - u^2}} = \sin^{-1}\frac{u}{a} + C$

17. $\displaystyle\int \frac{du}{a^2 + u^2} = \frac{1}{a}\tan^{-1}\frac{u}{a} + C$

18. $\displaystyle\int \frac{du}{u\sqrt{u^2 - a^2}} = \frac{1}{a}\sec^{-1}\frac{u}{a} + C$

19. $\displaystyle\int \frac{du}{a^2 - u^2} = \frac{1}{2a}\ln\left|\frac{u+a}{u-a}\right| + C$

20. $\displaystyle\int \frac{du}{u^2 - a^2} = \frac{1}{2a}\ln\left|\frac{u-a}{u+a}\right| + C$

21. $\displaystyle\int \sin^2 u\,du = \tfrac{1}{2}u - \tfrac{1}{4}\sin 2u + C$

22. $\displaystyle\int \cos^2 u\,du = \tfrac{1}{2}u + \tfrac{1}{4}\sin 2u + C$

23. $\displaystyle\int \tan^2 u\,du = \tan u - u + C$

24. $\displaystyle\int \cot^2 u\,du = -\cot u - u + C$

25. $\displaystyle\int \sin^3 u\,du = -\tfrac{1}{3}(2 + \sin^2 u)\cos u + C$

26. $\displaystyle\int \cos^3 u\,du = \tfrac{1}{3}(2 + \cos^2 u)\sin u + C$

27. $\displaystyle\int \tan^3 u\,du = \tfrac{1}{2}\tan^2 u + \ln|\cos u| + C$

28. $\displaystyle\int \cot^3 u\,du = -\tfrac{1}{2}\cot^2 u - \ln|\sin u| + C$

29. $\displaystyle\int \sec^3 u\,du = \tfrac{1}{2}\sec u \tan u + \tfrac{1}{2}\ln|\sec u + \tan u| + C$

30. $\displaystyle\int \csc^3 u\,du = -\tfrac{1}{2}\csc u \cot u + \tfrac{1}{2}\ln|\csc u - \cot u| + C$

31. $\displaystyle\int \sin^n u\,du = -\frac{1}{n}\sin^{n-1}u \cos u + \frac{n-1}{n}\int \sin^{n-2}u\,du$

32. $\displaystyle\int \cos^n u\,du = \frac{1}{n}\cos^{n-1}u \sin u + \frac{n-1}{n}\int \cos^{n-2}u\,du$

33. $\displaystyle\int \tan^n u\,du = \frac{1}{n-1}\tan^{n-1}u - \int \tan^{n-2}u\,du$

34. $\displaystyle\int \cot^n u\,du = \frac{-1}{n-1}\cot^{n-1}u - \int \cot^{n-2}u\,du$

35. $\displaystyle\int \sec^n u\,du = \frac{1}{n-1}\tan u \sec^{n-2}u + \frac{n-2}{n-1}\int \sec^{n-2}u\,du$

36. $\displaystyle\int \csc^n u\,du = \frac{-1}{n-1}\cot u \csc^{n-2}u + \frac{n-2}{n-1}\int \csc^{n-2}u\,du$

37. $\displaystyle\int \sin au \sin bu\,du = \frac{\sin(a-b)u}{2(a-b)} - \frac{\sin(a+b)u}{2(a+b)} + C$

38. $\displaystyle\int \cos au \cos bu\,du = \frac{\sin(a-b)u}{2(a-b)} + \frac{\sin(a+b)u}{2(a+b)} + C$

39. $\displaystyle\int \sin au \cos bu\,du = -\frac{\cos(a-b)u}{2(a-b)} - \frac{\cos(a+b)u}{2(a+b)} + C$

40. $\displaystyle\int u \sin u\,du = \sin u - u\cos u + C$

41. $\displaystyle\int u \cos u\,du = \cos u + u \sin u + C$

42. $\displaystyle\int u^n \sin u\,du = -u^n \cos u + n\int u^{n-1}\cos u\,du$

43. $\displaystyle\int u^n \cos u\,du = u^n \sin u - n\int u^{n-1}\sin u\,du$

44. $\displaystyle\int \sin^n u \cos^m u\,du = -\frac{\sin^{n-1}u \cos^{m+1}u}{n+m} + \frac{n-1}{n+m}\int \sin^{n-2}u \cos^m u\,du = \frac{\sin^{n+1}u \cos^{m-1}u}{n+m} + \frac{m-1}{n+m}\int \sin^n u \cos^{m-2}u\,du$